D1134696

Chemical Biomarkers in Aquatic Ecosystems

Chemical Biomarkers in Aquatic Ecosystems

THOMAS S. BIANCHI AND ELIZABETH A. CANUEL

PRINCETON UNIVERSITY PRESS
PRINCETON AND OXFORD

Published by Princeton University Press, 41 William Street,
Princeton, New Jersey 08540
In the United Kingdom: Princeton University Press, 6 Oxford Street,
Woodstock, Oxfordshire OX20 1TW

press.princeton.edu

Library of Congress Cataloging-in-Publication Data

Bianchi, Thomas S.
Chemical biomarkers in aquatic ecosystems/Thomas S. Bianchi and Elizabeth A. Canuel.
p. cm.
Includes bibliographical references and index.
ISBN 978-0-691-13414-7 (cloth : alk. paper)
1. Aquatic ecology. 2. Biochemical markers. I. Canuel, Elizabeth A., 1959– II. Title.
QH541.5.W3B53 2011
577.6—dc22 2010029921

British Library Cataloging-in-Publication Data is available

This book has been composed in Sabon and Din Pro
Printed on acid-free paper. ∞

Typeset by S R Nova Pvt Ltd, Bangalore, India
Printed in the United States of America

10 9 8 7 6 5 4 3 2 1

We dedicate this book to our spouses, Jo Ann Bianchi and Emmett Duffy, and our children, Christopher Bianchi and Conor Duffy, for their unending support and patience, without which we could never have completed this book.

Several current characteristics of organic geochemical research serve to limit advances toward a deeper understanding of global biogeochemical cycles. For the most part, these problems are shared by all organic geochemists (author included), but are not unique to our guild. The first of these is that geochemical research is decidedly provincial. That is, functionally similar processes occurring in weathering rocks, soils, lakes, rivers and the ocean tend to be studied by different types of geochemists with dissimilar backgrounds, terminologies and methodologies. This geographic specialization has limited the scope of both our fieldwork and insight, and has resulted in lost research and interpretive opportunities. . . . At present, there is insufficient information to obtain unique solutions at the global scale, even for bulk carbon. The main constraint is that meaningful rate constants for in situ degradation within key environment are not known.

— John I. Hedges, 1992

Contents

Preface

Due to the complexity of organic matter sources in aquatic systems, the application of chemical biomarkers has become widespread in limnology and oceanography. We define *biomarker molecules* in this book as compounds that characterize certain biotic sources and selectively retain their source information, even after stages of decomposition and diagenesis (after Meyers, 2003). Hence, the term biomarker molecule should not be confused with the terms commonly used by ecotoxicologists or molecular biologists (see Timbrell, 1998; Wilson and Suk, 2002; Decaprio, 2006, and references therein). Early definitions of biological markers, or simply *biomarkers* or *chemical biomarkers*, described these compounds as molecular fossils derived from formerly living organisms (Eglinton et al., 1964; Eglinton and Calvin, 1967).

More recently, biomarkers have been defined as complex organic compounds composed of carbon, hydrogen, and other elements, which occur in sediments, rocks, and crude oils, and show little to no change in their chemical structure from their precursor molecules that once existed in living organisms (Hunt et al., 1996; Peters et al., 2005; Gaines et al., 2009). Biomarkers have also been defined recently as lipid components found within *bitumen* in coal as well as *kerogens* that provide an unambiguous link to biological precursor compounds, because of their stability through *diagenesis* and *catagenesis* (Killops and Killops, 2005; Gaines et al., 2009). Additionally, the term *geolipid* has commonly been used to describe decay-resistant biomarkers in sediments because lipids are typically recalcitrant compared with other biochemical components of organic matter, making them more long-lived in the sedimentary record (Meyers, 1997). The first convincing linkage made between geolipids and living organisms was made by Treibs (1934), when a correlation was made between chlorophyll *a* in photosynthetic organisms and *porphyrins* in petroleum. This important discovery essentially marked the beginning of organic geochemistry.

Chemical biomarkers have provided numerous insights about present and past aspects of Earth history, including (1) the food and energy sources available to microbes and higher organisms, (2) microbial chemotaxonomy, (3) sources of fossil fuels, and (4) the evolution of life on Earth. A greater knowledge of the biogeochemical cycling of organic matter, which involves the transformation, fate, and transport of chemical substances, is critical in understanding the effects of natural and human-induced environmental changes—from a regional and global context. Approaching aquatic sciences from a biogeochemical perspective requires a fundamental understanding in the biochemistry of organic matter. Our motivation for writing this book is that many of the books on chemical biomarkers are edited volumes, too diffuse and difficult to use as textbooks for advanced undergraduate and graduate courses. We also wanted to develop a book with examples of the applications of biomarkers to a range of aquatic environments, including riverine, estuarine, and oceanic ecosystems.

This book is designed for advanced undergraduate and graduate level classes in aquatic and marine biogeochemistry, organic geochemistry, and global ecosystem dynamics. Prerequisites for such a course may include introductory courses in inorganic and organic chemistry, oceanography and/or limnology, and environmental and/or ecosystem ecology. The book should also prove to be a valuable resource for researchers in limnological, estuarine, marine, and environmental sciences. The basic organization of the book is derived from classes we have taught in global biogeochemistry, organic geochemistry, and marine/estuarine biogeochemistry. It is structured around individual classes of biomarkers, with each chapter describing the structure and synthesis of each class of compounds, their reactivity, and examples of specific applications for each class of biomarkers.

In chapter 1, we provide a general background on the synthesis of chemical biomarkers and their association with key metabolic pathways in organisms, as related to differences in cellular structure and function across the three domains of life. We discuss photosynthesis, the dominant pathway by which biomass is synthesized. Additionally, we present information about chemoautotrophic and microbial heterotrophic processes. The holistic view of biosynthetic pathways of chemical biomarkers here also provides a roadmap for other chapters in the book, where more specific details on chemical pathways are presented for each of the respective biomarker groups. While other organic geochemistry books have generally introduced the concepts of chemical biomarkers in the context of physical and chemical gradients found in natural ecosystems (e.g., anaerobic, aerobic), we begin by examining biosynthetic pathways at the cellular level of differentiation.

Chapter 2 provides a brief historical account of the success and limitations of using chemical biomarkers in aquatic ecosystems. This chapter also introduces the general concepts of chemical biomarkers as they relate to global biogeochemical cycling. The application of chemical biomarkers in modern and/or ancient ecosystems is largely a function of the inherent structure and stability of the molecule, as well as the physicochemical environment of the system wherein it exists. In some cases, redox changes in sediments have allowed for greater preservation of biomarker compounds; in well-defined laminated sediments, for example, a strong case can be made for paleo-reconstruction of past organic matter composition sources. However, many of the labile chemical biomarkers may be lost or transformed within minutes to hours of being released from the cell from processes such as bacterial and/or metazoan grazing, cell lysis, and photochemical breakdown, just to name a few. The role of trophic effects versus large-scale physiochemical gradients in preserving or destroying the integrity of chemical biomarkers varies greatly across different ecosystems. These effects will be discussed as they relate to aquatic systems such as lakes, estuaries, and oceans.

In chapter 3, we discuss the basic principles surrounding the application of stable isotopes in natural ecosystems, which are based on variations in the relative abundance of lighter isotopes from chemical rather than nuclear processes. Due to faster reaction kinetics of the lighter isotope of an element, reaction products in nature can be enriched in the lighter isotope. These fractionation processes can be complex, as discussed below, but have proven to be useful in determining geothermometry and paleoclimatology, as well as sources of organic matter in ecological studies. The most common stable isotopes used in oceanic and estuarine studies are ^{18}O, ^{2}H, ^{13}C, ^{15}N, and ^{34}S. The preference for using such isotopes is related to their low atomic mass, significant mass differences in isotopes, covalent character in bonding, multiple oxidations states, and sufficient abundance of the rare isotope.

Living plants and animals in the biosphere contain a constant level of ^{14}C, but when they die there is no further exchange with the atmosphere and the activity of ^{14}C decreases with a half-life of 5730 ± 40 yr; this provides the basis for establishing the age of archeological objects and fossil remains. The assumptions associated with dating materials are (1) that the initial activity of ^{14}C in plants and animals is a known constant and is independent of geographic location, and (2) that the sample has not been contaminated with modern ^{14}C. The application of ^{14}C measurements in organic carbon cycling studies has been extensive in oceanic environments. While only a few studies using such techniques in organic carbon cycling were applied early on in coastal and estuarine regions, there has been a considerable increase in recent years, with a general paradigm emerging in river/estuarine systems whereby dissolved organic carbon (DOC) is typically more ^{14}C-enriched (or younger) than particulate organic carbon (POC).

Recently developed methods, such as automated preparative capillary gas chromatography (PCGC), now allow for separation of target compounds for ^{14}C analysis based on acceleration mass spectrometry (AMS). In very general terms, compound-specific isotope analysis (CSIA) allows for accurate determination of ages of compounds that are specific to a particular source (e.g., phytoplankton) within the heterogeneous matrix of other carbon compounds commonly found in sediments (e.g., terrigenous sources). Similarly, the application of CSIA to stable isotopic work has also proven to be useful in distinguishing between different types of organic carbon sources in estuarine systems. Future work in this area along with the application of multiple isotopic mixing models, which in some cases include δD, in aquatic systems is discussed in this chapter. In recent years, application of CSIA has contributed new insights to the field of chemical biomarkers. Measurements of $\delta^{13}C$, $\delta^{15}N$, and δD stable isotope composition of lipid and amino acid biomarkers are covered within individual chapters about specific compound classes.

In chapter 4, we provide a background on the important role technology has played in the study of chemical biomarkers and the many advances in the field that have resulted from the development of new analytical tools. The reader is introduced to some of the classic analytical tools used in organic geochemistry, including gas chromatography–mass spectrometry (GC-MS), pyrolysis GC-MS, direct temperature-resolved MS (DT-MS), CSIA, high-performance liquid chromatography (HPLC), and nuclear magnetic resonance (NMR) spectroscopy. Additionally, characterization of dissolved organic matter (DOM) and chromophoric DOM (CDOM) by fluorescence, use of pulsed amperometric detector (PAD) detectors in the analysis of sugars, and capillary electrophoresis are introduced. Recent advances in the following areas are covered: (1) analysis of polar organic compounds utilizing liquid chromatography mass spectrometry (LC-MS), (2) multidimensional NMR, and (3) Fourier transform ion cyclotron resonance MS (FT-ICR-MS).

Future work will include discussion of the limitations of scaling-up from compound-level measurements to bulk carbon pools. Additionally, the challenges of moving from the laboratory benchtop to larger spatial and temporal scales, particularly as advances are made in observing/remote sensing platforms, are discussed.

Chapter 5 covers *carbohydrates*, the most abundant class of biopolymers on Earth and significant components of water column particulate organic matter (POM) and DOM in aquatic environments. Carbohydrates are important structural and storage molecules and are critical in the metabolism of terrestrial and aquatic organisms. Carbohydrates can be further divided into *monosaccharides* (simple sugars), *disaccharides* (two covalently linked monosaccharides), *oligosaccharides* (a few covalently linked monosaccharides), and *polysaccharides* (polymers made up of several mono- and disaccharide units). In phytoplankton, carbohydrates serve as important reservoirs of energy, structural support, and cellular signaling components. Carbohydrates make up approximately 20 to 40% of the cellular biomass in phytoplankton and 75% of the weight of vascular plants. *Minor sugars*, such as *acidic sugars*, *amino sugars*, and *O-methyl sugars*, tend to be more source-specific than major sugars and can potentially provide further information on the biogeochemical cycling of carbohydrates. As discussed in chapter 3, the application of CSIA on monosaccharides may prove useful in distinguishing between different types of organic carbon sources in aquatic systems.

In chapter 6, we discuss *proteins*, which make up approximately 50% of organic matter and contain about 85% of the organic N in marine organisms. Peptides and proteins comprise an important fraction of the POC (13–37%) and particulate organic nitrogen (PON) (30–81%), as well as dissolved organic nitrogen (DON) (5–20%) and DOC (3–4%) in oceanic and coastal waters. In sediments, proteins account for approximately 7 to 25% of organic carbon and an estimated 30 to 90% of total N.

Amino acids are the basic building blocks of proteins. This class of compounds is essential to all organisms and represents one of the most important components in the organic nitrogen cycle. Amino acids represent one of the most labile pools of organic carbon and nitrogen. The typical mole percentage of protein amino acids in different organisms shows considerable uniformity in compositional abundance. In aquatic systems, amino acids have typically been analyzed in characteristic dissolved and particulate fractions. For example, the typical pools of amino acids commonly measured in coastal systems are divided into the *total hydrolyzable amino acids* (THAA) in water column POM and sedimentary organic matter (SOM), and *dissolved free* and *combined amino acids* (DFAA and DCAA) in DOM.

Another process that can account for the occurrence of D-amino acids in organisms is *racemization*, which involves the conversion of L-amino acids to their mirror-image D form. Estuarine invertebrates have been found to have D-amino acids in their tissues as a result of this process. Rates of racemization of amino acids in shell materials, calibrated against radiocarbon measurements, have also been used as an index for historical reconstruction of coastal erosion. Future work on amino acids and proteins as biomarkers is discussed in the context of C and N isotopic analyses of D- and L-amino acids, as well as advances in the area of proteomics.

Chapter 7 examines *nucleic acids*, polymers of the nucleotides ribonucleic acid (RNA) and deoxyribonucleic acid (DNA), which act as the templates for protein synthesis. High levels of nucleic acids in microbes contribute to their elevated nitrogen and phosphorus contents. In recent years, isotopic signatures of nucleic acids have provided new insights about the sources of carbon-supporting bacterial activity.

This chapter discusses recent efforts to bridge the fields of organic geochemistry and molecular ecology. The coupling of biomarker information to molecular (genetic) data has the potential to provide new insights about specific sediment microbial communities and their effects on sediment organic matter. Recent studies have provided information about the evolutionary basis of biosynthetic pathways, which influence the capabilities of microorganisms to utilize specific substrates and the synthesis of unique biomarkers. Such efforts demonstrate the impact that microorganisms have on organic geochemistry.

Future directions will discuss attempts to characterize isotopic signatures of DNA. Progress is also being made in the application of proteomics to more fully characterize organic matter. While many of the major advances in biochemistry have been in protein chemistry, limitations remain in applying these methods to aquatic systems.

Chapter 8 covers *fatty acids*, the building blocks of lipids, which represent a significant fraction of the total lipid pool in aquatic organisms. We explore how chain length and levels of unsaturation (number of double bonds) have been shown to be correlated to decomposition, indicating a pre- and postdepositional selective loss of short-chain and polyunsaturated fatty acids. In contrast, saturated fatty acids are more stable and typically increase in relative proportion to total fatty acids with increasing sediment depth. *Polyunsaturated fatty acids* (PUFAs) are predominantly used as proxies for the presence of "fresh" algal sources, although some PUFAs also occur in vascular plants and deep-sea bacteria. Thus, these biomarkers represent a very diverse group of compounds present in aquatic systems.

This chapter discusses the numerous applications of fatty acid biomarkers to identifying the sources of organic matter in lakes, rivers, estuaries, and marine ecosystems. Fatty acid biomarkers have been applied to ultrafiltered dissolved organic matter (UDOM), POM, and sediments. Additionally, fatty acids have been used in ecological studies examining the dietary requirements of aquatic organisms and trophic relationships (e.g., work by Tenore and Marsh on effects of fatty acid composition on growth/reproduction of *Capitella* sp.) (Tenore and Chesney, 1985; Marsh and Tenore, 1990). Fatty acids have also been widely used in chemotaxonomic studies aimed at understanding microbial community structure. Recent studies have documented the influence of essential fatty acids in the transfer of algal organic matter to higher trophic levels (e.g., Müller-Navarra and colleagues) and trophic upgrading of food quality by protozoans and heterotrophic protists (e.g., work by Klein Breteler et al. and Veloza et al.).

In chapter 9 we discuss several classes of cyclic *isoprenoids* and their respective derivatives, which have proven to be important biomarkers that can be used to estimate algal and vascular plant contributions as well as diagenetic proxies. Sterols are a group of cyclic alcohols (typically between C_{26} and C_{30}) that vary structurally in the number, stereochemistry, and position of double bonds as well as methyl- and ethyl-group substitutions on the side chain. Sterols are biosynthesized from isoprene units using the mevalonate pathway and are classified as triterpenes (i.e., consisting of six isoprene units). Marine organisms such as phytoplankton and zooplankton tend to be dominated by C_{27} and C_{28} sterols. Vascular plants have been shown to have a relatively high abundance of C_{29} sterols, such as 24-ethylcholest-5-en-3β-ol (*sitosterol*) ($C_{29}\Delta^{5}$) and 24-ethylcholest-5,22E-dien-3β-ol (stigmasterol, $C_{29}\Delta^{5,22}$), as well as the C_{28} sterol 24-methylcholest-5-en-3β-ol (*campesterol*) ($C_{28}\Delta^{5}$). Finally, we show that although C_{29} sterols are considered to be the dominant sterols found in vascular plants, these compounds may be present in certain epibenthic cyanobacteria and phytoplanktonic species, indicating that these compounds represent a very diverse and powerful suite of chemical biomarkers in aquatic ecosystems.

Di- and *triterpenoids* are useful biomarkers for determining the sources and transformations of natural organic matter in aquatic ecosystems. Like sterols, these terpenoids are formed from isoprene units using the mevalonate pathway. The sources and diagenetic formation of molecules like retene, chrysene, picene, and perylene are discussed. Many of these compounds are natural products derived from plants, while others are formed from diagenesis. *Tetraterpenoids* (e.g., *carotenoids*) are covered in chapter 13.

Recent applications of $\delta^{13}C$ and δD to steroid and triterpenoid biomarker compounds will continue to help resolve complex mixtures of organic matter in natural systems. Compound-specific isotope analysis provides an opportunity to better understand the sources of these compounds in the environment and extend their application to ecological and paleoecological studies investigating changes in environmental and climatic conditions.

In chapter 10 we examine hydrocarbons present in the environment and derived from natural and anthropogenic sources. The abundance of hydrocarbons derived from anthropogenic sources (*petroleum hydrocarbons*) has increased significantly in aquatic systems since the industrial revolution (this addressed in hapter 14). Natural oil seeps and erosion of *bitumen* deposits can also contribute to hydrocarbon abundance and composition in systems. These petroleum hydrocarbons can be distinguished from biological hydrocarbons by their absence of odd-carbon chain lengths commonly found in biological hydrocarbons and the greater structural diversity found in petroleum hydrocarbons.

This chapter focuses on naturally produced hydrocarbons. We provide examples of how *aliphatic* and *isoprenoid hydrocarbons* have been successfully used to distinguish between algal, bacterial, and terrigenous vascular plant sources of carbon in aquatic systems. We discuss how pristine and phytane are formed from phytol under oxic vs. anoxic conditions, respectively. We also introduce the reader to highly branched isoprenoids and their use as algal biomarkers.

Chapter 11 focuses on several classes of polar lipids, including *alkenones*, which are di-, tri-, and tetra-unsaturated long-chain ketones. These compounds are produced by a restricted number of species of prymnesiophyte algae (coccolithophorid alga *Emiliania huxleyi*), living over a wide temperature range (2–29 °C). Prymnesiophytes are able to live under different temperature regimes because they are able to regulate the degree of unsaturation of these compounds; as ambient water temperature decreases, unsaturation increases. Long-chain ketones are more stable than most unsaturated lipids and can survive diagenesis. Because of these properties, alkenones have been used widely as paleothermometers.

Paleoclimate studies of continental environments have been hampered by the lack of a useful temperature proxy. *Glycerol dialkyl glycerol tetraethers* (GDGTs) occur ubiquitously, including sites where alkenones are not produced due to the absence/low abundance of alkenone-producing algae. The TEX_{86} index, based on the number of cyclopentane rings in the GDGTs, provides a useful paleotemperature index for lakes and other sites where alkenones are not produced.

The analysis of intact polar molecules is becoming more widespread with the advent of LC-MS techniques. In some cases, these compounds are proving to be more useful for identifying microbial communities associated with soils and sediments (e.g., *intact polar lipids*), than the more traditional approaches that involve hydrolysis of polar compounds and analysis of their subcomponents (e.g., phospholipid linked fatty acids [PLFA]). This compound class shows great promise for identifying microbes associated with soils and marsh sediments and the delivery of soil-derived organic matter to coastal regions (e.g., BIT index).

In chapter 12 we examine the primary *photosynthetic pigments* used in absorbing photosynthetically active radiation (PAR), which incluide *chlorophylls*, *carotenoids*, and *phycobilins*—with chlorophyll representing the dominant photosynthetic pigment. Although a greater amount of chlorophyll is found on land, 75% of the annual global turnover (ca. 10^9 Mg) occurs in oceans, lakes, and rivers/estuaries. All of the light-harvesting pigments are bound to proteins making up distinct carotenoid and chlorophyll–protein complexes. In this chapter, we examine the chemistry and application of these very important chemical biomarkers and discuss their limitations in aquatic systems. The matrix factorization program CHEMical TAXonomy (CHEMTAX) was introduced to calculate the relative abundance of major algal groups based on concentrations of diagnostic pigments and is also discussed.

The combined method of selective photopigment HPLC separation coupled with in-line flow scintillation counting using the chlorophyll *a* radiolabeling (^{14}C) technique can also provide information on phytoplankton growth rates under different environmental conditions. Fossil pigments have also been shown to be useful paleotracers of algal and bacterial communities.

As discussed in chapter 3, the application of CSIA to plant pigments may prove useful in distinguishing between different types of organic carbon sources in aquatic systems. We also address the application of satellite imagery in determining chlorophyll concentrations in natural waters, as well as new ideas for developing in situ HPLC systems for real-time data gathering in coastal observatories.

In chapter 13 we examine lignin, which has proven to be a useful chemical biomarker for tracing vascular-plant inputs to aquatic systems. Cellulose, hemicellulose, and lignin generally make up >75% of the biomass of woody plant materials. Lignins are a group of macromolecular heteropolymers (600–1000 kDa) found in the cell wall of vascular plants that are made up of phenylpropanoid units. The *shikimic acid pathway*, which is common in plants, bacteria, and fungi, is the pathway for synthesis of aromatic amino acids (e.g., tryptophan, phenylalanine, and tyrosine), thereby providing the parent compounds for the synthesis of the phenylpropanoid units in lignins. Specifically, the primary building blocks for lignins are the following monolignols: *p*-coumaryl alcohol; coniferyl alcohol; and sinapyl alcohol. Oxidation of lignin using the CuO oxidation method yields eleven dominant phenolic monomers, which can be separated into four families: *p*-hydroxyl, vanillyl (V), syringyl (S), and cinnamyl (C) phenols.

Cutins and suberins, which are lipid polymers in vascular plant tissues and serve as a protective layer (cuticle) and as cell wall components of cork cells, respectively, are also examined in this chapter. This chapter describes how cutins have been shown to be an effective biomarker for vascular plants in aquatic systems. When cutin is oxidized, using the CuO method commonly used for lignin analyses, a series of fatty acids are produced that can be divided into three groups: C_{16} hydroxy acids, C_{18} hydroxy acids, and C_n hydroxyl acids. Since cutins are not found in wood, they are similar to the *p*-hydroxyl lignin-derived cinnamyl phenols (e.g., *trans-p*-coumaric and ferulic acids) and serve as biomarkers of nonwoody vascular plant tissues.

We further explain how different analytical methods for lignin detection have been developed for use in aquatic systems. There are new microwave extraction techniques as well as applications of CSIA of lignin monomers using stable isotopes and radiocarbon. Finally, new techniques in multidimensional

NMR have allowed for better elucidation of aromatic structures potentially derived from lignins in natural waters.

Chapter 14 examines the application of anthropogenic compounds as biomarkers. Since World War II, human activities have introduced a wide array of compounds into the environment, including insecticides such as dichloro-diphenyl-trichloroethane (DDT) and pesticides, halocarbons (chlorofluorocarbons), sewage products (coprostanol), and polycyclic aromatic hydrocarbons (PAHs). This chapter introduces structural features of these compounds, their distribution and transformation in the environment, and their potential use(s) as tracers. A focus of this chapter is to present examples of how relationships between anthropogenic markers and biomarkers can be used to provide information about the sources, delivery, and fate of natural organic matter in aquatic ecosystems.

This chapter introduces various *emerging contaminants* (personal care pharmaceutical products (PCPPs), caffeine, and flame retardants) and their potential use as tracers for anthropogenic organic matter in aquatic ecosystems. We describe how $\delta^{13}C$, stable isotopes of Cl and Br, and radiocarbon can be used to apportion sources of organic contaminants (e.g., PAHs and PCBs).

Acknowledgments

Over the past two years it has taken to write this book, many people have helped along the way, and we are eternally grateful for their input. We thank our friends and colleagues for sharing new and unpublished work with us as well as their support during the writing and editing phases. Their positive comments and encouragement were a great help. In particular, E. Canuel would like to acknowledge the importance of the Gordon Research Conference on Organic Geochemistry for providing a stimulating, intellectual climate in which new ideas can be exchanged freely and warm friendships developed and nurtured. We would also like to thank our colleagues, postdocs, students, and family members who reviewed individual chapters of this book, including Jo Ann Bianchi, Emmett Duffy, Amber Hardison, Yuehan Lu, Christie Pondell, and Stephanie Salisbury. We also appreciate the dedicated and heroic efforts of Erin Ferer, Emily Jayne, and Stephanie Salisbury for their contributions to the glossary.

Throughout our careers, our curiosity and research interests in organic geochemistry have evolved through interactions with our students and colleagues. E. Canuel would like to acknowledge collaborations with past and present students, including Krisa Arzayus, Amber Hardison, Emily Jayne, Elizabeth Lerberg, Leigh Mccallister, John Pohlman, Christie Pondell, Stephanie Salisbury, Sarah Schillawski, Amanda Spivak, Craig Tobias, and Andy Zimmerman. T. Bianchi would like to thank the following past and present students who contributed to development of this book: Nianhong Chen, Shuiwang Duan, Richard Smith, and Kathyrn Schreiner. These interactions have encouraged us to venture into new areas of research and provided a constant stream of ideas and intellectual stimulation. We are indebted to our parents and mentors, who showed us the way and kept us on the path to becoming successful scientists. Sadly, T. Bianchi's father, Thomas Bianchi, passed away in 2008; while he never understood the life of his son as an academic, his blue-collar work ethic and kindness to others are traits that will be remembered for many years to come. E. Canuel thanks Stuart Wakeham for his friendship and intellectual input, and the many opportunities he has provided to her over 20+ years. T. Bianchi thanks many of his close friends for their intellectual discussions over the years, such as Mead Allison, Mark Baskaran, James Bauer, Robert Cook, Michael Dagg, Rodger Dawson, Ragnar Elmgren, Tim Filley, Patrick Hatcher, Franco Marcantonio, Mark Marvin-DiPasquale, Sid Mitra, Brent McKee, Hans Pearl, Rodney Powell, Eric Roden, Peter Santschi, Pichan Sawangwong, and the late Robert Wetzel. T. Bianchi acknowledges his friendship and collaborations with John Morse who passed away in 2009 after a long and distinguished career—he will be missed by his family and his many geochemical colleagues around the world. Finally, we also thank our muse, John Hedges. John's insights, friendship, and generosity continue to inspire us and serve as a role model for our personal and professional interactions with students and colleagues.

We thank Ingrid Gnerlich and the team at Princeton University Press for their patience and dedication to this project. We are also grateful to the two anonymous reviewers who provided many useful comments that contributed to the improvement of the book. We also are grateful for continued support from the National Science Foundation.

1. Metabolic Synthesis

1.1 Background

In this chapter, we begin by providing a brief background on the classification of organisms. We then provide a general background on the synthesis of chemical biomarkers and their association with key metabolic pathways in organisms, as they relate to differences in cellular structure and function across the three systematic domains of life. We also discuss photosynthesis, the dominant pathway by which biomass is synthesized, and provide information about chemoautotrophic and microbial heterotrophic processes. This holistic view of biosynthetic pathways of chemical biomarkers provides a roadmap for other chapters in this book, where more specific details on chemical pathways are presented for each of the respective classes of biomarkers. While other excellent books in the area of organic geochemistry have effectively introduced the concepts of chemical biomarkers in the context of physical and chemical gradients found in natural ecosystems (e.g., anaerobic, aerobic) (Killops and Killops, 2005; Peters et al., 2005), we begin by first examining biosynthetic pathways at the cellular level of differentiation. We believe that an understanding of the general biosynthetic pathways of these chemical biomarkers is critical when examining the complexity of rate-controlling processes that determine their production and fate in aquatic systems. We also provide the major features of different cell membrane structures, since these membranes play an important role in the transport of simple molecules in and out of the cell and influence the preservation of chemical biomarkers in sediments.

1.2 Classification of Organisms

The taxonomic classification of all living organisms is shown in fig. 1.1. The former five-Kingdom classification system consisted of the following phyla: Animalia, Plantae, Fungi, Protista, and Bacteria. It was replaced by a three-domain system—primarily derived from the *phylogenetic* analysis of base sequences of nucleic acids from rRNA (Woese et al., 1990) (fig. 1.1). These domains can be further divided into *heterotrophs* (e.g., animals and fungi) and *autotrophs* (e.g., vascular plants and algae),

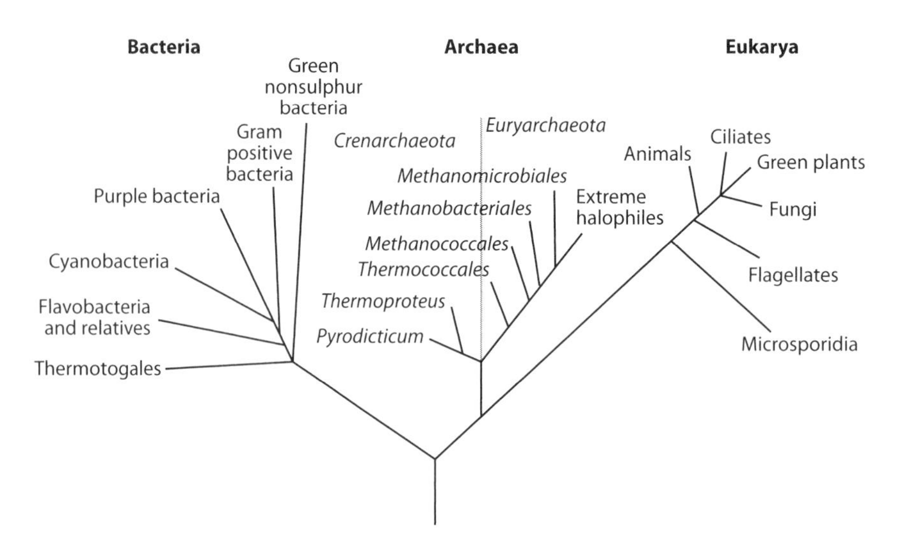

Figure 1.1. The three-domain system derived from the phylogenetic analysis of base sequences of nucleic acids from rRNA. Phylogenetically the Archaea fall into two distinct groups as shown by the gray line: the methanogens (and their relatives) and the thermophiles. (Adapted from Woese et al.,1990.)

which are either *prokaryotes* (unicellular organisms that do not possess a nuclear membranes [e.g., just a nucleoid, or DNA in the form of chromosomes] or *eukaryotes* (unicellular and multicellular organisms with nuclear membranes and DNA in the form of chromosomes) (fig. 1.2). The bacteria (eubacteria) and archaea (archaebacteria), both prokaryotes, represent important microbial groups and are involved in many of the biogeochemical cycling processes of aquatic ecosystems.

Photosynthesis is the dominant process by which organic matter is synthesized. During photosynthesis, organisms synthesize simple carbohydrates (carbon fixation) using light energy, water and an electron donor. The earliest photosynthetic organisms are thought to have been *anoxygenic*, likely using hydrogen, sulfur, or organic compounds as electron donors. Fossils of these organisms date to 3.4 billion years (3.4×10^9 years or 3.4 Gyr) before present (BP). Oxygenic photosynthesis evolved later; the first oxygenic photosynthetic organisms were likely the cyanobacteria, which became important around 2.1 Gyr BP (Brocks et al., 1999). As oxygen accumulated in the atmosphere through the photosynthetic activity of cyanobacteria (Schopf and Packer, 1987), life on Earth needed to quickly adapt. In fact, it is believed that a transfer of bacterial genes was responsible for the development of the first eukaryotic cell (*Gypania spiralis*), estimated, based on the fossil record, to be 2.1 Gyr old (Han and Runnegar, 1992). When a cell consumed aerobic (oxygen-using) bacteria, it was able to survive in the newly oxygenated world. Today, the aerobic bacteria have evolved to include mitochondria, which help the cell convert food into energy. Hence, the appearance of mitochondria and chloroplasts are believed to have evolved in eukaryotes as *endosymbionts* (Thorington and Margulis, 1981). The current theory of endosymbionts is based on the concept that an earlier combination of prokaryotes, once thought to be living together as symbionts, resulted in the origin of eukaryotes (Schenk et al., 1997; Peters et al., 2005). This theory suggests that some of the organelles in eukaryotes (e.g., mitochondria, kinetosomes, hydrogenosomes, and plastids) may have begun as symbionts. For example, it is thought that mitochondria and chloroplasts may have evolved from aerobic nonphotosynthetic bacteria and photosynthetic cyanobacteria, respectively. Endosymbiotic theory also explains the presence of eubacterial genes within eukaryotic organisms (Palenik, 2002). The important energy-transforming

Eukaryote

Nucleolus
Mitochondria
Nucleus
Cell membrane
DNA
Cytoplasm
Ribosomes
Endoplasmic reticulum
10–100 μm

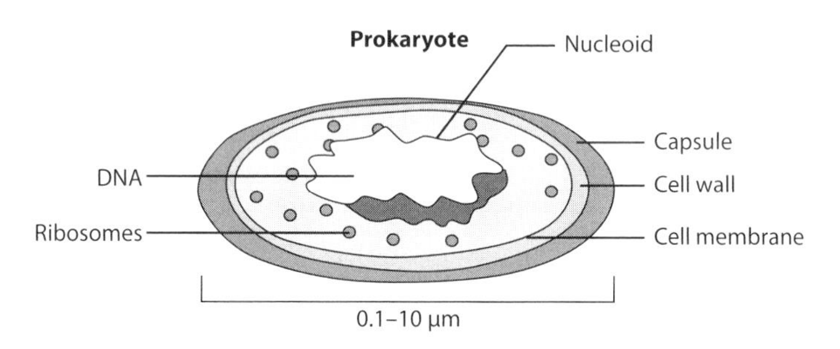

Figure 1.2. The basic cellular design of Eukaryotes, unicellular and multicellular organisms with nuclear membranes and DNA in the form of chromosomes, and Prokaryotes, unicellular organisms that do not possess a nuclear membrane (e.g., just a nucleoid, or DNA in the form of chromosomes).

processes that directly or indirectly occur in association with these organelles, such as photosynthesis, *glycolysis*, the *Calvin cycle*, and the *citric acid cycle*, are discussed in more detail later in this chapter.

While the origins of life on early Earth remain controversial, experimental evidence for the possible evolution of early life began back with the work of Miller and Urey (Miller, 1953; Miller and Urey, 1959). The basic premise of this work, which has remained central in many current studies, is that simple organic compounds, including amino acids, formed after a spark was applied to a flask containing constituents thought to be present in the Earth's early atmosphere (e.g., methane, ammonia, carbon dioxide, and water). It is thought that these simple organic compounds provided the "seeds" for the synthesis of more complex prebiotic organic compounds (fig. 1.3). In addition to the in situ formation of these simple molecules on Earth (based on the aforementioned lab experiments in the 1950s), extraterrestrial sources, such as comets, meteorites, and interstellar particles, have also been posited as possible sources for seeding early Earth with these simple molecules. There is now strong evidence for the presence of these prebiotic compounds in these extraterrestrial sources (Engel and Macko, 1986; Galimov, 2006; and references therein). One particular theory focuses on the notion that the synthesis of adenosine triphosphate (ATP) was most critical in the early stages of prebiotic evolution (Galimov, 2001, 2004). The hydrolysis of ATP to adenosine diphosphate (ADP) is critical in the ordering and assembly of more complex molecules. For example, the formation of peptides from amino acids, and nucleic acids from nucleotides, is inherently linked to the ATP molecule (fig. 1.3). Another important step in prebiotic chemical evolution of life was the development of a molecule that allowed for genetic coding in primitive organisms, and transfer ribonucleic acid (tRNA). Many scientists now suspect that all life diverged from a common ancestor relatively soon after life began (figs. 1.1 and 1.3). Based on DNA sequencing, it is believed that millions of years after the evolution

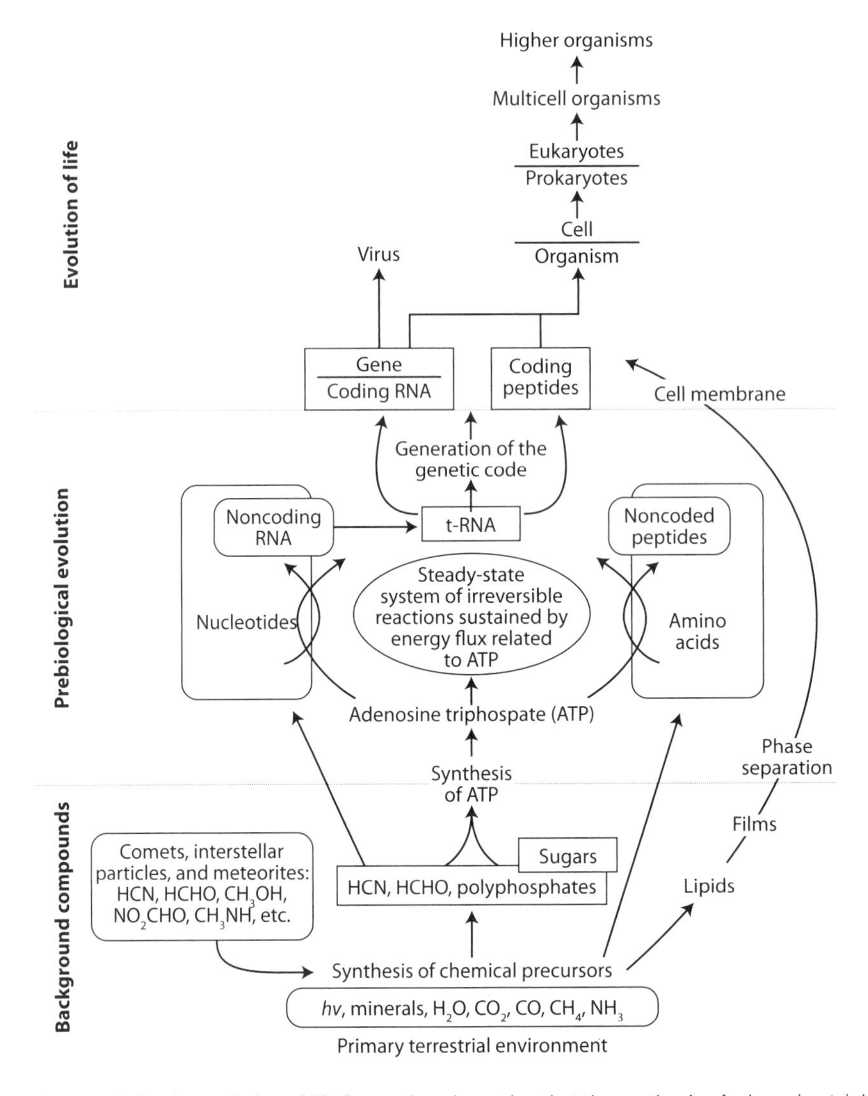

Figure 1.3. A scenario for the evolution of life focused on the notion that the synthesis of adenosine triphosphate (ATP) was most critical in the early stages of prebiotic evolution. (Adapted from Galimov, 2001, 2004.)

of archaebacteria and eubacteria, the ancestors of eukaryotes split off from the archaebacteria. The Earth is at least 4.6 Gyr old (Sogin, 2000), but it was the microbial organisms that dominated for the first 70 to 90 percent of Earth's history (Woese, 1981; Woese et al., 1990). While there has been considerable debate about the composition of the atmosphere of early Earth, it is now generally accepted that it was a reducing environment (Sagan and Chyba, 1997; Galimov, 2005). Kump et al. (2010) state the atmosphere is now thought to have been composed primarily of N_2 and CO_2. Banded iron formations (BIF) first appear in sediments deposited 3 Gyr BP, during the early history of the Earth. These formations include layers of iron oxides, either magnetite or hematite, alternating with iron-poor layers of shale and chert. It is thought that these iron oxide formations were formed in seawater from the reaction between oxygen produced during photosynthesis by cyanobacteria with dissolved reduced iron. The subsequent disappearance of banded iron formations in the geologic record approximately 1.8 Gyr BP is believed to have resulted following a phase of rising oxygen levels in the

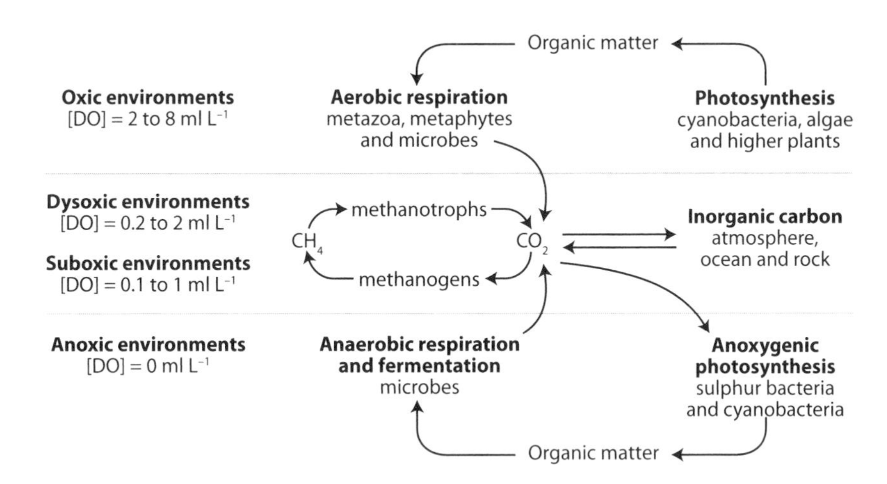

Figure 1.4. Schematic depicting the redox conditions during the mid-Proterozoic (ca. 1.8 to 0.8 Gyr ago) in the oceans, which are believed to have been considerably more anaerobic (no oxygen) and dysaerobic (low oxygen) than the oceans of today. (Adapted from Arnold et al., 2004.)

atmosphere that began about 2.4 to 2.3 Gyr BP (Farquhar et al., 2000; Bekker et al., 2004). Other recent work has suggested that even during the mid-Proterozoic (ca. 1.8 to 0.8 Gyr ago) the oceans were either *anaerobic* (no oxygen) or *dysaerobic* (low oxygen) compared to the oceans of today (Arnold et al., 2004) (fig. 1.4). The increase in oxygen in the oceans from this time period to the present was due to increased production by photoautotrophic organisms, which transfer atmospheric CO_2 and/or dissolved inorganic C (e.g., HCO_3^-) into biomass using photosynthesis; many of these early photoautotrophs were likely cyanobacteria (Brocks et al., 1999).

Archaebacteria were originally thought to live only in extreme environments (e.g., high temperatures, pH extremes, and high radiation levels), but have recently been found in a variety of habitats (Delong and Pace, 2001; Giovannoni and Stingl, 2005; Delong, 2006; Ingalls et al., 2006). Examples of archaebacteria include anaerobic methanogens and halophilic bacteria, and consist of organisms living both in cold (e.g., Antarctica) and hot (e.g., springs of Yellowstone Park [USA]) environments. The fact that these organisms can be found in extreme environments is consistent with the ancient heritage of this domain. Early Earth was likely a very hot environment, with many active volcanoes and an atmosphere composed mostly of nitrogen, methane, ammonia, carbon dioxide, and water—with little to no oxygen present. It is believed that the archaebacteria, and in some cases bacteria, evolved under these conditions, allowing them to live in harsh conditions today. For example, *thermophiles* live at high temperatures, of which the present record is 121°C (Kashefi and Lovley, 2003). In contrast, no known eukaryote can survive over 60°C. Another example is the *psychrophiles*, which live in extremely cold temperatures—there is one species in the Antarctic that grows best at 4°C. As a group, these hard-living archaebacteria are called *extremophiles*. The archaebacteria include other kinds of extremophiles, such as *acidophiles*, which live at pH levels as low as 1. *Alkaliphiles* thrive at high pH levels, while *halophiles* live in very salty environments. It should be noted that there are also alkaliphilic, acidophilic, and halophilic eukaryotes, and that not all archaebacteria are extremophiles. It has also been suggested that the thermophilic archaebacteria that live around deep-sea volcanic vents may represent the earliest life on Earth (Reysenbach et al., 2000). These thermophilic archaea harvest their energy very efficiently from chemicals (e.g., H_2, CO_2, and O_2) found at the vents using a process called *chemosynthesis*. These organisms are not greatly impacted by surface environmental changes. As such, thermophilic organisms living around deep-sea volcanic vents may have been the only organisms able to survive the large, frequent meteor impacts of Earth's early years.

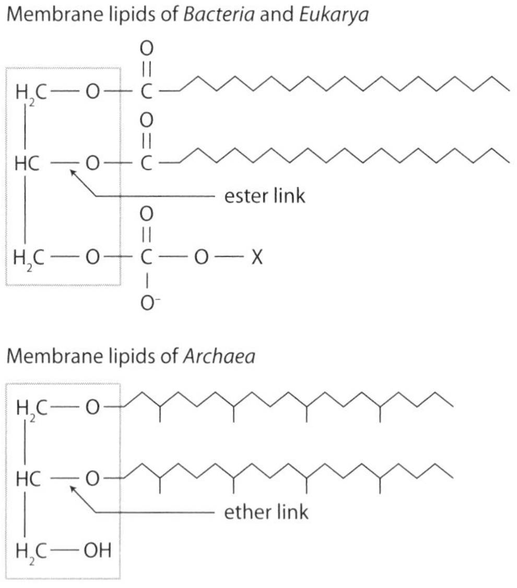

Membrane lipids of *Bacteria* and *Eukarya*

ester link

Membrane lipids of *Archaea*

ether link

Figure 1.5. Differences in the composition of the membrane lipids in bacteria and eukarya. Membrane lipids in bacteria are composed of ester linkages, while membrane lipids in Archaea, have ether linkages. Glycerol (H_2OC–CHO–CH_2O) is identified in the gray boxes.

Archaea are believed to be the most primitive forms of life on Earth, as reflected by their name, which is derived from *archae* meaning "ancient" (Woese, 1981; Kates et al., 1993; Brock et al., 1994; Sogin, 2000). Archaea are divided into two main phyla: the Euryarchaeota and Crenarchaeota. Like bacteria, archaebacteria have rod, spiral, and marble-like shapes. Phylogenetic trees also show relationships between archaea and eukarya and some have argued that the archaebacteria and eukaryotes arose from specialized eubacteria. One of the distinguishing features of the archaea is the composition of the lipids comprising their cell membranes; membrane lipids are associated with an ester linkage in bacteria, while those in archaebacteria are associated with an ether linkage (fig. 1.5). This is described further below and in subsequent chapters (e.g., chapter 8). The most striking chemical difference between archaea and other living cells is their cell membrane. There are four fundamental differences between the archaeal membrane and those of other cells: (1) *chirality* of glycerol, (2) *ether linkage*, (3) *isoprenoid* chains, and (4) branching of side chains. These are discussed in more detail below. In other words, archaebacteria build the same structures as other organisms, but they build them from different chemical components. For example, the cell walls of all bacteria contain *peptidoglycan*, while the cell walls of archaea are composed of surface-layer-proteins. Also, archaebacteria do not produce cell walls of cellulose (as do plants) or chitin (as do fungi); thus, the cell wall of archaebacteria is chemically distinct.

Glycerol is an important building block of lipids. This compound has three carbon atoms, each with a hydroxyl (–OH) group attached (fig. 1.5). When considering the chirality of glycerol, we need to begin with the basic unit from which cell membranes are built in bacteria and eukaryotes—*phospholipids*. Phospholipids are composed of a molecule of glycerol with fatty acids esterified to the C-1 and C-2 positions of the glycerol and a phosphate group attached to C-3 position by an *ester* bond (fig. 1.5). In the cell membrane, the glycerol and phosphate end of the molecules are at the surface of the membrane, with the long chains in the middle (fig. 1.6). This layering provides an effective chemical barrier around the cell, which helps to maintain chemical equilibrium. In archaea, the stereochemistry of the glycerol is the reverse of that found in bacteria and eukaryotes (i.e., the

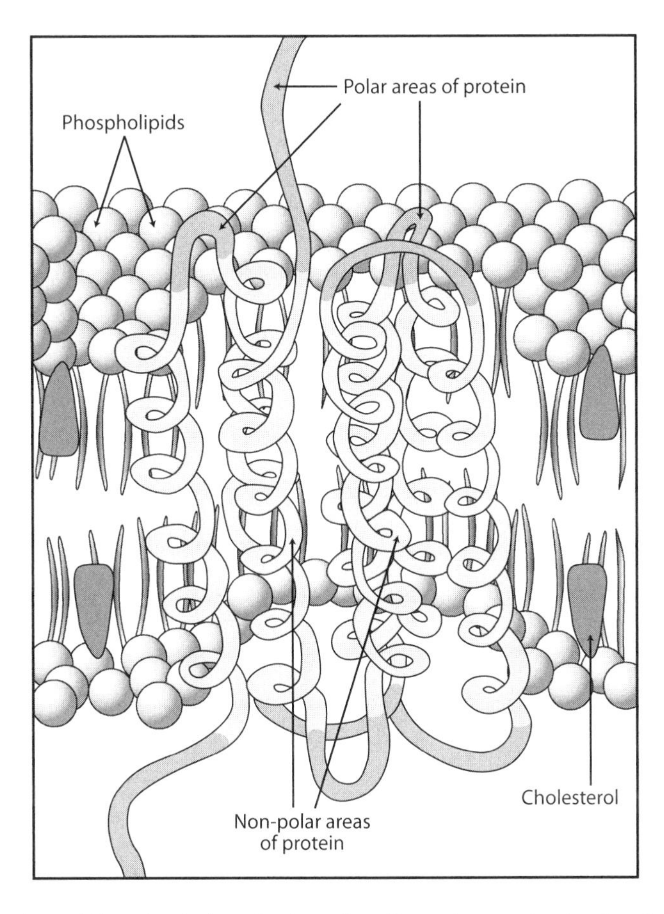

Figure 1.6. The structure of the eubacteria and eukaryotic cell membrane, showing the glycerol and phosphate end of the molecules at the surface of the membrane, with the long chains in the middle.

glycerol is a *stereoisomer* of the glycerol present in phospholipids), suggesting that archaea synthesize glycerol by a different biosynthetic pathway than bacteria and eukaryotes. Stereoisomers are optical isomers of one other, meaning there are two possible forms of the molecule and they are mirror images of each other. While eubacteria and eukaryotes have *dextrorotary* (D) D-glycerol in their membranes, archaeans have *levorotary* (L) L-glycerol. Chemical components of the cell have to be built by *enzymes*, and the "handedness" (chirality) of the molecule is determined by the shape of the enzymes.

In most organisms, side chains added to the glycerol are bonded using an *ester linkage*. Esters are acids where one of the hydroxyl (–OH) groups has been replaced by an O-alkyl group (fig. 1.5). The general formula for esters is: R^1–C(=O)–O–R^2. In the case of phospholipids, R^1 is the carboxyl carbon (C-1) on the fatty acid side chain and R^2 is the C-1 or C-2 position in glycerol. By contrast, membrane lipids in archaea are bound using an ether linkage, in which an oxygen atom is connected to two alkyl groups (general formula is R–O–R′). This causes the chemical properties of membrane lipids in archaea to differ from the membrane lipids of other organisms.

The side chains in the phospholipids of eubacteria and eukaryotes are *fatty acids* of usually 16 to 18 carbon atoms (fig. 1.5) (Killops and Killops, 2005; Peters et al., 2005) (see chapter 8 for more details about fatty acids). Archaebacteria do not use fatty acids to build their membrane phospholipids. Instead, they have side chains of 20 carbon atoms built from *isoprene* (2-methylbuta-1,3-diene or C_5H_8)—a C_5 compound, which forms the building block for a class of compounds called *terpenes*.

By definition, terpenes are a class of compounds constructed by connecting isoprene molecules together $(C_5H_8)_n$, where n is the number of linked isoprene units (see chapters 9 and 12 for more details).

The membrane lipids of archaea also differ from those of eubacteria or eukaryotes, because they have side chains off the main isoprene structure (fig. 1.5) (Kates et al., 1993). This results in some interesting properties in archaeal membranes. For example, isoprene side chains can be joined together, allowing side chains of the membrane lipids to join together, or become joined with side chains of other compounds on the other side of the membrane. No other group of organisms can form such transmembrane lipids. Another interesting property of the side chains is their ability to form carbon rings. This happens when one of the side branches curls around and bonds with another atom down the chain to make a five-carbon ring. These carbon rings are thought to provide structural stability to the membrane, which may allow archaebacteria to be more tolerant of high temperatures. They may work in the same way that cholesterol (another terpene) does in eukaryotic cells to stabilize membranes (fig. 1.6).

Cellular membranes of eukaryotes have diverse functions in the different regions and organelles of a cell (Brock et al., 1994). Membranes are vital because they separate the cell from the outside world. They also separate compartments inside the cell to protect important processes and events. In water, phospholipids are *amphipathic*: the polar head groups are attracted to water (*hydrophilic*) and the tails are *hydrophobic* or oriented away from water, creating the classic lipid bilayer (fig. 1.6). The tail groups orient against one another, forming a membrane with hydrophilic heads on both ends. This allows liposomes or small lipid vesicles to form, which can then transport materials across the cell membrane. Various *micelle*s, such as spherical micelles, also allow for a stable configuration for amphipathic lipids that have a conical shape, such as fatty acids. As mentioned earlier, cholesterol is an important constituent of cell membranes. It has a rigid ring system, has a short, branched hydrocarbon tail, and is largely hydrophobic. However, it has one polar group, a hydroxyl group, making it amphipathic. Cholesterol inserts into lipid bilayer membranes with its hydroxyl group oriented toward the aqueous phase and its hydrophobic ring system adjacent to fatty acid tails of phospholipids. The hydroxyl group of cholesterol forms hydrogen bonds with polar phospholipid head groups. It is thought that cholesterol contributes to membrane fluidity by hindering the packing together of phospholipids.

1.3 Photosynthesis and Respiration

The sum of all biochemical processes in organisms is called *metabolism*, which can further be divided into *catabolic* and *anabolic* pathways. These processes are responsible for the formation of many of the chemical biomarker compounds discussed in this book, which occur through an intermediary metabolism via glycolysis and the citric acid cycle (Voet and Voet, 2004). Kossel (1891) first noted the distinction between primary and secondary metabolism. Primary metabolism, also called basic metabolism, includes all the pathways and products that are essential for the cell itself. Secondary metabolism produces molecules that are not necessarily important for the survival of the cell itself, but are important for the whole organism. To understand biochemical and molecular biological processes such as *metabolism*, *differentiation*, *growth*, and *inheritance*, knowledge of the molecules involved is imperative. Cell-specific molecules are generated via a number of intermediate products from simple precursors, while others are broken down or rearranged. During evolution, pathways have been developed and maintained that produce functional molecules needed by the cell or the organism.

The chemical reactions of oxygenic photosynthesis (primary production) and oxidation (respiration or decomposition) of organic matter are described in equation 1.1:

$$\begin{array}{r} \leftarrow \text{respiration} \\ CO_2 + H_2O + \text{photons(light energy)} \rightarrow CH_2O + O_2 \\ \rightarrow \text{photosynthesis} \qquad\qquad\qquad\qquad\qquad \end{array} \tag{1.1}$$

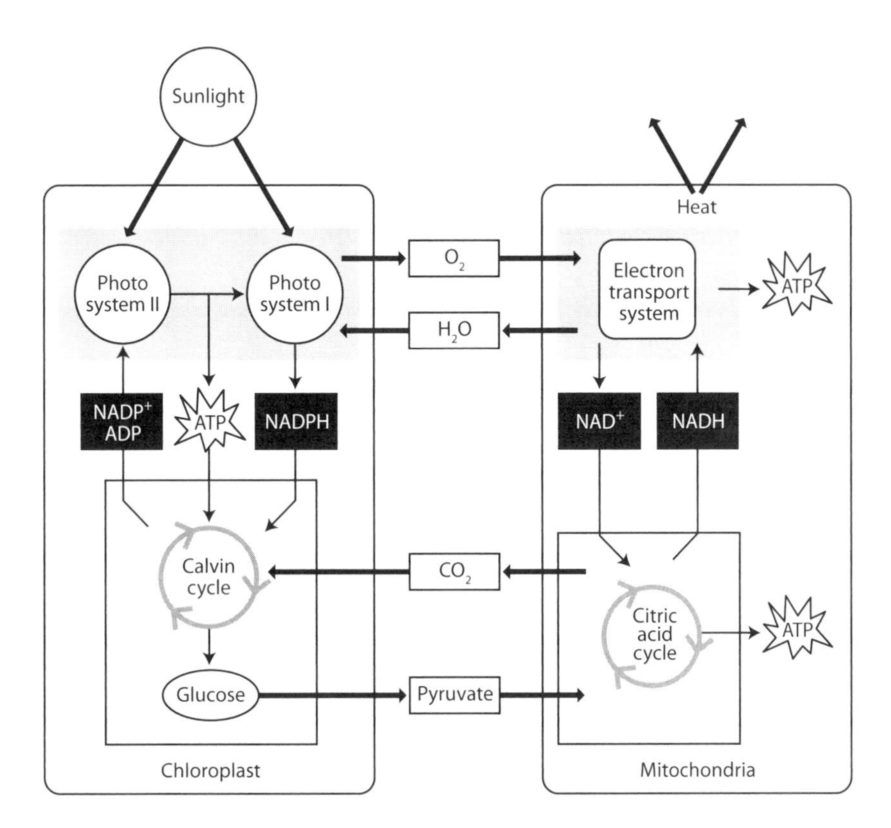

Figure 1.7. The pathways of photosynthesis and respiration, which, in part, produce oxygen and carbon dioxide, respectively, and occur in the chloroplasts and mitochondria.

The end products of electron transport during photosynthesis are NADH or NADPH and ATP, which can be used for carbon or nitrogen fixation and for intermediary metabolism. An important aspect of photosynthesis is the integration of carbon dioxide into organic compounds, called carbon fixation. (Ehleringer and Monson, 1993; Ehleringer et al., 2005). The pathways of photosynthesis and respiration, which, in part, produce oxygen and carbon dioxide, respectively, occur in the chloroplasts and mitochondria (fig. 1.7). If we look more closely at photosynthesis we find that photosystems I and II are located in the chloroplasts (figs. 1.7 and 1.8) (see chapter 12 for more details). The chlorophylls, also found in the chloroplast, serve as the principal antenna pigments for capturing sunlight in the photosystems (described later in this chapter). In higher plants, assimilation tissues are those tissues that are made from chloroplast-containing cells and are able to perform photosynthesis (fig. 1.9). The leaves of higher plants are by far the most important production centers—if you disregard unicellular aquatic algae. Leaves usually consist of the following three tissues: the *mesophyll*, *epidermis*, and *vascular tissues*. The mesophyll is a *parenchyma tissue* that is an important location for the reduction of carbon dioxide, which enters through the *stomata* in the epidermis in the carbon-fixation reactions of the Calvin cycle (Monson, 1989, and references therein). The Calvin cycle is the assimilatory path that is involved in all autotrophic carbon fixation, both photosynthetic and chemosynthetic.

Finally, it important to note that while the previous section was focused on oxygenic photosynthesis, bacteria are also capable of performing anoxygenic photosynthesis. In general, the four types of bacteria that perform this process are the purple, green sulfur, green-sliding, and gram-positive bacteria (Brocks et al., 1999). Most of the bacteria that perform anoxygenic photosynthesis live in environments where oxygen is in low supply as is discussed in later chapters. In anoxygenic photosynthesis, chemical species such as hydrogen sulfide and nitrite substitute for water as the electron donor.

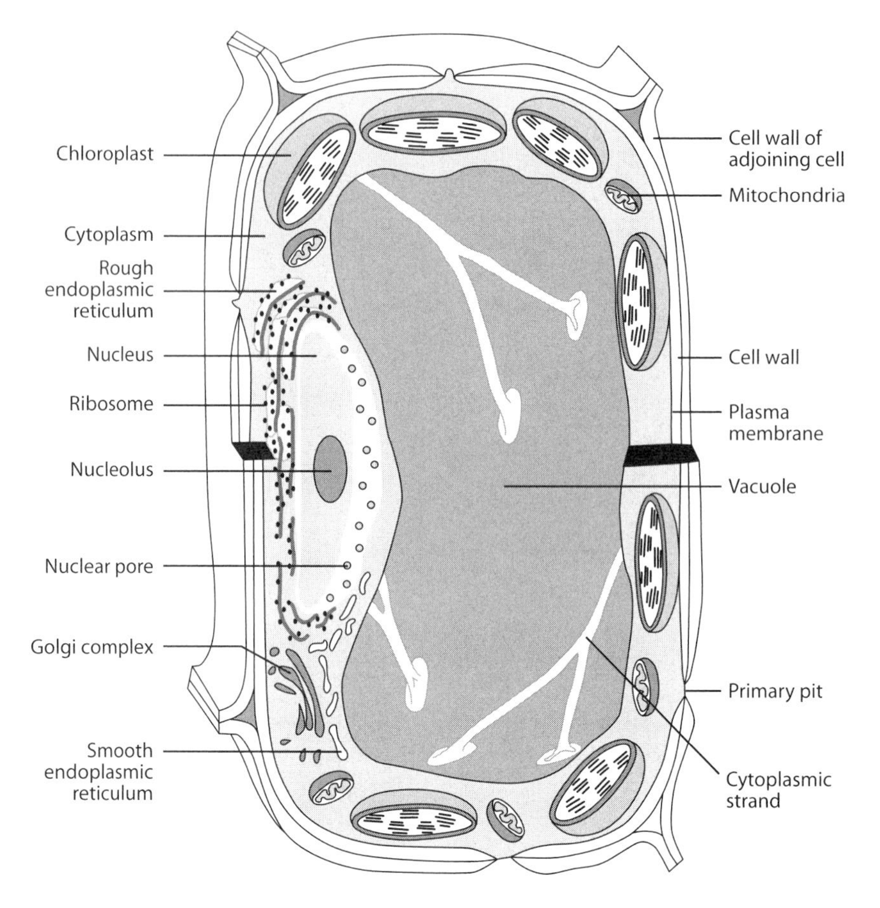

Figure 1.8. A plant cell and the associated organelles illustrating the chlorophylls (found in the chloroplast), which serve as the principal antenna pigments for capturing sunlight in the photosystems.

1.3.1 C_3, C_4, and CAM Pathways

The Calvin cycle (or Calvin-Benson cycle) was discovered by Melvin Calvin and his colleagues in the 1950s (fig. 1.9). It involves a series of biochemical reactions that take place in the chloroplasts of photosynthetic organisms (Bassham et al., 1950). These reactions are referred to as *dark* reactions because they are independent of light. Photosynthetic carbon fixation can occur by three pathways: C_3, C_4, and crassulacean acid metabolism (CAM), with C_3 photosynthesis being the most typical (Monson, 1989). In the next few paragraphs we discuss the biochemistry of the aforementioned three pathways in more detail; this will provide the foundation for understanding the biological fractionation of stable isotopes of carbon in plants using these different metabolic pathways, which is discussed in later chapters.

The C_3 carbon-fixation pathway is a process that converts carbon dioxide and ribulose-1,5-bisphosphate (RuBP, a 5-carbon sugar) into two molecules of the three-carbon compound 3-phosphoglycerate (PGA); hence, the name C_3 pathway (fig. 1.9). The reaction is catalyzed by the enzyme ribulose-1,5-biphosphate carboxylase/oxygenase (RuBisCO), as shown in equation 1.2:

$$6CO_2 + 6RuBP \xrightarrow{RuBisCO} 12\ 3\text{-phosphoglycerate (PGA)} \qquad (1.2)$$

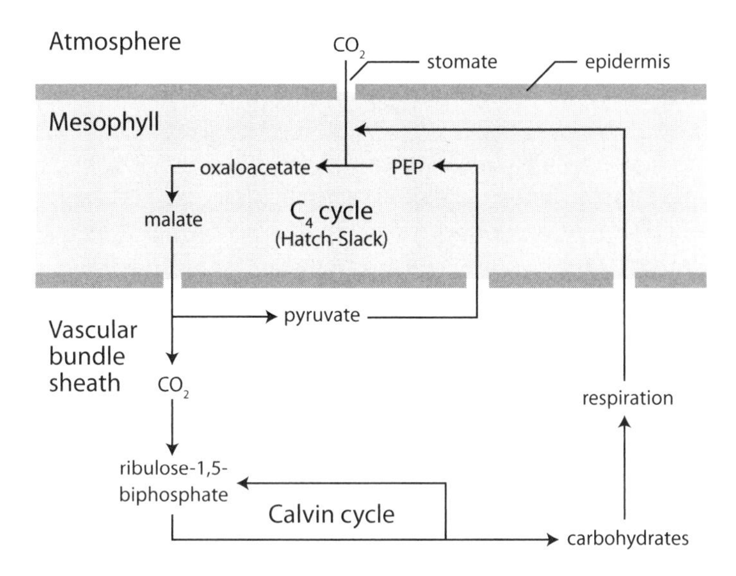

Figure 1.9. Diagram showing the assimilation tissues of higher plants that are made from chloroplast-containing cells and are able to perform photosynthesis in C_3 and C_4 plants. Although shown together here in one diagram, these pathways occur separately in plants.

This reaction occurs in all plants as the first step of the Calvin cycle (Swain, 1966; Ehleringer and Monson, 1993). In C_3 plants, one molecule of CO_2 is converted to PGA, which can subsequently be used to build larger molecules such as glucose. C_3 plants typically live where sunlight and temperatures are moderate, CO_2 concentrations are approximately 200 ppm or higher, and groundwater is abundant. C_3 plants, which are believed to have originated during the Mesozoic and Paleozoic eras, represent approximately 95% of Earth's plant biomass. In contrast, C_4 plants evolved considerably later, during the Cenozoic era, and did not become abundant until the Miocene (23 to 5 million years BP). Finally, it should be noted that the source of inorganic carbon and the enzyme utilized in equation 1.2 are different in algae and land plants. For example, land plants utilize atmospheric CO_2 (as shown in equation 1.2). In contrast, HCO_3^- is the common source of inorganic carbon used for photosynthesis in aquatic algae, where the enzyme *carbonic anhydrase* is used for uptake (Badger and Price, 1994).

In the C_4 pathway, also known as the *Hatch-Slack pathway*, carbon dioxide binds to phosphoenolpyruvic acid (PEP) in *mesophyll cells* to produce a four-carbon compound, oxaloacetate, followed by the production of malate (fig. 1.9) (Monson, 1989), as shown in equation 1.3:

$$\mathrm{PEP} + \mathrm{CO_2} \xrightarrow{\text{PEP carboxylase}} \text{oxaloacetate} \qquad (1.3)$$

In the vascular bundle sheath, Kranz cells split off CO_2 from malate, which is then fed into the Calvin cycle. The pyruvate is transported back into the mesophyll cells (active transport) and is phosphorylated to PEP with the aid of adenosine triphosphate. So, in C_4 plants, CO_2 is drawn from malate into this reaction, rather than directly from the atmosphere as in the C_3 pathway. It is important to note that PEP carboxylase (the enzyme involved in the reaction between PEP and CO_2 that yields oxaloacetate) has a higher affinity for CO_2 than RuBisCO, the enzyme used in C_3 plants. Thus, C_4 plants are dominant at relatively low concentrations of CO_2.

Plants using *crassulacean acid metabolism* (CAM) generally live in arid environments (Ransom and Thomas, 1960; Monson, 1989). The chemical reaction of the CO_2 accumulation is similar to that of C_4 plants; however, CO_2 fixation and its assimilation are separated by time, rather than spatially, as they are in C_4 plants. In C_4 plants, carbon fixation occurs when the stomata are opened for the uptake of CO_2 , but there can be large losses of water when the stomata are open. As a result, CAM plants have developed (or evolved) a mechanism to avoid water loss whereby CO_2 is taken up during the night and stored until photosynthesis occurs during daylight hours. Like C_4 plants, CO_2 reacts with PEP to form oxaloacetate, which is then reduced to form malate (fig. 1.9). Malate (and isocitrate) is then stored in the vacuoles until it is subsequently used during the daytime for photosynthesis.

Not surprisingly, CAM plants are adapted for and occur mainly in arid regions. In fact, both C_4 and CAM photosynthesis have adapted to arid conditions because they result in better *water use efficiency* (WUE) (Bacon, 2004). Moreover, CAM plants can modify their metabolism to save energy and water during harsh environmental periods. Just as CAM plants are adapted for dry conditions, C_4 plants are adapted for special conditions. C_4 plants photosynthesize faster under high heat and light conditions than C_3 plants because they use an extra biochemical pathway and their special anatomy reduces *photorespiration* (Ehleringer and Monson, 1993). From an ecological perspective, C_4 taxa tend to be uncommon in tropical regions (0–20° latitude), where dense tropical forests typically shade out C_4 grasses. C_4 taxa are more common moving away from the tropics and peak in savanna regions—with their abundances diminishing generally between 30 and 40° latitude.

1.3.2 Glycolysis and the Krebs Cycle

Oxidation of glucose is known as glycolysis. During this process, glucose is oxidized to either lactate or pyruvate. Under aerobic conditions, the dominant product in most tissues is pyruvate and the pathway is known as aerobic glycolysis. When oxygen is depleted, as, for instance, during prolonged vigorous activity, the dominant glycolytic product in many higher animal tissues is lactate and the process is known as anaerobic glycolysis. Anaerobic glycolysis occurs in the cytoplasm of the cell. It takes a six-carbon sugar (glucose), splits it into two molecules of the three-carbon sugar (glyceraldehyde), and then rearranges the atoms to produce lactate (fig. 1.10). This rearrangement of atoms provides energy to make ATP, which is the principal energy "currency" in the cell. It is also important to note that there is an overall conservation of atoms in this process. Hence, there are six carbon, six oxygen, and twelve hydrogen atoms at start and finish.

The primary purpose of acetyl co-enzyme A (*acetyl-CoA*) is to convey the carbon atoms within the group to the *Krebs cycle* or *citric acid cycle* (CAC), where they can be oxidized for energy production (fig. 1.11). Chemically, acetyl-CoA is the *thioester* formed when *coenzyme A* (a thiol) reacts with acetic acid (an *acyl* group carrier). Acetyl-CoA is produced during the second step of aerobic cellular respiration through the process of pyruvate *decarboxylation*, which occurs in the matrix of the *mitochondria* (figs. 1.7 and 1.11) (Lehninger et al., 1993). Acetyl-CoA then enters the CAC. In animals, acetyl-CoA is central to the balance between carbohydrate and fat metabolism (see chapters 5 and 8 for more on carbohydrate and fatty acid synthesis). Normally, acetyl-CoA from fatty acid metabolism feeds CAC, contributing to the cell's energy supply.

Organic molecules can be divided into simple molecules (e.g., amino acids, fatty acids, monosaccharides), which are the building blocks of macromolecules. Macromolecules are grouped into four classes of biochemicals: polysaccharides, proteins, lipids, and nucleic acids. As we proceed throughout the book we will continue to explore the basics of organic molecules, which, in part, require knowledge of the structure of *saturated* and *unsaturated aliphatic hydrocarbons*, *alcohols*, *aldehydes*, *carboxylic acids*, and *esters*. Similarly, the structures of *sugars* and their biosynthetic precursors, the *aromatic hydrocarbons* and compounds with heterocyclic rings, in addition to *amino acids* and *lipids* will be discussed in more detail in chapters 5, 6, 8, 9, 10, and 11. Both lipid metabolism and carbohydrate metabolism rely heavily on acetyl-CoA. Acetyl-CoA is also an important molecule in metabolism and is used in many biochemical reactions in primary metabolism (fig. 1.12). Similarly, there are interactions

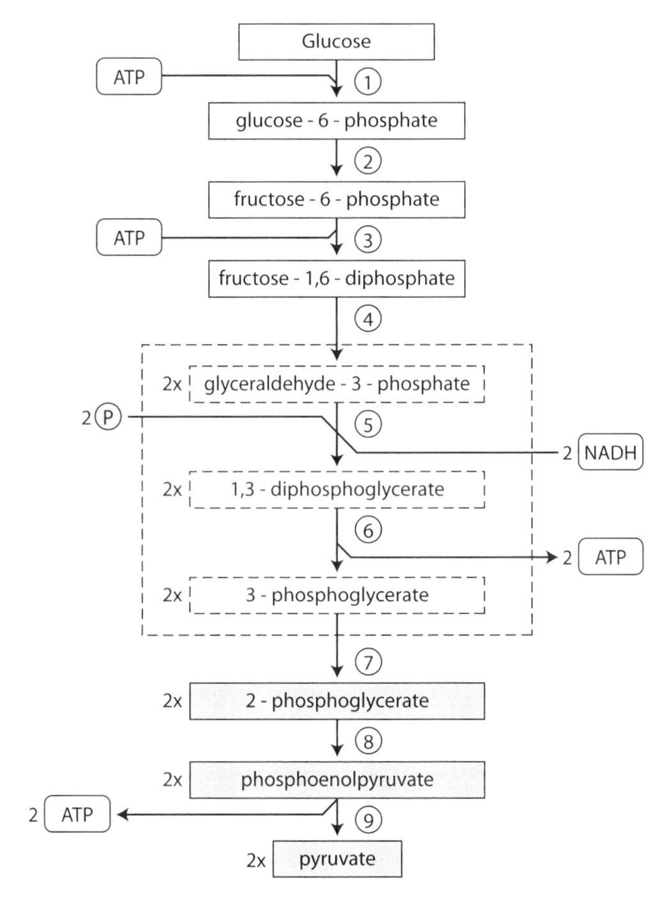

Figure 1.10. Pathway for the formation of pyruvate, via glycolysis, in the cytoplasm of cells.

between amino acids with acetyl-CoA and the citric acid cycle in the context of *protein* metabolism (more details in chapter 6). Fatty acid synthesis begins with acetyl-CoA and continues with the addition of two carbons units (see chapters 8 through 11 for more details). However, this is not the case with carbohydrate metabolism. Instead, carbohydrate synthesis is largely controlled by the availability of $NADP^+$, which acts as an electron acceptor of glucose 6-phosphate to 6-phosphogluconolactone, in the early stages of the 'pentose shunt' (Lehninger et al., 1993).

Thus far, we have been discussing the synthesis of organic compounds during primary metabolism, which comprises all pathways necessary for the survival of the cells. In contrast, secondary metabolism yields products that are usually synthesized only in special, differentiated cells and are not necessary for the cells themselves, but may be useful for the plant as a whole (fig. 1.13) (Atsatt and O'Dowd, 1976; Bell and Charlwood, 1980). Examples include flower pigments and scents or stabilizing elements (Harborne et al., 1975). Many of these compounds are not used as chemical biomarkers in aquatic research but are shown here to simply illustrate their relationship with the biomarkers discussed in this book. Plant cells, in particular, produce a significant amount of secondary products (Bell and Charlwood, 1980). Many of these compounds are highly toxic and are often stored in specific organelles, such as *vacuoles*. These storage functions serve, in some cases, as mechanisms for detoxification as well as reservoirs of important nitrogen-rich molecules. Many of these secondary compounds are found in specific plant organs, often in just one type of cell (and, there again, only in a certain compartment), and are often generated only during a specific developmental period of the

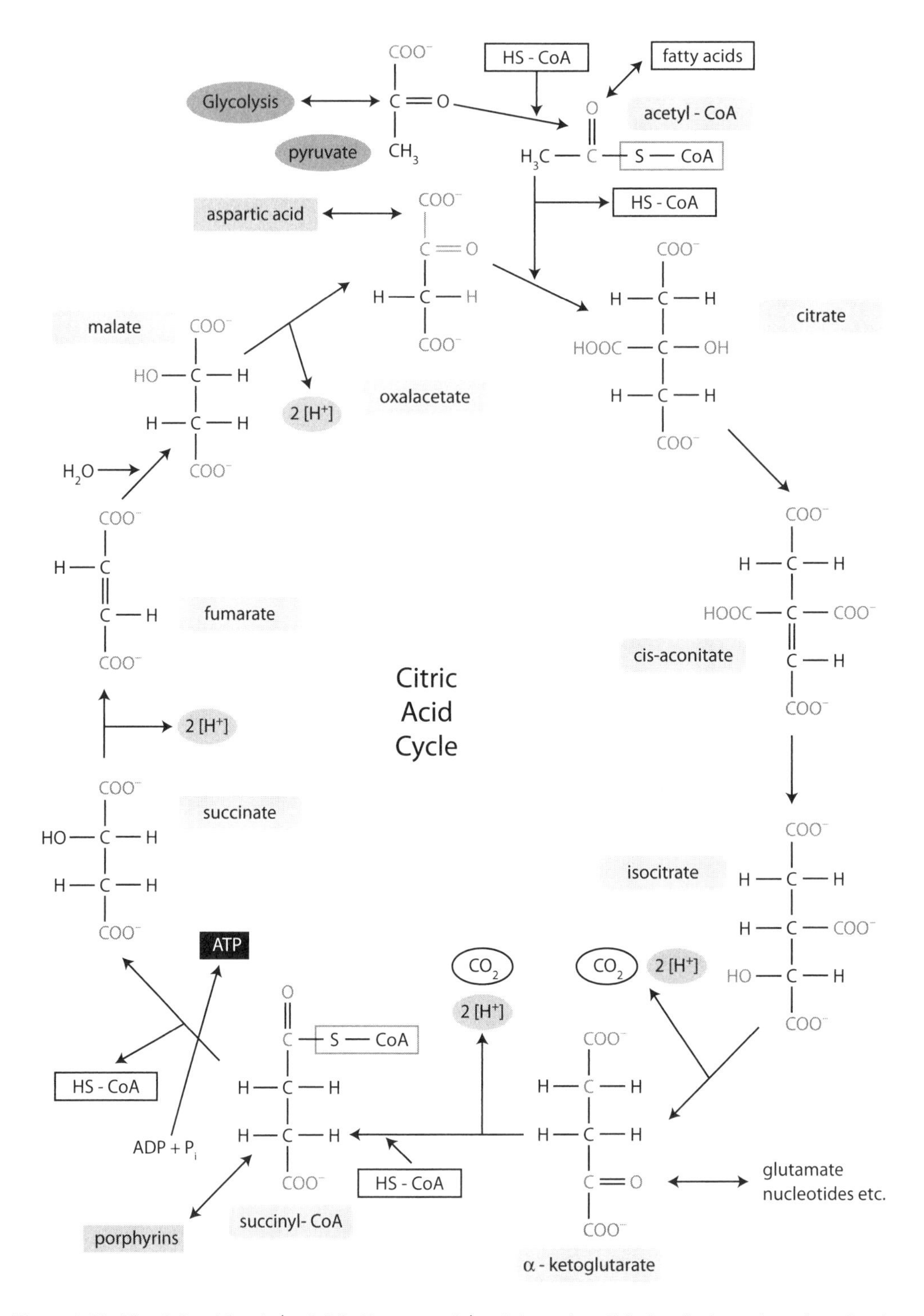

Figure 1.11. The citric acid cycle (or Calvin-Benson cycle) and the series of biochemical reactions that take place in the chloroplasts of photosynthetic organisms.

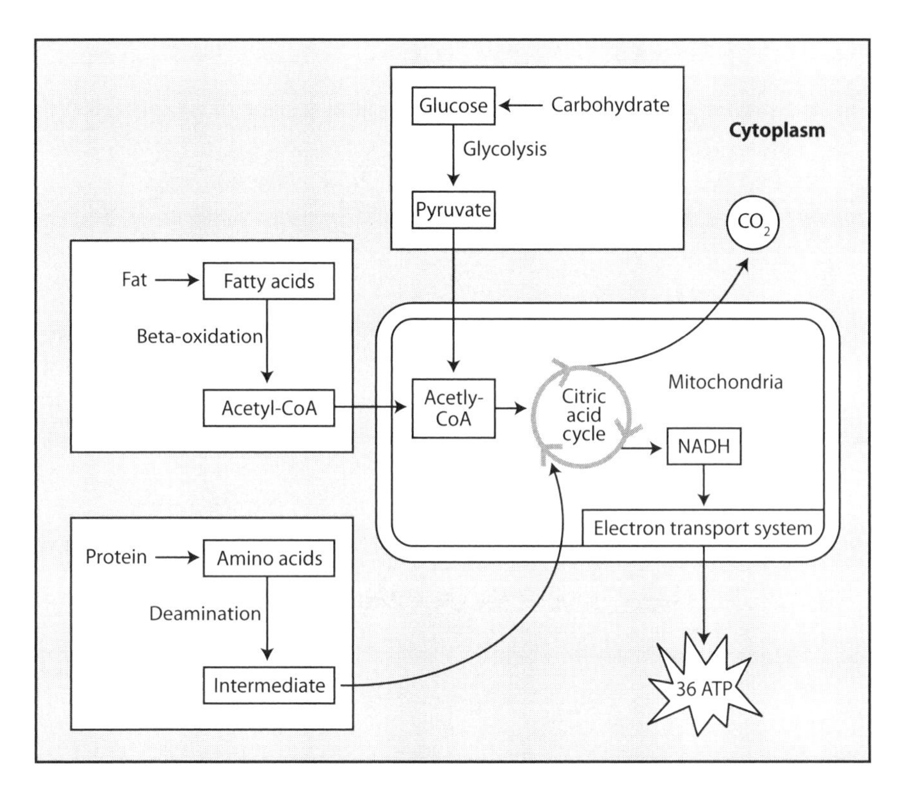

Figure 1.12. Some key pathways where acetyl-CoA represents an important molecule in metabolism of many biochemical reactions in primary metabolism.

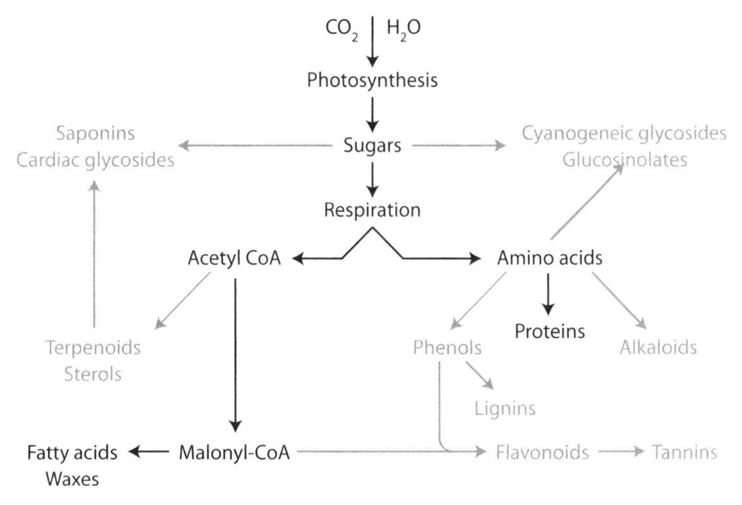

Figure 1.13. Primary metabolism, which comprises all pathways necessary for the survival of the cells, contrasts with secondary metabolism, which yields products that are usually synthesized only in special differentiated plant cells. Secondary processes are shaded.

plant. In some cases, the presence of secondary compounds in plants is clearly adaptive for reproductive purposes, such as flower pigments, or to provide structural integrity, such as lignins and cutins. Some secondary compounds inhibit the existence of competing species within an ecological niche. These molecules are not necessary for the growth and reproduction of a plant, but may serve a role in herbivore deterrence or the production of *phytoalexins*, which combat infective bacteria and fungi (Atsatt and O'Dowd, 1976; Bowers et al., 1986). In general, secondary compounds are often involved in key interactions between plants and their abiotic and biotic environments (Facchini et al., 2000).

1.3.3 Secondary Metabolism

In this section we begin to examine some of the basic pathways and products of secondary metabolism produced by plant cells (each of these groups of compounds is discussed in more detail in the later chapters). The pathways of secondary metabolism are inherently linked with the key processes (e.g., photosynthesis and respiration) and products of primary metabolism (e.g., sugars, fatty acids, and amino acids) (fig. 1.13). The major classes of secondary compounds in plants are the *saponins, cardiac* and *cyanogenic glycosides, terpenoids, sterols, phenols, phenylpropanoids, alkaloids, flavonoids*, and *tannins* (Mabry et al., 1970; Edwards and Gatehouse, 1999) (see chapters 5, and 8–14 for more details). The general pathway for creating secondary metabolites occurs through a branch-point enzyme, which regulates primary metabolism with secondary metabolism (Edwards and Gatehouse, 1999). For example, reactions that oxidize compounds are commonly catalyzed by dioxygenases, *heme*-containing enzymes that utilize oxygen and α-ketoglutarate in oxidation reactions, and release CO_2 and succinate. Another common step of secondary metabolite biosynthesis is methylation of potentially reactive carboxylic acid, amino, and hydroxyl groups that can spontaneously interact and form products that are undesirable to plants (Edwards and Gatehouse, 1999). An important methylating agent in secondary metabolism is *S*-adenosyl-L-methionine (Facchini et al., 1998, 2000), but this is only one of the many *methyltransferases* involved in methylating reactive groups.

We begin with the formation of two classes of secondary compounds, the *saponins* and *cardiac/cyanogenic glycosides*, which form in association with sugars produced during photosynthesis (fig. 1.13). These compounds are also formed via respiration through amino acid and terpenoid metabolism (Seigler, 1975). A few plant species have the ability to produce cyanides and cyanogenic glycosides, which are strong cytotoxins and competitive inhibitors for the iron atom associated with the heme group. Plant cells detoxify these strong cytotoxins by glycosylation, linking them via a beta-glycosidic bond to sugar residues (usually glucose) (Banthorpe and Charlwood, 1980; Luckner, 1984). *Glucosinolates* are anions that occur only in the cells of a limited number of *dicotyledonous* plant families (e.g., horseradish, radish, and mustards). Saponins are glycosides with a distinctive foaming characteristic, and consist of a polycyclic aglycone that is either a *choline* steroid or *triterpenoid*—attached via C_3 and an ether bond to a sugar side chain (Mabry et al., 1970).

Polymeric isoprene derivatives are a large family of substances that includes *steroids* and terpenoids (see chapter 9 for more details). These compounds are built from C_5 units of isoprene (2-methylbuta-1,3-diene). When two isoprene units combine, monoterpenoids are formed (C_{10}), while sesquiterpenoids are formed from three isoprene units and diterpenoids contain four isoprene units. Several thousand related compounds of this type have been isolated and characterized from different plant groups.

Steroids belong to the chemical family triterpenoids, a group of compounds generated by the polymerization of six isoprene units. Steroids typically have four fused rings—three six-membered and one five-membered—and range from 26 to 30 carbon atoms. Steroids occur widely in plants (both gymnosperms and angiosperms), as well as in fungi and animals. In plants, steroids are biosynthesized from cycloartenol, while in animals and fungi steroids are synthesized from lanosterol.

Carotenoids are tetraterpenoids, very common both in plant and animals, and contain 40 carbon atoms constructed from eight isoprene units (Banthorpe and Charlwood, 1980) (see chapter 12 for more details). They are products formed from the subsequent hydration, dehydration, ring formation, shifting of double bonds and/or methyl groups, chain elongation or shortening, and the subsequent incorporation of oxygen into a noncyclic $C_{40}H_{56}$ compound. Carotenoids can be further classified into carotenes (non-oxygen-containing) and xanthophylls (oxygen-containing).

Plant phenols are biosynthesized by several different routes and thus constitute a heterogeneous group (from a metabolic point of view). Although the presence of a hydroxyl group (–OH) makes them similar to alcohols, they are classified separately because the –OH group is not bonded to a saturated carbon atom. Phenols are also acidic due to the ease with which the –OH group dissociates. They are easily oxidized and can form polymers (dark aggregates), commonly observed in cut or dying plants. The starting product of the biosynthesis of most phenolic compounds is *shikimate* (Herrmann and Weaver, 1999). Two basic pathways are involved: the shikimic acid and malonic acid pathways (Herrmann and Weaver, 1999). Other derivatives include the phenylpropanol, lower-molecular-weight compounds such as coumarins, cinnamic acid, sinapinic acid, and the coniferyl alcohols; these substances and their derivatives are also intermediates in the biosynthesis of *lignin*. The malonic acid pathway, although an important source of phenolic secondary products in fungi and bacteria, is of less significance in higher plants. Animals, which cannot synthesize *phenylalanine*, *tyrosine*, and *tryptophan*, must acquire these essential nutrients from their diets.

Alkaloids are a group of nitrogen-containing bases. While the large majority of alkaloids are produced from amino acids, a few (e.g., caffeine) are derived from *purines* or *pyrimidines* (see chapter 6 for more details). Many of these compounds are used in *allelopathy*, the mutual influence of plant secretions on interactions among algae and higher plants. For example, allelopathic substances may damage the germination, growth, and development of other plants (Mabry et al., 1970; Harborne, 1977). Similarly, animals (particularly insects) have co-evolved defense strategies against the insecticide effects of some secondary plant products (Atsatt and O'Dowd, 1976; Bell and Charlwood, 1980; Bowers et al., 1986). Alkaloids are usually generated by the decarboxylation of amino acids or by the transamination of aldehydes. The biosynthesis of alkaloids derived from different amino acids requires different branch-point enzymes. In opiate biosynthesis, the branch-point enzyme is *tyrosine/dopa decarboxylase* (TYDC), which converts L-tyrosine into tyramine, as well as *dopa* into *dopamine* (Facchini and De Luca, 1995; Facchini et al., 1998). Aliphatic amines, which serve as insect attractants, are often produced during *anthesis*, the opening of a flower, or the formation of the fruiting body of certain fungi (e.g., the stinkhorn). A good example of insect attractants is the aliphatic–aromatic amines. Among the di- and polyamines are putrescine ($NH_2(CH_2)_4NH_2$), spermidine ($NH_2(CH2)_3NH(CH_2)_4NH_2$), and spermine ($NH_2(CH_2)_3NH(CH_2)_4NH(CH_2)_3NH_2$) (Pant and Ragostri, 1979; Facchini et al., 1998).

Flavonoids are polyphenolic compounds in plants that serve both as ultraviolet protection (Dakora, 1995) and in mediating plant–microbe interactions (Kosslak et al., 1987). While some classes of flavonoids, such as the *flavonones*, are colorless, other classes, *anthocyanes*, provide pigmentation in flowers (e.g., red and yellow) and other plant tissues. The basic structure of flavonoids is derived from the C_{15} body of flavone. Malonyl-CoA decarboxylase is an enzyme associated with *malonyl-CoA decarboxylase deficiency* (Harborne et al., 1975). It catalyzes the conversion of malonyl-CoA to CO_2 and to some degree, reverses the action of acetyl-CoA carboxylase. Flavonoids differ from other phenolic substances by the degree of oxidation of their central pyran ring and their biological properties.

Tannins are plant polyphenols that bind and precipitate proteins, and are usually divided into hydrolyzable and condensed tannins (Haslam, 1989; Waterman and Mole, 1994). Hydrolyzable tannins contain a polyol-carbohydrate (usually D-glucose), where the –OH groups of the carbohydrate are partially or totally esterified with phenolic groups such as gallic acid or ellagic acid. These tannins are hydrolyzed by weak acids or bases to produce carbohydrate and phenolic acids. Condensed tannins, also known as *proanthocyanidins*, are polymers of 2–50 (or more) flavonoid units that are joined by carbon–carbon bonds, which are not susceptible to cleavage by hydrolysis. While hydrolyzable tannins and most condensed tannins are water soluble, some very large condensed tannins are insoluble.

1.4 Summary

The earliest photosynthetic organisms are believed to have been anoxygenic, and have been dated back to 3.4 billion years BP. The first oxygenic photosynthetic organisms were likely the cyanobacteria, which became important around 2.1 billion years ago. As oxygen accumulated in the atmosphere through the photosynthetic activity of cyanobacteria, other oxygenic forms evolved, with the first eukaryotic cells evolving around 2.1 billion years ago. The sum of all biochemical processes in organisms is called metabolism, which involves both catabolic and anabolic pathways. Heterotrophic metabolism is the process by which organic compounds are either transformed or respired, thereby influencing the fate of organic compounds in the environment. These metabolic processes are responsible for the formation of many of the chemical biomarker compounds discussed in this book, which occur through an intermediary metabolism via glycolysis and the citric acid cycle. As life continues to evolve, new biosynthetic pathways are continually being developed, yielding organic compounds of increasing complexity and diversity.

2. Chemical Biomarker Applications to Ecology and Paleoecology

2.1 Background

In this chapter, we provide a brief historical account of the successes and limitations of using chemical biomarkers in aquatic ecosystems. We also introduce the general concepts of chemical biomarkers as they relate to global biogeochemical cycling. The application of chemical biomarkers in modern and/or ancient ecosystems is largely a function of the inherent structure and stability of the molecule in question, as well as the physicochemical environment and depositional conditions of the study system. In some cases, where preservation of biomarker compounds is enhanced, such as laminated sediments or environments with decreased oxygen exposure times, a strong case can be made for paleoreconstruction of past organic matter composition sources. However, it is important to keep in mind that even under the best scenarios, many of the most labile chemical biomarkers may be transformed or decomposed soon after senescence or death from processes such as microbial and/or metazoan grazing, viral infection and cell lysis, and photochemical breakdown. The role of trophic processes versus large-scale physiochemical gradients in preserving or destroying the integrity of sediment records of chemical biomarkers varies greatly across different ecosystems and will be discussed in relation to aquatic systems such as lakes, estuaries, and oceans.

2.2 Ecological Chemotaxonomy

Knowledge of the sources, reactivity, and fate of organic matter are critical for understanding the role of aquatic systems in global biogeochemical cycles (Bianchi and Canuel, 2001; Hedges and Keil, 1995; Wakeham and Canuel, 2006). Determining the relative contributions of organic matter fueling biogeochemical processes in the water column and sediments remains a significant challenge due to the wide diversity of organic matter sources across aquatic systems. Temporal and spatial variability in organic matter delivery adds further to the complexity of understanding these environments. In recent years significant improvements have been made in our ability to distinguish between organic

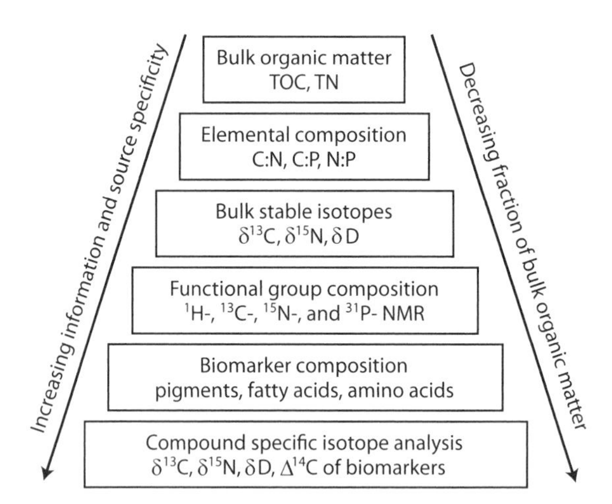

Figure 2.1. Schematic illustrating how approaches for characterizing organic matter differ in their source specificity (left) and the proportion of organic matter they represent (right).

matter sources in aquatic systems using tools such as elemental, isotopic (bulk and compound specific), and chemical biomarker methods. These tools differ in their ability both to identify specific sources of organic matter and to represent bulk organic matter (fig. 2.1). This chapter provides a general overview of the strengths and weaknesses of different approaches used for characterizing organic matter sources in aquatic environments.

2.3 Bulk Organic Matter Techniques

The abundance and ratios of elements important in biological cycles (e.g., C, H, N, O, S, and P) provide the basic foundation of information about organic matter sources and cycling. For example, concentrations of total organic carbon (TOC) provide the most important indicator of organic matter, since approximately 50% of most organic matter is composed of C. When bulk TOC measurements are combined with additional elemental information, as in the case of atomic C to N ratios ($(C:N)_a$ or mol : mol), basic source information can be inferred about algal and vascular plant sources (see review in Meyers, 1997). The broad range of $(C:N)_a$ ratios across divergent sources of organic matter in the *biosphere* demonstrate how such a ratio can provide an initial proxy for determining source information (table 2.1). Differences in $(C:N)_a$ ratios for vascular plants (>17) and microalgae (5 to 7) are largely due to differences in contributions from structural components. A large fraction of the organic matter in vascular plants is composed of carbon-rich biochemicals, such as carbohydrates and lignin. The most abundant carbohydrates supporting this high carbon content in vascular plants are structural *polysaccharides*, such as *cellulose, hemicellulose*, and *pectin* (Aspinall, 1970). In contrast, algae tend to be protein-rich and carbohydrate-poor because of both an absence of these structural components and higher contributions (relative to biomass) of protein and nucleic acids (fig. 2.2).

However, these generalizations do not hold for all plants. A recent survey of the elemental composition of a variety of plant types in San Francisco Bay (USA) showed that C : N ratios were highly variable, ranging from 4.3 to 196 (mol C mol^{-1} N), with the highest ratios measured in terrestrial and marsh vascular plants and lower ratios in aquatic plants (Cloern et al., 2002). Despite the general observation that the C : N ratios of terrestrial and marsh vascular plants were higher than for aquatic plants, C : N ratios ranged from minima of 10–20 to maxima of 40 to >100, suggesting large variability at the species level in biochemical composition. Seasonal variations in C : N ratios for individual plants

Table 2.1
$[C:N]_a$ for marine and terrestrial sources of organic matter

Name	$[C:N]_a$	*References*
Marine sources		
Brown macroalgae	12 to 23	Goñi and Hedges (1995)
Green macroalgae	8 to 15	Goñi and Hedges (1995)
Red macroalgae	9 to 10	Goñi and Hedges (1995)
Seagrasses		
Phytoplankton	7	Goñi and Hedges (1995)
Zooplankton	4 to 6	Goñi and Hedges (1995)
Gram-positive bacteria	6	Goñi and Hedges (1995)
Gram-negative bacteria	4	Goñi and Hedges (1995)
Oceanic particulate matter	4.5 to 10	Sterner and Elser (2002)
Terrigenous sources		
Grass	10	Emerson and Hedges (2008)
Tree leaves	10	Emerson and Hedges (2008)
Mangrove leaves (green)	49	Benner et al. (1990)
Mangrove leaves (senescent)	96	Benner et al. (1990)
Emergent vascular plants	12 to 108	Cloern et al. (2002)
C_3 saltmarsh plants	11 to 169	Cloern et al. (2002)
C_4 saltmarsh plants	12 to 65	Cloern et al. (2002)
Terrestrial woody	13 to 89	Cloern et al. (2002)

Plankton vs. Higher Plants

Plankton: mainly proteins (up to 50% or more)
variable amounts of lipid (5-25 %)
variable amounts of carbohydrates (<40%)
C/N ratio typically 6-7

Higher plants: mainly cellulose (30-50%)
lignin (15-25%)
protein (10%; generally <3%)
C/N from ~20 to as high as 100 (woody terrestrial)

Figure 2.2. Comparison of biochemical composition of plankton and vascular plants.

were also observed, suggesting shifts in biochemical composition during annual cycles or contributions from microbes colonizing the plants. At this stage, it is clear that $C:N_a$ ratios should be used cautiously as a proxy for tracing organic matter sources and, when used, should be coupled with stable isotopes or molecular biomarker techniques.

Uncertainty associated with using $(C:N)_a$ ratios as proxies for organic matter sources can derive from decomposition processes and remineralization. Selective remineralization of N, due to N limitation in marine systems, can result in artificially high $(C:N)_a$ ratios and contribute to misidentification of organic matter sources. Alternatively, decomposition processes can reduce $(C:N)_a$ ratios. Contributions of bacterial and/or fungal biomass during colonization of aging vascular plant detritus can represent a significant fraction of the total N pool (due to the typically low $(C:N)_a$

ratios [e.g., 3 to 4] found in bacteria and other microbes) (Tenore et al., 1982; Rice and Hanson, 1984). As a result, the bulk $(C:N)_a$ ratio of decaying vascular plant detritus may decrease during decomposition. Additionally, artifacts from the standard procedure of removing carbonate carbon when measuring TOC can also alter $(C:N)_a$ ratios because the residual N may represent contributions from both organic and inorganic N (Meyers, 2003). Since residual N is made up of both inorganic and organic N fractions, the N used in C:N ratios is defined as total nitrogen (TN). In most cases inorganic N is a relatively small fraction of TN associated with particulate and sedimentary organic matter. However, in particulate matter and sediments having low contributions of organic matter (e.g., $<0.3\%$), the relative importance of this residual inorganic N can be significant, resulting in underestimates of C:N ratios (Meyers, 2003). Specifically, this results from adsorption of NH_4^+ on low organic content sediments. Sorption processes can also influence the distribution of dissolved organic nitrogen and amino acids in soils and sediments because basic amino acids are preferentially sorbed to alumninosilicate clay minerals, while nonprotein amino acids remain in the dissolved phase (Aufdenkampe et al., 2001).

Stable isotopes provide complementary information to $(C:N)_a$ ratios and are often used to identify organic matter sources in aquatic systems. Isotopic mixing models have been used to evaluate sources of dissolved inorganic nutrients (C, N, S) (Day et al., 1989; Fry, 2006) and particulate and dissolved organic matter (POM and DOM) (Raymond and Bauer, 2001a, b; Gordon and Goñi, 2003; McCallister et al., 2004) in aquatic ecosystems. However, due to overlaps in the stable isotopic signatures of source materials in estuarine and coastal ecosystems, it can be difficult to identify organic matter sources when using single and dual bulk isotopes in complex systems (Cloern et al., 2002). More recently, new approaches using end-member mixing models that utilize multiple isotopic tracers coupled with chemical biomarker measurements have proven useful for elucidating organic matter sources in complex systems. In chapter 3 we describe simple conservative mixing models that attempt to define sources of inorganic nutrients, followed then by examples of models that utilize a multiple isotopic tracer and/or chemical biomarker approach to determine sources of organic matter.

One of the fundamental problems with these mixing models is that they assume that decomposition processes in the water column and sediments do not change the isotopic signature of organic compounds reflecting end-member source inputs. In cases where the decomposition products are known, specific compounds can be used as indices of decay—more is provided on these biomarkers in chapter 3. Other work using compound-specific isotope analysis (CSIA), as described in chapter 3, has shown that isotopic fractionations for lipid biomarker compounds vary under different redox regimes (Sun et al., 2004). More work is needed to understand the effects of decomposition processes on biomarkers if we are to incorporate molecular-based isotopic studies in efforts to determine sources of organic matter in aquatic systems.

Nuclear magnetic resonance (NMR) spectroscopy is a powerful and theoretically complex analytical tool that can be used to characterize bulk organic matter. We describe some of the basic principles of NMR in chapter 4 before discussing its application in aquatic systems. Proton (^1H) and ^{13}C-NMR have been the most common NMR tools for the nondestructive determination of functional groups in complex biopolymers in plants, soils/sediments, and DOM in aquatic ecosystems (Schnitzer and Preston, 1986; Hatcher, 1987; Orem and Hatcher, 1987; Benner et al., 1992; Hedges et al., 1992, 2002; Mopper et al., 2007). The application of ^{31}P NMR (Ingall et al., 1990; Hupfer et al., 1995; Nanny and Minear, 1997; Clark et al., 1998) and ^{15}N NMR (Almendros et al., 1991; Knicker and Ludemann, 1995; Knicker, 2000, 2002; Aluwihare et al., 2005) have also been useful in characterizing organic and inorganic pools of P and N in natural systems, respectively. When ^{13}C NMR was used to examine the sediments of Mangrove Lake, Bermuda (a small brackish to saline lake), the dominant carbon functional groups in organic matter were identified as aliphatic (30 ppm), *methoxy* (lignin) (56 ppm), *alkoxy* (polysaccharides) (72 ppm), aromatic/*olefinic* (130 ppm), and *carboxylic/amide* (175 ppm) (fig. 2.3) (Zang and Hatcher, 2002). The high aliphatic and amide

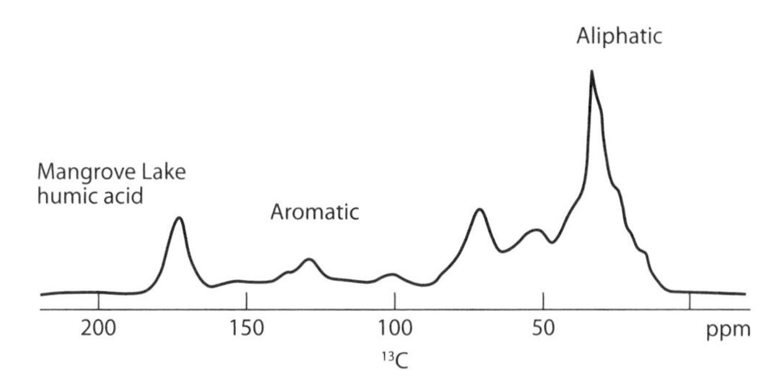

Figure 2.3. ^{13}C NMR spectrum of humic acids isolated from the Everglades, FL (USA). (Adapted from Zang and Hatcher, 2002.)

peaks at 30 and 175 ppm, respectively, are indicative of *sapropel* algal sources (Zang and Hatcher, 2002, and references therein). Using ^{15}N NMR, the high protein-rich algal source in these lake sediments was further corroborated by a large amide-type peak at 256 ppm (fig. 2.3) (Knicker, 2001; Zang and Hatcher, 2002). More recently, two-dimensional (2D) ^{15}N ^{13}C NMR has been used to study the fate of protein in experimentally degraded algae (Zang et al., 2001). Further details on the application and interpretation of NMR techniques are discussed in chapter 4.

2.4 Chemical Biomarkers: Applications and Limitations

Due to the complexity of organic matter sources in aquatic ecosystems and the aforementioned problems associated with using bulk measurements to constrain them, the application of chemical biomarkers, as defined in chapter 1, has become widespread in aquatic research (Hedges, 1992; Bianchi and Canuel, 2001; Bianchi, 2007). Applications of chemical biomarkers to historical (e.g., anthropogenic effects) and paleo (e.g., Earth system) timescales depend on a wide range of processes. If we look at the dominant processes believed to control the fate of plant pigments in aquatic systems through the water column and sediments, as an example, we see processes that span from days to millennia (fig. 2.4) (Leavitt, 1993). We use this example because many of these processes are applicable to the chemical biomarkers discussed in this book. For example, heterotrophic processing (and transformation) will influence most algal biomarkers. This may occur through the action of metazoans and/or by bacterial/archaebacterial decomposition processes over an ecological timeframe (e.g., hours to weeks). Similarly, while fig. 2.4 focuses on POM, biomarkers in DOM will also, and more dramatically, be altered by photochemical and microbial processes over a period of hours to days.

In the final stage of postdepositional alteration prior to burial, chemical biomarkers can be altered by biotic and abiotic processes, resulting in oxidation, reduction, saturation (e.g., hydrogenation), and other chemical transformations. While many of the biotic controls occur via bacterial/archaebacterial processes, many of the abiotic are linked to redox conditions. For example, one of the key processes affecting *hydrogenation*, a process that controls the conversion of chemical biomarkers to fossil biomarkers in anoxic sediments, occurs through inorganic sulfur species (e.g., H_2S), as demonstrated in the conversion of *β-carotene* to *β-carotane* (Hebting et al., 2006). Knowledge of these types of chemical transformations in sediments allows chemical biomarkers (particularly lipids) to be used as fossil biomarkers over periods of centuries to millennia (Olcott, 2007).

The limitations of chemical biomarkers as paleo-indicators are primarily based on the depositional environment, availability of laminated/varved sediments, chemical stability of the biomarker, and redox conditions. For example, *lipids* tend to be more decay resistant than other biochemical constituents of

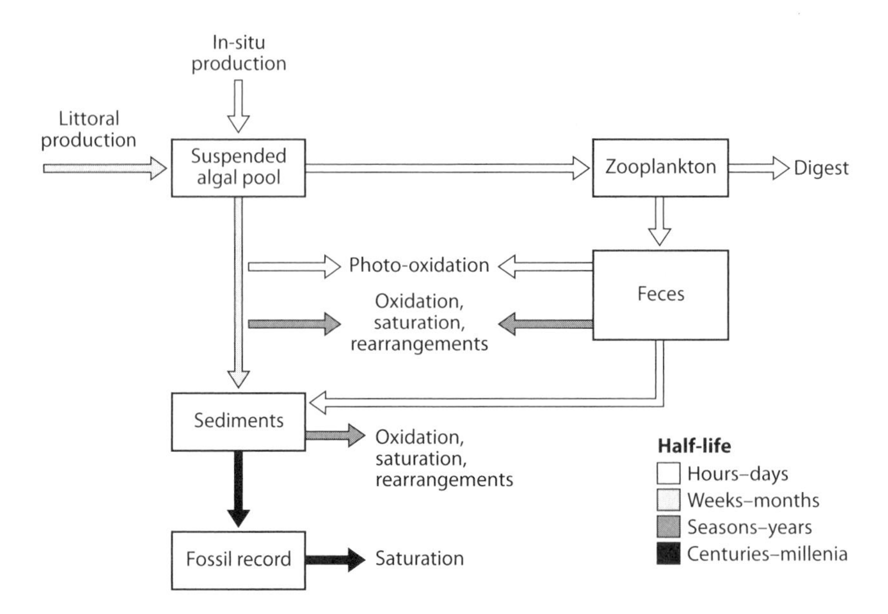

Figure 2.4. Processes believed to control the fate of plant pigments in aquatic systems through the water column and sediments. As an example, we illustrate processes that span from days to millennia. (Adapted from Leavitt, 1993.)

organic matter—hence, the name geolipid. Harvey et al. (1995a) showed that under both oxic and anoxic conditions approximately 33% of the lipids in cyanobacterial (*Synechococcus* sp.) detritus remained after a 93-day incubation period—a significantly higher percentage than for protein or carbohydrates. It was further concluded that all cellular components of both diatom and cyanobacterial detrital material decomposed more rapidly under oxic conditions, which supports the importance of the redox-controlling mechanism in preserving organic matter.

Laboratory experiments using estuarine sediments have shown greater loss of sterols and fatty acids under oxic versus anoxic conditions (Sun et al., 1997; Sun and Wakeham, 1998). Harvey et al. (1995) proposed that the relatively longer turnover time of lipids can possibly be explained by the following two mechanisms: (1) incorporation of free sulfide into lipids—particularly under anoxic conditions as shown in sediments (Kohnen et al., 1992; Russell et al., 1997), and (2) differences in structure noting fewer oxygen functionalized atoms compared to other biochemical components. A more recent study examining sediments from Long Island Sound estuary (USA) proposed a mechanism involving the frequency of oscillating redox conditions (Sun et al., 2002a, b). It was demonstrated that the rate and pathway of degradation of ^{13}C-labeled lipids were a function of the frequency of oxic to anoxic oscillations. Lipid degradation rates increased significantly with increasing oscillation and overall oxygen exposure time, but in some cases the rates were linear and in others they were exponential. A recent field study in the York River estuary (USA) showed that rates of sterol decomposition were higher in sediments influenced by physical mixing (Arzayus and Canuel, 2005). Results from these studies support findings that oscillating redox conditions (Aller, 1998) and *oxygen exposure time* (Hartnett et al., 1998) are important controlling variables for organic matter preservation in estuarine systems. The selective nature of decomposition processes over changing redox conditions, and the fact that certain biochemical constituents can decompose over timescales of a few days, reflect the importance of using short-lived biomarker compounds in understanding diagenetic processes in shallow dynamic aquatic systems, such as estuaries (Bianchi, 2007). Other experimental work has shown rapid decay of unsaturated sterols under oxic conditions, suggesting that chemical structure

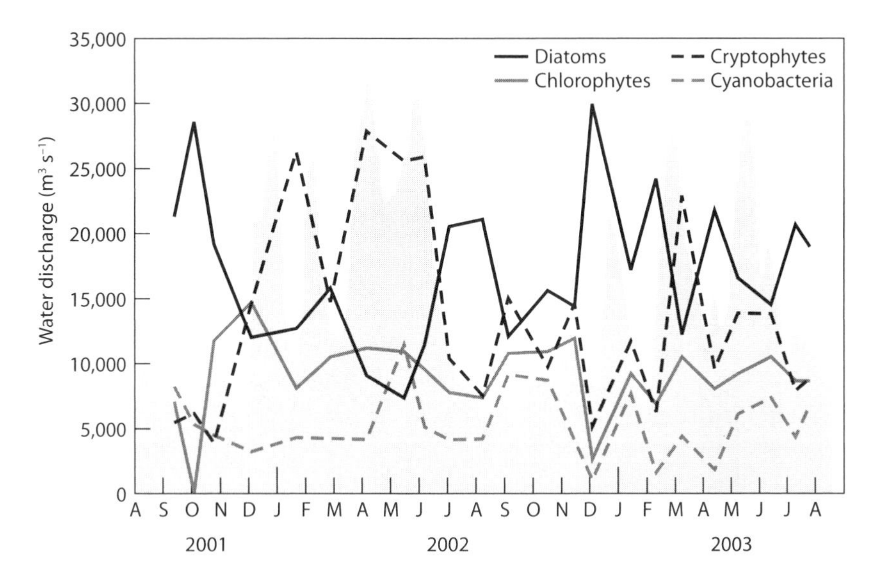

Figure 2.5. Patterns of phytoplankton abundance, based on plant pigment biomarkers, in the Mississippi River (USA). (Adapted from Duan and Bianchi, 2006.)

alone (e.g., unsaturation) may not always be a good predictor of the ease with which organic matter is decomposed (Harvey and Macko, 1997; Wakeham and Canuel, 2006).

Seasonal inventories of sedimentary *chloropigments* indicated that there are differences in timing between maximum *chlorophyll* concentrations in the water column and *pheopigment* inventories in sediments from Long Island Sound estuary (USA) (Sun et al., 1994). These disparities likely reflect differences in the growth patterns of *phytoplankton* (winter to early spring) and their export to sediments via production of *zooplankton* fecal pellets. These results also emphasize the importance of benthic–pelagic coupling in shallow estuarine systems. Differential decay of plant pigment biomarkers can provide additional information on the importance of pre- and postdepositional decay processes and decoupling between *benthic* and *pelagic* processes. For example, chlorophyll-decay rate constants in both *anoxic* and *oxic* sediments, with and without macrofauna, have been shown to be approximately $0.07\,d^{-1}$ (Bianchi and Findlay, 1991; Leavitt and Carpenter, 1990; Sun et al., 1994; Bianchi et al., 2000). Pigments "bound" to structural compounds such as lignin or surface *waxes* in higher plants have slower decay rate constants than similar pigments from nonvascular sources (Webster and Benfield, 1986; Bianchi and Findlay, 1991). Other work has shown the importance of pigment decay in the free versus bound state in estuarine sediments (Sun et al., 1994).

2.5 Application of Chemical Biomarkers to Modern and Paleo-Aquatic Ecosystems

Many aquatic systems contain diverse organic matter sources and the application of different chemical biomarkers has provided a useful tool for tracing their source and environmental fate. Some systems are more complex, such as estuaries, where there is a broad spectrum of subenvironments (e.g., *terrestrial, littoral, planktonic, benthic*) that contribute to organic matter inputs (Bianchi, 2007, and references therein), while others are relatively simple, such as the open ocean (e.g., planktonic) (Hurd and Spencer, 1991; Emerson and Hedges, 2009; and references therein). Plant pigments are one class of biomarkers that have been widely used as tracers for phytoplankton sources in aquatic ecosystems. For example, recent work in the Mississippi River (USA) has shown the patterns of different phytoplankton to be positively and negatively associated with river discharge conditions (fig. 2.5) (Duan and Bianchi, 2006).

Table 2.2
Assumptions involved in five basic applications with biomarkers (B) and associated organic matter (AOM)

Application type	*Assumption*
I. Detection of source	1. B properly identified 2. B source(s) known and unique
II. Determination of relative biomarker levels	1. All the above 2. B concentrations quantified with proportional accuracy 3. Relative concentrations of B diagenetically unchanged
III. Determination of relative concentration of AOM	1. All the above 2. B/AOM is constant in source (or mixed to uniformity) 3. B/AOM is not diagenetically al tered in situ
IV. Quantification of absolute concentration of AOM	1. All the above 2. B/AOM in natural sample(s) is known
V. Quantification of the paleoproduction of AOM	1. All the above 2. B/AOM is constant in source over a long time period 3. Efficiency of physical input of B and AOM is unchanged 4. B/AOM of source is not diagenetically altered

Source: Hedges & Prahl (1993).

In this study, different *carotenoids* were used to track the abundance of cryptophytes (cryptoxanthin), diatoms (e.g., fucoxanthin, diadinoxanthin), cyanobacteria (e.g., zeaxanthin), and chlorophytes (e.g., chlorophyll *b*, lutein). Additional details about the application of plant pigments are provided in chapter 12.

Lipid biomarker compounds such as fatty acids and sterols have also been widely used in studies aimed at elucidating organic matter sources in aquatic systems. A benefit of lipid biomarkers is that these compounds provide the ability to trace organic matter from phytoplankton, zooplankton, bacteria, and vascular plants simultaneously. These proxies have been used widely in studies of organic matter sources in lakes, rivers, estuaries, and oceanic ecosystems (see chapters 8 through 10). Lipids are useful molecular biomarkers for several reasons: (1) these compounds occur in unique distributions in different organic matter sources; (2) structural features (number and position of double bonds, functional group composition, etc.) provide information about their source; (3) differences in structural features result in a range in reactivity for different lipid biomarkers ranging from phospholipid linked fatty acids, which turn over soon after a cell dies, to hydrocarbons, which persist in the geologic record; (4) lipid biomarkers provide a diagnostic tool for differentiating between *autochthonous* (algae) and *allochthonous* (vascular plants) sources of organic matter in coastal ecosystems; and (5) lipid biomarkers can also provide information about nonbiogenic sources of organic matter, such as weathering of ancient sediments or inputs from fossil fuels.

Despite these strengths, lipids and other biomarkers have inherent limitations. One of these is that they are best applied to studies where a qualitative assessment of organic matter sources is needed (Hedges and Prahl, 1993). Another limitation is that many biomarkers are present at trace levels relative to total organic matter and may not represent bulk organic matter. Generally, biomarkers are easily applied to questions related to organic matter sources provided that particular compound(s) can be identified and the source of the compound(s) is both known and unique (table 2.2) (Hedges and Prahl, 1993). However, if the goal is to determine relative biomarker levels, it is necessary not only to meet the assumptions described above, but to also demonstrate the ability to quantify the concentration of the compound with proportional accuracy and that the relative concentrations of the compound are not changed by diagenetic alterations. As the goal increases from qualitative to quantitative determination of sources, more assumptions must be met (table 2.1). Additionally, as the quantification of source is extended from modern depositional environments to the longer timescales (e.g., paleo-record), assumptions about the source as well as its input and alteration must be met (see next section). In many cases, it is impossible to meet these assumptions, which limit the application of biomarkers solely to qualitative assessments of organic matter sources.

2.6 Paleo-Reconstruction

Laminated sediment chronologies of organic and inorganic P, N, C, and biogenic Si have been used to study long-term trends in eutrophication in lacustrine systems (Conley et al., 1993, and references therein). In particular, chemical biomarkers have been utilized as effective paleo-tracers of historical change in phytoplankton assemblages in *lacustrine* systems, where laminated sediments provide an excellent environment for preservation of chemical biomarkers (Watts and Maxwell, 1977). For example, biogenic silica, total organic carbon (TOC), and photosynthetic pigments were used to understand changes in phytoplankton community structure in dated sediment cores from Lake Baikal (Russia) representing the transition from glacial to postglacial conditions (Tani et al., 2002). Fossil pigments have also been used in paleolimnology to better understand historical changes in bacterial community composition, trophic levels, redox change, lake acidification, and past UV radiation (Leavitt and Hodgson, 2001). Additionally, lipid biomarker compounds have been widely used in paleolimnological studies (Meyers and Ishiwatari, 1993; Meyers, 1997, 2003). For example, stable isotopes and hydrocarbon biomarkers were used to determine changes in the dominant sources of organic matter inputs to Mud Lake, Florida (USA), a modern limestone sinkhole lake (Filley et al., 2001). These authors used downcore changes in *n-alkane* and *alcohol* biomarkers as organic matter indicators for the following sources: *cyanobacteria* (7- and 8-methylheptadecane), terrestrial/cuticular sources (C_{27}–C_{31}; C_{26}-ol to C_{30}-ol), mixed phototrophic sources (C_{23}–C_{25}; C_{22}-ol to C_{25}-ol; *phytols*) and phytoplankton (C_{17}; C_{22}-ol to C_{25}-ol) (fig. 2.6). If we examine the chronological changes they observed using $\delta^{13}C$ signatures of individual compounds (e.g., compound-specific isotope analysis), we find a trend that supports the chemical biomarker data: a shift to increasing dominance of phytoplankton at about 2500 ^{14}C yr BP (fig. 2.6). Thus, we can see how changes in organic matter sources, caused by increases in the Florida water table over time, are recorded in the composition of chemical biomarkers in the sediments of this lacustrine system. More details on the application of CSIA as well as *n*-alkanes and alcohols are provided in chapters 3 and 10, respectively.

Unlike lacustrine systems, *estuaries* typically have highly variable physical conditions, extreme spatial gradients, and bioturbation, which may limit the application of chemical biomarkers (Nixon, 1988). However, in spite of these limitations, past studies have shown increased preservation of total organic carbon in estuaries during the 20th century (Bianchi et al., 2000, 2002; Tunnicliffe, 2000; Zimmerman and Canuel, 2000, 2002). In fact, recent work has compared long-term ecological data (e.g., phytoplankton biomass composition, nutrients, and benthos) and plant pigments preserved in laminated sediments of the Himmerfjarden Bay (Baltic proper) (Bianchi et al., 2002). This bay has been monitored for the effects of municipal wastewater and runoff from agricultural land on coastal *eutrophication* since the 1980s (Elmgren and Larsson, 2001). This work essentially concluded that despite high-resolution sampling and freeze-core sampling of sediment layers (annual varves), there was no correlation between pigments preserved in sediments and phytoplankton biomass (Bianchi et al., 2002). However, when sedimentary pigments were averaged over longer time intervals (5 yr), averages of annual diatom biomass were positively correlated to downcore concentrations of fucoxanthin. This work demonstrated that under conditions favorable for preservation in sediments (e.g., presence of laminations), plant pigments could be used to trace past changes in phytoplankton in estuaries. More recent work has also shown significant variability in the effectiveness of using plant pigments for

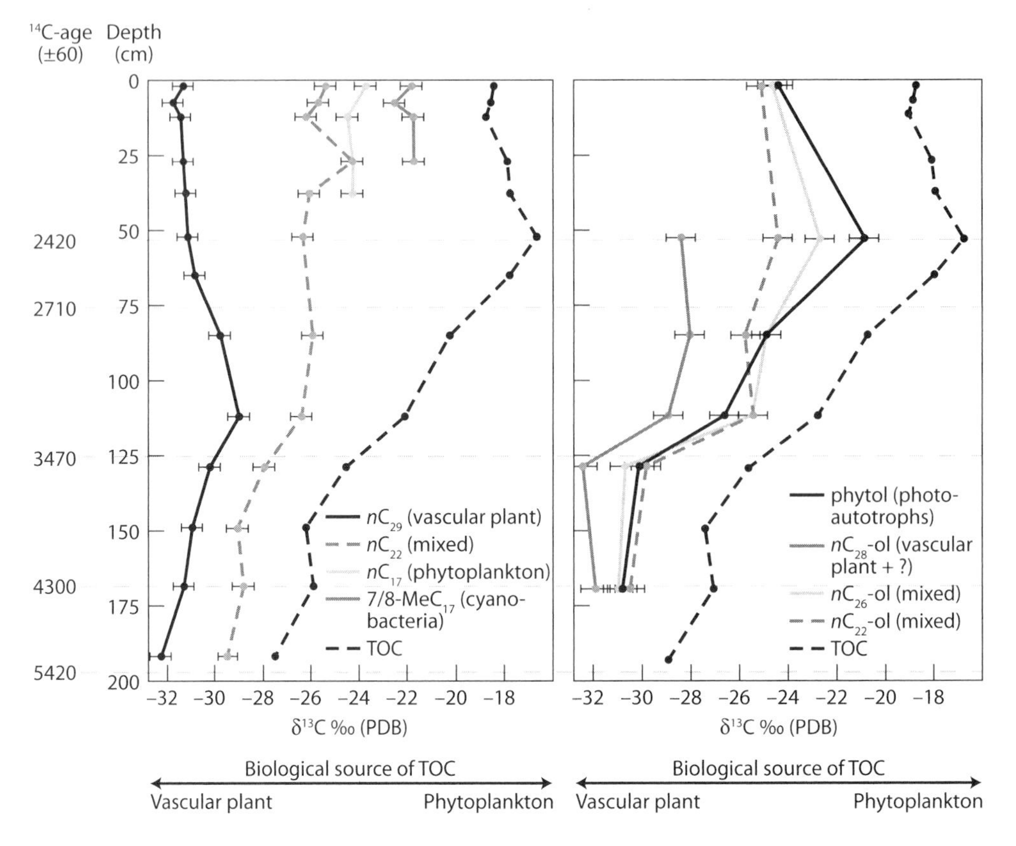

Figure 2.6. Downcore changes in *n*-alkane and alcohol biomarkers as organic matter indicators for the following sources: cyanobacteria, terrestrial/cuticular sources, mixed phototrophic sources, and phytoplankton. (Adapted from Filley et al., 2001.)

ecological reconstruction in four European estuaries (Reuss et al., 2005). Other work has shown that large cyanobacterial blooms in the Baltic Sea may be a natural feature of this system that has existed for thousands of years. For example, recently, it was demonstrated, using downcore variations in cyanobacterial pigment biomarkers and $\delta^{15}N$ values, that nitrogen-fixing blooms are nearly as old as the brackish water phase of the Baltic Sea, dating back to 7000 yr BP—soon after the last freshwater phase of the Baltic, the Ancylus Lake, changed into the final brackish water phase, the Litorina Sea (Bianchi et al., 2002).

2.7 Summary

In this chapter, we have illustrated some of the benefits and limitations of various approaches for tracing the sources of organic matter over modern, historical and paleo-timescales. It is important to be aware of the inherent trade-offs and assumptions in applying biomarker methods to aquatic environments, particularly in studies examining organic matter sources over historical or geological timescales. While bulk parameters such as $(C:N)_a$ and stable isotopes ($\delta^{13}C$, $\delta^{15}N$) represent a larger fraction of sediment organic matter, these proxies are limited in their source specificity. In contrast, biomarkers are generally more source-specific but are often present at low concentrations and thus may not be representative of bulk organic matter. Future studies should consider ways to move from qualitative applications of biomarkers to those where specific sources of sediment organic matter can be quantified. This requires improved understanding of the stability of biomarkers under a range of depositional conditions. Likewise, by using a combination of tools, it may be possible to obtain more reliable source information from bulk pools. Approaches that combine stable and radiocarbon isotopic signatures or utilize isotopes and NMR spectroscopy may prove useful in this regard. These approaches will be discussed further in subsequent chapters.

3. Stable Isotopes and Radiocarbon

3.1 Background

Isotopes are widely used tools in the fields of ecology, geochemistry, limnology and oceanography, providing information about the sources and cycling of (1) bulk organic matter pools (dissolved, colloidal, particulate, and sedimentary), (2) biochemical classes, and (3) individual biomarker compounds. Stable isotope studies in aquatic geochemistry have incorporated the use of a variety of light, biogenic elements, including hydrogen, carbon, nitrogen, oxygen, and sulfur. The use of stable isotopes has contributed novel insights about food web structure and organic matter sources over a variety of timescales, including contemporary, historical, and geological. Stable isotopes can also be used to assess the strength or magnitude of the processes that cause isotopes to fractionate during the synthesis and recycling of organic matter. In addition to stable isotopes, recent advances in accelerator mass spectrometry have provided the ability to measure radiocarbon (the radioactive isotope of carbon) in bulk carbon pools and individual compounds present in a variety of environmental matrices, providing an additional tool for elucidating the sources of organic matter. Radiocarbon measurements have been used to provide information about the ages and sources of organic matter across a range of aquatic environments as well as the ages of carbon supporting heterotrophic organisms. This chapter provides background information about the use of stable isotopes and radiocarbon and presents examples of the application of these isotopes to a variety of research questions in aquatic science.

3.1.1 Background Information and Isotope Vocabulary

The three primary subatomic particles are protons, electrons, and neutrons. Protons and neutrons are similar in mass (1.6726×10^{-24}g and $1.6749543 \times 10^{-24}$g, respectively) but protons are positively charged and neutrons have no charge. In contrast, electrons are considerably smaller (9.109534×10^{-28}g) and carry a negative charge. An element is defined according to the number of protons (*atomic number* or Z) in its nucleus. While the number of protons is always the same, the number of neutrons

Table 3.1
Isotopic abundances and relative atomic masses of stable isotopes of elements used in organic geochemistry

Symbol	*Atomic number*	*Mass number*	*Abundance* (%)	*Atomic weight* ($^{12}C = 12$)
H	1	1	99.985	1.007825
D	1	2	0.015	2.014
C	6	12	98.89	12
		13	1.11	13.00335
N	7	14	99.63	14.00407
		15	0.37	15.00011
O	8	16	99.759	15.99491
		17	0.037	16.99914
		18	0.204	17.99916
S	16	32	95.0	31.97207
		33	0.76	32.97146
		34	4.22	33.96786

Source: Adapted from Sharp (2007).

(*neutron number* or N) may vary. The *mass number* (A) is the number of neutrons plus protons in a nucleus ($A = Z + N$). *Isotopes* are different forms of an element that have the same Z value but a different N. Carbon, for example, has 6 protons (hence, atomic number = 6) but can have 6, 7, or 8 neutrons (atomic mass 12, 13, or 14, respectively). As a result, carbon has two stable isotopes (^{12}C and ^{13}C) and one radioactive isotope (^{14}C). The behavior of different isotopes of an element is essentially, but not exactly, the same, and it is the subtle differences in behavior that are capitalized upon in the application of stable isotopes. For example, isotopes will react at different rates due to slight differences in their atomic masses, with the "lighter" isotope reacting faster and to a greater extent. As a result, reaction products will be enriched in the lighter isotope. The extent to which the lighter isotope will be enriched will depend on the reaction mechanism, the degree to which the reaction has proceeded, the isotopic composition of the reactants, and environmental factors such as temperature and pressure.

3.1.2 Stable Isotopes

The fundamental principle surrounding the application of stable isotopes in natural ecosystems is based on variations in the relative abundance of lighter isotopes from chemical rather than nuclear processes (Hoefs, 1980, 2004; Clayton, 2003; Fry, 2006; Sharp, 2007). Reaction products in nature are generally enriched in the lighter isotope due to the faster reaction kinetics of the lighter isotope of an element. These fractionation processes can be complex, as discussed below, but have proven to be useful to a variety of applications, such as geothermometry and paleoclimatology, as well as for identifying sources of organic matter in ecological studies. The most common stable isotopes used in lacustrine, oceanic, and estuarine studies are ^{18}O, ^{2}H, ^{13}C, ^{15}N, and ^{34}S. The preference for using such isotopes is related to their low atomic mass, significant mass differences in isotopes, covalent character in bonding, multiple oxidation states, and sufficient abundance of the rare isotope.

The average relative abundances of isotopes in the Earth's crust, oceans, and atmosphere, commonly expressed as stable isotope ratios, are shown in table 3.1. Small differences in the ratios of a particular element in natural samples can be detected using *mass spectrometry*, but, they cannot be determined with high *precision* and *accuracy* (Nier, 1947). The solution to this problem is to measure isotope ratios in a sample concurrently with a standard or reference material; however, this still does not

Table 3.2
Internationally accepted stable isotope standards

Element	*Standard*	*Abbreviation*
H	Vienna Standard Mean Ocean Water	VSMOW
C	Vienna PDB	VPDB
N	Atmospheric N_2	AIR
O	Vienna Standard Mean Ocean Water	VSMOW
	Vienna PDB	VPDB
S	Triolite (FeS) from Cañon Diablo Triolite	CDT

provide adequate precision and accuracy for the absolute abundance of each isotope. The equation used to describe this relative difference or delta (δ) value is as follows:

$$\delta = [(R_{\text{sample}} - R_{\text{standard}})/R_{\text{standard}}] \times 1000 \tag{3.1}$$

where R = the ratio of the heavy to the light stable isotope, and delta (δ) is expressed in units of per mil or parts per thousand (‰). The isotope composition of a sample is considered enriched when the ratio of the heavy isotope to the light isotope is more abundant in the sample relative to the standard, resulting in positive delta values. Conversely, delta values will be negative, or depleted, when the ratio of the heavy isotope to the light isotope in the sample is lower than the standard. Finally, the delta value will be 0 when the sample and the standard have the same isotopic composition. A range of standard materials has been developed with the following criteria: materials that are homogeneous in composition, readily available, and easy to prepare and measure, and that have isotopic composition in the middle of the natural range of materials that will be analyzed. A list of the international standards typically used for stable isotopic measurements of O, C, H, N, and S is provided in table 3.2 (Hoefs, 1980; Faure, 1986; Fry, 2006; Sharp, 2007). The isotopic composition of a sample is measured relative to a standard using isotope ratio mass spectrometry (Hayes, 1983; Hayes et al., 1987; Boutton, 1991). Details about this methodology are provided in chapter 4.

3.1.3 Basic Principles of Radioactivity

Radioactivity is defined as the spontaneous adjustment of nuclei of unstable nuclides to a more stable state. Radiation (e.g., *alpha, beta*, and *gamma rays*) is released in different forms as a direct result of changes in the nuclei of these nuclides. There are two general rules concerning the stability of nuclides, the *symmetry* and *Oddo-Harkins* rules (Hoefs, 2004). The symmetry rule states that a stable nuclide with a low atomic number has approximately the same number of neutrons and protons, whereby the N/Z ratio is near unity. However, when the stable nuclide has more than 20 protons or neutrons the N/Z ratio is always greater than unity. The maximum value of N/Z is about 1.5 for the heaviest nuclides; this occurs because of electrostatic *Coulomb repulsion* of positively charged protons with increasing Z. The Oddo-Harkins rule simply states that nuclides with even atomic numbers are more abundant than nuclides with odd numbers. So, instability in nuclei is generally caused by having an inappropriate number of neutrons relative to the number of protons. Some of the pathways by which a nucleus can spontaneously transform are as follows: (1) *alpha decay*, or loss of an alpha particle (nucleus of a helium-4 atom) from the nucleus, which results in a decrease in the atomic number by two (2 protons) and the mass number by four units (2 protons and 2 neutrons); (2) *beta (negatron) decay*, which occurs when a neutron changes to a proton and a *negatron* (negatively charged electron) is emitted—thereby increasing the atomic number by one unit; (3) emission of a *positron* (positively charged electron), which results in a proton becoming a neutron and a decrease in the atomic number by one unit; and (4) *electron capture*, where a proton is changed to a neutron after combining with the captured extranuclear electron (from the K shell)—the atomic number is decreased by one unit.

Experimentally measured rates of decay of radioactive atoms show that decay is *first order*, where the number of atoms decomposing in a unit of time is proportional to the number present. This is expressed by the following equation (Faure, 1986):

$$dN/dt = -\lambda N \tag{3.2}$$

where

N = number of unchanged atoms at time t
λ = decay constant

If we rearrange and integrate equation 3.1, from $t = 0$ to t and from N_0 to N we get

$$-\int dN/N = \lambda \int dt$$

$$-\ln N = \lambda t + C \tag{3.3}$$

where

$\ln N$ = is the natural logarithm of N
C = constant of integration

When $N = N_0$ and $t = 0$, equation 3.3 can be modified to

$$C = -\ln N_0 \tag{3.4}$$

We can then substitute into equation 3.3 to obtain the following equations:

$$-\ln N = \lambda t - \ln N_0 \tag{3.5}$$

$$\ln N - \ln N_0 = -\lambda t$$

$$\ln N/N_0 = -\lambda t$$

$$N/N_0 = e^{-\lambda t}$$

$$N = N_0 e^{-\lambda t} \tag{3.6}$$

In addition to λ, another term used for characterizing rate of decay is *half-life* ($t_{1/2}$), the time required for half of the initial number of atoms to decay. If we substitute $t = t_{1/2}$ and $N = N_0/2$ into equation 3.6 we get the following equation:

$$N_0/2 = N_0 e^{-\lambda t}_{1/2}$$

$$\ln 1/2 = -\lambda t_{1/2}$$

$$\ln 2 = \lambda t_{1/2}$$

$$t_{1/2} = \ln 2/\lambda = 0.693/\lambda \tag{3.7}$$

Finally, the term used to examine the average life expectancy, or *mean-life* (τ), of a radioactive atom is expressed as follows:

$$\tau = 1/\lambda \tag{3.8}$$

On the assumption that the radioactive parent atom produces a stable radiogenic daughter (D^*), which is zero at $t = 0$, we can describe the number of daughter atoms at any time with the following equation:

$$D^* = N_0 - N \tag{3.9}$$

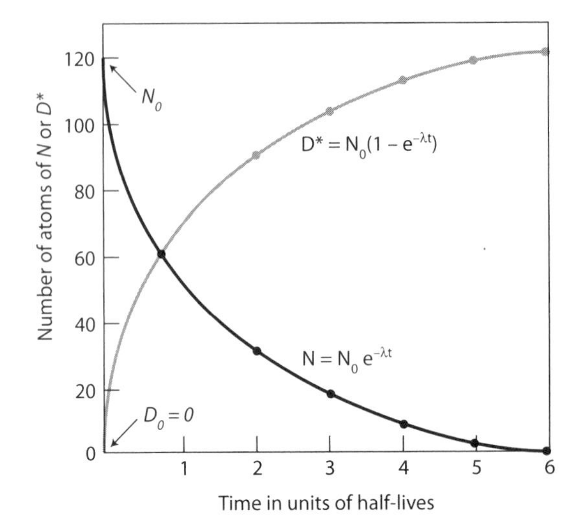

Figure 3.1. The hypothetical decay of a radionuclide (N) to a stable radiogenic daughter (D^*). The successive loss of atoms of N from radioactive decay is followed by proportional increase in the daughter atoms. (Adapted from Faure, 1986.)

This assumes no addition or loss of daughter atoms and that any loss of the parent atoms is due to radioactive decay. If we then substitute equation 3.6 into 3.9 we get the following:

$$D^* = N_0 - N_0 e^{-\lambda t}$$

$$D^* = N_0(1 - e^{-\lambda t}) \tag{3.10}$$

With further substitution we get

$$D^* = Ne^{\lambda t} - N = N(e^{\lambda t} - 1) \tag{3.11}$$

Figure 3.1 illustrates the hypothetical decay of a radionuclide (N) to a stable radiogenic daughter (D^*); the successive loss of atoms of N from radioactive decay are followed by proportional increase in the daughter atoms. Assuming the total number of daughter atoms includes those formed radiogenically, plus those initially present at $t = 0$, the total number of daughter atoms present in a system (D) is

$$D = D_0 + D^* \tag{3.12}$$

and with further substitution from equation 3.11,

$$D^* = N(e^{\lambda t} - 1)$$

$$D = D_0 = N(e^{\lambda t} - 1) \tag{3.13}$$

The radioactivity of a sample is measured as the number of disintegrations per minute (dpm). Since it is not practical to describe radioactivity by the number of atoms remaining (N) in the aforementioned equations, the observed *specific activity* (A) of a sample is usually described by the following equation:

$$A = \lambda N \tag{3.14}$$

The specific activity (A) of a radionuclide represents the observed counting rate in a sample. Radionuclides are useful in evaluating processes over timescales that are four to five times their half-lives, where beyond that time period there is typically only 1% or less of the nuclide remaining and it cannot be measured effectively. For example, ^{210}Pb has a half-life of 22.3 yr and a useful life of

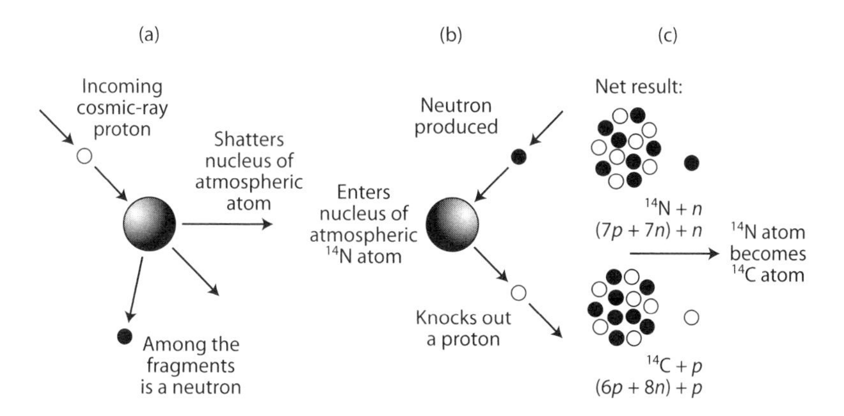

Figure 3.2. The reaction between neutrons and nitrogen ($^{14}N + n \rightarrow {}^{14}C + p$) is the dominant mechanism for the formation of ^{14}C in the atmosphere. (Adapted from Broecker and Peng, 1982.)

approximately 112 yr; thus, dating sediment or rocks older than about 140 yr is beyond the range of this nuclide.

3.1.4 Radiocarbon

The existence of radiocarbon (^{14}C) was not realized until 1934, when an unknown radionuclide was formed during exposure of nitrogen to neutrons in a cloud chamber (Kurie, 1934). In 1940, Martin Kamen confirmed the existence of ^{14}C when he prepared a measurable quantity of ^{14}C. Over the next few decades more details on the production rate of ^{14}C in the atmosphere and possible applications for dating archeological samples continued (Anderson et al., 1947; Arnold and Libby, 1949, 1951; Kamen, 1963; Ralph, 1971; Libby, 1982). In recent years, following the development of accelerator mass spectrometry and its ability to measure smaller amounts of carbon, radiocarbon has become a useful tool in biogeochemical studies (McNichol and Aluiwhare, 2007).

Similar to the other cosmogenic radionuclides, ^{14}C is produced by the reaction of cosmic rays with atmospheric atoms such as N_2, O_2, and others to produce broken pieces of nuclei, which are called *spallation* products (Suess, 1958, 1968). Some of these spallation products are neutrons that can also interact with atmospheric atoms to produce additional products, which include ^{14}C and other radionuclides (^{3}H, ^{10}Be, ^{26}Al, ^{36}Cl, ^{39}Ar, and ^{81}Kr) (fig. 3.2) (Broecker and Peng, 1982). Thus, the reaction between neutrons and nitrogen ($^{14}N + n \rightarrow {}^{14}C + p$) is the dominant mechanism for the formation of ^{14}C in the atmosphere. Once formed, ^{14}C decay occurs by the following reaction, and has a half-life of 5730 ± 40 yr:

$$^{14}C \rightarrow {}^{14}N + \beta^{-} + \text{neutrino} \tag{3.15}$$

The free ^{14}C atoms formed in the atmosphere become oxidized and form $^{14}CO_2$, which is rapidly mixed throughout the atmosphere (Libby, 1952). This ^{14}C then becomes incorporated into other reservoirs, such as the biosphere, as plants fix carbon during the process of photosynthesis; the exchange between the atmosphere and the surface ocean is estimated to take approximately five years (Broecker and Peng, 1982).

Living plants and animals in the biosphere contain a constant level of ^{14}C, but, when they die there is no further exchange with the atmosphere and the activity of ^{14}C decreases with a half-life of 5730 ± 40 yr; this provides the basis for establishing the age of archeological objects and fossil remains. The assumptions associated with dating materials are (1) that the initial activity of ^{14}C in

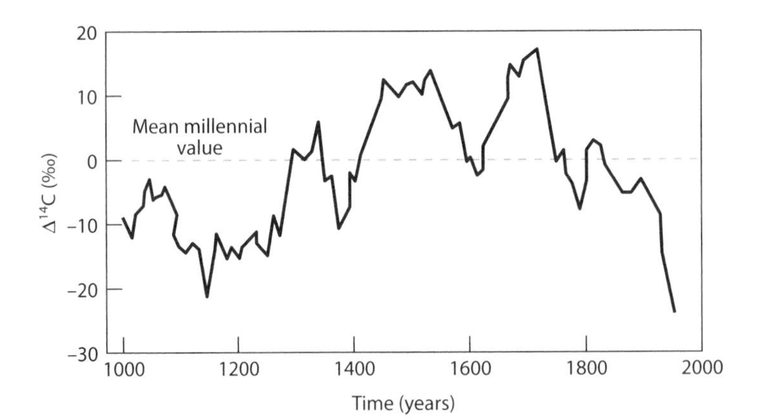

Figure 3.3. Variations in the atmospheric ratio of $^{14}C/C$ in CO_2 (expressed in $\Delta^{14}C$ notation) over the last 1000 years. (Adapted from Stuiver and Quay, 1981.)

plants and animals is a known constant and is independent of geographic location; and (2) that the sample has not been contaminated with modern ^{14}C (Faure, 1986). Unfortunately, measurements of ^{14}C in wood samples dated by dendrochronology indicate that there have been changes in the initial ^{14}C content over time (Anderson and Libby, 1951). Variations in the atmospheric ratio of $^{14}C/C$ in CO_2 (expressed in Δ notation, see below) over the last 1000 years are shown in fig. 3.3 (Stuiver and Quay, 1981). This variation is believed to result from a combination of the following factors: (1) changes in cosmic-ray flux from activities in the Sun; (2) changes in the Earth's magnetic field; and (3) changes in the reservoir of carbon on Earth (Faure, 1986). In addition to natural variations in ^{14}C, anthropogenic activities have both decreased and increased (by almost double) the atmospheric content of ^{14}C; this is due to combustion of fossil fuels over the past 100 years and nuclear weapons (i.e., "bomb fallout"), respectively. Variations in atmospheric ^{14}C content over the past 1000 years are quite evident. The dilution effect from inputs of fossil fuel combustion are clear after about 1850 in the early stages of the Industrial Revolution, and is referred to as the *Suess effect*. The dilution stems from the fact that fossil carbon has been isolated from exchange with the atmosphere for a long time period so that the ^{14}C signal has decayed, making it ^{14}C-free. Conversely, anthropogenically derived increases in the atmospheric content of ^{14}C are from nuclear bomb testing in the 1950s and early 1960s, whereby atmospheric ^{14}C nearly doubled over a short period of time. Ironically, this spike serves as a useful tracer, since it has been decreasing steadily since the *Test-Ban Treaty* of 1962. As a result, any carbon reservoir with a $\Delta^{14}C > 0$ ppt has clearly been affected by *bomb* ^{14}C (Eglinton and Pearson, 2001). Thus, the difference between pre- and postbomb $\Delta^{14}C$ values can be used to better understand biogeochemical processes that exchange with the atmosphere within years to decades compared to those that show no exchange (Eglinton and Pearson, 2001; McNichol and Aluiwhare, 2007). Finally, two other earlier times of anomalously high values of ^{14}C occurred ca. 1500 and 1710. These periods are referred to as the *de Vries effect*, and the causes remain unknown.

The basic equation used for determining $\Delta^{14}C$ is defined as

$$\Delta^{14}C = [(^{14}C/C)_{sample} - (^{14}C/C)_{standard}]/[(^{14}C/C)_{standard}] \times 1000 - IF \tag{3.16}$$

Like stable isotopes, the $\Delta^{14}C$ of a sample is measured relative to a standard to improve the accuracy and precision of accelerator mass spectrometry measurements (Elmore and Phillips, 1987). Multiplying the ratio by 1000 results in the delta ($\Delta^{14}C$) values having units of parts per thousand, also known as per mil (‰). For standards, it is necessary to use wood from trees harvested before about 1850 preindustrial to avoid the *Suess effect*. The standard value for preindustrialized atmospheric CO_2 is

13.56 dpm g^{-1} or $^{14}C/C$ equals 1.176×10^{-12} (Broecker and Peng, 1982). A term correcting for the effects of isotopic fractionation (IF) is also subtracted out of this equation. Isotopes are fractionated due to physical and chemical reactions. Enzymatic processes occurring during photosynthesis, for example, causes variations in the abundance of carbon isotopes (^{12}C, ^{13}C, and ^{14}C) in plants using different carbon-fixation processes (Faure, 1986). The National Bureau of Standards currently provides an oxalic acid ^{14}C standard that is used for this correction, but, there have been a number of challenges associated with development of this standard (Craig, 1954, 1961; Stuiver and Polach, 1977).

As described by Eglinton and Pearson (2001), the $\Delta^{14}C$ equation used for geochemical samples is as follows:

$$\Delta^{14}C = [\,f_m(e^{\lambda(y-x)}/e^{\lambda(y-1950)}) - 1] \times 1000 \tag{3.17}$$

where

f_m = fraction modern, which is expressed as the ratio $^{14}C/^{12}C_{sample}/^{14}C/^{12}C_{standard}$, similar to conventional terms for stable isotopic ratios described in equation 3.1 (e.g., $R = [^{13}C/^{12}C]$); $\lambda = 1/8267(y^{-1})(= t_{1/2}/\ln 2)$;
y = year of measurement;
x = year of sample formation (=deposition year when known from other radionuclide measurements)

The true half-life of ^{14}C should be used when making decay-related corrections (Eglinton and Pearson, 2001; McNichol and Aluwihare, 2007). Finally, the radiocarbon age (^{14}C yr) can be determined using the Libby half-life of 5568 yr as shown in the following equation:

$$\text{Age} = -8033\ln(f_m) \tag{3.18}$$

3.2 Stable Isotope and Radiocarbon Fractionation and Mixing

3.2.1 Isotopic Fractionation

Isotopic fractionation can occur from both *kinetic* and *equilibrium* effects. In simple and intuitive terms, physical processes affecting *kinetic fractionation* are largely due to the higher energy and more rapid diffusion and/or phase changes (e.g., evaporation) associated with the lighter isotope of a particular element. These fractionations tend to be associated with unidirectional processes (e.g., evaporation, diffusion, and dissociation reactions) and most biological (enzymatic) reactions. The energy of a molecule can be described by electronic, nuclear spin, translational, rotational, and vibrational properties (Faure, 1986; Fry, 2006; Hoefs, 2004). Biologically mediated reactions involving enzymes, for example, kinetically fractionate isotopes, resulting in significant differences in the isotopic composition of substrates (or reactants) and products. Assuming that the substrate is not limiting to a reaction and the isotope ratio of the product is measured within a short period of time, the fractionation factor (α) can be defined as follows:

$$\alpha = R_p/R_s \tag{3.19}$$

where

R_p = isotope ratio for the product
R_s = isotope ratio for the substrate (or reactant)

Biologically mediated kinetic fractionation will lead to a preference for the lighter isotope in the reaction products and a more negative delta value. This is due both to the fact that the lighter isotope has a higher average velocity as well as the lower dissociation energies and stabilities associated with bonds involving molecules with lighter isotopes. It is important to point out that organic matter tends to have low $^{13}C/^{12}C$ ratios, reflecting a preference for the ^{12}C isotope during carbon fixation. Kinetic isotope effects due to differences in dissociation energies can also be quite large in bacterial reactions. It is also important to note that kinetic isotope effects are described as primary, if the sensitivity to isotopic substitution exists at the position at which chemical bond changes during reaction, or secondary, when the sensitivity to isotopic substitution occurs at an atomic position not directly involved in the reaction itself.

Kinetic isotope fractionations typically occur in reactions that are unidirectional, where the reaction rates actually depend on the isotopic composition of the substrates and products. In an open system, the reactant is never significantly depleted; therefore, the degree of fractionation is constant over time and over the course of the process, provided other conditions also stay constant. In contrast, in a closed system, depletion of the reactant results in the heavier isotope becoming consumed to a greater extent than in the open system. If all the reactant is eventually consumed, the accumulated product will have the same ratio as the initial reactant. Therefore, in a closed or even partially closed system, isotopic ratios of [intermediate] products may not be predictable solely on the basis of the fractionation factor. In a closed system, the isotopic ratio of the substrate is related to the amount of substrate that has not been used (Mariotti et al., 1981); this is described by the *Rayleigh* equation:

$$R_s/R_{s0} = f^{(\alpha-1)} \tag{3.20}$$

where

R_{s0} = isotope ratio of the substrate at time zero
R_s = isotope ratio of the substrate at a particular time point
f = fraction of unreacted substrate at a particular time point

Equilibrium (thermodynamic) fractionation is due to physical processes and describes the distribution of isotopes of an element among phases of a molecule at equilibrium. In this case, there is isotope exchange, but no net chemical reaction.

An example of equilibrium isotope fractionation is the concentration of heavy isotopes of oxygen in liquid water, relative to water vapor:

$$H_2{}^{18}O(g) + H_2{}^{16}O(\ell) \rightleftharpoons H_2{}^{16}O(g) + H_2{}^{18}O(\ell) \tag{3.21}$$

At 20°C, the equilibrium fractionation factor for this reaction is

$$\alpha = \frac{(^{18}O/^{16}O)_{liquid}}{(^{18}O/^{16}O)_{gas}} = 1.0098 \tag{3.22}$$

Equilibrium fractionation is a type of mass-dependent isotope fractionation, while mass-independent fractionation is usually assumed to be a nonequilibrium process. The equilibrium fraction can be defined as it relates to kinetic isotope effects:

$$\alpha_{eq} = k_2/k_1 \tag{3.23}$$

where, k_1 and k_2 are the equilibrium rate constants for the heavy and light isotopes, respectively.

3.2.2 Carbon Stable Isotopes

Stable carbon isotopes are commonly used to distinguish between allochthonous versus autochthonous sources of organic carbon. One of the most important pieces of information that can be gathered from this information is the separation of C_3 and C_4 plant inputs (Peterson and Fry, 1989; Goñi et al., 1998; Bianchi et al., 2002; Cloern et al., 2002). Other pioneering work on the application of stable carbon isotopes in estuaries showed that these isotopes can be used to understand *trophic interactions* in food chain dynamics (Fry and Parker, 1979; Fry and Sherr, 1984). Since ^{13}C moves in a predictable way, with a 1 to 2‰ enrichment of ^{13}C for each step in a predator/prey interaction, the food source can be estimated from stable isotope analysis of tissues isolated from the predator (Parsons and Chen, 1995). Organic carbon is composed of a heterogeneous mixture of organic compounds having different ^{13}C values due to their biosynthetic pathway. Bulk $\delta^{13}C$ values represent the average isotopic composition of organic matter but subfractions of organic matter exhibit a range in isotopic signatures (e.g., protein and carbohydrate tend to have higher (more positive) $\delta^{13}C$ values while lipid and lignin generally have lower (more negative) $\delta^{13}C$ values (Deines, 1980; Hayes, 1983; Schouten et al., 1998).

Isotopic signatures are also complicated by differences in the susceptibility of different pools of organic matter to diagenesis. On average, polysaccharides and proteins are less decay resistant (more labile) than other biochemical classes (e.g., lipid and lignin). Since polysaccharides and proteins tend to be more ^{13}C-enriched than lipids (Deines, 1980; Hayes, 1983; Schouten et al., 1998), microbial decomposition of organic matter selectively removes the more labile ^{13}C-enriched components, resulting in the depletion of $\delta^{13}C$ in the remaining lignin-rich residue (Benner et al., 1987). Clearly, these general trends get considerably more complicated when factoring in different food sources (e.g., terrestrial versus aquatic) as well as size and physiology of the organisms involved (Incze et al., 1982; Hughes and Sherr, 1983; Goering et al., 1990). Isotopic fractionation distinguished on the basis of fractionation through photosynthetic pathways or trophic interactions provides the basis for discrimination of carbon sources (Fogel and Cifuentes, 1993; Hayes, 1993). Specifically, the carbon isotopic composition of naturally synthesized compounds is controlled by (1) carbon sources, (2) isotopic effects during assimilation processes in producer organisms, (3) isotope effects during metabolism and biosynthesis, and (4) cellular carbon budgets. The classic work by Park and Epstein (1960) and others (O'Leary, 1981; Fogel and Cifuentes, 1993) established that the enzyme ribulose 1,5-bisphosphate carboxylase (RuBP carboxylase) controls fractionation of carbon isotopes during photosynthesis in plants.

The C_3 pathway involves the fixation of atmospheric CO_2 or dissolved CO_2 (or seawater HCO_3^-) and results in production of one three-carbon compound (3-phosphoglycerate) from the enzymatic conversion of ribulose 1,5-bisphosphate. Through extensive lab studies, it was established that the isotopic fractionation between photosynthetic carbon ($\delta^{13}C = -27‰$) and atmospheric CO_2 ($\delta^{13}C = -7‰$) in C_3 land plants is approximately $-20‰$ (Stuiver, 1978; Guy et al., 1987; O'Leary, 1988). The fractionation is the same for C_3 aquatic plants, but photosynthetic biomass is enriched since dissolved CO_2 or HCO_3^- is used for carbon fixation ($\delta^{13}C = \sim 0‰$). Based on this fraction information, the following model for carbon isotopic fractionation in C_3 land plants was established:

$$\Delta = a + (c_i/c_a)(b - a) \tag{3.24}$$

where

$\Delta =$ isotopic fractionation
$a =$ isotope effects from diffusion ($-4.4‰$) (O'Leary, 1988)
$b =$ combined isotope effect from RuBP and phosphoenolpyruvate (PEP) carboxylases ($-27‰$)
$c_i/c_a =$ ratio of CO_2 inside the plant to atmospheric CO_2

The model assumes that the c_i/c_a ratio is important in determining the carbon isotopic composition of plant tissues (Farquhar et al., 1982, 1989; Guy et al., 1986). In general, when an unlimited supply of CO_2 is available to the plant, enzymatic fractionation will dominate; however, when CO_2 is limiting, fractionation during diffusion will predominate (Fogel and Cifuentes, 1993). These models show that carbon fixation in C_3 land plants is controlled by a dynamic process between atmospheric CO_2 availability—which is largely controlled by stomatal conductance—and enzymatic fractionation during photosynthesis. Since opening stomata for CO_2 also allows water to be lost from plant tissues to the atmosphere, plants using the C_3 photosynthetic pathway have had to evolve a balance between carbon uptake and water loss in a diversity of environments.

Over time other photosynthetic systems evolved that allowed for more efficient carbon uptake in plants (C_4 and crassulacean acid metabolism [CAM]) without significant water loss in extreme environments (Ehleringer et al., 1991). In C_4 plants (e.g., corn, prairie grasses, and *Spartina* spp. marsh grasses), the first step in carbon dioxide fixation results in the formation of the four-carbon compound oxaloacetate; this phase of fixation is catalyzed by the enzyme phosphoenolpyruvate (PEP) carboxylase. The fractionation during C_4 fixation is considerably smaller (~2.2‰) than that occurring with RuBP (O'Leary, 1988), making C_4 plants more enriched in ^{13}C (δ = ~8 to 18‰) (Smith and Epstein, 1971; O'Leary, 1981). Differences in the efficiency with which CO_2 is converted to new plant material in C_4 plants, via the Calvin cycle in bundle-sheath cells of vascular plants, further minimize carbon fractionation (Fogel and Cifuentes, 1993). A modified fractionation model for carbon isotopes in C_4 plants, as developed by Farquhar (1983), is as follows:

$$\Delta = a + (b_4 + b_3\Phi - a)(c_i/c_a) \tag{3.25}$$

where

b_4 = isotopic effects from diffusion into bundle - sheath cells
b_3 = isotopic fractionation during carboxylation (−2 to −4‰) (O'Leary, 1988)
Φ = leakiness of the plant to CO_2

Unlike C_3 or C_4 land plants, diffusion of CO_2 limits photosynthesis in aquatic plants. Consequently, algae have adapted a mechanism that allows them to actively transport or "pump" either CO_2 or HCO_3^- across their membrane, thereby accumulating DIC (dissolved inorganic carbon) in the cell (Lucas and Berry, 1985). Although carbon fixation is through RuBP carboxylase for both algae and C_3 higher plants, their isotopic signatures differ because much of the stored CO_2 pool does not leave the cell in algae. Thus, the rationale for using carbon isotopes to distinguish between aquatic and land plant inputs to estuaries is based on the idea that phytoplankton generally use HCO_3^- as a carbon source for photosynthesis ($\delta^{13}C = 0$‰), as opposed to land plants which use CO_2 from the atmosphere ($\delta^{13}C = -7$‰) (Degens et al., 1968; O'Leary, 1981). This results in algae being more enriched in ^{13}C (hence, having more positive $\delta^{13}C$ values) than terrestrial plants (table 3.3). The model for carbon fractionation in aquatic plants is as follows:

$$\Delta = d + b_3(F_3/F_1) \tag{3.26}$$

where

d = equilibrium isotope effect between CO_2 and HCO_3^-
b_3 = isotope fractionation associated with carboxylation
F_3/F_1 = ratio of CO_2 leaking out of the cell to the amount inside the cell

The carbon isotopic composition of phytoplankton has been shown to be strongly affected by pCO_2 of surface waters in the ocean (Rau et al., 1989, 1992). Moreover, fractionation of carbon isotopes by phytoplankton is also correlated with cell growth rate, cell size, cell membrane permeability, and CO_2 (aq) (Laws et al., 1995; Rau et al., 1997).

In addition to fractionations occurring during carbon uptake, isotopic fractions occur during the biosynthesis of different biochemicals. This results in differences in the isotopic signature of different

Table 3.3
Published ranges of isotope values of potential organic matter sources to estuaries

Source	$\delta^{13}C$ (‰)	$\delta^{15}N$ (‰)	$\Delta^{14}C$ (‰)	*Reference*
Terrigenous (vascular plant)	−26 to −30	−2 to +2		Fry and Sherr (1984); Deegan and Garritt (1997)
Terrigenous soils (surface) forest litter	−23 to −27	2.6 to 6.4	+152 to +310	Cloem et al. (2002); Richter et al. (1999)
Freshwater phytoplankton	−24 to −30	5 to 8		Anderson and Arthur (1983); Sigleo and Macko (1985)
Marine, estuarine phytoplankton	−18 to −24	6 to 9		Fry and Sherr (1984); Currin et al. (1995)
C-4 salt marsh plants	−12 to −14	3 to 7		Fry and Sherr (1984); Currin et al. (1995)
Benthic microalgae	−12 to −18	0 to 5		Currin et al. (1995)
C-3 freshwater, brackish marsh plants	−23 to −26	3.5 to 5.5		Fry and Sherr (1984); Sullivan and Moncreiff (1990)

classes of organic matter (e.g., protein, carbohydrate, and lipid) as well as individual compounds. The ^{13}C content of biomolecules depends on (1) the ^{13}C content of carbon source utilized, (2) isotope effects associated with carbon assimilation, (3) isotope effects associated with metabolism and biosynthesis, and (4) cellular carbon budgets at each branch point during biosynthesis. Compound-specific isotope analysis (CSIA) provides the ability to analyze the stable isotopic composition of individual biomarkers, thereby enhancing our ability to identify the sources of organic matter in environmental samples and the metabolic pathways influencing organic matter in marine and aquatic systems (Macko et al., 1986, 1990; Freeman et al., 1990; Hayes, 1993). Hayes (2001) provides a thorough review of biosynthetic processes and their influence on carbon and hydrogen isotopes at the biochemical level. CSIA was first applied to $\delta^{13}C$ analysis of hydrocarbons but has subsequently been extended to other biogenic elements (e.g., nitrogen and hydrogen) and a variety of biochemicals (amino acids, carbohydrates, lipids, and lignin phenols). Examples of the broad application of CSIA across a variety of organic biomarkers are included in later chapters of this book.

3.2.3 Nitrogen Stable Isotopes

Like carbon, nitrogen isotopes ($\delta^{15}N$) are susceptible to both equilibrium and kinetic fractionations. Isotopic discriminations between inorganic nitrogen sources and organic nitrogen produced by photosynthetic organisms are the result of fractionation during the primary assimilation of inorganic nitrogen (e.g., N_2, NO_3^-, NO_2^-, NO_4^+). These inorganic nitrogen species can have a wide range of $\delta^{15}N$, depending on the signatures of the original source pools, the degree to which they are recycled by microbial processes, and the enzymatic pathways by which inorganic nitrogen is fixed into organic matter during photosynthesis. Microbial processes such as mineralization, nitrogen fixation, assimilation, nitrification, and denitrification influence the $\delta^{15}N$ composition of inorganic and organic nitrogen species (Sharp, 2007). In addition, anthropogenic activities such as application of organic and inorganic fertilizers also influence the $\delta^{15}N$ composition of environmental samples. As a result, the $\delta^{15}N$ composition of primary producers and heterotrophic organisms can vary considerably and unpredictably.

Since the geochemical cycling of nitrogen is more complex than that of carbon due to both the large number of species contributing to the nitrogen pool (inorganic and organic forms) and the numerous processes nitrogen undergoes in the environment, it is difficult to use N isotopes alone to identify sources of organic matter. Instead, nitrogen-stable isotopes ($\delta^{15}N$) are often used in combination with carbon or sulfur isotopes (Peterson et al., 1985; Currin et al., 1995; Cloern et al., 2002). However, $\delta^{15}N$ isotopes afford an additional use as tracers for elucidating trophic relationships between organisms. The basis for this application is that $\delta^{15}N$ is enriched by 2 to 3 per mil with each increase in trophic level, due to preferential excretion of the lighter isotope. As a result, stable isotopes of nitrogen are widely used in food web studies to investigate the trophic relationships between organisms.

3.2.4 Isotopic Mixing Models

In many aquatic systems, the sources of carbon and nitrogen consist of complex mixtures, which vary in their $\delta^{13}C$ and $\delta^{15}N$ composition. Isotope mixing models provide a useful tool for deciphering the sources of carbon or nitrogen in such systems (Fry, 2006), particularly when there are a small number of potential sources and they are isotopically distinct. Isotope mixing models have been widely used in ecological studies, many in estuarine or coastal settings with complex sources (Peterson and Fry, 1987; Peterson and Howarth, 1987; Currin et al., 1995; Wainright et al., 2000). To illustrate the use of isotope mixing models, we provide the following example of a simple, two end-member model. An estuarine organism has a $\delta^{13}C$ value of −16 per mil, suggesting it likely assimilates a mixture of two different food sources (e.g., source-1 and source-2), each with a distinct isotopic signature (−22 and −12 per mil, respectively). The relative contributions of these food sources to the organism's diet can be calculated using the following equations:

$$\delta^{13}C_{organism} = (f_{source-1} \times \delta^{13}C_{source-1}) + (f_{source-2} \times \delta^{13}C_{source-2}) \tag{3.27}$$

and

$$f_{source-1} + f_{source-2} = f_{organism} = 1 \tag{3.28}$$

In this case, since we are solving for both $f_{source-1}$ and $f_{source-2}$, one variable can be expressed in terms of the other:

$$f_{source-1} = 1 - f_{source-2} \tag{3.29}$$

Equation 3.26 then becomes

$$\delta^{13}C_{organism} = [(1 - f_{source-2}) \times \delta^{13}C_{source-1}] + (f_{source-2} \times \delta^{13}C_{source-2}) \tag{3.30}$$

We can then solve for $f_{source-2}$ using the known values of $\delta^{13}C_{organism}$, $\delta^{13}C_{source-1}$, and $\delta^{13}C_{source-2}$. In this case the solution to the mixing model suggests that the organism's biomass is composed of 60% source-2 (or $f_{source-2} = 0.6$) and 40% source-1 (or $f_{source-1} = 0.4$).

Although isotope mixing models have been widely applied in ecological and oceanographic studies, they work best when applied to systems with a small number of sources, each being isotopically distinct. An assumption of these models is that the mixing between two sources is linear. As the number of sources increases, it becomes impossible to obtain a unique solution to these models. For example, if a third source with an intermediate isotopic signature of −16 per mil was added to the example above, it would be impossible to obtain a unique solution as illustrated by the fact that the equation could be solved with a ratio of source-1:source-2:source-3 of 40:60:0 or 0:0:100. In complicated situations where multiple sources are involved, multiple stable isotopes or a combination of stable isotope and biomarker measurements may be useful for resolving the relative contribution of different sources (see chapter 8).

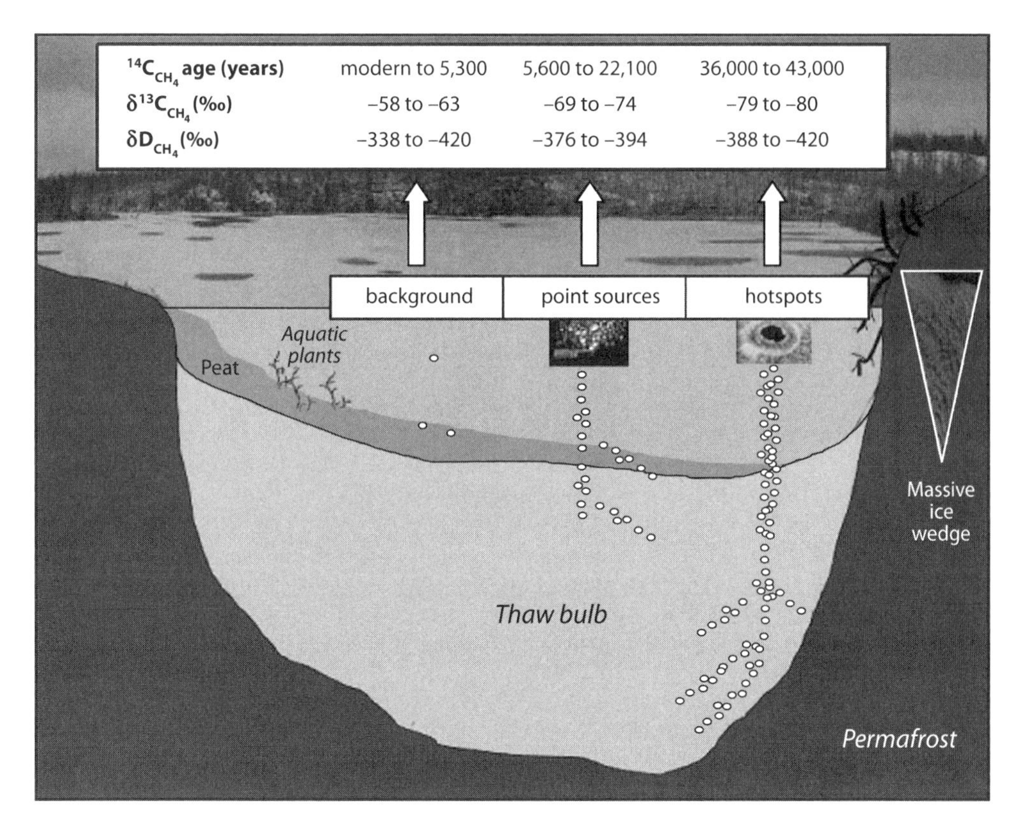

Figure 3.4. CH_4 emissions from a range of arctic and boreal lakes in Siberia and Alaska using $\delta^{13}C$, δD, and $\Delta^{14}C$ isotopic ratios. (Adapted from Walter et al. 2008.)

3.3 Applications: Stable Isotopes and Radiocarbon as Biomarkers

3.3.1 Stable Isotopes and Radiocarbon Biomarkers in Lakes

Past work has shown that isotopic signatures of CH_4 can be used to construct isotope mass balance models to better estimate the source and sinks of atmospheric CH_4 (Hein et al., 1997; Bousquet et al., 2006). Methane emissions from lakes and wetlands have shown high temporal variability with *ebullition* being a major contributor, particularly in lakes (Casper et al., 2000; Walter et al., 2006). Stable hydrogen and carbon isotopes can be used to distinguish between bacterial, thermogenic, and biomass burning sources of CH_4 (Whiticar et al., 1986). Similarly, ^{14}C ratios of CH_4 have been used to estimate the relative importance of fossil carbon natural gas seepage and coal mining processes (Wahlen et al., 1989). Most of the isotopic studies in lakes have been conducted in North America and Europe (Martens et al., 1986; Wahlen et al., 1989). More recently, a study was conducted on CH_4 emissions from a range of arctic and boreal lakes in Siberia and Alaska using $\delta^{13}C$, δD, and $\Delta^{14}C$ isotopic ratios (Walter et al., 2008). The overall objective was to determine the role of ebullition as related to *thermokarst* erosion in global atmospheric CH_4 processes. This study concluded that the isotopic compositions, concentrations and fluxes could be categorized into three distinct zones (background, point sources, and hot spots) (fig. 3.4). The major pathway for CH_4 production from point sources was from CO_2 reduction, while general background sources of CH_4 were from acetate fermentation. Results from this study also revealed that CH_4 from ebullition had an isotopically distinct

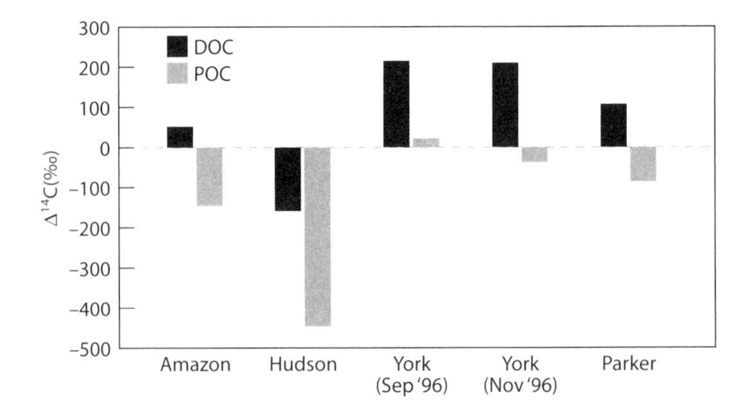

Figure 3.5. The general paradigm emerging in river/estuarine systems is that DOC tends to be more ^{14}C-enriched (or younger) than POC. (Adapted from Raymond and Bauer, 2001a.)

signature ($\delta^{13}CH_4$ = −58‰, ^{14}C age = modern) relative to CH_4 from wetlands ($\delta^{13}CH_4$ = −70‰, ^{14}C age = 16,500 yr BP) in northern latitudes. To date, this process has been ignored in many of the inverse models used for modeling CH_4 cycling in high latitudes (Walter et al., 2008).

3.3.2 Stable Isotopes and Radiocarbon Biomarkers along the River–Estuarine Continuum

Carbon and nitrogen isotopes have been used successfully in estuarine and coastal systems to trace source inputs of terrestrial and aquatic organic matter (Peterson et al., 1985; Cifuentes et al., 1988; Horrigan et al., 1990; Westerhausen et al., 1993) as well as sewage and nutrients (Voss and Struck, 1997; Caraco et al., 1998; Holmes et al., 2000; Hughes et al., 2000). Additionally, stable isotopes have been widely used in food web studies in salt marshes and other coastal habitats. These studies are based on the premise that organisms retain the isotopic signals of the foods they assimilate as demonstrated in the classic work of DeNiro and Epstein (1978, 1981a, b). While the basic assumption of this application of isotopes remains, subsequent studies have shown the complexities associated with distinguishing stable isotopic signatures due to seasonal changes in food resources (Fogel et al., 1992; Simenstad et al., 1993; Currin et al., 1995; Fry, 2006), effects of isotopic shifts between trophic levels (Fry and Sherr, 1984; Peterson et al., 1986), and decomposition effects on isotopic composition during early diagenesis in litter and sediments (DeNiro and Epstein, 1978, 1981a, b; Benner et al., 1987; Jasper and Hayes, 1990; Montoya, 1994; Sachs 1997).

Until recently, ^{14}C measurements have not been widely used in studies investigating organic carbon cycling in coastal and estuarine regions (Spiker and Rubin, 1975; Hedges et al., 1986). However, there has been a considerable increase in the application of ^{14}C measurements, both alone and in combination with stable isotopes, in recent years (Santschi et al., 1995; Guo et al., 1996; Guo and Santschi, 1997; Cherrier et al., 1999; Mitra et al., 2000; Raymond and Bauer, 2001a–c; McCallister et al., 2004). The general paradigm emerging in river/estuarine systems is that dissolved organic carbon (DOC) tends to be more ^{14}C-enriched (or younger) than particulate organic carbon (POC) (fig. 3.5) (Raymond and Bauer, 2001a). This is because DOC is believed to be derived from fresh leachate of surface soil litter (Hedges et al., 1986; Raymond and Bauer, 2001a–c). Understanding the linkage between the carbon sources in drainage basin soils and river/estuarine systems is critical in determining the connection between terrestrial and aquatic systems. The residence time of organic carbon in soils has been examined using ^{14}C measurements (O'Brien, 1986; Trumbore et al., 1989; Schiff et al., 1990;

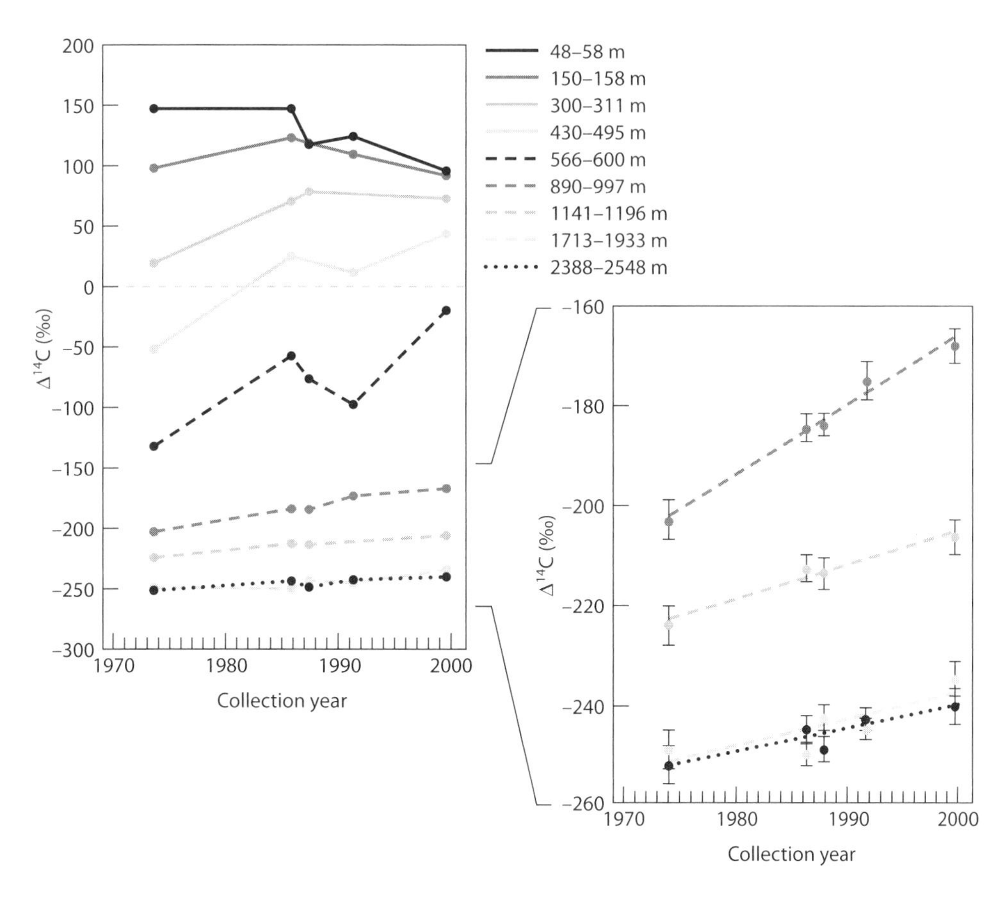

Figure 3.6. Bomb ^{14}C has increased in the deep waters of the North Central Pacific (NCP) over the period of 1973 to 2000. This trend was observed after reoccupation of a site in the NCP. (Adapted from Druffel et al., 2008.)

Trumbore, 2000). These studies support the concept that ^{14}C-enriched DOC from surface soils is exported to streams and is younger than the soil organic carbon from which it is derived.

Radiocarbon isotopes have also provided new insights about the organic matter supporting bacterial production. Until recently, aged organic matter was thought to be resistant to microbial attack. McCallister and colleagues showed that dissolved organic matter (DOM) leached from ancient deposits was utilized and oxidized by bacteria in the water. This finding opens an entirely new window in estuarine biogeochemistry. McCallister et al. (2004) were able to provide quantitative estimates of the "diets" of bacteria growing in situ in different reaches of two estuaries as a function of season. Bacteria in the mesohaline zone of the York River, Virginia (USA) in May 2000 were sustained mostly on OM derived from estuarine phytoplankton (83–87% of total), but this local source was supplemented by terrigenous DOM (13%) and small amounts of OM from a salt marsh (1–5%). In contrast, in the Hudson River (USA), up to 25% of carbon assimilated by bacteria originated from ~24,000 yr BP sedimentary deposits.

3.3.3 Stable Isotopes and Radiocarbon as Biomarkers in the Ocean

^{14}C measurements have been widely used in oceanic environments to gain insights about the age distribution of different pools of organic carbon (Williams and Gordon, 1970; Williams and Druffel,

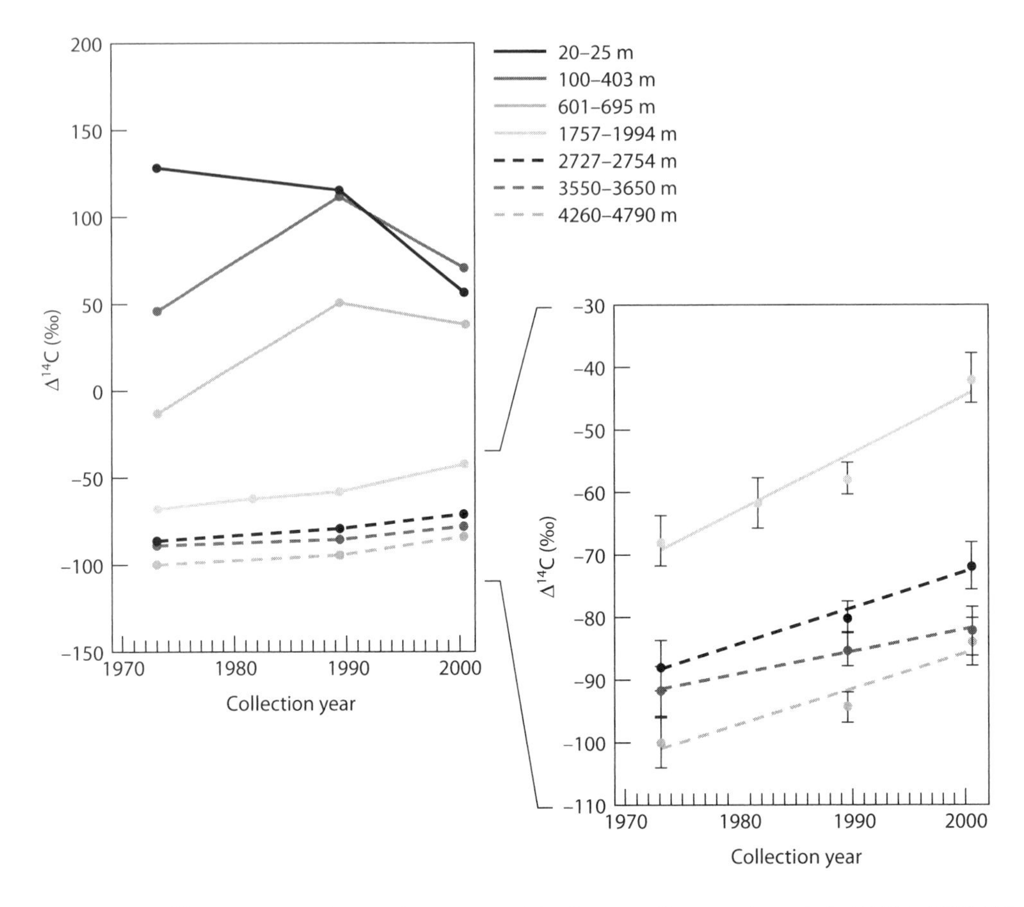

Figure 3.7. Bomb ^{14}C has increased in the deep waters of the Sargasso Sea over the period of 1973 to 2000. This trend was observed after reoccupation of a site in the SS. (Adapted from Druffel et al., 2008.)

1987; Bauer and Druffel, 1998; Bauer et al., 1998; Druffel et al., 1992; Bauer, 2002). For example, Broecker and Peng (1982) used the ^{14}C in CO_2 extracted from oceanic water to determine the age and circulation patterns between surface and bottom waters. Dissolved inorganic carbon in the oceans represents the largest exchangeable pool carbon (36,000 Gt C) on Earth (Druffel et al., 2008). Recent work has shown that after reoccupation at two sites in the north central Pacific (NCP) and Sargasso Sea (SS) there was an observed increase in $\Delta^{14}C$ in bottom waters (Druffel et al., 2008). In particular, bomb ^{14}C has continued to increase in the deep waters of the NCP and SS over the period of 1973 to 2000. In the NCP, the results suggest changes in deep circulation or the introduction of bomb ^{14}C to this depth. Increases in bomb ^{14}C to the deep SS from 1989 to 2000 likely reflect southward transport of North Atlantic deep water (NADW) to this site. (figs. 3.6 and 3.7). The decrease in mean $\delta^{13}C$ values of DIC within the upper 1000 m at both sites between the period of 1972 and 2000 suggested the presence of ^{13}C-depeleted atmospheric anthropogenic sources (fig. 3.8). Interestingly, it was concluded that remineralization of POC in surface waters could not supply enough bomb ^{14}C to deep waters at these sites. Moreover, it was concluded that the increased bomb ^{14}C SS bottom waters was from NADW, produced in the Norwegian and Greenland Sea, as shown with other tracers (Smethie et al., 1986; Ostlund and Rooth, 1990). However, the observed increases in bomb ^{14}C at the NCP were likely due to changes in deep water circulation patterns. Further work is needed to better understand these purported changes (Druffel et al., 2008).

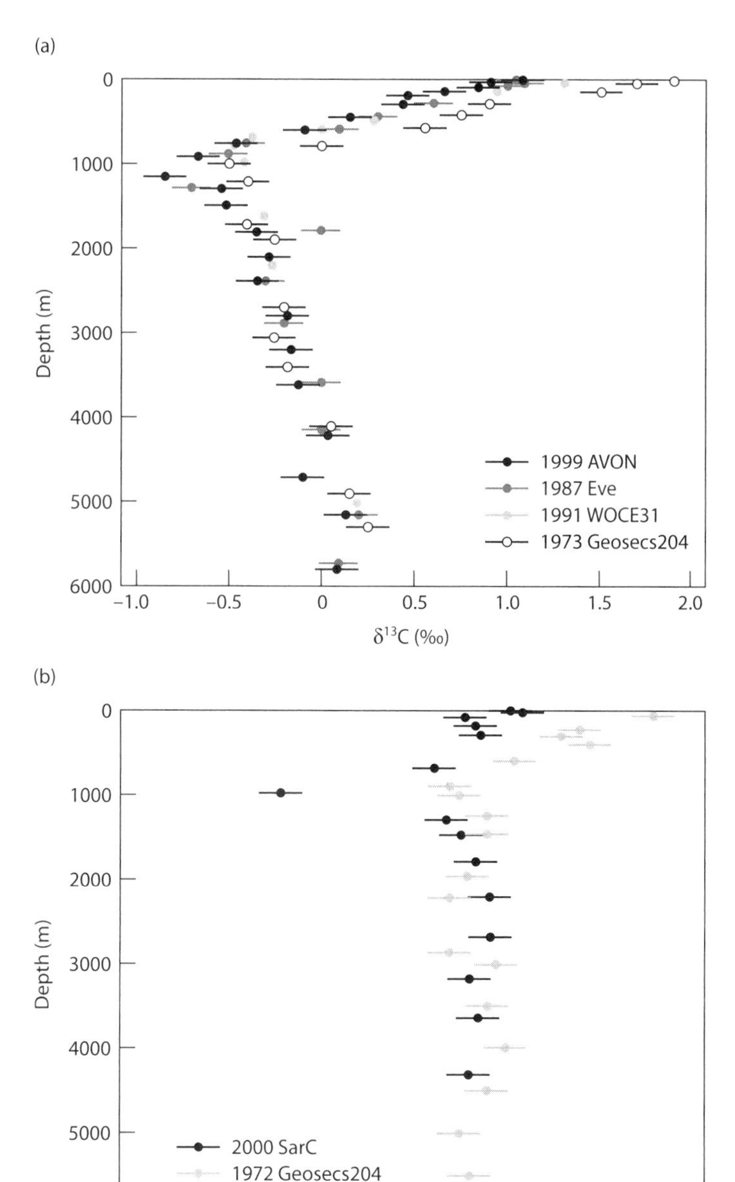

Figure 3.8. A decrease in mean $\delta^{13}C$ values of DIC within the upper 1000 m at two sites in the period between 1972 and 2000. Reoccupation of sites in the (a) North Central Pacific and (b) Sargasso Sea suggests the presence of ^{13}C-depeleted atmospheric anthropogenic sources. (Adapted from Druffel et al., 2008.)

3.4 Summary

In this chapter we introduced some of the basic principles of isotope geochemistry along with some examples of how stable isotopes and radiocarbon have been used in examining organic matter source inputs to aquatic ecosystems. Technological advances in the last decade or so have allowed for greater sensitivity with smaller samples, which shows great promise for further coupling of chemical biomarker and bulk isotopic techniques. In particular, CSIA has proven to be invaluable in providing more information on the specific sources and age of organic matter. The more exhaustive analytical approaches now being published that use dual-isotopic (e.g., ^{13}C and ^{14}C) tracers, CSIA, and multiple chemical biomarkers have allowed for greater insights in our understanding of the cycling, preservation, and sources of organic matter in aquatic ecosystems.

4. Analytical Chemical Methods and Instrumentation

4.1 Background

Sampling and analytical tools are at the heart of the field of organic geochemistry, and technology has advanced the field throughout its history. This chapter reviews some of the primary analytical methods used by organic geochemists as well as new and evolving tools that show promise for their application to aquatic organic geochemistry. The ability to detect organic compounds in environmental samples at lower concentrations, as well as separate and characterize compounds of higher mass and complexity, has increased dramatically in recent decades. Overall, advances in our ability to characterize organic matter in environmental samples have developed in three areas: (1) new sampling methods that allow either concentration or isolation of organic matter associated with a wider variety of matrices, (2) innovations in analytical chemistry that provide improved detection limits and/or allow the analysis of new compound classes, and (3) the adaptation and application of methods borrowed from fields outside environmental science.

While the specific details of the analytical techniques used in biogeochemical methods are not typically presented in most books in the fields of ecosystem ecology and geosciences, we feel it is critical to provide at least some foundation, albeit limited, on the instrumental approaches used in chemical biomarker analyses. To achieve this, we begin this chapter with a description of some of the relevant field sampling techniques for isolating dissolved organic matter (DOM) and particulate organic matter (POM) in the context of the continually expanding discipline of *separation sciences*, As described in the *Journal of Separation Science*, this field includes "all areas of chromatographic and electrophoretic separation methods... This includes fundamentals, instrumentation and applications, both in analytical and preparative mode." Before returning to advanced separation techniques involving the more sophisticated electrophoretic and chromatographic analytical approaches, we first describe how the bulk properties of POM and DOM are analyzed in the absence of separation science. We begin with spectroscopic analyses from the early simple spectro- and flourometric analyses to more advanced methods currently used in spectrofluorometry, and then on to nuclear magnetic resonance (NMR) spectroscopic applications. We then return to separation sciences with details on the electrophoretic and chromatographic instrumental analyses, from the simple spectrofluorometric detector approaches to more sophisticated instrumentation such as mass spectrometry. Finally, we conclude with stable and radiocarbon analyses, including both bulk and compound-specific approaches.

An important consideration in these analytical approaches is that the ability to see organic matter in terms of its discrete biochemical components decreases as it becomes increasingly altered (i.e., a larger fraction of organic matter becomes uncharacterizable over time). Many of the techniques described here are opening up analytical windows and providing new insights about the composition and fate of organic matter in the environment.

4.2 Analytical Processes for the Separation and Collection of Natural Samples

Generally speaking, the analysis of organic compounds associated with environmental samples involves multiple steps, including (1) sample collection, (2) extraction or isolation of the organic fractions of interest from the sample matrix, (3) separation of the organic fraction into its components, (4) compound detection, identification, and quantification, and (5) data analysis and interpretation. Prior to sampling, an experimental design should be developed that places the samples in a framework that will allow the data to be interpreted within an environmental context. External factors that influence the processes being studied should be considered in the sampling plan (e.g., temporal and spatial variability, biological activity, physical processes). Sampling should also take into consideration how to best represent the environment or process(es) being studied. As a result, factors such as environmental variability, replication, and data analysis methods should be built into the sampling plan. It is also important to avoid sampling methods that will introduce artifacts or bias that may influence the samples or interpretation of data evolving from them. This includes consideration of the types of materials used for the sampling devices, sampling efficiency, contamination, sorptive losses, as well as storage and preservation of samples.

4.2.1 DOM Separation

Organic geochemical studies are typically conducted on environmental samples associated with both dissolved and solid matrices (fig. 4.1). Traditional samples include suspended particles that are filtered or concentrated from water samples, sinking particles from sediment trap studies, soil and sediment samples, and fossil materials such as rocks, coal, and petroleum. A number of new analytical methods have been developed in recent decades for concentrating DOM, including resins (e.g., XAD), *ultrafiltration*, *solid-phase extraction,* and, most recently, *reverse osmosis–electrodialysis* (RO-ED). Following concentration, DOM can be further characterized by a variety of spectroscopic and chromatographic methods (Guo and Santschi, 1997; Minor et al., 2001, 2002; Guo et al., 2003; Simjouw et al., 2005). The analysis of DOM has advanced considerably in recent decades (see reviews by Hedges et al., 1993a; Hansell and Carlson, 2002; Mopper et al., 2007). However, until recently, characterization of marine DOM has been particularly challenging because the methods used to concentrate DOM also concentrate salts that interfere with analyses. An additional challenge is that many of the concentration methods isolate only a portion of the DOM and that portion may not be representative of the entire DOM pool. Ultrafiltration technology was introduced in the early 1990s and has since been widely applied to the concentration of DOM in aquatic samples (Bauer et al., 1992; Benner et al., 1992; Amon and Benner, 1996; Aluwihare et al., 1997, 2002; Loh et al., 2004; Repeta and Aluwihare 2006). Relative to other methods (e.g., XAD resins), ultrafiltration offers several advantages: (1) high surface area; (2) the ability to separate colloidal organic matter (COM) and truly

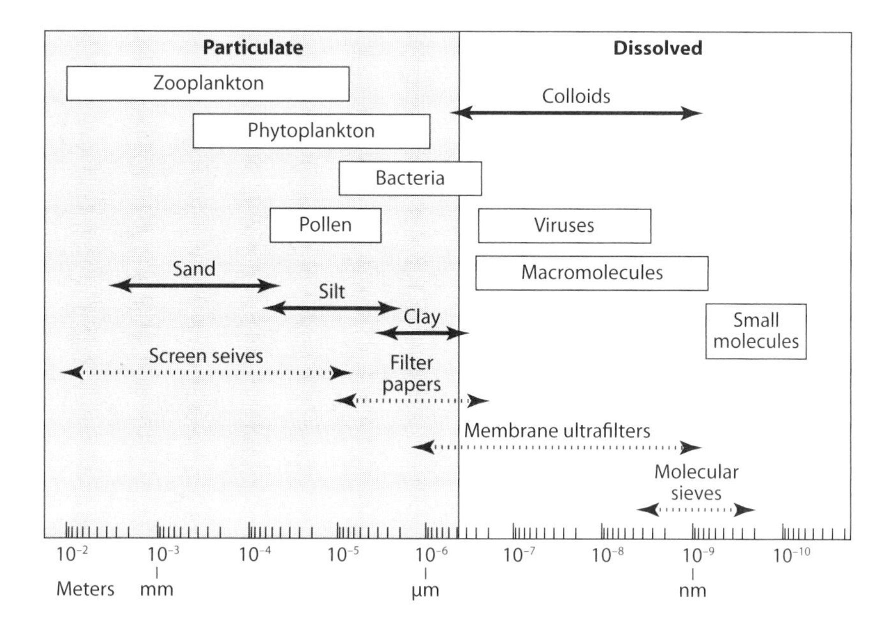

Figure 4.1. Sizes and separation methods for dissolved and particulate organic matter in seawater. (Adapted from Hansell and Carlson, 2002.)

dissolved organic matter; (3) higher flows, which facilitate concentration of larger volumes of DOM and COM for characterization; (4) no pretreatment, such as changes in pH, which likely causes fewer artifacts; and (5) higher yields (recoveries) of DOM than were typical of other methods (e.g., more than twice that provided by isolation of DOM using XAD resins).

More recently, the development of coupled RO-ED provides a promising method for circumventing some of the analytical challenges associated with DOM characterization (Vetter et al., 2007; Koprivnjak et al., 2009). Chloride and strong cations (Na^+, Ca^{2+}, Mg^{2+}, Sr^{2+}, and K^+) are removed from solution during electrodialysis by exchange through semi-permeable membranes aided by an electric field (Mopper et al., 2007). Cation exchange chromatography can then be used to remove low concentrations of residual cations and chloride can be removed by freeze-drying since it has a high vapor pressure. This method has several advantages over ultrafiltration and XAD resins including the following: (1) it does not require changes in pH, (2) the DOM does not interact with organic solvents or other eluting agents, (3) losses due to size fractionation are minimal, and (4) DOC recoveries are greater than 60% (Vetter et al., 2007). RO-ED yields a desalted, concentrated liquid sample from which low-ash natural organic matter can be obtained as a freeze-dried solid material. This material can then be further characterized by spectroscopic and chromatographic methods.

4.2.2 POM Separation

Advances have also been made in the collection and analysis of particulate organic matter (POM). Sediment traps have been the traditional method for collecting sinking particles and have been used widely in studies in both lacustrine and oceanic environments to characterize the flux and composition of sinking particulate matter (see chapters 7–10). However, sediment traps have limitations, and problems with lateral flow, incorporation of swimmers, and preservatives have been widely acknowledged (Lee et al., 1992; Hedges et al., 1993b; Wakeham et al., 1993; Buesseler et al., 2000, 2007). Recently, a new and improved class of sediment traps has been designed that addresses

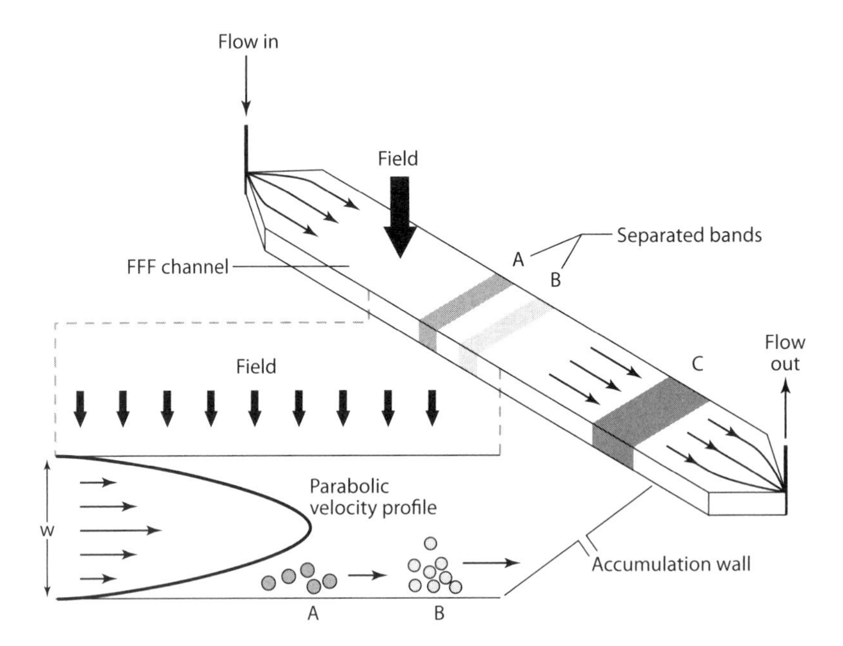

Figure 4.2. Diagram of field-flow fractionation (FFF) channel system, A,B, and C represent separated analyte components. (Adapted from Schure et al., 2000.)

some of these issues (Peterson et al., 2005). Specific modifications to the new sediment trap design include (1) collection of particles based on discrete particle settling-velocity ranges; (2) development of a large, free-floating NetTrap based on the design of a closing plankton net capable of collecting large amounts (similar to 1 g) of very fresh sinking particulate material in short time periods (24–36 h); and (3) an elutriation system to fractionate particles from the NetTrap into settling-velocity classes using countercurrents of varying speeds. This new generation of sediment traps has been tested recently in a variety of coastal and oceanic systems (Goutx et al., 2007; Lamborg et al., 2008).

Another area of current research has been focused on understanding the "architecture" of organic matter. A series of papers published in the 1990s hypothesized that the associations between organic matter and mineral phases controlled the preservation of organic matter in the environment (Keil et al., 1994, 1997; Mayer, 1994; Hedges and Keil, 1995; Ransom et al., 1998). In recent years, this has led to detailed characterizations of particulate matter and sediments, including size and density separation (Arnarson and Keil, 2001; Coppola et al., 2007). Additionally, these studies have utilized a variety of spectroscopic tools to better understand the associations between organic matter and mineral surfaces (Brandes et al., 2004, 2007; Steen et al., 2006; Yoon, 2009). Methods from the fields of material sciences and soil geochemistry have increasingly been employed to better understand mineral–organic matter interactions and to examine their role in organic matter preservation (Zimmerman et al., 2004a,b; Steen et al., 2006). While the role of sorptive preservation has not been completely answered, we now have a better understanding of the complex associations between organic matter and inorganic materials as well as how these associations influence enzyme activity and organic matter preservation.

4.2.2.1 Field-Flow Fractionation

Field-flow fractionation (FFF) represents a family of analytical techniques developed specifically for isolating and characterizing macromolecules, supramolecular assemblies, colloids, and particles (fig. 4.2) (Schure et al., 2000). This method separates both soluble and colloidal components over a

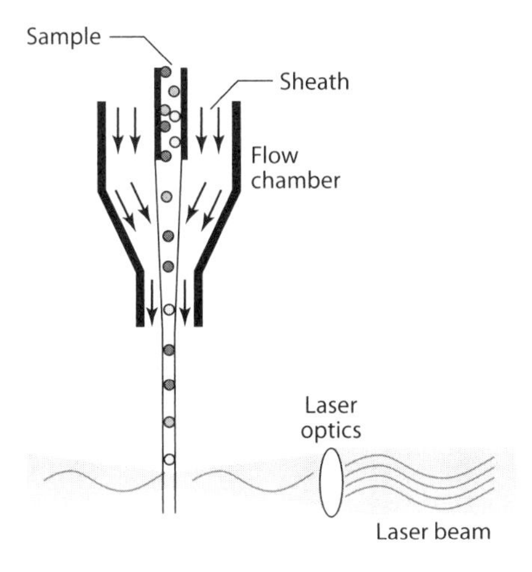

Figure 4.3. Flow cytometry schematic. (From *Current Protocols in Cytometry*, Unit 1.2 , p1.2.2.)

wide size range and is unmatched in its powers in the submicrometer range, with a large dynamic range from a few nanometers to a few micrometers (Schimpf et al., 2000; Messaud et al., 2009). It also has the benefit of rapid analysis, typically 10 to 30 min. FFF utilizes the combined effects of a laminar flow profile with an exponential concentration profile of analyte components (Messaud et al., 2009), and is undergoing increasingly widespread use in aquatic sciences. For example, FFF has provided a unique approach for the separation of marine colloidal matter by molecular size (Beckett and Hart, 1993; Hasselhov et al., 1996; Floge and Wells, 2007). FFF is a chromatographic style analytical technique that separates colloids as a function of their diffusion coefficient (Giddings, 1993). This allows for the ability to obtain a continuous size spectrum of colloidal matter that may then be analyzed by various in-line detectors (e.g., spectrophotometers, fluorometers) and other techniques (Floge and Wells, 2007).

4.2.2.2 Flow Cytometry

Flow cytometry was introduced into biological oceanography as a tool for rapidly and precisely estimating cell numbers of pico- and nano-phytoplankton (e.g., algal cells of less than 20 μm in size), within the ecological timescale (e.g., hours to days) of phytoplankton cycles (Legendre and Le Fevre, 1989; Legendre and Yentsch, 1989; Wietzorrek et al., 1994). Flow cytometers use the principle of hydrodynamic focusing for presenting cells to a laser (or any other light excitation source). The sample is injected into the center of a sheath flow. The combined flow is reduced in diameter, forcing the cell into the center of the stream, which allows the laser to detect one cell at a time. A schematic of the flow chamber and its relation to the laser beam in the sensing area is shown in fig. 4.3. This method has been used to complement epifluorescence microscopy by allowing higher-resolution sampling of cell composition and number along ship transects and during incubation experiments. Flow cytometry also allows for the quantification of cellular pigments (e.g., chlorophyll and phycobilins). This information is useful in determining how changes in light, nutrients, and other physical parameters affect phytoplankton physiology over ecological time. Flow cytometry has also been used to study ecophysiological aspects of pico- and nano-phytoplankton through the use of specific fluorescent dyes and/or fluorogenic substrates (Dorsey et al., 1989). Recently, there have been new developments that allow for the application of flow cytometric sorting and *direct temperature-resolved mass spectrometry* (DT-MS) to examine POC samples (Minor et al., 1999). Direct temperature-resolved mass spectrometry is a rapid (~2-min) measurement that provides molecular-level characterization across a wide range of compound classes, including both desorbable and pyrolyzable components.

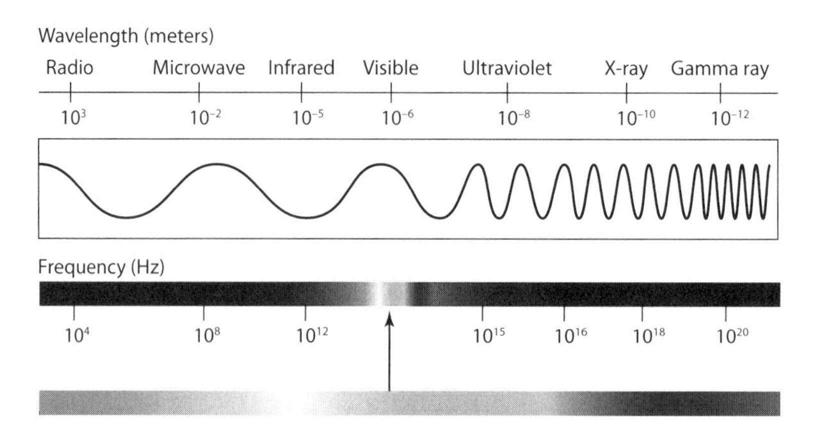

Figure 4.4. The electromagnetic spectrum.

Consequently, this technique along with flow cytometry bridges the gap between information available from bulk measurements, such as elemental analysis, and information derived from detailed but laborious compound-class-specific analyses. Finally, since flow cytometry has been shown to be useful because of its speed and quantitative measurements, it has been limited by the need to take discrete water samples for analysis onboard ship or in the laboratory. To solve this problem recent work has developed an automated flow cytometer (FlowCytobot) that can operate in situ and unattended (Olson et al., 2003).

4.3 Spectroscopic Techniques and Instrumentation

4.3.1 Basic Principles

While compounds typically exist at their lowest energy or ground state, certain electronic, vibrational, and rotational transitions can occur in these compounds upon interaction with the appropriate electromagnetic radiation. For example, once electrons reach their excited state in these compounds they can return to their ground state by emitting the energy they absorbed at the same or lower frequency or through heat loss, which may have no radiation relaxation. (McMahon, 2007). Certain molecules absorb *ultraviolet* (UV), *visible* (Vis), and/or near *infrared* (NIR) radiation. NIR can cause vibrations of molecules to increase, while absorption of UV and Vis can cause electrons to move to higher electronic orbitals. More dramatically, x-rays can cause bonds to actually break and can ionize molecular structures. In the *electromagnetic spectrum* (fig. 4.4), ranges of radiation for UV, Vis, and NIR are from 190 to 350 nm, 350 to 800 nm, and approximately 800 to 2500 nm, respectively (McMahon, 2007). In the UV-Vis absorption the concentrations of certain molecular species is a function of the *Beer-Lambert law* as shown here:

$$A_\lambda = \varepsilon_\lambda c l \tag{4.1}$$

where

A_λ = absorbance at a particular wavelength (λ)

ε_λ = extinction coefficient at a particular wavelength (λ)

c = concentration

l = pathlength

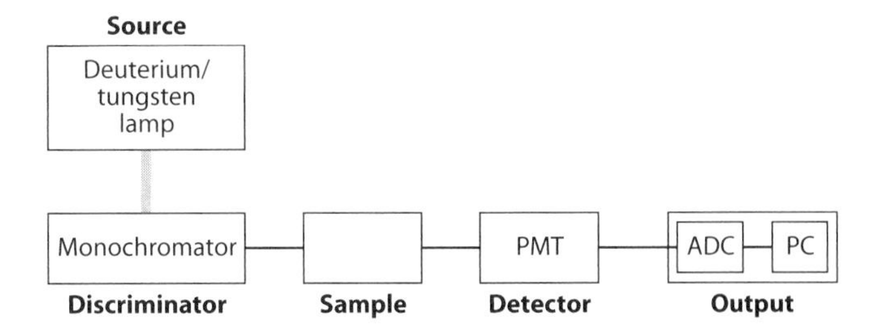

Figure 4.5. Diagram of a typical single-beam UV-Vis-NIR spectrophotometer (PMT = photomultiplier tube; ADC = analog to digital conversion; PC = personal computer). (Adapted from McMahon, 2007.)

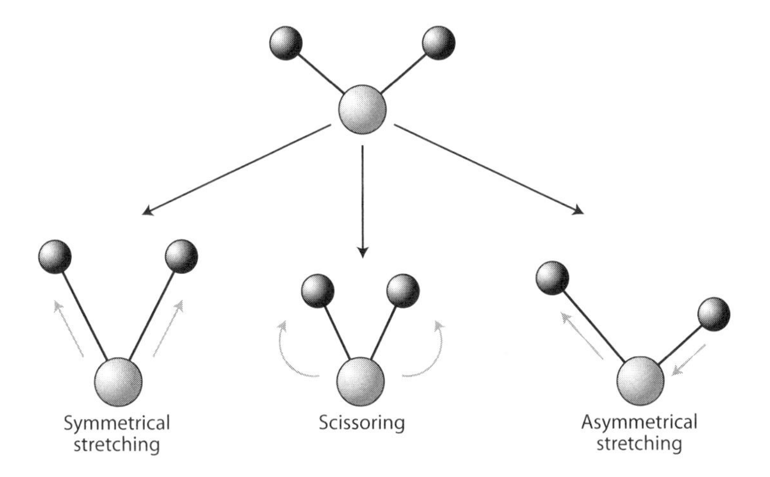

Figure 4.6. Fourier transform infrared (FT-IR) vibrational transitions.

A schematic of a typical spectrophotometer, which can be single or double beam, is shown in fig. 4.5. In this particular example we use a single-beam to illustrate the basic design. Single-beam spectrophotometers have the disadvantage of being unable to compensate for real-time fluctuations in system parameters, which could adversely affect the resulting data. A dual-beam spectrophotometer analyzes light emanating from a sample (e.g., by reflectance, transmission, or any other suitable reactance) in response to a sample being illuminated by a monochromatic light beam of known wavelengths. This single monochromatic source beam is split into dual beams, one beam for illumination of the sample and the other for a reference. This allows a dual-beam spectrophotometer to provide a reference signal that, when compared with the sample signal, minimizes the effects of real-time fluctuations.

The operating principles of IR spectroscopy are based on vibrational techniques. These techniques are useful for determining structural information about molecules as well as details on functional groups and their orientation with respect to isomerization. Here we refer to infrared (IR) as the midrange of the electromagnetic spectrum, which covers 2500 to 25,000 nm. When molecules absorb IR there are polarity changes in the molecule, which result in vibrational and rotational transitions of the molecule (McMahon, 2007). Some examples of the transitional changes resulting from IR absorption are illustrated in fig. 4.6. Many of the modern instruments are known as dispersive instruments and have been replaced by *Fourier transform infrared* (FT-IR) spectrometers. The FT-IR instrument uses an interferometer, rather than a monochronometer, for obtaining a spectrum, which allows for improvements in signal-to-noise ratios, speed, and simultaneous measurements across all

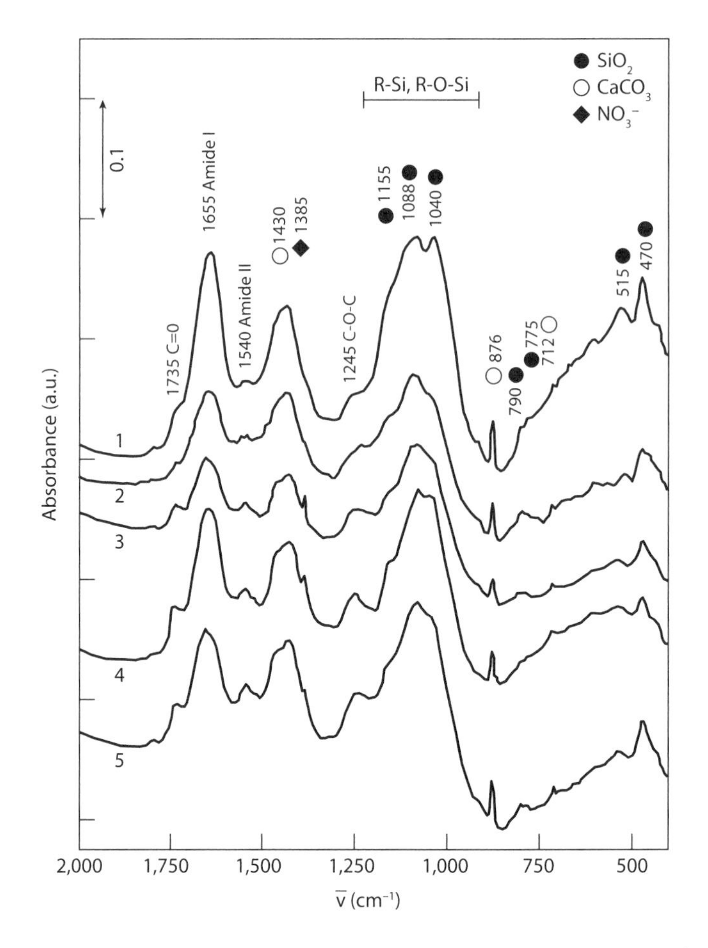

Figure 4.7. FT-IR spectra of macroaggregates in KBr pellets (over the range 2000–400 cm^{-1}) collected on June 13, 2000 (1); July 16, 1997 (2); July 31, 1997 (3); August 13, 1997 (4); and September 4, 1997 (5). (Adapted from Kovac et al., 2002.)

wavelengths, which develop when light distorts the electron density in a molecule (see details on Raman effects in McMahon [2007]). FT-IR has been used frequently in aquatic systems to characterize both DOM and POM. For example, the FT-IR spectra of macroaggregates collected from the Adriatic Sea in 1997 and 2000 (fig. 4.7) (Kovac et al., 2002) generally showed absorptions in the range 2800 to 3600 cm^{-1}, which are believed to correspond with hydrogen bonded OH and N–H groups and weaker absorptions in the range 2800–3000 cm^{-1} originating from various aliphatic components. In the range 400–1800 cm^{-1}, numerous bands were noted and attributed to the vibrations of organic (proteins and polysaccharides) and inorganic components (carbonates and silicates) believed to be largely derived from diatoms.

4.3.2 Spectrofluorometry

As mentioned earlier, chlorophyll *a* concentrations have been used routinely for estimating standing stock and productivity of photoautotrophic cells in freshwater and marine environments. Spectrophotometric analysis of chlorophyll pigments was developed in the 1930s and 1940s (Weber et al., 1986). Richards and Thompson (1952) first introduced a trichromatic technique that was supposed to measure chlorophylls *a*, *b*, and *c*; this technique used a set of equations that attempted to correct for interferences from other chlorophylls at the maximum absorption wavelength for each (Richards and Thompson, 1952). Since the 1960s, fluorometric methods have generally been used for analyses in the field. However, many of the early fluorometers could not unambiguously distinguish fluorescence band overlaps when detected with wide bandpass filters (Trees et al., 1985). Moreover, such fluorometers generally underestimated chlorophyll *b* and overestimated chlorophyll *a* when chlorophyll *c* co-occurred in high concentrations (Bianchi et al., 1995; Jeffrey et al., 1997). Many of the more recently developed spectrofluorometers have improved considerably and are better able to deal with these interference problems (e.g., Dandonneau and Niang, 2007). Moreover, in vivo spectrofluorometric analyses over large distances in the field still prove useful for rapid results, relative to the more time-consuming, albeit accurate method of high-performance liquid chromatography (HPLC) (Jeffrey et al., 1997, and references therein).

Another application of fluorescence spectroscopy in aquatic waters has involved analyses of DOM. In particular, 3-dimensional excitation–emission matrix (EEM) spectroscopy offers the ability, through nondestructive and highly sensitive means, to examine the types and origin of DOM in different aquatic systems (e.g., Mopper and Schultz, 1993; Coble, 1996, 2007; Cory and McKnight, 2005; Kowalczuk et al., 2005; Murphy et al., 2008). Fluorescence spectroscopy of DOM in aquatic systems typically involves scanning and recording a number of different emission spectra (e.g., 260–700 nm) and excitation wavelengths (e.g., between 250 and 410 nm, at increments of approximately 10 nm) (Parlianti et al., 2000). More recently, fluorescence EEM spectra (EEMs) have been analyzed using parallel factor analysis (PARAFAC), whereby the fluorescence signal for DOM is decomposed, using a multiway method, into unique fluorescent groups that are related in abundance to DOM precursor materials (Stedmon et al., 2003; Stedmon and Markager, 2005). Fluorescence EEMs are usually collected in a given natural system and then fit to a preexisting PARAFAC model data, because despite an attempt to provide a tutorial for building a PARAFAC model (Stedmon and Bro, 2008), component selection and other problems still exist when developing a new model (Cory and McKnight, 2005). Consequently, many researchers have resorted to using preexisting PARAFAC models to fit their EEM data (Cory and McNight, 2005); however, fitting EEMs to a PARAFAC model developed for different aquatic environment may result in an inaccurate interpretation of EEM data. On the positive side, a recent study reported that fitting EEMs to an existing PARAFAC model was fine as long as the DOM for the model and EEMs were from similar aquatic environments (Fellman et al., 2009).

4.3.3 Nuclear Magnetic Resonance (NMR) Spectroscopy

NMR spectroscopy is a powerful and theoretically complex analytical tool that can be used to characterize organic matter and elucidate the structures of organic compounds. We will now briefly describe some of the basic principles of NMR before discussing its application in estuarine systems. In the most general terms, many nuclei (e.g., ^{1}H, ^{13}C, ^{31}P, ^{15}N) have an intrinsic spin state that when placed in a strong magnetic field, simultaneously irradiated with electromagnetic (microwave) radiation, absorbs energy through a process called *magnetic resonance* (Solomons, 1980; Levitt, 2001). This resonance is detected by the instrument as an NMR signal. NMR uses the interaction between spin and strong magnetic fields using a radio-frequency (RF) radiation that allows simulation transitions between different spin states of the sample in a particular magnetic field (McCarthy et al., 2008). A simple illustration of an NMR spectrometer is shown in fig. 4.8. RF transmitters typically generate frequencies of a few MHz to almost 1 GHz, which are very effective at irradiating sample molecules. Most NMR instruments scan multiple magnetic field strengths that interact with the spinning nuclei that become aligned with the external magnetic field of the instrument. The microwave frequencies of liquid and solid-state NMR instruments generally range from 60 to 600 MHz-with the application

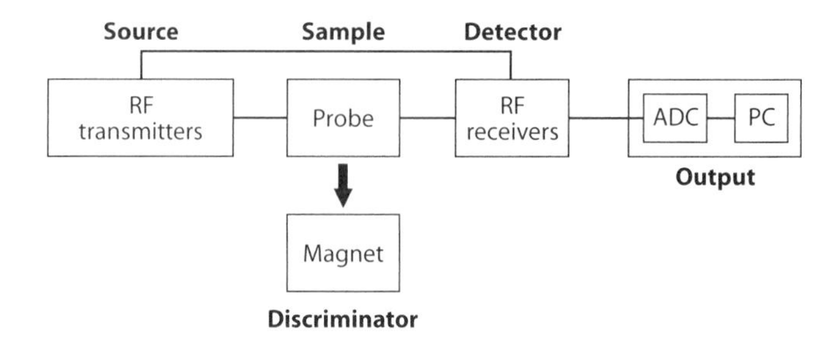

Figure 4.8. General design for a nuclear magnetic resonance (NMR) spectrometer. RF, radio frequency; PC, personal computer; ADC, analog-to-digital converter. (Adapted from McMahon, 2007.)

of higher frequency resulting in greater resolution. The actual magnetic field detected by the nucleus of a "naked" atom is not the same as one that has a different electronic environment; these differences provide the basis for determining the structure of a molecule. Differences in field strengths are caused by the *shielding* or *deshielding* of electrons associated with adjacent atoms, and this electron interference is referred to as a *chemical shift*. The chemical shift of a nucleus will be proportional to the external magnetic field and, since many spectrometers have different magnetic field strengths, it is common to report the shift independent of the external field strength. Chemical shifts are typically small (in the range of Hz) relative to field strength (range of MHz), so it has become convenient to express chemical shift in units of part per million (ppm) relative to an appropriate standard. Cross-polarization, magic-angle spinning (CPMAS) NMR is another commonly used method that takes molecules with high degrees of spin and imparts this spin to molecules that have none, making it extremely useful for getting signals on atoms that usually do not have a spin (Levitt, 2001).

Proton (^{1}H) and ^{13}C-NMR have been the most common NMR tools for the non-destructive determination of functional groups in complex biopolymers in plants, soils and sediments, and DOM in aquatic ecosystems (Schnitzer and Preston, 1986; Hatcher, 1987; Orem and Hatcher, 1987; Benner et al., 1992; Hedges et al., 1992, 2002). Although ^{13}C-NMR signals are generated in the same way as proton signals, the magnetic field strength is wider for the chemical shifts of ^{13}C nuclei. One of the key advantages in using ^{13}C-NMR for characterizing organic matter in natural systems is to be able to determine the relative abundances of carbon associated with the major functional groups. Although ^{13}C-NMR does not provide as much insight into the specific decay pathways of certain biopolymers, compared to the use of chemical biomarkers, it does provide a nondestructive approach that allows a greater understanding of a larger fraction of the bulk carbon. A sample CPMAS ^{13}C-NMR spectrum of colloidal material collected in the Gulf of Mexico is presented in fig. 4.9, showing a large peak between 60 and 90 ppm, indicative of carbohydrates (table 4.1)—most likely derived from phytoplankton exudates. Similarly, a sample spectrum using ^{1}H-NMR of a pure diatom culture shows a broad peak in the region of $\delta = 3.4$–5.0 ppm (CH) and the signals for $\delta = 5.0$–5.8 ppm, indicative of anomeric protons, which are assigned to the carbohydrates and proton resonances for organo-silicon compounds (Si–R, Si–O–R) at $d < 0.7$ ppm, both reflective of diatoms (fig. 4.10) (Kovac et al., 2002).

The application of ^{31}P-NMR (Ingall et al., 1990; Hupfer et al., 1995; Nanny and Minear, 1997; Clark et al., 1998) and ^{15}N-NMR (Alemandros et al., 1991; Knicker and Ludemann, 1995; Knicker, 2000) has been useful in characterizing organic and inorganic pools of P and N in natural systems, respectively. More recently, two-dimensional (2D) ^{15}N ^{13}C-NMR has been used to study the fate of protein in experimentally degraded algae (Zang et al., 2001; Mopper et al., 2007).

One area where ^{13}C-NMR first appeared as a useful tool for estuarine studies was its application in characterizing bulk chemical changes in decomposing wetland plant materials (Benner et al., 1990; Filip et al., 1991). For example, Benner et al. (1990) used four structural types of carbon (paraffinic,

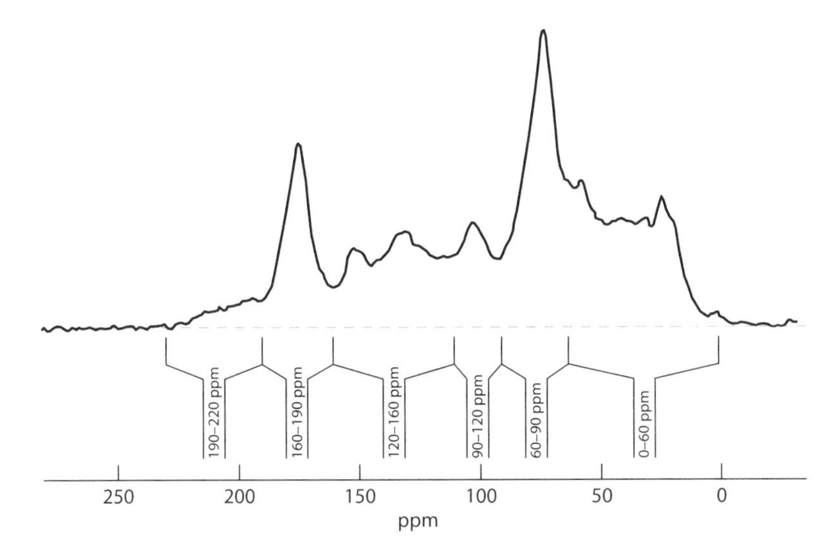

Figure 4.9. CPMAS ^{13}C-NMR spectrum of marine DOM showing the integration values (in % normalized to the total of 100% C) for the various spectral regions. The spectrum was obtained from ultrafiltered DOM (>1 kDa) from the Gulf of Mexico. (Unpublished data from Bianchi et al., Texas A&M University.)

Table 4.1
Chemical shift regions used for integration of peak areas in ^{13}C-NMR spectra of natural organic matter (NOM) and their respective assignments

Integration region (ppm)	*Identity*
0–45	Paraffinic carbons from lipids and biopolymers
45–60	Methoxyl, mainly from lignin, and amino groups
60–90	Carbohydrate carbons
90–120	Carbohydrate anomeric and proton-substituted aromatic carbons
120–140	Carbon-substituted aromatic carbons, mainly from lignin and nonhydrolyzable tannins
140–160	Oxygen-substituted aromatic carbons, mainly from lignin and hydrolyzable tannins
160–190	Carboxyl and aliphatic amide carbons from degraded lignin and fatty acids
190–220	Aldehyde and ketone carbons

Source: After Dria et al. (2002).

carbohydrate, aromatic, and carbonyl) to determine changes in bulk C mangrove leaves (*Rhizophora mangle*) during decomposition in tropical estuarine waters. This work showed a preferential loss of carbohydrates and preservation of paraffinic polymers. Similarly, ^{13}C-NMR analyses on DOM released from decomposing *Spartina alterniflora* showed preferential loss of carbohydrates from fresh tissues (Filip et al., 1991). Two-dimensional $^{15}N-^{13}C$ NMR analysis of the remains of the decomposed microalgae, labeled with ^{13}C and ^{15}N isotopes, indicated that proteins and highly aliphatic compounds were preserved over time (Zang et al., 2001), supporting the concept of encapsulation, where labile compounds are protected by macromolecular components, such as *algaenan* (Knicker and Hatcher, 1997; Nguyen and Harvey, 2001; Nguyen et al., 2003).

Characterization of DOM using ^{13}C-NMR indicates that river- and marsh-derived DOM is largely composed of aromatic carbon, reflective of lignin inputs from vascular plant materials (Lobartini et al., 1991; Hedges et al., 1992; Engelhaupt and Bianchi, 2001). Conversely, DOM in the Mississippi River

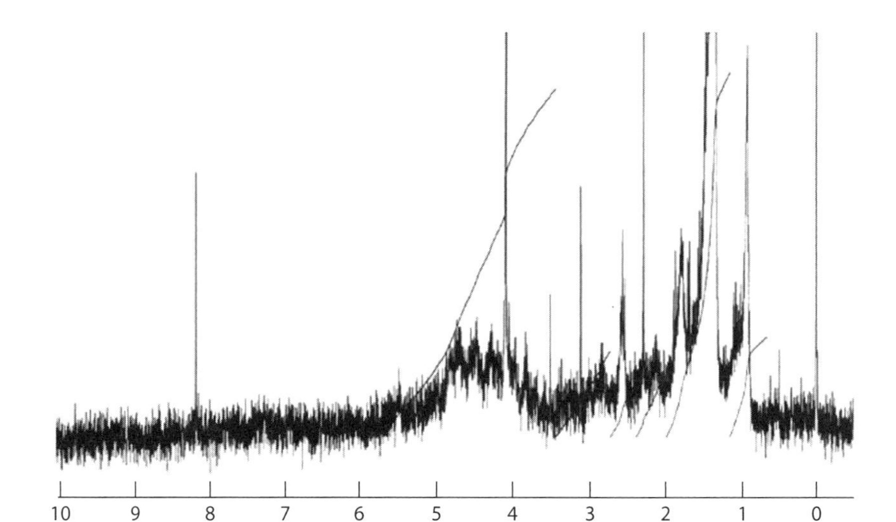

Figure 4.10. ^{1}H-NMR spectrum of the diatom culture *Skeletonema costatum*. Resonances of carbohydrates are at $\delta = 3.4–5.8$ ppm, of aliphatic components (R) are at $\delta = 0.9–1.8$ ppm, and of functional groups, such as ester and amide groups (COR), are at $\delta = 2.1–2.7$ ppm. (Adapted from Kovac et al., 2002.)

(USA) was found to be primarily composed of aliphatic versus aromatic compounds, and believed to be largely derived from freshwater diatoms growing under high-nutrient, low-light conditions (Bianchi et al., 2004). Other sources of DOM to estuaries that can significantly affect bulk carbon composition include sediment porewater fluxes and local runoff from soils. For example, ^{13}C NMR work indicated that under anoxic conditions porewaters were dominated by DOM primarily composed of carbohydrates and paraffinic structures derived from the decay of algal/bacterial cellulose and other unknown materials—with minimal degradation of lignin (Orem and Hatcher, 1987). Under aerobic conditions, porewater DOM had lower carbohydrate and higher abundance of aromatic structures. Recent work by Engelhaupt and Bianchi (2001) further supported that lignin was not degraded efficiently under low oxygen conditions in a tidal stream, adjacent to Lake Pontchartrain estuary (USA).

4.4 Advanced Separation Techniques and Instrumentation

4.4.1 Gas Chromatography and Gas Chromatography–Mass Spectrometry

Capillary gas chromatography (GC) and GC–mass spectrometry (GC-MS) have been the traditional primary tools used by organic geochemists to identify and quantify organic compounds in environmental samples. GC analysis separates a mixture of organic compounds by partitioning the compounds between a mobile phase (gas) and stationary phase (column) (fig. 4.11). Commercially available fused silica capillary columns, with dimensions of 0.20 to 0.25 mm i.d. and 30 to 60 m length, are used routinely for modern-day GC analysis. Typically, the stationary phase for these columns is chemically bonded to the silica, resulting in a larger thermal range and minimal column bleed. Applications involving GC are limited to compounds that can be volatilized either directly or following derivitatization. Although only a relatively small fraction (5 to 10%) of the organic compounds present in environmental samples are amenable to GC analysis, this method has broad application to many classes of organic contaminants and natural products (e.g., hydrocarbons, halogenated hydrocarbons,

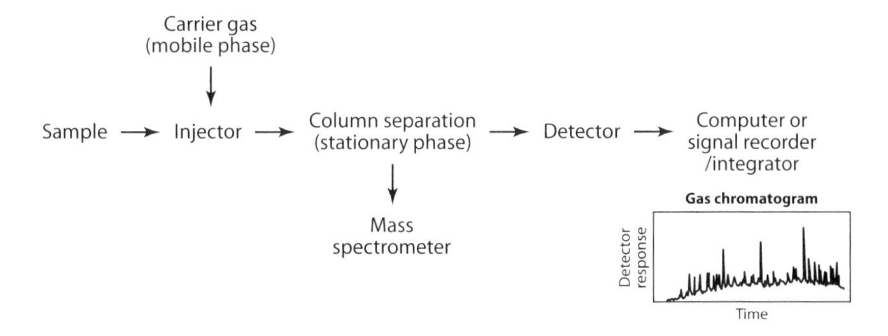

Figure 4.11. Schematic with major components of gas chromatograph (GC) showing effluent from column going to either a detector or a mass spectrometer.

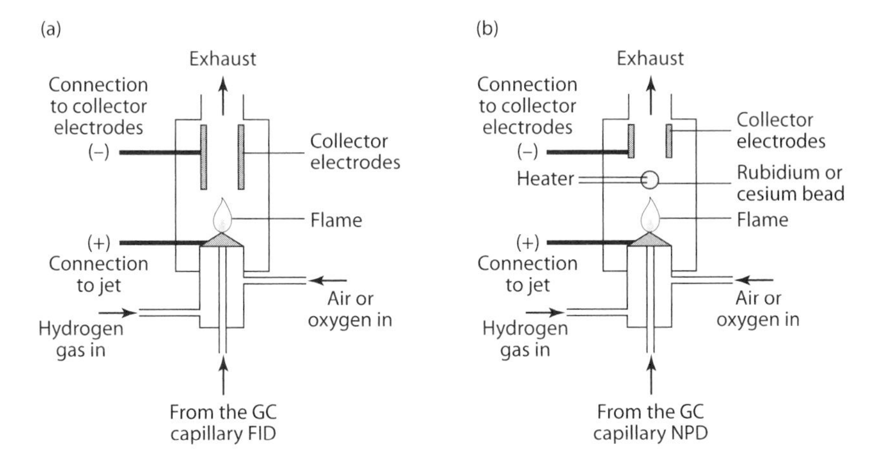

Figure 4.12. Schematics of (a) a flame ionization detector (FID) and (b) a nitrogen phosphorus detector (NPD). (Adapted from McMahon, 2007.)

pesticides, lignin phenols, fatty acids, and sterols). GC-MS combines the ability to resolve structurally similar compounds from one another (GC) with mass spectral information that can be used to reliably identify individual compounds. Computerized GC-MS is able to both detect and identify compounds using multiple pieces of information, including GC retention times, elution order and patterns, as well as MS fragmentation patterns. Peters and colleagues provide nice overviews of the theory behind GC and GC-MS as well as their application to the discipline of organic geochemistry in their books (Peters and Moldowan, 1993; Peters et al., 2005).

4.4.2 GC Detectors and Ionization Techniques

A wide variety of detectors have been used in combination with capillary GC for the analysis of environmental samples. The flame ionization detector (FID) is the most common detector used for GC analyses and is often considered the universal detector, since it responds to all organic compounds with similar levels of sensitivity (fig. 4.12a). The FID uses a hydrogen/air flame to decompose carbon-containing molecules. The nitrogen phosphorus detector (NPD) is more sensitive and selective for organic compounds that contain nitrogen and/or phosphorus (fig. 4.12b) (McMahon, 2007). FIDs are mass-sensitive detectors that offer the benefits of a high linear dynamic range (10^6), moderately

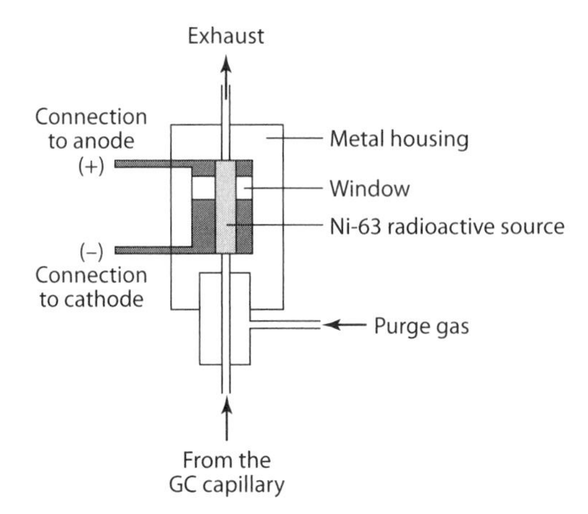

Figure 4.13. Schematic of an electron capture detector (ECD). (Adapted from McMahon, 2007.)

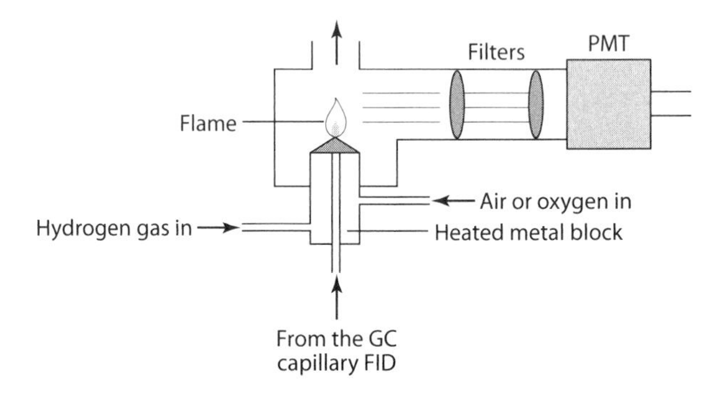

Figure 4.14. Schematic of a flame photometric detector (FPD). (Adapted from McMahon, 2007.)

high sensitivity and good stability. A drawback of FID is that it is destructive to samples. Effluent from the GC column is mixed with hydrogen and burned at the tip of a jet in an excess of air. When organic matter enters the detector, it is combusted in the flame, forming ions that are collected by an electrode positioned above the jet. The response of the FID (peak area) is proportional to the number of carbon atoms passing through the detector. Electron capture detectors (ECD) are also widely used in organic geochemistry (fig. 4.13). The ECD is a selective detector that responds to molecules with high electron affinity. It is particularly sensitive to halogenated compounds but can also be used for nitro compounds, highly oxygenated compounds, and some aromatic compounds. Unlike FID, ECD response is variable and its sensitivity can vary by several orders of magnitude, even for compounds that are structurally similar. It also has a relatively limited linear dynamic range. Another detector used commonly in GC analysis of organic compounds is the flame photometric detector (FPD), which is used for trace analysis of phosphorus- and sulfur-containing compounds. In contrast to the above detectors (fig. 4.14), the thermal conductivity detector (TCD) is not destructive to samples (fig. 4.15). For this reason, it is commonly used in preparative GC. The TCD responds to a variety of sample types, including gases, but is not widely used in trace analysis because of its low sensitivity.

The most important detector used by organic geochemists is the mass spectrometer (fig. 4.16). GC-MS analysis can be conducted in two modes: *electron impact* (EI) and *chemical ionization* (CI). In EI mode, effluent leaving the GC column is bombarded with electrons (usually at 70 eV). This causes fragmentation of the molecules, resulting in a molecular ion ($M + e^- \rightarrow M^{+\cdot} + 2e^-$) and additional

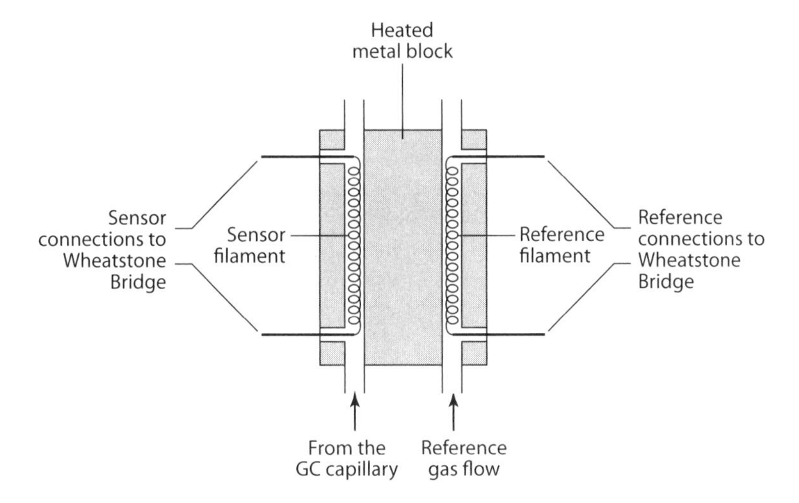

Figure 4.15. Schematic of a thermal conductivity detector (TCD). (Adapted from McMahon, 2007.)

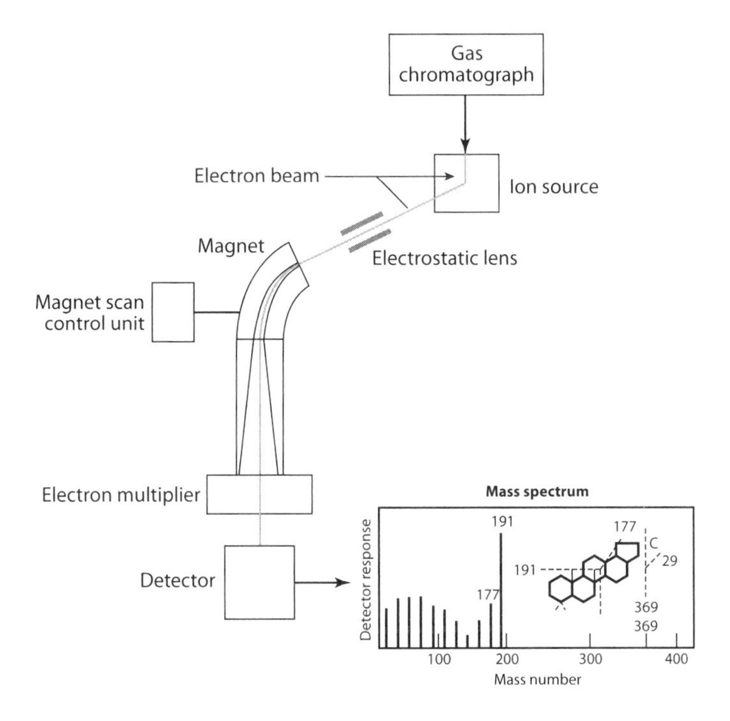

Figure 4.16. Schematic showing effluent leaving a gas chromatograph interfaced to a mass spectrometer (GC-MS). The major parts of the mass spectrometer are shown. GC-MS analysis provides the ability to identify compounds through their fragmentation patterns, as illustrated in the example mass spectrum shown above. (Adapted from Libes, 1992. An introduction to marine biochemistry. Witey, New York.)

fragments. While the molecular ion ($M^{+\cdot}$) provides information about the molecular weight of the molecule of interest, the fragmentation pattern (*mass spectrum*) provides a unique fingerprint based on the structure of a given molecule (fig. 4.16). A useful analogy for GC-MS is to imagine breaking a case of wine glasses where each glass breaks exactly the same way. Similarly, in GC-MS, a given molecule will fragment reproducibly each time it is analyzed, and it is the resulting fragmentation pattern that

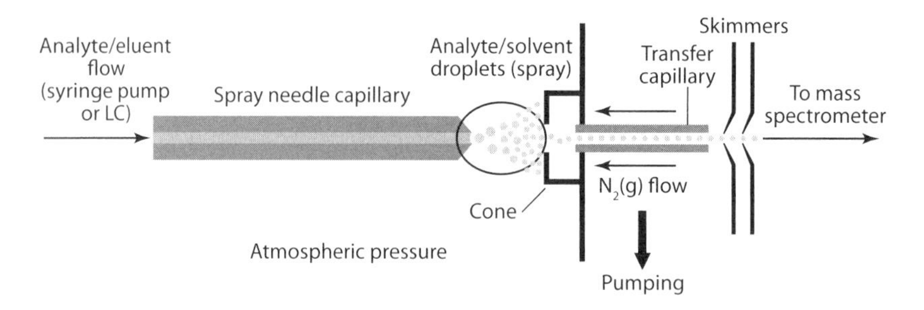

Figure 4.17. Schematic of electrospray ionization (ESI). (Adapted from McMahon, 2007.)

provides unique structural information. The ways by which a molecule fragments are largely controlled by differences in bond strength (e.g., C–C bonds cleave easier than C–H bonds because they are weaker) and the types of functional groups. GC-MS operated in EI mode has been used to identify a number of novel biomarker compounds present in environmental samples (Rowland and Robson, 1990; Sinninghe Damsté et al., 1999; Prahl et al., 2006; Wakeham et al., 2007).

GC-MS analyses can also be conducted in chemical ionization (CI) mode. In CI, a reagent gas (typically methane or ammonia) is introduced into the ion source, where it is ionized to yield primary and secondary ions. The compound to be analyzed is introduced from the GC column and reacts with the secondary ions in the source. Ions are formed by proton transfer or hydride abstraction, yielding principal ions (either $[M + 1]^+$ or $[M - 1]^+$). The lower energy associated with CI results in less fragmentation and an identifiable molecular ion. These ions are useful for determining the molecular weight of the compound of interest.

It is also useful to point out that GC-MS analyses can be conducted in either full-scan or selective ion monitoring (SIM) mode. In full-scan mode, a mass range (e.g., 50 to 600 amu) is scanned every few seconds, limiting the dwell times per mass unit. As a result, full scan requires about 1 nanomole of each compound (similar to FID). In SIM, only selected ions are scanned, allowing for detection to about 1 picomole. Since SIM scans fewer masses, dwell times are generally longer, providing greater sensitivity. However, SIM requires knowledge of retention times and fragmentation characteristics of the compounds being studied so that unique ions and response factors can be identified prior to analysis.

Electrospray ionization (ESI) takes place at atmospheric pressure and is considered to be a "soft" ionization technique, which makes it particularly useful for analyzing proteins, peptides, and oligonucloetides. This technique is very useful for liquid samples, as is discussed later in this chapter. As the eluent from a liquid chromatograph (LC) is introduced to the mass spectrometer it is nebulized with a gas at atmospheric pressure under a strong electrostatic field along with a heated drying gas (usually nitrogen) (fig. 4.17) (McMahon, 2007). This electrostatic field causes further dissociation of the molecules being analyzed. As the heated gas causes evaporation, the droplets from the spray get smaller and there is an enhancement in the charge of each droplet. Eventually the charges become so strong that the repulsive forces of the ions in each droplet are greater than the cohesive forces and the ions are desorbed and separated into the gas phase from the droplets. These ions are then allowed to pass through a capillary sampling port into the mass analyzer. Atmospheric pressure chemical ionization (APCI) is similar to ESI in that it also creates gas-phase ions from a liquid eluent. During APCI the eluent is vaporized, usually at temperatures that range between 250 and 400°C (fig. 4.18) (McMahon, 2007). The gas-phase solvent molecules are then ionized by electrons from a corona needle. The solvent ions can then transfer charge to the analyte molecules through chemical reactions (or chemical ionization). Once this occurs, the ions are passed through an opening to the mass analyzer. The gas-phase ionization of APCI is more effective than ESI when analyzing more polar molecules. Finally, atmospheric pressure photoionization (APPI) is a relatively new technique that converts the

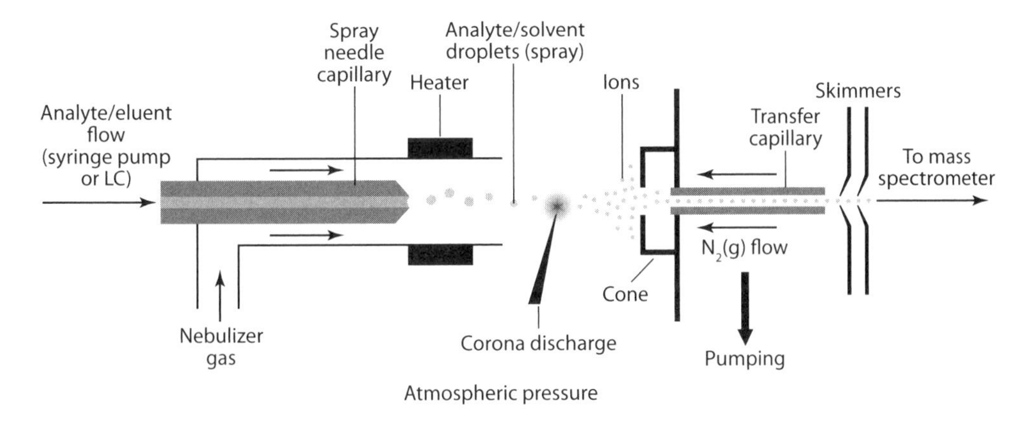

Figure 4.18. Schematic showing the major components of atmospheric pressure chemical ionization (APCI). (Adapted form McMahon, 2007.)

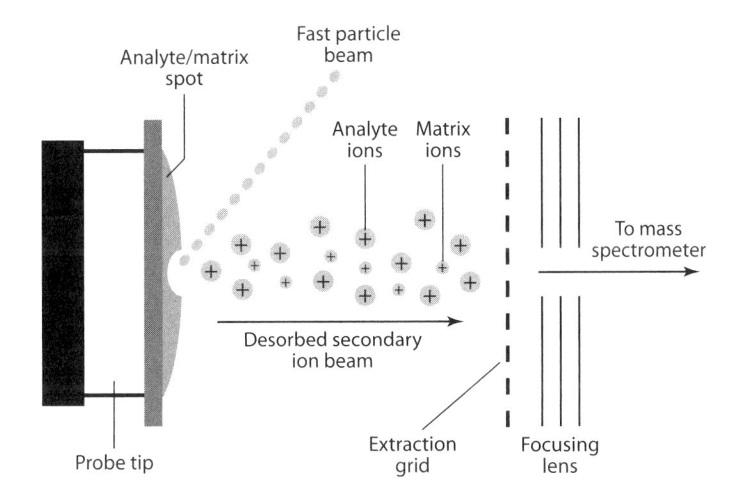

Figure 4.19. A schematic showing the major components of fast atom bombardment (FAB) ionization. (Adapted from McMahon, 2007.)

eluent into a vapor, which then allows a discharge lamp, which generates photons in a narrow window of ionization energies, to produce ions that can then pass into the mass analyzer (McMahon, 2007; Mopper et al., 2007).

Fast atom bombardment (FAB) is an ionization technique that allows for the bombardment of the analyte/matrix mixture at a particular spot with a fast particle beam (fig. 4.19) (McMahon, 2007). The matrix here is typically a small organic species that is used to keep the sample surface homogeneous at the end of the sample probe. The particle beam is typically a neutral inert gas (e.g., argon) at 4 to 10 keV, making the process a relatively soft ionization. The particle beam interacts with the analyte mixture, causing collisions and disruptions that result in the ejection (sputtering) of analyte ions. These ions are then captured and focused before entering the mass analyzer.

Matrix-assisted laser desorption ionization (MALDI) offers an ionization technique that can analyze molecules with high molecular masses (e.g., >10,000 Da). It is another soft ionization technique that also uses a matrix species, which mixes with the analyte; the role of the matrix species in this case is

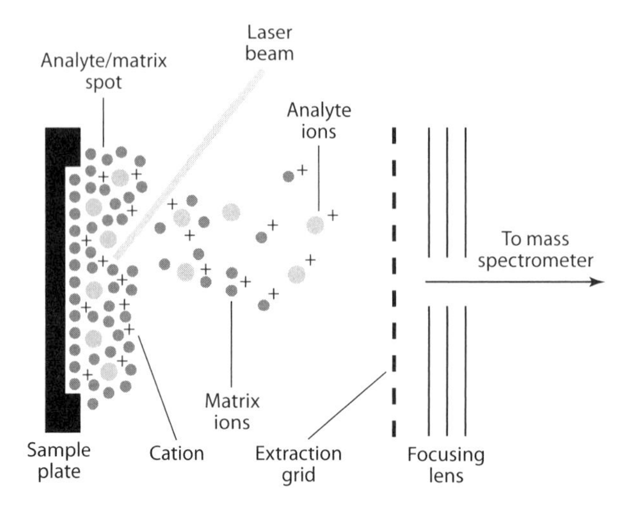

Figure 4.20. Schematic showing the major components of matrix-assisted laser desorption ionization (MALDI). (Adapted from McMahon, 2007.)

to create a mixture that strongly absorbs UV radiation (fig. 4.20) (McMahon, 2007). The mixture is placed on a dry sample plate and inserted into the source region where it is irradiated by a laser beam (e.g., nitrogen at 337 nm). The mixture absorbs the energy from the laser and then ions are ejected into the gas phase and focused into the mass analyzer.

4.4.3 Mass Analyzers

In mass spectrometry, the fragment ions must be focused into a concentrated beam in the mass analyzer. This is usually accomplished using either magnets or *quadrupole* rods (fig. 4.21). Mass spectrometers can utilize double-focusing or single-focusing magnetic systems. Double-focusing instruments use an electrostatic sector to focus the beam before entering, or after leaving, the magnetic sector to achieve high mass resolution (2000 or more). Single-focusing instruments do not have electrostatic focusing, resulting in lower mass resolution (~1000 or lower). Quadrupole instruments use direct current and radio-frequency fields operated on rods as a mass filter. The magnitude of the direct current and radio-frequency fields controls which ions reach the detector, causing other ions to collide with the rods.

The *quadrupole ion trap* mass spectrometer, commonly referred to simply as an ion trap, is one of the most common analyzers in labs today. An ion trap detector is composed of a circular ring electrode and two caps that collectively form the chamber (fig. 4.22) (McMahon, 2007). As ions enter the chamber they become "trapped" by an electromagnetic field, which is created by a 3D quadrupole; an additional field can also be applied to selectively eject ions from the trap. The entire process of trapping, storing, manipulating, and ejecting ions is performed in a continuous cycle. One of the key advantages of an ion trap is its compact size and ability to trap and accumulate ions, which allows for greater signal-to-noise ratios when making measurements.

In a *time-of-flight* (TOF) mass analyzer, a consistent electromagnetic force is applied to all ions at the same time, causing them to accelerate down a flight tube (fig. 4.23) (McMahon, 2007). The lighter ions travel faster and reach the detector quicker, making the mass-to-charge ratios a function of their arrival times. Some advantages of TOF mass analyzers are that they have a broad mass range and are very accurate in mass measurements. TOF mass analyzers are commonly set up in conjunction with MALDI.

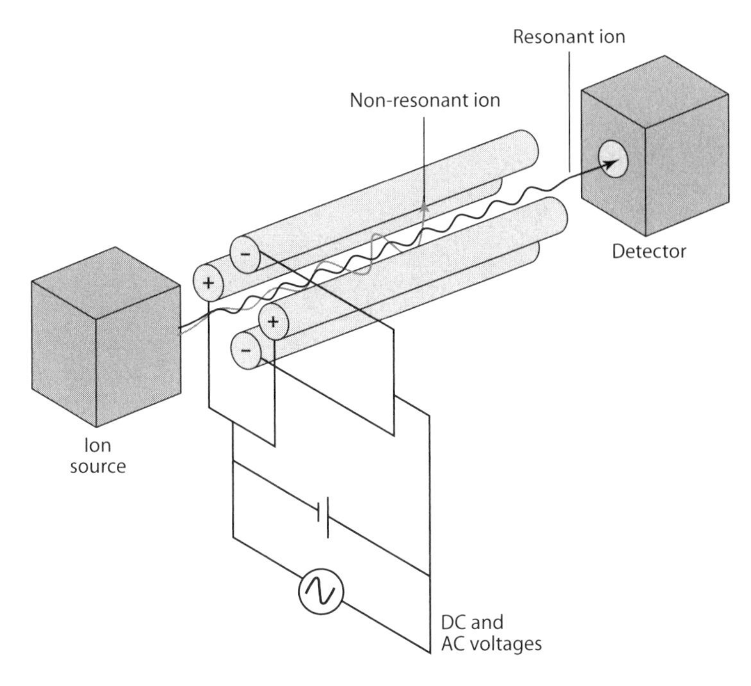

Figure 4.21. A schematic of a quadrupole mass analyzer showing the four parallel rods arranged in a square that are used for mass separations. (Adapted from McMahon, 2007.)

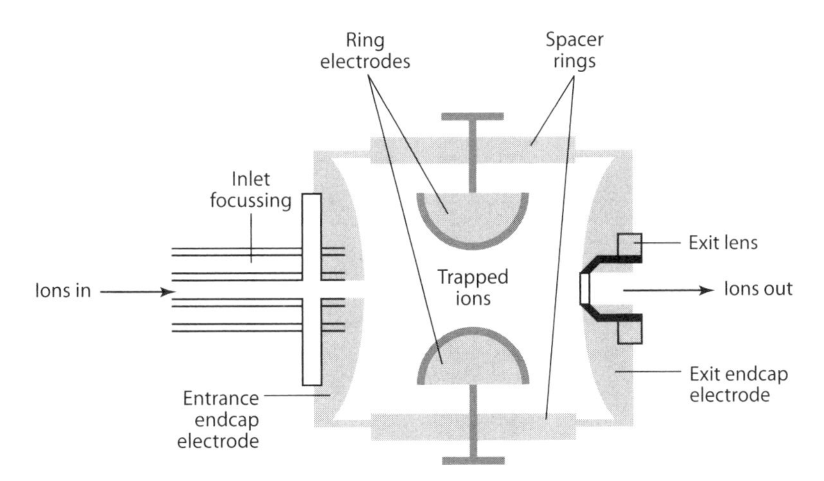

Figure 4.22. Schematic showing the major components of an ion trap mass analyzer. (Adapted from McMahon, 2007.)

Fourier transform ion cyclotron resonance (FT-ICR) mass spectrometry is another trapping analyzer like the ion trap. In this case, ions enter the trap and are trapped by circular orbits or strong electrical and magnetic fields; these orbits pass receiver plates that can detect their frequencies (fig. 4.24) (McMahon, 2007). These ions are then exposed to a radio-frequency electrical field, which generates time-dependent current. This current is then converted by Fourier transform into orbital frequencies that correspond to the mass-to-charge ratios of the ions. For example, the frequency of motion of an ion is inversely proportional to its mass. These mass analyzers have a wide mass range and are

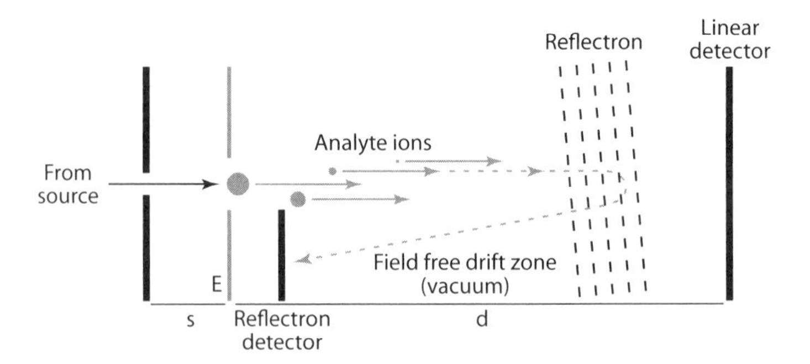

Figure 4.23. A schematic showing the major features of a time-of-flight (TOF) mass analyzer. (Adapted from McMahon, 2007.)

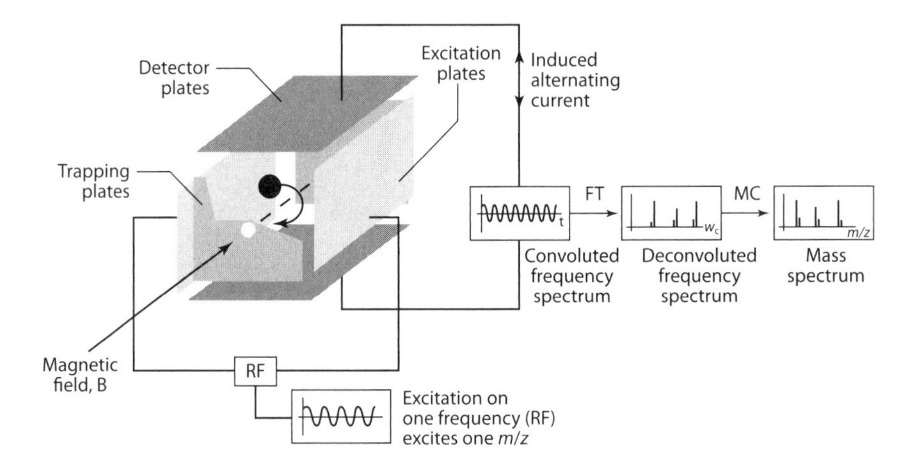

Figure 4.24. A schematic showing the major features of a Fourier transform ion cyclotron resonance (FT-ICR) mass analyzer. (Adapted from McMahon, 2007.)

very sensitive. This approach is unique in that the ions are observed without separation or collection, but by the absorption and emission of radio-frequency energy at their characteristic (mass-dependent) cyclotron frequencies as they experience cyclotron motion in a strong magnetic field (Dunbar and McMahon, 1998).

Although FT-ICR is the most expensive of all mass analyzers we can see the power of using such resolution with natural samples. For example, the APPI FT-ICR and ESI FT-ICR mass spectra of an aquatic DOM extract from Lake Drummond, Dismal Swamp, Virginia (USA) are shown here (fig. 4.25) (Mopper et al., 2007). The mass error between the theoretical formula weights and measured masses is <500 ppb, and the resolving power is >450,000 at 500 Da. Further analysis reveals that the entire APPI spectrum reveals the presence of a diverse suite of aromatic biomolecules, as well as novel arene and azarene compounds (e.g., a homologous series of polycyclic aromatic hydrocarbons PAHs extending to $C_{60}H_{16}$). Further details on how to examine such spectra can be found in Asamoto (1991) and Mopper et al. (2007). After exact elemental formulas have been calculated, further data analysis is required to understand the vast data sets that comprise FT-ICR-MS of a natural DOM sample. For example, Kendrick mass analysis, which was originally used to identify series of compounds with identical chemical backbones with differing numbers of $-CH_2$ groups (Kendrick, 1963), has been adapted to separate polar petroleum compounds (Hughey et al., 2001) and compounds within DOM (Kujawinski et al., 2002; Stenson et al., 2003; Hertkorn et al., 2006). Basically, the measured mass

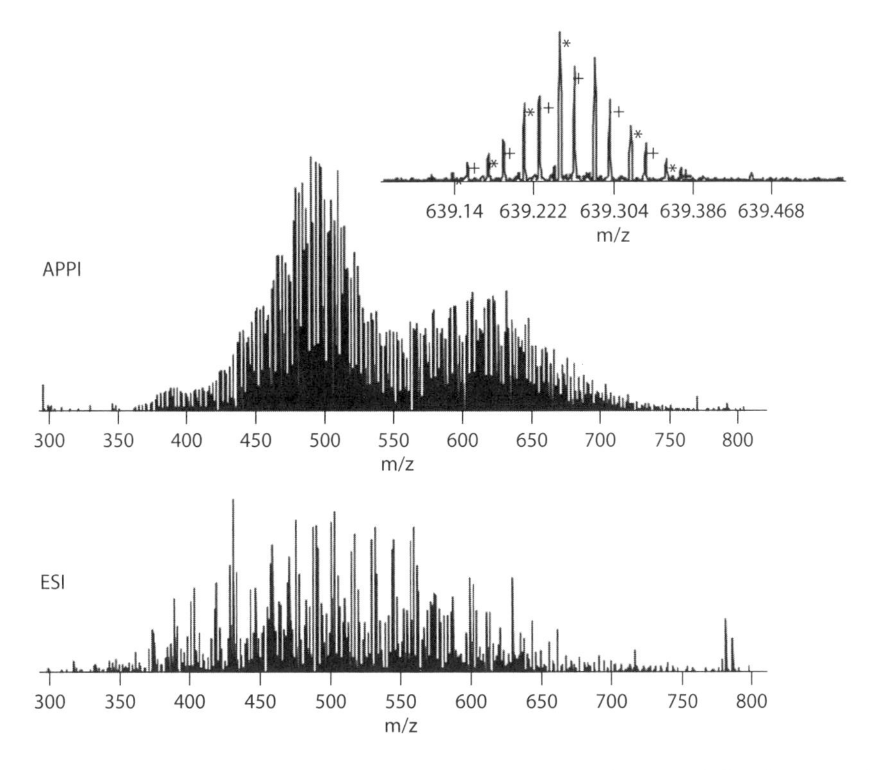

Figure 4.25. Negative-ion ESI and APPI-FT-ICR mass spectra of Lake Drummond dissolved organic matter (DOM). The inset shows a single nominal mass region (m/z 639), with the asterisks indicating peaks from direct ionization of DOM and the plus signs indicating peaks from chemical ionization via a dopant (toluene). The APPI spectrum shows many more peaks than the ESI spectrum due to the ionization of a wider range of molecules. (Adapted from Mopper et al., 2007.)

is converted to a *Kendrick mass*, where the mass of $-CH_2$ is defined as 14.000 Da, rather than the IUPAC mass of 14.01565 Da. Another calculation yields the Kendrick mass defect (KMD) (IUPAC mass - Kendrick mass), which is used as a constant for compounds with identical chemical backbones yet different numbers of $-CH_2$ groups (Mopper et al., 2007). The application of *van Krevelen diagrams* (van Krevelen, 1950), which are two-dimensional scatter plots of H : C ratios on the y-axis versus O : C ratios on the x-axis, have also been used in FT-ICR analyses (Kim et al., 2003). These diagrams allow for examination of possible reaction pathways and a qualitative analysis of compound classes as shown in fig. 4.26 (data from Mopper et al. [2007], and references therein). Van Krevelen (1950) first used these diagrams to better understand the reaction processes of plant molecules in humification and coal formation. Using the H : C and and O : C ratios, principal reactions such as decarboxylation, dehydration, dehydrogenation, hydrogenation, and oxidation can be represented by straight lines. Using this approach, complicated mass spectra can be visualized in two ways: (1) as possible reaction pathways and (2) as qualitative analyses of major classes of compounds that comprise the spectra.

4.4.4 Multidimensional GC ($GC \times GC$) Analysis

In complex samples, peaks may not be resolved from one another (i.e., co-eluting peaks), resulting in mass spectra for the sum of the peaks that compromises the ability to reliably identify individual peaks. In recent years, multidimensional GC (GC × GC) analysis has been applied to complex environmental

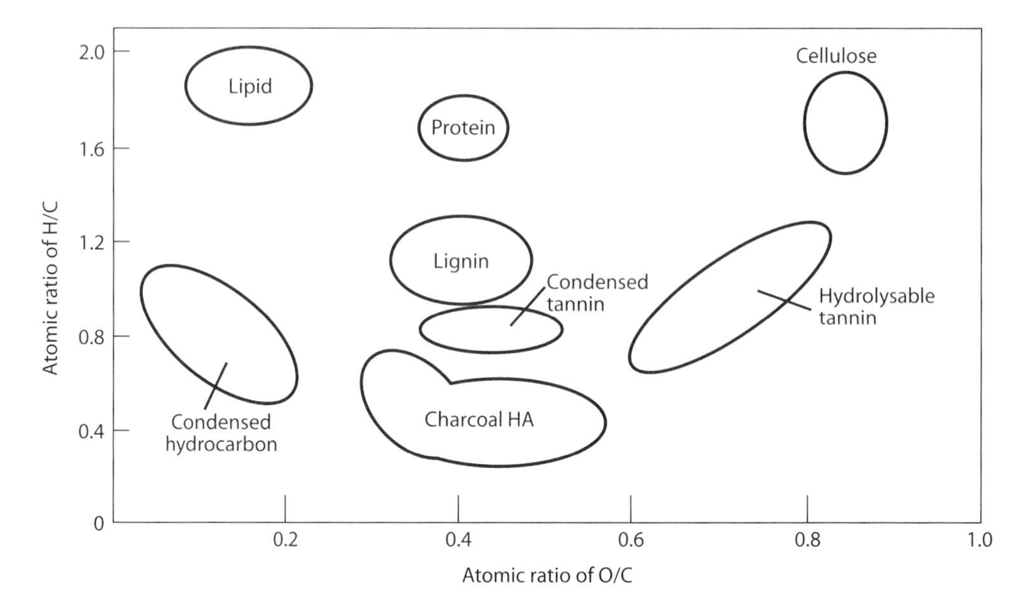

Figure 4.26. The van Krevelen diagram is a data analysis technique where the elemental composition (H:C vs. O:C) of a sample is evaluated to gain insight about contributions from different biochemicals and the maturity of the sample. Positions where major biomolecular components occur on a van Krevelen diagram are provided. (Adapted from Mopper et al., 2007.)

samples where enhanced separation is needed (Frysinger et al., 2002; Mondello et al., 2008; Cortes et al., 2009). In GC × GC, samples are analyzed on two sequential columns connected in series, with a transfer device, or modulator, located at the head of the second column. Retention on each column is independent (e.g., the primary column is often a conventional column, while a short microbore capillary is used for the secondary column) and specific to a particular compound. As a result, it is common to define parameters of the GC × GC analysis separately for each column. This methodology shows promise for its ability to deconvolute complex extracts from marine sediments. Recent studies by Frysinger et al. (2002) and Mondello et al. (2008), for example, identified a wide variety of chemical contaminants in sediment extracts (figs. 4.27 and 4.28).

GC × GC chromatograms are composed of thousands of peaks, making compound identifications challenging (fig. 4.28). Thus, multidimensional GC is often combined with MS detection. Initial efforts used GC × GC analysis in combination with quadropole MS (qMS), but the limited data acquisition speeds of qMS provided limited success. Subsequently, TOF-MS was demonstrated to be more suitable for interfacing with GC × GC analyses (Mondello et al., 2008). In recent years, multidimensional GC-MS has proven particularly useful for resolving complex mixtures in environmental samples, such as those associated with oil residues (Reddy et al., 2002; Frysinger et al., 2003; Nelson et al., 2006; Wardlaw et al., 2008), petrochemicals (Blomberg et al., 1997; von Muhlen et al., 2006), and toxicants (Marriott et al., 2003; Skoczynska et al., 2008).

4.4.5 GCMS-MS Analysis

GCMS-MS (or tandem mass spectrometry) operates on the principle that organic molecules (parent compound) ionized in the ion source break down into smaller charged fragments (daughters), with some of these daughter ions characteristic of their parent molecules. GCMS-MS allows one to

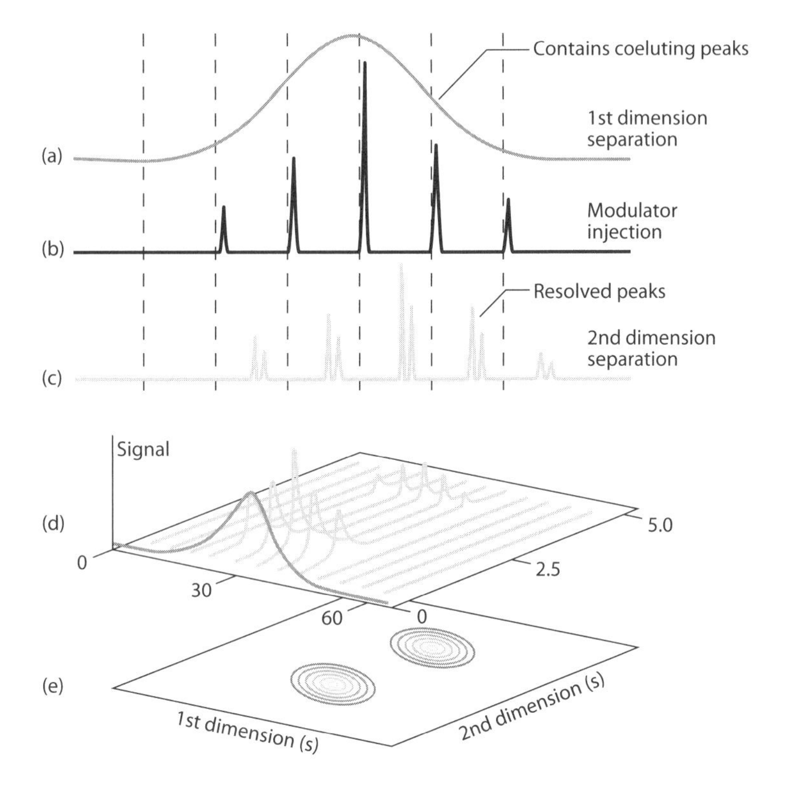

Figure 4.27. Conceptual figure illustrating comprehensive two-dimensional GC. (Adapted from Frysinger et al., 2002.)

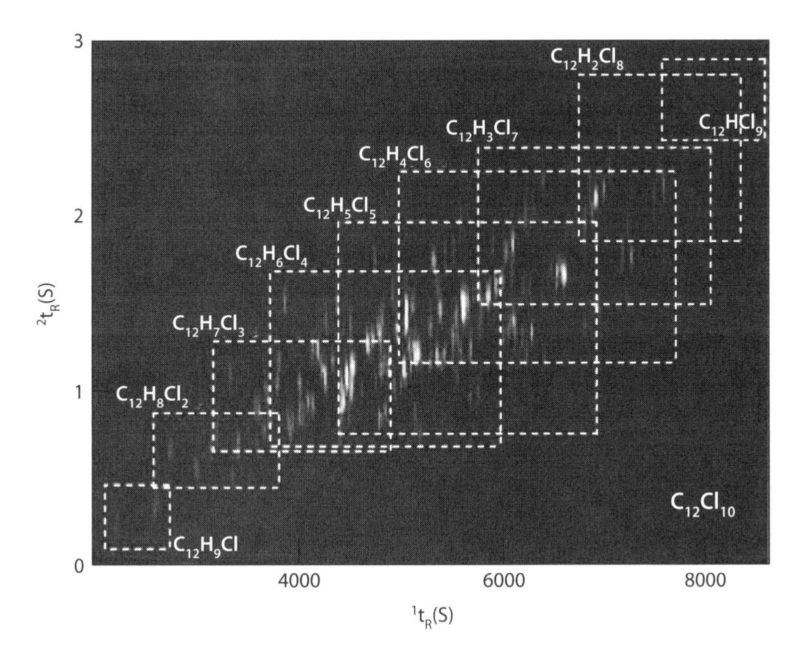

Figure 4.28. GC × GC-TOF-MS chromatogram of the 209 PCB congeners using an HT-8/BPX-50 column. (Adapted from Mondello et al., 2008.)

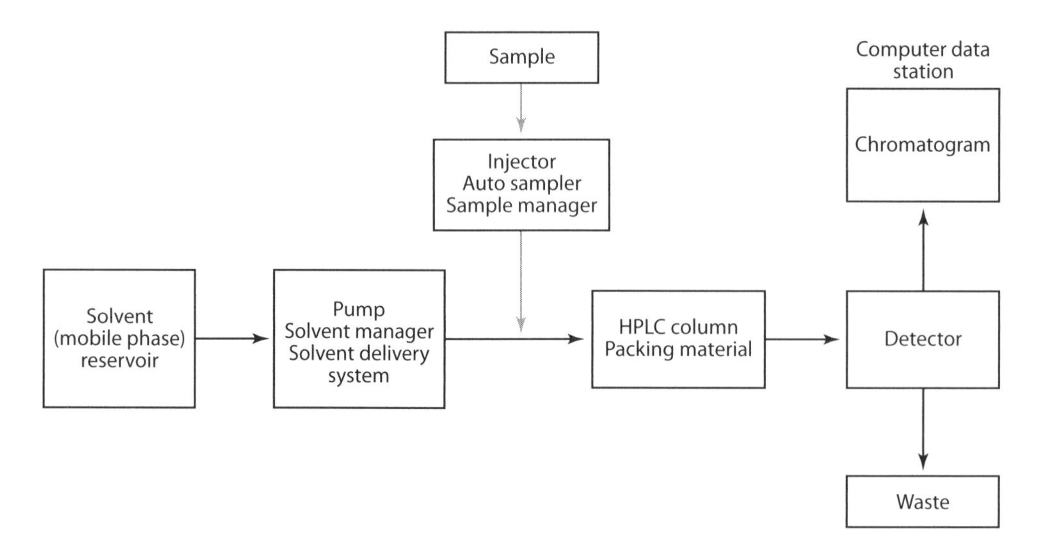

Figure 4.29. Schematic showing the major components of a typical high-performance liquid chromatography (HPLC) system.

determine the parents of selected daughters (Peters et al., 2005). Tandem mass spectrometry is typically conducted using triple-quadrupole or multisector mass spectrometers with three quadrupoles connected in series: the parent quadrupole (Q1), the collision cell quadrupole (Q2), and the daughter quadrupole (Q3). However, tandem mass spectrometry can also combine magnetic and electrostatic sectors with collision cell quadrupoles. Hybrid instruments are another option; in this case, a magnetic sector instrument is combined with collision cell and mass filtering quadrupoles. Tandem mass spectrometry is particularly useful for the analysis of complex samples since it has a high level of specificity. GCMS-MS also has high signal-to-noise ratios, typically orders of magnitude better than those for conventional GC-MS in SIM mode (fig. 4.28).

4.5 High-Performance Liquid Chromatography

4.5.1 High-Performance Liquid Chromatography

Although HPLC is less sensitive than GC, it offers a broader spectrum of applications. The analyte is transferred through the stationary phase of a column (either packed or capillary) with the mobile phase, whereby the components of a mixture are separated on the column. If the column is polar (e.g., silica based) and the mobile phases are nonpolar, the HPLC method is considered to be *normal phase, NP-HPLC*; the opposite conditions in mobile and stationary phases are classified as *reversed phase, RP-HPLC*. Most of the applications today are RP-HPLC using a stationary phase that is a C_8 or C_{18} bonded phase. A typical HPLC instrument consists of the components shown in fig. 4.29. Pumping rates typically range from nL to mL min $^{-1}$, with column dimensions of 250×4.6 mm i.d., 3- to 10-μm particle diameters, and pressures that operate up to ca. 3000 psi. There is a reservoir that holds the solvents or mobile phases and a high-pressure pump or delivery system that largely controls the flow rate and pressure, which interfaces with the size and composition of the stationary phase in controlling peak retention times. The detector is usually a *multiwavelength UV* (sensitivity around 0.1–1 ng) in tandem with *multiwavelength fluorescence* detection (sensitivity 0.001–0.01 ng). For other applications, such as analysis of sugars, *pulsed amperometric detectors* (PAD) are used. Their sensitivity

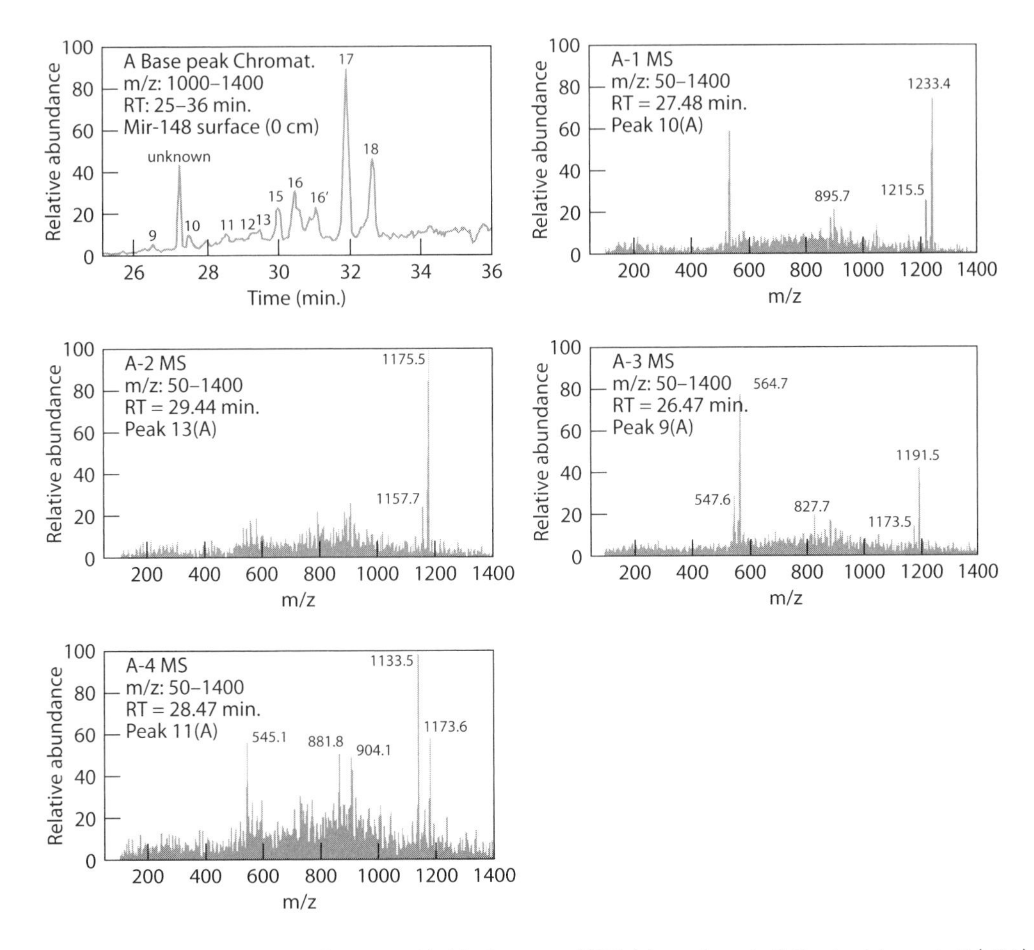

Figure 4.30. HPLC-MS spectra of carotenoid chlorin esters (CCEs) in sediment off the Louisiana shelf (USA). Base peak chromatograms (*m/z* confined to the 1000 to 1400 range) and mass spectrum of CCEs. (Adapted from Chen et al., 2003.)

is in the range of 0.01 to 1 ng. In the case of plant pigments, *photodiode array detectors* (PDA) allow for a very preliminary identification of peaks based on the fact that carotenoids do not fluoresce, while chlorophylls show fluorescence and absorbance (see a sample HPLC chromatogram of pigments in chapter 12). The PDA uses a broad spectrum of light sources that passes through a sensing cell, which is then dispersed by a polychromator that allows the light to be distributed through an array of diodes (McMahon, 2007). Thus, with PDA, the availability of one wavelength or a variety of wavelengths can be used to perform spectroscopic scanning to obtain a UV-Vis spectrum for each peak.

4.5.2 HPLC-MS

HPLC-MS utilizes the separating power of LC with the detection capabilities of MS. Similar to GC-MS, there are choices for ionization techniques as well as mass analyzers (see section 4.4 above). The typical ionization techniques used in HPLC-MS are ESI and APCI, while the mass analyzers are usually quadrupole or triple quadrupole, ion trap, and TOF. An example of the application of HPLC-MS to coastal systems is provided in fig. 4.30 (Chen et al., 2003). These sample spectra show identification

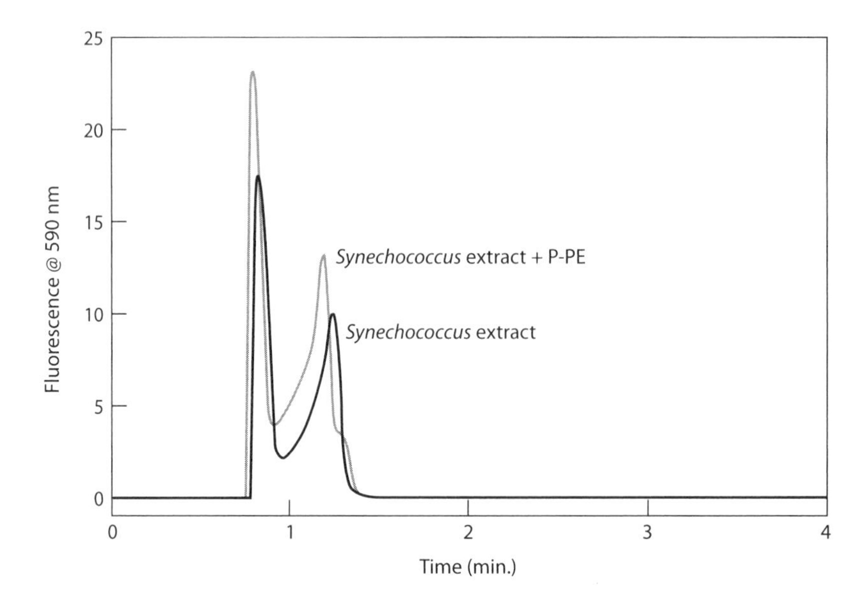

Figure 4.31. Electropherograms of phycoerythrin extracted from *Synechoccus*. The extract was spiked with reference-phycoerythrin (R-PE) to yield a sample with a final concentration of 1.04×10^{-9} M (w/added R-PE). (Adapted from Viskari et al., 2001.)

of carotenoid chlorin esters (CCEs) in sediments off the Louisiana shelf (USA). CCEs are one of the major and most stable decay products of chlorophyll *a* found in these sediments. Many advances are being made with LC-MS, and even more recent and exciting work is being done with HPLC-NMR (McMahon, 2007).

4.5.3 Capillary Electrophoresis

The fundamental principles of capillary electrophoresis (CE) are based on an electric field being applied to a solution of charged species in a capillary, whereby the ions separate from each other as they move toward the electrode of opposite charge. The electrophoretic separation is based on differences in the mobility of different solutes under such conditions. The general components of CE are a power supply, injector, capillary, and detector. The detection of the components can be selective or universal, depending on the detector. The source is composed of an eluent buffer and power supply; most power supplies provide from −30 to +30 kV with currents that range from 200 to 300 mA. Separation begins when the buffer and analytes migrate through the capillary with their own mobility under the influences of electroosmotic flow (Kujawinski et al., 2004). The response of the detector is plotted in an electropherogram with the order of elution set up with positive ion, neutral species, and then negative ions (McMahon, 2007). New techniques using laser-induced fluorescence detection have proven useful for analyzing phycobilin pigments in algae, as shown in fig. 4.31 (Viskari et al., 2001). Finally, capillary gel electrophoresis is particularly good for the separation of macromolecules. New techniques are expanding into CE-NMR as well (see detail in McMahon, 2007).

4.6 Stable Isotope and Radiocarbon Analyses

The most effective means of measuring isotopic abundances is the use of mass spectrometric techniques. The general principle of a mass spectrometer is to separate charged atoms and ions based on their

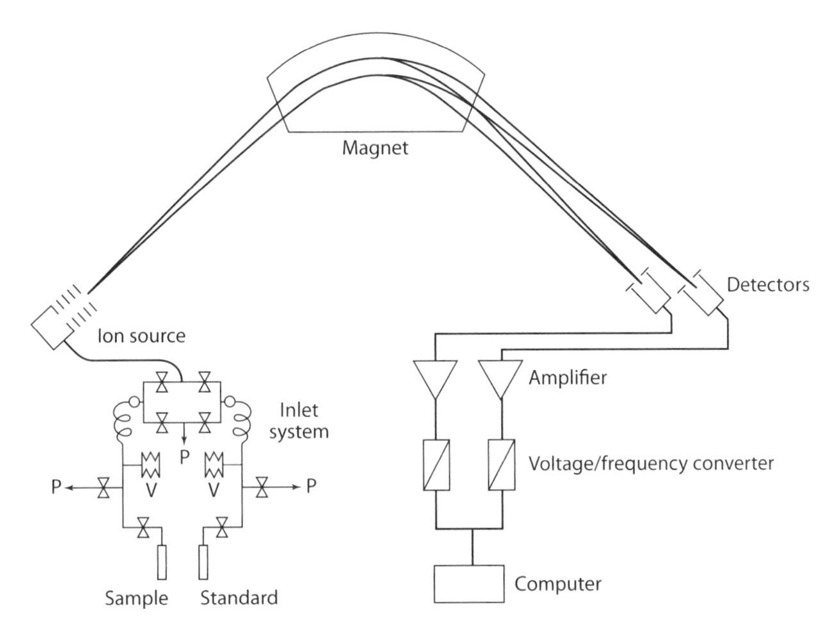

Figure 4.32. Schematic showing major components of an isotope ratio mass spectrometer. P, pumping system; V, variable volume. (Adapted from Hoefs, 1997.)

masses as well as their movements under different magnetic and/or electric fields (Hoefs, 2004). The mass spectrometer consists of the following four components: (1) inlet system, (2) ion source, (3) mass analyzer, and (4) ion detector (fig. 4.32). The first inlet system used was a dual-inlet system, first introduced by Nier (1947), which allowed for the rapid analysis of two gases (sample and standard) using a changeover valve within a matter of seconds. The *ion source* is where ions are generated by a tungsten or rhenium filament, after which they are accelerated and focused into a narrow beam; the efficiency of ionization determines the sensitivity of the mass spectrometer (Hoefs, 2004). The *mass analyzer* receives the beam of ions from the source and separates them according to their mass/charge (m/z), as they pass through a magnetic field and are deflected into circular paths. After passing through the magnetic field the separated ions are collected in an *ion detector* where the electrical impulse is converted and then fed into an amplifier (Hoefs, 2004). In the mid 1980s, the dual-inlet approach was modified to a continuous-flow isotope ratio monitoring mass spectrometer where a trace gas is analyzed in a carrier phase gas using chromatographic techniques. This allows the isotope ratio mass spectrometer to be interfaced with either an elemental analyzer for analysis of the isotopic composition of elements or a gas chromatograph for determining the isotopic composition of specific compounds by GC– isotope ratio mass spectrometry (GC-IRMS) (fig. 4.33). This development in the 1980s allowed for much smaller samples to be analyzed at a much faster rate with high sensitivity. Other microanalytical techniques now include laser techniques (Crowe et al., 1990) as well as *secondary ion mass spectrometry* (SIMS) (Valley and Graham, 1993). For more details on mass spectrometry refer to books by Hoefs (2004), Fry (2006), and Sharp (2006).

The development of accelerator mass spectrometry (AMS) (Bennett et al., 1977; Muller, 1977; Nelson et al., 1977) has made a revolutionary contribution to the measurement of cosmogenic radionuclides such as ^{14}C. Cosmogenic radionuclides have extremely low concentrations with isotope ratios, in the range of 10^{-12} to 10^{-15}. In AMS, these small concentrations are measured by counting individual atoms, which is realized by a combination of ion beam and detection techniques developed in nuclear physics. Analyses using ^{14}C-AMS utilize specialized methods for accurately determining $\Delta^{14}C$ in samples that contain micrograms rather than milligrams of carbon (Eglinton and Pearson, 2001;

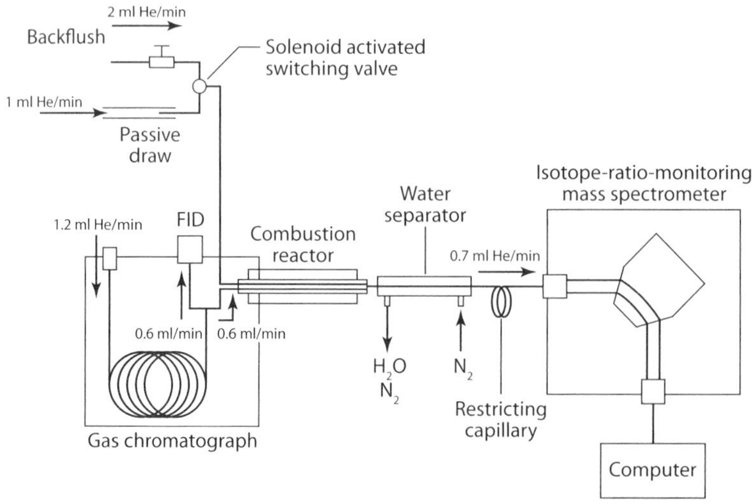

Figure 4.33. Schematic showing major components of gas chromatograph coupled to isotope ratio mass spectrometer (GC-IRMS) showing in-line combustion reactor and water separator. (Adapted from Hayes et al., 1990.)

McNichol and Aluiwhare, 2007). The analytical precision is 0.5% (age uncertainty of 40 yr BP) for sub-recent material; the detection limit allows a resolution back to ca. 50,000 years.

4.7 Compound-Specific Isotope Analysis

The heterogeneity of organic matrices in soils, sediments, and dissolved and suspended particulate materials complicates ^{14}C age-based determinations. There is considerable uncertainty in organic carbon measurements in fractions that have ^{14}C values less than contemporary. In particular, organic matter in aquatic systems is composed of biochemical classes (e.g., carbohydrates, proteins, lipids, and nucleic acids), in addition to more complex polymeric macromolecules, particularly those in sediments and those associated with humic substances, that vary in age. Bulk stable isotope (e.g., ^{13}C) and ^{14}C measurements mask the problems of heterogeneous samples that comprise a mixture of modern and old carbon; different mixtures of these components can result in very different age determinations. To address this problem, compound-specific stable isotope analysis (CSIA) (see more details in the next section) was first performed on single biomarker compounds (e.g., hydrocarbons) to reduce the variability of source pools and identify the sources of individual compounds or groups of compounds in complex mixtures (Freeman et al., 1990; Hayes et al., 1990). This was followed by the development of methods for compound-specific radiocarbon analysis (CSRA), in samples with a carbon content as small as $< 100\,\mu g$ (Eglinton et al., 1996, 1997; Pearson et al., 1998). These techniques involve automated preparative capillary gas chromatography (PCGC), and now allow for separation of target compounds for ^{14}C analysis based on AMS (Eglinton et al., 1996, 1997; McNichol et al., 2000; Eglinton and Pearson, 2001; Ingalls and Pearson, 2005).

As mentioned in chapter 3, separation of prebomb from postbomb $\Delta^{14}C$ values can prove useful in understanding the rates of exchange between the atmosphere and aquatic metabolic processes. This is illustrated by a study that investigated the $\Delta^{14}C$ of lipids from the sediments off Santa Monica, California (USA) (fig. 4.33) (Eglinton and Pearson, 2001). Results from this study showed that the sources of carbon for these compounds were derived from production in the euphotic zone. The

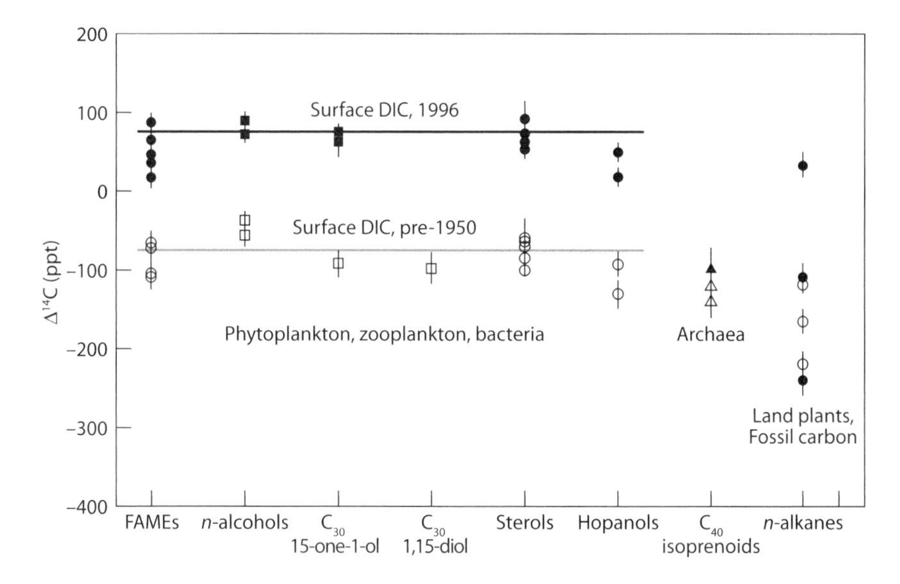

Figure 4.34. The $\Delta^{14}C$ of lipids from the sediments off Santa Monica, California (USA); open symbols are pre-1950 samples and filled symbols are post-1950 samples; DIC, dissolved inorganic carbon. The majority of lipids in prebomb sediments have $\Delta^{14}C$ values equal to $\Delta^{14}C$ DIC from that time (gray line), while most of the postbomb $\Delta^{14}C$ values equal $\Delta^{14}C$ DIC from today (black line). (Adapted from Eglinton and Pearson, 2001.)

majority of lipids in prebomb sediments have $\Delta^{14}C$ values equal to $\Delta^{14}C$ DIC from that time (gray line), while most of the postbomb $\Delta^{14}C$ values equal $\Delta^{14}C$ dissolved inorganic carbon (DIC) from today (black line). There are many other interesting distinctions among these compounds further discussed by Eglinton and Pearson (2001), thus corroborating the utility of CSIA and CSRA techniques. Recent work has also utilized methods such as high-performance liquid chromatography (Aluiwhare et al., 2002; Quan and Repeta, 2005) and even HPLC-MS (Ingalls et al., 2004) for isolation of compounds. In very general terms, CSRA and CSIA allow for accurate determination of ages and sources of compounds within the heterogeneous matrix of other carbon compounds commonly found in sediments (e.g., terrigenous versus marine sources).

The production of biomass can be viewed as occurring in two distinct steps: (1) assimilation of carbon and (2) biosynthesis of cellular components. As a result, the ^{13}C content of biomolecules depends on the ^{13}C content of the carbon source utilized, isotope effects associated with carbon assimilation, isotope effects associated with metabolism and biosynthesis, and cellular carbon budgets at each branch point. The first isotopic analyses of biomarkers involved HPLC separation and collection of individual compounds (Abelson and Hoering, 1961). Solvent from each fraction was evaporated and the residue combusted to CO_2 and N_2, which were subsequently purified cryogenically and analyzed by stable isotope mass spectrometers. While successful, there were several limitations to this approach, including (1) sample size (usually required mg quantities); (2) the labor-intensive nature of these measurements; and (3) in addition to establishing separation techniques to provide sufficient material for isotope analysis, an analytical scheme must produce little or no isotopic fractionation.

The specific challenges to isotopic analysis of individual compounds remained the inlet system and sensitivity for several decades. Finally, new methods were developed that allowed the gas chromatograph to be directly interfaced to an isotope ratio mass spectrometer via an in-line combustion oven and gas purification system (fig. 4.34). Hayes and colleagues provided significant advances to this technology by developing an instrument that combined a combustion interface with a small-volume (20-μL) removal of water from the gas stream by diffusion through a selective membrane, an isotope-ratio mass spectrometer optimized for compound-specific application, and highly specialized software (Freeman et al., 1989; Hayes et al., 1990). As peaks elute from the column in the gas chromatograph, they are immediately combusted to CO_2 and the ratio of the mass-45 ion current to the mass-44 ion current is measured instantaneously. Ion currents at masses 44, 45, and 46 are corrected for electronic offsets, contributions from column bleed, and contributions due to oxygen-17. The isotopic compositions of individual compounds are then determined by integrating across the width of each peak and by comparison to isotopic standards co-injected with the samples.

Following this advance, compound-specific isotopic analyses have been applied to a spectrum of organic compound classes, as discussed throughout this book (e.g., lipids, amino acids, monosaccharides, and lignin-phenols) (Engel et al., 1994; Goñi and Eglinton, 1996; Macko et al., 1997). Compound-specific analyses have also been extended to analysis of the stable isotopes of elements other than carbon, including hydrogen, sulfur, and nitrogen. Recently, with the development of preparative capillary gas chromatography, radiocarbon isotopic analyses of individual compounds can also be conducted (see chapters 8–10 and 14).

4.8 Summary

Analytical methodology has played an important role in advancing the field of organic geochemistry throughout its history and will continue to do so. This chapter highlights some recent advances in sample collection and analytical chemistry. More work is needed to characterize a larger fraction of the organic matter present as DOM and to better characterize the architecture of organic matter associated with particles and sediments. New approaches are also needed to further probe the uncharacterizable fraction of organic matter, since a large portion of sedimentary organic matter is unrecognizable as biochemical classes with traditional methods. Recent advances in LC-MS have expanded the classes of compounds than can be analyzed and allowed the analysis of larger, polar compounds. These tools were developed largely in the biomedical, pharmaceutical, and proteomics fields; organic geochemists should continue to stay abreast of analytical advances in these and other fields.

5. Carbohydrates: Neutral and Minor Sugars

5.1 Background

Carbohydrates are important structural and storage molecules and are critical in the metabolism of terrestrial and aquatic organisms (Aspinall, 1970). The general chemical formula for carbohydrates is $(CH_2O)_n$. These compounds can be defined more specifically as polyhydroxyl aldehydes and *ketones*—or compounds that can be hydrolyzed to them. Carbohydrates can be further divided into monosaccharides (simple sugars), disaccharides (two covalently linked monosaccharides), oligosaccharides (a few, usually 3 to 10, covalently linked monosaccharides), and polysaccharides (polymers made up of chains of mono- and disaccharides). Based on the number of carbon atoms, monosaccharides (e.g., 3, 4, 5, and 6) are commonly termed trioses, tetroses, pentoses, and hexoses, respectively. Carbohydrates have chiral centers (usually more than one), making possible the occurrence of *enantiomers* (mirror images), whereby n chiral centers allows for the possibility of 2^n sterioisomers. For example, glucose has two stereoisomers: one reflects plane-polarized light to the right (dextrorotatory [D]) and the other reflects it to the left (levorotatory [L]), as shown in the traditional *Fisher projection* (fig. 5.1). Almost all naturally occurring monosaccharides occur in the D configuration; examples of some of these *aldoses* and *ketoses* are shown in figs. 5.1 and 5.2. Pentoses and hexoses can become cyclic when the aldehyde or keto group reacts with a hydroxyl group on one of the distal carbons. In glucose, for example, C-1 reacts with the C-5 forming a *pyranose ring*, as shown in a typical *Haworth projection* (fig. 5.3). This results in the formation of two *diasteromers* in equilibrium with each other around the asymmetric center at C-1, called *anomers;* when the OH group is oriented above or below the plane of the ring at the anomeric C-1 its position is referred to as β or α, respectively. Similarly, fructose can form a 5-C ring called a *furanose* when the C-2 keto group reacts with the hydroxyl on C-5 (fig. 5.2).

Cellulose, a highly abundant polysaccharide found in plant cell walls, is composed of long, linear chains of glucose units connected as *1,4′–β-glycoside linkages* (fig. 5.4). These β-linkages enhance intrachain hydrogen bonding as well as *van der Waals* interactions, resulting in straight, rigid chains of *microfibrils*, making cellulose very difficult to digest by many heterotrophs (Voet and Voet, 2004).

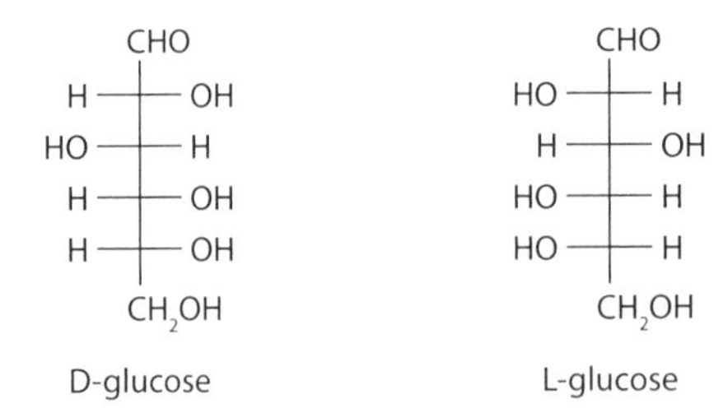

Figure 5.1. Glucose has two stereoisomers—one that reflects plane-polarized light to the right (dextrorotatory [D]) and the other reflects it to the left (levorotatory [L])—as shown in the traditional Fisher projection.

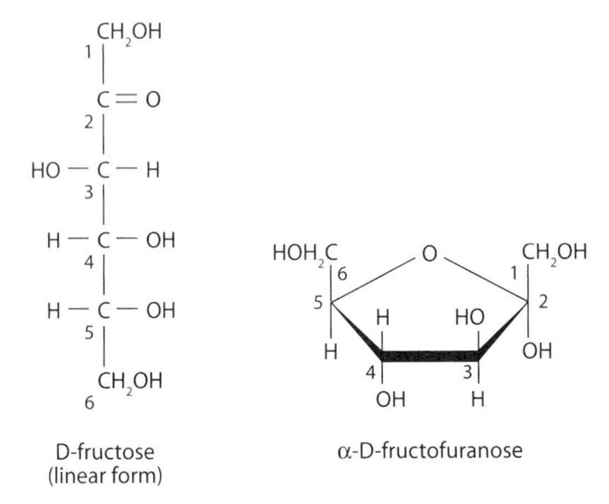

Figure 5.2. Fructose can form a five-carbon ring called a furanose when the C-2 keto group reacts with the hydroxyl groups at C-5.

In phytoplankton, carbohydrates serve as important reservoirs of energy, structural support, and cellular signaling components (Lee, 1980; Bishop and Jennings, 1982). Carbohydrates make up approximately 20 to 40% of the cellular biomass in phytoplankton (Parsons et al., 1984) and 75% of the weight in vascular plants (Aspinall, 1983). Structural polysaccharides in vascular plants, such as *α-cellulose, hemicellulose, and pectin*, are dominant constituents of plant biomass, with cellulose being the most abundant biopolymer (Aspinall, 1983; Boschker et al., 1995). Certain carbohydrates can also represent significant components of other structural (e.g., lignins) and secondary compounds (e.g., tannins) found in plants (Zucker, 1983). Carbohydrates serve as precursors for the formation of humic material (e.g., kerogen), a dominant form of organic matter in soils and sediments (Nissenbaum and Kaplan, 1972; Yamaoka, 1983). Extracellular polysaccharides are important for binding microbes to surfaces, construction of invertebrate feeding tubes and fecal pellets, and in microfilm surfaces (Fazio et al., 1982). For example, carbohydrates changed the mechanism of erodibility of sediments in a tidal flat in the Westerschelde estuary (Netherlands) from individual rolling grains to clumps of resuspended materials (Lucas et al., 2003). Similarly, carbohydrates associated with biofilms altered the properties of cohesive sediments (Tolhurst et al., 2008). Fibrillar material enriched in polysaccharides has been shown to be an important component of colloidal organic matter in aquatic systems (Buffle and Leppard, 1995; Santschi et al., 1998). Carbohydrates are also important in the formation of aggregates such as "*marine snow*" (Alldredge et al., 1993; Passow, 2002).

Polysaccharides can further be divided into the dominant, or major, sugars (e.g., those found in most organic matter), as well as other more source-specific forms, the minor sugars. While there have

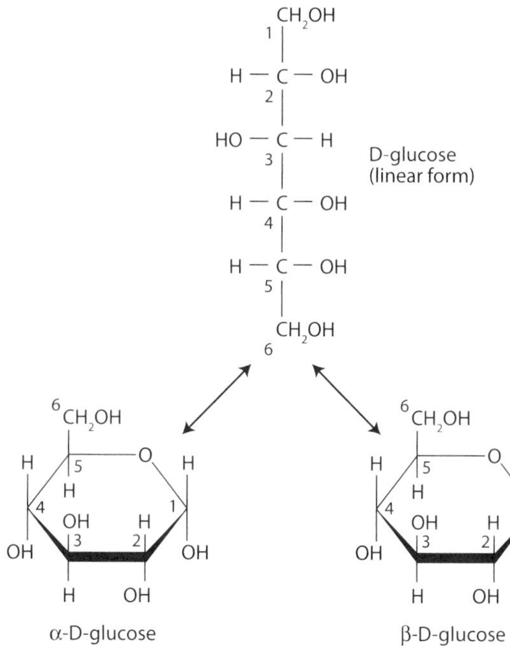

Figure 5.3. In glucose, C-1 reacts with the C-5, forming a pyranose ring, as shown in a typical Haworth projection.

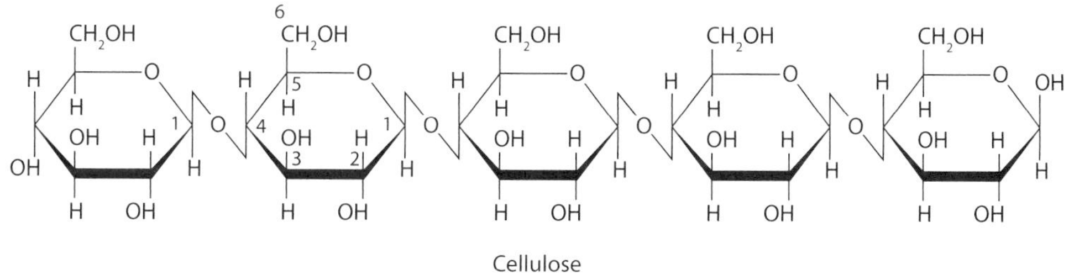

Figure 5.4. Cellulose, a highly abundant polysaccharide found in plant cell walls, is composed of long, linear chains of glucose units connected by 1,4′-β-glycoside linkages.

been problems with using some of these carbohydrate classes as effective biomarkers of organic matter in aquatic systems (due to a lack of source specificity), there remains considerable interest in their application as biomarkers because they are the dominant molecular constituents of organic matter.

Considerable attention has been given to measuring the relative abundances of the major sugars (Mopper, 1977; Cowie and Hedges, 1984b; Hamilton and Hedges, 1988). The major sugars can be further classified as unsubstituted aldoses and ketoses. The aldoses, or *neutral sugars* (rhamnose, fucose, lyxose, ribose, arabinose, xylose, mannose, galactose, and glucose) (fig. 5.5), are the monomeric units of the dominant structural polysaccharides of vascular plants (e.g., cellulose and hemicelluloses). For example, cellulose is composed exclusively of glucose monomers, making it the most abundant neutral sugar in vascular plants. Since neutral sugars make up the majority of organic matter in the biosphere, molecular measurements of neutral sugars in natural organic matter are commonly thought of as total carbohydrates (Opsahl and Benner, 1999). Minor sugars, such as acidic sugars, amino sugars, and O-methyl sugars tend to be more source-specific than major sugars and can potentially provide information useful for understanding the biogeochemical cycling of carbohydrates (Mopper

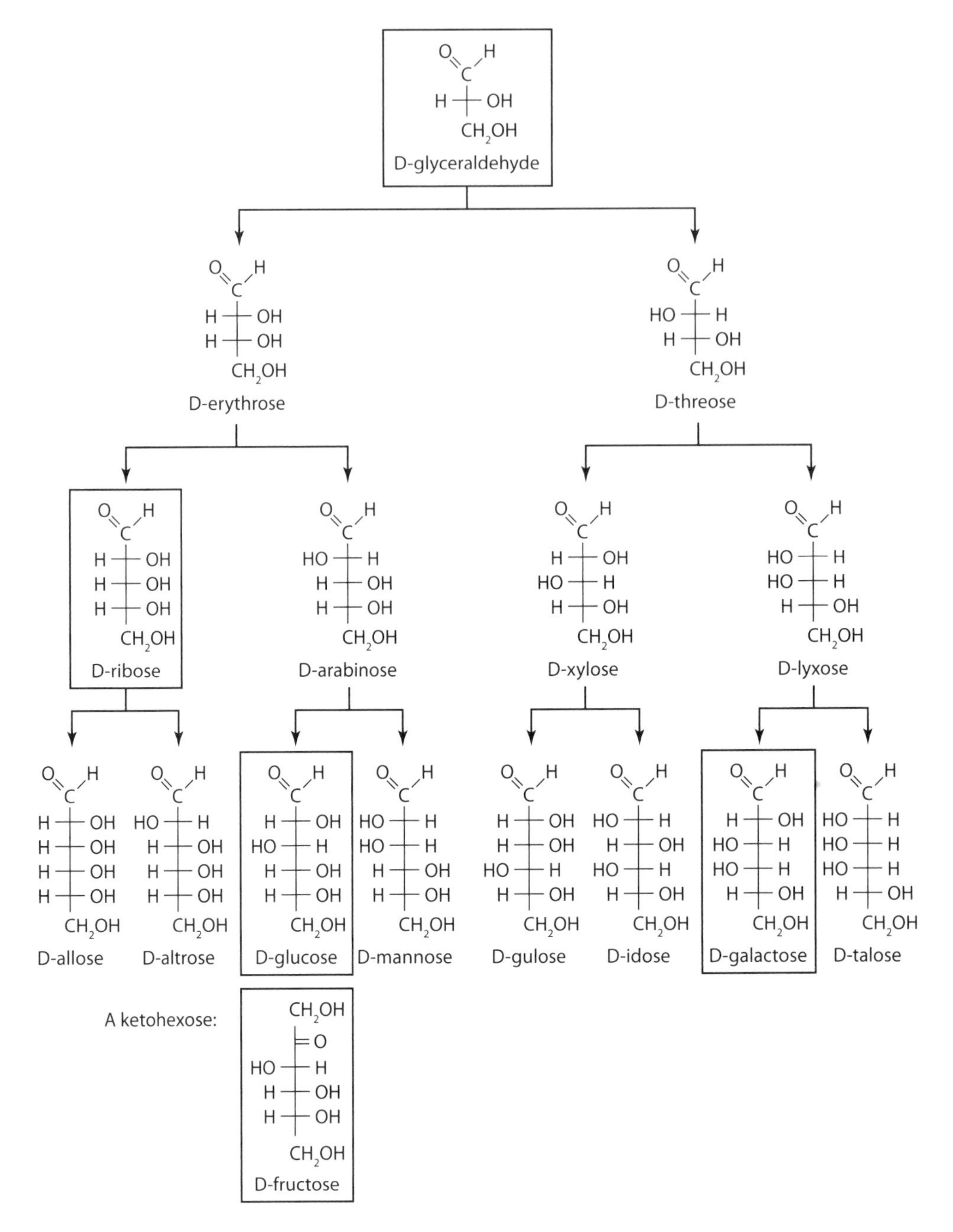

Figure 5.5. The major sugars can be further classified as unsubstituted aldoses and ketoses. The aldoses, or neutral sugars, include rhamnose, fucose, lyxose, ribose, arabinose, xylose, mannose, galactose, and glucose.

and Larsson, 1978; Klok et al., 1984a,b; Moers and Larter, 1993; Bergamaschi et al., 1999). In the case of amino sugars, such as glucosamine, an amino group substitutes for one of the hydroxyl groups and in some cases may be acetylated as in α-D-*N*-acetylglucosamine (fig. 5.6). Minor sugars exist in a diversity of forms and differ from simple sugars by the substitution of one or more hydroxyl groups with an alternative functional group (Aspinall, 1983). The contribution of minor sugars expressed as a

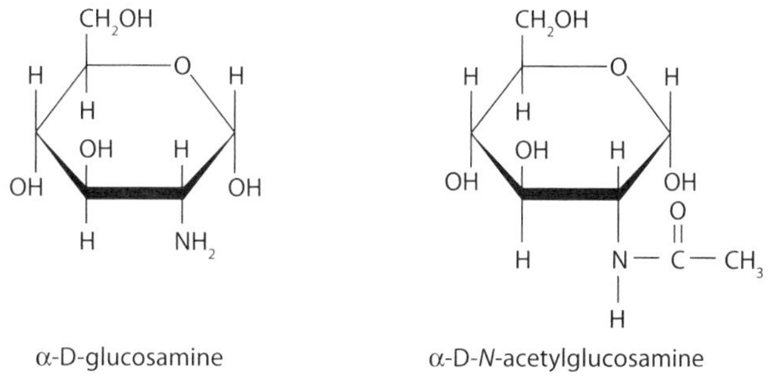

CH$_2$OH, H, O, H, H, OH, H, OH, OH, H, NH$_2$

α-D-glucosamine

CH$_2$OH, H, O, H, H, OH, H, OH, OH, O, H, N— C— CH$_3$, H

α-D-*N*-acetylglucosamine

Figure 5.6. In the amino sugar glucosamine an amino group substitutes for one of the hydroxyl groups and in some cases may be acetylated as in α-D-*N*-acetylglucosamine.

percentage of total carbohydrates is shown for select sources of organic matter (fig. 5.7) (Bergamaschi et al., 1999). Since the monomers that comprise carbohydrates are similar across many sources of organic matter, and many are not geochemically stable in the environment, individual sugars are not generally used as biomarkers without additional information. Acidic sugars, such as *uronic acids,* are consituents of bacterial biomass (Uhlinger and White, 1983) and have been used as proxies for bacterial sources of dissolved and particulate organic matter (DOM and POM) (Benner and Kaiser, 2003). Uronic acids are important in a variety of marine sources; they make up the extracellular polymers used by microbes to bind to surfaces, are used by invertebrates to make feeding nets, and play an important role in the pelletization of fecal pellets (Fazio et al., 1982). Uronic acids are also important components of hemicellulose and pectin (Aspinall, 1970, 1983) and are sometimes used as source indicators (biomarkers) for angiosperms and gymnosperms in terrestrial and coastal systems (Aspinall, 1970; Whistler and Richards, 1970; Aspnall, 1983). For example, *glucuronic* and *galacturonic acids* are the most abundant uronic acids in vascular plants (Danishefsky and Abraham, 1970; Bergamaschi et al., 1999), in contrast to those most abundant in bacteria. Cyanobacteria, prymnesiophytes, and heterotrophic bacteria have been purported to be producers of acid polysaccharides in the upper water column of coastal waters (Hung et al., 2003b).

The concentration range of total uronic acids in estuarine and coastal waters has been shown to be 3.2 to 6.5 μM-C, accounting for approximately 7.9 to 12.3% of the total carbohydrate content, respectively (Hung and Santschi, 2001). In general, the dominant marine sources of acid polysaccharides are phytoplankton, prokaryotes, and macroalgae (Biddanda and Benner, 1997; Decho and Herndl, 1995; Hung et al., 2003b). Similarly, O-methyl sugars can be found in soils and sediments (Mopper and Larsson, 1978; Ogner, 1980; Klok et al., 1984a,b; Bergamaschi et al., 1999), bacteria (Kenne and Lindberg, 1983), vascular plants (Stephen, 1983), and algae (Painter, 1983). In certain sources of organic matter, such as eelgrass (*Zostera marina*), minor sugars can represent a greater fraction of the total carbohydrate pool than neutral sugars (fig. 5.7). Certain O-methyl (O-Me) sugars and amino sugars may have potential as source-specific indicators of filamentous cyanobacteria (e.g., 2/4/-O-Me-xylose, 3- and 4-O-Me-fucose, 3- and 4-O-Me-arabinose, 2/4-O-Me-ribose, and 4-O-Me-glucose) and coccoid cyanobacteria and/or fermenting, SO_4^{-2}-reducing and methanogenic bacteria (e.g., glucosamine, galactosamine, 2-and 4-O-Me-rhamnose, 2/5-O-Me-galactose, 3/4-O-Me-mannose) (Moers and Larter, 1993).

5.2 Biosynthesis of Neutral and Minor Sugars

Before discussing the details of neutral and minor sugar formation, it should be noted that while we will focus on the photoautotrophic pathway, there are other net carbon fixation pathways used

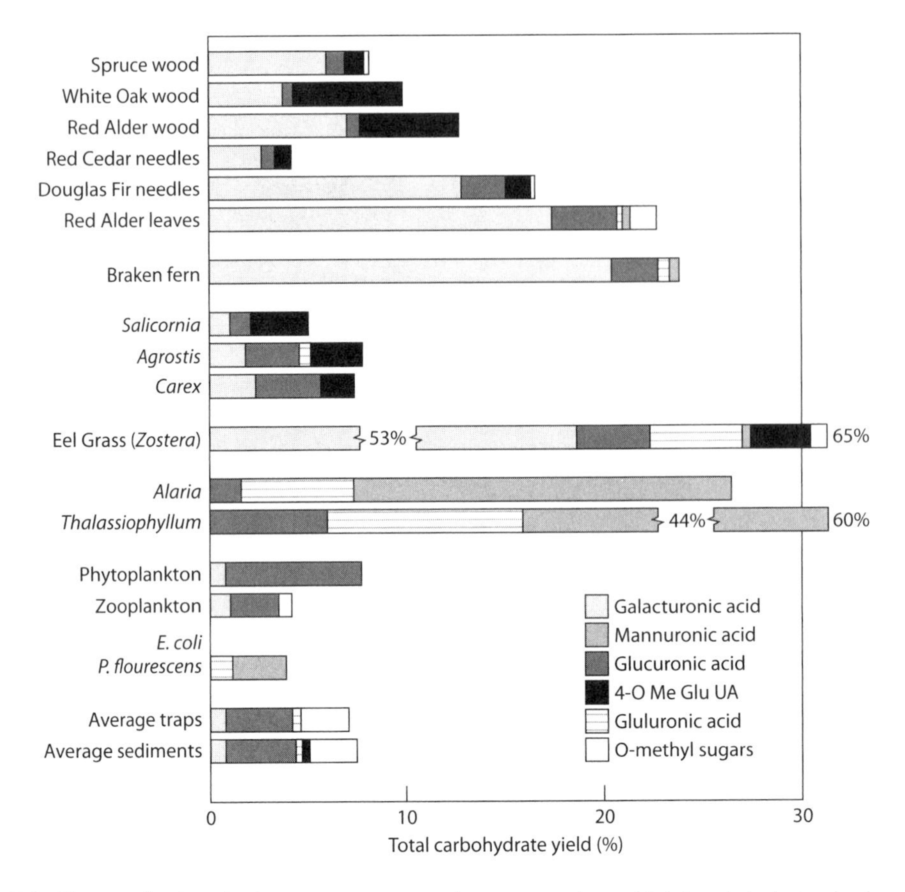

Figure 5.7. The contribution of minor sugars expressed as a percentage of total carbohydrates is shown for a variety of organic matter sources. (Adapted from Bergamaschi et al., 1999.)

by Bacteria and Archaea to synthesize biomass from inorganic carbon (Hayes, 2001, and reference therein). For our purposes, we will focus on sugar biosynthesis as related to the dominant C_3 pathway in plants. As discussed in chapter 1, the C_3 carbon-fixation pathway is a process by which carbon dioxide and ribulose -1,5-bisphosphate (RuBP, a 5-carbon sugar) are converted into two molecules of the three-carbon compound 3-phosphoglycerate (PGA) (see fig. 1.9 in chapter 1). The reaction is catalyzed by the enzyme ribulose-1,5-biphosphate carboxylase/oxygenase (RuBisCO), as shown in equation 1.2 (chapter 1):

$$6CO_2 + 6RuBp \xrightarrow{RuBisCO} 12\ 3\text{- phosphoglycerate (PGA)}$$

In the initial phases of simple sugar synthesis, which occurs in the chloroplasts, PGA is reduced by hydrogen, whereby nicotinamide adenine dinucledtide phosphate (NADPH) serves as the hydride donor. The biosynthetic pathway for hexoses (e.g., fucose and glucose) is shown in fig. 5.8. The regeneration of PGA from consumption here is through an extensive series of *transketolase* and *aldolase*, reactions in the initial stages, and some *isomerase* and *epimerase* reactions in the final stages, as summarized in equation 5.1:

$$\begin{aligned}&\text{fructose 6 phosphate} + \text{2-glyceraldehyde 3-dihydroxyacetone} +\\&\text{phosphate} + 3\ \text{ATP} + H_2O \rightarrow \text{3-ribulose 1, 5-bisphosphate} + 3\ \text{ADP} + P_i\end{aligned} \quad (5.1)$$

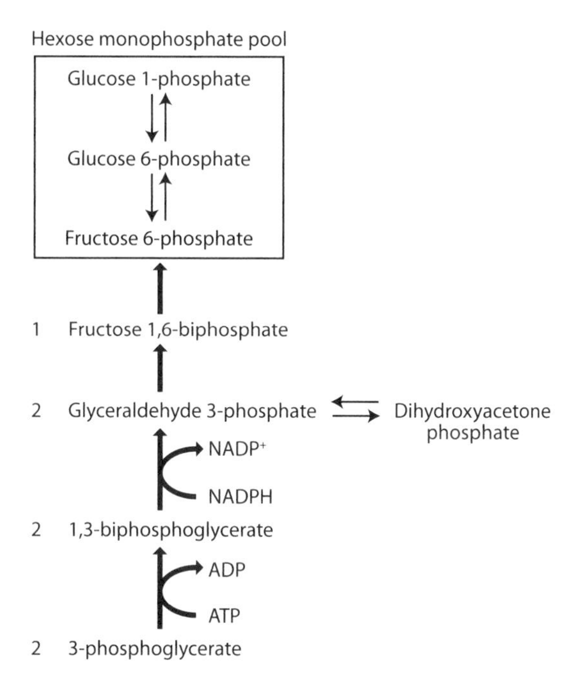

Figure 5.8. The biosynthetic pathway for hexoses (e.g., fucose and glucose).

When examining the net balanced equation (equation 5.2) for the Calvin cycle (described in chapter 1, fig. 1.10), there is a need for the replacement of NADPH molecules:

$$6CO_2 + 18ATP + 12NADPH + 12H_2O \rightarrow$$
$$C_6H_{12}O_6 + 18H_2O + 18P_i + 12NADP^+ + 6H^+ \qquad (5.2)$$

While discussing the biosynthesis of sugars, it is necessary to briefly discuss the heterotrophic processing of *photosynthate* as it relates to the regeneration of NADPH and pentoses through the oxidation of hexoses. The key pathway here is the *pentose phosphate pathway* (PPP), as shown in equation 5.3:

$$C_6H_{12}O_6 + 2NADP^+ + H_2O \rightarrow$$
$$\text{ribulose 1,5-bisphosphate} + 2NADPH + 2H^+ + CO_2 \qquad (5.3)$$

The PPP occurs both outside and inside the chloroplasts, and can be oxidative (phase I) and non-oxidative (phase 2) (fig. 5.9). Moreover, the PPP and glycolysis (see chapter 1, fig. 1.11) are linked through the enzymes transketolase and transaldolase. Specifically, when the need for NADPH is greater than for ribulose 5-phosphate, ribulose 5-phosphate is converted into glyceraldehyde 3-phosphate and fructose 6-phosphate.

5.3 Analysis of Neutral and Minor Sugars

Many of the early techniques used for measuring total dissolved carbohydrates in natural waters involved concentration procedures that resulted in fractionated samples of uncharacterized carbohydrates (Handa, 1970; Liebezeit et al., 1980; Duursma and Dawson, 1981; Ittekkot et al., 1981). The current accepted spectrophotometric method for measurements of bulk carbohydrates in natural waters is the 3-methyl-2-benzothiazolinone hydrazone hydrochloride (MBTH) method, first developed

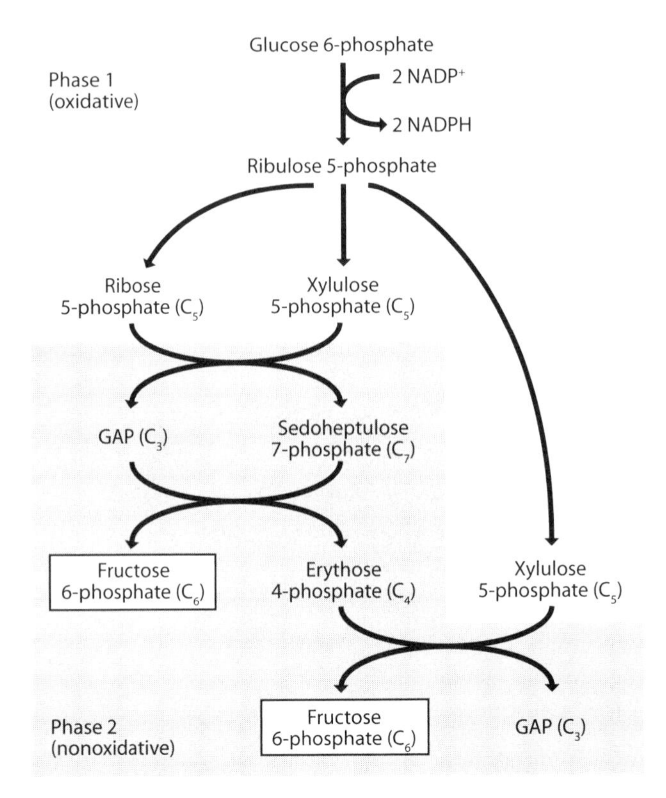

Figure 5.9. Reaction pathways for phases 1 and 2 of the pentose phosphate pathway (PPP).

by Burney and Sieburth (1977) and Johnson and Sieburth (1977), and more recently modified by Pakulski and Benner (1992). In general, dried seawater samples (via SpeedVac) are first pretreated with 12 M H_2SO_4 for 2 h at 25°C. They are then hydrolyzed in 1.2 M H_2SO_4 for 3 h at 100°C in sealed ampoules to break down polysaccharides, or monomeric sugars that contain polymeric natural substances, into monosaccharides (aldoses). After the samples are cooled in ice, they are neutralized (with NaOH) and further oxidized to *alditols*. Alditol groups are then further oxidized to formaldehyde. The formaldehyde can then be quantified spectrophotometrically after periodic acid oxidation of monosaccharide-derived alditols with MBTH. However, this method does not allow for separation of individual sugars, which is critical for a better understanding of the origin and diagenetic characteristics of organic matter. More recently, total carbohydrates have been determined using the modified spectrophotometric method of Hung and Santschi (2000), which oxidizes free reduced sugars with 2,4,6-trippyridyl-*s*-tri-azine (TPTZ), followed by analysis of the products using spectrophotometry (Myklestad, 1997).

While quantifying total carbohydrate pools provides insights about the contribution of carbohydrates to total carbon pools, individual sugars provide source-specific information. Another technique used for separating neutral sugars is gas chromatography (GC) and GC-mass spectrometry (GC-MS) (Modzeles et al., 1971; Cowie and Hedges, 1984b; Ochiai and Nakajima, 1988). In the method described by Cowie and Hedges (1984b), samples are typically treated for 2h with 72 wt% H_2SO_4 at room temperature and then hydrolyzed for 3 h at 100°C in 1.2 M H_2SO_4. The final anomeric mixtures are then equilibrated with lithium perchlorate in pyridine prior to GC-MS analysis as trimethylsilyl (TMS) derivatives. Samples are injected onto a fused silica (SE-30) capillary column, and separated using a method where the temperature is held at 140°C for 4 min and then increases at a rate of

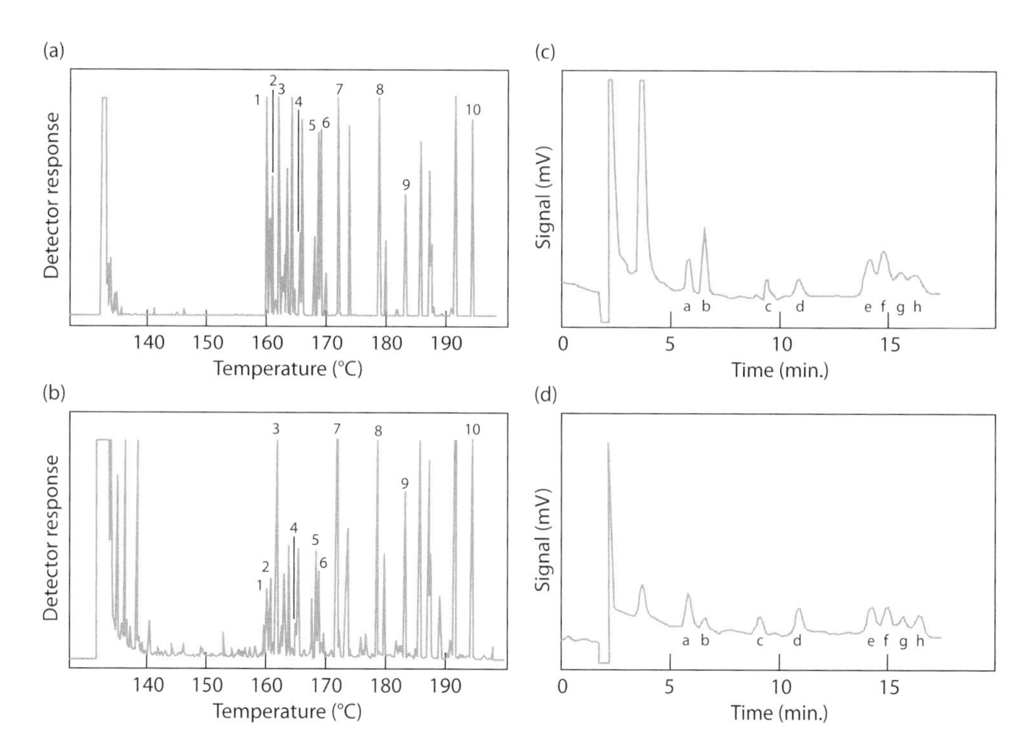

Figure 5.10. (a) Gas chromatogram of standard neutral TMS–sugar mixture. (b) Gas chromatogram of samples from Dabob Bay (USA) sediment (0–1 cm). Numbered peaks for both chromatograms a and b are as follows: 1, lyxose; 2, arabinose; 3, rhamnose; 4, ribose; 5, xylose; 6, fucose; 7, (internal std.); 8, mannose; 9, galactose; 10, glucose. (Adapted from Cowie and Hedges 1984b.) (c) High-performance liquid chromatography (HPLC) chromatograms of hydrolyzed water (unfiltered) 2° S from the equatorial Pacific. (d) Chromatogram from a standard containing 20 nM concentrations of the following individual sugars: a, fucose; b, deoxyribose; c, rhamnose; d, arabinose; e, galactose; f, glucose; g, mannose; h, xylose. (Adapted from Skoog and Benner, 1997.)

4°C min^{-1} for an additional 4 min. Detection is achieved with a flame ionization detector. Adonitol is used as an internal standard with good separation of neutral sugars (fig. 5.10a,b). The detection limit for this GC-MS method is about 0.1 ng for individual sugars with a precision of $\leq$ 5% (based on sample mean deviation) (Cowie and Hedges, 1984b).

When comparing colorimetric assays (Burney and Sieburth, 1977; Johnson and Sieburth, 1977) to gas chromatography (Cowie and Hedges, 1984b; Ochiai and Nakajima, 1988; Medeiros and Simoneit, 2007) and liquid chromatography (Mopper, 1977; Ittekkot et al., 1982; Wicks et al., 1991), it is important to note that the colorimetric methods lack compositional information, while the GC method requires derivatization steps that can be time-consuming. The next advance in carbohydrate analyses was the development of a method that combined high-performance liquid chromatography and pulsed amperometric detection (HPLC-PAD) (Wicks et al., 1991; Jorgensen and Jensen, 1994; Mopper et al., 1992; Gremm and Kaplan, 1997); this method provided both accuracy and sensitivity. More recently, HPLC-PAD methods have been modified for analyzing aldoses by H_2SO_4 hydrolysis and high performance anion exchange chromatography with PAD (HPAEC-PAD) (Skoog and Benner, 1997; Kaiser and Benner, 2000). In general, dried seawater samples (via SpeedVac) are first pretreated with 12 M H_2SO_4, sonicated, and then stirred until salts are dissolved. Samples are then hydrolyzed in 1.2 M H_2SO_4 for 3 h at 100°C in sealed ampoules. After cooling, the samples are neutralized ($CaCO_3$ or $Ba(OH)_2$) (Mopper et al., 1992; Pakulski and Benner, 1992), and then passed through anion

(AG2-X8, 20-50 mesh, Bio-Rad) and cation (AG 50W-X8, 100-200 mesh, Bio-Rad) exchange resins. Samples are then analyzed using ion chromatography, run isocratically using 28 mmol L^{-1}NaOH on a PSA-10 column system and PAD (Rocklin and Pohl, 1983; Johnson and Lacourse, 1990), with a working gold and a Ag/AgCl reference electrode. This method allows for the separation of the following seven aldoses: fucose, rhamnose, arabinose, galactose, glucose, mannose, and xylose (fig. 5.10c,d). Deoxyribose is usually added as an internal standard and the precision for this method is approximately $\leq$5% (Skoog and Benner, 1997).

The minor sugars, uronic acids, have been shown to represent a significant fraction of extracellular *acylheteropolysaccharides* (APS), which are secreted by bacteria and algae (Leppard,1993; Hung and Santschi, 2001; Hung et al., 2003b). Although GC and HPLC have been used extensively to analyze uronic acids in plant tissues, foods, and plankton, these methods are not sensitive enough to use in seawater (Mopper, 1977; Fazio et al., 1982; Walters and Hedges, 1988). A technique was recently developed for measuring total uronic acids by spectrophotometry after separation and preconcentration (Hung and Santschi, 2001). Briefly, the uronic acids are separated by cation exchange and preconcentrated by lyophilization. Final concentrations of uronic acids are based on a sulfamate/*m*-hydroxydiphenyl assay in concentrated sulfuric acid. The detection limit for uronic acids is 0.08 μM and the precision was $\leq$10%.

5.4 Applications

5.4.1 Neutral and Minor Sugars as Biomarkers in Lakes

Carbohydrates are the most abundant class of biopolymers on Earth (Aspinall, 1983) and consequently represent significant components of water column POM and DOM in aquatic environments (Cowie and Hedges, 1984a,b; Arnosti et al., 1994; Aluwihare et al., 1997; Bergamaschi et al., 1999; Opsahl and Benner, 1999; Burdige et al., 2000; Hung et al., 2003a, b). Early work suggested that fucose, which is rarely found in vascular plants and is more abundant in phytoplankton and bacteria, may be a good source indicator for aquatic organic matter (Aspinall,1970,1983; Percival, 1970). Although there has been considerable work on the composition and abundance of carbohydrates in seawater (Burney and Sieburth, 1977; Mopper et al., 1992; Rich et al., 1996; Borch and Kirchmann, 1997; Skoog and Benner, 1997), less work has been conducted in lacustrine systems (Wicks et al., 1991; Boschker et al., 1995; Tranvik and Jorgensen, 1995; Wilkinson et al., 1997). However, a recent study examined the molecular characterization of carbohydrates in dissolved organic carbon (DOC) of Antarctic lakes on Signy Island with different trophic status: Heywood Lake (*eutrophic*), Light Lake (*oligotrophic to mesotrophic*), and two oligotrophic lakes—Sombre and Moss lakes (Quayle and Convey, 2006). In general, higher concentrations of bulk DOC, high-molecular-weight DOC, and hydrolyzable neutral sugars reflected the high production of Heywood Lake (fig. 5.11). The high nutrient inputs to this system were due to seepage flow from an adjacent fur seal colony.

Hydrolyzable neutral sugar yields have commonly been used as an index of organic matter degradation, with the higher yields indicating “fresher” material (Cowie and Hedges, 1994). When examining the hydrolyzable neutral sugar yields as related to trophic status of these lakes there was no clear relationship, other than the “freshness” of material associated with the high productivity in Heywood Lake (fig. 5.11) (Quayle and Convey, 2006). These high yields were also considerably higher than those found in other aquatic ecosystems. Finally, most of the hydrolyzable neutral sugars in these lakes were dominated by glucose, especially in Heywood Lake (Quayle and Convey, 2006). Glucose has been shown to be linked with the cellular materials of phytoplankton, particularly diatoms (Skoog and Benner, 1997), which also have a high abundance of arabinose in their cell walls—another neutral sugar found to be relatively abundant in these lakes.

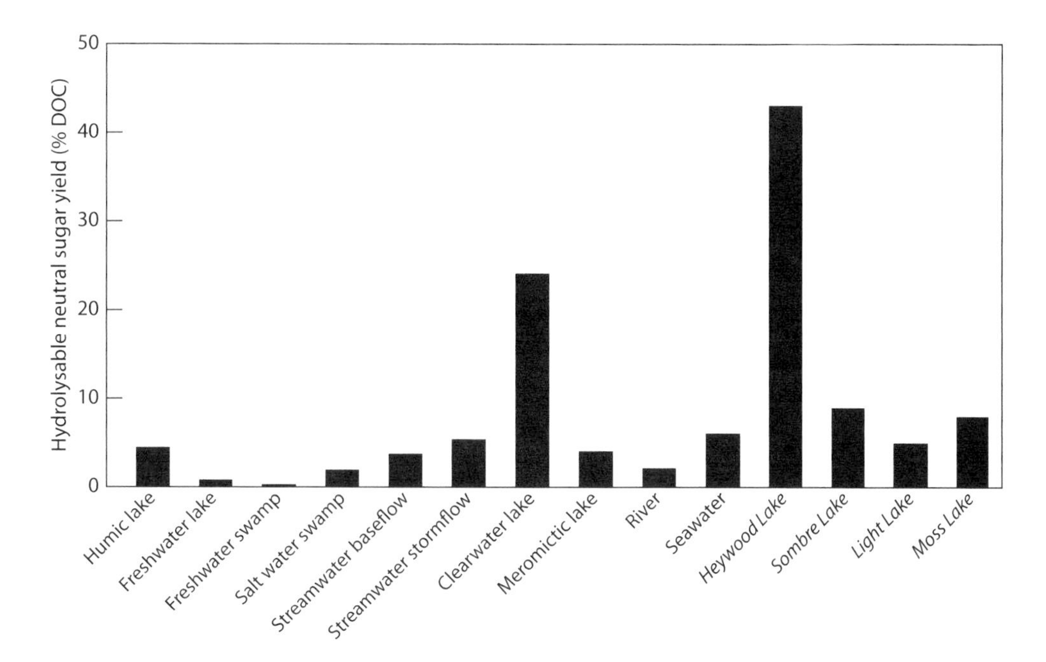

Figure 5.11. Concentrations of neutral sugars across a variety of environments, concentrations of bulk DOC, high-molecular-weight DOC, and hydrolyzable neutral sugars are higher in Heywood Lake (Australia) reflecting higher primary production. (Adapted from Quayle and Convey, 2006.)

5.4.2 Neutral and Minor Sugars as Biomarkers along the River–Estuarine Continuum

Considerable variability in concentrations of total carbohydrates, monosaccharides, and minor sugars (uronic acids) (normalized to total DOC) have been observed in rivers (table 5.1). However, when examining this variability we need to remember that different techniques were used in these studies. Measurements made with spectrophotometry are generally in good agreement with one another (Buffle, 1990; Paez-Osuna et al., 1998; Hung and Santschi, 2001; Hung et al., 2001; Gueguen et al., 2004). However, results obtained using spectrophotometric methods were generally higher than those using chromatographic methods, since a large fraction of the carbohydrate-like material in aquatic environments has yet to be identified using chromatographic techniques (Benner and Kaiser, 2003). Other work characterizing carbohydrates in Arctic rivers (Ob and Yenisei rivers) has shown that the molecular composition of neutral sugars in *ultrafiltered dissolved organic matter* (UDOM; $< 0.2\,\mu m$ to $>1\,kDa$) was very similar and not useful for distinguishing terrestrial versus marine sources of DOM in the Arctic Ocean (fig. 5.12) (Amon and Benner, 2003).

Previous work in a hypersaline estuarine lagoon suggested that fucose, ribose, mannose, and galactose were good source indicators for bacteria and cyanobacteria (Moers and Larter, 1993). Similarly, xylose or mannose:xylose ratios have been proposed as indicators for angiosperm versus gymnosperm plant tissues, while arabinose + galactose versus lyxose + arabinose provides a useful index for separating woody versus nonwoody vascular plant tissues (Cowie and Hedges, 1984a,b). Additionally, arabinose has been found in association with mangrove detritus in lagoonal sediments (Moers and Larter, 1993), but has also been linked with the surface microlayers of brackish and marine waters (Compiano et al., 1993) and is believed to be derived from phytoplankton sources (Ittekkot et al., 1982). In general, all organisms contain the same complement of simple sugars, although the relative abundance may change as a function of source (Cowie and Hedges, 1984a). Thus, the relative

Table 5.1
Relative concentrations of bulk carbohydrates (TCHO), colloidal polysaccharides (CCHO ≤ 1 kDa), total monosaccharides (MCHO), and uronic acids (URA), normalized to dissolved organic carbon (DOC) in dissolved and colloidal samples from fresh waters

Location sample type	*TCHO (C%)*	*CCHO (C%)*	*MCHO (C%)*	*URA (C%)*	*References*
Williamson River, USA (D)			2		Sweet and Perdue (1982)
Chena River, USA (C, D)	5.6	1.8			Guo et al. (2003)
Mississippi River, USA (C)		8.5			Guo et al. (2003)
Amazon River, South America (C)			4		Hedges et al. (1994)
Yukon River, USA (D)	22–27				Guéguen et al. (2004)
Lake Bret, Switzerland (D)	14–24				Wilkinson et al. (1997)
River (D)	6–24				Buffle (1990)
Stream, USA (D)			6–8		Cheng and Kaplan (2001)
Mississippi River, USA (C)			1–4		Benner and Opsahl (2001)
Trinity River, USA (CD)	17	15		2	Hung et al. (2001)
Mississippi River, USA (C)	6–8				Bianchi et al. (2004)
Santa Ana River, USA (D)	5–13				Ding et al. (1999)
Delaware River, USA (C)		15			Repeta et al. (2002)
Mississippi River, USA (C)		9			Repeta et al. (2002)
Lake Superior, USA (C)		30			Repeta et al. (2002)
Culiacan River, Mexico (D)	17				Paez-Osuna et al. (1998)
Natural waters, UK (D)	5–31				Boult et al. (2001)
Trinity River, USA (C)	9–36	—	8–31	1–5	This study

Note: D, dissolved; C, colloid (>1 kDa); TCHO, the summation of 5–7 monosaccharides, glucose, galactose, rhamnose, fucose, arabinose, xylose, and mannose.

lack of source specificity in neutral sugar composition and differential stability of individual sugars (Hedges et al., 1988; Macko et al., 1989; Cowie et al., 1995) has in some cases compromised their effectiveness as biomarkers. For example, sediment trap and sediment samples in an anoxic fjord (Saanich Inlet, Canada) indicated that glucose, lyxose, and mannose in vascular plant detrital particles were lost in the upper sediments (Hamilton and Hedges, 1988). Conversely, rhamnose and fucose were produced within the fjord and were likely derived from bacteria known to be abundant in these deoxy

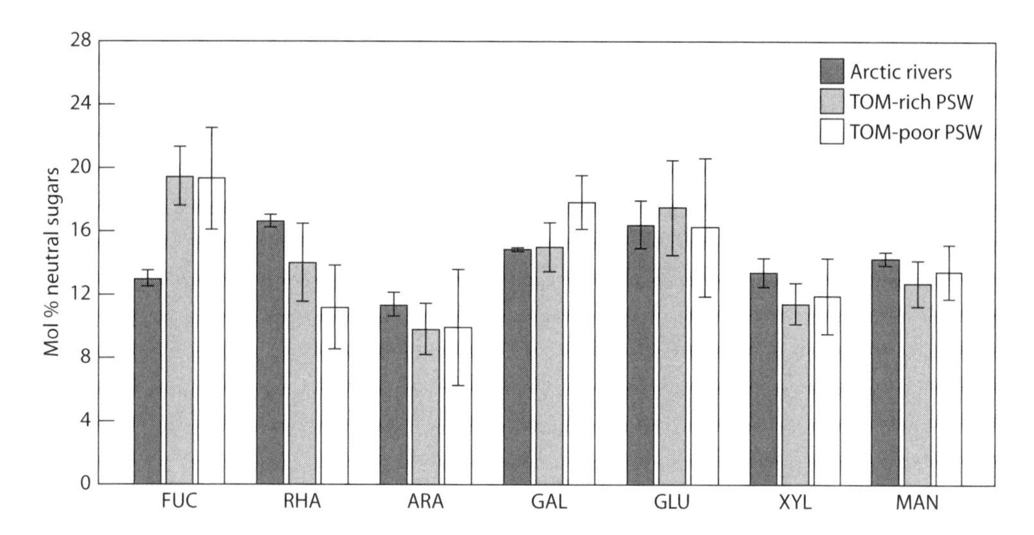

Figure 5.12. Carbohydrates in Arctic rivers (the Ob and Yenisei) showing the molecular composition of neutral sugars in ultrafiltered dissolved organic matter (UDOM; $< 02\,\mu$m to > 1 kDa). (Adapted from Amon and Benner, 2003.)

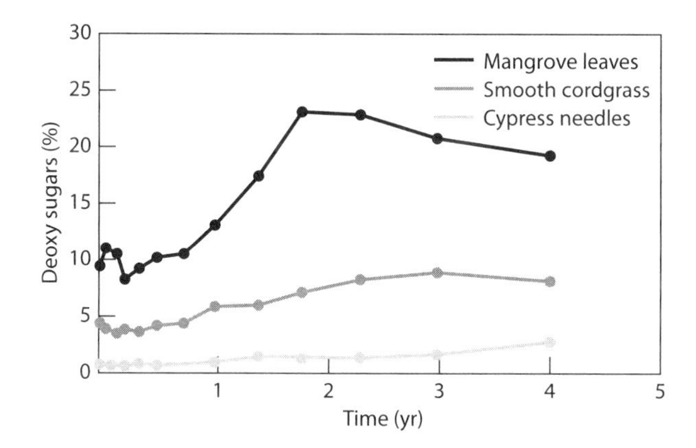

Figure 5.13. Results from long-term (4-yr) incubation experiment that examined changes in the carbohydrate abundance and composition of three different vascular plant tissues (mangrove leaves [*Avicennia germinans*], cypress needles [*Taxodium distichum*], and smooth cordgrass [*Spartina alterniflora*]). While glucose and xylose were degraded over time, deoxy sugars increased in relative abundance, indicating the importance of contributions from microbial biomass with increasing age of detritus. (Adapted from Opsahl and Benner, 1999.)

sugars (Cowie and Hedges, 1984b). A long-term (4-yr) incubation experiment that examined changes in the carbohydrate abundance and composition of five different vascular plant tissues (mangrove leaves and wood [*Avicennia germinans*], cypress needles and wood [*Taxodium distichum*], and smooth cordgrass [*Spartina alterniflora*]) showed that while glucose and xylose were degraded over time, deoxy sugars increased in relative abundance, which is indicative of the importance of contributions from microbial biomass with increasing age of detritus (Opsahl and Benner, 1999) (fig. 5.13).

Similarities in dissolved combined neutral sugar (DCNS) composition have been attributed to the production of refractory heteropolysaccharides through bacterial processing of organic matter (Kirchmann and Borch, 2003). Other work has shown high concentrations of acylheteropolysaccharides (APS) in high–molecular-weight (HMW) DOM from different aquatic systems believed to be derived

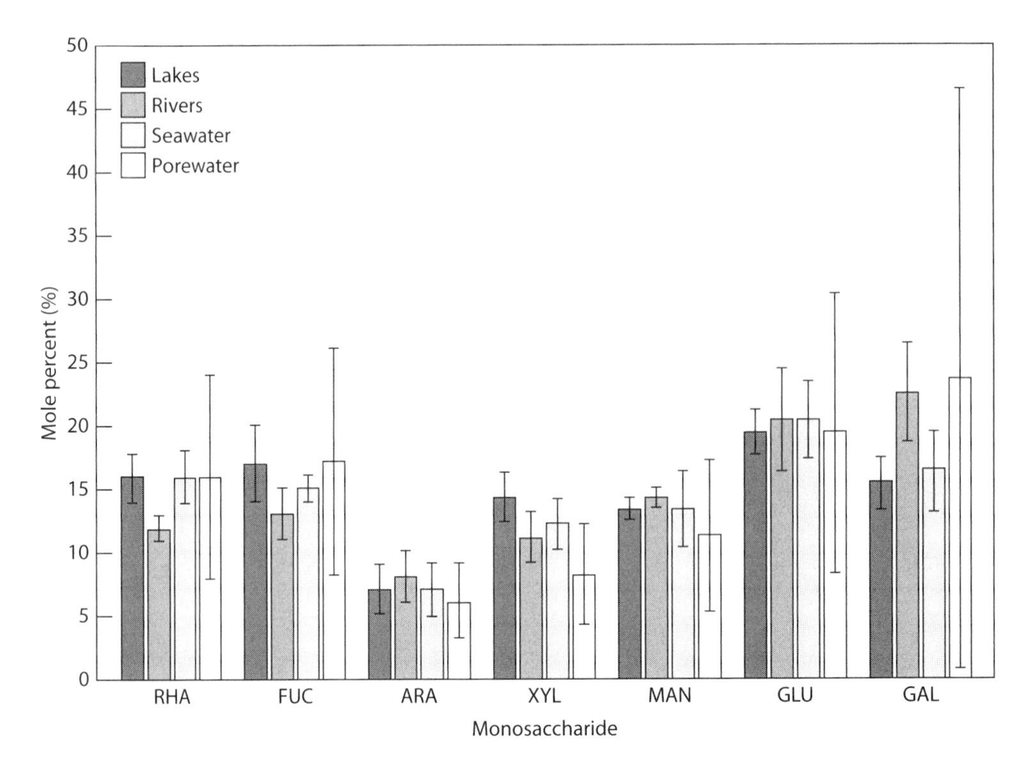

Figure 5.14. The relative abundance of seven major neutral sugars (rhamnose, fucose, arabinose, xylose, mannose, glucose, and galactose) in HMW DOM from different aquatic ecosystems. (Adapted from Repeta et al., 2002.)

from freshwater phytoplankton (Repeta et al., 2002). In fact, the relative abundance of seven major neutral sugars (rhamnose, fucose, arabinose, xylose, mannose, glucose, and galactose) in HMW DOM was found to be similar and the relative amounts in samples from different aquatic systems were nearly the same (fig. 5.14).

Although marine and freshwater microalgae are both capable of producing APS (Aluwihare et al., 1997; Repeta et al., 2002), controls on the overall cycling of APS in aquatic systems remain largely unknown. Spectroscopic and molecular data indicate that APS in HMW DOM across divergent systems is very similar (Aluwihare et al., 1997). Multivariate analyses using direct temperature-resolved mass spectrometry (DT-MS) showed that HMW DOM (>1D) from Chesapeake Bay (USA) and Oosterschelde estuary (Netherlands) was largely composed of amino sugars, deoxysugars, and O-methyl sugars—suggestive of bacterial processing (Minor et al., 2001, 2002). These studies further support the hypothesis that similarities in lipid composition or "signatures" found in estuarine DOM (Zou et al., 2004) were ultimately determined by bacterial processing along the estuarine gradient. These molecular studies also support the notion that much of the DOC in estuaries is composed of both labile and refractory components (Raymond and Bauer, 2000, 2001; Cauwet, 2002; Bianchi et al., 2004)—with much of the labile fraction degraded before being exported to the ocean.

In sediments, carbohydrates have been estimated to be approximately 10 to 20% of the total organic matter (Hamilton and Hedges, 1988; Cowie et al., 1982; Martens et al., 1992). Like the water column, controls on the turnover and extracellular hydrolysis of polysaccharides to monosaccharides are not well understood for the dissolved and particulate carbohydrate fractions in sediments (Lyons et al., 1979; Boschker et al., 1995; Arnosti and Holmer, 1999). Recent work has shown that neutral sugars represent 30 to 50% of the dissolved carbohydrates in Chesapeake Bay porewaters (fig. 5.15) (Burdige

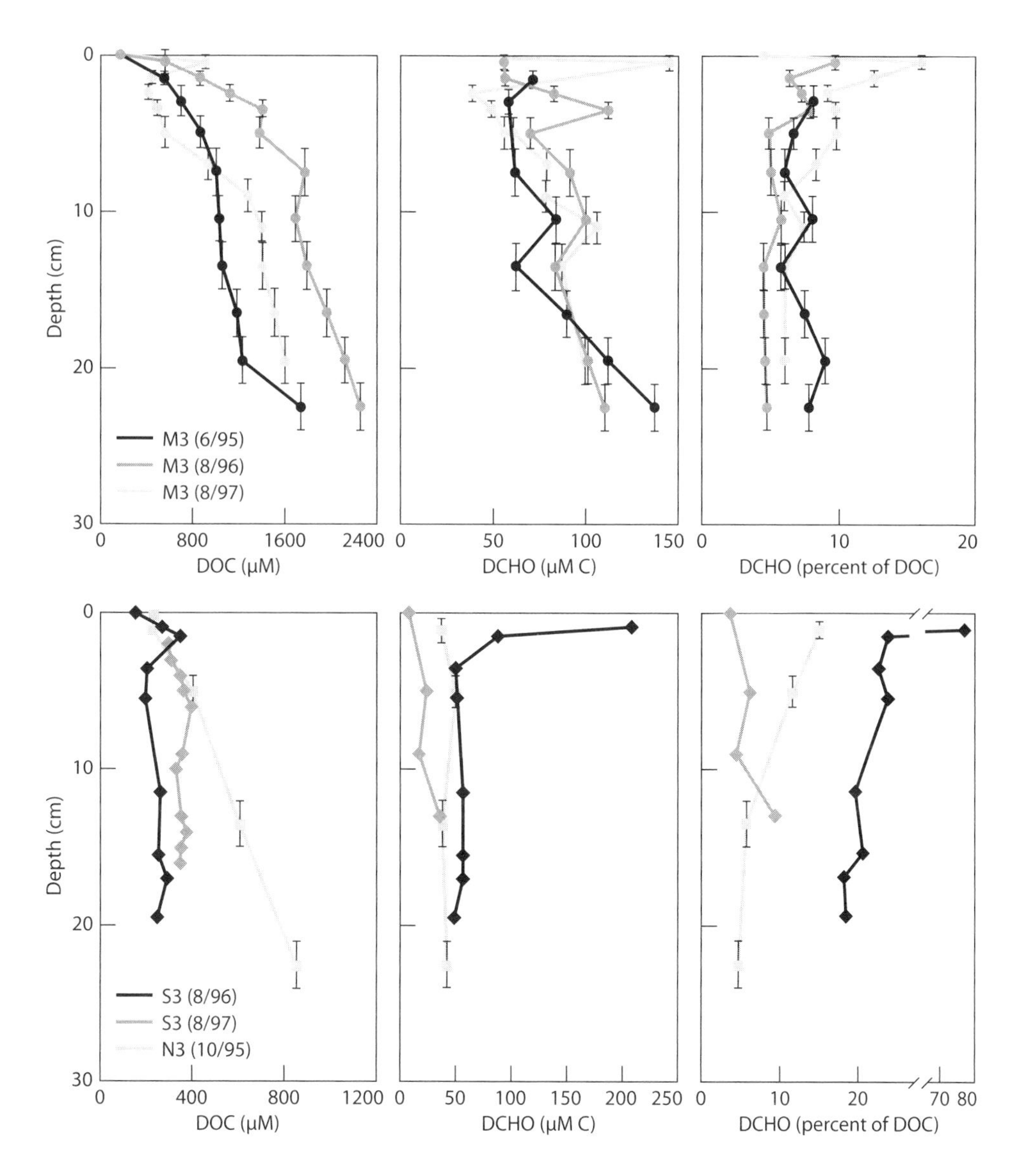

Figure 5.15. Porewater profiles of DOC and dissolved carbohydrates (DCHO) at three sites in Chesapeake Bay. Neutral sugars represented 30 to 50% of the dissolved carbohydrates in Chesapeake Bay (USA) porewaters. (Adapted from Burdige et al., 2000.)

et al., 2000). The percent abundance of neutral sugars in porewaters ranged from glucose at the high end (28%) to rhamnose (6%) and arabinose (7%) at the low end (Burdige et al., 2000); these results agree with the relative percentages of *dissolved combined neutral sugars* (DCNS) found in the water column of Delaware Bay (USA) (Kirchmann and Borch, 2003). However, concentrations of particulate and dissolved carbohydrates in Chesapeake Bay sediments appeared to be decoupled, with more of the dissolved carbohydrate fraction found in the HMW DOM of porewaters. Such differences between particulate and dissolved carbohydrate fractions have been reported in other studies (Arnosti and Holmer, 1999). Recent work based on enzymatic hydrolysis rates showed that 40 to 62% of the monosaccharides in the surface waters of Chesapeake Bay were consumed rapidly by microbes, with lower percentages in shelf waters (Steen et al., 2008). A relatively higher abundance of dissolved carbohydrates in the HMW DOM fraction may reflect an accumulation of HMW intermediates generated during POC remineralization in sediments (Burdige et al., 2000). Although carbohydrates are considered to be a significant fraction of the DOC pool in both oceanic (20–30%) (Pakulski and Benner, 1994) and estuarine waters (9–24%) (Senior and Chevolot, 1991; Murrell and Hollibaugh, 2000; Hung and Santschi, 2001), little is known about the uptake dynamics and general cycling of carbohydrates in estuaries. The turnover of polysaccharides (e.g., xylan, laminaran, pullulan, fucoidan) is also poorly understood, with recent work suggesting that molecular size may not be as important a variable in controlling the extracellular hydrolysis of these macromolecular compounds in coastal systems as previously thought (Arnosti and Repeta, 1994; Keith and Arnosti, 2001). While we do know that neutral sugars are a dominant component (40 to 50%) of the total carbohydrates in the water column (Borch and Kirchmann, 1997; Minor et al., 2001), little is known about differential uptake of these monosaccharides by bacteria (Rich et al., 1997; Kirchman, 2003; Kirchmann and Borch, 2003). Many of the abundant sugars (e.g., glucose and rhamnose) typically occur as DCNS (Kirchmann and Borch 2003). Other studies have shown remarkable temporal and spatial stability in the composition of DCNS in the Delaware Bay estuary—despite significant variations in POC and DOC (Kirchmann and Borch, 2003). Glucose (23 mol%) and arabinose (6 mol%) were the most and least abundant compounds in DCNS. The relative abundance of neutral sugars, specifically glucose (Hernes et al., 1996), is typically viewed as an index of organic matter lability (Amon et al., 2001; Amon and Benner, 2003). Thus, it is remarkable that the composition of DCNS in Delaware Bay estuary did not show significant seasonal variation and was similar to the composition found in the open ocean (Borch and Kirchmann, 1997; Skoog and Benner, 1997; Burdige et al., 2000).

5.4.3 Neutral and Minor Sugars as Biomarkers in the Ocean

As stated earlier, carbohydrates are one of the most important sources of carbon to bacterioplankton and represent one of the best-characterized components (ca. 25%) of marine DOM (e.g., Pakulski and Benner, 1994; Borch and Kirchmann, 1997; Skoog and Benner, 1997). Based on the negative correlation between polysaccharides and monosaccharides at depths above the oxygen minimum in the Pacific Ocean, the Gulf of Mexico, and Gerlache Straight (Antarctica), it was concluded that monosaccharides are, in fact, derived from the hydrolysis of polysaccharides (Pakulski and Benner, 1994). This is consistent with the general hypothesis that polysaccharides, largely derived from phytoplankton in surface waters, are highly available to microbes that degrade these larger molecules into the smaller monosaccharides (Ittekkot et al., 1981). More specifically, when examining the compositions and concentrations of aldoses in POM and HMW DOM from surface water (2 m) and the component that was removed between surface and deep waters in the Equatorial Pacific, glucose and galactose were the dominant sugars in both fractions (Skoog and Benner, 1997) (fig. 5.16). Moreover, the fraction of carbohydrates removed from surface waters to depth was essentially the same, suggesting that there was no preferential biological removal of specific sugars in the POM or HMW DOM pools.

While past work has shown that inorganic nutrients are also capable of enhancing bacterial growth (Wheeler and Kirchman, 1986; Horrigan et al., 1988), the interplay between nutrients and carbohydrate uptake was ignored for many years. Only recently has work focused on bacterial production as related glucose uptake in the North Pacific gyre (Skoog et al., 2002). This work showed that glucose-supported bacterial production (BP) was strongly dependent on the availability of nitrogen and phosphorous (fig. 5.17). In particular, glucose represented a larger fraction of the bacterial carbon demand in nutrient-rich waters (sub-Arctic gyre) than in nutrient-poor water (North Pacific gyre). The glucose uptakes rates in this study ranged from <0.01 to $4.2\,nM\,d^{-1}$, which were lower than rates in

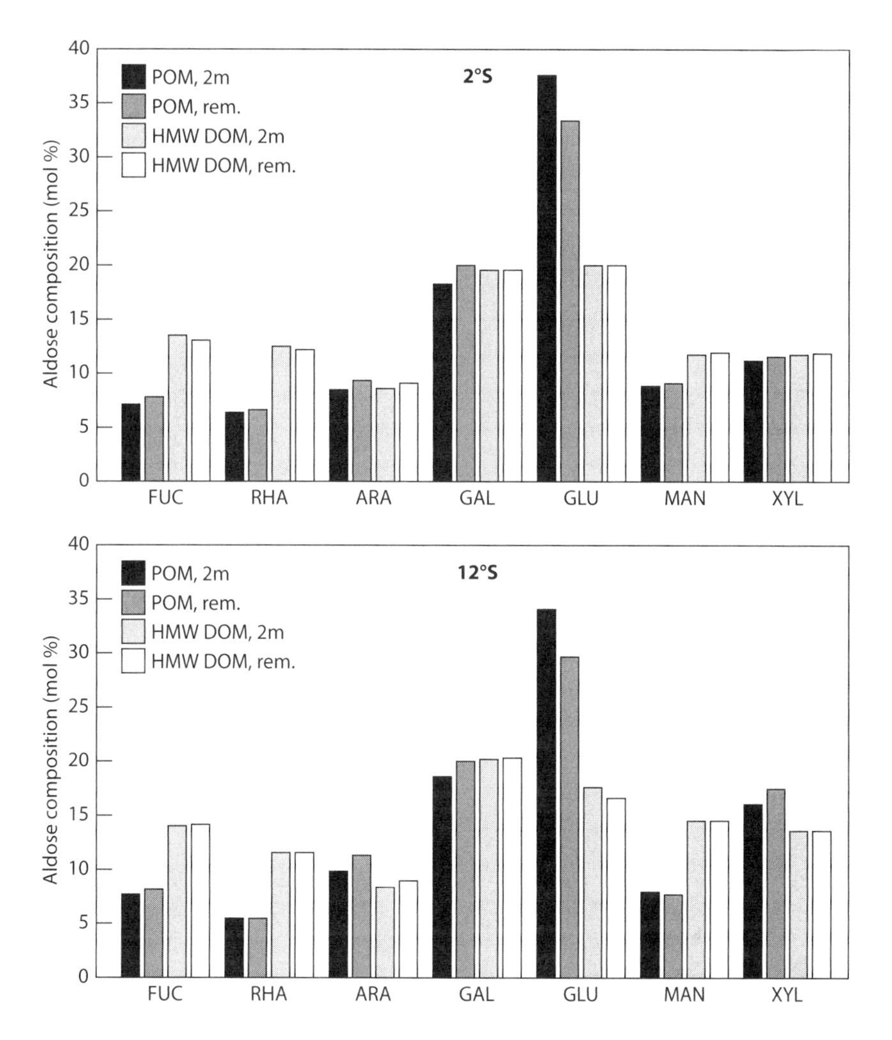

Figure 5.16. The compositions and concentrations of aldoses in POM and HMW DOM from surface water (2 m) and deep waters (rem) in the Equatorial Pacific; glucose (GLU) and galactose (GAL) were the dominant sugars in both fractions. (Adapted from Skoog and Benner, 1997.)

the tropical Pacific (3 to 148 nN d^{-1}) (Rich et al., 1996), and higher than those in the Gulf of Mexico (< 1 to 3 nM d^{-1}) (Skoog et al., 1999).

Although there has been some work on carbohydrate abundance and composition in estuarine porewaters (e.g., Burdige et al., 2000), little is known about their abundance in *hydrothermal fluids*. Much of the work in hydrothermal fluids has shown the abundance of hydrocarbons (Simoneit et al., 1996). However, a recent study analyzed total dissolved neutral aldoses (TDNA) in Vulcano, one of the 7 active islands in the Aeolian archipeligo of the north coast of Sicily, Italy (Skoog et al., 2007). This work nicely complemented other studies analyzing low-molecular-weight organic acids (Amend et al., 1998) and amino acids (Svensson et al., 2004) in vent fluids from Vulcano. Skoog et al. (2007) analyzed TDNA in three different geological environments (*sediment seeps*, *submarine vents*, and hydrothermal wells). While all three environments had considerable variability in concentrations of TDNA, most had enough labile aldose to support microbial growth. However, the mole fraction abundance of fucose + rhamnose, biomarkers for algal sources (Mopper, 1977; Zhou et al., 1998), indicated that the primary

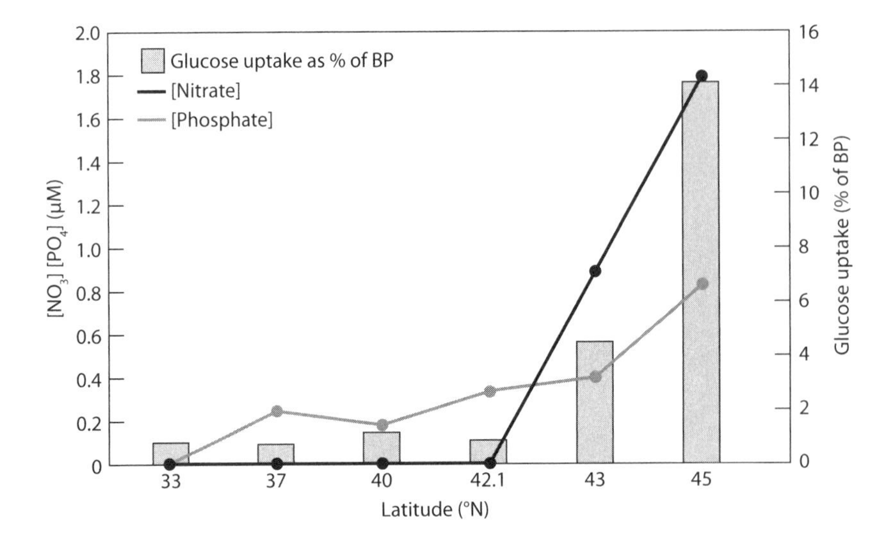

Figure 5.17. Bacterial production (BP) expressed as glucose uptake in the North Pacific gyre. Concentration of dissolved nitrate and phosphate are shown for comparison. (Adapted from Skoog et al., 2002.)

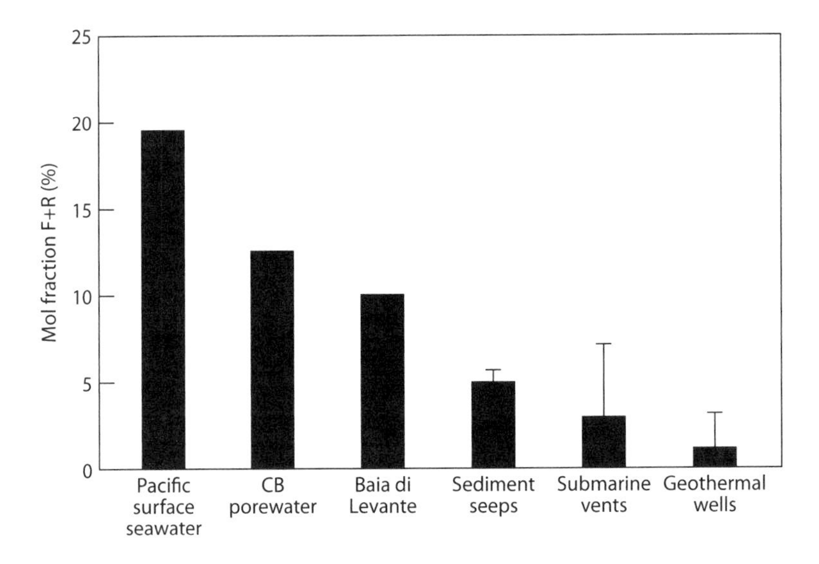

Figure 5.18. Concentrations of total dissolved neutral aldoses Fucose (F) + Rhamnose (R) from different aquatic and geological environments (sediment seeps, submarine vents, and hydrothermal wells). (Adapted from Skoog et al., 2007.)

source of aldoses in the Vulcano hydrothermal samples was not algal in origin (Skoog et al., 2007) (fig. 5.18).

5.5 Summary

In this chapter we focused on some of the most abundant molecules on the planet, the carbohydrates. These molecules exhibit a range in reactivity and are thus important on long-term (geological) and short-term (ecological) time intervals. Carbohydrates are particularly important in the context of trophic dynamics with the larger carbon cycle in terms of ecological energy transfer across different trophic levels. Although these compounds are ubiquitous in aquatic environments and in many cases do not provide unique, class-specific differentiation of plant/algal source materials, we have provided examples where such cases have been demonstrated. The techniques for measuring carbohydrate compounds continue to improve and should provide greater applications of separating out important more details in energy exchange processes in aquatic systems. Finally, we have shown that these compounds in many cases provide the "glue" for organic cells to stick to surface as well as a matrix for larger dissolved organic polysaccharide molecules (e.g., APS) to be formed in natural waters. Further work is needed on understanding the role of the polymers in global carbon cycling.

6. Proteins: Amino Acids and Amines

6.1 Background

Proteins make up approximately 50% of organic matter (Romankevich, 1984) and contain about 85% of the organic N in marine organisms (Billen, 1984). Peptides and proteins comprise an important fraction of the particulate organic carbon (POC) (13 to 37%) and nitrogen (PON) (30 to 81%) (Cowie et al., 1992; Nguyen and Harvey, 1994; Van Mooy et al., 2002), as well as dissolved organic nitrogen (DON) (5–20%) and carbon (DOC) (3–4%) in oceanic and coastal waters (Sharp, 1983). Other estimates show that acid-hydrolyzable, proteinaceous material accounts for as much as 50% of oceanic PON (Tanoue, 1992). In sediments, proteins account for approximately 7 to 25% of organic carbon (Degens, 1977; Burdige and Martens, 1988; Keil et al., 1998) and are estimated to comprise 30 to 90% of total N (Henrichs et al., 1984; Burdige and Martens, 1988; Haugen and Lichtentaler, 1991; Cowie et al., 1992). Other labile fractions of organic N include *amino sugars*, *polyamines*, *aliphatic amines*, *purines*, and *pyrimidines*.

Proteins are made of approximately 20 α-amino acids, which can be divided into different categories based on their functional groups (fig. 6.1). An *α-amino acid* is composed of an amino group, a carboxyl group, an H atom, and a distinctive R group (side chain) that is bonded to a C atom—this C atom is the α-carbon because it is adjacent to the carboxylic group. Since amino acids contain both basic (–NH_2) and acidic (–COOH) groups they are considered to be *amphoteric*. In a dry state, amino acids are dipolar ions, where the carboxyl group exists as a carboxylate ion (–COO^-) and the amine group exists as an ammonium group (–NH_3^+); these dipolar ions are called *zwitterions*. In aqueous solutions, equilibrium exists between the dipolar ion and the anionic and cationic forms. Thus, the dominant form of an amino acid is strongly dependent on the pH of the solution; the acidic strength of these different functional groups is defined by its pK_a. Under highly acidic and basic conditions amino acids occur as cations and anions, respectively. At intermediate pH the zwitterion is at its highest concentration, with the cationic and anionic forms in equilibrium; this is called the *isoelectric point*. Peptide bonds are formed from condensation reactions between the carboxyl group of one amino acid and the amine group of another (fig. 6.2). As a result of the asymmetric α-carbon, each amino acid exists as two *enantiomers*. Amino acids in living systems are almost exclusively L-isomers.

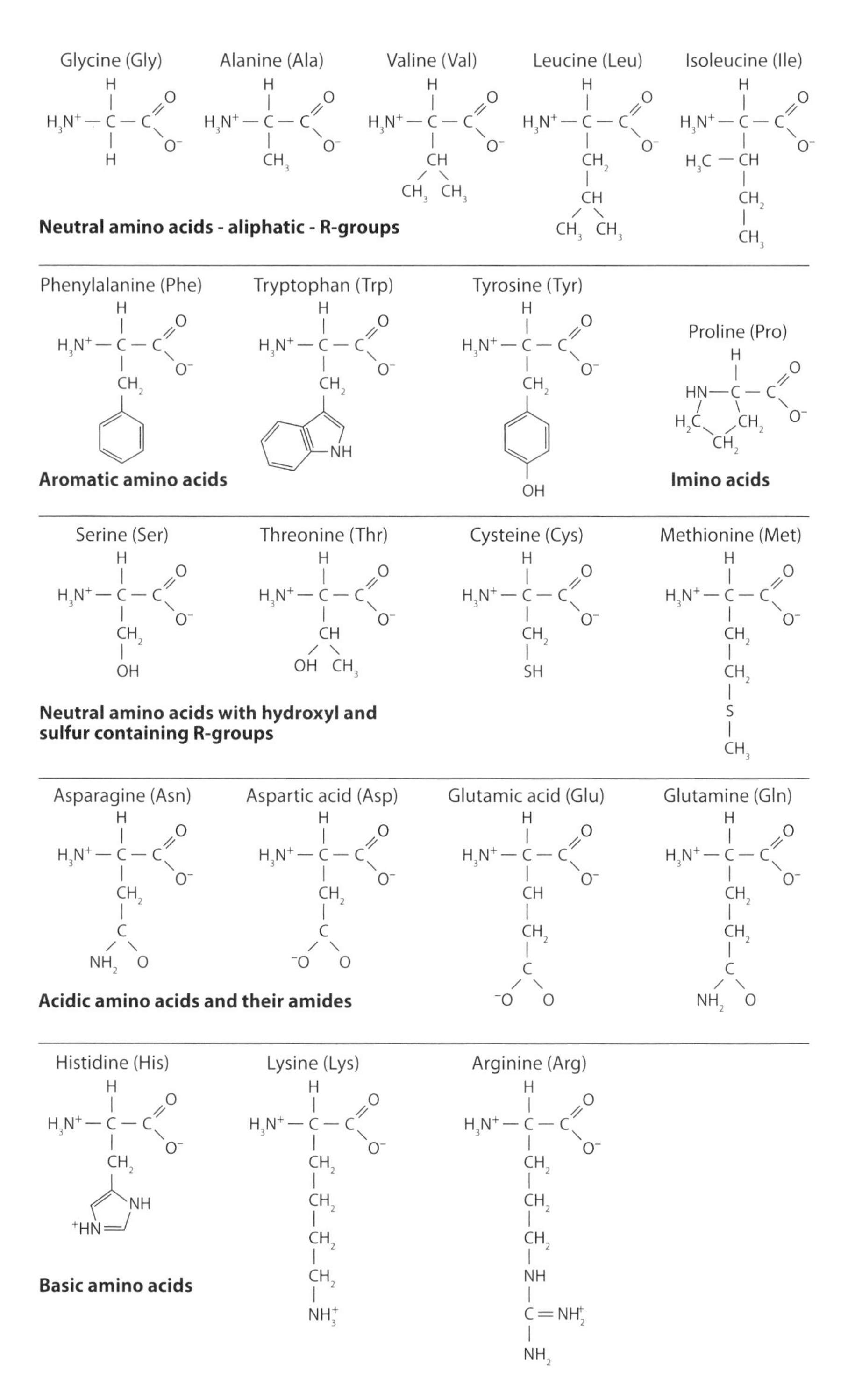

Figure 6.1. Proteins are made of approximately 20 α-amino acids, which can be divided into different categories (e.g., neutral, aromatic, acidic) based on their functional groups.

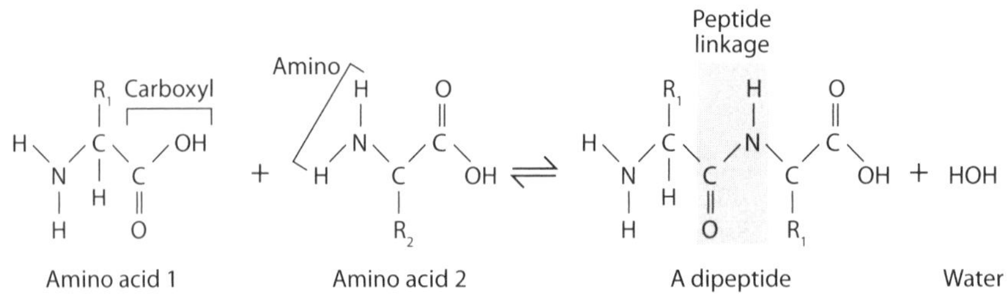

Figure 6.2. Peptide bonds are formed from condensation reactions between the carboxyl group of one amino acid and the amine group of another.

However, the *peptidoglycan* layer in bacterial cell walls has a high content of D-amino acids. Recent work has shown that the four major D-amino acids in bacterioplankton are alanine, serine, aspartic acid, and glutamic acid, and that under low carbon conditions bacteria may rely on the uptake of D-amino acids (Perez et al., 2003). Consequently, a high D/L ratio in dissolved organic matter (DOM) may indicate contributions from bacterial biomass and/or high bacterial cycling of this material. Recently, the D/L ratio has been successfully used as an index of the diagenetic state of DON (Preston, 1987; Dittmar et al., 2001; Dittmar, 2004). Another process that can account for the occurrence of D-amino acids in organisms is *racemization,* which involves the conversion of L-amino acids to their mirror-image D-form. Estuarine invertebrates have been found to have D-amino acids in their tissues as a result of this process (Preston, 1987; Preston et al., 1997). Racemization is an abiotic process whereby L-amino acids present in living organisms are converted to D-amino acids over time following an organism's death. Since the rate of racemization can be time-calibrated, the conversion of L-amino acids to D-amino acids provides a dating tool. Rates of racemization of amino acids in shell, bone, teeth, microfossils, and other biological materials, calibrated against radiocarbon measurements, have been used as a dating tool (e.g., as an index for historical reconstruction of coastal erosion [Goodfriend and Rollins, 1998]).

Amino acids are essential to all organisms and consequently represent one of the most important components in the organic nitrogen cycle. Of the 20 amino acids normally found in proteins, 11 can be synthesized by animals, referred to as *nonessential* amino acids (alanine, arginine, asparagine, aspartic acid, cysteine, glutamine, glutamic acid, glycine, proline, serine, and tyrosine), and the other 9, which must be obtained through dietary means, are referred to as *essential amino acids* (histidine, isoleucine, leucine, lysine, methionine, phenylalanine, threonine, tryptophan, and valine)—typically synthesized by plants, fungi, and bacteria. The typical mole percentage of protein amino acids in different organisms shows considerable uniformity in compositional abundance (fig. 6.3) (Cowie et al., 1992). Hence, it has been argued that differences in amino acid composition in the water column and/or sediments are likely to arise from decomposition rather than source differences (Dauwe and Middelburg, 1998; Dauwe et al., 1999). In fact, the following degradation index (DI) was derived, using *principal components analysis* (PCA), including a diversity of samples ranging from fresh phytoplankton to highly degraded turbidite sediment samples:

$$\mathrm{DI} = \sum_i [(\mathrm{var}_i - \mathrm{avg\,var}_i/\mathrm{std\,var}_i)\mathrm{loading}_i] \tag{6.1}$$

where

var_i = mole percent (mol%) of amino acid

$\mathrm{avg\,var}_i$ = mean amino acid mol%

$\mathrm{std\,var}_i$ = standard deviation of amino acid mol%

$\mathrm{loading}_i$ = PCA-derived loading of amino acid_i

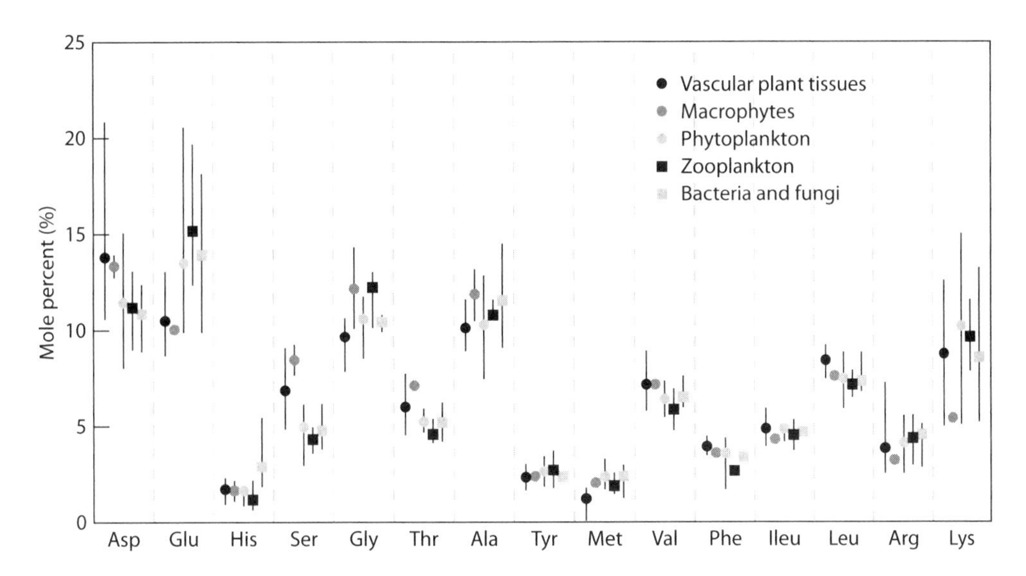

Figure 6.3. The typical mole percentage of protein amino acids in different organisms shows considerable uniformity in compositional abundance. (Adapted from Cowie et al., 1992.)

The more negative the DI values, the more degraded the sample, with positive DI values indicative of fresh materials. Elevated mole percentages (e.g., >1%) of *nonprotein amino acids* (NPAA) have also been used as an index of organic matter degradation (Lee and Bada, 1977; Whelan and Emeis, 1992; Cowie and Hedges, 1994; Keil et al., 1998). For example, amino acids such as β-alanine, γ-aminobutyric acid, and ornithine are produced enzymatically during decay processes from precursors such as aspartic acid, glutamic acid, and arginine, respectively. β-Aminoglutaric is another NPAA found in marine porewaters (Henrichs and Farrington, 1979, 1987), with glutamic acid as its possible precursor (Burdige, 1989). Another factor contributing to differences in mole percentages of amino acids between the water column and sediments is selective adsorption of different amino acids (e.g., basic versus acidic) on different clay minerals during pre- and postdepositional processes (Keil et al., 2000; Ding and Henrichs, 2002). Finally, recent work has shown that traditional hydrolysis techniques used in amino acid analysis of sediments destroy most amide bonds, thereby underestimating the contribution of amino acids to the total sedimentary nitrogen pool (Nunn and Keil, 2005, 2006).

In oceanic and coastal waters, proteins are believed to be degraded by hydrolysis reactions whereby other smaller components, such as peptides and free amino acids, are produced (Billen, 1984; Hoppe, 1991). Although many higher consumers can hydrolyze proteins in their food resources internally, bacteria must rely on external hydrolysis (via *ecto-* or *extracellular enzymes*) of proteins outside the cell (Payne, 1980). Extracellular hydrolysis allows for smaller molecules to be transported across the cell membrane; molecules typically larger than 600 Da cannot be transported across microbial cell membranes (Nikaido and Vaara, 1985). The digestibility of protein in different food resources has been examined using nonspecific *proteolytic enzymes* (e.g., protease-k), which can separate large polypeptides from smaller oligo and monomers (Mayer et al., 1986, 1995). This yields another category of groups of amino acids commonly referred to as enzymatically hydrolyzable amino acids (EHAA), which can be used to determine what essential amino acids are limiting to organisms (Dauwe et al., 1999). Only recently have we begun to understand the role of peptides in the cycling of amino acids in coastal systems. It appears that not all peptides are capable of being hydrolyzed and that peptides with greater than two amino acids are hydrolyzed more rapidly than *dipeptides* (Pantoja and Lee, 1999). Extracellular hydrolysis of peptides and amino acids has also been known to occur by certain phytoplankton species (Palenik and Morel, 1991). During the extracellular oxidation of amino acids, NH_4^+, an important source of dissolved inorganic nitrogen (DIN) for primary producers, is liberated and then becomes available for uptake. Recent work has shown that approximately 33% of the NH_4^+ uptake by the pelagophyte, *Aureococcus anophagefferens*, was provided by amino acid oxidation in Quantuck Bay estuary (USA) (Mulholland et al., 2002). This agrees with earlier work that documented elevated amino acid oxidation rates in Long Island estuaries (USA) during phytoplankton bloom events (Pantoja and Lee, 1994; Mulholland et al., 1998). In general, amino acids have typically been analyzed in dissolved and particulate fractions. Pools of amino acids commonly measured in coastal systems include the *total hydrolyzable amino acids* (THAA) in water column POM (North, 1975; Cowie et al., 1992; Van Mooy et al., 2002) and sedimentary organic matter (SOM) (Cowie et al., 1992), and *dissolved free* and *combined amino acids* (DFAA and DCAA) in DOM (Mopper and Lindroth, 1982; Coffin, 1989; Keil and Kirchman, 1991a,b). Despite the importance of living biomass being the source of amino acids, oceanic POM contains relatively small amounts of living biomass (e.g., living biomass: detritus = 1 : 10) (Volkman and Tanoue, 2002). The cellular proteinaceous material that has been transformed to detrital components has been referred to as particulate combined amino acids (PCAA) (Volkman and Tanoue, 2002; Saijo and Tanoue, 2005).

Amino acids are generally considered to be more labile than bulk C or N in the water column and sediments and have been shown to be useful indicators for the delivery of "fresh" organic matter to surface sediments (Henrichs and Farrington, 1987; Burdige and Martens, 1988). In fact, 90% of the primary amino acids and N production was found to be remineralized during predepositional processes, particularly herbivorous grazing by zooplankton (Cowie et al., 1992). The major fraction of amino acids in sediments occurs in the solid phase, with very small fractions in the free dissolved state (1–10%) (Henrichs et al., 1984). Low concentrations of DFAA in pore waters may also reflect high turnover rates (Christensen and Blackburn, 1980; Jorgensen, 1987) since microorganisms preferentially utilize DFAA (Coffin, 1989). A considerable amount of the THAA fraction exists in resistant forms bound to humic compounds and mineral matrices (Hedges and Hare, 1987; Alberts et al., 1992). Thus, differential degradation among amino acids appears to be controlled by differences in functionality and intracellular compartmentalization. For example, aromatic amino acids (e.g., tyrosine and phenylalanine), along with glutamic acid and arginine, decay more rapidly than some of the more refractory acidic species, such as glycine, serine, alanine, and lysine (Burdige and Martens, 1988; Cowie et al., 1992). Serine, glycine, and threonine have been shown to be concentrated into the siliceous exoskeletons of diatom cell walls, which may "shield" them from bacterial decay processes (Kroger et al., 1999; Ingalls et al., 2003). Enhanced preservation of amino acids may also occur in carbonate sediments due to adsorption and incorporation into calcitic exoskeletons (King and Hare, 1972; Constantz and Weiner, 1988). Conversely, preferential adsorption of basic amino acids to fine particulate matter in river and estuarine systems occurs because of the attraction between the net positive charge of the amine group and the negative charge of the aluminosilicate clay minerals (Gibbs, 1967; Rosenfeld, 1979; Hedges et al., 2000; Aufdenkampe et al., 2001). Molecular shape/structure (Huheey, 1983) and weight (higher van der Waals attractive forces with increasing weight) also affect the adsorption of amines (Wang and Lee, 1993). Similarly, preferential adsorption of amino acids to clay particles can result in mixtures with other surface amino acids (Henrichs and Sugai, 1993; Wang and Lee, 1993, 1995) or reactions with other organic molecules (e.g., *melanoidin polymers* are produced from condensation reactions of glucose with acid, neutral, and basic amino acids) (Hedges, 1978).

Redox conditions have also proven to be an important controlling factor in determining the decay rates and selective utilization of amino acids. Recent work found preferential utilization of nitrogen-rich amino acids over the non-amino acid fraction of POM in suboxic waters (Van Mooy et al., 2002). This selective utilization of the amino acid fraction was believed to be responsible for supporting water column denitrification processes. Other work in shallow coastal waters has demonstrated an oxygen

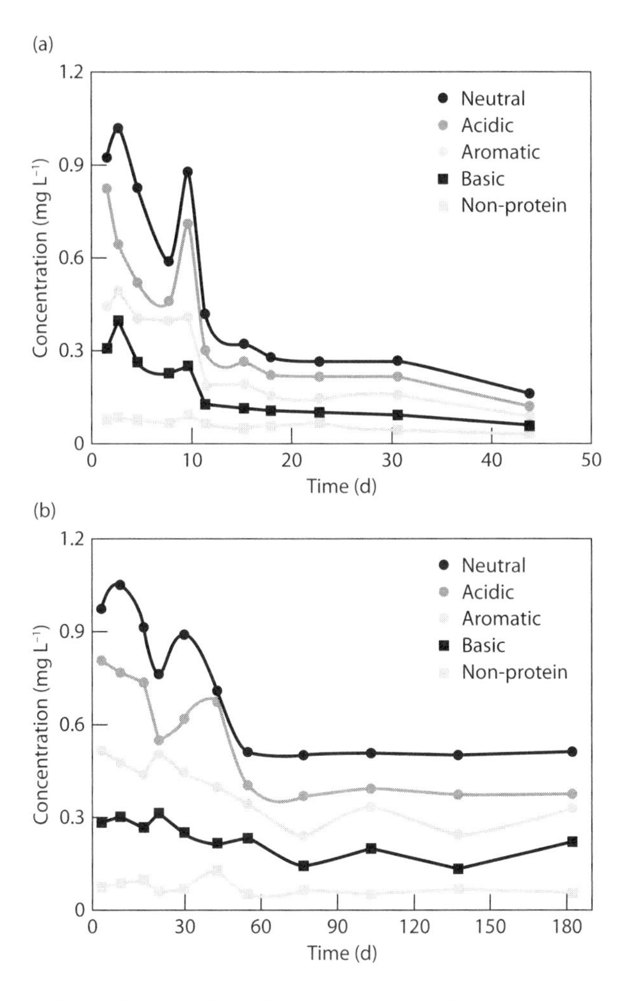

Figure 6.4. Using an experimental approach, phytoplankton protein (shows as neutral, acidic, aromatic, and basic amino acids) was preferentially utilized under (a) oxic conditions (8–65% preserved) relative to (b) anoxic conditions (15–95% preserved). (Adapted from Nguyen and Harvey, 1997.)

effect on the decomposition of amino acids (Haugen and Lichtentaler, 1991; Cowie et al., 1992). Using an experimental approach, phytoplankton protein was found to be preferentially preserved under anoxic (15–95%) compared to oxic (8–65%) conditions (Nguyen and Harvey, 1997). However, there appeared to be only minor selective loss of amino acids within the THAA pool based on differences in functionality (fig. 6.4). Patterns of minimal change in amino acid composition with sediment depth have also been observed in both oxic and anoxic coastal sediments (Rosenfeld, 1979; Henrichs and Farrington, 1987). This suggests that the two dominant groups of amino acids commonly found in phytoplankton (acid and neutral amino acids) are utilized with similar efficiency during early diagenesis. Alternatively, the lack of change in amino acid composition with sediment depth could reflect in situ addition of amino acids during organic matter processing by microbes (Keil et al., 2000). However, Nguyen and Harvey (1997) observed that there was selective preservation of glycine and serine during their experiment examining the decay of diatom material, further supporting the notion that these amino acids are more refractory because of their association with the silica matrix in diatom cell walls. Glycine is also believed to be the most abundant amino acid typically found in sediments because it is thought to have minimal nutritional value, perhaps due to the absence of a side chain (–H), which may be selectively "avoided" by enzymes (Keil et al., 2000; Nunn and Keil, 2005, 2006). Similarly, serine and threonine have been shown to be preserved in the silica matrix of diatom frustules (Ingalls et al., 2003), allowing for selective preservation in sediments (Nunn and Keil, 2006).

While much of the focus to date on the composition of DON has been the amino acid fraction, as described in the aforementioned paragraphs, there are other naturally occurring classes of dissolved free primary amines, such as *aliphatic amines* and *hexosamines*, produced from metabolic and hydrolytic processes in aquatic systems that have largely been ignored (Aminot and Kerouel, 2006). This total pool can be referred to as total dissolved free primary amines (TDFPA), operationally defined as the response of an amine-sensitive method that is analytically standardized against glycine (Aminot and Kerouel, 2006).

Finally, we would like to briefly introduce a group of compounds called *mycosporine-like amino acids* (MAAs) that are ubiquitous in aquatic and many other organisms and have ultraviolet radiation (UVR, 290–400 nm) absorbing qualities (see review by Shick and Dunlop, 2002). These compounds have their maximum absorbance in the 310- to 360-nm range, with a high molar absorbtivity ($\varepsilon = 28,100\text{–}50,000\ M^{-1}\ cm^{-1}$) for UVA and UVB. Early observations showed for the first time the presence of these compounds in marine (Wittenberg, 1960), and more recently in freshwater (Sommaruga and Garcia-Pichel, 1999) organisms. However, more detailed work on the natural product chemistry of these compounds showed that these compounds were similar to a family of metabolites found in terrestrial fungi (fig. 6.5) (Shick and Dunlop, 2002). Evidence for the importance of MAAs as "sunscreens" in aquatic organisms continues to grow (Tartarotti et al., 2004; Whitehead and Hedges, 2005).

6.2 Proteins: Amino Acid and Amine Biosynthesis

The first step in the biosynthesis of amino acids is acquiring N from the natural environment. In the case of plants and algae, NO_3^- and/or NH_4^+ are preferentially taken up as nutrients, whereby the acquired N can then be used in the biosynthesis of amino acids. Another pathway for the biological uptake of N is through biological *N_2 fixation* (BNF). BNF is a process performed by prokaryotic (both heterotrophic and phototrophic) organisms (sometimes occurring symbiotically with other organisms), resulting in the enzyme-catalyzed reduction of N_2 to NH_3 or NH_4^+, or organic nitrogen compounds (Jaffe, 2000). Organisms that directly fix N_2 are also called *diazotrophs*. The high activation energy required to break the triple bond makes BNF an energetically expensive process for organisms, thereby restricting BNF to select groups of organisms. BNF is an important process in aquatic systems (Howarth et al., 1988a; Howarth and Cole, 1988b); some of the heterotrophic and phototrophic bacteria capable of BNF in coastal and estuarine systems provide the essential N for amino acid synthesis. In the case of NO_3^- uptake, the first step toward amino acid synthesis occurs through the reduction of NO_3^- to NO_2^-, using *nitrate reductase*, in combination with nicotinamide adenosine dinucleotide (NADH) + H^+—key products of glycolysis (see chapter 1). The next reaction involves further reduction of NO_2^- to NH_4^+ using the enzyme *nitrite reductase* and reduced *ferrodoxin*, which acts as an electron donor—all as part of the electron transport chain involving nicotinamide dinucleotide phosphate (NADP) and flavin adenine dinucleotide (FAD) (see chapter 1).

When examining the citric acid cycle (fig. 6.6) (Campbell and Shawn, 2007), we find the important nitrogen-free carbon compounds (e.g., oxaloacetate and α-ketoglutarate) serve as important intermediates for amino acid biosynthesis. In particular, α-ketoglutarate is aminated to enable glutamate to be produced as the primary mechanism for assimilating N into amino acids. Perhaps the most common next step, using Chlorophytes as a model, is reductive amination of alpha-ketoglutarate catalyzed by the $NADP^+$-dependent *glutamate dehydrogenase* in combination with NH_4^+ to produce

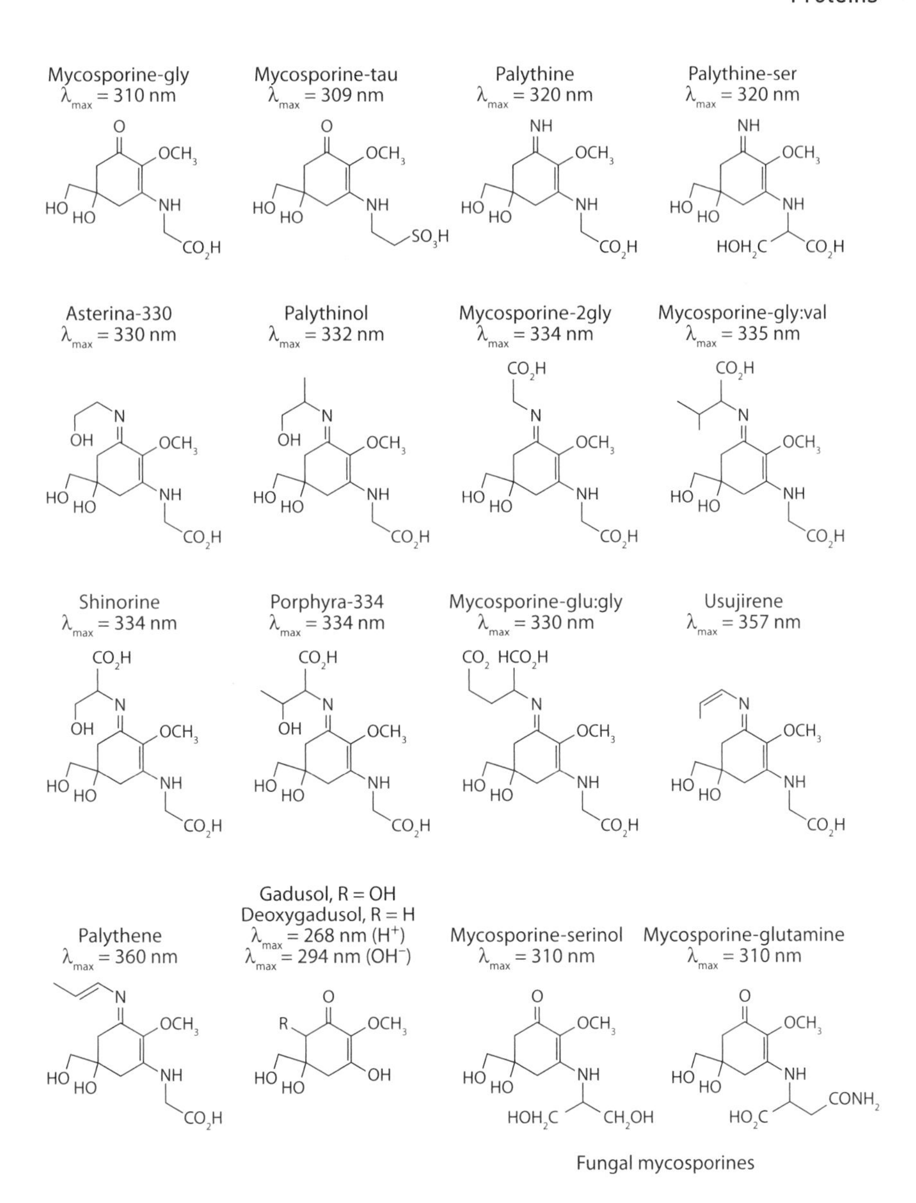

Figure 6.5. Mycosporine-like amino acids (MAAs) are ubiquitous in aquatic and many other organisms. The natural products chemistry of these compounds is similar to a family of metabolites found in terrestrial fungi. (Adapted Shick and Dunlop, 2002.)

glutamate, or glutamic acid (fig. 6.7a). This reaction occurs in the chloroplasts and defines glutamate as the first amino acid produced, in a group of similar amino acids called the glutamate family, described below. It should also be noted that through another pathway, which also occurs in the chloroplasts, glutamate binds with another NH_4^+ to form a second amino acid, glutamine (fig. 6.7b).

Although the biosynthetic pathways of amino acids are diverse, the carbon-based compounds from which they are built are intermediates in the glycolysis, pentose phosphate, and citric acid cycles (fig. 6.6) (Campbell and Shawn, 2007). Moreover, it is common practice to further group amino acids into the following 6 families based on their metabolic precursors: (1) glutamate family (glutamate, glutamine, proline, arginine, and ornithine); (2) pyruvate family (alanine, valine, and

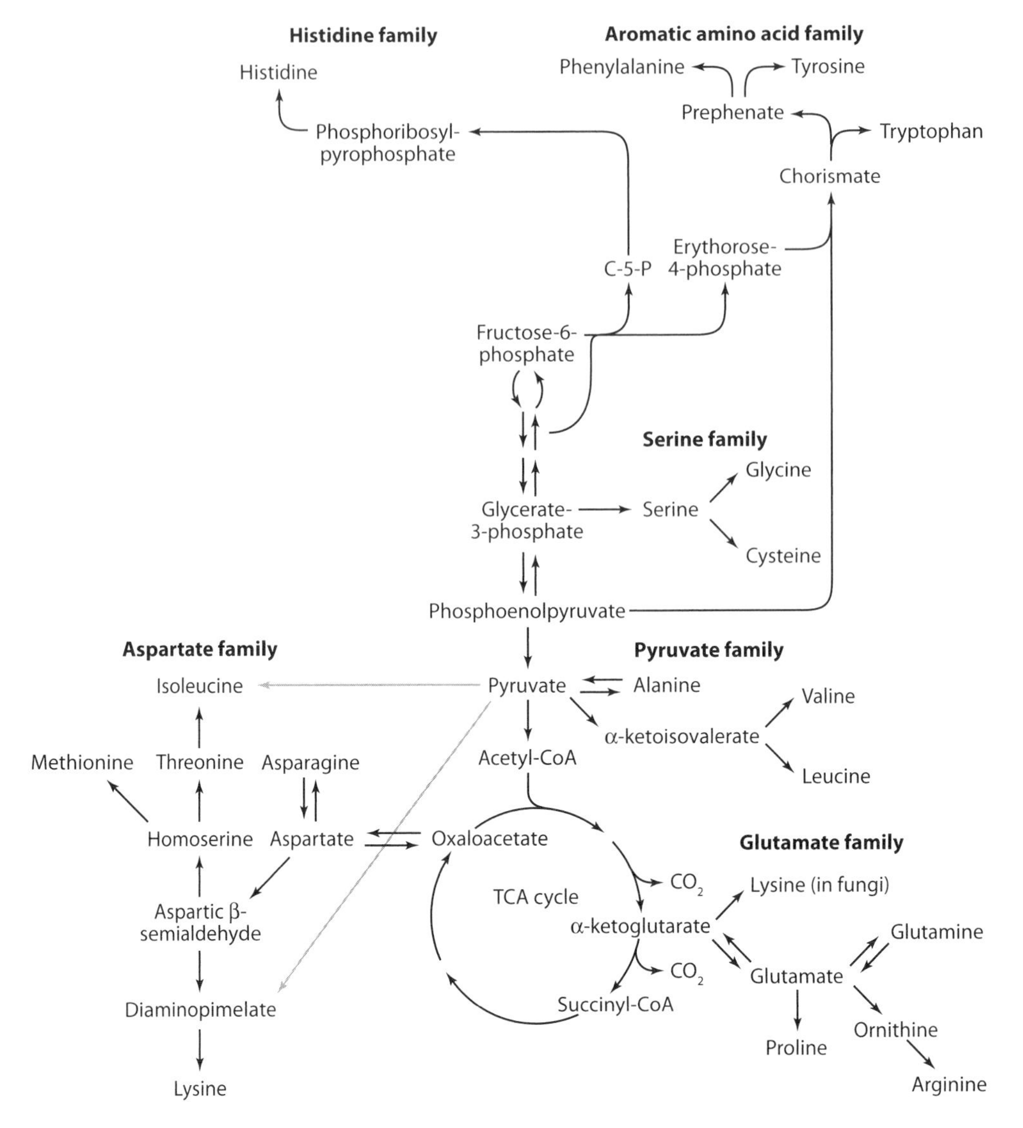

Figure 6.6. The biosynthetic pathways of amino acids are diverse; the carbon-based compounds from which they are built are intermediates in the glycolysis, pentose phosphate, and citric acid cycles. (Adapted from Campbell and Shawn, 2007.)

leucine; (3) aspartate family (apartate, asparagine, lysine, threonine, isoleucine, and methionine); (4) serine family (serine, glycine, and cysteine); (5) aromatic family (phenylalanine, tyrosine, and tryptophan); and (6) histidine family (fig. 6.6). We now provide a brief summary of each of these 6 biosynthetic pathways, used by bacteria, fungi, and plants, beginning from the point of post N uptake and glutamate and glutamine (key amino group donors) formation, described above.

The name of the glutamate family (glutamate, glutamine, proline, arginine, and ornithine) derives from the convergence and metabolism of these amino acids on glutamate. Following the synthesis of glutamate and glutamine, described above, phosphorylation of glutamate occurs by adenosine triphosphate (ATP) with the enzyme *γ-glutamyl kinase*, which is then followed by reduction to glutamate-*γ* semialdehyde, which spontaneously cyclizes to pyrrolidine-5-carboxylate via the loss of

Catalyzed by glutamate dehydrogenase

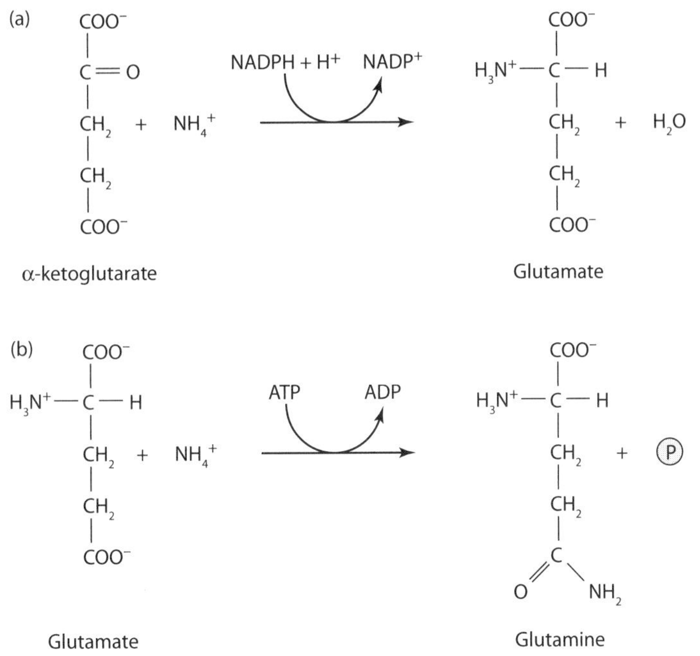

(a) $NADPH + H^+$ $NADP^+$

α-ketoglutarate + NH_4^+ → Glutamate + H_2O

(b) ATP ADP

Glutamate + NH_4^+ → Glutamine + Ⓟ

Figure 6.7. (a) The nitrogen-free carbon compound (α-ketoglutarate) serves as an important intermediate for amino acid biosynthesis in Chlorophytes. (b) Subsequently, glutamate can bind with another NH_4^+ to form a second amino acid, glutamine.

Catalyzed by γ-glutamyl kinase

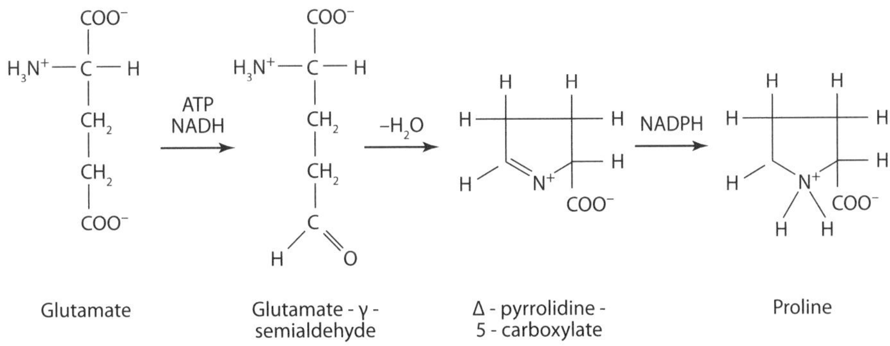

Glutamate → (ATP, NADH) → Glutamate - γ - semialdehyde → ($-H_2O$) → Δ - pyrrolidine - 5 - carboxylate → (NADPH) → Proline

Figure 6.8. Phosphorylation of glutamate occurs by reaction with ATP, which is then followed by reduction to glutamate-γ-semialdehyde, which spontaneously cyclizes to pyrrolidine-5-carboxylate via the loss of H_2O and no enzyme required to produce proline. (Adapted from Campbell and Shawn, 2007.)

H_2O and no enzyme is required (fig. 6.8) (Campbell and Shawn, 2007). Proline contains no primary but a secondary amino group and is therefore actually an *alpha*-imino acid, although it is often referred to as an amino acid. Basically, the *α-carboxyl* group on the glutamate reacts with the ATP to produce an *acyl phosphate*. The formation of the semialdehyde also requires the presence of either NADP or NADPH. The pyrrolidine-5-carboxylate is then reduced by NADPH to proline. However, this step after the formation of a semialdehyde is a branching point, whereby in one case proline is formed, while in another it leads to ornithine and arginine. In this case glutamate-5-semialdehyde is transaminated to ornithine, whereby glutamate acts as an amino group donor. Arginine is an important intermediate in ornithine synthesis that can be generated in plant cells. It serves as an acceptor of *carbamyl phosphate*, which results in the splitting off of a phosphate group, resulting in the formation of the next intermediate, citrulline (Campbell and Shawn, 2007).

The *pyruvate family* (alanine, valine, and leucine) derives its name from pyruvate, the starting compound in the biosynthesis reactions. Alanine is an amino acid derivative of pyruvate, and valine and leucine are synthesized independently of each other by elongation of the pyruvate chain, interconversion, and subsequent transamination. These are branched-chain amino acids. Due to the numerous steps involved in the biosynthesis of these amino acids, we will introduce just the initial and final steps of the reactions (fig. 6.6) (Campbell and Shawn, 2007). In general, the first step common to the synthesis of all 3 amino acids is as follows:

$$\text{pyruvate} + \text{thiamine pyrophosphate (TPP)} \xrightarrow{\text{(catalyzed by acetolactate synthase)}} \text{hydroxyethyl-TPP} \qquad (6.2)$$

Once hydroxyethyl-TPP is formed it can react with another pyruvate molecule to form *α-acetolactate*, allowing valine and isoleucine to be formed, or it may react with *α-ketobutyrate*, which leads to the formation of isoleucine. After a few other steps, there is a branching point at the formation of *α-ketoisovalerate,* at which one pathway leads to valine, and the other to leucine. The final steps of the biosynthesis of these three amino acids involve the transfer of an amino group from glutamate to the corresponding *α-ketoacid* of each of the three branched-chain amino acids. Once again we see the importance of glutamate in amino acid biosynthesis.

The *aspartate family* (apartate, asparagine, threonine, isoleucine, lysine, and methionine) derives its name from the compound from which it is derived, aspartate, which is formed from glutamate (fig. 6.9) (Azevedo et al., 1997). Aspartate is generated by the following transamination reactions and results in the addition of a further amino group:

$$\text{glutamate} + \text{oxaloacetate} \rightarrow \alpha\text{-ketoglutarate} + \text{aspartate and/or asparagine} \qquad (6.3)$$

This reaction can occur through the fixation of an NH_4^+, similar to that previously described for the synthesis of glutamine, or it can occur by the transfer of one amino group from glutamine to aspartate, as shown in the following reactions:

$$NH_4^+ + \text{aspartate} + \text{ATP} \rightarrow \text{asparagine} + \text{ADP} + P_i \qquad (6.4)$$

$$\text{glutamine} + \text{aspartate} + \text{ATP} \rightarrow \text{glutamate} + \text{asparagine} + \text{AMP} + \text{PP} \qquad (6.5)$$

The biosynthesis of threonine begins with the first enzyme, aspartate kinase, which catalyzes the conversion of aspartate to β-aspartyl phosphate (fig. 6.9). The next step involves the conversion of β-aspartyl phosphate to aspartate semialdehyde via aspartate-semialdehyde dehydrogenase, which is then converted to homoserine in the presence of homoserine dehydrogenase, NADH, and NADPH. Dihydrodipicolinate synthase is the first enzyme at a branch point for lysine synthesis, whereby it catalyzes the condensation of pyruvate and aspartate semialdehyde into dihydrodipicolinate (fig. 6.9). Homoserine kinase is the next important enzyme that catalyzes the formation of threonine, isoleucine, and methionine. The first step in the breakdown of cysteine is the conversion of methionine to *S*-adenosylmethionine (SAM) through the transfer of an adenosyl group from an ATP to the S atom of methionine. SAM is an important molecule used in one-carbon transfers, and further discussion on this is provided in the next section on the serine family. The next branch point that leads to the synthesis of methionine involves the enzyme cystathione synthase, which catalyses the synthesis of cystathione from O-phosphohomoserine and cysteine (fig. 6.9). In the continuation of threonine

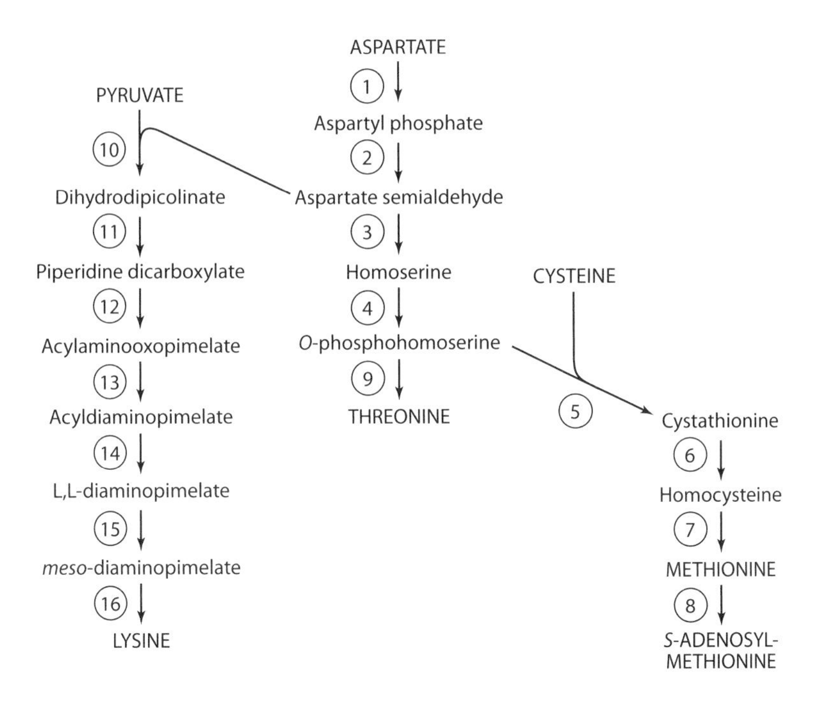

Figure 6.9. The aspartate family (aspartate, asparagine, threonine, isoleucine, lysine, and methionine) derives its name from the initial formation of aspartate from glutamate. The numbers refer to the sequence of steps leading to the synthesis of lysine, noting some of the products produced along the way. (Adapted from Azevedo et al., 1997.)

biosynthesis, homoserine is then converted to *O*-phosphohomoserine by homoserine kinase in the presence of ATP; *O*-phosphohomoserine is finally converted to threonine via threonine synthase. It should also be noted that the regulatory properties of homoserine kinase in these reactions vary widely across plants and microbes (Azevedo et al., 1997).

In the *serine family* (serine, glycine, and cysteine) we begin with the synthesis of serine, which is formed in a two-step reaction from 3-phosphoglycerate, an intermediate in glycolysis (fig. 6.10) (Campbell and Shawn, 2007). The first step is an oxidation to 3-phosphohydroxypyruvate and the second is a transamination to 3-phosphoserine, with glutamate as an amino group donor. The final step in the formation of serine occurs through the dephosphorylation of 3-phosphoserine. Another pathway for the synthesis of serine begins with the dephosphorylation of 3-phosphoglycerate, with the formation of hydroxyproline, which is then converted to serine by transamination. Before we discuss the formation of glycine, we need to describe the notation of an important carrier of one-carbon functional groups (e.g., methyl, formyl, methylene, formimino, and methenyl). Tetrahydrofolate or tetrahydrofolic acid (THF) is composed of pteridine, *p*-aminobenzoate, and glutamate (fig. 6.11). The one-carbon group is bonded to the N-5 and/or N-10 nitrogen atom (denoted N^5 and N^{10}) in THF. Now, serine is converted to glycine in the following reaction:

$$\begin{gathered}\text{(enzyme : serine hydroxymethyltransferase)}\\ \text{serine} + \text{THF} \rightarrow \text{glycine} + N^5, N^{10}\text{-methylene} - \text{THF}\end{gathered} \tag{6.6}$$

The final step in producing glycine involves the following condensation reaction:

$$\begin{gathered}\text{(enzyme : glycine synthase, with the addition of NADH)}\\ N^5, N^{10}\text{-methylene-THF} + CO_2 + NH_4^+ + \rightarrow \text{glycine}\end{gathered} \tag{6.7}$$

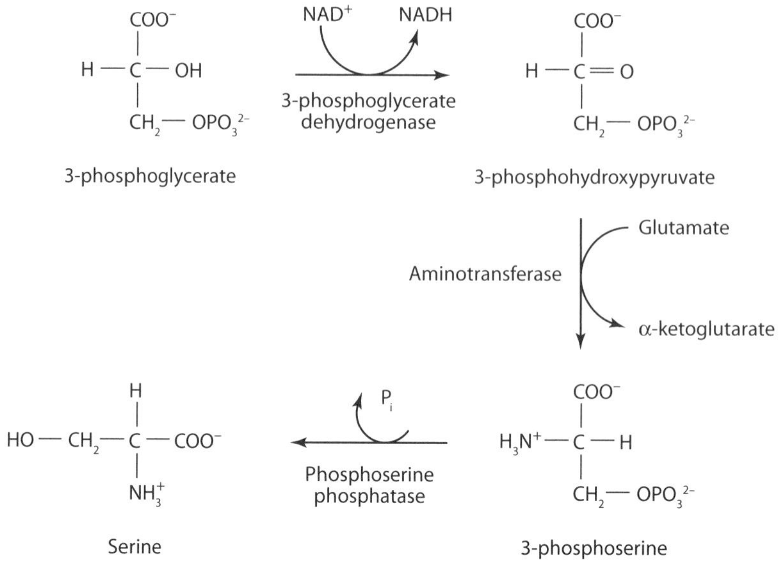

Figure 6.10. Serine is synthesized from 3-phosphoglycerate, an intermediate in glycolysis, in a two-step reaction. (Adapted from Campbell and Shawn, 2007.)

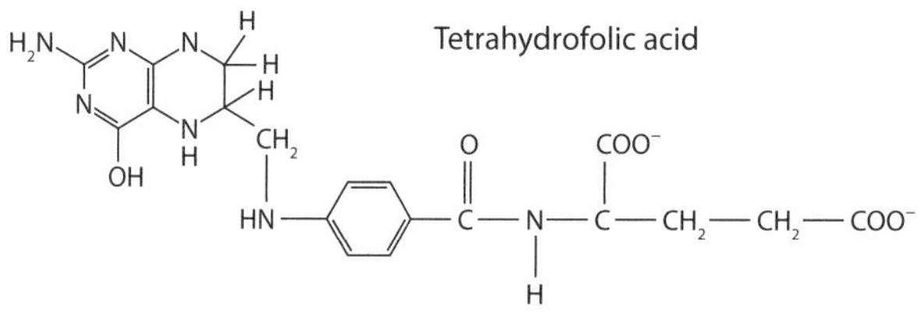

Figure 6.11. Tetrahydrofolate, or tetrahydrofolic acid, is an important carrier of one-carbon functional groups (e.g., methyl, formyl, methylene, formimino, and methenyl) and is composed of pteridine, *p*-aminobenzoate, and glutamate.

Another path for the conversion of serine to glycine occurs through the loss of the $-CH_2OH$ side-chain of serine, whereby the one-carbon group unit becomes bonded to THF. Cysteine, the last amino acid in this group, is synthesized from serine and homocysteine, a methionine breakdown product (fig. 6.12a) (Campbell and Shawn, 2007). One of the key differences here is that N^5-methyl-THF is not used because of the higher energetic demand needed in these reactions. As mentioned previously, the first step in the breakdown of cysteine is the conversion of methionine to SAM through the transfer of an adenosyl group from an ATP to the S atom of methionine. The methyl group of the *S*-adenosylmethionine is transferred to phosphatidylethanolamine to create *S*-adenosylhomocysteine, which is finally hydrolyzed to homocysteine (fig. 6.12b). As mentioned earlier, the methyl group transfer is made using SAM, which is more efficient due to the positive charge on the adjacent S atom, which activates the methyl group of methionine group. SAM is also used to transfer methyl groups in the methylation of DNA. Homocysteine combines with serine in a condensation reaction to produce cystathionine, which is then deaminated and finally cleaved to produce α-ketobutyrate and cysteine (fig. 6.12c).

Biosynthesis of amino acids in the aromatic family (phenylalanine, tyrosine, and tryptophan) begins with the compounds phosphoenolpyruvate and erythrose-4-phosphate. Shikimate and chorismate are important intermediates in this pathway, whereby the former is an important link between the aromatic

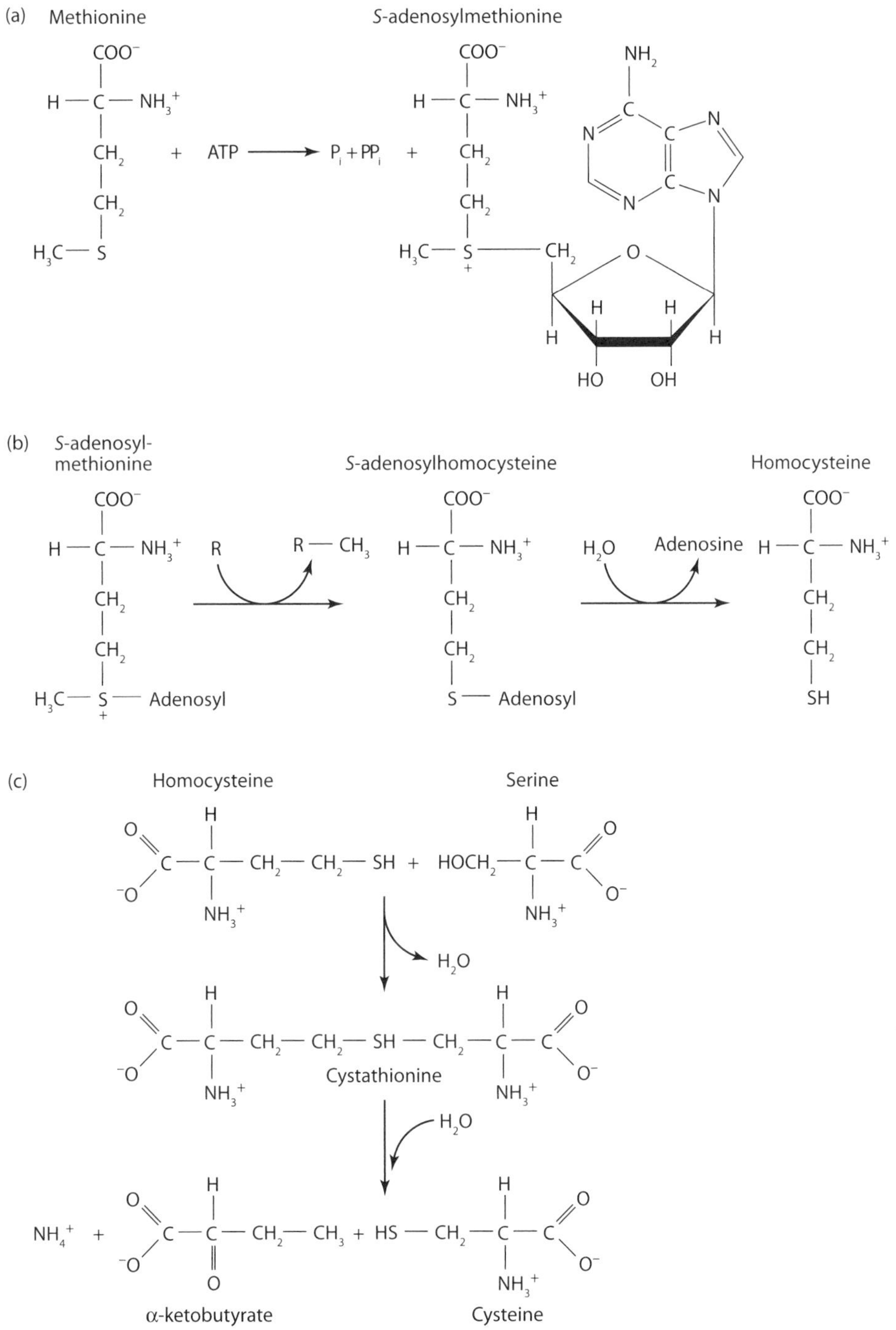

Figure 6.12. (a) Cysteine synthesis begins with the breakdown of methionine to *S*-adenosylmethionine, (b) The methyl group of the *S*-adenosylmethionine is transferred to phosphatidylethanolamine to create *S*-adenosylhomocysteine, which is hydrolyzed to homocysteine. (c) Homocysteine combines with serine in a condensation reaction to produce cystathionine, which is then deaminated and cleaved to produce α-ketobutyrate and cysteine. (Adapted from Campbell and Shawn, 2007.)

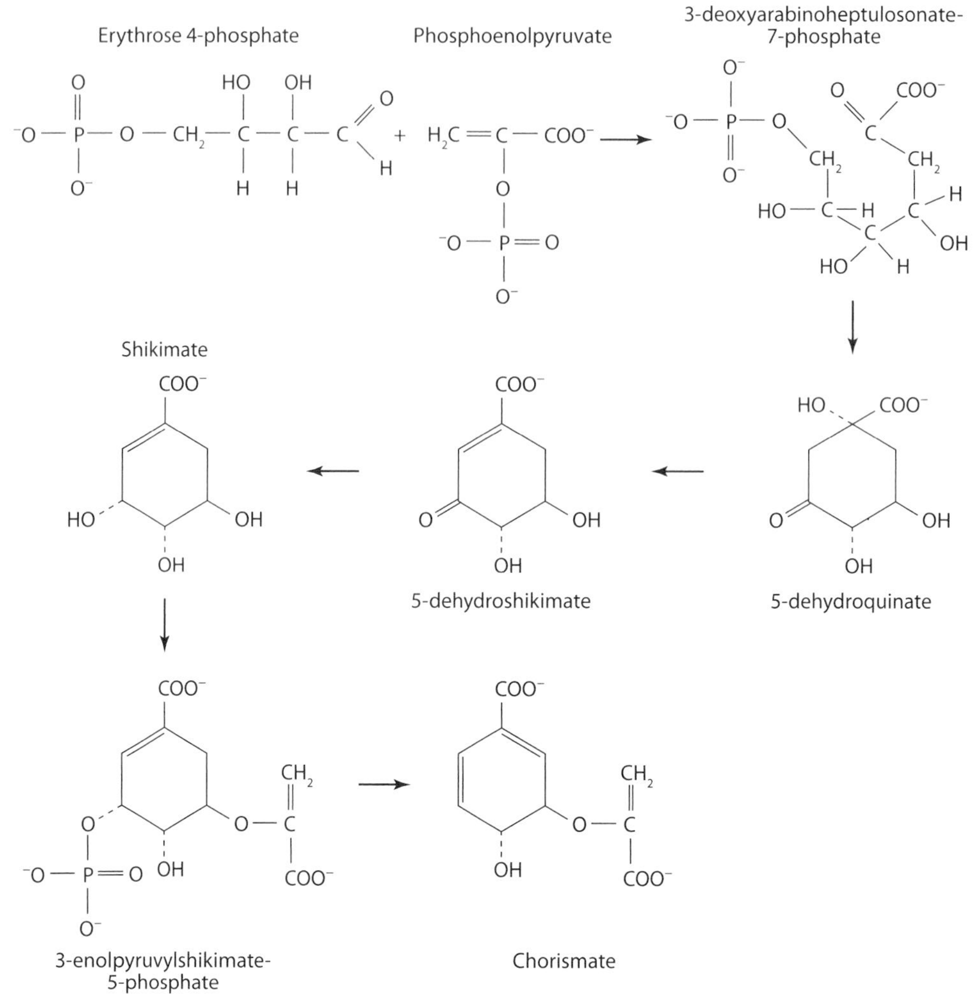

Figure 6.13. Biosynthesis of amino acids in the aromatic family (phenylalanine, tyrosine, and tryptophan) begins with the compounds phosphoenolpyruvate and erythrose-4-phosphate. (Adapted from Campbell and Shawn, 2007.)

amino acids and the biosynthesis of the phenylpropanoid building blocks of lignin (chapter 13), and the latter is the starting compound for the different pathways that lead to the aromatic amino acids (fig. 6.13) (Campbell and Shawn, 2007). It should first be noted that there are several pathways for the synthesis of the aromatic amino acids, which we will not cover completely. If we begin with chorismate, a mutase first converts chorismate to prephenate, which serves as an intermediate in the formation of tyrosine and phenylalanine (fig. 6.14a) (Campbell and Shawn, 2007). The intermediates from decarboxylation and dehydration and/or oxidative decarboxylation are phenylpyruvate and *p*-hydroxyphenylpyruvate, respectively. The only difference between these two amino acids is that the *para*-carbon on the benzene ring of tyrosine is hydroxylated. In another pathway, which leads to the formation of tryptophan, chorismate obtains an amino group from the side chain of glutamine to form anthranilate (fig. 6.14b). Once again, it is important to note that glutamine is important as an amino donor for a number of other amino acid biosynthetic pathways. Phosphoribosylpyrophosphate (PRPP) condenses with anthranilate to form the moiety phosphoribosylanthranilate, then rearranges

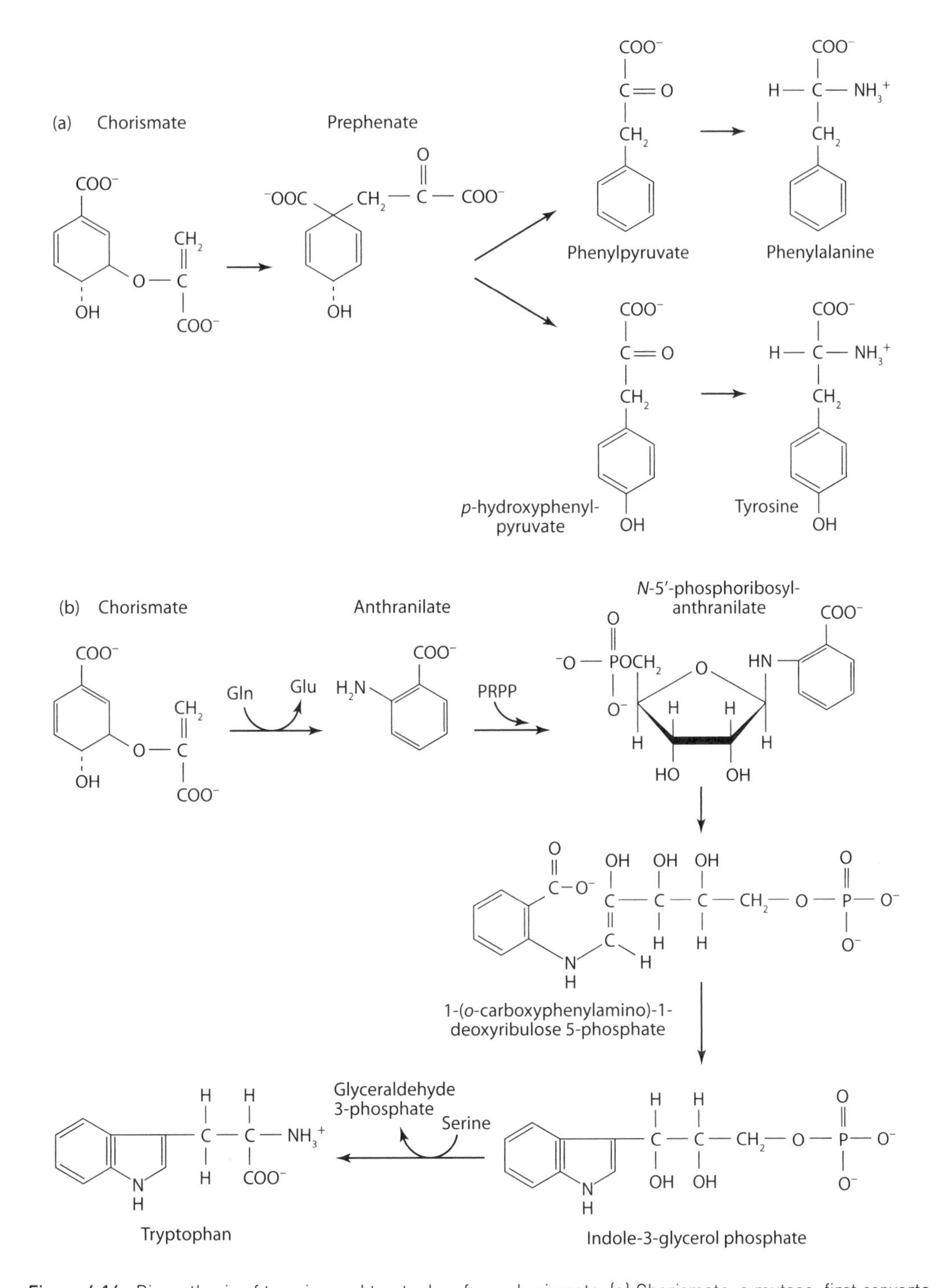

Figure 6.14. Biosynthesis of tyrosine and tryptophan from chorismate. (a) Chorismate, a mutase, first converts chorismate to *prephenate*, which serves as an intermediate in the formation of tyrosine and phenylalanine. (b) A second pathway involving chorismate leads to the formation of tryptophan, where chorismate obtains an amino group from the side chain of glutamine to form anthranilate. (Adapted from Campbell and Shawn, 2007.)

to form the intermediate, 1-(*o*-carboxyphenylamino)-1-deoxyribulose 5-phosphate. This intermediate undergoes dehydration to form indole-3-glycerol phosphate, which combines with serine to form tryptophan. PRPP is also important in the synthesis of histidine, discussed next, as well as in purines and pyrimidines.

Histidine is in its own family because its biosynthesis is uniquely and fundamentally linked to the pathways of *nucleotide* formation. In the first step of histidine synthesis, PRPP condenses with ATP to form the purine, *N*-5′-phosphoribosyl ATP (fig. 6.15). Subsequent reactions result in the hydrolysis of the pyrophosphate, which then condenses out. At this time, glutamine again plays a role as an amino group donor, assisting in the formation of 5-aminoimidazole-4-carboxamide ribonucleotide, which also serves as an important intermediate in purine biosynthesis. Further on in this pathway, we see that the immediate precursor of tryptophan is an indole, resulting in an indole ring that is characteristic of its structure.

6.3 Analysis of Proteins: Amino Acids and Amines

Chromatographic methods have been the most successful techniques used for analyzing amino acids in natural waters, with many of the earlier techniques involving paper chromatography (Palmork et al., 1963; Rittenberg et al., 1963; Degens et al., 1964; Starikov and Korzhiko, 1969), followed by thin-layer and ion chromatography (see reviews by Dawson and Pritchard, 1978; Dawson and Liebezeit, 1981). The next key advances were made with application of high-performance liquid chromatography (HPLC) techniques, which occurred mostly in the 1970s to the early 1980s (Riley and Segar, 1970; Seiler, 1977; Lindroth and Mopper, 1979; Dawson and Liebezeit, 1981; Jorgensen et al., 1981; Lee, 1982; Mopper and Lindroth, 1982). Many of these HPLC methods, which used fluorescence detectors, achieved detection capability down to the subpicomolar to nanomolar range. Some of the early techniques required time-consuming stages, such as desalting and preconcentration, as well as a delay in the time of analysis since ship-board capability was not possible; all of this commonly resulted in losses and/or contamination of DFAA (Garrasi et al., 1979). Interestingly, one of the earliest preconcentration procedures used ligand-exchange chromatography on Cu-Chelex 100 (Siegel and Degens, 1966).

The first methods that utilized post-column derivatization for analysis of amino acids in seawater, without the use of desalting or preconcentration, were those of Garrasi et al. (1979) and Lindroth and Mopper (1979). However, the Garrasi et al. (1979) method involved ion-exchange chromatography and use of an amino acid analyzer, which were more costly to maintain and relatively more complex than the chromatographic technique of Lindroth and Mopper (1979), which has become the standard method for seawater analysis. This method uses *o*-phthaldialdehyde (OPA) and the thiol (2-mercaptoethanol) to derivatize amino acids and other primary amines into highly fluorescent isoindolyl derivatives, based on the reaction chemistry shown in fig. 6.16. A chiral thiol compound is needed to form diastereomeric derivatives, which can then be separated by HPLC; more details on different methods used to separate amino acid enantiomers using different mobile and stationary phases as well as derivatizing agents can be found in Bhushan and Joshi (1993). A sample chromatogram from Mopper and Dawson (1986) illustrates the resolution of amino acid separation obtained using a slightly modified version of the method developed by Lindroth and Mopper (1979), for the analysis of DFAA in seawater (fig. 6.17). The reaction of OPA with amino acids (with the exception of the imino acids proline and hydroxyproline) in alkaline media and the presence of 2-mecaptoetahnol to produce a strong fluorescence was first described by Roth (1971). Fluorogenic detection of amino acids with OPA was also shown to be more effective than the *ninhydrin reaction*, and 5 to 10 times more sensitive than *fluorescamine* (Benson and Hare, 1975). Primary amines can also be analyzed by derivatization with OPA and fluorescamine, in the absence of heating (Josefsson et al., 1977). While there are a

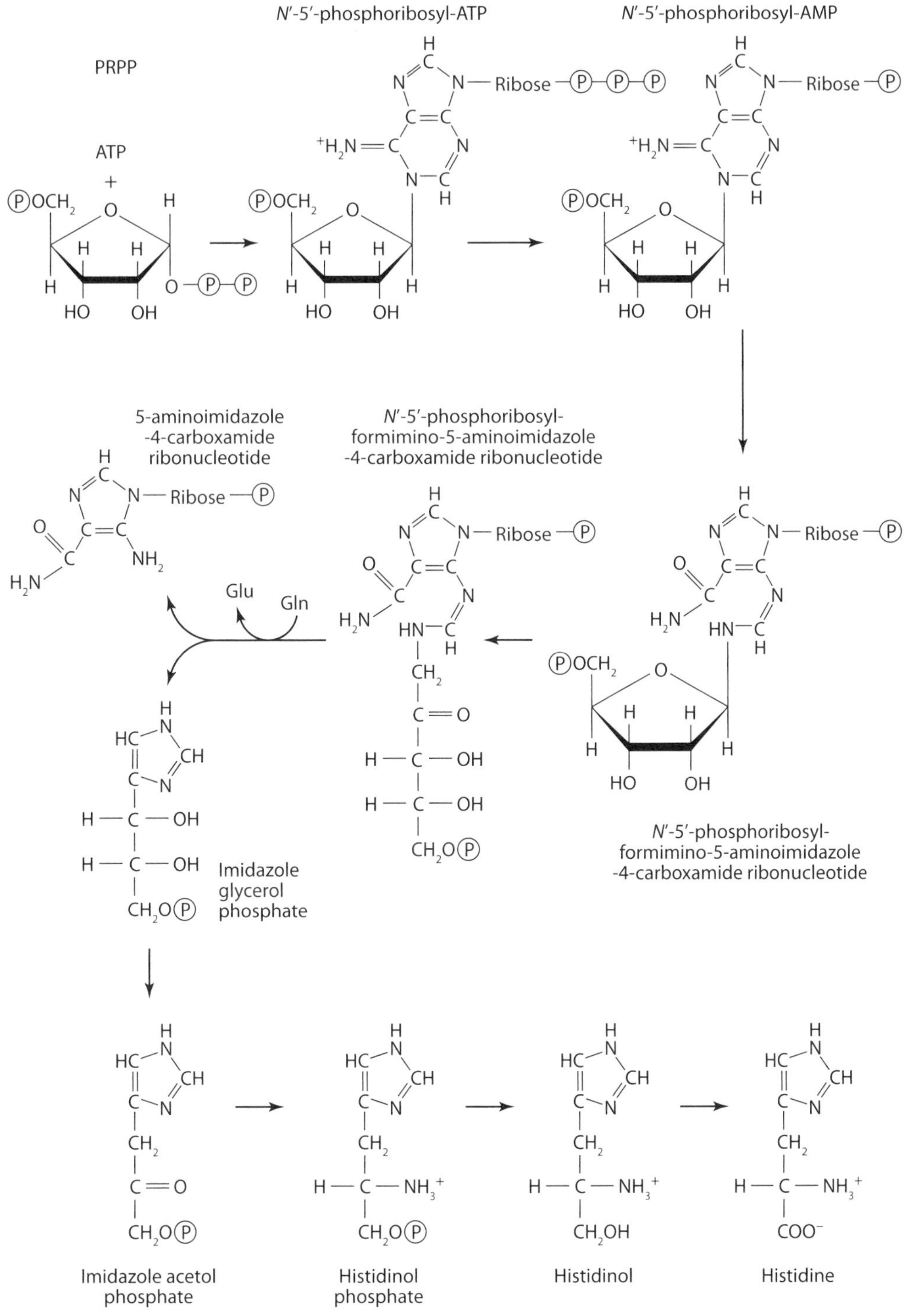

PRPP

ATP

+

N′-5′-phosphoribosyl-ATP

N′-5′-phosphoribosyl-AMP

N′-5′-phosphoribosyl-formimino-5-aminoimidazole -4-carboxamide ribonucleotide

N′-5′-phosphoribosyl-formimino-5-aminoimidazole -4-carboxamide ribonucleotide

5-aminoimidazole -4-carboxamide ribonucleotide

Glu

Gln

Imidazole glycerol phosphate

Imidazole acetol phosphate

Histidinol phosphate

Histidinol

Histidine

Figure 6.15. Histidine is in its own family because its biosynthesis is uniquely and fundamentally linked to the pathways of nucleotide formation. (Adapted from Campbell and Shawn, 2007.)

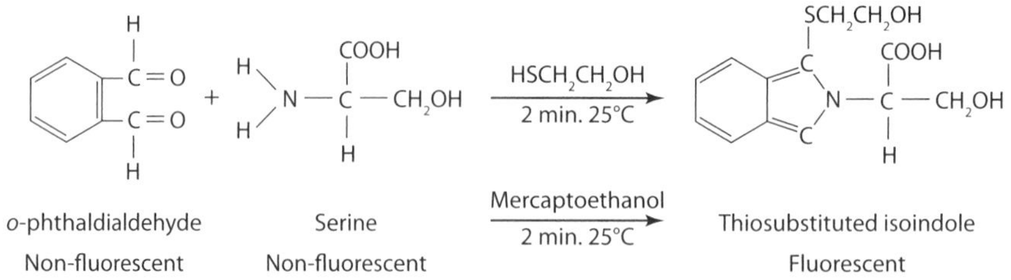

$$\text{o-phthaldialdehyde (non-fluorescent)} + \text{Serine (non-fluorescent)} \xrightarrow[\text{2 min. 25°C}]{HSCH_2CH_2OH} \text{Thiosubstituted isoindole (fluorescent)}$$

Mercaptoethanol / 2 min. 25°C

o-phthaldialdehyde Non-fluorescent — Serine Non-fluorescent — Thiosubstituted isoindole Fluorescent

Figure 6.16. The chemical reaction for the use of *o*-phthaldialdehyde (OPA) and the thiol (2-mercaptoethanol) to derivatize amino acids and other primary amines into highly fluorescent isoindolyl derivatives. (Adapted from Lindroth and Mopper, 1979.)

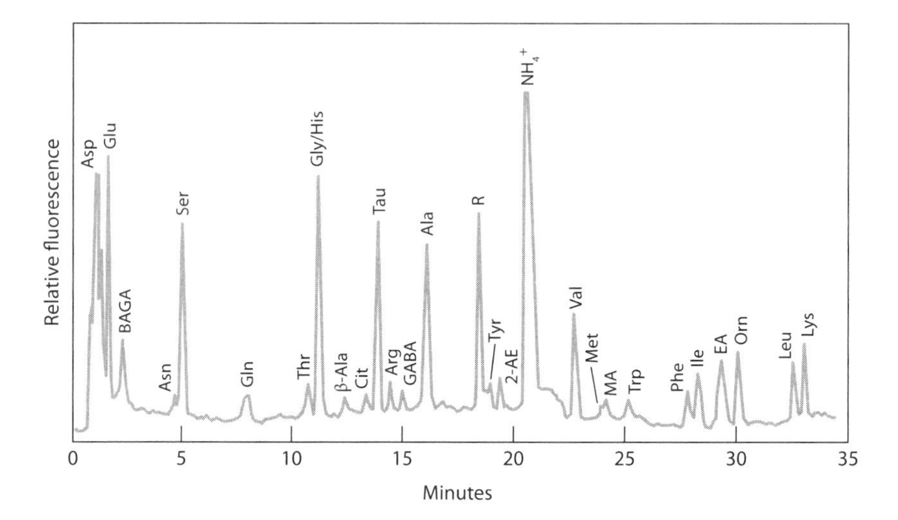

Figure 6.17. A sample chromatogram from Mopper and Dawson (1986) illustrating the resolution of amino acid separation obtained using a slightly modified version of Lindroth and Mopper (1979) for the analysis of dissolved free amino acids in seawater.

number of older modifications of the Lindroth and Mopper (1979) method (Jones et al., 1981; Cowie et al., 1992; Sugai and Henrichs, 1992), the more recent changes in the method are typically used now (e.g., Lee et al., 2000; Duan and Bianchi, 2007). Finally, vapor-phase hydrolysis is another efficient approach for hydrolyzing amino acid samples (Tsugita et al., 1987; Keil and Kirchman, 1991a, b).

Although a few studies have measured TDFAA, the methods generally used include the manual approach using fluorescamine (North, 1975), and the automated continuous flow method using OPA (Josefsson et al., 1977). Others have used a modified version of the North (1975) method with fluorescamine (Jorgensen, 1979; Jorgensen et al., 1980; Sellner and Nealley, 1997). When using fluorescamine and OPA, ammonium is considered to be the most problematic compound that interferes with fluorescence measurements (Aminot and Kerouel, 2006). An HPLC method was also developed that allowed for the separation of ammonium and primary amines in seawater (Gardner and St. John, 1991). Finally, based on the recent review of TDFAA methods (Aminot and Kerouel, 2006), other recommendations were made for a new protocol using the more sensitive OPA (Dawson and Liebezeit, 1981, 1983; Liebezeit and Behrends, 1999).

Hydrolysis is likely the most important decomposition process influencing proteins and peptides associated with particles in aquatic systems (Smith et al., 1992). Unfortunately, the ability to measure the hydrolysis of proteins in aquatic systems has proven difficult (Pantoja et al., 1997). Moreover, the general paradigm of protein degradation to peptides and amino acids in the ocean (Billen, 1984; Hoppe, 1991) has generally ignored the size class of amino acid sequences with a molecular weight of <6000 Da as important intermediates in the overall degradation of organic matter (Pantoja and Lee, 1999). While many larger organisms can degrade proteins internally, bacteria use extracellular enzymes to hydrolyze proteins and peptides to smaller substrates (e.g., <600 Da) that can be transported across the microbial cell membrane (Nikaido and Vaara, 1985). Although some work has examined proteolytic activity in seawater and sediments (Hollibaugh and Azam, 1983; Hoppe, 1983; Somville and Billen, 1983), the actual hydrolytic enzymes involved have not been examined. Other work has shown that fluorogenic substrates can be used to examine peptide bond hydrolysis. The classic example is that of an aminopeptidase (N-terminal) cleavage of the amide bond in leucine-methylcoumarinylamide (Leu-MCA), which results in a fluorescent product (Kanaoka et al., 1977). This technique has been used to measure hydrolysis rates in oceanic waters and sediments (Rheinheimer et al., 1989; Boetius and Lochte, 1994). Tritiated amino acids have also been used to label peptides and hydrolysis of proteins in sediments (Ding and Henrichs, 2002). Perhaps the most novel advance was the ability to study the extracellular hydrolysis rates in aquatic systems (Pantoja et al., 1997; Pantoja and Lee, 1999). In general, this method uses fluorescent peptide, which affects the structure and size of peptides on substrates (derivatives of trialanine, tetraalanine, leucyl-alanine, etc.) that are synthesized by linking a peptide to a fluorescent anhydride (Lucifer Yellow anhydride) via condensation, based on the procedures described by Stewart (1981) and Pantoja et al. (1997). The fluorescent substrates are then separated by HPLC using a gradient of K_2HPO_4 and methanol, with fluorescence detection at 424 and 550 nm excitation and emission, respectively (Pantoja et al., 1997). Finally, it was recently shown in a comparison of six different extraction techniques (0.5 N NaOH, 0.1 N NaOH, Triton X-100, hot water, NH_4HCO_3, and HF) of intact peptides and proteins in sediments) that there were significant differences in amino acid yields (Nunn and Keil, 2006), supportive of the idea that a large fraction of sedimentary nitrogen is protected within an organic matrix (Keil et al., 2000).

As mentioned earlier, the dominant and most well-defined pool of amino acids in the ocean is DCAA (Coffin 1989; Keil and Kirchman, 1991a,b), but the larger protein and longer peptide sources of these smaller peptides are not well understood (Tanoue et al., 1995). Early work showed that selective utilization of amino acids in the euphotic zone by heterotrophic processes (Lee, 1982; Ittekkot et al., 1984), as well as rapid removal at the sediment–water interface (Henrichs et al., 1984; Burdige and Martens, 1988), results in the accumulation (ca. 30 to 40%) of longer peptide proteinaceous material in sediments (Nunn and Keil, 2005). Thus, the ability to measure peptides in sediments is particularly important for understanding the nitrogen cycle in sediments (Mayer, 1986). The analysis of these peptides, using the reaction of Coomassie blue dye with proteins (Bradford, 1976) in natural aquatic systems and lab cultures, has made significant advances from the more basic colorimetric methods (Setchell, 1981; Mayer et al., 1986; Long and Azam, 1996) to more sophisticated two-dimensional (2D) electrophoretic approaches (Nguyen and Harvey, 1998; Saijo and Tanoue, 2004). This application of more recent 2D work is commonly referred to *proteomics*, which examines the size, structure, and function of proteins in the context of the entire *proteome*. Early work showed that the reaction of Coomassie blue dye with proteins (Bradford, 1976) was effective for analyzing only large polypeptides, typically in the range of 10 to 25 amino acid residues (Sedmak and Grossberg, 1977; Righetti and Chillemi, 1978). Other work showed that this method was also sensitive to colored humic materials (Setchell, 1981). Further work utilized an enzymatic approach whereby the proteinaceous material could be separated from the humic material (Mayer et al., 1986). In this method the analysis of proteins, operationally defined as peptide residues larger than 7 to 15 amino acids, was confined to residues that could be extracted from sediments by NaOH, previously defined as EHAA. Details of this method are shown in (fig. 6.18); for sediments, the detection limits are in the range of 0.05 to 0.5 mg protein g sediment^{-1}, with protein recovery of about 80%, and the precision is from ±15 to 30% (Mayer et al., 1986).

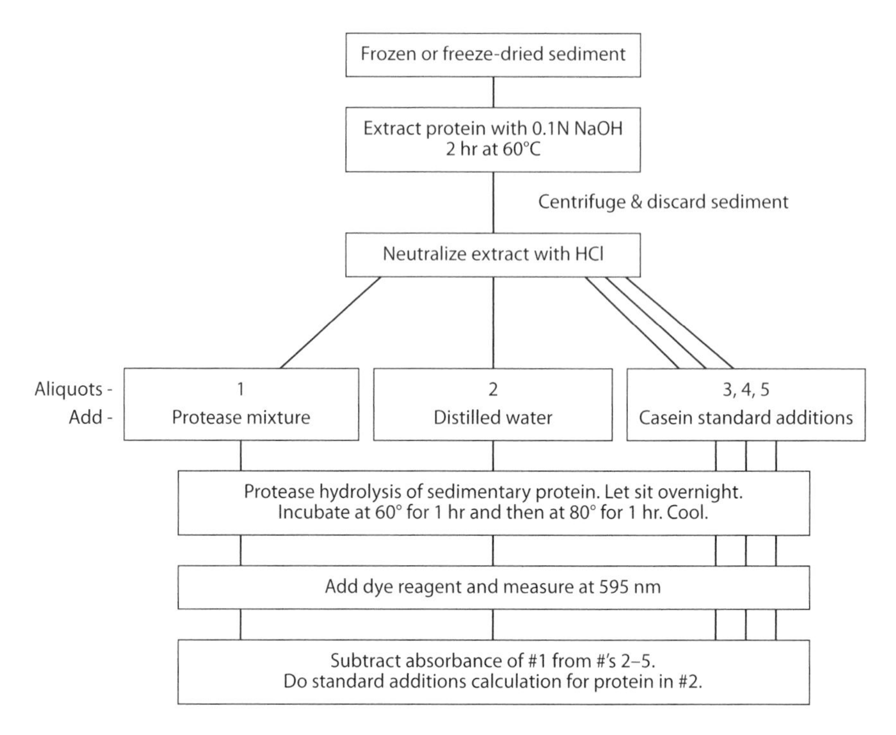

Figure 6.18. Flowchart showing an enzymatic approach that separates proteinaceous material from humic material. (Adapted from Mayer et al., 1986.)

More recently, one-dimensional sodium dodecyl sulfate–polyacrylamide gel electrophoresis (1D SDS-PAGE) was used to separate proteins into background and specific proteins in oceanic waters (Tanoue, 1992, 1996; Tanoue et al., 1996). Background proteins in particulate organic matter (POM) from oligotrophic surface waters were not separated by 1D SDS-PAGE, presumably because of their wide range of molecular masses and low concentration levels (Tanoue, 1996). Conversely, specific proteins, composed of a smaller range of molecular weights, were resolved and superimposed over the background proteins in these same surface waters. However, specific proteins were not recognizable in POM from more productive surface waters because of the high levels of background proteins associated with high phytoplankton biomass. Thus, it was clear that a technique requiring higher resolution was needed for separating these proteins. The proteomic method of two-dimensional electrophoresis (2DE) has been widely used as an effective method for separating complex proteins (Laemmli, 1970; O'Farrell, 1975; Oakley et al., 1980; Bjellqvist et al., 1982; Ramagli and Rodriguez, 1985; Matsudaira, 1987). The basic premise here is that proteins are separated based on their isoelectric points using isoelectric focusing in the first dimension and molecular mass in the second dimension by SDS-PAGE. Until the recent application of 2DE to particulate proteins in Pacific surface waters (Saijo and Tanoue, 2004), this technique had only been applied to protein separation in foram shells (Robbins and Brew, 1990) and diatom cultures (Nguyen and Harvey, 1998).

6.4 Applications

6.4.1 Proteins: Amino Acids and Amines as Biomarkers in Lakes

The earliest work on amino acids in lakes was conducted with an attempt to better understand the overall controls on system productivity (Hellebust, 1965; Gocke, 1970). Some of the first work using

HPLC methods to measure DFAA in lakes was conducted by Jorgensen (1987). In this study, three Danish lakes (two eutrophic [Frederiksborg Slotsso and Hylke] and one oligotrophic [Almind]) were sampled every three days during several weeks in an attempt to examine the relative importance of DFAA to the metabolism of bacterioplankton. The most abundant DFAA were serine, glycine, alanine, and ornithine; other DFAA such as aspartic acid, threonine, valine, phenylalanine, and lysine were also high at certain times (fig. 6.19). The overall conclusions from this study were that DFAA showed large fluctuations in concentration in all three lakes (78 to 3672 nM), with the largest in Lake Almind—the oligotrophic system. However, only the eutrophic systems had diel trends in DFAA concentrations, with positive correlations between bacterial and phytoplankton production. Overall, phytoplankton production in both eutrophic lakes was important in controlling the abundance and assimilation of DFAA. More recently, it was also shown, using PCA analyses, that amino acid cycling in a set of German lakes in the region around Berlin was largely controlled by the time of day they were sampled. This, again, supports the idea that the diel cycles of phytoplankton are key in controlling the DFAA pool in lacustrine systems.

6.4.2 Proteins: Amino Acids and Amines as Biomarkers along the River–Estuarine Continuum

Amino acids represent an important portion of the labile fraction of riverine organic matter (Ittekkot and Zhang, 1989; Spitzy and Ittekkot, 1991). Thus, a greater understanding of factors controlling amino acid abundance and composition provides a powerful tool to study the sources and biogeochemical cycling of labile organic matter in rivers (Ittekkot and Arain, 1986; Hedges et al., 1994; Aufdenkampe et al., 2001). In a recent study, the spatial and temporal variability of particulate and dissolved amino acids was compared in a large (the lower Mississippi River) and small river (the Pearl River) system in the United States (Duan and Bianchi, 2007). The Mississippi River is one of the largest regulated rivers (levees and dams) in the world, and represents an important source of organic matter and nutrients to the northern Gulf of Mexico (Trefry et al., 1994; Guo et al., 1999). In contrast, the Pearl River is a less-disturbed, small, black-water river (3rd-order stream), in Mississippi and southwest Louisiana that drains the same lower southeast region of the Mississippi River flood plain. Differences between the particulate amino acid (PAA) and dissolved amino acid (DAA) fractions were attributed, in part, to selective partitioning of amino acids between water and suspended sediments, as demonstrated previously in the Amazon River (South America) (Aufdenkampe et al., 2001). It was also concluded that biological factors, such as zooplankton feeding and cell lysis of phytoplankton, contributed to changes in PAA and DAA in these two very different river systems. In fact, progressive enrichment in % nonprotein amino acids from living biota of riverine and marine POM to high-molecular-weight-dissolved organic matter (HMW DOM, $<0.2\,\mu$m and >1 kDa), with a decrease in the amino acid degradation index, suggested that microbial degradation during downstream transport was reflected in the lower Mississippi River signature of amino acids (fig. 6.20).

In general, DCAA concentrations are higher than DFAA in estuaries and may represent as much as 13% of the total DON (Keil and Kirchman, 1991a,b). Furthermore, DCAA can support approximately 50% (Keil and Kirchman, 1991a,b, 1993) and 25% (Middelboe and Kirchman, 1995) of the bacterial N and C demand, respectively, in estuaries. While most DFAA are derived from the heterotrophic breakdown of peptides and proteins in POM and DOM, they can be formed when NH_4^+ reacts with some carboxylic acids in the presence of NADPH (reduced form) (De Stefano et al., 2000). For example, glutamic acid can be formed from the reaction of NH_4^+ with α–ketoglutaric acid (De Stefano et al., 2000). Independent of their source, DFAA have clearly been shown to be important sources of C and N to microbial communities in estuarine and coastal systems (Crawford et al., 1974; Dawson and Gocke, 1978; Keil and Kirchman, 1991a,b; Middelboe and Kirchman, 1995). In fact, the extremely

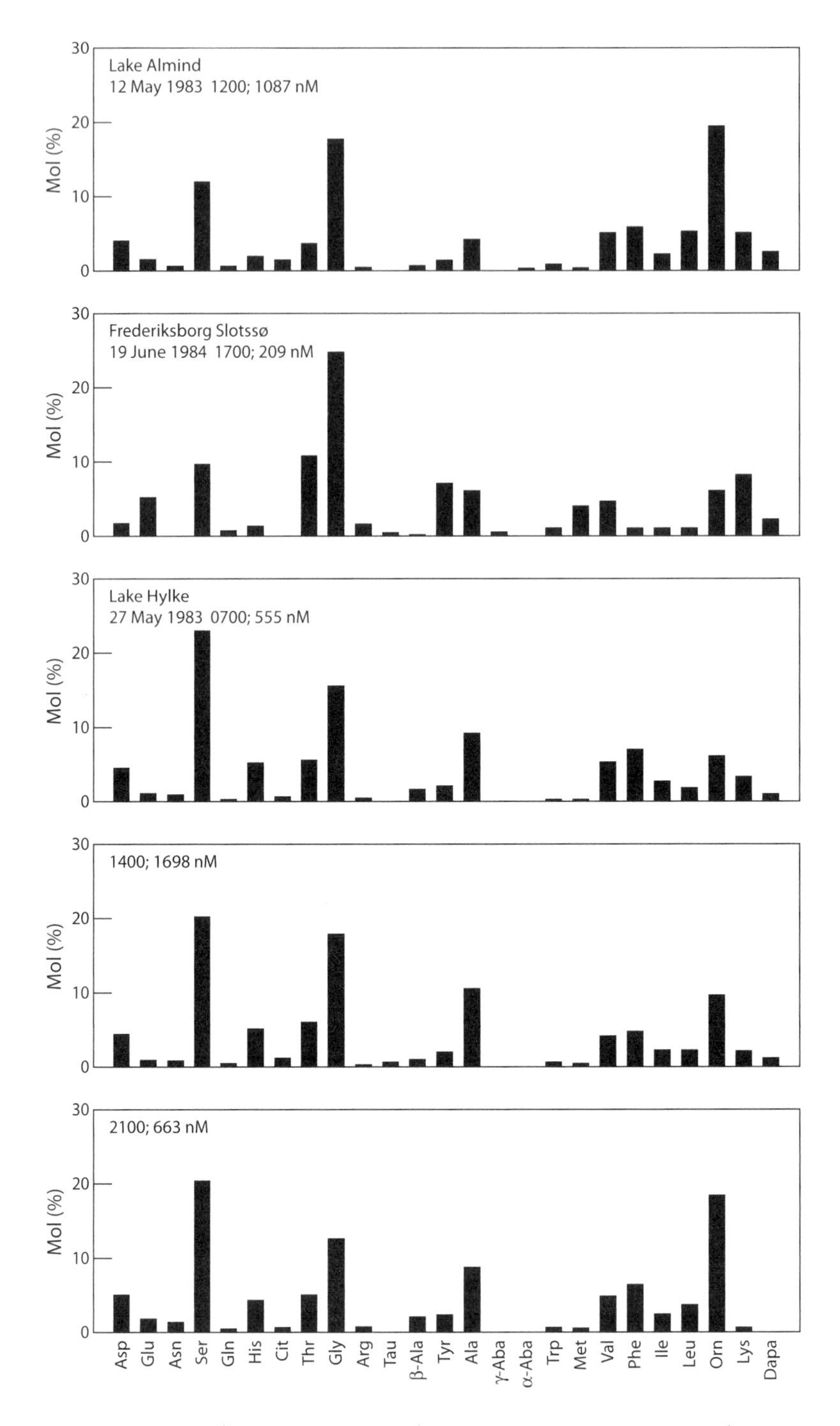

Figure 6.19. Three Danish lakes (two eutrophic lakes (Frederiksborg Slotsso and Hylke) and one oligotrophic (Almind) were sampled every three days during several weeks in an attempt to examine the relative importance of dissolved free amino acids in the metabolism of bacterioplankton. The most abundant dissolved free amino acids were serine, glycine, alanine, and ornithine, while others, such as aspartic acid, threonine, valine, phenylalanine, and lysine, were also high at certain times. (Adapted from Jorgensen, 1987.)

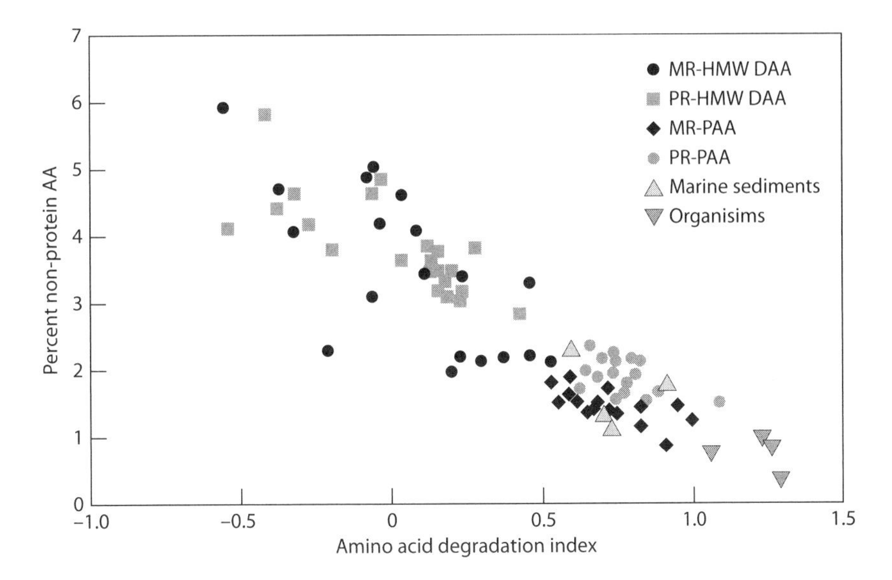

Figure 6.20. Relationship between nonprotein amino acids and amino acid degradation index for organisms, marine sediments, riverine particulate amino acids (PAA), and high-molecular-weight dissolved amino acids (HMW-DAA; $<0.2\,\mu m$ and >1 kDa). The lower amino acid degradation index in the Mississippi River suggests microbial processing as material is transported downstream. MR, Mississippi River, LA (USA); PR, Pearl River (USA).

rapid (minutes) cycling of amino acids observed in coastal waters in the Hudson River (USA) (Mitchell et al., 1985) and Chesapeake Bay (USA) (Fuhrman, 1990) estuaries suggests strong coupling between DFAA sources (e.g., phytoplankton) and bacteria.

The biogeochemical cycling of dissolved amino acids in organic-rich, anoxic porewaters has been shown to be controlled by the following "internal" transformations: (1) conversion of sedimentary amino acids to DFAA; (2) microbial remineralization (e.g., the amino acids serve as electron donors or for direct uptake by sulfate-reducing bacteria) (Hanson and Gardner, 1978; Gardner and Hanson, 1979); and (3) uptake or reincorporation into larger molecules (e.g., geopolymerization) in sediments (Burdige and Martens, 1988, 1990; Burdige, 2002). DFAA in porewaters are considered to be intermediates of biotic and abiotic processing (Burdige and Martens, 1988). During predepositional settlement of POM to the sediment surface, organic matter may be completely mineralized to inorganic nutrients or may be transformed to DOM. Postdepositional decay of organic matter results in selective loss of low-molecular-weight compounds, such as DFAA and simple sugars, with increasing depth in the sediments (Henrichs et al., 1984)—as shown in DFAA profiles for Cape Lookout Bight (USA) sediments (fig. 6.21) (Burdige and Martens, 1990). The highest concentrations of DFAA in porewaters are typically in the top few centimeters of surface sediments (20 to 60 μM), compared to lower asymptotic values at depth (2 to 5 μM) and in overlying waters ($<1\,\mu$M). The dominant DFAA in these porewaters were nonprotein AA (β-aminoglutaric acid, δ-aminovaleric, and β-alanine), which were primarily derived from fermentation processes. The high concentration gradient of DFAA between the overlying waters and surface porewaters results in active fluxes across the sediment–water interface that are mediated by biological and diffusive processes. For Cape Lookout Bight, the highest estimated effluxes of total DFAA were found in summer months, with annual rates that ranged from 0.02 to 0.09 mol total DFAA $m^{-2}\,yr^{-1}$ or 52 to 257 $\mu mol\,m^{-2}\,d^{-1}$ (Burdige and Martens, 1990). These fluxes are considerably higher than those found in other nearshore environments, such as the Gullmar Fjord (Sweden) (ca. 18 $\mu mol\,m^{-2}\,d^{-1}$) and the Skagerrak (northeastern North Sea) (-20 to 13 $\mu mol\,m^{-2}\,d^{-1}$)

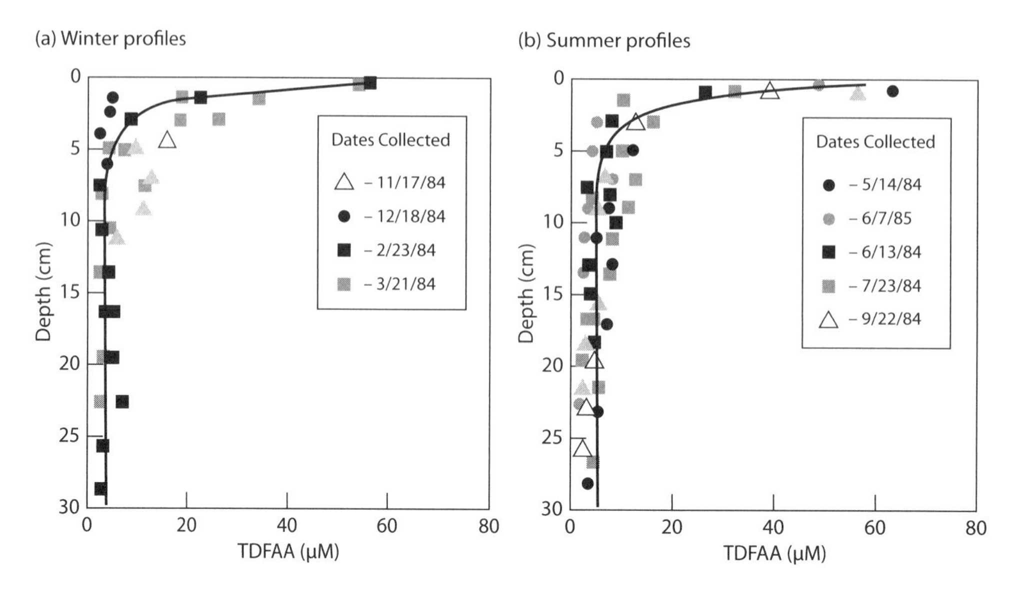

Figure 6.21. Postdepositional decay of organic matter results in selective loss of low-molecular-weight compounds, such as dissolved free amino acids (DFAA) and simple sugars (not shown here), with increasing depth in the sediments—as shown in total DFAA (TDFAA) profiles for Cape Lookout Bight (USA) sediments. (Adapted from Burdige and Martens, 1990.)

(Landen and Hall, 2000). Although bacterial uptake of DFAA in surface sediments has been shown to decrease DFAA fluxes across the sediment–water interface (Jorgensen, 1984), the likely explanation for such high rates at the Cape Lookout site was significantly greater amounts of sedimentary organic matter. Finally, the interaction between the DFAA and DCAA pools in porewaters is likely to be quite important in understanding carbon preservation (Burdige, 2002). Unfortunately, relatively few studies have examined DCAA in porewaters (Caughey, 1982; Colombo et al., 1998; Lomstein et al., 1998; Pantoja and Lee, 1999)—with preliminary results indicating that the DCAA pool is 1 to 4 times the size of the DFAA pool.

6.4.3 Proteins: Amino Acids and Amines as Biomarkers in the Ocean

As described earlier, particulate combined AAs (PCAA) are an important component of oceanic DOM that is fundamentally linked with many of the major biogeochemical cycles in the ocean (Wakeham, 1999; Volkman and Tanoue, 2002). Recent work has examined the chemical composition of amino acids in POM as related to the formation of PCAA in Pacific Ocean surface waters, using 1D and 2D gel electrophoresis (Saijo and Tanoue, 2005). In the application of gel electrophoresis, the stainable and nonstainable materials represent different peptide chain lengths. For example, anything less than or greater than 6 to 8 amino acid residues was found to be nonstainable and stainable, respectively (Saijo and Tanoue, 2005). The stainable and nonstainable materials were found to be similar in POM from Pacific surface waters (fig. 6.22). In both forms (stainable and nonstainable), glycine, valine, isoleucine, threonine, serine, and aspartic acid were depleted relative to POM. Conversely, proline, lysine, and arginine were relatively enriched compared to bulk POM. The acid and nonstainable material made up 31–42% and 49–63% of all the fractions analyzed, respectively, with proteins making up less than 2% of the PCAA. The presence of stainable material was interpreted as intermediate degradation of high- to lower-molecular-weight products during the decomposition of POM in the water column.

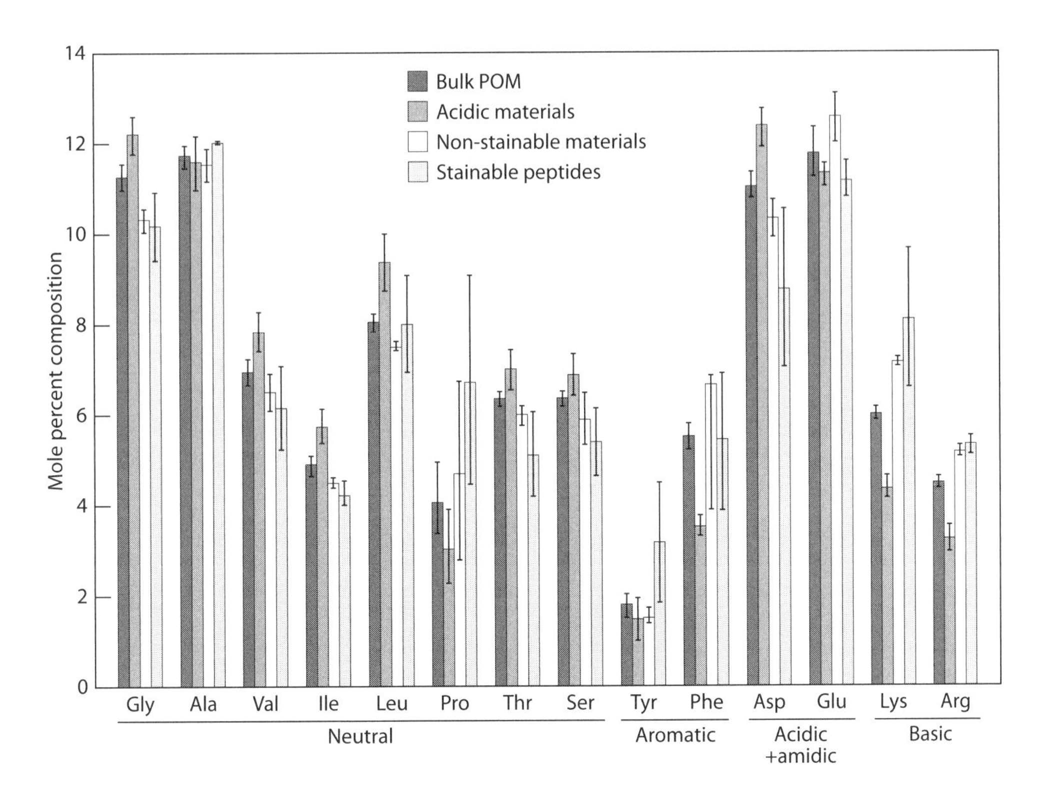

Figure 6.22. Composition of amino acids in particulate organic matter (POM) as related to the formation of particulate combined amino acids (PCAA) in Pacific Ocean surface waters, using 1D and 2D gel electrophoresis. In the application of gel electrophoresis, the stainable and nonstainable materials represent different peptide chain lengths. (Adapted from Saijo and Tanoue, 2005.)

These data support the size reactivity model, whereby larger, more labile material is decomposed into smaller, more refractory material during decomposition (Amon and Benner, 1996). The potential sources of acidic materials were likely from alanine and glutamic acid, which are abundant amino acids in peptidoglycan (Schleifier and Kandler, 1972) and/or perhaps remnants of *glycoproteins* (Winterbum and Phelps, 1972).

As particles in the ocean sink, POM is transported to deep waters and is generally converted to smaller, more refractory material. The ballasting that aids in the transport of some of these particles in sinking fecal pellets is provided by biogenic minerals (Honjo et al., 2000). The biominerals calcite and aragonite have a glycoprotein template that is important for determining the morphology of these minerals. Proteins that are enriched in glycine, serine, and threonine are associated with biogenic opal, while proteins rich in aspartic acid are found in $CaCO_3$ (Constantz and Weiner, 1988). The relative importance of calcifying versus silicifying plankton can be determined by the relative abundance and composition of amino acids (Ittekkot et al., 1984; Gupta and Kawahata, 2000). Thus, mineral-bound amino acids may be useful indicators of the source and diagenetic state of sinking particles in the ocean (Lee et al., 2000). Based on this premise, recent work examined the concentrations and fluxes of amino acids in plankton in the Pacific sector of the Southern Ocean (Ingalls et al., 2003). The average amount of THAA in net plankton, sediment trap material, and sediments is shown in (fig. 6.23). The THAA, which included both non-mineral-bound amino acids and calcium-bound amino acids, were enriched in glycine and serine, reflective of diatom sources. The biomineral-bound amino acids (CaTHAA and SiTHAA) were a small percentage of the total THAA, with SiTHAA becoming a larger component with

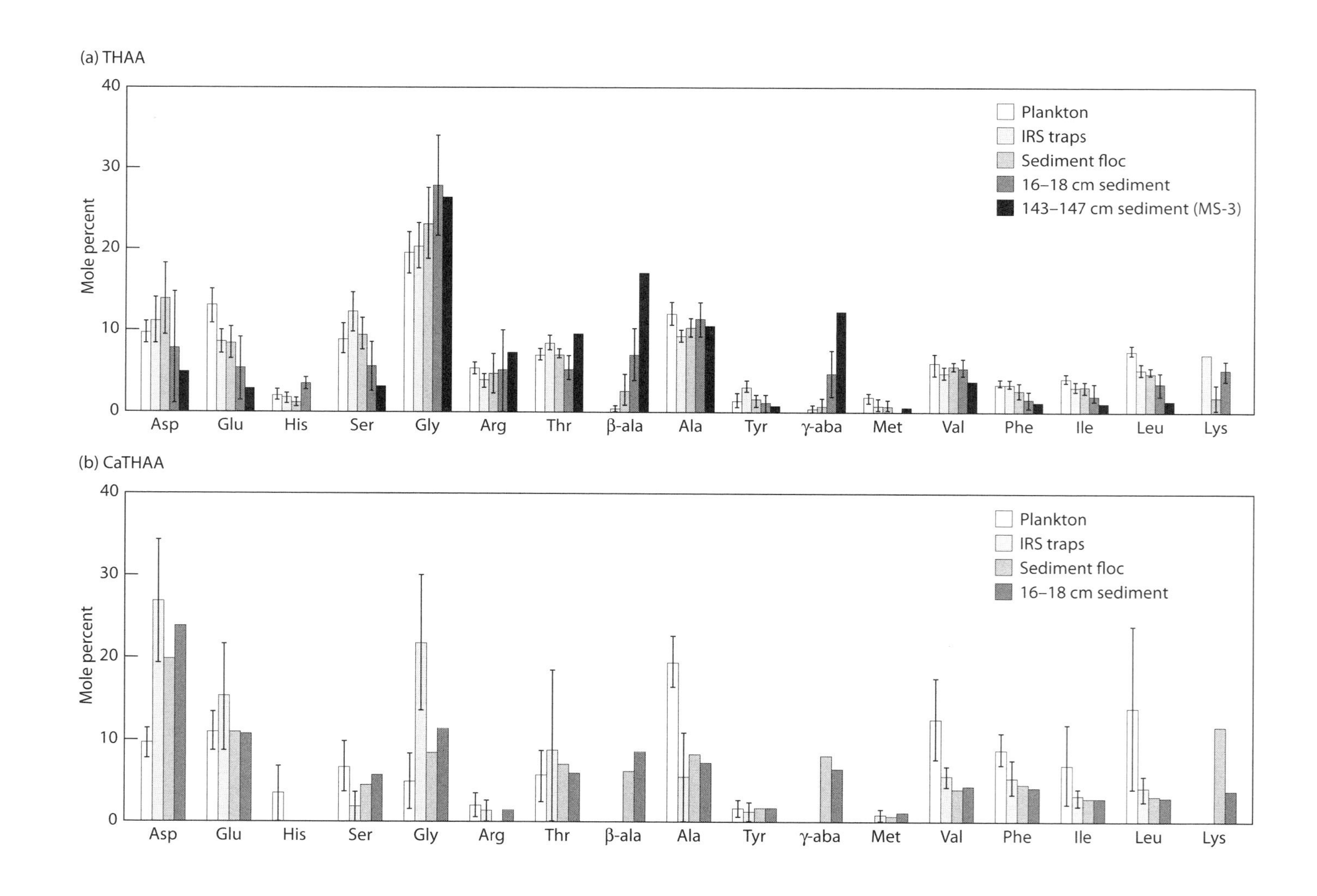
(a) THAA
Mole percent
40
30
20
10
0
Plankton
IRS traps
Sediment floc
16–18 cm sediment
143–147 cm sediment (MS-3)
Asp
Glu
His
Ser
Gly
Arg
Thr
β-ala
Ala
Tyr
γ-aba
Met
Val
Phe
Ile
Leu
Lys
(b) CaTHAA
Mole percent
40
30
20
10
0
Plankton
IRS traps
Sediment floc
16–18 cm sediment
Asp
Glu
His
Ser
Gly
Arg
Thr
β-ala
Ala
Tyr
γ-aba
Met
Val
Phe
Ile
Leu
Lys

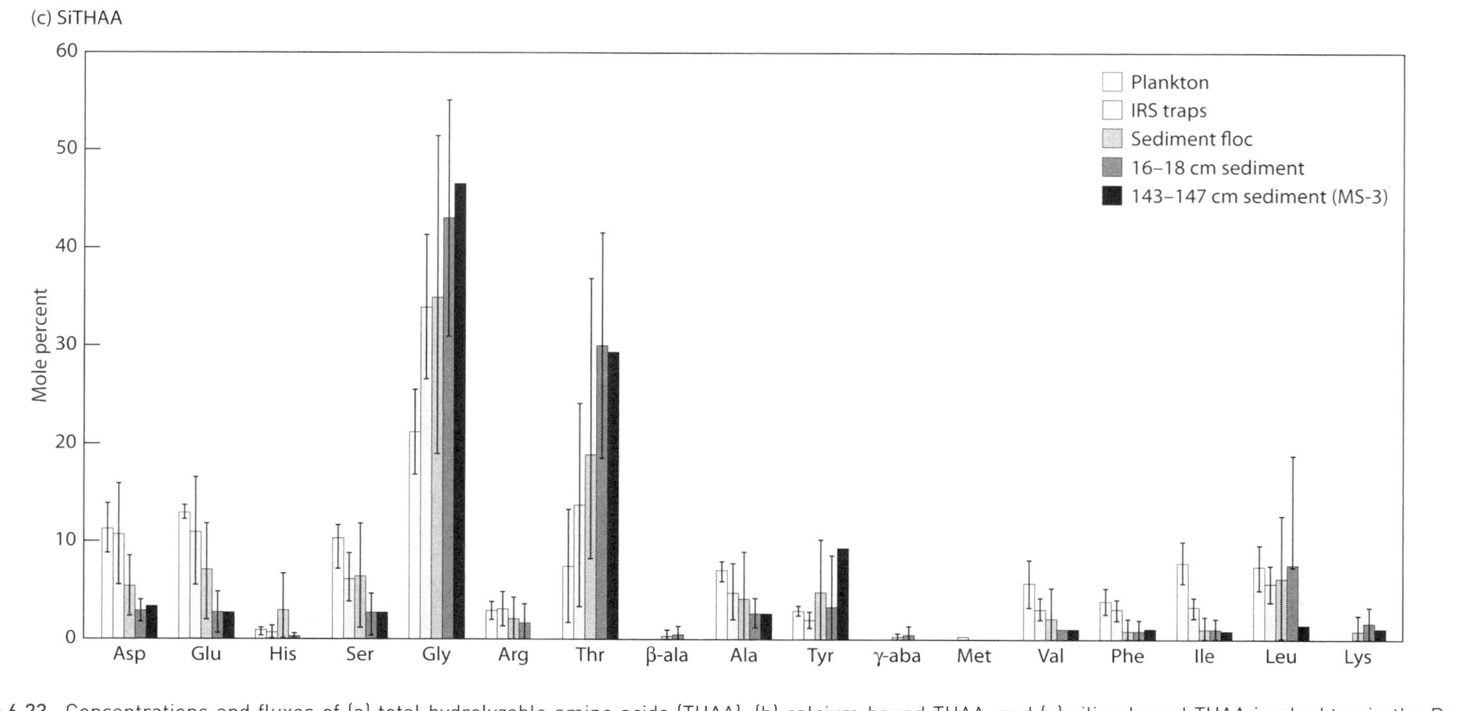

Figure 6.23. Concentrations and fluxes of (a) total hydrolyzable amino acids (THAA), (b) calcium bound THAA, and (c) silica-bound THAA in plankton in the Pacific sector of the Southern Ocean. The average total hydrolyzable amino acids (THAA) in net plankton, sediment trap material, and sediments are shown here. (Adapted from Ingalls et al., 2003.)

depth due to selective preservation relative to CaTHAA. Sediments beneath the Antarctic Circumpolar Current were enriched in biomineral-bound amino acids due to the high diatom productivity in that region.

6.5 Summary

Proteins make up approximately 50% of organic matter and contain about 85% of the N in marine organisms. Nevertheless, N remains an important limiting element for biological systems. The focus on the organic N pool in the early years of aquatic chemistry was on colorimetric techniques that examined very crude estimates of proteins or the smaller amino acid pools. Since the DFAA pools cycle so quickly it has been difficult to link these pools with biogeochemical processes over ecological timescales. Recent developments involving proteomics provide new and exciting methods that allow for the examination of the function and structure of proteins. More advanced techniques using HPLC and HPLC-MS have also provided better approaches to studying these rapidly cycling amino pools. The past use of amino acids, amines, and proteins as chemical biomarkers has been somewhat restricted by their lack of source specificity relative to other source-specific organic compounds (e.g., lipids). However, these aforementioned techniques now allow for further examination of one of the more important dissolved and particulate pools in aquatic systems.

7. Nucleic Acids and Molecular Tools

7.1 Background

In this chapter, we examine molecular tools based on *nucleic acids,* polymers of the nucleotides ribonucleic acid (RNA) and deoxyribonucleic acid (DNA). Topics include the biogeochemical significance of nucleic acids as a pool of organic nitrogen and phosphorus, as well as recent work characterizing the stable and radiocarbon isotopic signatures of nucleic acids to identify the sources and age of organic matter supporting heterotrophic bacteria. In addition to a discussion of nucleic acids as a class of biochemicals, we will discuss the application of molecular genetic information to organic geochemistry. The recent combination of biomarker information and molecular genetic data provides new insights about aquatic microbial communities and their influence on organic matter decomposition. Recent studies utilizing molecular tools have identified new and previously uncultured microbes, and clarified the evolutionary history of biosynthetic pathways, the capabilities of microorganisms to utilize specific substrates, and the synthesis of unique biomarkers. These tools provide new perspectives on the diversity of microbial communities and *phylogenetic* relationships between microbial groups as well as the evolutionary significance behind many of the proxies used in organic geochemistry.

7.2 Introduction

Nucleic acids are the fourth most abundant class of biochemicals, following carbohydrates, proteins, and lipids. This class of macromolecules is composed of polymers of *nucleotides* whose composition and arrangement convey genetic information. The macromolecule deoxyribonucleic acid is made up of two strands, oriented in a double helix, where the bases on opposite strands are linked together by hydrogen bonds. DNA contains the genetic information used by cells to encode protein synthesis and provides information about the evolutionary relationships between organisms. In contrast, ribonucleic acid is generally a single-stranded macromolecule with a shorter chain of nucleotides than DNA. There are three types of RNA—messenger RNA (mRNA), transfer RNA (tRNA), and ribosomal RNA (rRNA)—which together facilitate transmission and expression of genetic information from DNA to protein.

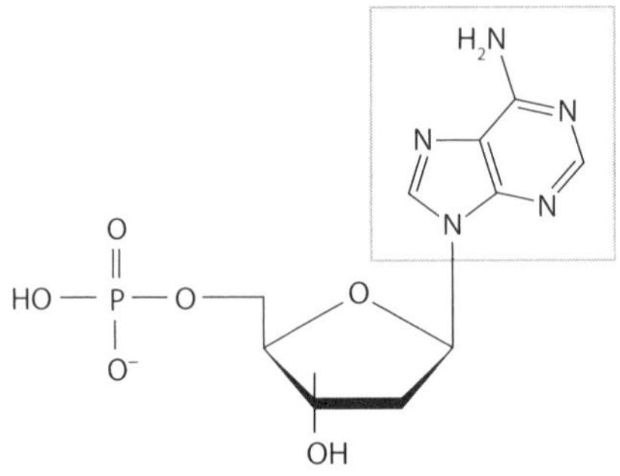

Deoxyadenosine monophosphate

Figure 7.1. Example showing subcomponents of nucleotides. All nucleotides include a purine or pyrimidine base (shown in box), a pentose sugar, and a phosphate group.

Messenger RNA carries information from DNA to structures called *ribosomes*, and there provides the template for translating the amino acid sequences involved in the synthesis of proteins. Ribosomes are subcellular structures composed of a combination of proteins and ribosomal RNAs (rRNA), which together "read" messenger RNAs and translate the information into proteins. Transfer RNAs transport amino acids to the ribosomal site of protein synthesis during translation and add them to the building polypeptide (protein) chain. RNA also serves as the genome, i.e., genetic code, of most viruses.

All nucleotides are composed of three units: (1) a nitrogen-containing base, either a *purine* or *pyrimidine*, (2) a pentose sugar moiety, and (3) a phosphate group, with the base bonded to the sugars at the C-1 position (fig. 7.1). These nucleotide units are bonded together in long chains by phosphodiester linkages between the pentose sugars to form the DNA or RNA. DNA contains four bases: the pyrimidines cytosine and thymine and the purines adenine and guanine, often abbreviated by the letters C, T, A, and G, respectively. In RNA, uracil (an unmethylated form of thymine) replaces thymine as the base corresponding to adenine. An additional difference between DNA and RNA is the sugar units involved: DNA contains deoxyribose, while RNA contains ribose. As a result, RNA is generally less stable than DNA, since the presence of an additional hydroxyl group attached to the pentose ring makes it more susceptible to hydrolysis.

Relative to the other classes of biochemicals, nucleic acids contain a high proportion of nitrogen and phosphorus. Both DNA and RNA are built in proportions of one phosphate per sugar per nitrogenous base. On average, the elemental composition of purines and pyrimidines is 39.1% N and 43.1% C (Sterner and Elser, 2002). When the elemental composition of these bases is combined with that of the phosphate and sugar groups, nucleic acids are 32.7% C, 14.5% N, and 8.7% P, yielding a C:N:P stoichiometry of 9.5:3.7:1. Interestingly, nucleic acids are similar to proteins in nitrogen content (14.5 versus 17%), but differ in phosphorus content, with proteins being P-poor while nucleic acids are P-rich. Since DNA is generally a small fraction of biomass, its stoichiometry has little influence on the overall C:N:P composition of organisms. However, levels of RNA vary significantly among taxa and within individual organisms as a function of growth, development, and physiology. For example, RNA can make up 15% or more of the biomass of some metazoans and 40% of the biomass of some microbes (Sterner and Elser, 2002). Because RNA is both P-rich and much more abundant than DNA in biomass, RNA content can influence the C:P and N:P ratios of organisms.

7.3 DNA in Aquatic Environments

Nucleic acids are present in a variety of forms in aquatic environments, including (1) as intracellular DNA associated with living organisms, (2) as protein-encapsulated DNA associated with viruses,

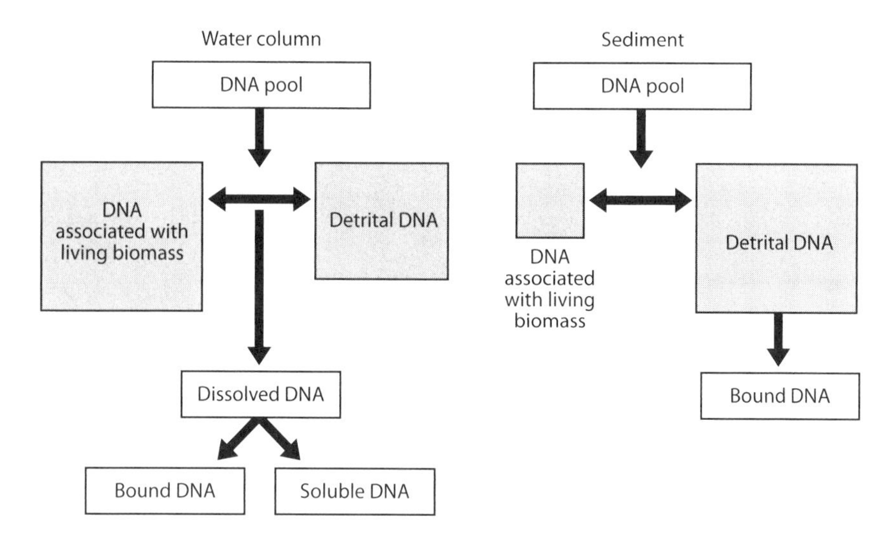

Figure 7.2. Schematic showing DNA pools in the water column and sediments. (Adapted from Danovaro et al., 2006.)

(3) as free (soluble) DNA, and (4) absorbed to detritus and mineral particles (Danovaro et al., 2006). Living biomass is the ultimate source of all nucleic acids but a portion of the nucleic acids present in the environment derive from extracellular sources formed by exudation and excretion from living cells, losses resulting from grazing, release following cell death and lysis, viral lysis of cells, and sorption/desorption of dissolved DNA from suspended particulate matter. While the pool of DNA associated with living biomass is larger than that associated with detrital sources in the water column, the situation is reversed in sediments where most (~90%) of the DNA is associated with detritus (fig. 7.2). In the water column, the dissolved pool of DNA is primarily extracellular, with viruses contributing about 20%. Concentrations of dissolved DNA are highest in estuaries and decrease with distance offshore (Deflaun et al., 1987; Paul et al., 1991a,b; Brum, 2005) (table 7.1). In contrast to the dissolved phase, the particulate DNA pool is primarily in the biomass of prokaryotes in the picoplankton size range (0.2 to 1.0 μm) (Danovaro et al., 2006).

Nucleic acids can be important sources of labile organic phosphorus and nitrogen in aquatic ecosystems. The dominant mechanism for the recycling of extracellular DNA involves the enzyme DNase, which regenerates phosphorus and supplies organic nitrogen and phosphorus as well as nucleosides and nucleobases for bacterial metabolism. In the water column, extracellular DNA may supply ~50 and 10% of the daily P and N requirements, respectively, for bacterioplankton, with this pool potentially supplying a greater portion of nutrient requirements in P-depleted regions (Danovaro et al., 2006). In deep-sea sediments, extracellular DNA contributes 4, 7, and 47% of the daily C, N, and P demand for prokaryotes (Dell'anno and Danovaro, 2005). However, the half-life for extracellular DNA is generally longer in sediments than in the water column, with some preservation of DNA occurring over geological timescales (Coolen et al., 2006; Coolen and Overmann, 2007). This may reflect the fact that enzymatic degradation of extracellular DNA decreases when DNA is bound to macromolecules or inorganic particles, as is typical in sediments (Romanowski et al., 1991).

Several recent studies have documented the cycling and turnover of extracellular DNA in the water column and sediments of various aquatic environments (Jorgensen and Jacobsen, 1996; Brum, 2005; Dell'anno et al., 2005; Reitzel et al., 2006; Corinaldesi et al., 2007a) (table 7.1). For example, dissolved extracellular DNA can serve as a sole source of phosphorus, as well as carbon and energy, for metal-reducing bacteria of the genus *Shewanella* (Pinchuk et al., 2008). Particularly in deep-sea

Table 7.1
Concentration and turnover of dissolved DNA in aquatic environments

Location	*Site characteristic*	*DNA concentrations*	*Turnover time*	*References*
Station ALOHA	Oligotrophic ocean			
5 to 100 m		1.08 to 1.27 ng L^{-1}	9.6 to 24 h	Brum et al. (2004); Brum et al. (2005)
150 to 500 m		0.21 to 0.61 ng L^{-1}		
Kaneohe Bay, HI (USA)	Coastal waters	3.41 ± 0.09 ngL^{-1}		Brum et al. (2004)
Roskilde Fjord, New Zealand	Fjord waters	2 to 5 μgL^{-1}	6 to 21 h	Jorgensen and Jacobsen (1996)
	Fjord waters + nutrient	2 to 11 μgL^{-1}	5 to 89 h	
Eastern Mediterranean Sea	Deep-sea sediments	55.5 ± 8.7 μg DNA g^{-1}	n.d.	Dell'Anno et al. (2005)
	Open-slope sediments	12.0 ± 3.8 μg DNA g^{-1}	n.d.	
Mediterranean Sea	Coastal sediments	9.8 ± 2.6 μg g^{-1}	0.35 yr	Corinaldesi et al. (2007)
	Deep-sea sediments	19.8 ± 0.6 μg DNA g^{-1}	1.2 yr	
Gulf of Mexico	Offshore–oligotrophic	0.2 to 1.9μgL^{-1}	n.d.	DeFlaun et al. (1987)
	Nearshore–estuarine	10 to 19 μgL^{-1}		
Gulf of Mexico	Offshore–oligotrophic	4.6 μg^{-2}	25.5 to 28 h	Paul et al. (1989)
Bayboro Harbor, FL (USA)	Estuarine	9. 39 to 11.6 μgL^{-1}	6. 4 to 11.0	Paul et al. (1989)
Crystal River, FL (USA)	Freshwater spring	$1.43 \pm 1.1$$\mu$g$L^{1}$	9.76 ± 3.1h	Paul et al. (1989)
Medard Reservoir, FL (USA)	Freshwater	11.9 ± 8.9	10.8 ± 3.9	Paul et al. (1989)

ecosystems, recent studies have also highlighted the potential importance of viruses in the cycling of DNA, (Dell'anno and Danovaro, 2005; Corinaldesi et al., 2007b; Danovaro et al., 2008). Danovaro and colleagues showed that mortality of prokaryotes by viral lysis increases with water depth (fig. 7.3), and described the "viral shunt," a process that releases carbon during lysis of prokaryotic cells and enhances turnover of nitrogen and phosphorus. This mechanism is thought to cause mortality in benthic prokaryotes, reducing competition for resources, while simultaneously generating labile materials that stimulate production in uninfected deep-sea prokaryotes.

7.4 Analysis of Nucleic Acids

Traditional methods of DNA analysis involve prefiltration of water samples to remove larger particulates, followed by filtration onto a micropore membrane filter (0.2 μm) and lysing of the cells retained on the filter (Fuhrman et al., 1988). DNA is then concentrated using ethanolic (or cetyltrimethylammonium bromide; CTAB) precipitation and quantified using the fluorescent dye Hoechst 33258. A problem with this methodology is that it often requires large volumes in order to concentrate sufficient material and that materials other than DNA can be isolated, which can

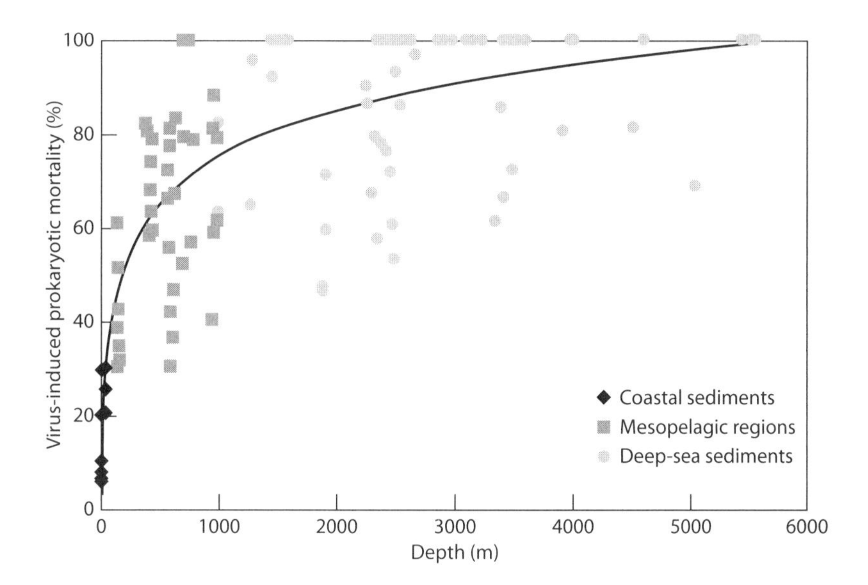

Figure 7.3. Depth-related distribution of virus-induced mortality of prokaryotes showing statistically significant relationship, $y = 14.46 \ln(x) - 24.98$, $n = 119$, $r^2 = .686$ ($p < .001$). Data include coastal sediments (diamonds), mesopelagic regions (squares), and deep-sea sediments (circles). (Adapted from Danovaro et al., 2008.)

interfere with subsequent analyses. More recently, Kirchman and colleagues developed an alternative approach, "filter PCR," which consists of filtering small volumes of water (as small as 25 μL) through polycarbonate filters, followed by amplification using sections of the filter directly in the polymerase chain reaction (PCR) (Kirchman et al., 2001). Once the DNA is amplified sufficiently, genome libraries can be obtained using specific genes (e.g., 16S rRNA) to provide insights about the composition and diversity of the aquatic microbial community (Giovannoni et al., 1990; Glockner et al., 2003; Palenik et al., 2003; Armbrust et al., 2004; Culley et al., 2006). An alternative approach involves the visualization of DNA by fluorescence in situ hybridization (FISH). FISH allows identification of particular DNA or RNA sequences within the cell and can thus identify organisms with particular gene sequences in microscopic preparations. The DNA probe is labeled with a radioactive or fluorescent probe, which can be viewed microscopically or by flow cytometry. A detailed description of the extensive molecular methods used in genetic analysis is beyond the scope of this chapter. Instead, we refer readers to recent books and review articles for additional information about molecular approaches (Munn, 2004; Dupont et al., 2007; Stepanauskas and Sieracki, 2007; Kirchman, 2008).

Compared with samples from water column or biomass, DNA extraction is more complicated in sediments where cells can be associated with macromolecules (e.g., humic substances) or mineral particles, and a larger proportion of the pool is in extracellular forms. Separation of DNA from sediments is typically accomplished using either physical (e.g., sonication) or chemical means (e.g., sodium dodecyl sulfate, SDS). As discussed above, it is important to be able to separate intracellular and extracellular nucleic acids since these pools provide different information and differ in their ecological significance. Intracellular DNA provides a reservoir of genetic information about the microbial community, whereas extracellular forms may be both important sources of nitrogen and phosphorus, as well as exogenous nucleotides that can be recycled by bacteria (Danovaro et al., 2006). Recently, Dell'Anno and colleagues (2002) developed a nuclease-based method for extraction of extracellular DNA in marine sediments. They found that extracellular DNA made up a significant but variable portion of the total DNA pool in sediments (from <10 to >70%) and that the base composition of extracellular DNA changed with depth in the upper 15 cm of sediments, suggesting differences

Table 7.2
Stable and radiocarbon composition of nucleic acids

Site	$\delta^{13}C$ *(per mil)*	$\delta^{15}N$ *(per mil)*	$\Delta^{14}C$ *(per mil)*	*References*
Gulf Breeze, FL (USA)	−22.6 to −21.1			Coffin et al. (1990)
Perdido Bay, FL (USA)	−27.5 to −26.4			Coffin et al. (1990)
Range Point, FL (USA)	−21.4 to −20.5			Coffin et al. (1990)
York River, VA (USA)	−21.2 to −29.4	+5.5 to +17.3	−35 to +234	McCallister et al. (2004)
Hudson River, NY (USA)	−25.0 to −28.4	+0.9 to +8.7	−153 to +16	McCallister et al. (2004)
Santa Rosa Sound, FL (USA)	−20.8 to −21.0		+108 to +132	Cherrier et al. (1999)
Sta M Pacific Ocean	−20.7 to −21.3		−13 to −61	Cherrier et al. (1999)

in reactivity. This method may have application to the characterization of dissolved and particulate organic matter since extracellular DNA is a ubiquitous component of these pools as well.

7.5 Isotopic Signatures of Nucleic Acids

Several studies have examined the stable carbon ($\delta^{13}C$) and radiocarbon ($\Delta^{14}C$) signatures of nucleic acids as a tool for assessing the sources and age of organic matter supporting heterotrophic bacteria (table 7.2). Initial efforts documented fidelity between the $\delta^{13}C$ signatures of nucleic acids isolated from heterotrophic bacteria and their substrates in both laboratory and field studies (Coffin et al., 1990; Coffin and Cifuentes, 1999). Subsequent studies using $\delta^{13}C$ signatures of nucleic acids have documented that estuarine bacteria use dissolved organic matter originating from a spectrum of sources, including phytoplankton (Coffin et al., 1990), terrigenous organic matter (Coffin and Cifuentes, 1999), marsh-derived organic matter (Creach et al., 1999; McCallister et al., 2004), and light hydrocarbons from oil seeps (Kelley et al., 1998). Complementary to these efforts, $\delta^{13}C$ and $\delta^{15}N$ signatures of DNA isolated from soils have been measured recently (Schwartz et al., 2007). The isotopic composition of DNA differed significantly across soil samples with $\delta^{13}C$ and $\delta^{15}N$ signatures of DNA related to $\delta^{13}C$ and $\delta^{15}N$ signatures of soil, suggesting that isotopic compositions of nucleic acids in soil microbes were strongly influenced by the soil organic matter supporting them. This study also found that DNA was enriched in ^{13}C relative to soil, indicating that soil microbes fractionated C during assimilation or preferentially used ^{13}C-enriched substrates. However, DNA enrichment in ^{15}N relative to soil was not consistently observed, suggesting that microorganisms do not fractionate N or preferentially use N substrates at different rates across ecosystems.

Recent work has expanded the isotopic analysis of nucleic acids to multicellular organisms (Jahren et al., 2004). Analyses of nucleic acids isolated from twelve higher plant species that spanned the full phylogenetic diversity of seed plants showed that $\delta^{13}C$ signatures of plant nucleic acids were enriched in ^{13}C relative to bulk plant tissue by a constant value, 1.39 per mil. Results from this study differed from those conducted with the microorganisms discussed above, where minimal fractionations were found. Because the isotopic enrichment was constant across higher plants, it may be possible to use $\delta^{13}C$ signatures of nucleic acids as a proxy for plants. This study further suggests the potential for using fossil plant DNA as a paleoenvironmental proxy.

Although isotopic analysis of nucleic acids provides information about the entire microbial community, often the goal is to obtain $\delta^{13}C$ and $\delta^{15}N$ signatures for selected phylogenetic groups. This has required the development of new analytical approaches that focus on the DNA or RNA isolated from specific groups of organisms (Pearson et al., 2004b; Sessions et al., 2005). In some cases, moreover, it is beneficial to know the phylogenetic identity and metabolic activity of individual cells in complex microbial communities. Stable isotope probing (SIP) involves introducing a stable isotope-labeled substrate into a microbial community and following the fate of the substrate by extracting diagnostic molecular species such as nucleic acids from the community and determining whether specific molecules have incorporated the isotope (Kreuzer-Martin, 2007). Building on this approach, a new method has been developed that combines rRNA-based in situ hybridization with stable isotope imaging based on nanometer-scale, secondary-ion mass spectrometry (NanoSIMS) (Behrens et al., 2008). A second method combines fluorescence in situ hybridization with stable-isotope Raman spectroscopy for simultaneous cultivation-independent identification and determination of C-13 incorporation into microbial cells (Huang et al., 2007). Together, these approaches show tremendous potential for elucidating the isotopic signatures (natural abundance or tracer levels) of specific microbial groups for application to both ecological and paleoenvironmental studies.

Finally, recent studies have examined the radiocarbon composition of nucleic acids to enable characterization of the ages of carbon assimilated by bacteria (Cherrier et al., 1999; McCallister et al., 2004). To date, radiocarbon studies of nucleic acids have been conducted in estuarine, coastal, and open ocean environments (table 7.2). Contrary to the dogma that bacteria utilize only labile, recently produced organic matter, radiocarbon ages of bacterial nucleic acids indicate that bacteria assimilate organic matter ranging in age from modern to older (fossil) components.

7.6 Molecular Characterization of Contemporary Microbial Communities

Genetic analyses of the DNA sequences of microbes have revolutionized our understanding of both the taxonomic and functional diversity of aquatic microbes in the last decade or so. Molecular methods have revealed a diverse community of microorganisms with varied metabolic capabilities, new species or unexpected distributions of existing species, and novel pathways of organic matter cycling (Giovannoni et al., 1990; Delong, 1992; Fuhrman et al., 1992; Karner et al., 2001; Delong and Karl, 2005). Molecular genetic approaches contributed to the recognition of Archaea as the third fundamental domain of life on par with the Eubacteria and Eukaryotes (see chapter 1) and revealed the tremendous diversity of aquatic microbes. In oceanic ecosystems, new groups of Archaea and their broad distribution and metabolic capabilities have been revealed through phylogenetic surveys using the 16S rRNA gene. This gene is especially useful because it is present across prokaryotes, eukaryotes, and Archaea and contains both variable regions that are diagnostic for specific groups of organisms as well as highly conserved regions, making it a useful tool for understanding microbial diversity. Originally, Archaea were thought to be restricted to extreme environments, such as hot springs or deep-sea hydrothermal vents, where they are represented by methanogens, sulfate reducers, and extreme thermophiles, as well as in highly saline landlocked seas represented by halophiles. This view began to change when Fuhrman and colleagues first discovered 16S rRNA sequences from a previously undescribed group of Archaea at depths of 100 and 500 m in the Pacific Ocean (Fuhrman et al., 1992). Subsequently, several new species of Archaea have been discovered and found to be dominant in marine microbial communities. The abundance and diversity of Archaea have now been observed in a broad range of environments, including oxygenated coastal surface waters of North America (Delong, 1992), the mesopelagic zone of the Pacific Ocean (Karner et al., 2001), estuarine environments (Vieira et al., 2007), surface sediments (Schippers and Neretin, 2006; Wilms et al., 2006), and deep subsurface sediments (Sorensen and Teske, 2006; Teske, 2006; Lipp et al., 2008).

In addition to highlighting their diversity and broad distribution, molecular studies have also identified the unexpected metabolic capabilities of Archaea. In a recent study, the abundance of the gene encoding for the archaeal ammonia monooxygenase alfa subunit (*amoA*) was correlated with a decline in ammonium concentrations and with the abundance of Crenarchaeota (Wuchter et al., 2006). This study showed that crenarchaeotal *amoA* copy numbers were 1 to 3 orders of magnitude higher than those of bacterial *amoA* in the upper 1000 m of the Atlantic Ocean, where most of the ammonium regeneration and oxidation takes place, suggesting an important role for Archaea in oceanic nitrification. Additionally, molecular studies have documented the diversity and widespread occurrence of Archaea with the ability to oxidize ammonia (Francis et al., 2005; Lam et al., 2007). Ammonia-oxidizing Archaea (AOA) are ubiquitous throughout the world's oceans, including at the base of the euphotic zone, suboxic water columns, and estuarine and coastal sediments (Francis et al., 2007). In the water column of the Black Sea, for example, ammonia-oxidizing crenarchaea and proteobacteria were both shown to be important nitrifiers and were coupled to anaerobic ammonium oxidation (anammox) through both indirect and direct processes, respectively. Through N-incubation experiments and modeled calculations, it was shown that ammonia-oxidizing crenarchaea and proteobacteria each supplied about half of the nitrite required by anammox (Lam et al., 2007). Together, these studies highlight the previously unforeseen importance of Archaea to the global nitrogen and carbon cycles as revealed through molecular studies (see more about nitrogen cycling in section below).

Molecular methods have also clarified the process of anaerobic oxidation of methane (AOM) (Kaneko et al., 1996) and its role in ocean biogeochemistry. AOM is likely responsible for consuming much of the methane produced in marine sediments and could represent a significant sink, consuming 5 to 20% of the total annual methane flux to the atmosphere (Valentine and Reeburgh, 2000). Although biogeochemical evidence for the process has been documented in the form of isotopically depleted lipid biomarker compounds unique to Archaea, an organism(s) responsible for this process has never been isolated. Laboratory and field-based studies hypothesize that AOM may be mediated by a consortium of methanogenic and sulfate-reducing bacteria (Hoehler et al., 1994).

In an attempt to identify the organisms responsible for AOM, Hinrichs et al. (1999) analyzed sequences of 16S rRNA from a methane seep in the Eel River basin off the coast of California (USA) (Hinrichs et al., 1999). They identified a complex community of microbes, including anaerobic bacteria, sulfate-reducing bacteria, and Archaea. The Archaea consisted primarily of a new group (ANME-1) related to the methanogens *Methanomicrobiales* and *Methanosarcinales* (fig. 7.4). However, the authors were unable to resolve the mechanism by which AOM operated in these sediments.

Molecular methods have also been used to document fine-scale microspatial analysis of cell associations. Orphan et al. (2002) used fluorescent in situ hybridization (FISH) to examine the associations between organisms responsible for methane oxidation in anoxic cold seep sediments (Orphan et al., 2002). This study demonstrated that AOM involves at least two groups of Archaea (ANME-1 and ANME-2) and that it can be carried out by single cells, aggregates of monospecific cells, or multispecies consortia (fig. 7.5). Results from this study also showed a range of associations between ANME-1 and ANME-2 and sulfate-reducing bacteria. ANME-1 occasionally associated with bacteria in a poorly organized way. In contrast, the associations between ANME-2 and bacteria were typically structured (fig. 7.5). Subsequently, a third group of Archaea (ANME-3) has been identified, further illustrating the diversity of organisms involved in AOM.

Molecular methods have provided evidence for unexpected associations between different metabolisms, such as anaerobic oxidation of methane coupled to denitrification of nitrate (N-DAMO) (Raghoebarsing et al., 2006). Using specific gene probes and fluorescence in situ hybridization (FISH), Raghoebarsing et al. (2006) documented an association between Archaea and bacterial cells. They observed that the Archaea were clustered inside a matrix of bacterial cells, with a ratio of 8 bacterial cells to 1 Archaea cell. Growth rates of the Archaea were slow, with doubling times of several weeks. The nitrite-dependent AOM rate was 140 mmol CH_4 per g protein h^{-1}, corresponding to approximately 0.4 fmol $CH_4 cell^{-1} d^{-1}$for the Archaea in an enrichment culture. These rates are similar to those determined for the archaeal partner involved in sulfate-dependent AOM (0.7 fmol CH_4 per cell per day) (Nauhaus et al., 2005). The observation of related species of these Archaea in several locations

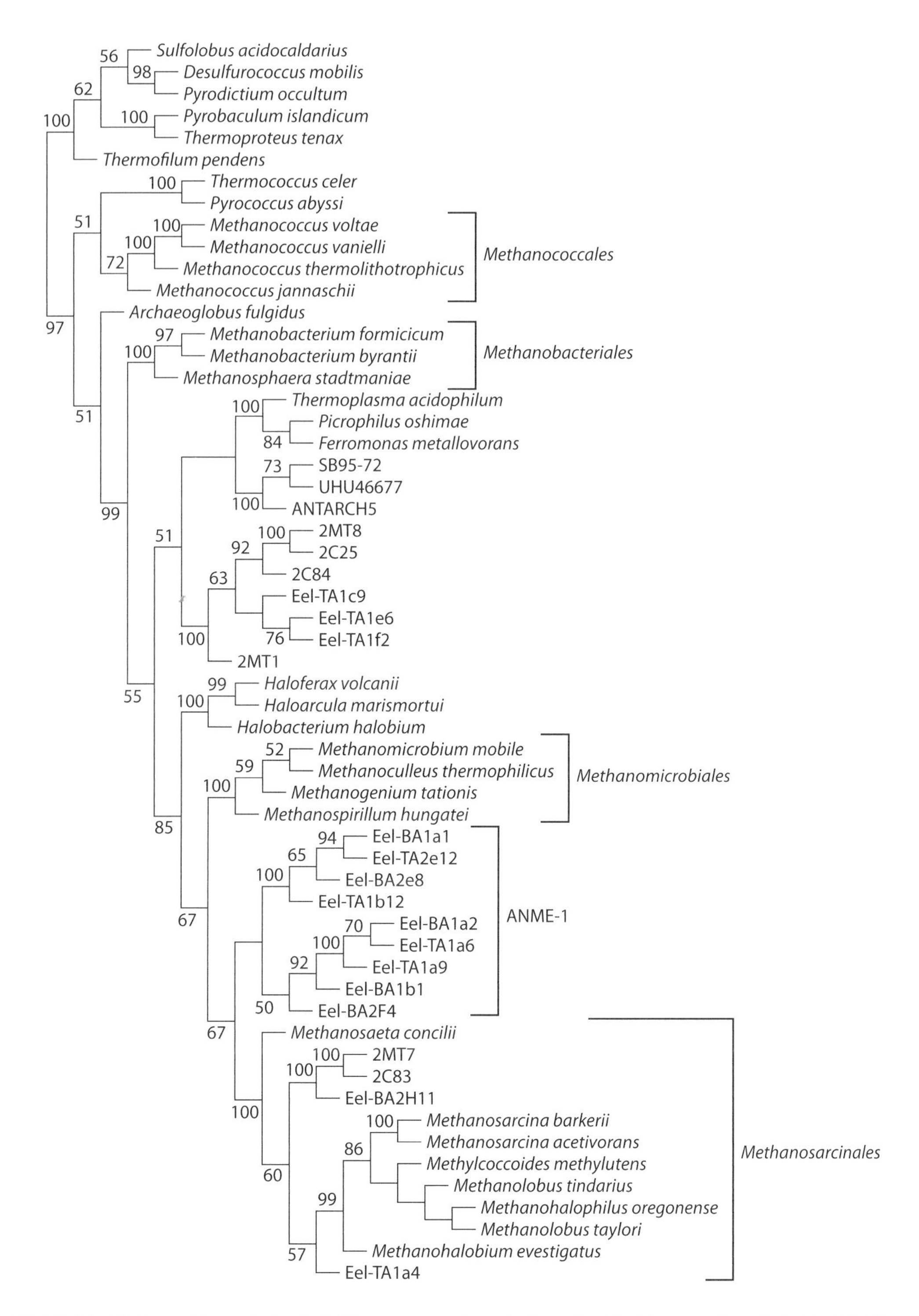

Figure 7.4. Phylogenetic analysis of rRNA sequences from Archaea isolated from sediments in the Eel River basin (USA). (Adapted from Hinrichs et al., 1999.)

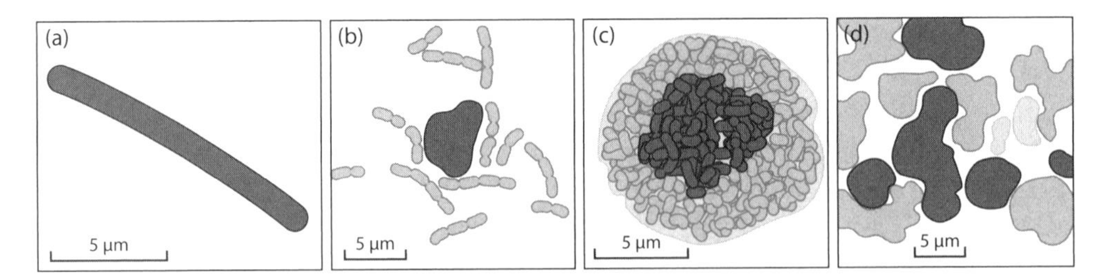

Figure 7.5. Photos showing examples individual cells and cell aggregates of ANME-1 and ANME-2 Archaea and their associations with bacteria. Samples were isolated from Eel River basin (USA) sediments and the cells were visualized using fluorescently labeled oligomer probes. The bar represents 5 μm. (a) ANME-1 archael filament stained with Cy-3-labeled probe. (b) Illustration showing association between ANME-1 Archaea rods (light gray) and sulfate-reducing bacteria, *Desulfosarcina* sp. (dark gray). (c) Illustration showing layered aggregate with ANME-2 Archaea (dark gray core) surrounded by *Desulfosarcina* sp. (light gray). (d) Loosely aggregated microcolonies of ANME-2 (dark gray) and *Desulfosarcina* sp. (light gray). (Adapted from Orphan et al., 2002.)

suggests that N-DAMO may be of greater importance to biogeochemical processes than previously thought. With advances in molecular biology, additional groups of organisms with unique metabolic capabilities will likely continue to be discovered, leading to a better understanding of biogeochemical processes and the decomposition of organic matter.

7.7 Molecular Methods: Insights about the Nitrogen Cycle

The aquatic nitrogen cycle includes a number of microbially mediated processes that are carried out by a complex community of organisms (fig. 7.6). In recent years, molecular studies have played an important role in revealing new insights about the microorganisms and processes involved in the nitrogen cycle (Jetten, 2008). In addition to the discovery of anaerobic ammonium oxidation (annamox) (Strous et al., 1999) and its widespread occurrence (Francis et al., 2007), molecular studies have uncovered new metabolic pathways, such as nitrite-dependent anaerobic methane oxidation (N-DAMO) (Raghoebarsing et al., 2006), unexpected diversity in N-fixing organisms (Zehr et al., 2007), and the role of Bacteria and Archaea in aerobic ammonia oxidation (Koenneke et al., 2005; Zehr et al., 2007; Prosser and Nicol, 2008). Genome sequencing has demonstrated the diversity of species and metabolic capabilities of organisms involved in the nitrogen cycle (Strous et al., 2006; Arp et al., 2007). We refer readers to Jetten (2008) and Zehr and Ward (2002), who provide excellent reviews of recent advances in the application of molecular methods to understanding the nitrogen cycle.

In addition to using DNA to identify the organisms involved in nitrogen cycling, a number of molecular studies have focused on the use of functional genes in order to target specific processes (Zehr and Ward, 2002) (table 7.3). To date, a number of genes involved in denitrification have been isolated, including nitrate reductase (narG and napA), nitrite reductase (nirS and nirK), and nitrous oxide reductase (nosZ) (Scala and Kerkhof, 1998; Braker et al., 2000; Casciotti and Ward, 2001) and gene probes for nitrite and nitrate assimilation are now available (Wang et al., 2000; Allen et al., 2001). Genes have also been identified for a variety of nitrogen cycling processes, including the nitrogenase gene (nifH) used in nitrogen fixation (Zehr and McReynolds, 1989), the ammonia monooxygenase gene (amoA) used in anammox (Rotthauwe et al., 1997; Alzerreca et al., 1999), and the urease (Strous et al., 1999) gene involved in the metabolism of organic nitrogen (Collier et al., 1999). Gene probes can be used in microarrays to measure gene expression in the environment (Ward et al., 2007; Bulow et al., 2008) and to provide quantitative information about the composition of the microbial community

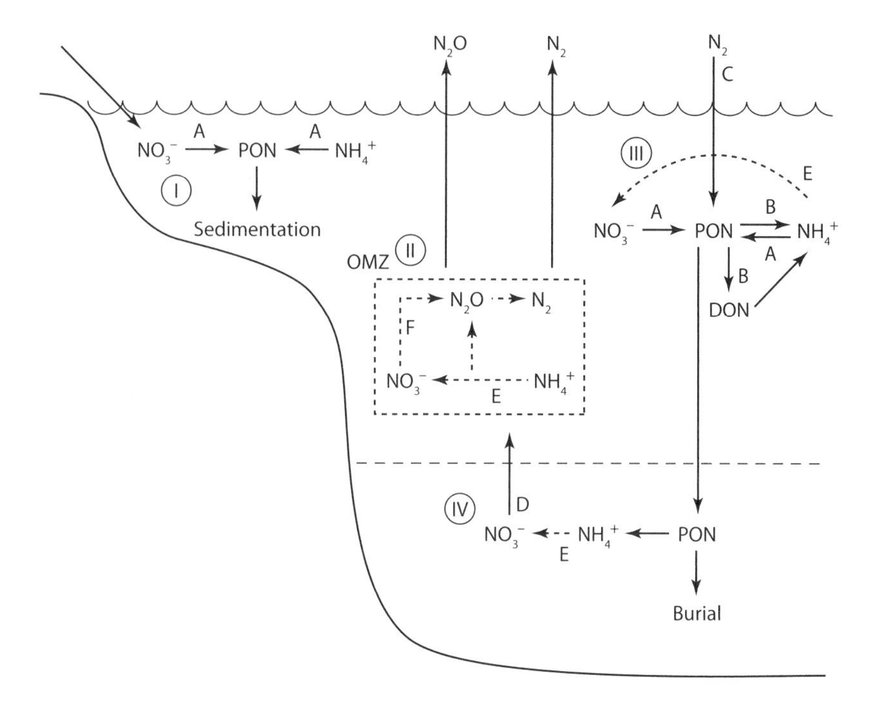

Figure 7.6. Nitrogen cycling processes in coastal shelf and upwelling environments (I), oxygen minimum zones (II), surface waters of the open ocean (III), and deep water (IV). PON, particulate organic nitrogen. Dashed lines indicate transformations involving multiple steps. Pathways: A, DIN assimilation; B, ammonium regeneration; C, nitrogen fixation; D, nitrate diffusion/advection from deep water; E, nitrification; F, denitrification. (Adapted from Zehr and Ward, 2002.)

and its response(s) to changing environmental conditions with high resolution and throughput. Using genes for nitrate reductase (NR) and ribulose-1,5-bisphosphate carboxylase/oxygenase (RuBisCO), the microarray approach was recently adopted to characterize nitrogen and carbon assimilation by phytoplankton in the English Channel (Ward, 2008), showing a shift from the dominance of diatoms during the spring bloom to haptophytes and flagellates later in the summer. The nitrate reductase gene provided complementary information, identifying two uncultivated diatom phylotypes as the most abundant (DNA) and active (NR gene expression) organisms.

7.8 Biosynthesis of Biomarkers: Phylogenetic Relationships and Evolution

Molecular genetic analyses of nucleic acids can contribute to our understanding of the evolution of the microbial community and its influence on Earth processes over geological timescales (Brocks and Pearson, 2005; Volkman, 2005). Several studies have linked the occurrence of particular biomarkers, including *n*-alkanes, hopanes, steranes, and steroids, to the appearance of particular organisms or metabolic pathways (fig. 7.7). The discovery of 2α-methylhopanes in the rock record, for example, provides evidence for the presence of cyanobacteria long before the earliest morphological fossils, suggesting the onset of photosynthesis by 2.77 Ga (Brocks et al., 1999). Additionally, biomarkers document the presence of microaerophilic bacteria (3α-methylhopanes), the appearance of the eukaryotic cell (C_{27} to C_{30} steranes), and other microbes (*n*-alkanes and isoprenoids) during the late Archean (Brocks et al., 1999, 2003). However, interpretation of the fossil record is often based on the known synthesis of biomarkers in contemporary ecosystems and pathways, for the biosynthesis of these

Table 7.3
Examples of genes isolated for nitrogen-cycling microbes

Process	*Gene*	*Protein*	*References*
N_2 fixation	nifHDK	Nitrogenase	Zehr and McReynolds (1989)
Nitrite assimilation	nir	Nitrite reductase	Wang and Post (2000)
Nitrate assimilation	narB, nasA	Assimilatory nitrate reductase	Allen et al. (2001)
Ammonium assimilation	glnA	Glutamine synthetase	Kramer et al. (1996)
Nitrate respiration	nirS	Nitrate reductase	Braker et al. (2000);
and denitrification	nirK	Nitrate reductase	Casciotti and Ward (2001); Scala and Kerkhof (1998)
	norB	Nitric oxide reductase	
	nosZ	Nitrous oxide reductase	
Organic N metabolism	ure	Urease	Collier et al. (1999)
Ammonium oxidation /nitrification	amo	Ammonia monooxygenase	Alzerreca et al. (1999); Rotthauwe et al. (1997)
nitrogen regulation (cyanobacteria)	ntcA	Nitrogen regulatory protein	Lindell et al. (1998)

Source: Zehr and Word (2002).

compounds may have changed over time. Dinosteranes are generally ascribed to dinoflagellate sources since these compounds are diagenetic products of the biomarker dinosterol. However, dinosteranes, found in Archean rocks, for example, likely derived from Eukarya, since dinoflagellates did not emerge until the early Cambrian (Moldowan and Talyzina, 2009). This finding suggests that dinosteranes may have a source independent of dinoflagellates (Volkman, 2005). Thus, approaches that combine biomarker and isotopic evidence with morphological fossils and molecular information enhance our confidence in interpreting the geologic record.

The source specificity of sterols and other triterpenoids and the evolution of the pathways for their biosynthesis has been a particular source of controversy (Pearson et al., 2004a; Volkman, 2005). The discovery of C_{27} to C_{30} steranes in the rock record has been used to infer the presence of eukaryotes (Brocks et al., 1999), but other studies have suggested that cyanobacteria may also synthesize sterols, adding uncertainty to this interpretation. The recent sequencing of the complete genome for several species of cyanobacteria (Kaneko et al., 1996; Kaneko and Tabata, 1997) now provides evidence that sterol synthesis by this group is unlikely. Although genomic studies of cyanobacteria reveal the presence of some genes resembling those required for sterol synthesis (e.g., Δ^{24}-sterol *C*-methyltransferase, sterol-*C*-methyltransferase), the complete set of genes needed for sterol synthesis was not present (Volkman, 2005). It now appears that previous reports of sterols and stanols in cyanobacteria may be due to contamination. On the other hand, genomes have been characterized for only a few species of cyanobacteria to date, so it remains possible that a few species may have this capability (Volkman, 2005).

Similar to cyanobacteria, the capability for sterol synthesis in bacteria has also been a subject of debate. A recent study investigating gene sequences for the proteobacterium *Methylococcus capsulatus* and the planctomycete *Gemmata obscuriglobus* provided the first evidence for sterol synthesis in bacteria (Pearson et al., 2004a). This study revealed a limited pathway for sterol synthesis in *Gemmata*, yielding lanosterol and its isomer, parkeol. Phylogenetic analysis suggests a common ancestor for sterol synthesis in bacteria and eukaryotes, and indicates that the pathway in *Gemmata obscuriglobus*

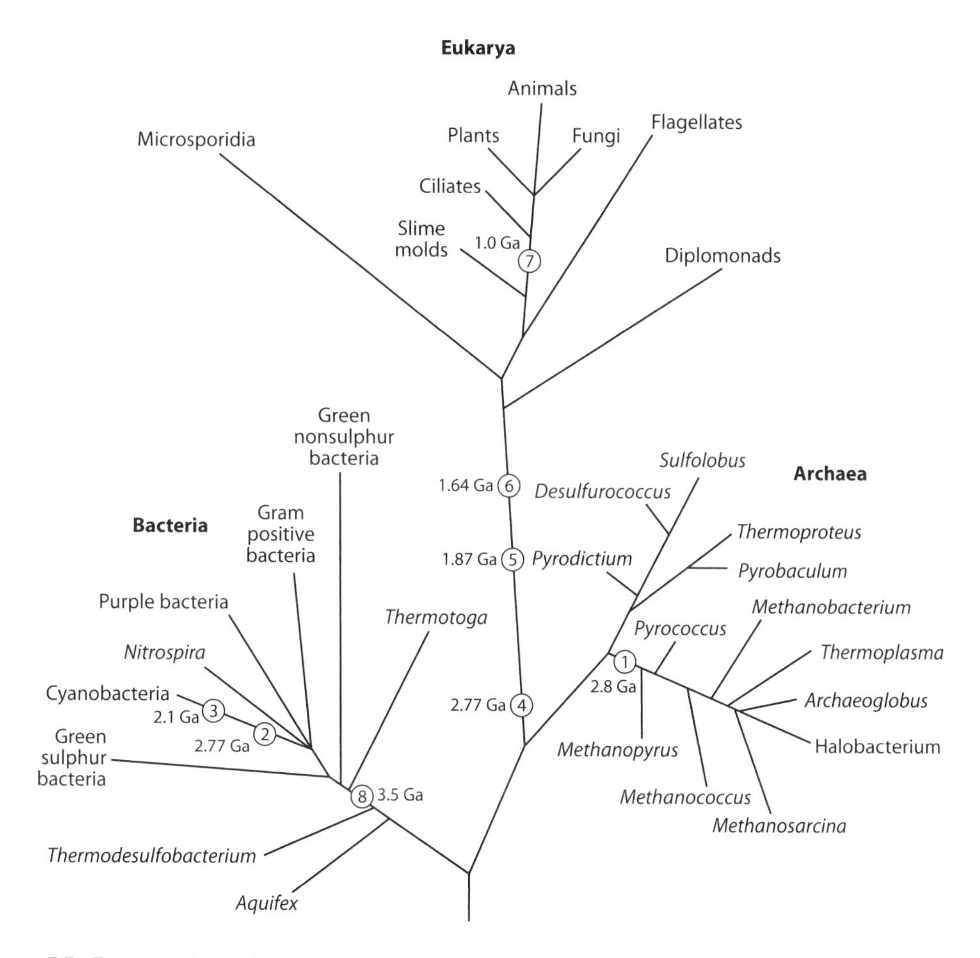

Figure 7.7. Example of tree of life with ages of branches based on biogeochemical and paleontological data. The numbers in circles correspond to (1) indirect evidence for methanogenic bacteria based on isotopic anomalies, (2) biomarker evidence (2α-methylhopanes) for cyanobacteria, (3) oldest fossil with morphology of cyanobacteria, (4) biomarker evidence (steranes) for Eukarya, (5) oldest fossils with eukaryotic morphology, (6) previous oldest sterane biomarkers, (7) oldest eukaryotic fossils linked to extant phylum (Rhodophyta), and (8) sulfur isotope evidence for sulfate-reducing bacteria. (Adapted from Brocks et al., 2003.)

may be related to an ancestral pathway for sterol biosynthesis in eukaryotes. To date, however, only eukaryotes have been documented to produce sterols with side-chain modifications at C-24, suggesting that the presence of fossil biomarkers (24-methylated and ethylated steranes) indicates the presence of eukaryotes.

Traditional approaches in microbial ecology have involved culturing microorganisms from environmental samples, but only a small portion of the community is culturable. The advent of molecular genetic approaches has revolutionized the field, allowing researchers to scan genomic databases to find organisms with genes for particular biomarkers or metabolic pathways. This approach was recently used to identify organisms capable of anaerobic biosynthesis of hopanoid biomarkers and revealed three species of bacteria with this capability (Fischer and Pearson, 2007). Genomic studies have also been used to investigate the distribution of hopanoid cyclases in the environment. To date, most of the organisms with this capability remain unknown, despite the widespread abundance of hopanoid compounds in the sediment record. Genes derived from known cyanobacterial sequences were not detected in the contemporary environments analyzed as part of this study and only a small portion of bacteria (~10%) appeared to have the genes for this capability (Pearson et al., 2007).

Molecular studies have also provided insights about the evolution of isoprenoid biosynthesis (Lange et al., 2000) and the function of isoprenoids (Bosak et al., 2008). Lange et al. (2000) investigated the distribution of the genes for isopentyl diphosphate (IPP), an intermediate involved in the biosynthesis of isoprenoids. IPP biosynthesis occurs by two pathways—the mevalonate pathway and the deoxyxylulose-5-phosphate (DXP) pathway. The phylogenetic distribution of genes for IPP suggests that the mevalonate pathway is the more ancestral pathway and is associated with archaebacteria, while the DXP pathways is associated with eubacteria. Lange et al. (2000) also provided evidence for plants acquiring the DXP pathway through lateral transfer from eubacteria. Molecular studies can also be used to investigate the function that specific biomarkers play and whether they may have evolved in response to environmental or metabolic conditions. A recent study, for example, identified the role of sporulenes, a class of tetracyclic isoprenoids found in spores of the bacterium *Bacillus subtilis* (Bosak et al., 2008). The authors conducted experiments using the gene sodF, which encodes for an enzyme that disproportionates toxic superoxide into oxygen and hydrogen peroxide. These experiments found that sporulanes increased the resistance of spores to a reactive oxygen species (H_2O_2), suggesting that polycyclic isoprenoids may provide a protective role. This finding broadens the traditional view that isoprenoids function by reinforcing or (rigidifying) cell membranes.

The application of molecular (DNA) information to the field of organic geochemistry is still at its frontier. As new information becomes available about the genes for the biosynthesis of particular biomarkers, molecular studies promise greatly to increase our ability to predict whether organisms have the capability to synthesize particular biomarkers, their evolutionary significance, and the function these compounds play in present ecosystems as well as the function they may have played in the geologic past.

7.9 Ancient DNA Preserved in Sediments

Remnants of DNA preserved in the sediment record can be used in combination with traditional biomarkers to provide information about the aquatic microbial community. Coolen and colleagues have used this approach in a variety of settings, including a meromictic salt lake (Mahoney Lake, British Columbia, Canada) (Coolen and Overmann, 1998), Ace Lake in Antarctica (Coolen et al., 2004a,b), and the Black Sea (Coolen et al., 2006). Using gene sequences of 16S rRNA, Coolen and Overmann (1998) documented four species of purple sulfur bacteria (Chromatiaceae) in Holocene sediments collected from Mahoney Lake. DNA from one species of Chromatiaceae, which is abundant in the chemocline of the present lake (*Amoebobacter purpureus ML1*), was identified in 7 of the 10 sediment horizons analyzed, with the oldest sediments deposited 9100 years BP. Interestingly, concentrations of *A. purpureus* DNA did not correlate with okenone or bacteriophaeophytin a (biomarkers typically used to identify Chromatiaceae) (fig. 7.8), although there was a correlation between total DNA and okenone ($r^2 = .968; p < .001$). The ratio of DNA to okenone was also considerably lower in the sediment layers than in intact cells. Together, these data suggest that decomposition of DNA was higher than for hydrocarbon biomarkers, even under the anoxic conditions present in Mahoney Lake.

Coolen and colleagues also combined DNA and biomarkers in their reconstruction of the microbial community associated with Ace Lake, Antarctica. Ace Lake was a freshwater lake that became saline during the Holocene when a connection to the sea was established (marine inlet period). It was subsequently isolated again due to isostatic rebound. Following its isolation, the lake became stratified when meltwater was introduced, eventually leading to the development of anoxia in its bottom waters (Coolen et al., 2004a,b). Data from fossil 18S rDNA and lipid biomarkers showed changes in phytoplankton community structure reflecting changes in lake chemistry and salinity. While

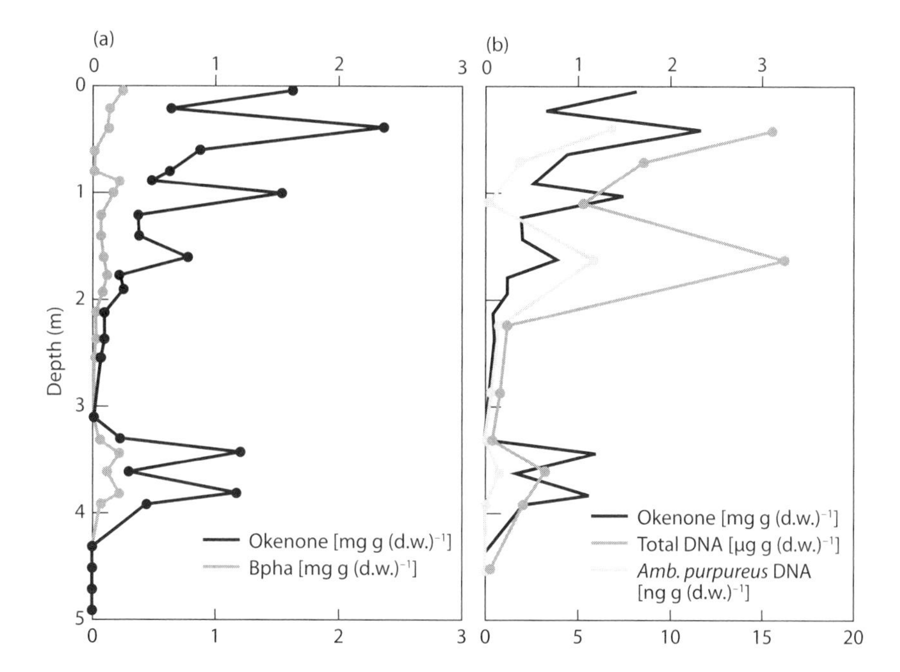

Figure 7.8. (a) Profiles of okenone and bacteriophaeophytin *a* (Bph *a*), biomarkers for *Chromatiaceae*, in Holocene sediments collected from Mahoney Lake, Canada. (b) Total DNA and DNA from *Amoebobacter purpureus ML1* are compared to okenone concentration profiles. Total DNA was correlated to okenone, while DNA from *A. purpureus ML1* was not. The upper axis corresponds to okenone and the lower axis to total DNA and *A. purpureus* DNA. (Adapted from Coolen and Overmann, 1998.)

diatoms and haptophytes are not abundant in the present lake system, DNA data showed that these organisms were present in horizons deposited when the lake was a marine inlet (Coolen et al., 2004b). The changing environmental conditions in Ace Lake also contributed to changes in the heterotrophic community. Lipids specific to methanogenic Archaea (^{13}C depleted $\Delta^{8(14)}$-sterols) and 16S rRNA genes both indicated an active methane cycle in the lake over the last 3000 years and suggested that the methanogens were introduced to the lake when it became a marine inlet (9400 yr BP). Methanogens were also dominant in the deepest layers of the cores, representing the time period when the system was a freshwater lake. These studies document the value of combining lipid biomarkers with genetic studies for paleoenvironmental reconstructions.

A similar approach was used to investigate the species of haptophytes responsible for unsaturated C_{37} alkenones preserved in sediments of the Black Sea (Coolen et al., 2006). Previous studies have used alkenones to reconstruct sea surface temperatures (SST) under the assumption that the alkenones derived from the well-characterized coccolithophorid, *Emiliani huxleyi*. However, other haptophytes also synthesize alkenones and the relationships between levels of unsaturation and temperature vary for different species. To validate the interpretation of the SST data based on alkenone biomarkers, Coolen et al. (2006) used 18S rDNA to examine what species of haptophytes were present in Holocene sediments collected from the Black Sea and whether *E. huxleyi* was the dominant species. The dominant alkenones in unit I sediments were C_{37} and C_{38} methyl and ethyl ketones with 2 to 3 double bonds, with concentrations decreasing to below levels of detection in unit II sediments (fig. 7.9). The abundance of 18S rDNA derived from *E. huxleyi* followed some of the major patterns in the alkenone profiles, but the correlation was weak ($r^2 = .13$). DNA from *E. huxleyi* comprised a small fraction of the total DNA from Eukarya ($< 0.8\%$), but sequences from other ancient haptophytes were not found, suggesting that the SST calibration based on *E. huxleyi* is valid over the past 3600 years. Interestingly,

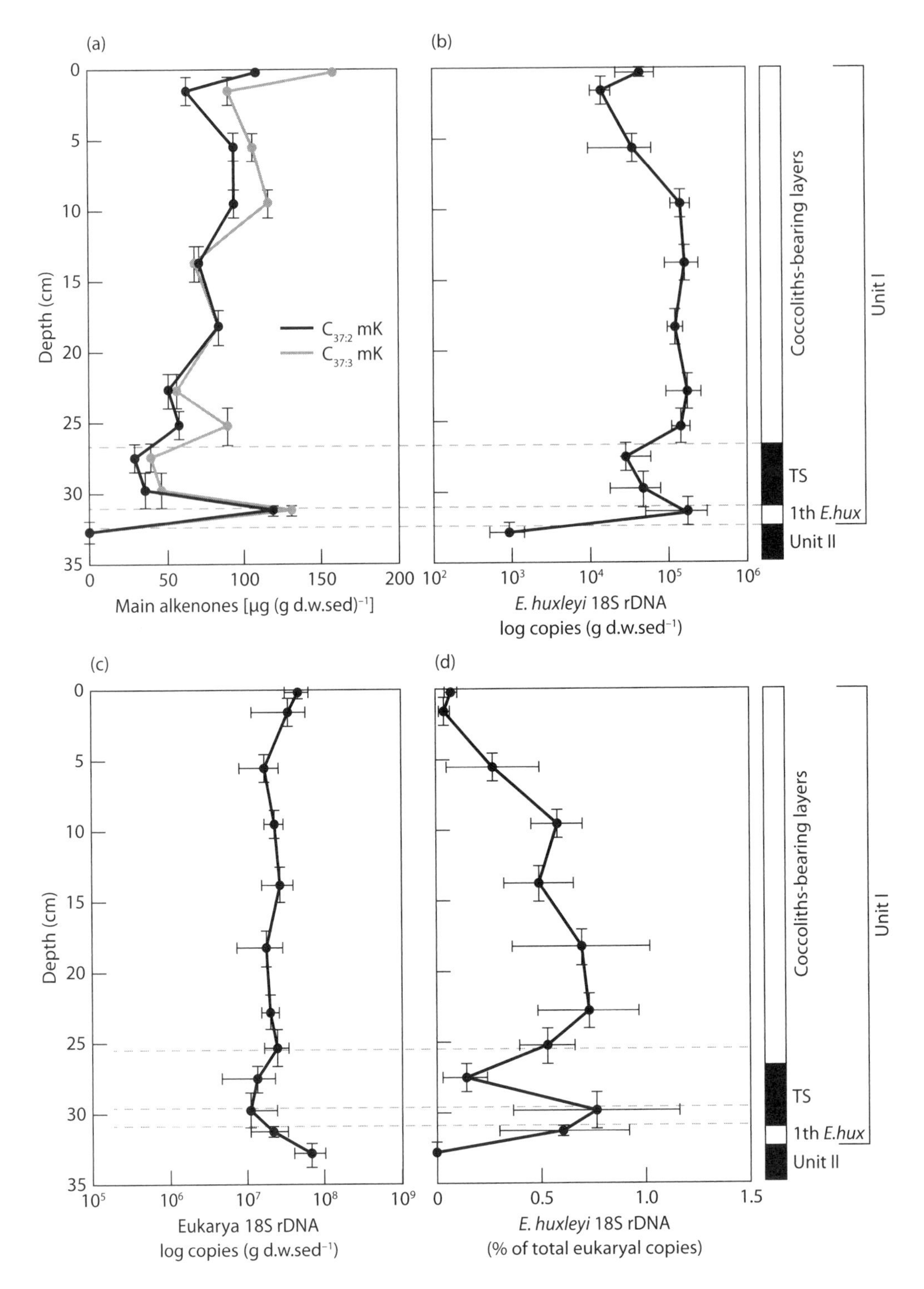

Figure 7.9. Sediment core profiles from the Black Sea showing concentrations of C_{37} and C_{38} alkenones (a), 18S rDNA from *E. huxleyi* expressed relative to g dry wt sediment (b), 18S rDNA from Eukarya (c), and 18S rDNA from *E. huxleyi* expressed relative to total DNA from Eukarya (d). (Adapted from Coolen et al., 2006.)

dinoflagellates (*Gymnodinium* spp.) were the most abundant source of fossil DNA, similar to what is observed during summer phytoplankton blooms in the modern Black Sea.

The isolation and interpretation of DNA from ancient sediments is still in its early stages and not without controversy. While recent work claimed to have isolated fossil DNA from Cretaceous shales with an age of approximately 112 Myr BP (Inagaki et al., 2005), other work has contended that genetic signals from such small fragments of fossil DNA (e.g., ca. 900 base pairs long) were not valid, based on a resampling of the shales within this region, using both molecular fossil and DNA information (Sinninghe Damsté and Coolen, 2006). Nevertheless, overall, the body of work by Coolen and colleagues does indicate the value of combining fossil DNA and biomarkers approaches in paleoecological and paleoenvironmental studies.

7.10 Summary

Future research in organic geochemistry will benefit from molecular explorations of the phylogenetic relationships between organisms to yield a better understanding of the metabolism of different microbes and the biosynthesis of different classes of biomarker compounds. Previous studies have examined the phylogenetic distribution of biomarkers by screening or surveying organisms in order to understand how widely biomarkers were distributed across different microbial groups. Increasingly, organic geochemical studies will utilize molecular phylogeny to understand the ancestry of biomarkers, providing new insights about the origins of biosynthetic pathways and evolutionary relationships between organisms and biosynthetic pathways. These approaches will provide answers to long-standing questions, such as (1) How does the diversity of biomarkers compare and contrast to the modern microbial community? (2) What is the evolutionary, ecological, and environmental basis for lipid biomarkers preserved in the rock record? and (3) How do environmental conditions and metabolic processes influence the diversity of lipid biomarker compounds? In this way, organic geochemists will be able to generate a "biomarker tree of life" (Brocks and Pearson, 2005).

8. Lipids: Fatty Acids

8.1 Background

Lipids are operationally defined as all substances produced by organisms that are effectively insoluble in water but extractable by solvents (e.g., chloroform, hexane, toluene, and acetone). This broad definition includes a wide range of compounds, such as *pigments, fats, waxes, steroids,* and *phospholipids*. Alternatively, lipids may be more narrowly defined as fats, waxes, steroids, and phospholipids. This second definition relates specifically to their biochemical function in terms of energy storage rather than their *hydrophobicity*, which allows their extraction into organic solvents.

Fatty acids are among the most versatile classes of lipid *biomarker* compounds and this class of biomarkers has been applied to a wide variety of environmental studies, including characterization of organic matter sources in aquatic systems, analysis of the nutritional value of organic matter, trophic studies, and microbial ecology. Characteristics of fatty acids that contribute to their usefulness as biomarkers include (1) their unique distributions in various organisms (archaebacteria, eubacteria, microalgae, higher plants), (2) distinct structural features indicative of their sources, and (3) the fact that individual compounds and fatty acid classes exhibit a spectrum in reactivity. This range in reactivity largely derives from variations in functional group composition and carbon chain length observed within this class of biomarker compounds (Canuel and Martens, 1996; Sun et al., 1997). Fatty acids exist in free forms (e.g., free fatty acids), bound forms (e.g., through ester linkages in lipid classes such as wax esters, *triacylglycerols*, and phospholipids) (fig. 8.1), and in combination with other biochemical classes, such as glycolipids (*macromolecules* formed by combination of lipids with *carbohydrates*) and *lipoproteins* (macromolecules formed by the combination of lipids with proteins). Free fatty acids occur in low abundances, whereas most fatty acids are present in esterified forms in neutral (e.g., triacylglycerols and wax esters) and polar lipids (e.g., phospholipids). These lipid classes vary in biochemical function (e.g., membrane lipids, energy storage) and lipid stores change in response to physiological factors such as growth and reproduction (Cavaletto and Gardner, 1998). Lipid compound classes and their fatty acid constituents can also change in response to environmental factors (e.g., light and nutrient availability).

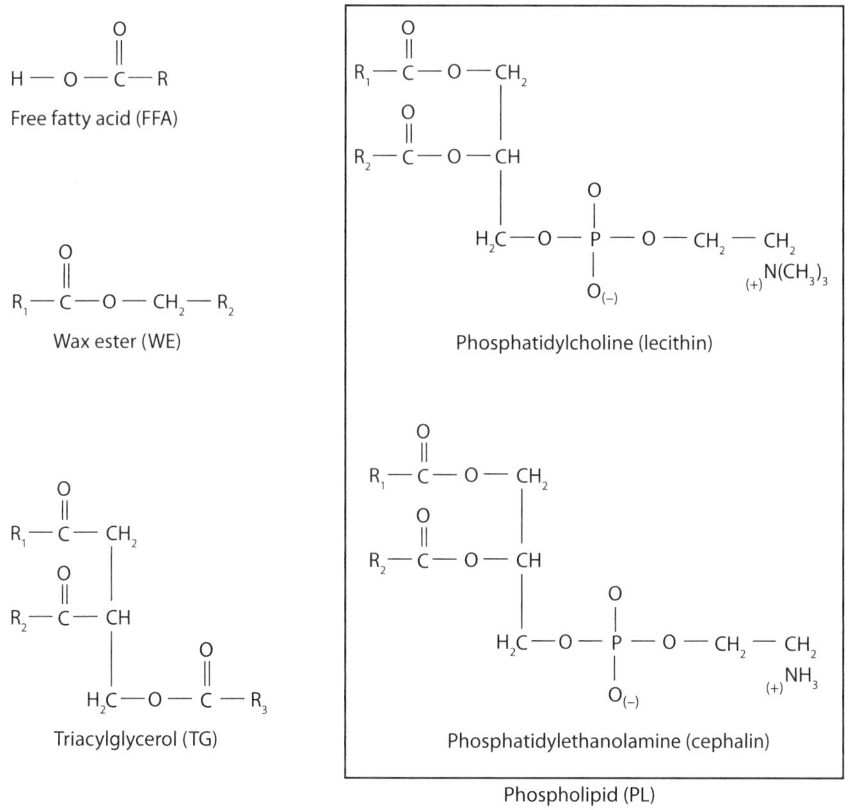

Free fatty acid (FFA)

Wax ester (WE)

Triacylglycerol (TG)

Phosphatidylcholine (lecithin)

Phosphatidylethanolamine (cephalin)

Phospholipid (PL)

Figure 8.1. Free fatty acids and lipid subclasses composed of fatty acids.

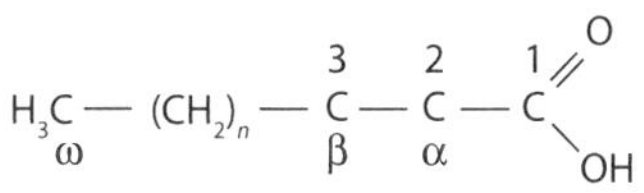

$$H_3C-(CH_2)_n-C-C-C(=O)OH$$

Figure 8.2. The "Δ" notation is used for numbering of fatty acids from the C1, or carboxyl (COOH), end of the carbon chain while the ω notation is used for numbering from the aliphatic end of the molecule.

Fatty acids are described using the nomenclature: $A : B\omega C$, where A = number of carbon atoms, B = number of double bonds, and C = position of double bonds. The omega (ω) notation is used when numbering the position of the double bonds relative to the aliphatic end of the fatty acid (fig. 8.2). The delta (Δ) notation is used when the numbering begins at the carboxylic end. When describing fatty acids, CX is used to denote the carbon atom at the X position using the numbering scheme presented in figure 8.2, while C_x is used to describe the carbon chain length. Fatty acids can also often have methyl groups forming branches along the carbon chain; *iso* fatty acids have a methyl group at the $n - 1$ position, and *anteiso* fatty acids have a methyl group at the $n - 2$ position using the ω notation.

Glycerides are *esters* of the alcohol glycerol. The glycerol molecule has three hydroxyl (OH) groups, allowing it to react with one, two, or three carboxylic acids to form monoglycerides, diglycerides, and triglycerides (also called triacylglycerols) (fig. 8.1). Triglycerides (triacylglycerols) typically function as energy reserves; C_{12} to C_{36} fatty acids comprise glycerides, with C_{16}, C_{18}, and C_{20} carboxylic acids being most common. In animals, C_{16} and C_{18} saturated (no double bonds) fatty acids dominate, while plants often contain C_{18} to C_{22} fatty acids with one or more double bonds. Microalgae are often composed of highly unsaturated fatty acids with 4 to 6 double bonds (termed polyunsaturated fatty acids or PUFA). Overall, the greater the number of double bonds, the lower the melting points. Hence, animal fats tend to be solids, while plant fats occur as liquids (oils).

Another class of fatty acid-containing lipids is the wax esters (or waxes). While triacylglycerols are the major energy reserve in most vascular plants and algae, some marine animals (e.g., copepods and other zooplankton) store energy in the form of wax esters. These compounds are composed of a long-chain fatty acid esterified to a long-chain alcohol (fig. 8.1). Generally, the fatty acid and alcohols are similar in chain lengths (Wakeham, 1982, 1985).

Phospholipids are composed of fatty acids and function mainly as membrane lipids (fig. 8.1). Phospholipids include a head that is hydrophilic and nonpolar tails that are hydrophobic. In water, phospholipids form a bilayer, where the hydrophobic tails line up against each other, forming a membrane with hydrophilic heads extending out of the water. This creates *liposomes*, small lipid vesicles that can be used to transport materials into living organisms across the cell membrane. Phospholipids may be built on two backbones: the glycerol backbone (glycerophospholipid or phosphoglyceride) or the *sphingosine* backbone (*sphingomyelin*). In phosphoglycerides, the carboxyl group of each fatty acid is esterified to the hydroxyl groups on C_1 and C_2 of the glycerol molecules and the phosphate group is attached to C_3 by an ester link (fig. 8.1). Sphingosine, an amino alcohol formed from palmitate and serine, forms the backbone of sphingomyelin.

Another class of cell membrane components is the glycolipids, which include compounds such as monogalactosyldiacylglycerol (MGDG), digalactosyldiacylglycerol (DGDG), and sulfoquinovosyldiacylglycerol (SQDG), and other more complex compounds (Volkman, 2006). Generally, glycolipids are less polar than most phospholipids and are typically isolated by chromatography on diethylaminoethyl (DEAE)–cellulose and silica gel columns. There are few reports of glycolipids in environmental samples due to their rapid hydrolysis in the environment. In one study, intact glycolipids were found in the sediments of Ace Lake, in Antarctica (Sinninghe Damsté et al., 2001). Two novel compounds were identified in this study, docosanyl 3-*O*-methyl-*a*-rhamnopyranoside and docosanyl 3-*O*-methylxylopyranoside. Additionally, a major glycolipid was identified following a bloom of halophilic archaea in the Dean Sea (Oren and Gurevich, 1993). Highly unsaturated C_{18} fatty acids ($18{:}4\omega 3$ and $18{:}5\omega 3$) have been recovered from the glycolipid fraction associated with the chloroplast of dinoflagellates (Leblond and Chapman, 2000). In 12 of 16 strains of dinoflagellates, over 90% of $18 : 5\omega 3$ were associated with the glycolipid fraction.

8.2 Fatty Acid Biosynthesis

Fatty acids generally have even-numbered carbon atoms because they are formed from acetyl (C_2) units. The first step in the synthesis of fatty acids is the formation of acetyl coenzyme A (acetyl-CoA) from glucose (fig. 8.3). CO_2 is added to one molecule of acetyl-CoA to form malonyl-CoA, while the acetyl group on another molecule of acetyl-CoA is transferred to the enzyme, fatty acid synthetase. The malonyl unit (C_3) combines with the enzyme-bound acetyl unit, producing a butyryl (C_4) group, resulting in the loss of CO_2. Palmitic acid (16:0) is constructed after six further steps involving subsequent addition of malonyl, decarboxylation, dehydration, and reduction. Longer chain lengths are built from palmitate using chain-elongation enzymes. One such chain-elongation system, found in the mitochondria of animals and plants, uses acetyl-CoA rather than malonyl-CoA as the source of two carbon units. Another is found in the microsomal fraction of mammalian cells and uses malonyl-CoA as the source of the two carbon units and produces CoA-bound intermediates (Erwin, 1973). Desaturation of these fatty acids (e.g., by dehydrogenation) results in the synthesis of unsaturated fatty acids.

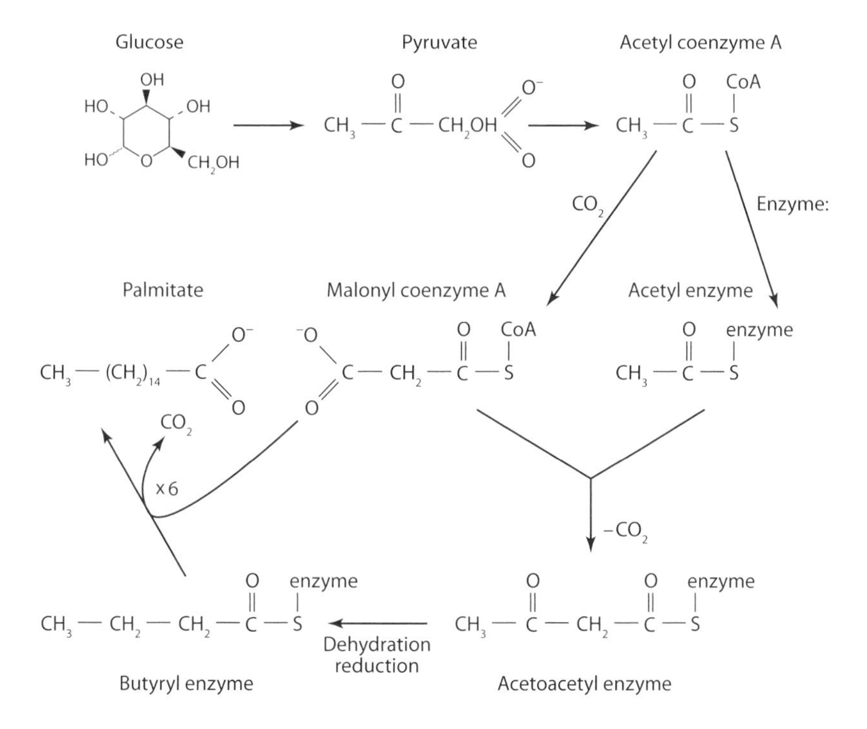

Figure 8.3. Fatty acid biosynthesis in plants and animals. (Adapted from Killops and Killops, 2005.)

Bacteria are more versatile than other organisms in the substrates they can use for fatty acid synthesis. For example, some bacteria can utilize the amino acid L-valine or L-leucine to synthesize iso-branched fatty acids or L-isoleucine to synthesize anteiso-branched fatty acids (fig. 8.4). Iso-branched fatty acids have a methyl group at the $n-1$ position, while anteiso-branched fatty acids have a methyl group at the $n-2$ position, using the ω notation. Additionally, odd-numbered fatty acids can form when propionate is used as a substrate by bacteria.

Structural features (e.g., number and position of double bonds) and functional group composition (hydroxyl groups, cycloalkyl groups, and methyl branching) are important in determining the ability to use fatty acids as proxies for different organic matter sources (table 8.1). The 2- and 3-hydroxy fatty acids (α- and β-hydroxy fatty acids) occur in a wide variety of organisms and have been identified in recent sediments in both lacustrine and marine environments (Cranwell, 1982; Volkman, 2006). These compounds are products of the α- and β-oxidation of monocarboxylic acids. 2-Hydroxy fatty acids in the range C_{12}–C_{28} occur widely in both the free and bound phases and have both higher plant and microbial sources. In yeasts, 2-hydroxy acids are intermediates in fatty acid synthesis. The 3-hydroxy carboxylic acids are generally enriched over 2-hydroxy carboxylic acids in the bound phase. Normal and branched 3-hydroxy acids in the C_{10}–C_{18} range are structural components of cell membranes in gram-negative bacteria. The dominant 3-hydroxy fatty acid is generally C_{14} but iso- and anteiso-C_{15} and C_{17} 3-hydroxy acids have been shown to dominate in the bacterium *Desulfovibrio desulfericans* (Volkman, 2006).

ω-Hydroxy acids are found in the cuticular waxes of plants and occur as bound constituents in the biopolymers of cutin (C_{16} and C_{18}) and suberin (C_{16}–C_{24}) (Kolattukudy, 1980; Cranwell, 1982; Volkman, 2006) (more details provided in chapter 13). An additional class of hydroxy acids is the (ω-1)-hydroxy acids, but their origin is still unknown. Several microorganisms are able to use a preterminal hydroxylation mechanism for modification of alkane and fatty acid substrates. Long-chain

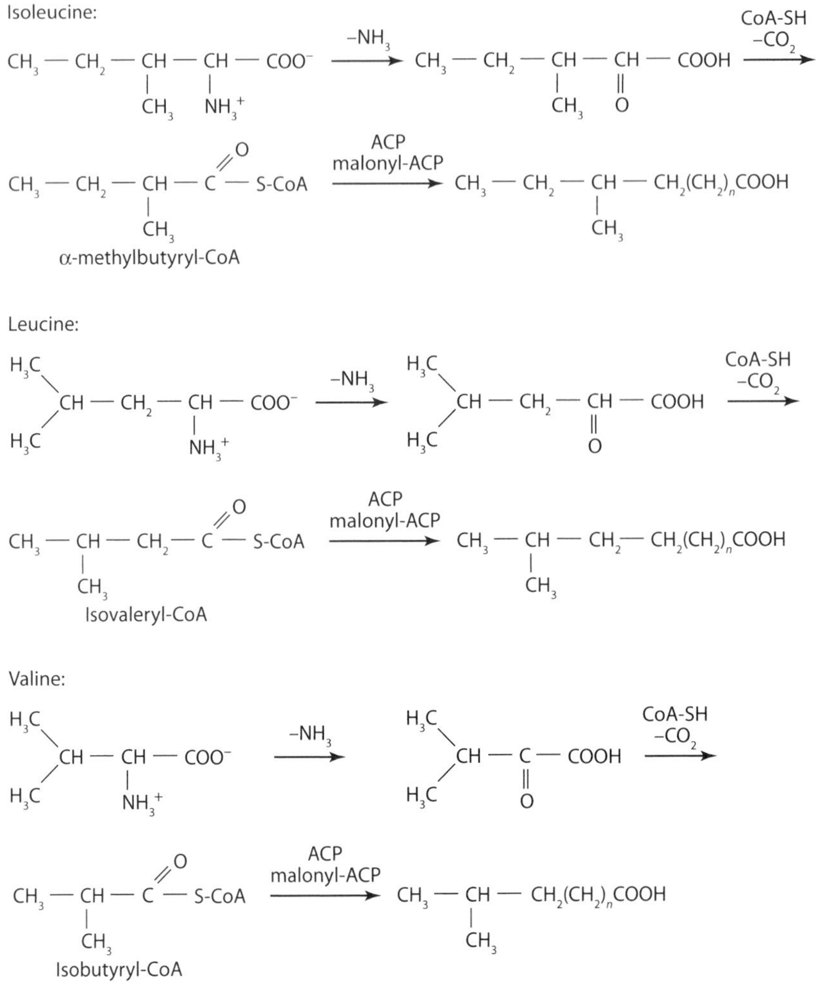

Isoleucine:

$$CH_3-CH_2-CH(CH_3)-CH(NH_3^+)-COO^- \xrightarrow{-NH_3} CH_3-CH_2-CH(CH_3)-C(=O)-COOH \xrightarrow[-CO_2]{CoA\text{-}SH}$$

$$\underset{\alpha\text{-methylbutyryl-CoA}}{CH_3-CH_2-CH(CH_3)-C(=O)-S\text{-}CoA} \xrightarrow[\text{malonyl-ACP}]{ACP} CH_3-CH_2-CH(CH_3)-CH_2(CH_2)_nCOOH$$

Leucine:

$$(H_3C)_2CH-CH_2-CH(NH_3^+)-COO^- \xrightarrow{-NH_3} (H_3C)_2CH-CH_2-C(=O)-COOH \xrightarrow[-CO_2]{CoA\text{-}SH}$$

$$\underset{\text{Isovaleryl-CoA}}{CH_3-CH(CH_3)-CH_2-C(=O)-S\text{-}CoA} \xrightarrow[\text{malonyl-ACP}]{ACP} CH_3-CH(CH_3)-CH_2-CH_2(CH_2)_nCOOH$$

Valine:

$$(H_3C)_2CH-CH(NH_3^+)-COO^- \xrightarrow{-NH_3} (H_3C)_2CH-C(=O)-COOH \xrightarrow[-CO_2]{CoA\text{-}SH}$$

$$\underset{\text{Isobutyryl-CoA}}{CH_3-CH(CH_3)-C(=O)-S\text{-}CoA} \xrightarrow[\text{malonyl-ACP}]{ACP} CH_3-CH(CH_3)-CH_2(CH_2)_nCOOH$$

Figure 8.4. Biosynthesis of iso- and anteiso- branched fatty acids in bacteria. Branched fatty acids result from the use of amino acids (e.g., isoleucine, leucine, and valine), in place of acetate, as the starting substrate. (Adapted from Gillan and Johns, 1986.)

(ω-1)-hydroxy fatty acids have been found in methane-utilizing bacteria and cyanobacteria (De Leeuw et al., 1992; Skerratt et al., 1992).

8.3 Analysis of Fatty Acids

Fatty acids are typically extracted from a sample matrix (e.g., particulate matter, sediment, ultrafiltered dissolved organic matter, or animal and plant tissues) using a combination of organic solvents, generally with a portion of the solvent mixture including a polar solvent such as acetone or methanol (e.g., chloroform and methanol, 2:1, v:v) (Bligh and Dyer, 1959; Smedes and Askland, 1999). A wide variety of extraction methods can be used, including soxhlet extraction, sonication, microwave extraction, or accelerated solvent extraction (ASE) (e.g., MacNaughton et al., 1997; Poerschmann and Carlson, 2006). Often, total fatty acids are analyzed following *saponification* (base hydrolysis) of the

Table 8.1
Structural features of fatty acids and their use as source indicators

Feature	*Source*
Short chain (14:0, 16:0, 18:0)	Nonspecific (marine and terrigenous; bacteria, plants, and animals)
Long chain (24:0, 26:0, 28:0)	Higher (vascular) plants
Monounsaturated short chain (16:1 and 18:1)	Generally, nonspecific However, isomers may be source-specific
18:1ω7 (*cis*-vaccenic) is abundant in, but not exclusive to, bacteria 18:1ω9 (oleic acid) animals, higher plants and algae 16:1ω7: bacteria 16:1ω9: algal	
Polyunsaturated	
20:4, 20:5, 22:4, 22:5, 22:6	Marine and aquatic phytoplankton, zooplankton and fish
18:2 and 18:3	Higher plants and algae
Methylbranched chains (iso and anteiso)	Bacteria
Internally branched and cycloalkyl	
10-methyl isomers of 16:0 and 18:0	Fungi and bacteria
cyclopropyl 17:0 and 19:0	Eubacteriia
Hydroxy	
α-hydroxy (2-hydroxy) β-hydroxy (3-hydroxy)	Intermediates in oxidation of monocarboxylic acids Mostly bacteria, some cyanobacteria (predominantly C_{14}–C_{16}) and some microalgae
C_{15} and C_{17} β-hydroxy	Bacteria
C_{26} to C_{30} β-hydroxy	Eustigmatophytes
Diacids	
α, ω diacids	Higher plants Bacterial oxidation of other compounds

lipid extract. However, additional information can be gained if fatty acids associated with specific lipid subclasses (e.g., wax esters, triacylglycerols, and phospholipids) are separated chromatographically and their constituent fatty acids analyzed by individual compound class (fig. 8.5). In this case, the lipid extract is first separated into compound classes using adsorptive chromatography and the lipid subclasses of interest are then saponified (Muri and Wakeham, 2006; Wakeham et al., 2006). During saponification, ester bonds are cleaved (hydrolyzed) under basic conditions to form an alcohol and the salt of a carboxylic acid, as illustrated in equation 8.1:

$$\begin{gathered}\mathrm{CH_2{-}OOC{-}R{-}CH{-}OOC{-}R{-}CH_2{-}OOC{-}R\ (fat) + 3NaOH\ (or\ KOH)} \\ \mathrm{+Heat} \rightarrow \\ \mathrm{CH_2{-}OH{-}CH{-}OH{-}CH_2{-}OH\ (glycerol) + 3R{-}CO_2{-}Na\ (salt\ of\ carboxylic\ acid)}\end{gathered} \tag{8.1}$$

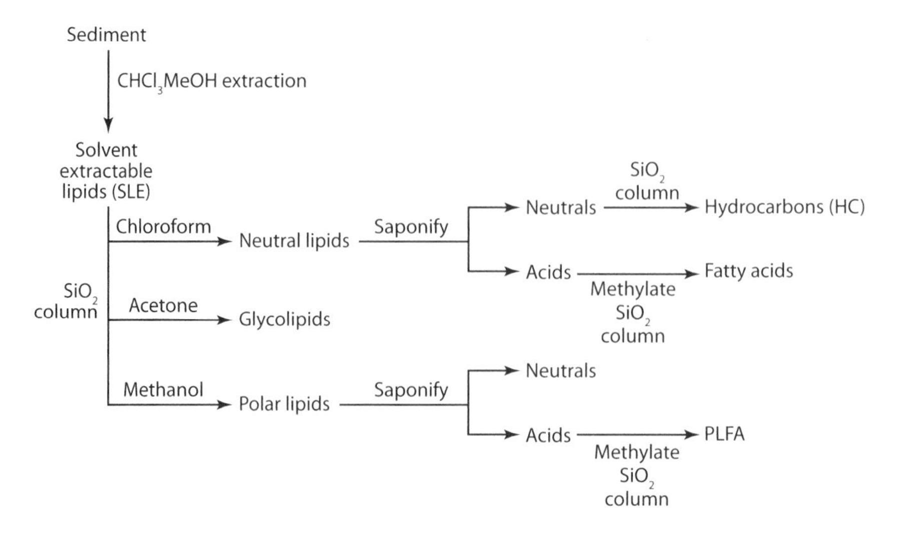

Figure 8.5. Schematic showing separation of different lipid classes by polarity and subsequent saponification and methylation of neutral lipids and carboxylic acids. (Adapted from Wakeham et al., 2006.)

Following saponification, fatty acids are generally converted to methyl *esters* for subsequent analysis by gas chromatography. Methylation can be accomplished using a variety of methods, including BF_3 in methanol (Grob, 1977; Metcalfe et al., 1966), trimethylchlorosilane (TMCS)/methanol, diazomethane, and mild alkaline methanolysis (Whyte, 1988). Fatty acid methyl esters (FAME) are typically analyzed by gas chromatography–flame ionization detection using a polar column. Double-bond positions are generally determined using dimethyl disulfide adducts (DMDS) and gas chromatography–mass spectrometry (Nichols et al., 1986). Double bonds can also be identified by electron impact gas chromatography–mass spectrometry (EI GC-MS) of picolinyl esters or 4,4-dimethyloxazoline (DMOX) derivatives (Christie, 1998).

In addition to the information provided by fatty acid biomarker compounds alone, complementary information may be provided by determining the stable isotopic composition of individual fatty acids using compound-specific isotope analysis (CSIA) (more details are provided in chapter 3) (Canuel et al., 1997; Boschker et al., 1998; Boschker and Middelburg, 2002; Bouillon and Boschker, 2005). Stable isotopic signatures of fatty acids provide additional insights about the sources of organic matter and the cycling of this compound class. Both natural abundance of stable isotopes and tracer studies have provided a number of new insights to the fields of microbial ecology and biogeochemistry in recent years. Additionally, recent studies have provided information about the radiocarbon ($\Delta^{14}C$) composition of fatty acids. Radiocarbon signatures can also provide additional information about the sources and age of biomarkers (Pearson et al., 2001; Wakeham et al., 2006; Mollenhauer and Eglinton, 2007).

8.4 Applications

8.4.1 Fatty Acids as Biomarkers

Fatty acids are among the most versatile classes of lipid biomarker compounds and have been used widely to trace sources of organic matter in aquatic environments. This is because individual compounds, and groups of compounds, can be attributed to microalgal, bacterial, fungal, marine faunal, and higher plant sources, providing a powerful tool for elucidating sources of organic matter sources (Volkman et al., 1998). More than 500 fatty acids have been identified in plants and microorganisms, but the most abundant compounds are considerably fewer in number. While fatty acids can provide qualitative information about the dominant sources of organic matter, it is difficult to extrapolate quantitative contributions of organic matter from specific sources because lipid content varies with tissue type, changes in life cycle, and environmental conditions. Higher plants and microbes, for example, may change the proportion of unsaturated fatty acids in response to changes in temperature in order to maintain membrane fluidity (Yano et al., 1997). Plants and animals also alter their lipid content during their life cycle in response to morphological and physiological changes. Further work is clearly needed to better understand how such changes may affect the chemistry of lipids (and other biomarkers) across temporal and spatial gradients in highly variable environments.

Two of the most common fatty acids in most organisms are palmitic acid (16:0) and stearic acid (18:0). The dominant fatty acids in higher plants include even-numbered saturated fatty acids (C_{12}–C_{18}), 18:1ω9, and two polyunsaturated fatty acids, 18:2ω2 and 18:2ω3. However, these compounds are present in other organisms and do not provide a source-specific tracer for identifying contributions of organic matter derived solely from higher plants. The long-chain saturated fatty acids (C_{24}–C_{36}) are more specific for higher plants but can be present in microalgae at trace levels. Total fatty acids (the fatty acids yielded following the saponification of a total lipid extract) have been used as biomarkers in a number of aquatic systems. A review of these applications follows.

8.4.2 Fatty Acids as Biomarkers in Lakes

Biomarker studies in lakes have largely focused on three applications of fatty acid biomarkers: (1) understanding the transformation and decomposition of aquatic organic matter as it sinks through the water column through studies of suspended and/or sinking particulate matter; (2) utilizing sediment cores to understand organic matter diagenesis; and (3) *paleolimnological* studies. Much of the classic work on the organic geochemistry of lacustrine systems was done by Cranwell and colleagues in the 1970s and 1980s (Cranwell, 1982). Subsequently, Meyers and colleagues conducted numerous studies in the Laurentian Great Lakes (USA) utilizing fatty acid biomarkers (Meyers et al., 1984; Meyers and Eadie, 1993; Meyers, 2003). These studies showed that the abundance and composition of fatty acids change with increasing water column depth. Fatty acid composition of surface water samples reflected organic matter contributions from both aeolian and aquatic sources, while deeper water samples primarily reflected aquatic contributions. These studies also revealed the influence of microbial degradation with increasing water depth. Similar observations were made in Lake Haruna, Japan, where autochthonous fatty acids (C_{12}–C_{19} saturated fatty acids and polyunsaturated C_{18} fatty acids) decreased relative to total organic carbon with increasing water depth (Kawamura et al., 1987). In contrast, saturated C_{20}–C_{30} fatty acids of terrigenous vascular plant origin, remained constant with depth. These results reflect selective degradation of algal fatty acids during microbial processing of settling organic matter.

The preservation of organic matter produced within lakes, as well as that delivered from the surrounding watershed, provides a record of changes in lakes due to both human impacts and natural processes. While ocean sediments provide information about global processes such as changes in climate and responses to orbital cycles, lake sediments provide the best record of continental climate histories and information about how local and regional climates have changed (Meyers, 2003). Additionally, lake sediment records often have higher temporal resolution, providing details about climate change that may not be available from oceanic sediments because of lower rates of sediment accumulation (as discussed in chapter 2). Studies utilizing fatty acid biomarkers have focused on using the sediment record to understand past environmental conditions, climate change, and eutrophication in lake ecosystems (Meyers and Ishiwatari, 1993; Meyers, 2003). Fatty acid chain length has been used to record changes in the aquatic versus terrigenous sources. For example, eutrophic lakes tend to be dominated by aquatic sources (e.g., Lake Haruna and Lake Suwa in Japan) and are enriched in short-chain fatty acids. In contrast, oligotrophic lakes (e.g., Lake Motosu) generally have higher proportions of long- to short-chain fatty acids (Meyers and Ishiwatari, 1993).

One tool for assessing changes in sources is the ratio of terrigenous to aquatic fatty acids (TAR_{FA}) (Meyers, 1997). This ratio is calculated as

$$TAR_{FA} = \frac{C_{24} + C_{26} + C_{28}}{C_{12} + C_{14} + C_{16}} \quad (8.2)$$

where, C_{12}–C_{28} are saturated fatty acids with carbon chain lengths of 12 to 28. TAR_{FA} values of 1 suggest equal contributions of terrigenous (long-chain) and aquatic (short-chain) sources of organic matter. However, care must be used in employing this ratio because the susceptibility of short- and long-chain fatty acids to decomposition processes is not the same (Haddad et al., 1992; Canuel and Martens, 1996). Another index used to examine source changes is the carbon preference index (CPI), which provides a numerical representation of how well biological chain-length information is preserved in sediments. In fresh lipid material, even-numbered chain lengths dominate the fatty acid distributions. To gain additional insights from CPI, fatty acids can be separated into lower-molecular-weight (C_{12}–C_{18}) and higher-molecular-weight (C_{22}–C_{32}) ranges (CPI_L and CPI_H, respectively). Plants generally have high values for CPI_L (between 12 and 100), while CPI_H values tend to be lower (between 0.9 and 8) due to the presence of long-chain, odd-numbered acids in some land plants (Meyers and Ishiwatari, 1993). Bacterial processing of organic matter generally reduces CPI values because bacteria generally have low CPI values. An additional tool used to interpret profiles of fatty acids in sediments is the composition diversity index (CDI) (Matsuda and Koyama, 1977). Higher values of CDI indicate a greater diversity of sources, thereby providing information about changes in organic matter sources in lakes due to eutrophication and microbial contributions.

Generally, because of their susceptibility to diagenesis, the application of fatty acid biomarkers to paleolimnological studies must be performed cautiously, and in concert with other proxies. For example, past work showed that fatty acid and alkanol biomarkers in a 0.5-m-long core from Coburn Mountain Pond, Maine (USA) were sensitive to diagenetic alteration, while *n*-alkane biomarkers were not (Ho and Meyers, 1994). Recent studies investigating the preservation of fatty acid biomarkers within lakes have further demonstrated biases due to selective degradation, indicating the need to be cautious in the application of these proxies for source apportionment in some lake sediments (Muri et al., 2004; Muri and Wakeham, 2006). Despite these limitations, fatty acid biomarkers have still proven useful in elucidating changes in the contribution of microbial lipids and short-term changes in sources of organic matter in numerous lacustrine systems (see reviews by Meyers et al., 1980; Meyers and Eadie, 1993; Meyers and Ishiwatari, 1993; Meyers, 1997, 2003).

A very recent area of study involves the measurement of stable carbon ($\delta^{13}C$) and hydrogen (δD) isotopic signatures of fatty acid biomarkers in lakes (Tenzer et al., 1999; Chikaraishi and Naraoka, 2005). Carbon isotopic signatures ($\delta^{13}C$) of biomarkers provide additional insights about carbon sources, particularly when a compound derives from two sources that are isotopically distinct (Canuel et al., 1997). Additional information is provided through the hydrogen isotopic signatures that relates to hydrologic processes, such as water masses and humidity on the Earth's surface. Compound-specific, dual-isotope (carbon and hydrogen) analysis of biomarkers can supply more detailed geochemical information, and technical advances in this area show promise for contributing to our further understanding of the sources and dynamics of organic matter in lacustrine environments.

8.4.3 Fatty Acids as Biomarkers along the River–Estuarine Continuum

Estuaries are characterized by numerous sources of organic matter, including materials of both autochthonous and allochthonous origin (Canuel et al., 1995; Canuel, 2001). Water column primary production is often high in estuarine ecosystems due to the availability of light and inorganic nutrients. Additionally, because estuaries tend to be relatively shallow and light often penetrates to the *benthos*, these systems are often characterized by high benthic primary production. As a result, autochthonous sources may include several sources (e.g., phytoplankton, seagrasses, macroalgae, and benthic microalgae), ranging in their lability and potential to support production at higher trophic levels (Bianchi, 2007, and references therein). Because of their proximity to land and linkages to the surrounding watershed by rivers, allochthonous sources of organic matter are also important. Allochthonous sources include vascular plant tissues and *detritus* from marshes and upland sources as well as soil-derived organic matter. Additionally, estuaries are in close proximity to human populations and their activities, resulting in contributions of organic matter from *anthropogenic* sources. These may include sewage and pharmaceutical compounds, organic pollutants, as well as the products of industrialization and urbanization. Lastly, it is important to note that organic matter from these different sources (primary production, terrigenous, and anthropogenic), in most cases, does not pass through estuaries passively. Residence time of the estuary and reactivity of the organic matter will largely determine if material is passively transported. Estuaries are important regions of organic matter processing; thus, products of photochemical and heterotrophic processes and bacterial biomass contribute additionally to the organic matter pools within these systems (McCallister et al., 2006). This can increase the challenge of tracing the sources and fate of organic matter along the estuarine salinity continuum.

The application of fatty acid biomarkers to estuarine systems has provided numerous insights not available through studies of bulk properties of organic matter (e.g., C:N ratios, $\delta^{13}C_{TOC}$, chlorophyll *a*, protein:carbohydrate ratios, etc.). This is largely due to the inability of these methods to resolve sources due to overlapping signatures. Saliot and coworkers were among the first to use fatty acid biomarkers to trace the sources of organic matter along the estuarine salinity gradient (Saliot et al., 1988, 2002; Bigot et al., 1989; Scribe et al., 1991; Derieux et al., 1998; Pinturier-Geiss et al., 2001; Brinis et al., 2004). Building on this work, several investigators have used fatty acid biomarkers to examine the dynamics of organic matter composition within estuaries. These studies have incorporated measures of fatty acids into studies examining the effects of processes varying at tidal, seasonal, and interannual timescales. Canuel and coworkers have shown that the concentrations, sources, and lability of fatty acids differed during low-flow and high-flow periods both in San Francisco Bay and Chesapeake Bay (USA) (Canuel, 2001). These studies show the importance of seasonal changes in hydrology and primary production in controlling POM composition. Several studies have focused on the spring phytoplankton bloom (Mayzaud et al., 1989; Canuel, 2001; Ramos et al., 2003). This seasonal event often supplies much of the organic matter to animal communities and may play an important role in the timing of reproduction and recruitment. The effects of physical processes on fatty acid composition at shorter timescales have demonstrated variations in the fatty acid composition of suspended particulate organic matter (POM) over tidal timescales (Bodineau et al., 1998). Recent studies utilizing fatty acid biomarkers have identified the *estuarine turbidity maximum* (ETM) as an important zone for physical and biological processing of organic matter (Bodineau et al., 1998; David et al., 2006). Additional studies have utilized fatty acid biomarkers to trace the sources and fate of terrigenous and anthropogenic organic matter along the estuarine continuum (Quemeneur and Marty, 1992; Saliot et al., 2002; Countway et al., 2003).

Surface sediments provide a more integrated view of organic matter sources than POM, which provides more of a snapshot of organic matter composition at a specific time or place. A number of studies investigating the composition of surface sediments in estuaries have utilized fatty acid biomarkers (Shi et al., 2001; Zimmerman and Canuel, 2001; Hu et al., 2006). As mentioned above, estuaries are influenced by human activities due to their proximity to land and are linked to activities in the adjacent watershed through rivers. As a result, sediment records collected from estuaries can provide a useful tool for examining the effects of human activities on estuarine ecosystems. Fatty acids have been used in studies investigating the affects of changes in land-use and other human activities on organic matter inputs to estuaries (Zimmerman and Canuel, 2000, 2002).

Studies of estuarine sediments have also been used to examine questions related to organic matter diagenesis. Several studies have utilized laboratory experiments to examine and quantify rates of fatty acid degradation (Sun et al., 1997, 2002; Grossi et al., 2003; Sun and Dai, 2005) (table 8.2). These studies have focused largely on the effects of sediment oxygenation, physical mixing, and *bioturbation* on the decomposition of fatty acids. Additional insights about the reactivity of fatty acids in sediments have been provided through field studies (Haddad et al., 1992; Canuel and Martens, 1996; Arzayus and Canuel, 2005). Together, these studies have shown that decomposition rates decrease with decreasing levels of functionality and that increased exposure to oxygen promotes decomposition of more refractory compounds (e.g., saturated, long-chain fatty acids). However, additional effort is needed in this area, particularly in examining the effects of benthic communities on fatty acid decomposition, rather than single species. Future experimental studies aimed at understanding the decomposition of organic compounds should consider ways to better represent the complexities of the physical environment and biological communities.

With recent advances in *ultrafiltration*, a number of studies have examined the composition of ultrafiltered dissolved organic matter (UDOM) utilizing fatty acid biomarker compounds. Mannino and Harvey (1999) examined the composition of POM and UDOM in surface waters of Delaware Bay. The UDOM fraction had a lower diversity of fatty acid ompounds than the POM and was dominated by saturated fatty acids (C_{14}, C_{16}, and C_{18}), with unsaturated and branched fatty acids present at low levels. This study also showed that patterns of fatty acid abundance and composition differed for the POM and dissolved pools. When expressed on an OC normalized basis, fatty acids associated with POM were most abundant at the coastal ocean station, while fatty acid concentrations in the high-molecular-weight DOM pool tended to be higher in the midestuary at the chlorophyll maximum (fig. 8.6). Zou et al. (2004) compared to the lipid biomarker composition of high-molecular-weight DOM (HMW DOM, >1000 Da) across multiple estuarine systems. Fatty acid composition was dominated by saturated C_{14}–C_{18} fatty acids (68 to 80% of total fatty acids). Bacterial fatty acids, including 18:1ω7, normal odd-number C_{15}–C_{17} fatty acids, and iso- and anteiso-branched C_{15} and C_{17} fatty acids were present in all samples, suggesting contributions from bacterial membrane components and bacterial processing of organic matter.

More recent studies have examined fatty acid composition of UDOM and POM within the context of estuarine dynamics by designing studies around the major biological and hydrological forcings (Loh et al., 2006; McCallister et al., 2006). Like Mannino and Harvey (1999), these studies found differences in the composition and abundance of fatty acids associated with UDOM and POM. Loh et al. (2006) compared differences in POM and UDOM at two sites in the Chesapeake Bay (USA) representing the freshwater and marine end members during periods of high and low freshwater input. Although fatty acids representing autochthonous sources dominated, long-chain fatty acids, likely of higher plant origin, were elevated at the river site during high flow. A similar study conducted by McCallister et al. (2006) examined POM and HMW DOM at three sites in the York River Estuary, Virginia (USA), during high- and low-flow conditions. During high flow, HMW DOM was dominated by fatty acids representing a mixture of terrigenous higher plant and bacterial sources, while POM was dominated by fatty acids of phytoplankton/zooplankton origin (fig. 8.7). During low flow, bacterial fatty acids dominated the HMW DOM collected from mesohaline and high-salinity regions, while POM was dominated by plankton sources in these regions. Both the HMW DOM and POM collected from the riverine freshwater site were enriched in fatty acids representing terrigenous vascular plant sources.

Table 8.2
Turnover times for fatty acids in marine/estuarine sediments

Location	*Depth*	*Bottom water*	*Sediment interface*	*Depositional environment*	*τ (days)*	*References*
Carteau Bay, Mediterranean Sea	Surface sediments	oxic	oxic	bioturbated	21 to 76	Grossi et al. (2003)
Long Island Sound, NY, USA	Surface sediments	oxic	oxic	bioturbated	11 to 17	Sun and Wakeham (1999)
					8 to 50	Sun et al. (1997)
York River Estuary, VA, USA	33–84 cm	oxic	oxic	bioturbated	Algal FA: 25×10^3	Arzayus and Canuel (2004)
	~30–40 years in age				Higher plant FA: n.s.	
York River Estuary, VA, USA	41–54 cm	oxic	oxic	physically mixed	Algal FA: 17×10^3	Arzayus and Canuel (2004)
	~30–40 years in age				Higher plant FA: 12×10^3	
Cape Lookout Bight, NC, USA	0–1 cm	oxic	hypoxic/ anoxic seasonally	nonbioturbated	SCFA:19 to 33 SCFA:19 to 33 LCFA: n.s. BrFA: 22 to 200 MUFA: 17 to 28 PUFA: 19 to 26	Canuel and Martens (1996)
Cape Lookout Bight, NC, USA	0–10 cm	oxic	hypoxic/ anoxic seasonally	nonbioturbated	SCFA: 19 to 250 LCFA: 125 to 250 BrFA: 26 to 200 MUFA: 17 to 200 PUFA: 19 to 143	Canuel and Martens (1996)
Cape Lookout Bight, NC, USA	0–100 cm	oxic	hypoxic/ anoxic seasonally	nonbioturbated	SCFA: 250 LCFA: 2500	Haddad et al. (1992)
Black Sea	Surface sediments	anoxic	anoxic	nonbioturbated	6500 to 14,000	Sun and Wakeham (1994)

8.4.4 Fatty Acids as Biomarkers in the Ocean

A major focus of biomarker studies in oceanic systems has been on understanding the sources and transformations of organic matter as it settles through the water column and during early diagenesis (see reviews by (Lee and Wakeham, 1988; Wakeham and Lee, 1993; Lee et al., 2004). Much of the most recent research has taken place under the auspices of international programs such as the Joint Global

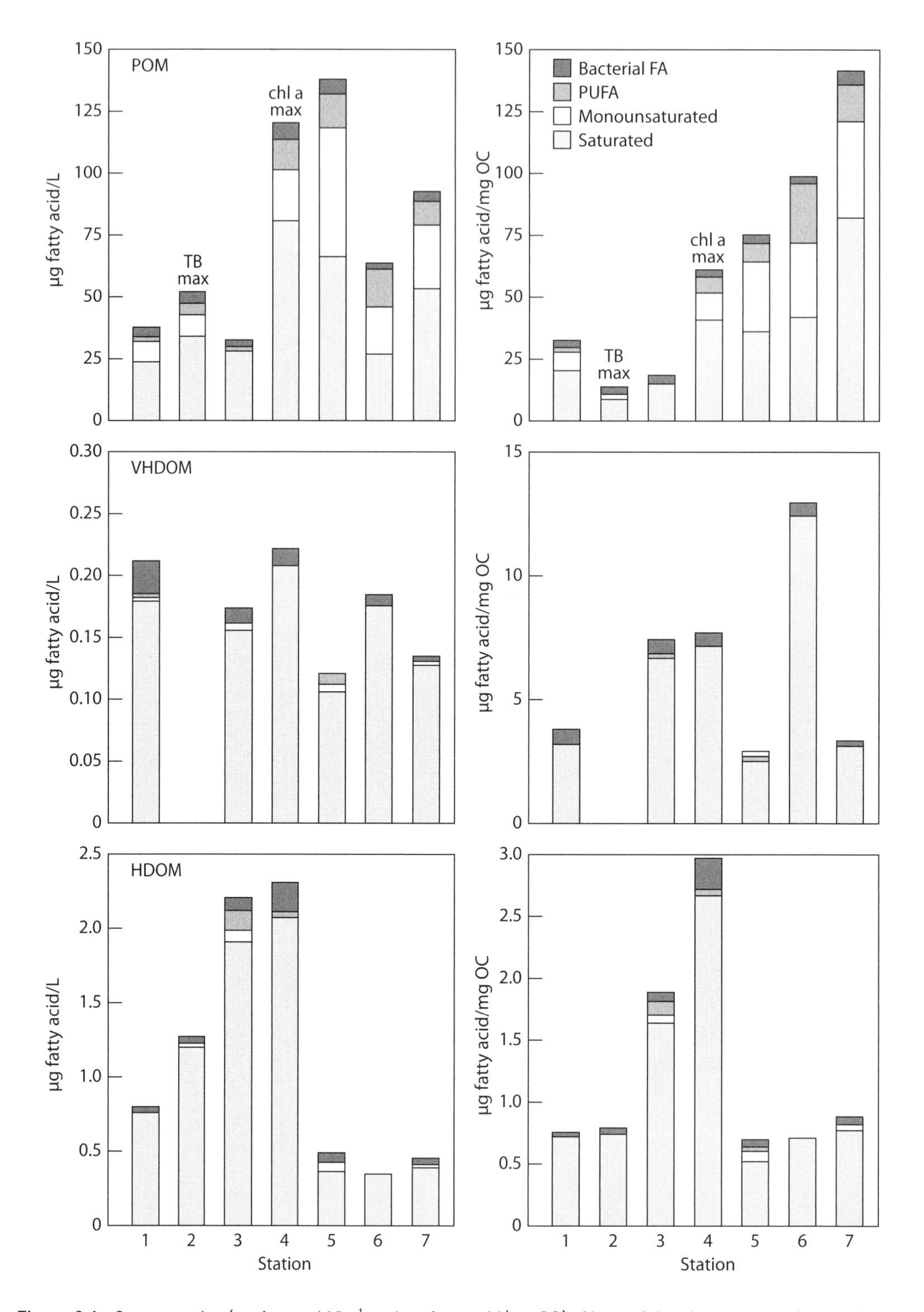

Figure 8.6. Concentration (μg fatty acid L^{-1} and μg fatty acid/mg OC) of bacterial, polyunsaturated, monounsaturated, and saturated fatty acids in particulate organic matter (POM) and two size classes of dissolved organic matter (DOM) along the salinity gradient of the Delaware River estuary (USA). (Adapted from Mannino and Harvey, 1999.)

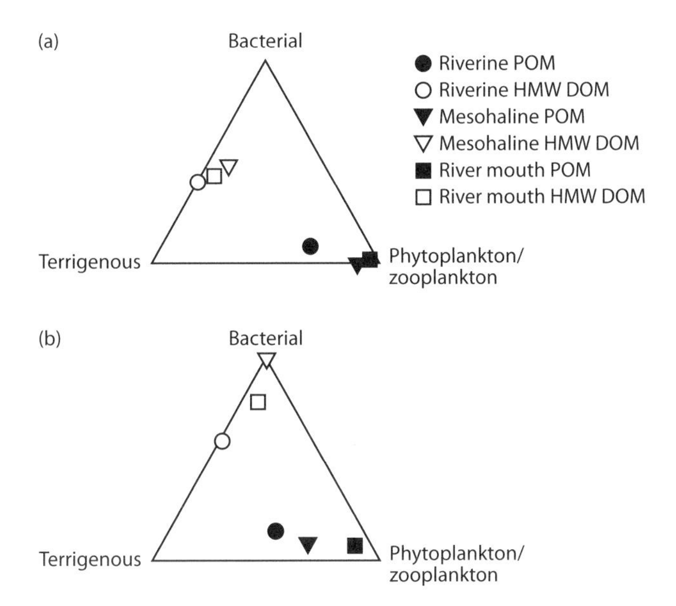

Figure 8.7. Relative abundance of fatty acids representing bacterial, terrigenous, and plankton sources in POM and high-molecular-weight DOM along the York River estuary, VA (USA). (Adapted from McCallister et al., 2006.)

Ocean Flux Study (JGOFS), MedFlux, and PECHE (Production and Export of Carbon: Controls by Heterotrophs, supported by the Centre National de la Recherche Scientifique, France). In the equatorial Pacific Ocean, Wakeham and colleagues (1997) examined fluxes of compounds representing each of the major classes of biochemicals and found: (1) plankton-derived organic matter decreasing exponentially with increasing water depth; (2) elevated contributions from bacteria near the sediment surface; and (3) selective preservation of organic compounds from bacteria, phytoplankton, and terrigenous plants within the sediments (Wakeham et al., 1997). These general trends are consistent with those observed in other regions of the oceans (Wakeham, 1995; Wakeham et al., 2002; Goutx et al., 2007).

Consistent with these findings, fatty acids were classified into four groups reflecting their sources and susceptibility to decomposition: (1) compounds characterized by maxima in surface waters; (2) compounds that were enriched within the water column; (3) compounds enriched in surficial sediments; and (4) compounds with their highest abundance within sediments (10–12 cm) (fig. 8.8). Surface water POM was dominated by fatty acids from phytoplankton sources (e.g., 20:5ω3 and 22:6ω3 polyunsaturated fatty acids) in the central equatorial Pacific. Fatty acids within the water column were dominated by compounds of phytoplankton origin that are resistant to decomposition as well as compounds from zooplankton (e.g., oleic acid, 18:1ω9). Bacterial fatty acids were enriched at the sediment surface (0–5 cm), including *cis*-vaccenic acid (18:1ω7 and iso- and anteiso-C_{15} and C_{17} fatty acids). Compounds resistant to microbial degradation were most abundant deeper within the sediments (10–12 cm), including high-molecular-weight, straight-chain fatty acids (C_{24} to C_30).

Additional studies have examined the fatty acid composition of POM in water columns characterized by low dissolved oxygen concentrations or anoxia (e.g., Black Sea and Cariaco Trench). The goal of these studies was to examine changes in organic matter composition across the oxic–anoxic boundary. Fatty acids from anaerobic organisms dominated the anoxic water column of the Black Sea (130 to 400 m). Branched fatty acids, including iso- and anteiso-branched C_{15} and C_{17} fatty acids, iso-16:0, and 10-methylhexadecanoic acid, comprised 30% of the fatty acid distribution within this zone (fig. 8.9) (Wakeham and Beier, 1991; Wakeham, 1995). These fatty acids are common in marine

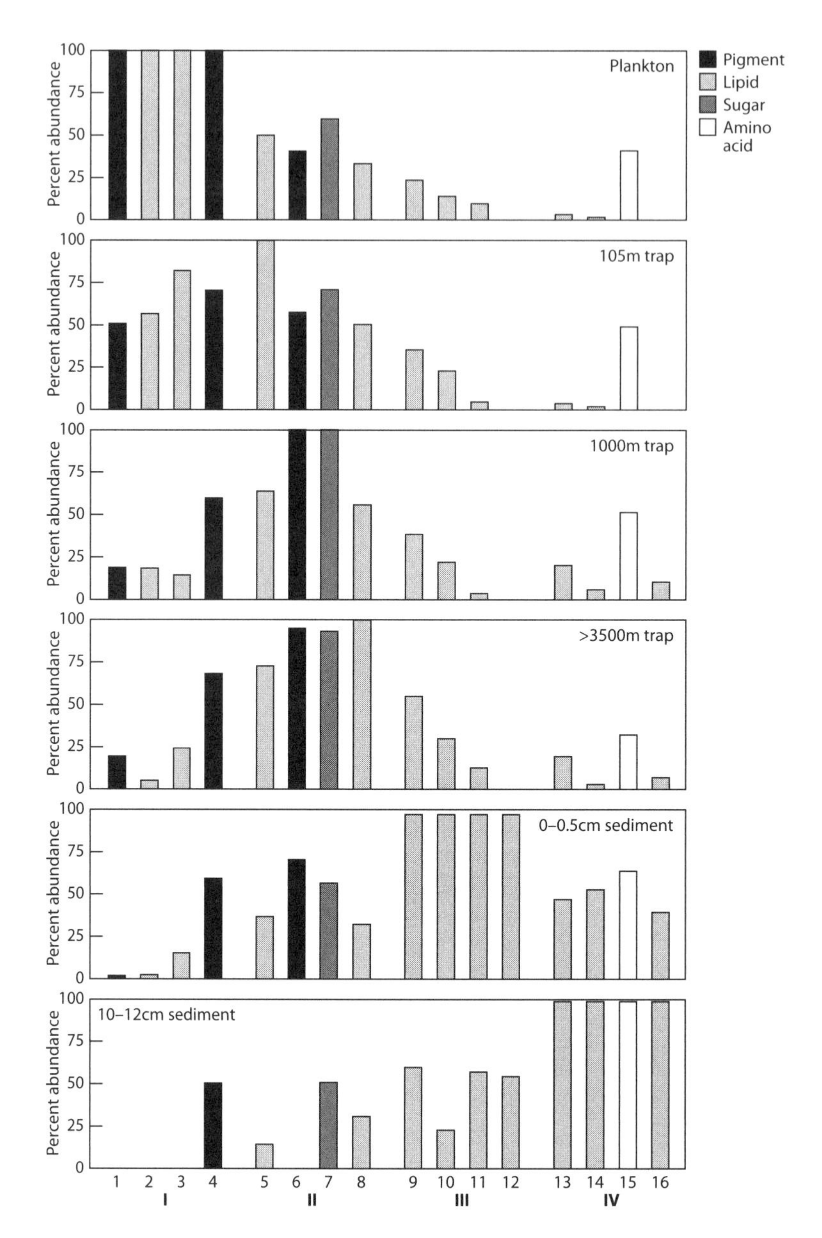

Figure 8.8. Percent abundance of different biochemicals associated with water-column POM and surface sediments collected from the central equatorial Pacific. Surface water POM was dominated by fatty acids from phytoplankton sources (e.g., 20:5ω3 and 22:6ω3 polyunsaturated fatty acids). Within the water column, fatty acids were dominated by compounds of phytoplankton origin that are resistant to decomposition as well as compounds from zooplankton (e.g., oleic acid, 18:1ω9). Bacterial fatty acids were enriched at the sediment surface (0–5 cm), including *cis*-vaccenic acid (18:1ω7 and iso- and anteiso-C_{15} and C_{17} fatty acids). Compounds resistant to microbial degradation were most abundant deeper within the sediments (10–12 cm), including high molecular weight, straight-chain fatty acids (C_{24} to C_{30}). Compound identities: 1, chlorophyll *a*; 2, 22:6ω3 polyunsaturated fatty acid; 3, 20:5ω3 polyunsaturated fatty acid; 4, glucose; 5, 24-methylcholesta-5,24(28)-dien-3β-ol; 6, phaeophorbide α; 7, galactose; 8, oleic acid (18:1ω9); 9, cis-vaccenic acid (18:1ω7); 10, anteiso-15:0 fatty acid; 11, 24-ethylcholest-5-en-3β-ol; 12, bisnorhopane; 13, $C_{24} + C_{26} + C_{30}$ fatty acids; 14, $C_{37} + C_{38}$ methyl and ethyl, di- and tri-unsaturated alkenones; 15, glycine; and 16, squalene. Groupings I–IV are based on compound behavior and geochemical stability in the environment (see text). (Adapted from Wakeham et al., 1997.)

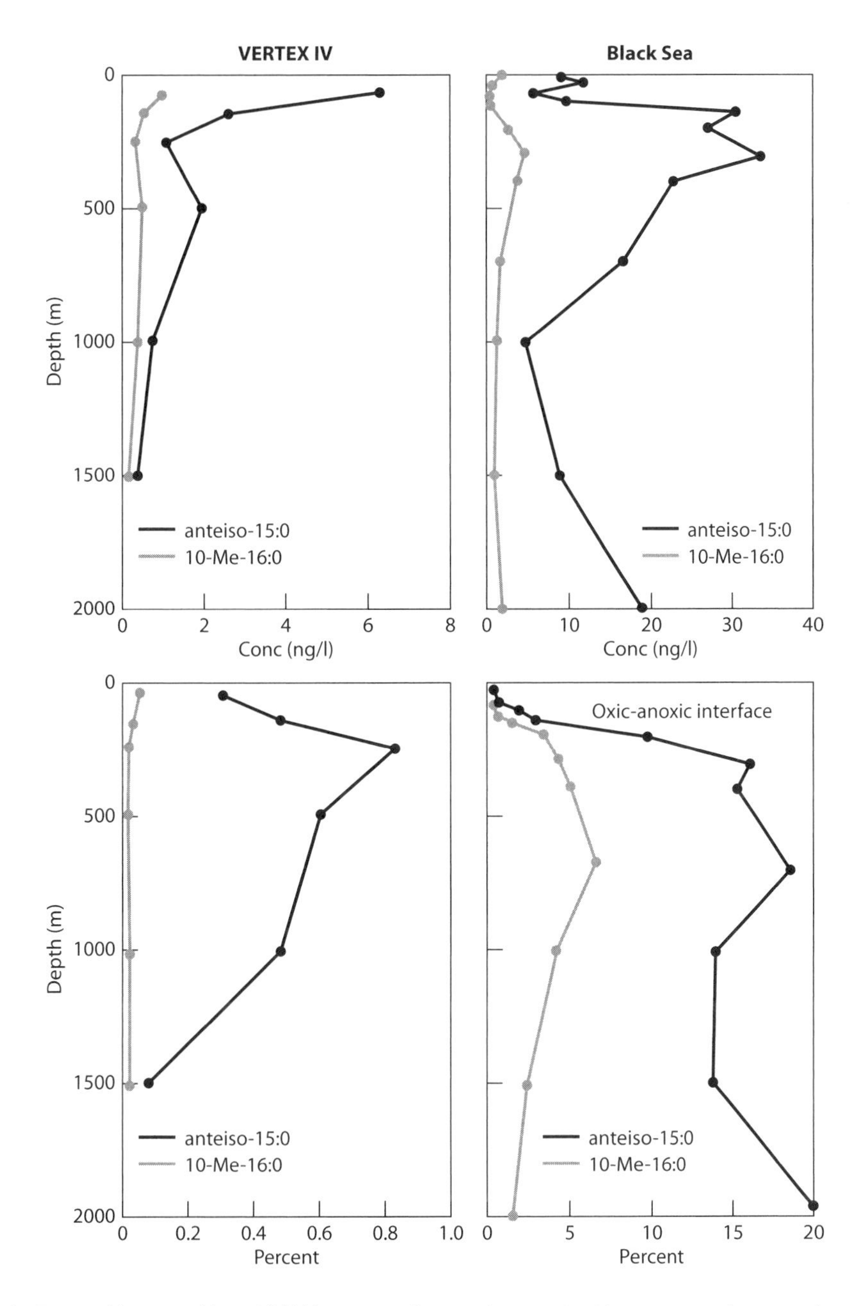

Figure 8.9. Fatty acid composition of POM in water columns characterized by oxygenated water column (VERTEX IV) vs. low dissolved oxygen concentrations or anoxia (Black Sea). Fatty acids from anaerobic organisms dominated the anoxic water column of the Black Sea (130 to 400 m). Branched fatty acids including iso- and anteiso- branched C_{15} and C_{17} fatty acids, iso-16:0, and 10-methylhexadecanoic acid comprised 30% of the fatty acid distribution within this zone. (Adapted from Wakeham and Beier, 1991.)

sulfate-reducing bacteria (Kaneda, 1991; Parkes et al., 1993), which are abundant in the anoxic zone of the Black Sea. Monoenoic fatty acids (16:1ω7 and 18:1ω7) also increased in relative abundance within the anoxic zone. Likely sources for these fatty acids include sulfate-reducing bacteria or anaerobic photosynthetic bacteria, such as the green sulfur bacteria *Chlorobium* spp. Similar findings were obtained for studies in the Cariaco Trench, where the oxic–anoxic interface occurs between 260 and 300 m water depth. In the Cariaco Trench, branched C_{15} fatty acids increased below the oxic–anoxic interface, suggesting contributions from microbial activity in the anoxic zone. Remarkably, anteiso-C_{15} fatty acid was the second most abundant fatty acid within the anoxic water column, comprising approximately 10% of the total fatty acids.

Additional studies have characterized fatty acid composition of surface sediments in the ocean. However, similar to estuarine and lacustrine systems, care must be used in interpreting fatty acid compositions in oceanic sediments because of the effects of postdepositional processes (e.g., diagenesis, bioturbation). Additionally, transport processes may influence organic compounds differentially, depending on the phases with which they are associated. As resuspended material is transported, labile components may be preferentially removed, influencing the composition of materials that are eventually deposited. Advection can also control the composition of sedimentary organic matter if specific pools of organic matter are associated with inorganic (mineral) phases that differ in their density or propensity to be transported (see discussion about effects of differential transport on fatty acids below as well as the effects on alkenones discussed in chapter 11). As described above, selective degradation results in preservation of only the most decay-resistant compounds in subsurface sediments (Wakeham et al., 1997). Thus, the absence of labile compounds may be due to their preferential removal by heterotrophic processes rather than the absence of contributions of organic matter from a particular source. Additionally, labile compounds may be more reflective of inputs from the overlying water column, while more refractory compounds may reflect contributions of organic matter from vertical and lateral sources (Mollenhauer and Eglinton, 2007).

In recent years, the application of compound-specific $\Delta^{14}C$ analysis has provided new insights about the sources of organic matter as well as diagenetic and sedimentological controls on organic matter composition (Pearson et al., 2001; Mollenhauer and Eglinton, 2007). Pearson and colleagues measured ^{14}C signatures of lipid compounds present in prebomb and postbomb horizons of sediment cores collected from Santa Monica Basin (SMB) and surface sediments collected from Santa Barbara Basin (SBB) (USA). $\Delta^{14}C$ signatures of C_{16} and C_{18} fatty acids in prebomb and postbomb sediment (-76 ± 14‰ for $C_{16:0}$ and $+86 \pm 11$‰ for $C_{16:0}$ and $C_{18:0}$, respectively) were consistent with prebomb and postbomb surfacewater $\Delta^{14}C_{DIC}$ (-82‰ and $+71$‰, respectively) (table 8.3). The $\Delta^{14}C$ signatures, along with similar $\delta^{13}C$ signatures, indicate a common source or one that is unvarying in its relative contributions from different sources, consistent with surface water production. $\Delta^{14}C$ values for bacterial fatty acids (iso- and anteiso-C_{15}) and *n*-15:0 fatty acid ranged from -113 ± 14‰ (anteiso-C_{15}) in prebomb sediments and between $+32$ and $+44$‰ for postbomb horizons (table 8.3). In both cases, values were depleted relative to $\Delta^{14}C_{DIC}$, suggesting that these fatty acids are not derived entirely from bacterial heterotrophy. Long-chain fatty acids are generally attributed to higher plant sources (table 8.1). Interestingly, $\Delta^{14}C$ signatures for $C_{26:0}$ fatty acid were depleted relative to the signature for $C_{24:0}$ fatty acid in both prebomb and postbomb sediments of SBB and SMB, (table 8.3). One explanation for this observation is that the $C_{26:0}$ fatty acid was influenced by contributions from pre-aged terrestrial carbon to a greater extent.

Mollenhauer and Eglinton (2007) further investigated the $\delta^{13}C$ and $\Delta^{14}C$ signatures of lipid biomarker compounds in SBB and SMB sediments (Mollenhauer and Eglinton, 2007). Specifically, they examined the isotopic signatures of fatty acids in sediments representing different depositional regimes (flank vs. depocenter) and under contrasting bottom water oxygen concentrations. These data showed the preferential loss of short-chain fatty acids ($C_{14:0}$, $C_{16:0}$, and $C_{18:0}$) relative to long-chain ($>C_{24:0}$) fatty acids. They also showed that labile compounds (e.g., short-chain fatty acids,) were enriched in radiocarbon relative to more recalcitrant compounds fatty acids ($>C_{24:0}$) (fig. 8.10). This suggests that a larger proportion of the refractory compounds may be supplied from pre-aged material.

Table 8.3
Compound-specific $\Delta^{14}C$ and $\delta^{13}C$ data from Santa Barbara Basin and Santa Monica Basin sediments (USA)

	Postbomb			*Prebomb*		
Fatty acid	*Horizon (cm)*	$\delta^{13}C$ (‰)[a]	$\Delta^{14}C$ (‰)[b]	*Horizon (cm)*	$\delta^{13}C$ (‰)	$\Delta^{14}C$ (‰)
i-15:0	0–0.75	−23.2 (0.1)	+33 (15)	4.5–5.5	−25.9 (0.2)	—
a-15:0	0–0.75	−24.1 (0.2)	+32 (12)	4.5–5.5	−30.6 (0.6) −24.6*	−113(14)
15:0	0–0.75	−24.0 (0.2)	+44 (13)	4.5–5.5	−26.5 (0.6)	—
16:0	0–0.75	−25.0 (0.3)	+86 (11)	4.5–5.5	−26.3 (0.2)	−76 (14)
18 : 1ω7	0–0.75	−23.2 (0.3) −27.2*	+64 (11)	4.5–5.5	−26.4 (0.2)	—
18:0	0–0.75	−24.1 (0.2)	+83 (14)	4.5–5.5	−26.3 (0.5)	—
24:0	0–0.75	−26.4 (0.2)	+62 (14)	4.5–5.5	−26.3 (0.2)	−69 (15)
26:0	0–0.75	−26.8 (0.4)	+14 (13)	4.5–5.5	−26.7 (0.2)	−108 (19)

[a] $\delta^{13}C$ values based on irm-GC-MS measurements unless noted with asterisk. Values marked with asterisk were determined by off-line irMS splits of CO_2.
[b] Average of two $\Delta^{14}C$ values.
Source: Pearson et al. (2001).

They also showed that labile organic compounds were preferentially preserved in depocenters than in flank sediments, as evidenced by "younger" ages for biomarkers. These data suggest the importance of considering transport processes, depositional environment, and diagenesis when interpreting biomarker distributions in sediments.

8.5 Fatty Acids as Tools for Understanding Trophic Relationships

Several approaches have been utilized for understanding the trophic relationships between aquatic organisms including the analysis of gut contents (Napolitano, 1999) and stable isotopes (Peterson and Fry, 1987; Michener, 1994). Additionally, lipid biomarker compounds may be used to trace the flow of energy in marine and freshwater systems. The premise for these studies is that the biomarkers can be linked to known sources and that the compounds retain their basic structure after ingestion.

There are numerous examples of studies where lipid biomarker compounds have been used to understand the marine food web (Dalsgaard et al., 2003). Lipids are biochemical components present in all organisms. These biochemicals can be used both as tracers of organic matter (food) source, trophic level, and physiology. Lipid compounds can be taxon-specific, particularly in lower trophic levels (autotrophs and bacteria), where most lipids are synthesized de novo. In contrast, higher-trophic-level organisms obtain most of their lipids from dietary sources. One class of lipids, the fatty acids, has proven particularly useful in discerning trophic relationships. Fatty acids have been used in three ways to study food webs. First, by examining fatty acid composition in predators, qualitative information about spatial or temporal variations in diets both among and within individuals or populations can be obtained (Iverson et al., 2004; Budge et al., 2006). The second application of fatty acids to trophic studies relies on individual compounds or *biomarkers*—that is, unique fatty acids that can be linked to a single origin or prey species. This approach can be difficult to apply successfully because many fatty acids are ubiquitous in a wide range of marine and terrigenous organisms. However, unusual levels of particular fatty acids or ratios of fatty acids may indicate the importance or dominance of some organisms in the diet (Dalsgaard et al., 2003). The third application of fatty acids involves

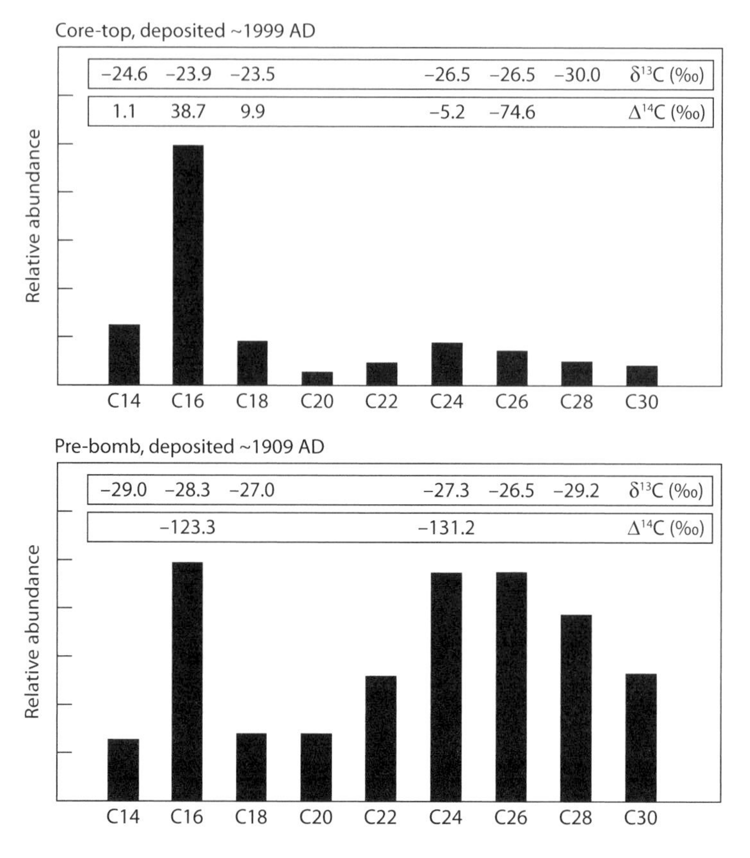

Figure 8.10. Relative abundance of fatty acids normalized to C_{16} in sediments collected from Santa Monica and Santa Barbara basins, CA (USA) (top). Stable isotopic composition and radiocarbon age of fatty acids (bottom). (Adapted from Mollenhauer and Eglinton, 2007.)

quantitative estimate of diet from fatty acid signatures of predator and prey. This approach is the most challenging and requires knowledge of the fatty acid composition of all important prey species, as well as information about within-species variability in fatty acid composition for both prey and predator (Budge et al., 2006). Quantitative fatty acid signature analysis (QFASA) is a new method of estimating predator diets that uses a statistical model to compute the most likely combination of fatty acid signatures for prey that matches the fatty acid composition in the predator, with consideration for predator metabolism (Iverson et al., 2004).

A new area of active research involves investigating the role of specific algal fatty acids in supporting zooplankton nutrition. These studies are based on the premise that the efficiency with which biomass and energy are transferred through the food web, thereby supporting the production of higher trophic levels (such as fish), is related to the abundance of specific ω3-highly unsaturated fatty acids (ω3-HUFA). Feeding experiments have shown that variation in the nutritional quality of various algal species can be explained in part by their fatty acid content (Park et al., 2002; Jones, 2005; Klein Bretelar et al., 2005). Fatty acids important to marine animal nutrition include eicosapentaenoic acid (EPA; 20:5ω3), docosahexaenoic acid (DHA; 22:6ω3), and linoleic acid (18:3ω6) (Ahlgren et al., 1990; Müller-Navarra, 1995a,b; Weers, 1997). These HUFA can only be obtained directly from an animal's diet or following conversion of dietary components (Brett and Müller-Navarra, 1997; Müller-Navarra, 2004). HUFA with double bonds at the ω3 and ω6 positions are not likely synthesized by animals due

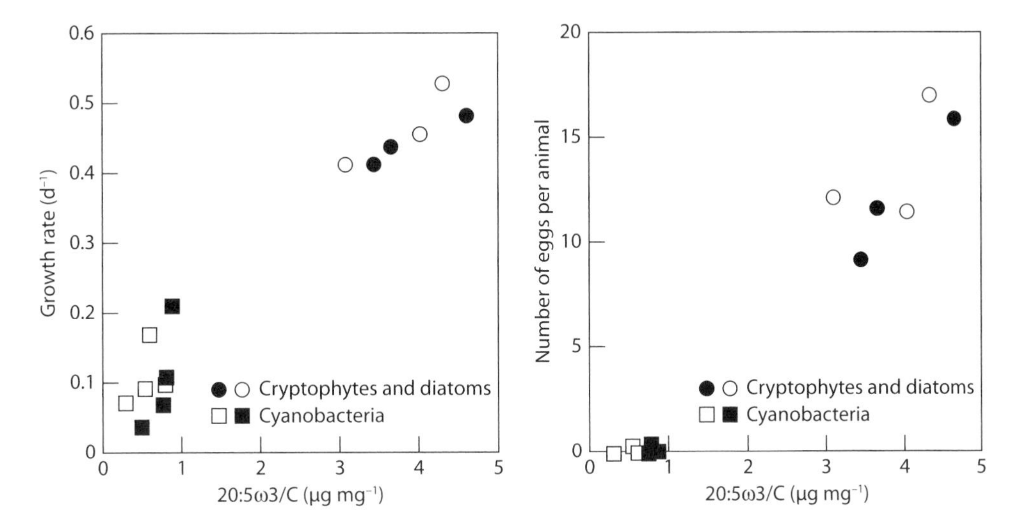

Figure 8.11. Results from laboratory experiment showing relationship between (left) *Daphnia* growth rates and (right) egg production with 20:5ω3 fatty acid. The circles correspond to diatoms and cryptophytes, while the squares correspond to cyanobacteria. Open symbols refer to the < 30-μm fraction and closed symbols to total particulate matter. (Adapted from Müller-Navarra et al., 2000.)

to limitations in the positions where double bonds can be inserted by animal fatty acid synthetases (Canuel et al., 1995).

Recent work has demonstrated relationships between HUFA and zooplankton (*Daphnia*) growth and fecundity (fig. 8.11) (Müller-Navarra et al., 2000). Similarly, it has also been shown that the chemical composition of seston and microalgae can influence copepod egg production (Jonasdottir, 1994; Jonasdottir et al., 1995). Since algal species differ in their fatty acid composition, the dominance of algal species enriched in HUFA (e.g., crypophytes, diatoms), may have a positive influence on transfer of energy to higher trophic levels (zooplankton), while the dominance of species depleted in HUFA (e.g., cyanobacteria) may result in accumulation of algal organic matter or processing by microbes rather than transfer to higher trophic levels. However, zooplankton do not generally feed on mono-specific algal diets and one should use caution when extrapolating from laboratory experiments where simple diets are used. A recent study showed that cyanobacteria, when fed alone may have low food quality, but cyanobacteria used in combination with other algal sources may supplement the diet of the calanoid copepod, *Acartia tonsa* (Schmidt and Jonasdottir, 1997). Additionally, it is important to keep in mind that animals live as members of a community, and food quality may change in response to processing within the food web. Recent studies of planktonic food webs have shown that some flagellates and ciliates may increase the PUFA content of the their food ("trophic upgrading") (Klein Bretelar et al., 1999; Bec et al., 2003; Park et al., 2003; Veloza et al., 2006), thereby increasing the nutritional value of microalgae.

The role fatty acids play in determining the nutritional value of organic matter to benthic communities has been investigated to a lesser extent. In a study in San Francisco Bay (USA), the fatty acid composition of tissues of the Asian clam, *Potamocorbula amurensis,* showed that this organism feeds selectively on microalgae along the estuarine salinity continuum (Canuel et al., 1995). C_{20} and C_{22} polyunsaturated were the most abundant fatty acids present in tissues of *P. amurensis* in the southern San Francisco Bay during the spring phytoplankton bloom, as well as during baseline conditions in northern San Francisco Bay (fig. 8.12). In another study, the fatty acid DHA (22:6ω3), was shown to limit the ability of larvae of the zebra mussel *Dreissena polymorpha,* to convert C_{18}

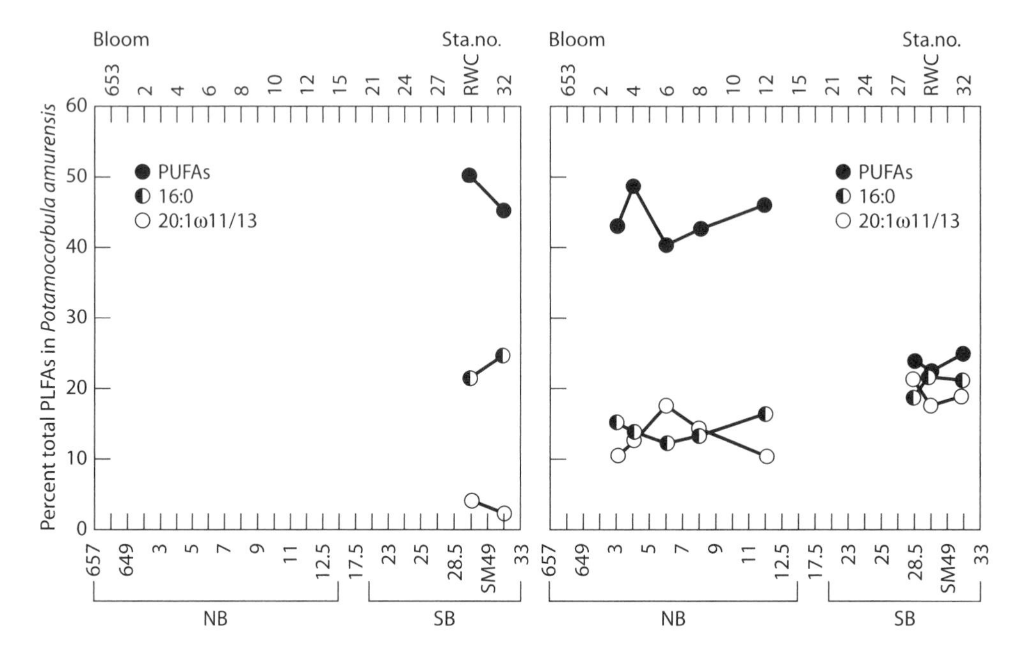

Figure 8.12. Fatty acid composition of the Asian clam, *P. amurensis*, along the salinity gradient of San Francisco Bay (SFB), CA (USA). Results show that polyunsaturated fatty acids (PUFAs) dominate the FA composition of animals from southern SFB during the Spring bloom and in northern SFB during baseline (nonbloom) conditions. These data suggest selective feeding of microalgae by the clam. (Adapted from Canuel et al.,1995.)

PUFA into long-chained (ω3) PUFA (Wacker et al., 2002). This study also showed that a reduction in the rates for elongation and desaturation of EPA to DHA limited growth of larvae. An additional new avenue for the application of fatty acid biomarkers to trophic studies involves investigating the influence of animal diversity on sediment organic matter composition (Canuel et al., 2007; Spivack et al., 2007). In nature, animals do not live as single species but as members of a community. Thus, it is important to understand not only the effects individual species have on sediment biogeochemistry, but also how diverse communities of interacting species influence sediment geochemical processes. In recent mesocosm experiments, Canuel and coworkers showed that changes in epibenthic foodweb composition influenced both the total abundance ($\sum$FA) and composition of fatty acid biomarkers (fig. 8.13). These experiments tested the effects of grazer diversity and food-chain length on the quantity and quality of accumulated sediment organic matter in experimental eelgrass (*Zostera marina*) mesocosms (Canuel et al., 2007; Spivak et al., 2007).

Results from one of these experiments showed that species composition and richness of grazers strongly affected the relative abundance (% of total FA) of short-chain (C_{12}, C_{14}, C_{16}) FA (Canuel et al., 2007). Grazer species composition, but not richness, influenced the relative abundance of branched (iso- and anteiso-C_{13}, C_{15}, and C_{17}), polyunsaturated (C_{18}, C_{20}, and C_{22}), and long-chain (C_{22}–C_{30}) fatty acids. Increasing food-chain length by addition of predatory blue crabs (*Callinectes sapidus*) resulted in a trophic cascade, increasing algal biomass and accumulation of algal organic matter in sediments, enhancing the quality of sediment organic matter. Concomitantly, the relative proportion of bacterial branched FA increased. Overall, the identity and number of epibenthic consumers strongly influenced the accumulation and composition of sediment organic matter and the pathways by which it is processed.

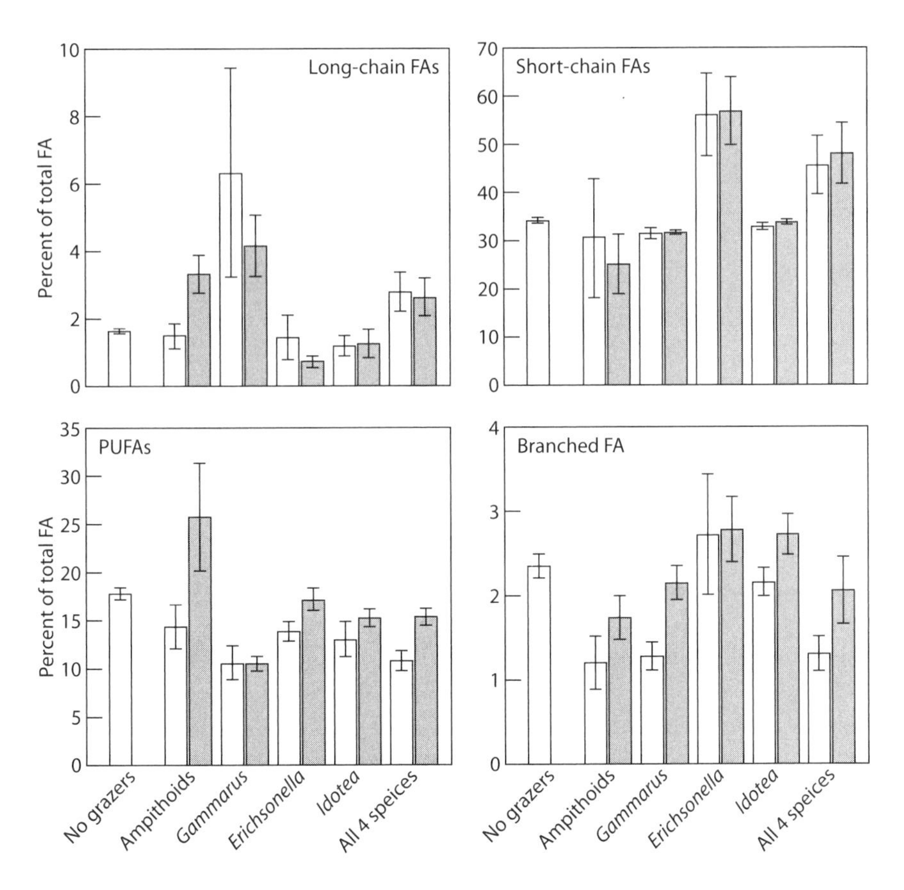

Figure 8.13. Mesocosm experiments show that epifaunal grazer species influence fatty acid composition ($p < .01$). Data from this experiment provide evidence for a trophic cascade, where the presence of crabs increases the accumulation of labile algal material (PUFA); this results in increased bacterial biomass (branched FA) ($p < .01$). Open bars = no crabs; filled bars = with crabs. (Adapted from Canuel et al., 2007.)

8.6 Phospholipid-Linked Fatty Acids (PLFA) as Chemotaxonomic Markers

Environmental studies in the fields of microbial ecology and biogeochemistry require information about the species richness and distribution of the microbial community. Traditionally, this information has been obtained through microscopic counts, but recent advances in *chemotaxonomy* have expanded our ability to obtain quantitative information about the microbial community. Chemotaxonomic approaches offer several advantages over microscopic analyses, since microscopic counts require a trained microscopist and are time-consuming. Also, samples for microscopic analyses generally have to be stored for some time before analysis, requiring that samples be preserved, or fixed, and not all species hold up to fixation. In contrast, microbial membrane lipids, specifically phospholipids-linked fatty acids (PLFA), are present in both the bacterial and microalgal communities, providing a more complete picture of the microbial community, and PLFA analyses can be easily applied to a variety of matrices, including water, sediment, and soil (Guckert et al., 1985; White, 1994; Dijkman and Kromkamp, 2006). Additionally, PLFA provide a useful tool for characterizing microbial communities for the following reasons: (1) they are found in all living organisms; (2) they decompose within weeks following cell death, thereby providing information about the viable or recently viable community; (3) PLFA composition differs between members of the microbial community, providing a tool for assessing microbial community composition; and (4) specific PLFA can provide additional information about the nutritional status of the microbial populations.

PLFA analyses have provided numerous insights about the microbial communities associated with soils, sediments, and the water column of marine and freshwater ecosystems. Because microorganisms have unique PLFA signatures (fig. 8.14), PLFA provide chemotaxonomic information about the communities associated with soils and sediments. In terrigenous environments, PLFA have been used to characterize soil biota (Bull et al., 2000; Cifuentes and Salata, 2001; Crossman et al., 2005) as well as the responses of soil microbes to various perturbations, such as drought, changes in land use, and contaminants (Khan and Scullion, 2000; Aries et al., 2001; Slater et al., 2006; Wakeham et al., 2006). Using ^{13}C incorporation into PLFA, Bull and colleagues showed that the oxidation of atmospheric methane in a soil from the United Kingdom was mediated by a new methanotrophic organism, similar in PLFA composition to type II methanotrophs (Bull et al., 2000). Studies in soils have also examined the interactions between the aboveground plant community and soil microbes. Zak et al. (2003) examined whether plant diversity influenced the soil microbial communities using PLFA. In this study, the authors showed that soil microbial community composition changed in response to increased plant diversity (Zak et al., 2003). Recently, PLFA have also been used to study how soil microbial communities respond to restoration efforts (McKinley, 2005; Bossio et al., 2006). In a restoration study in the Sacramento–San Joaquin Delta of California (USA), Bossio and colleagues (2006) used PLFA analyses to show that active microbial communities of restored wetlands differed from those of agricultural fields. This study demonstrated that flooding regime and aeration were the primary factors controlling microbial community composition.

Over the past decade or more, PLFA have been widely used to examine the microbial community structure in estuarine and marine sediments. Rajendran and colleagues used PLFA to characterize sediment microbial composition in several coastal bays differing in anthropogenic impacts (Rajendran et al., 1992, 1993, 1997). These studies revealed differences in microbial community structure in the sediments that reflected differences in the extent to which the sites were perturbed, indicating that PLFA composition could be used to predict the extent of pollution and/or eutrophication. Similar work was conducted in the Mediterranean Sea, investigating the responses of sediment microbial composition to levels of hydrocarbon contamination (Polymenakou et al., 2006).

An additional advantage of PLFA analysis is that it is often difficult, if not impossible, to separate the microbial community from environmental matrices such as POM, sediments, and soils. Such approaches are needed for studies measuring the incorporation of stable or radiocarbon isotopes into microbial pools, for understanding the sources of organic matter supporting bacterial production, and for discriminating between contributions of organic matter from viable organisms versus detritus. Extraction of PLFA from a variety of sample matrices provides a tool for quantifying the contribution of microbial biomass and examining its composition (White et al., 1979; Guckert et al., 1985) as well as for isolating viable biomass for subsequent isotopic analysis. Boschker et al. (1998) pioneered this approach by demonstrating that ^{13}C-labeled substrates were incorporated into PLFA and could therefore be used to link biogeochemical processes to specific microbial groups. Additionally, $\delta^{13}C$ of PLFA has been used to study the fate of ^{13}C-labeled microphytobenthos and phytoplankton (Middelburg et al., 2000; Van Den Meersche et al., 2004).

Subsequent studies have employed natural abundance isotope ratios of PLFA to study organic matter sources utilized by bacteria (Cifuentes and Salata, 2001; Boschker and Middelburg, 2002). More recent studies analyzing the stable isotopic composition ($\delta^{13}C$) of PLFA have provided insights about the sources of organic matter supporting in situ bacterial production (Boschker et al., 2000; Bouillon et al., 2004; Bouillon and Boschker, 2005). These studies indicated that algal-derived carbon is often the dominant substrate for sediment bacteria across a range of sediments differing in TOC content and vegetation types.

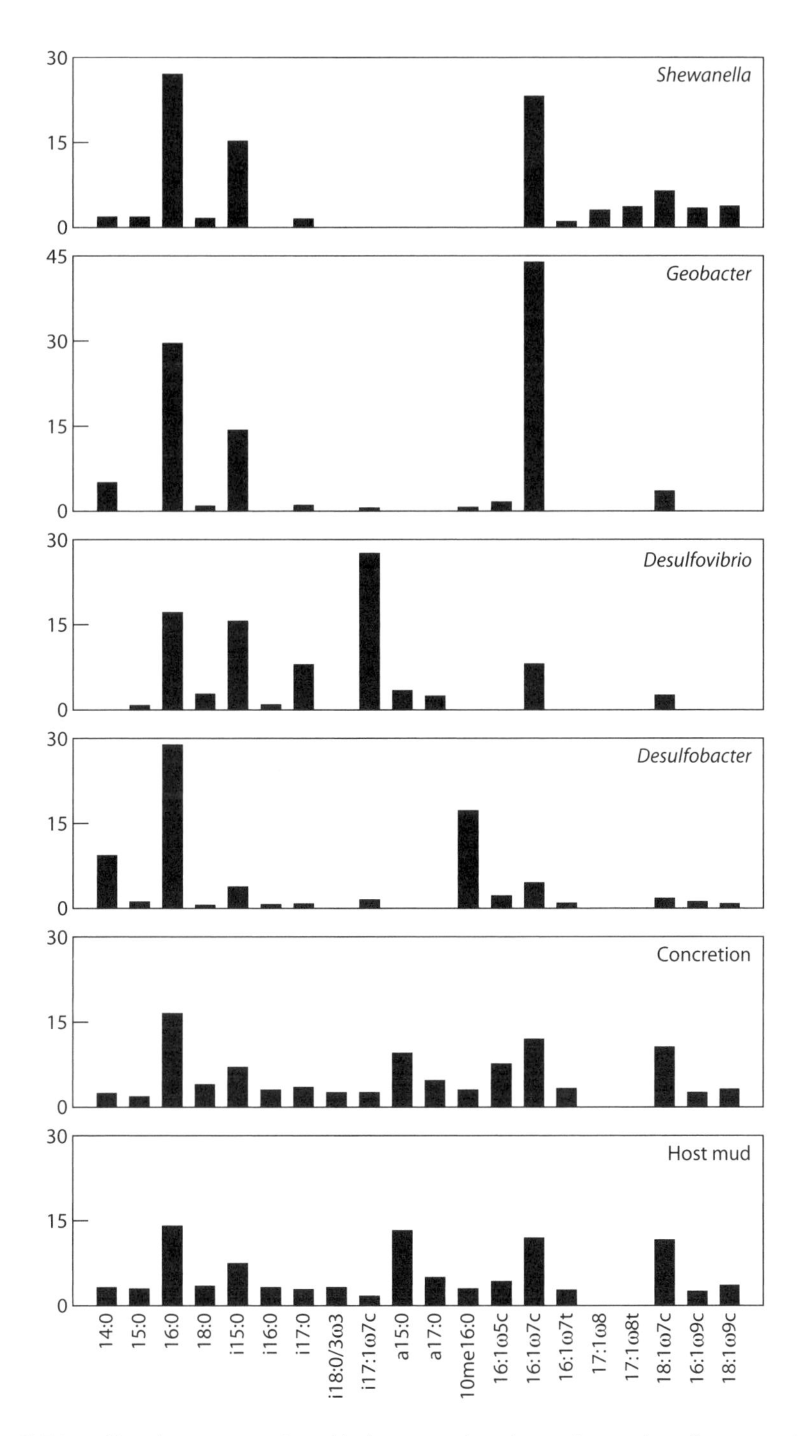

Figure 8.14. PLFA profiles demonstrate that siderite concretions in a salt marsh sediment resulted from the activity of sulfate-reducing bacteria (SRB). The panels illustrate differences in PLFA composition for four groups of bacteria—two genera are sulfate-reducing bacteria (SRB) (*Desulfobacter* and *Desulfovibrio*) and two are dissimilatory Fe(III) reducers (*Geobacter* and *Shewanella*). The characteristic fatty acids for the SRB are 10-methyl-palmitic PLFA (10Me16:0) and iso-17-monenoic PLFA (17:1ω7c). These data suggest that *Desulfovibrio* is more abundant within the concretion than in the mud. (Adapted from by Coleman et al., 1993.)

Similarly, radiocarbon ($\Delta^{14}C$) analysis of PLFA has been used to investigate incorporation of fossil carbon into microbial biomass (Petsch et al., 2001; Slater et al., 2006; Wakeham et al., 2006). Using natural-abundance carbon-14 analysis of membrane lipids, Petsch et al. (2001) showed that a significant portion (74 to 94%) of the lipid carbon associated with cultures grown on shales derived from ancient organic matter. Similar methods have also been utilized to investigate whether microbes utilize fossil carbon from anthropogenic sources. Slater and colleagues examined whether microbes living in sediments contaminated during an oil spill in 1969 were actively degrading petroleum residues. This study revealed that sediment microbes utilized sediment organic matter rather than fossil organic matter, since the authors did not find evidence for incorporation of ^{14}C into PLFA. An implication of this finding is that hydrocarbon contamination at this site will likely persist indefinitely since it does not appear to be utilized by sediment microbes. In contrast, Wakeham and colleagues (2006) found that bacterial PLFAs in surface sediments at a contaminated site were depleted in ^{14}C, suggesting incorporation of petroleum-derived carbon into bacterial membrane lipids (Wakeham et al., 2006). A mass balance calculation indicated that 6 to 10% of the carbon in bacterial PLFAs at the contaminated site could derive from petroleum residues, suggesting that weathered petroleum may contain components of sufficient lability to support bacterial production in marsh sediments.

8.7 Summary

Fatty acids are one of the most versatile classes of lipid biomarkers due to the wide range in structural features (e.g., number of double bonds, functional group composition, and methyl branches). These features allow the fatty acids to be linked to a variety of sources (e.g., algae, bacteria, and higher plants) and to be useful over a range of timescales (e.g., ecological to paleo). In recent years, the application of compound-specific stable isotope and radiocarbon analyses has extended the applications of fatty acid biomarkers and allowed additional information to be retrieved from this class of compounds. As a result, new applications of this class of biomarkers are evolving and will likely become useful not only to organic geochemists but to microbial ecologists, paleoecologists, and environmental scientists from a broad array of disciplines.

9. Isoprenoid Lipids: Steroids, Hopanoids, and Triterpenoids

9.1 Background

Isoprenoids are a diverse class of naturally occurring organic compounds that are classified as lipids. This class of compounds has been the focus of studies in physiology, biochemistry, and natural products chemistry for 150 years (Patterson and Nes, 1991). Compounds within this class are derived from the five-carbon isoprene unit and many isoprenoids are multicyclic structures (fig. 9.1) that differ from one another both in their basic carbon skeletons as well as their functional group composition. Isoprenoids are classified according to both the number of isoprene units and the number of cyclic structures they contain (table 9.1; fig. 9.1). Isoprene units can be joined together in head-to-tail, tail-to-tail, and head-to-head configurations (fig. 9.2) and the linkages between isoprene units can be identified by the number of CH_2 groups between methyl branches (e.g., three CH_2 groups for head-to-tail, four CH_2 groups for tail-to-tail, and two CH_2 groups for head-to-head) (fig 9.2) (Hayes, 2001).

Subtle differences in functional group composition and in the number and positions of double bonds contribute to the versatility of this class of compounds and its use as a class of biomarkers. Isoprenoids are found in all classes of living organisms and are considered one of the most diverse classes of

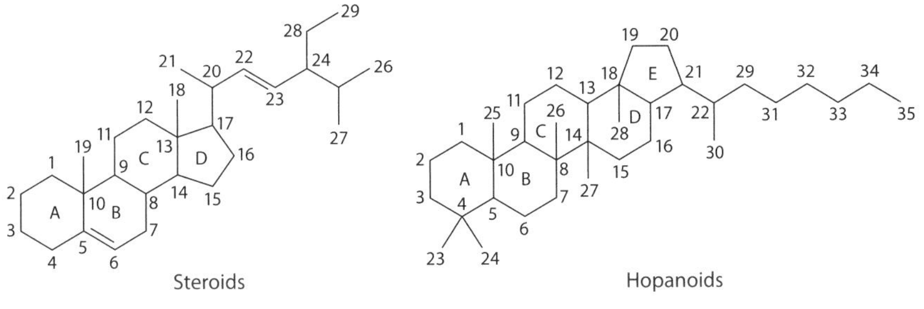

Figure 9.1. Steroid and hopanoid structures identifying each of the rings with a letter. The numbering of the carbon atoms for each of the skeletons is included. (Adapted from Killops and Killops, 2005.)

Table 9.1
Classification of terpenoids

Classification	*Number of isoprene units*
Monoterpenoids	2
Sesquiterpenoids	3
Diterpenoids	4
Sesterterpenoids	5
Triterpenoids	6
Tetraterpenoids	8
Polyterpenoids	>8

Source: Adapted from Killops and Killops (2005).

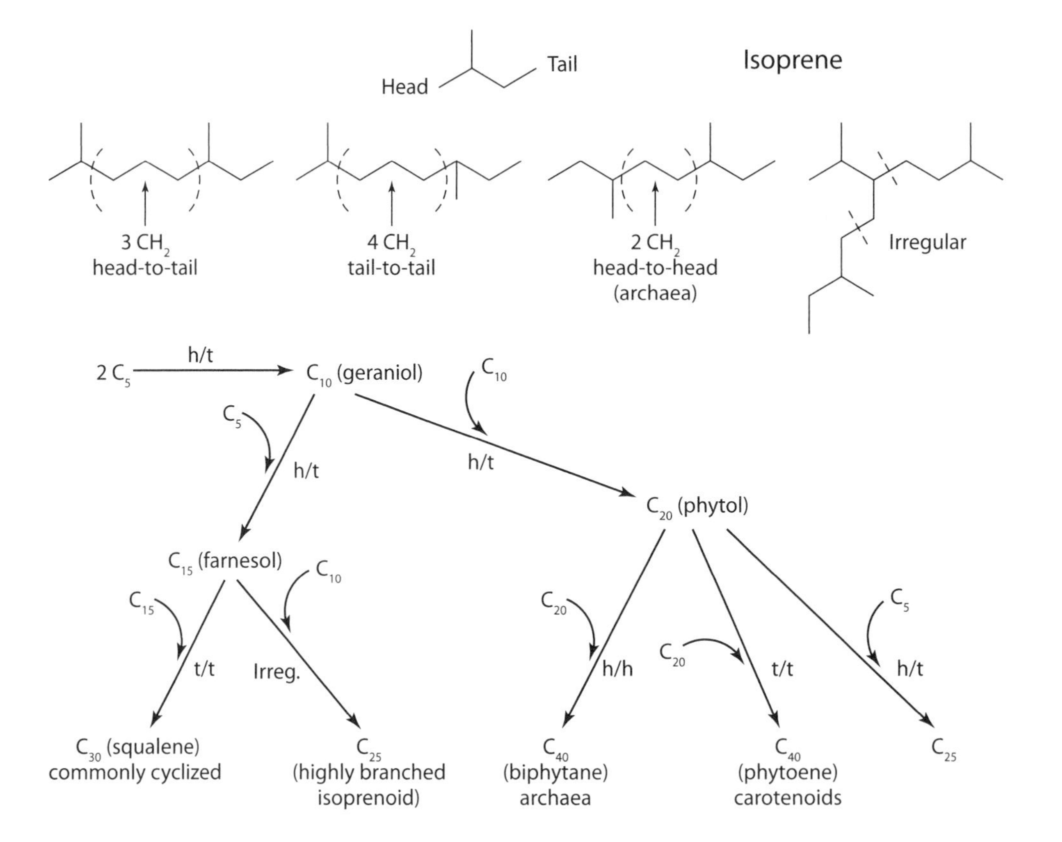

Figure 9.2. Structure of isoprene, the basic building block for all isoprenoids, and examples illustrating how isoprene units can be joined together by condensation reactions to form different isoprenoid compounds: head-to-tail (h/t), tail-to-tail (t/t), head-to-head (h/h), and irregular isoprenoids. (Adapted from Hayes, 2001.)

natural products (Sacchettini and Poulter, 1997). These compounds are essential components of the cell membrane in eukaryotic organisms and also occur in some prokaryotes. Isoprenoid hydrocarbons bound together by ether linkages are characteristic of the membrane lipids of Archaea see chapter 11. In contrast, eubacteria and eukaryotes synthesize cyclic isoprenoids, hopanoids, and steroids as lipid membrane constituents. These compounds are inserted between the glycolipids of the bilayer membranes with their polar ends in contact with the aqueous phase. The rigid nature of the ring structures of steroids and hopanoids limits the random motion of adjacent fatty acids chains closest to the polar head groups, while the aliphatic tail may interact with the distal portion of the adjacent fatty acid chains causing the bilayer to become more fluid (Peters et al., 2005a). In addition to the role of this class of compounds in controlling membrane fluidity and permeability, isoprenoids function as visual pigments, reproductive hormones, defensive agents, pheromones, signaling agents, and photoprotective agents (Sacchettini and Poulter, 1997).

Isoprenoids occur in both cyclic and acyclic forms. In this chapter, we focus on biomarker compounds belonging to the subclass, cyclic isoprenoids, which includes, *steroids*, *hopanoids*, and *triterpenoids*. Acyclic isoprenoids are covered in chapter 10. The cyclic isoprenoids have been widely used to understand Earth's history, the evolution of ancient organisms, and processes such as oxygenation of the Earth's early atmosphere (Brocks et al., 2003; Summons et al., 1999). They have also been widely used in studies examining the sources of contemporary organic matter associated with particulate organic matter and surface sediments (Volkman, 2006).

9.2 Biosynthesis of Terpenoids

Terpenoids (or isopentenoids) are defined as a class of cyclic biomarkers composed of isoprene subunits (C_5H_{10}). These compounds are synthesized from the universal C_5 building block isopentenyl diphosphate (IPP or IDP). IPP can be formed by two different pathways: the Bloch-Lynen (or mevalonate) pathway and the methylerythritol–phosphate (MEP) pathway (Hayes, 2001; Peters et al., 2005a; Volkman, 2005). In the Bloch-Lynen pathway, IPP is synthesized from three molecules of acetyl-CoA via mevalonate (the MVA pathway, see fig. 9.3). Alternatively, in some microorganisms, IPP can be synthesized via the MEP pathway, whereby pyruvate and glyceraldehyde form 1-deoxyxylulose 5-phosphate (DOXP) followed by intramolecular rearrangement and reduction to 2-C-methylerythritol 4-phosphate (MEP) (fig. 9.3). While animals, fungi, and Archaea use the MVA pathway, most eubacteria, some algae, protists, and land plants use the MEP pathway. In eukaryotes with chloroplasts, synthesis of IPP by the MEV pathway takes place in the nucleus and fluid within the cytoplasm, while the MEP pathway is used in the chloroplasts (table 9.2). Some organisms (e.g., some gram-positive eubacteria) have genes for both pathways, although both may not be activated (Peters et al., 2005a).

The condensation of IPP homologs leads to the synthesis of squalene, the precursor for steroid biosynthesis in plants and animals and hopanoid biosynthesis in eubacteria (fig. 9.4). The synthesis of hopanoids does not require molecular oxygen because the oxygen in the hopanoid skeleton originates from water. In contrast, many of the steps required for the biosynthesis of sterols require oxygen. For sterol synthesis, oxygen is required for squalene to be converted to squalene-2,3-oxide by the enzyme oxidosqualene cyclase. Subsequently, squalene-2,3-oxide is converted to either protosterolscycloartenol or lanosterol—by enzymatic cyclization. In plants, cycloartenol is the precursor for all steroids, while lanosterol is the precursor for steroid biosynthesis in animals (fig. 9.4). Details of the biosynthesis of sterols from cycloartenol (common in higher plants) and lanosterol (common in animals, protists, and many algae) are provided in Peters et al. (2005a). A thorough review of the current state of knowledge regarding the evolution of steroid and triterpenoid biosynthesis and their connection to oxygen in the environment is provided by Summons et al. (2006).

9.3 Sterol Biomarkers

Sterols are identified using the shorthand notation $C_x\Delta^y$, where x is the total number of carbon atoms and y indicates the positions of the double bonds (see numbering scheme for cholestane in fig. 9.1).

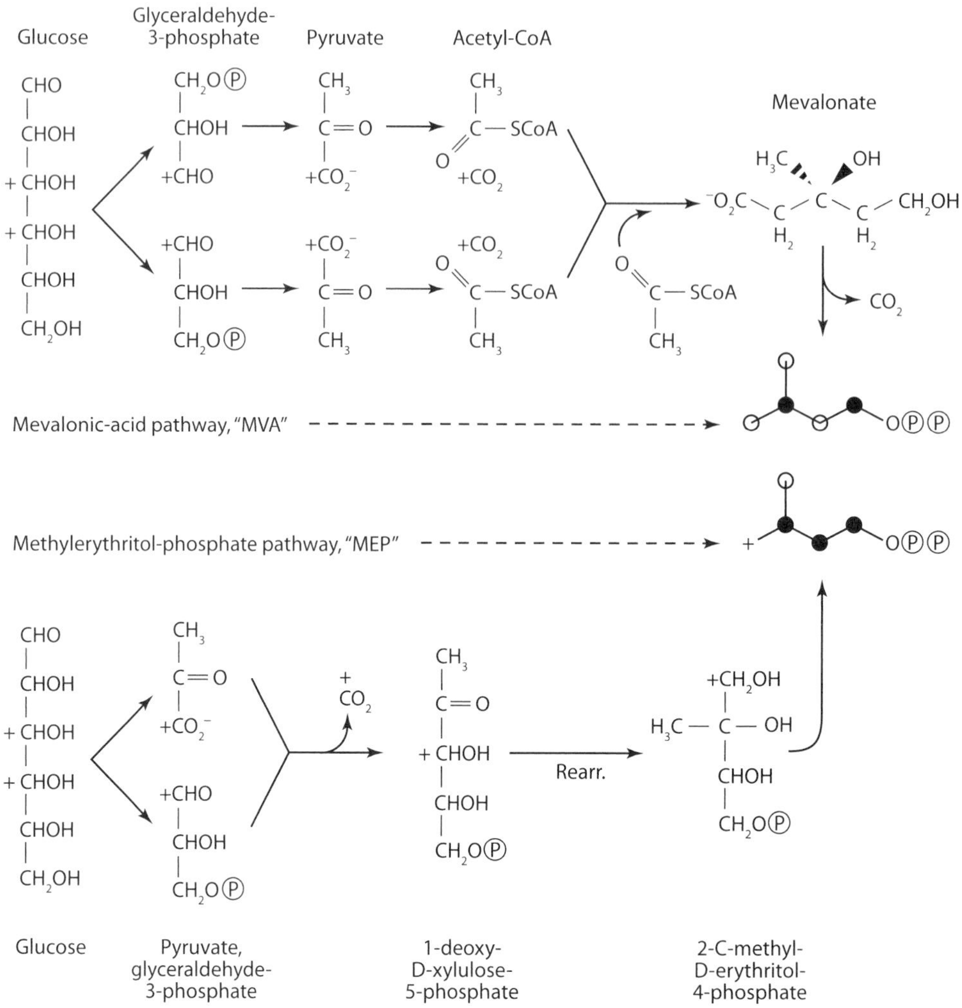

Figure 9.3. Figure showing the MVA and MEP pathways for synthesizing isopentenyl diphosphate, the basic building block for isoprenoids. (Adapted from Hayes, 2001.)

The position of the first (lowest-numbered) carbon atom involved in the double bond (identified as "en") and the carbon atom bonded to the hydroxy group are included in the naming (e.g., $C_{27}\Delta^5$ is named cholest-5-en-3β-ol). The second (higher-numbered) carbon atom involved in the double bond is given only when more than one possibility exists. In such cases, the number of the carbon atom is given in parentheses (e.g., $C_{28}\Delta^{5,24(28)}$ is the shorthand notation for 24-methylcholest-5,24(28)-dien-3β-ol). Many sterols also have alkyl side chains, often occurring at the C-24 position, which are represented in the naming scheme (e.g., C_{28} sterols typically have a methyl group at C-24, and C_{29} sterols typically have an ethyl group at C-24). In cases where a methyl group is missing from the basic C_{27} cholestane skeleton, the position of the missing carbon atom is indicated by "nor." The general name "stanol" is used to describe compounds with no unsaturated bonds (double bonds) in the ring portion of the skeleton (e.g., 5α (H)-stanols), while "stenol" refers to compounds with unsaturated bonds on one or more of the rings.

Table 9.2
Pathways used for the biosynthesis of isoprenoid lipids

Organism	*Pathway*		*References*[a]
Prokaryotes			Lange et al. (2000)
Bacteria			Boucher & Doolittle (2000)
Aquificales, Thermotogales	MEP		
Photosynthetic bacteria			
Chloroflexus	MVA		Rieder et al. (1998)
Chlorobium	MEP		Boucher & Doolittle (2000)
Gram-positive eubacteria			
Commonly	MEP		
Streptococcus, Staphylococcus	MVA		Boucher & Doolittle (2000)
Streptomyces	MEP & MVA		Seto et al. (1996)
Spirochaetes			Boucher & Doolittle (2000)
Borrelia burgdorferi	MVA		
Treponema pallidum	MEP		
Proteobacteria			
Commonly	MEP		
Myxococcus, Nannocystis	MVA		Kohl et al. (1983)
Cyanobacteria	MEP		Disch et al. (1998)
Archaea	MVA		Lange et al. (2000)
Eukaryotes			
Non-plastid-bearing	MVA		Lange et al. (2000)
Plastid-bearing	*Plastid*	*Cytosol*	
Chlorophyta[b]	MEP	MEP	Schwender et al. (2001)
Streptophyta[c]	MEP	MVA	Lichtenthaler et al. (1997)
Euglenoids	MVA	MVA	Lichtenthaler (1999)

[a] Where no reference is cited, either the assignment is based on generalizations introduced and supported in three major reviews—Lange et al. (2000), Boucher & Doolittle (2000), and Lichtenthaler (1999)—or it is based on work cited in those reviews, the first two of which are based on genetic analyses rather than labeling studies.
[b] Chlorophytes include the Trebouxiophyceae, Chlorophyceae, and Ulvophyceae.
[c] Streptophytes include higher plants and eukaryotic algae other than those named above.
Source: Hayes (2001).

Sterols are particularly useful biomarkers in aquatic systems because of the large diversity of compounds synthesized by microorganisms (table 9.3); (see review by Volkman et al., 1998). In particular, there is tremendous diversity in the sterol composition of microalgae, the dominant primary producers in aquatic environments. This is due to both the large number of classes of microalgae, genera, and species as well as the long evolutionary history of most classes of organisms. Some of these sterols are widespread, but others appear to be restricted to just a few algal classes. For example, 24-methylcholesta-5,24(28)-dien-3β-ol) is specific to some centric diatoms, such as *Thalassiosira* and *Skeletonema* (Barrett et al., 1995), while 24-methylcholesta-5,22E-dien-3β-ol occurs in some diatoms but is also present in some haptophytes and cryptophytes (Volkman, 1986; Volkman et al., 1998). In addition to sterol biomarkers for diatoms, other microalgae have unique sterols: for example, many species of green microalgae have sterols with double bonds at Δ^7, dinoflagellates biosynthesize 4-methyl sterols and the C_{30} sterol 4α, 23, 24-trimethyl-5α-cholest-22E-en-3β-ol, and 24-*n*-propylidenecholesterol is unique to some species of cryptophytes (Volkman, 2006). The "brown tide" algae *Aurecococcus* has also been shown to have 24-propylidenecholesterol. A thorough review of the sterol composition of microalgae is provided by Volkman et al. (1998). In addition to the use

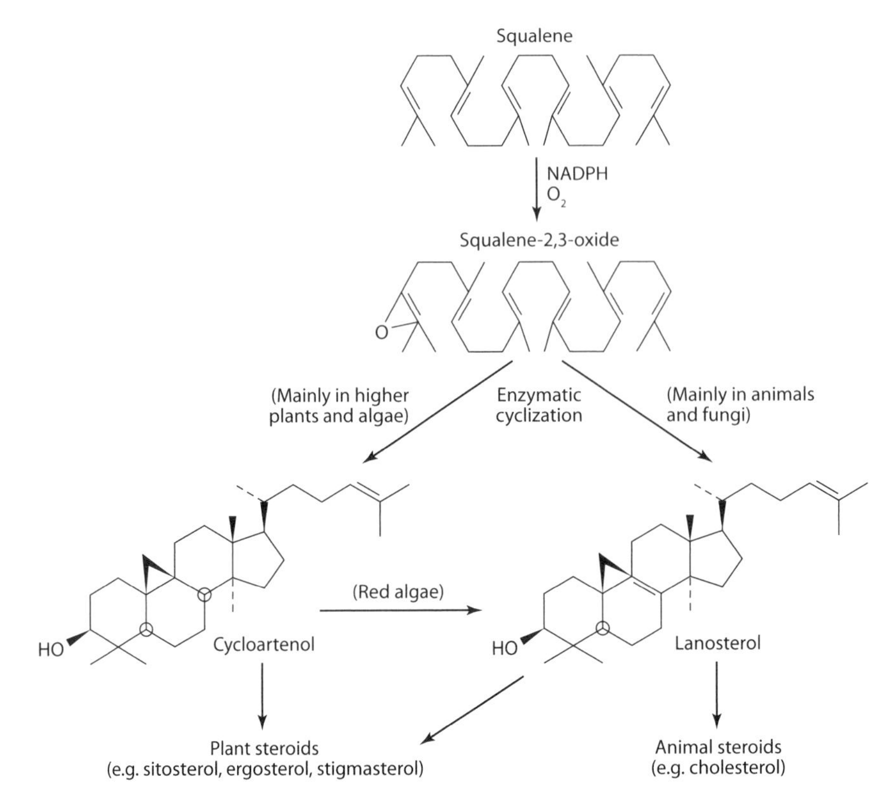

Figure 9.4. Synthesis of cycloartenol and lanosterol from squalene. Cycloartenol and lanosterol are the precursors for sterol synthesis in higher plants vs. animals, protists, and algae, respectively. (Adapted from Killops and Killops, 2005.)

of sterols to trace inputs of aquatic organic matter from microalgae, compounds from this class of biomarkers have been identified in methylotrophic bacteria, yeasts, and fungi (table 9.3) (Volkman, 2003).

In recent years, there has been some question about whether organisms such as cyanobacteria and prokaryotes are capable of synthesizing sterols (Volkman, 2003). Recently, molecular biologists have identified isoprenoid lipid synthesis genes, providing a tool for identifying whether organisms are capable of biosynthesizing specific biomarker compounds (Volkman, 2005). Another recent paper describes the presence of tetracyclic isoprenoids in the spores of the bacterium *Bacillus subtilis* (Bosak et al., 2008) (see chapter 7). Advances using molecular genetics methods will allow organic geochemists to use isoprenoid biomarkers to assign sources of organic matter with more confidence and increases the possibility of using molecular tools (rDNA) to identify microbial contributions from particular organisms in recently deposited sediments. Additionally, the presence of isoprenoid biomarkers in ancient sediments may provide information about the existence of specific biosynthetic pathways throughout Earth's history.

9.4 Analysis and Application of Sterol Biomarkers

Sterols are generally isolated from the total lipid extract by solid-phase chromatography and analyzed by gas chromatography following derivatization of the hydroxy group. The two common methods for

Table 9.3
Major sterols in microorganisms

Microorganism	*Major or common sterols*[a]
Microalgae	
Bacillariophyceae	$C_{28}\Delta^{5,22}$, $C_{28}\Delta^{5,24(28)}$, $C_{27}\Delta^{5}$, $C_{29}\Delta^{5}$, $C_{27}\Delta^{5,22}$
Bangiophyceae	$C_{27}\Delta^{5}$, $C_{27}\Delta^{5,22}$, $C_{28}\Delta^{7,22}$
Chlorophyceae	$C_{28}\Delta^{5}$, $C_{28}\Delta^{5,7,22}$, $C_{28}\Delta^{7,22}$, $C_{29}\Delta^{5,22}$, $C_{29}\Delta^{5}$
Chrysophyceae	$C_{29}\Delta^{5,22}$, $C_{29}\Delta^{5}$, $C_{28}\Delta^{5,22}$
Cryptophyceae	$C_{28}\Delta^{5,22}$
Dinophyceae	4Me-Δ^{0}, dinosterol, $C_{27}\Delta^{5}$, $C_{28}\Delta^{5,24(28)}$
Euglenophyceae	$C_{28}\Delta^{5,7,22}$, $C_{29}\Delta^{5}$, $C_{28}\Delta^{7}$, $C_{29}\Delta^{5,7}$, $C_{28}\Delta^{7,22}$
Eustigmatophyceae	$C_{27}\Delta^{5}$ (marine) or $C_{29}\Delta^{5}$ (freshwater)
Haptophyceae	$C_{28}\Delta^{5,22}$, $C_{27}\Delta^{5}$, $C_{29}\Delta^{5,22}$, $C_{29}\Delta^{5}$
Pelagophyceae	$C_{30}\Delta^{5,24(28)}$, $C_{29}\Delta^{5,22}$, $C_{29}\Delta^{5}$, $C_{28}\Delta^{5,24(28)}$
Prasinophyceae	$C_{28}\Delta^{5}$, $C_{28}\Delta^{5,24(28)}$, $C_{28}\Delta^{5}$
Raphidophyceae	$C_{29}\Delta^{5}$, $C_{28}\Delta^{5,24(28)}$
Rhodophyceae	$C_{27}\Delta^{5}$, $C_{27}\Delta^{5,22}$
Xanthophyceae	$C_{29}\Delta^{5}$, $C_{27}\Delta^{5}$
Cyanobacteria:	$C_{27}\Delta^{5}$, $C_{29}\Delta^{5}$, $C_{27}\Delta^{0}$, $C_{29}\Delta^{0}$ (evidence equivocal)
Methylotrophic bacteria	4Me-Δ^{8}
Other bacteria	$C_{27}\Delta^{5}$
Yeasts and fungi	$C_{28}\Delta^{5,7,22}$, $C_{28}\Delta^{7}$, $C_{28}\Delta^{7,24(28)}$
Thraustochytrids	$C_{27}\Delta^{5}$, $C_{29}\Delta^{5,22}$, $C_{28}\Delta^{5,22}$, $C_{29}\Delta^{5,7,22}$

[a]The nomenclature is $C_x\Delta^y$ when: x is the total number of carbon atoms and y indicates the positions of the double bonds, In general, C_{28} sterols have a methyl group at C-24, and C_{29} sterols have a 24-ethyl substituent. Table adapted from data in Volkman (1986), Jones et al. (1994), and Volkrnan et al. (1998).
Source: Volkman (2003).

derivatization are conversion to acetates using acetic anhydride and conversion to trimethylsilyl ethers (TMS-ethers) using bis(trimethylsilyl)trifluoroacetamide (Brooks et al., 1968). Derivatization to TMS-ethers provides diagnostic mass spectra (m/z 129 is diagnostic for TMS-sterols), but these derivatives are less stable than acetates. Identification of sterols as TMS-ethers is usually verified using gas chromatography–mass spectrometry, where specific fragments can be used to identify compounds. For example, sterols with Δ^5-unsaturations (double bonds) have a base peak at m/z 129 and fragmentation ions at M^+ -90 and M^+ -129 (Volkman, 2006). Saturated sterols (stanols) usually elute following their unsaturated precursor and are characterized by m/z 215 as the base peak or m/z 229 when the compound includes a 4-methyl group. There are several good compilations of retention time information and characteristic mass spectra for sterols and sterol degradation products (Wardroper, 1979; Peters et al., 2005a).

Sterols and their diagenetic products are common in aquatic systems and sediments, reflecting their contributions from multiple sources, including microalgae, higher plants, and animals. Additionally, sterols can be used to trace anthropogenic inputs through the use of compounds specific for sewage—coprostanol (5β-cholestan-3β-ol) and its epimer, epicoprostanol (5β-cholestan-3β-ol) (fig. 9.5; table 9.4) (Laureillard and Saliot, 1993; Seguel et al., 2001; Carreira et al., 2004). These compounds have also been used to examine changes in sewage inputs over time using sediment cores (Carreira et al., 2004) and as an archaeological tool (Bull et al., 2003). In cases where sewage sterols are present only at low concentrations, questions remain about their usefulness for tracing anthropogenic inputs (Sherblom et al., 1997). However, combining information about the distribution of fecal sterols with additional tracers for sewage and/or anthropogenic inputs can help resolve these issues

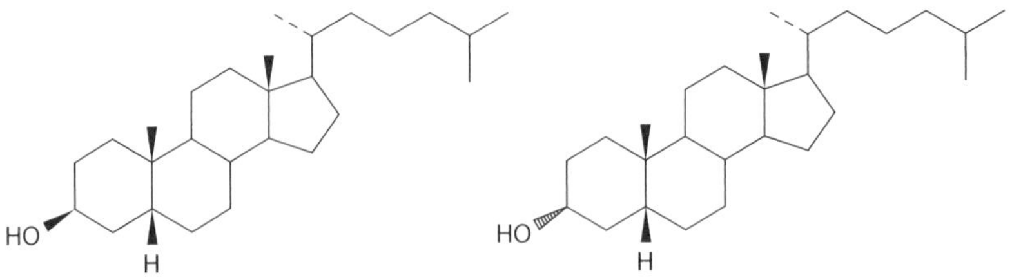

5β-cholestan-3β-ol (coprostanol)

5β-cholestan-3α-ol (epicoprostanol)

Figure 9.5. Structures of coprostanol and epicoprostanol, sterol biomarkers for sewage.

Table 9.4
Concentration of coprostanol in surface sediments of temperate and tropical estuaries

Location	*Sediment layer (cm)*[a]	*n*[b]	*Year of research*	*Concentration range* ($\mu g\,g^{-1}$)	*References*
Tamar Estuary, UK	0–3	8	1985	0.80–17.0	Readman et al. (1986)
Santa Monica Basin, USA	0–2	6	1985	0.50–5.10	Venkatesan and Kaplan (1990)
Barcelona, Spain	0–3	10	1986–87	1.0–390.0	Grimalt and Albaiges (1990)
Havana Bay, Cuba	0–3	2	1986–87	0.41–1.10	Grimalt et al. (1990)
Morlaix River Estuary, France	n.i.	10	1987–88	0.7-30.0	Quémcnéur and Marty (1992)
Narragansett Bay, USA	Superficial	25	1985–86	0.13–39.3	LeBlanc et al. (1992)
Venice Lagoon, Italy	0–5	25	1986	0.20–41.0	Sherwin et al. (1993)
Tokyo Bay, Japan	n.i.	—	1989	0.02–0.24	Chalaux et al. (1995)
SE coast of Taiwan	0–4	24	1992	<0.05–0.82	Jeng and Han (1996)
Hamilton Port, Lake Ontario, Canada	0–2	30	1992	0.11–147.0	Bachtiar et al. (1996)
Ria Formosa, Algarve, Portugal	n.i.	—	1994	0.10–41.8	Mudge and Bebianno (1997)
Bilbao Estuary, Spain	0–0.5	20	1995–96	2.20–293.0	González-Oreja and Saiz-Salinas (1998)
Capibaribe River Estuary, Brazil	0–3	10	1994	0.52–7.31	Fernandes et al. (1999)
Boston Harbor, USA	0–2	8	1988	0.26–12.0	Eganhouse and Sherblom (2001)
Guanabara Bay, Brazil	0–3	8	1996	0.33–40.0	Carriera et al. (2004)

[a] n.i. = not informed.
[b] *n* = number of samples collected.
Source: Carriera et al. (2004).

(Leeming et al., 1997). Additional complexities associated with the application of sewage-derived sterols include transformation of compounds into 5β(H)-stanols in highly reducing sediments (Volkman, 1986) and contributions from marine and domesticated mammals (Venkatesan and Santiago, 1989; Tyagi et al., 2008).

Studies in several environments have utilized sterols to trace contributions of organic matter from autochthonous and allochthonous sources. Sterols representing plankton and vascular plant sources have been used to identify contributions from aquatic and terrigenous sources in polar, temperate, and tropical systems (Yunker et al., 1995; Mudge and Norris, 1997; Belicka et al., 2004; Xu and Jaffè, 2007; Waterson and Canuel, 2008). Additionally, sterol biomarkers have been used to compare the composition and sources of particulate and dissolved organic matter obtained using ultrafiltration (Mannino and Harvey, 1999; Loh et al., 2006; McCallister et al., 2006). These studies generally show elevated concentrations of sterols from higher plants in the dissolved phases at the freshwater end member and decreasing concentrations with distance downstream (fig. 9.6). Contributions of sterols from plankton and algae are generally higher in the particulate than in the dissolved phases, with maxima located in the region of the chlorophyll maximum located downstream of the estuarine turbidity maximum.

Additionally, sterols have been used in surface sediments and sediment cores collected from lacustrine, estuarine, and marine environments to trace changes in organic matter inputs over time (Meyers and Ishiwatari, 1993; Meyers, 1997). Several studies have investigated sources of organic matter associated with suspended particles (Wakeham and Ertel, 1988; Wakeham and Canuel, 1990; Wakeham, 1995) and recently deposited sediments (Mudge and Norris, 1997; Zimmerman and Canuel, 2001; Volkman et al., 2007). Additionally, studies using sediment core records have revealed changes in organic matter delivery arising from processes such as eutrophication (Meyers and Ishiwatari, 1993; Zimmerman and Canuel, 2000; 2002), productivity and paleoproductivity (Hinrichs et al., 1999; Mèjanelle and Laureillard, 2008), and climate (Kennedy and Brassell, 1992).

The pathways for sterol diagenesis are well established (fig. 9.7), allowing application of both sterols and their decomposition products to identify the origins of organic matter. Studies utilizing sterol biomarkers range from those examining sources of organic matter in contemporary ecosystems (see above) to the use of diagenetic products of sterols in ancient sediments, rocks, and petroleum (Summons et al., 1987; Pratt et al., 1991; McCaffrey et al., 1994). Several studies have demonstrated changes in sterol composition resulting from early diagenesis in surface sediments (Conte et al., 1994; Harvey and Macko, 1997; Sun and Wakeham, 1998). Additional studies have shown evidence for the conversion of stenols to stanols across the oxic–anoxic interface of systems characterized by anoxic water columns (Wakeham and Ertel, 1988; Beier et al., 1991; Wakeham, 1995). The production of sterenes (steroidal alkenes) provides additional evidence for the transformation of biogenic sterols to the degradation products found in mature sediments (Wakeham et al., 1984).

Stable isotopes ($\delta^{13}C$ and δD) and radiocarbon ($\Delta^{14}C$) signatures of sterols have been used as complementary information for tracing sources of organic matter in the environment. In the case of carbon isotopes, these methods can be useful for differentiating between sources of particular biomarkers in cases where there are multiple sources for the compound and they are isotopically distinct. For example, $\delta^{13}C$ signatures of 24-ethylcholest-5-en-3β-ol in a sediment core collected off the coast of Japan revealed that this compound was derived from marine algae (-24.2 ± 1.1 per mil, on average) in sediments deposited since the last glacial period (10- to 73-cm section; 2–9 ka). However, in the last glacial stage (140- to 274-cm section; 15–29 ka), 24-ethylcholest-5-en-3β-ol most likely reflected contributions from terrestrial higher plants (-30.5 ± 0.9 per mil on average) (Matsumoto et al., 2001). An excellent review of the biosynthesis of sterols and isoprenoids and factors controlling fractionation of carbon and hydrogen at the molecular level is provided in Hayes (2001).

Compound-specific isotope analysis of sterols has been used to identify plankton and vascular plant sources in complex environments characterized by multiple sources (Canuel et al., 1997; Pancost et al., 1999; Shi et al., 2001). Additionally, radiocarbon isotopic signatures of sterols have been examined in recent studies (Pearson et al., 2000, 2001; Pearson, 2007). Data from these studies showed that the carbon source for the majority of the sterol biomarkers is marine primary production from the euphotic zone or subsequent heterotrophic processing of this recently produced biomass.

In recent years, Chikaraishi and colleagues have pioneered the application of combined $\delta^{13}C$ and δD isotopes of sterols for tracing sources of organic matter in the environment (Chikaraishi and

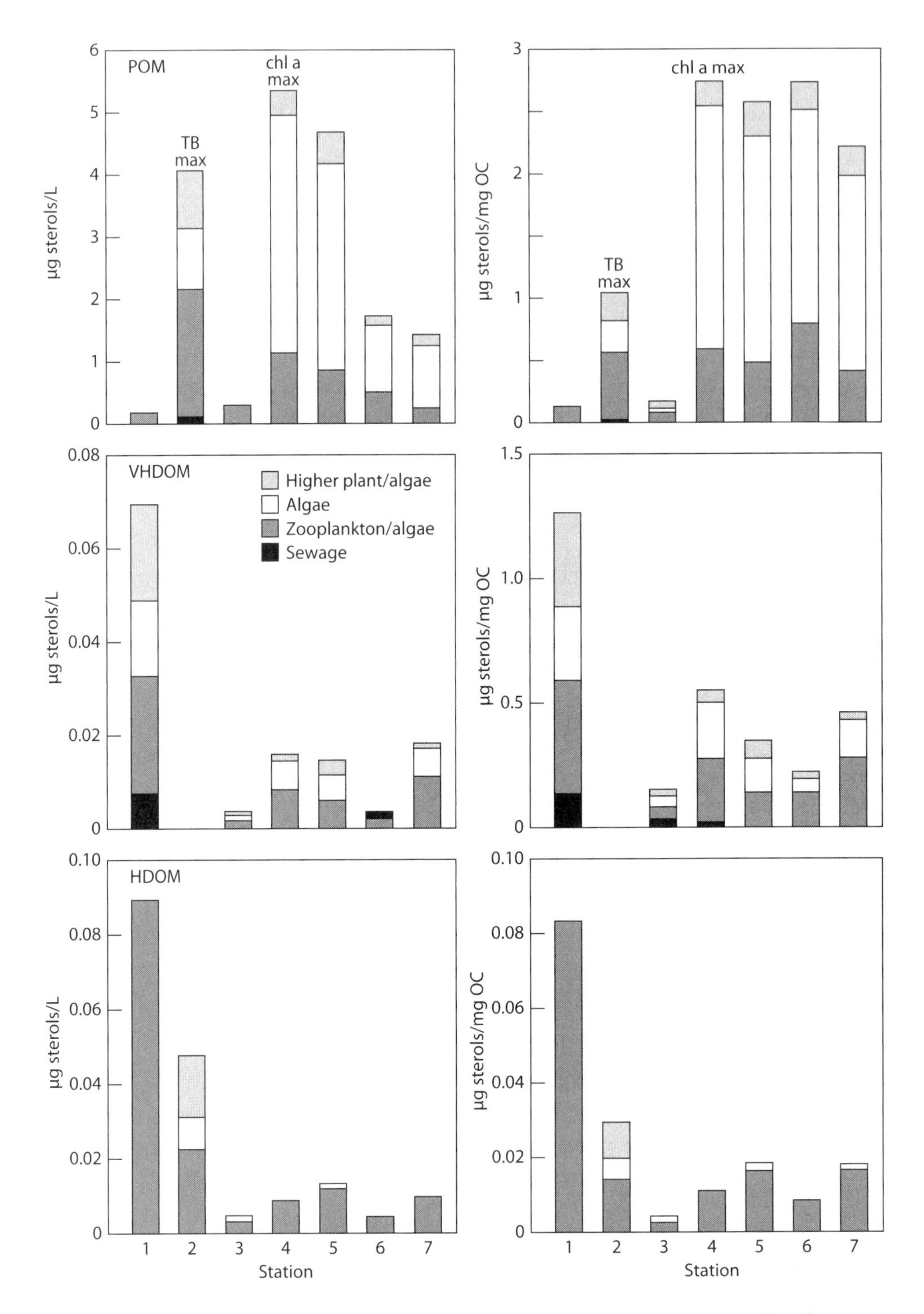

Figure 9.6. Changes in sterol abundance and composition of particulate organic matter (POM), very high-molecular-weight dissolved organic matter (VHDOM; 30 kDa to 0.2 μm), and high-molecular-weight DOM (HDOM; 1–30 kDa), along the Delaware Estuary (USA). (Adapted from Mannino and Harvey, 1999.)

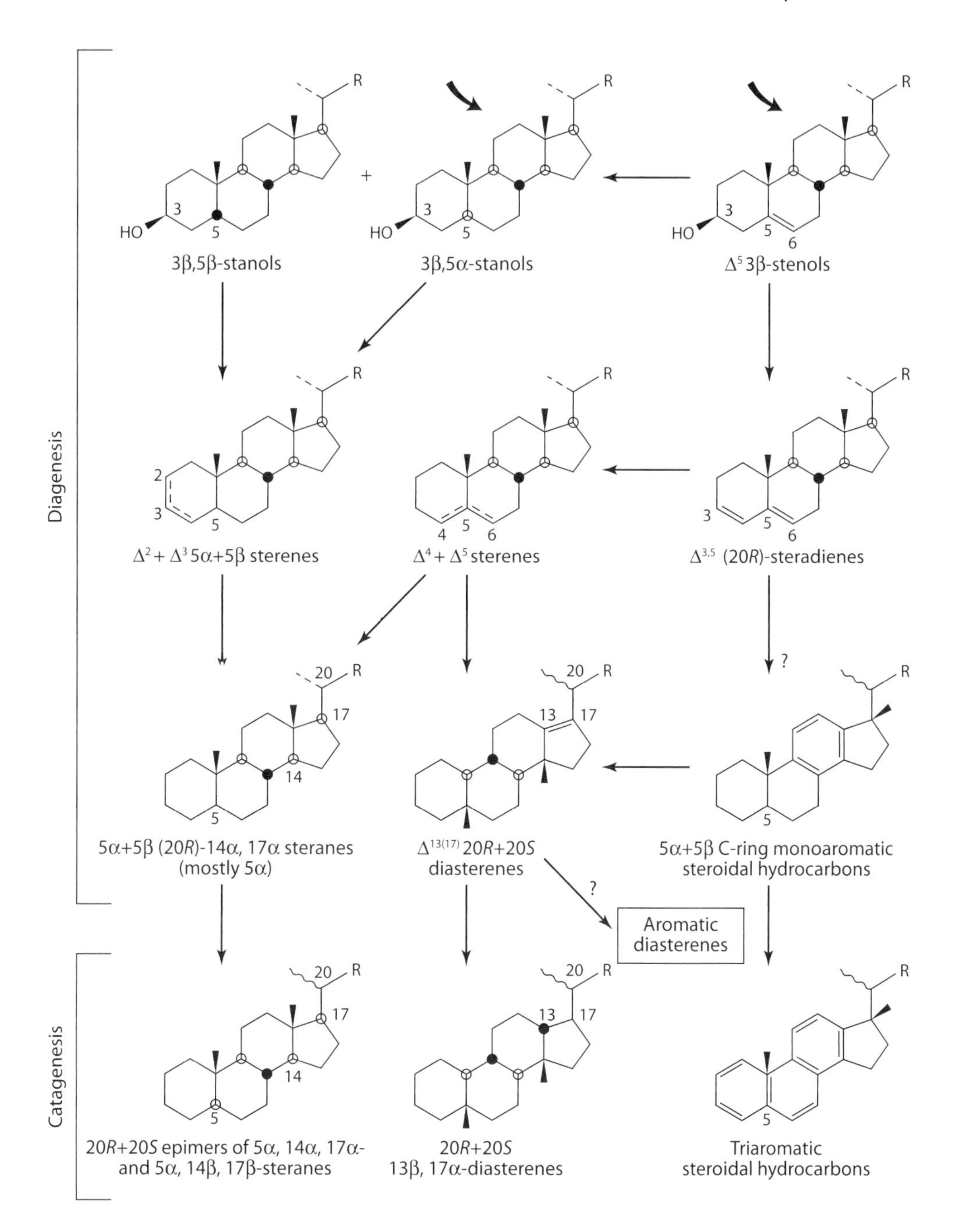

Figure 9.7. Conversion of stenols to stanols, stanols to sterenes, sterenes to steranes, and steranes to diasteranes and aromatic cyclic hydrocarbons during diagenesis and catagenesis. (Adapted from Killops and Killops, 2005.)

Naraoka, 2005; Chikaraishi, 2006). δD isotopic signatures offer a proxy for identifying the isotopic (D/H) signature of environmental water, providing additional paleoenvironmental information. Sauer and colleagues tested the efficacy of this method (Sauer et al., 2001). They found that the fractionation between sedimentary lipids and environmental water for algal sterols (24-methylcholest-3β-ol,

24-ethylcholest-5,22-dien-3β-ol, and 4,23,24-trimethylcholesterol, or dinosterol), was -201 ± 10 per mil in both marine and freshwater sites. Based on the consistency of these isotopic signatures, the authors concluded that hydrogen-isotopic analyses of algal sterols provided a reliable tool for reconstructing the isotopic signature (D/H) of environmental waters. Additional advances in the stable and radiocarbon isotope analysis of sterols holds considerable promise for elucidating present and past records of organic matter sources as well as paleoenvironment information.

9.5 Related Compounds: Steroid Ketones, Steryl Chlorin Esters, and Steryl Esters

Several classes of compounds related to sterols have been found in marine environments (Gagosian et al., 1980; Bayona et al., 1989; Volkman, 2006). Steroidal ketones may derive from oxidation of Δ^5-unsaturated sterols by microbial processes occurring in surface sediments during early diagenesis. These compounds may also reflect contributions from some microalgae such as dinoflagellates. The dinoflagellate *Scrippsiella*, for example, contains a diverse assemblage of steroidal ketones (Harvey et al., 1988).

Steryl chlorin esters (SCEs) can be formed through reaction of sterols with pyropheophorbide *a*, a degradation product of the pigment chlorophyll *a*. SCEs have been associated with both water column particulate matter and surface sediments in marine environments (King and Repeta, 1994; Goericke et al., 1999; Talbot et al., 1999; Riffè-Chalard et al., 2000). They have also been identified in lake sediments (Squier et al., 2002; Itoh et al., 2007). It is thought that these compounds are formed during zooplankton grazing (King and Wakeham, 1996; Dachs et al., 1998; Talbot et al., 2000; Chen et al., 2003a,b). Because SCEs are stable in the environment, they provide a mechanism for preserving phytoplankton biomarkers in sediments (Talbot et al., 1999; Volkman, 2006).

There are few reports of intact steryl esters in the environment. The best direct evidence for these compounds is their identification in sediment trap samples from the North Atlantic (Wakeham, 1982; Wakeham and Frew, 1982). The identified compounds were dominated by C_{16} and C_{18} fatty acids esterified to cholesterol, suggesting a zooplankton source (Volkman, 2006). Similarly, there are limited observations of steryl ethers in environmental samples (Schouten et al., 2000).

9.6 Hopanoids

Hopanoids (also sometimes called bacteriohopanoids) are a diverse class of natural products and geolipids (Ourisson and Albrecht, 1992), belonging to the general class of compounds triterpenoids. Triterpenoids are C_{30} isoprenoids characterized by either a 5-membered or 6-membered E-ring (hopanoids and triterpenoid alcohols, respectively; see hopanoid structure in fig 9.1). Bacteria synthesize hopanoids from squalene using a pathway similar to that used by other organisms to synthesize sterols from oxidosqualene (see section 9.2 above and fig. 9.8). As described in section 9.2, an important difference between steroid and triterpenoid biosynthesis is that the synthesis of triterpenoids does not require molecular oxygen (Summons et al., 2006). Hopanoids are characterized by diverse functional group composition and include alkenes, ketones, acids, and alcohols. Hopanoids can be analyzed easily using gas chromatography–molecular spectroscopy (GC-MS), and are readily identifiable by their m/z 191 mass fragment (or m/z 205 for hopanoids with a methyl group on the A ring or m/z 177 for demethylated hopanes).

Hopanoids reflect both direct contributions from bacterial biomass and are products of diagenesis. Interestingly, the hopanoid bacteriohopane was first identified in sediments. Based on this discovery and that of related compounds, investigators predicted that bacteriohopanoids derived from the biosphere, and their occurrence in bacteria was subsequently confirmed (Ourisson et al., 1979). Hopanoids are

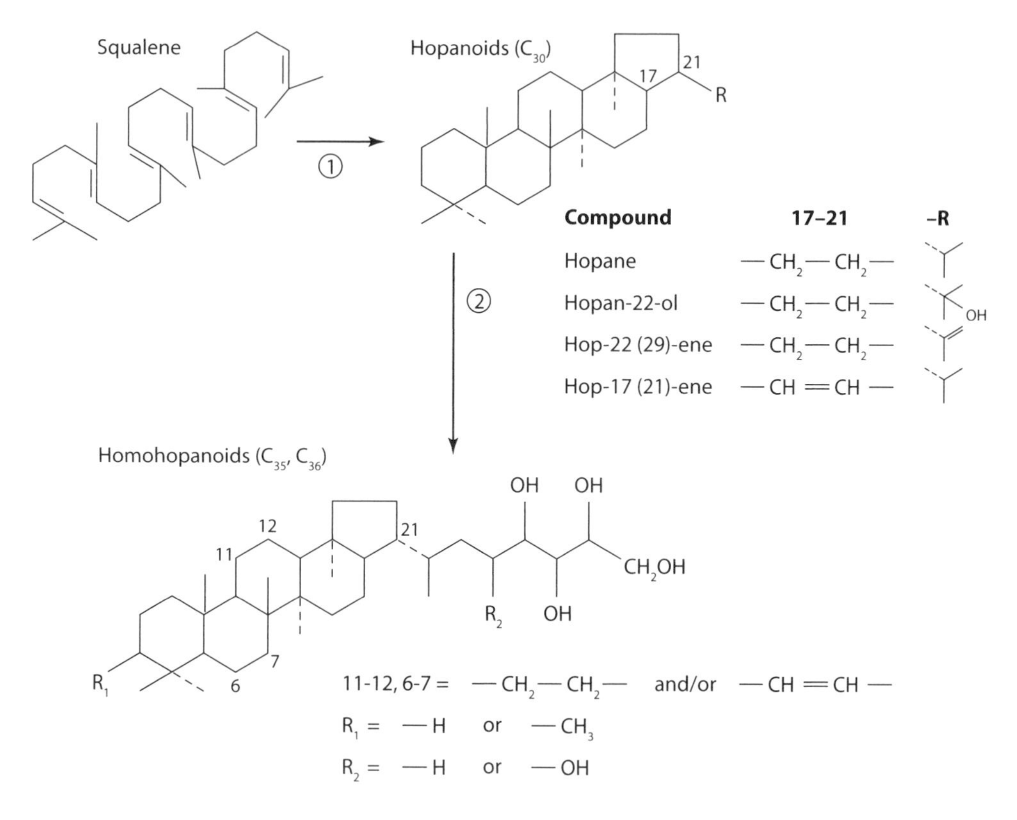

Figure 9.8. Synthesis of the bacterial biomarkers hopanoids and homohopanoids from squalene. (Adapted from Taylor, 1984.)

common cell membrane components in eubacteria and have been identified in both gram-positive and gram-negative organisms. Hopanoids often occur in strains with high G+C (guanine+cytosine), which may reflect the need to provide greater membrane stability in response to osmotic stress (Kannenberg and Poralla, 1999). These compounds have also been identified in cyanobacteria, methanotrophic bacteria, and members of the α-proteobacteria (Peters et al., 2005a). Talbot and colleagues identified bacteriohopanepolyols (BHPs) in cultured cyanobacteria (pure cultures and enrichment cultures) representing a variety of environments, including marine, freshwater, and hydrothermal systems (Talbot et al., 2008).

Hopanoids have been identified in recent and ancient sediments, rocks, and petroleum. 17β, 21β-Hopanols were identified in surface sediments collected as part of a study comparing several lakes in Spain representing a range in salinity and hydrological (shallow ephemeral to deep permanent) conditions. Hopanols comprised up to 4.5% total lipids, with the highest values found in deeper, seasonal, and permanent sites, likely representing input from bacterial activity (Pearson et al., 2007). Hopanols were also found in surface sediments collected along the salinity gradient in Chesapeake Bay (USA) (Zimmerman and Canuel, 2001). Recent data suggest that hopanol composition may vary in response to environmental conditions. Farrimond and colleagues recently found that the composition of bacteriohopanepolyols varies between depositional environments. While hexafunctionalized biohopanoids were most abundant in lake sediments, most likely due to contributions from type I methanotrophic bacteria (which produce dominantly hexafunctionalized biohopanoids), these compounds were minor constituents (2–6% of total biohopanoids) in marine environments where sulfate reduction is dominant (Farrimond et al., 2000).

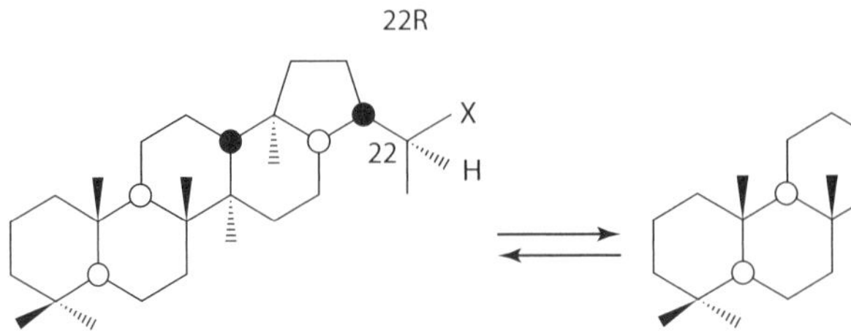

22R

22S

x = n-C_2H_5, n-C_3H_7, n-C_4H_9, n-C_5H_{11}, n-C_6H_{13}

Figure 9.9. Biological hopanes are synthesized with the R configuration at C-22 but are converted to equal abundances of R and S diasteromers over time. (Adapted from Peters et al., 2005.)

Hopanes, hydrocarbon derivatives of hopanoids, are well preserved in the sediment record and are the archetypal geolipids. For example, the presence of hydrocarbon products of the 2-methyl-bacteriohopanepolyols, biomarkers for cyanobacteria, have been used to constrain the age of the oldest cyanobacteria and the onset of oxygenic photosynthesis (Summons et al., 1999; Brocks et al., 2003). Hopanes are also used to identify source characteristics and maturity of petroleum. The ratio of C_{31}/C_{30} hopane, for example, can be used to identify marine vs. lacustrine depositional environments (Peters et al., 2005b). 28,30-Bisnorhopane (BNH) and 25,28,30-trisnorhopane (TNH) are typical of petroleum source rocks deposited under anoxic conditions. BNH is thought to derive from chemoautotrophic bacteria that live at the oxic–anoxic interface (Peters et al., 2005b). In addition, several pairs of isomers of steranes (alkane products of sterols) and hopanes are useful for assessing the maturity of petroleum. Because isomerization occurs preferentially at specific positions, ratios of these isomers can provide information about the thermal maturity of petroleum. For example, biological hopanes are synthesized with the R configuration at C-22, which is converted to a mixture of 22R and 22S diasteromers over time (fig. 9.9). Hence, ratios of 22S/(22S+22R) provide an index of maturity (fig. 9.10). These ratios can be useful tools for application in the petroleum geochemistry but also in identifying the sources of petroleum in the environment from spills or seeps.

9.7 Triterpenoids

Most triterpenoids with a 6-membered E ring (fig. 9.11) are of higher plant origin, where they function as resins. These compounds are characterized into three series: the oleanoids (e.g., β-amyrin), ursanoids (e.g., α-amyrin), and lupanoids (e.g., lupeol) (fig. 9.11). Like hopanoids, these compounds can have various functional groups and are found in the environment as alcohols, alkenes and alkanes. They can be readily identified by GC-MS using the m/z 191 fragment.

Recent studies have used triterpenoid alcohols as biomarkers for tracing higher plant inputs to mangrove environments (Koch et al., 2003) and tropical river–estuarine systems (Volkman et al., 2007). In the tropical Ord River (Australia), Volkman and colleagues found that triterpenoid alcohols were dominated by α-amyrin (urs-12-en-3β-ol), β-amyrin (olean-12-en-3β-ol), lupeol (lup-20(29)-en-3β-ol), and taraxerol (taraxer-14-en-3β-ol). Minor amounts of triterpenoid diols were detected in some samples, with the highest concentrations in marine sediments, and these compounds were rarely detected in the freshwater sediments. A study in northern Brazil noted the usefulness of triterpenoid alcohols for tracing organic matter derived from mangroves (Koch et al., 2003).

In addition to the application of this class of biomarkers for tracing inputs of terrigenous organic matter to contemporary sediments, diagenetic products of triterpenoid alcohols can be used to trace terrigenous sources of organic matter in the historical and geological past. Similar to steroids, these

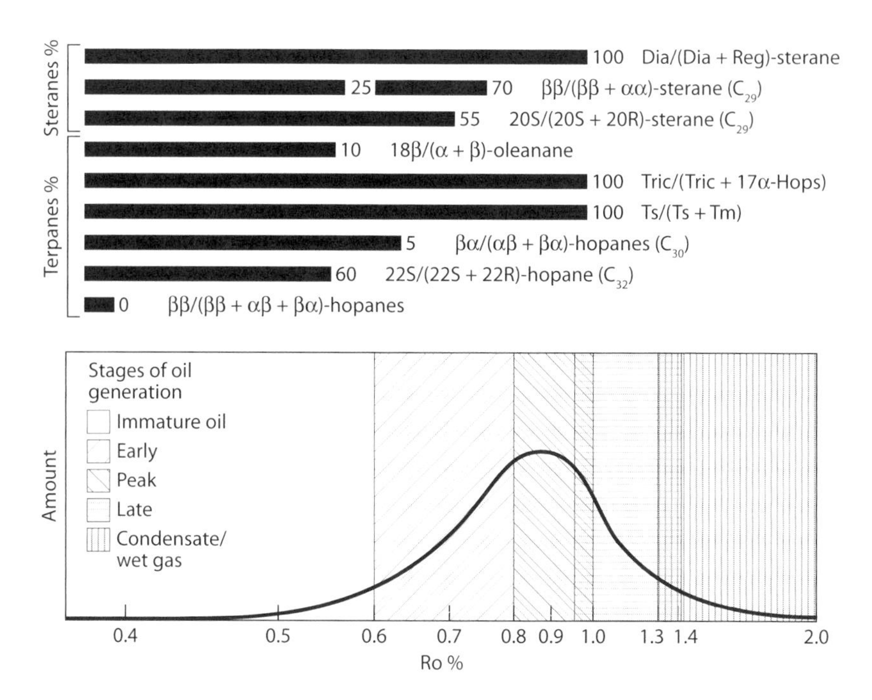

Figure 9.10. The upper panel provides ratios of isomers of steranes and hopanes useful for estimating the maturity of petroleum. The lower panel illustrates the stages in oil generation, beginning with immature oil and ending with condensate/wet gas, showing that the isomer ratios are appropriate for different stages of the maturation process. (Adapted from Peters et al., 2005.)

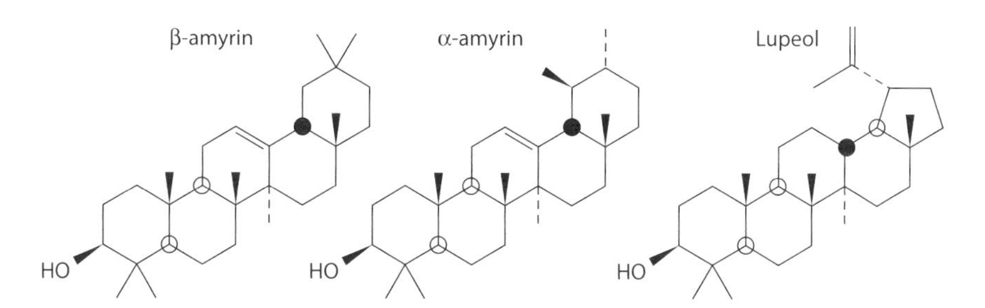

Figure 9.11. Structures of triterpenoids common to higher plants. The compounds were selected to provide examples of each of the three series of pentacyclic triterpenoids—the oleanoid (e.g, β-amyrin), ursanoid (e.g., α-amyrin), and lupanoid (e.g., lupeol). (Adapted from Killops and Killops, 2005.)

compounds undergo defunctionalization during early diagenesis and reduction and aromatization during later stages of diagenesis or catagenesis. Laboratory experiments and observations from sediment cores have contributed to our understanding of the pathways for diagenesis of triterpenols (ten Haven et al., 1992a,b). These studies suggest that both chemical and bacterial processes are likely important in the conversion of 3α- and 3β-triterpenols to Δ^2 -triterpenes by dehydration reactions (see proposed reaction scheme [Killops and Killops, 2005]). Studies such as those conducted by ten Haven and colleagues have provided insights about the decomposition and diagenetic products of triterpenoids. These and similar studies are important for interpreting the distribution of geolipids, which are well preserved in the sediment and rock records.

9.8 Summary

The cyclic isoprenoids and their diagenetic products provide a useful class of biomarkers for reconstructing inputs of organic matter from algal (steroids), bacterial (hopanoids), and vascular plant (steroids and triterpenoids) sources through paleolimnologic and paleoceanographic studies. Future efforts should focus on increasing our understanding of the diagenesis of these compound classes in order to assess the integrity and robustness of sediment core records. There are several excellent compilations of the sterol composition of microalgae and vascular plants. Additional focus should be place on developing similar libraries of the hopanol and triterpenol composition for a broader range of organisms. Through studies of contemporary environments, quantitative relationships between biomarkers preserved in surface sediments and contributions from their biological sources can be developed. Lastly, there is tremendous potential to employ compound-specific isotope analysis ($\delta^{13}C$, δD, ^{14}C) of the steroids, hopanoids, and triterpenoids to gain environmental and paleoenvironmental information locked into these structures during their biosynthesis.

10. Lipids: Hydrocarbons

In this chapter we focus on naturally produced hydrocarbons, while hydrocarbons derived from anthropogenic activities are covered later in this book (see chapter 14). Hydrocarbons are one of the most widely used classes of biomarkers and have been applied to geochemical studies conducted over a range of timescales, from contemporary ecosystems to ancient sediments and rocks. In this chapter, we discuss traditional biomarkers such as aliphatic and cyclic hydrocarbons, which have been successfully used to distinguish between algal, bacterial, and terrigenous vascular plant sources of carbon in aquatic systems. We also present several classes of isoprenoid hydrocarbon biomarkers, including highly branched isoprenoids and their use as algal biomarkers as well as isoprenoid biomarkers recently identified for archaea. We discuss some of the transformations hydrocarbons undergo during diagenesis and how environmental conditions influence these processes (e.g., aerobic and anaerobic biodegradation of hydrocarbons). Recent applications of stable and radiocarbon isotopes to hydrocarbon biomarkers are also presented in this chapter.

10.1 Background

Hydrocarbons are a class of organic compounds that consist of the elements carbon (C) and hydrogen (H). These compounds are composed of a backbone consisting of carbon atoms, also called a carbon skeleton, with hydrogen atoms attached. The carbon atoms are linked by *covalent bonds*, which are formed when electrons are shared by adjacent atoms. Carbon is unique in its ability to share electrons with other carbon atoms, resulting in large molecules dominated by carbon–carbon bonds (Peters et al., 2005). In fact, carbon is one of only two elements capable of forming bonds with itself under the environmental conditions on Earth, with the other being sulfur. Hydrocarbons are stable biomolecules that are the primary constituents of fossil fuels, including petroleum, coal, and natural gas. The most abundant hydrocarbon is methane.

Hydrocarbons may be *saturated* (no double bonds) or *unsaturated* (one or more double bonds) and may occur in a spectrum of configurations, including straight carbon chains (*n*-alkanes), with

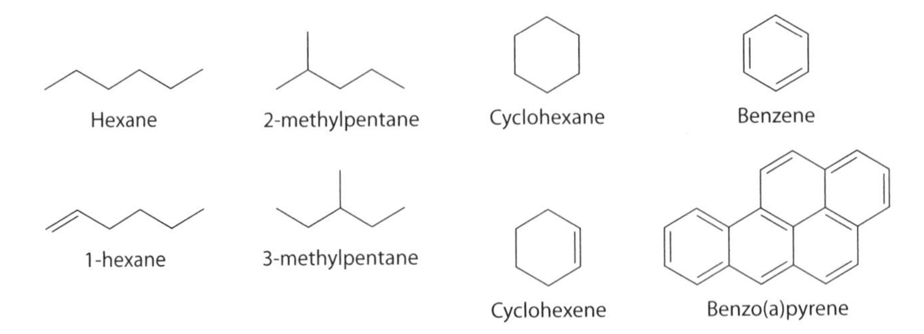

Figure 10.1. Examples of different hydrocarbon structures. The examples include an *n*-alkane (hexane), an *n*-alkene (1-hexene), branched alkanes (2-methylpentane and 3-methylpentane), a cyclic alkane (cyclohexane), a cyclic alkene (cyclohexene), a monoaromatic hydrocarbon (benzene), and a polycyclic aromatic hydrocarbon (benzo(a)pyrene).

Table 10.1
Examples of acyclic alkane homologues

Number of carbon atoms	*Name*	*Formula*
1	Methane	CH_4
2	Ethane	C_2H_6
3	Propane	C_3H_8
4	Butane	C_4H_{10}
5	Pentane	C_5H_{12}
6	Hexane	C_6H_{14}
7	Heptane	C_7H_{16}
8	Octane	C_8H_{18}
9	Nonane	C_9H_{20}
10	Decane	$C_{10}H_{22}$
20	Eicosane	$C_{20}H_{42}$
30	Triacontane	$C_{30}H_{62}$

Source: Adapted from Peters et al. (2005).

methyl or other hydrocarbon branches (e.g., isoprenoidal hydrocarbons), as well as in cyclic and aromatic structures (fig. 10.1). Saturated hydrocarbons *(n-alkanes* or aliphatic hydrocarbons) are represented by the general formula C_nH_{2n+2}. Unsaturated hydrocarbons with one or more double bonds are called *alkenes* (or *olefins*), and have the formula $C_nH_{2n+2-2z}$, where z is equal to the number of double bonds. Hydrocarbons with one double bond, for example, have the formula C_nH_{2n} for noncyclic or nonaromatic structures. Compounds with one or more triple bonds are called *alkynes*. The *conjugated* double bond is a particularly stable configuration, where double-bonded carbon atoms (C=C) alternate with single bonded atoms (C–C). Cyclic hydrocarbons can also be saturated or unsaturated. Aromatic hydrocarbons (*arenes*) have enhanced stability because of this bonding pattern.

Hydrocarbons often occur as a series of compounds in which each member differs from the next by a constant amount (*homologous series*). For example, the family of alkanes has the general formula C_nH_{2n+2} and forms a homologous series where the constant difference is a CH_2 group (table 10.1). Alkanes are generally biosynthesized from carboxylic acids by enzymatic *decarboxylation* as illustrated in the following equation:

$$\underset{\text{acid}}{CH_3(CH_2)_nCH_2COOH} \rightarrow \underset{\text{alkane}}{CH_3(CH_2)_nCH3} + CO_2 \tag{10.1}$$

Since fatty acids are generally dominated by even-numbered homologues, decarboxylation results in most naturally occurring hydrocarbons having an odd-numbered predominance. In fact, this is one way of distinguishing between natural hydrocarbons and those introduced to the environment through anthropogenic activities. Petroleum hydrocarbons, for example, typically lack the characteristic odd-carbon-chain-length dominance of biological hydrocarbons. This is due to both the random cleavage of alkyl chains in the kerogen matrix and the progressive loss of high odd-over-even components during petroleum formation (Killops and Killops, 2005). A second characteristic of petroleum hydrocarbons is the presence of a mixture of compounds that cannot be resolved by high-resolution capillary gas chromatography. These compounds are referred to as the *unresolved complex mixture* or UCM, and often appear as a "hump" in gas chromatograms for samples influenced by petroleum hydrocarbons (see fig. 10.11a).

In addition to straight-chained compounds, hydrocarbons often contain methyl branches both at the iso and anteiso positions as well as elsewhere along the carbon skeleton. Iso and anteiso-hydrocarbons are common in plant waxes, where they generally occur over the carbon range C_{25}–C_{31} with odd carbon number predominance. C_{10}-branched alkanes makeup another class of branched hydrocarbons; these compounds originate from the terpene constituents of essential oils synthesized by many higher plants. A third class of methyl-branched alkanes is the midchain branched hydrocarbons.

Branched hydrocarbons derived from *isoprene* units are common constituents in many bacteria or may be formed from algal organic matter during grazing and senescence (see chapter 9). These isoprenoid hydrocarbons occur in both cyclic and acyclic forms (see fig. 10.2). *Regular isoprenoids* are defined as acyclic, branched, saturated molecules with a methyl group on every fourth carbon atom, irrespective of the total number of carbon atoms. The arrangement implies a head-to-tail linkage of the isoprene groups (e.g., pristane (2,6,10,14-tetramethylpentadecane) and phytane (2,6,10,14-tetramethylhexadecane) fig. 10.2). Pristane and phytane are decomposition products of phytol, the side chain on *chlorophyll a*. Under oxidizing conditions, phytol is preferentially converted to pristene (phytol is first oxidized to phytenic acid, which is then decarboxylated to pristene). Pristene is subsequently reduced to pristane. However, under anoxic conditions, phytol is likely to undergo reduction and dehydration, resulting in the decomposition product phytane (see Killops and Killops (2005) for a more thorough description of these processes).

Acyclic 'irregular isoprenoids' are molecules with either a head-to-head or tail-to-tail linkage. Examples of compounds with a tail-to-tail linkage include 2,6,10,15,19-pentamethyleicosane (PME), lycopane (2,6,10,14,19,23,27,31-octamethyldotriacontane), and squalane (2,6,10,15,19,23-hexamethyltetracosane) (fig. 10.2). Irregular isoprenoids also include compounds with a head-to-head linkage, such as bis-phytane ($C_{40}H_{82}$) (fig. 10.2). Isoprenoids can also have cyclic structures. Examples of cyclic isoprenoidal hydrocarbons include sterenes and triterpenes, which are derived from bacterial transformation of steroids and triterpenoids during diagenesis (fig. 10.3).

Aromatic hydrocarbons are introduced to the environment through natural processes as well as through anthropogenic activities, such as petroleum spills and combustion of fossil fuels. One class of aromatic hydrocarbons with both natural and anthropogenic sources is the polycyclic aromatic hydrocarbons (PAH) (see benzo(*a*)pyrene in fig. 10.1). Although the abundance of this class of compounds in the environment largely reflects anthropogenic activities, some compounds are produced during natural combustion processes, such as forest fires, as well as during diagenesis (e.g., perylene). Although its precursor compound(s) remain unknown, perylene is thought to derive from microbial transformations of organic matter occurring during early diagenesis (Silliman et al., 1998, 2000, 2001). A second class of aromatic hydrocarbons derived from biogenic sources is the suite of compounds derived from phenanthrene. These compounds are formed by dehydrogenation of abietic acid, a component of pine resin. Once introduced to the environment, PAH can have a different fates, depending on their molecular weight and physical behavior, including abiotic weathering,

Head Tail
Isoprene

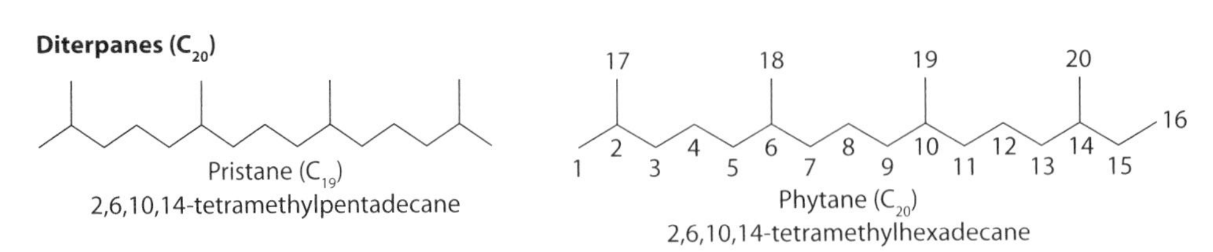

Sesterterpanes (C_{25})

Tail-to-tail linkage

2,6,10,15,19-pentamethyleicosane ($C_{25}H_{52}$)

Triterpanes (C_{30})

Tail-to-tail linkage

Squalane ($C_{30}H_{62}$)

Tetraterpanes (C_{40})

Tail-to-tail linkage

Lycopane ($C_{40}H_{82}$)

Tail-to-tail linkage

Perhydro-β-carotene

Head-to-head linkage

Bis-phytane ($C_{40}H_{82}$)

Head-to-tail linkage

Regular C_{40}-isoprenoid

Figure 10.2. Isoprene structure showing head and tail positions. Examples of isoprenoid hydrocarbons showing different linkages between isoprene units. (Adapted from Killops and Killops, 2005.)

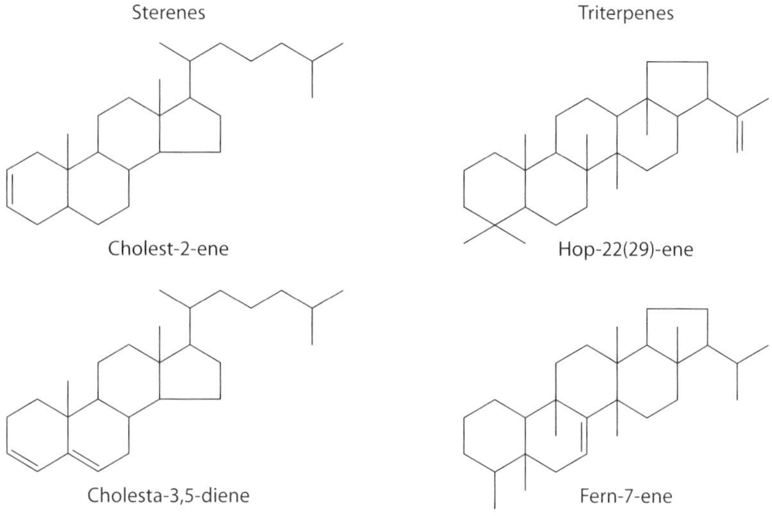

Sterenes

Triterpenes

Cholest-2-ene

Hop-22(29)-ene

Cholesta-3,5-diene

Fern-7-ene

Figure 10.3. Examples of cyclic isoprenoid hydrocarbons resulting from bacterial decomposition of sterols (sterenes) and triterpenols (terpenes).

photochemical decomposition, uptake by organisms and bioaccumulation, sediment deposition, and microbial degradation.

10.2 Analysis of Hydrocarbons

Like other classes of lipids (see chapters 8 and 9), hydrocarbons must first be separated from a sample matrix by extraction into organic solvent(s). Dichloromethane is often used to extract hydrocarbons, although other solvents and mixtures of solvents are also common (Meyers and Teranes, 2001; Volkman, 2006). It is also common to use solvents such as methanol and acetone in combination with nonpolar solvents such as dichloromethane, particularly when wet samples are extracted or when there is interest in analyzing both polar and nonpolar lipid biomarker compounds. In addition to 'extractable' compounds, hydrocarbons may occur in 'bound' forms. These compounds must be released from the matrix they are associated with using acid or base hydrolysis before they can be extracted into organic solvents.

Solvent extraction can be conducted in a number of ways, including shaking the sample with solvents, when sample sizes are relatively small (i.e., less than a few grams). Other extraction, methods include soxhlet extraction, sonication, microwave-assisted extraction, and accelerated solvent extraction. Once the sample is extracted, hydrocarbons are usually separated from other lipid classes chromatographically. Chromatographic methods used for this purpose include column chromatography, thin-layer chromatography, medium-pressure liquid chromatography, and high-pressure liquid chromatography (Meyers and Teranes, 2001). The hydrocarbon fraction is subsequently analyzed by capillary column gas chromatography, which can be used both to separate the mixture into individual compounds as well as to quantify each compound. Individual compounds are generally identified based on comparison of their *retention time* with the retention time of standards. It is important that the identification of each compound be subsequently verified by gas chromatography–mass spectrometry (GC-MS), since more than one compound can have the same retention time (e.g., branched and unsaturated hydrocarbons may elute at the same retention time) (Volkman et al., 1998). GC-MS may be used to both identify and quantify hydrocarbons, and several diagnostic ions can be used to reliably identify hydrocarbons associated with environmental samples.

10.3 Hydrocarbons as Biomarkers

10.3.1 Alkane Biomarkers

The dominant sources of natural hydrocarbons delivered to aquatic environments include autochthonous sources such as algae and bacteria as well as allochthonous inputs from terrigenous plants (Cranwell, 1982; Meyers and Ishiwatari, 1993; Meyers, 1997). A summary of hydrocarbon biomarkers representing these sources is provided in table 10.2. Hydrocarbons from bacterial sources occur in the form of open (noncyclic) chains, generally with carbon chain lengths between C_{10} and C_{30}. Several studies have attributed even-numbered homologues (C_{12} to C_{24}) to bacterial sources (Nishimura and Baker, 1986; Grimalt and Albaiges, 1987). Midchain, methyl-branched alkanes, often C_{18}, have also been associated with bacterial sources, while midchain methyl-branched C_{17} alkanes such as 7- and 8-methyl heptadecane are thought to derive from cyanobacteria (Shiea et al., 1990).

Algae, including phytoplankton, are usually characterized by low concentrations of hydrocarbons (3–5% of the lipid fraction). The hydrocarbon fraction of phytoplankton is generally dominated by saturated and unsaturated hydrocarbons having both straight and branched chains. Marine algae typically synthesize *n*-alkanes with chain lengths ranging from C_{14} to C_{32}. The ratio of even- to odd-numbered homologues is frequently close to one in marine algae with *n*-C_{15}, *n*-C_{17}, or *n*-C_{19} as the dominant alkanes (up to 90%). C_{31} and C_{33} alkanes, and alkenes are also common in many species of microalgae, as well as very-long-chain alkenes, such as diunsaturated C_{31}, tri- and tetraunsaturated C_{33}, and di- and triunsaturated C_{37} and C_{38} (table 10.2).

In contrast to bacteria and algae, terrestrial vascular plants are dominated by long-chain *n*-alkanes in the range C_{23} to C_{35}. In vascular plants, the distribution of *n*-alkanes is usually dominated by odd-numbered homologues with C_{27}, C_{29}, and C_{31} as the dominant compounds. The carbon number at which the *n*-alkane distribution maximizes (C_{max}) can sometimes be a diagnostic tracer for different plant sources. For example, some crops differ in their C_{max}, providing a tool for evaluating the sources of soils deposited to some aquatic systems (Rogge et al., 2007). Long-chain alkanes are associated with the waxy protective coatings on leaf cuticles and provide protection from infection, physical damage, and dessication (Eglinton and Hamilton, 1967). Interestingly, the cell walls of fungi are also characterized by long-chain *n*-alkanes, suggesting they play a protective function, similar to *n*-alkanes found in vascular plants.

Submerged macrophytes often have biomarker compositions intermediate between algae and terrestrial vascular plants (fig. 10.4a). Aquatic macrophytes are generally characterized by a dominance of mid-chain-length *n*-alkanes with odd-numbered homologues. Previous studies have shown that submerged and floating aquatic plants have *n*-alkane distributions that maximize at C_{21}, C_{23}, or C_{25} (Cranwell, 1984; Viso et al., 1993; Ficken et al., 2000). For example, *n*-alkanes of the seagrass *Zostera marina* range from C_{17} to C_{27}, with a maximum at C_{21} (Canuel et al., 1997).

The somewhat unique *n*-alkane composition of submerged/floating aquatic macrophytes, contributed to the development of the proxy, P_{aq}, for expressing relative contributions of these plants versus emergent and terrestrial plant input to lake sediments (Ficken et al., 2000). P_{aq} is calculated as follows:

$$P_{aq} = (C_{23} + C_{25})/(C_{23} + C_{25} + C_{29} + C_{31}) \qquad (10.2)$$

This proxy was recently used in combination with compound-specific stable carbon isotope analysis in the Florida Everglades (USA) to differentiate between organic matter inputs from submerged and emergent vegetation vs. terrestrial vegetation (Mead et al., 2005). Results from this study showed that P_{aq} values for the emergent/terrestrial plants were generally lower (0.13–0.51) than for freshwater/marine submerged vegetation (0.45–1.00) and that $\delta^{13}C$ values for the *n*-alkanes (C_{23} to C_{31})

Table 10.2
Examples of hydrocarbon biomarkers

Compound or proxy	*Source*	*References*
n-Alkanes with C_{15}, C_{17}, or C_{19} predominance	Algae	Cranwell (1982)
n-Alkanes with C_{21}, C_{23} or C_{25} predominance	Aquatic macrophytes	Ficken et al. (2000)
Long-chain *n*-alkanes with odd-even predominance with maxima at C_{27}, C_{29} or C_{31}	Terrigenous plants	Eglinton and Hamilton (1967) Cranwell (1982)
n-Alkanes C_{20}–C_{40} without odd–even predominance	Petroleum	Peters et al. (2005) and references therein
Unresolved complex mixture (UCM)	Petroleum	Peters et al. (2005) and references therein
IP25	Sea ice diatoms	Belt et al. (2007)
P_{aq}	Aquatic macrophytes	Ficken et al. (2000)
$TAR_{HC} = (C_{27} + C_{29} + C_{31})/ C_{15} + C_{17} + C_{19})$	Terrigenous to aquatic ratio	Meyers (1997)
3,6,9,12,15,18-Heneidosahexaene (n-$C_{21:6}$)	Diatoms	Lee and Loeblich (1971)
C_{31} and C_{33} *n*-alkanes and *n*-alkenes	Plankton	Freeman et al. (1994)
Highly branched isoprenoids (C_{20}, C_{25}, C_{30})	Diatoms	Wraige et al. (1997, 1999) Sinninghe Damsté et al. (1999); Belt et al. (2001)
Very-long-chain alkenes (e.g., diunsaturated C_{31}, tri- and tetraunsaturated C_{33}, di- and tri-unsaturated C_{37} and C_{38})	Microalgae	Volkman (2005) and references therein
Pristane	Zooplankton processing of chlorophyll *a*	Blumer et al. (1964)
	Erosion of sedimentary rocks	Meyers (2003)
Phytane	Methanogenic bacteria	Risatti et al. (1984)
Diploptene	Chemoautotrophs	Freeman et al. (1994)
2,6,10,15,19-Pentamethyleicosane (PME)	Methanogens; thermoacidophiles	Tornabene and Langworthy (1979); Brassell et al. (1981)
2,6,10,14,19,23,27-Octamethyldotriacontane (lycopane)	Methanogens Photoautotrophic organisms	Tornabene and Langworthy (1979); Brassell et al. (1981); Wakeham et al. (1993); Freeman et al. (1994)
C_{30} isoprenoids (e.g., squalane, squalene)	Methanogens	Volkman and Maxwell (1986)
2,6,10,15,19,23-Hexamethyltetracospentaene (tetrahydrosqualene)	Methanogens	Tornabene and Langworthy (1979); Tornabene et al. (1979); Risatti et al. (1984)
Sterenes	Microbial dehydration of sterols	Wakeham et al. (1984); Wakeham (1989)
Hop-9(11)-ene	Photosynthetic bacterium, *Rhodommicrobium vannielli*	Howard (1980)

Table 10.2
Continued

Compound or proxy	*Source*	*References*
Hop-22(29)-ene	Prokaryotes	Wakeham et al. (1990)
Hop-22(29)-ene and fernenes	Ferns	Ageta and Arai (1984)
Lupenes and oleananes	Higher plants	Brassell and Eglinton (1986); Moldowan et al. (1994)
Lycopane/C_{31} alkane ratio	Paleoxicity proxy	Sinninghe Damsté et al. (2003)

were distinct for terrestrial/emergent and freshwater/marine submerged plants. This information was then used to characterize sources of vegetation deposited in regions dominated by emergent/submerged freshwater vegetation, terrestrial-dominated estuarine sites, and a region in the coastal zone dominated by submerged marine vegetation (fig. 10.4b, c).

Since higher plants synthesize hydrocarbons with a strong predominance of odd-numbered over even-numbered *n*-alkanes (often by a factor of 10 or more), it is useful to use the odd-over-even predominance to evaluate contributions of organic matter from vascular plants. The *carbon preference index* (CPI) has been developed as a proxy for evaluating potential contributions from terrigenous plants to sediments and source rocks. CPI is calculated as the ratio, by weight, of odd to even molecules in the range C_{24} to C_{34} centered around the $C_{29}H_{60}$ *n*-alkane (Bray and Evans, 1961). CPI is calculated as follows:

$$\begin{aligned} CPI = 0.5[(C_{25} + C_{27} + C_{29} + C_{31} + C_{33}/C_{24} + C_{26} + C_{28} + C_{30} + C_{32}) \\ +(C_{25} + C_{27} + C_{29} + C_{31} + C_{33}/C_{26} + C_{28} + C_{30} + C_{32} + C_{34})] \end{aligned} \quad (10.3)$$

Since this equation does not always calculate a CPI = 1 when there is no apparent carbon number preference, a second method for calculating CPI has been developed (Marzi et al., 1993):

$$\begin{aligned} CPI2 = 70.5[(C_{25} + C_{27} + C_{29} + C_{31})/(C_{26} + C_{28} + C_{30} + C_{32}) \\ +(C_{27} + C_{29} + C_{31} + C_{33})/(C_{26} + C_{28} + C_{30} + C_{32})] \end{aligned} \quad (10.4)$$

Higher plant contributions to recent sediments are usually characterized by CPI values much higher than 1. With increasing maturity (e.g., petroleum), CPI values usually approach 1. It is unusual to have CPI values less than 1, although this is sometimes observed with shorter chain homologues ($<C_{23}$) and in some depositional settings. In these cases, hydrocarbons usually derive directly from microbial synthesis rather than reduction of fatty acids or alcohols. The CPI for petroleum is also sometimes calculated as $2C_{29}/(C_{28} + C_{30})$ (Philippi, 1965). However, this method is limited in its reliability for calculating CPI since it is based on only three compounds.

Although CPI calculations often provide a useful index for identifying whether hydrocarbon sources are predominantly biogenic vs. petrogenic, it is important to be cognizant of some of the limitations of this proxy. For example, the overall trend of weight percentage vs. chain length may be variable for different organic matter sources, resulting in CPI values that differ from the expected range. A second consideration in using the CPI calculation is that the percentages of compounds $C_{24}H_{50}$ through $C_{34}H_{70}$ may decrease rapidly with increasing chain length, while the ratio of odd to even *n*-alkane weight percentages changes by only a small absolute amount (e.g., from 1 to 1.5 or 2). The numerator and denominator of the Bray and Evans equation are dominated by the few *n*-alkanes near $C_{24}H_{50}$, causing the CPI to be close to unity. Because samples often do not contain all the alkanes from $C_{24}H_{50}$ through $C_{34}H_{70}$, the weight percentages of a smaller number of compounds are often used in computing CPI. As a result, other ranges of *n*-alkanes may be used to determine whether there is an odd-to-even predominance (OEP) (Scanlan and Smith, 1970).

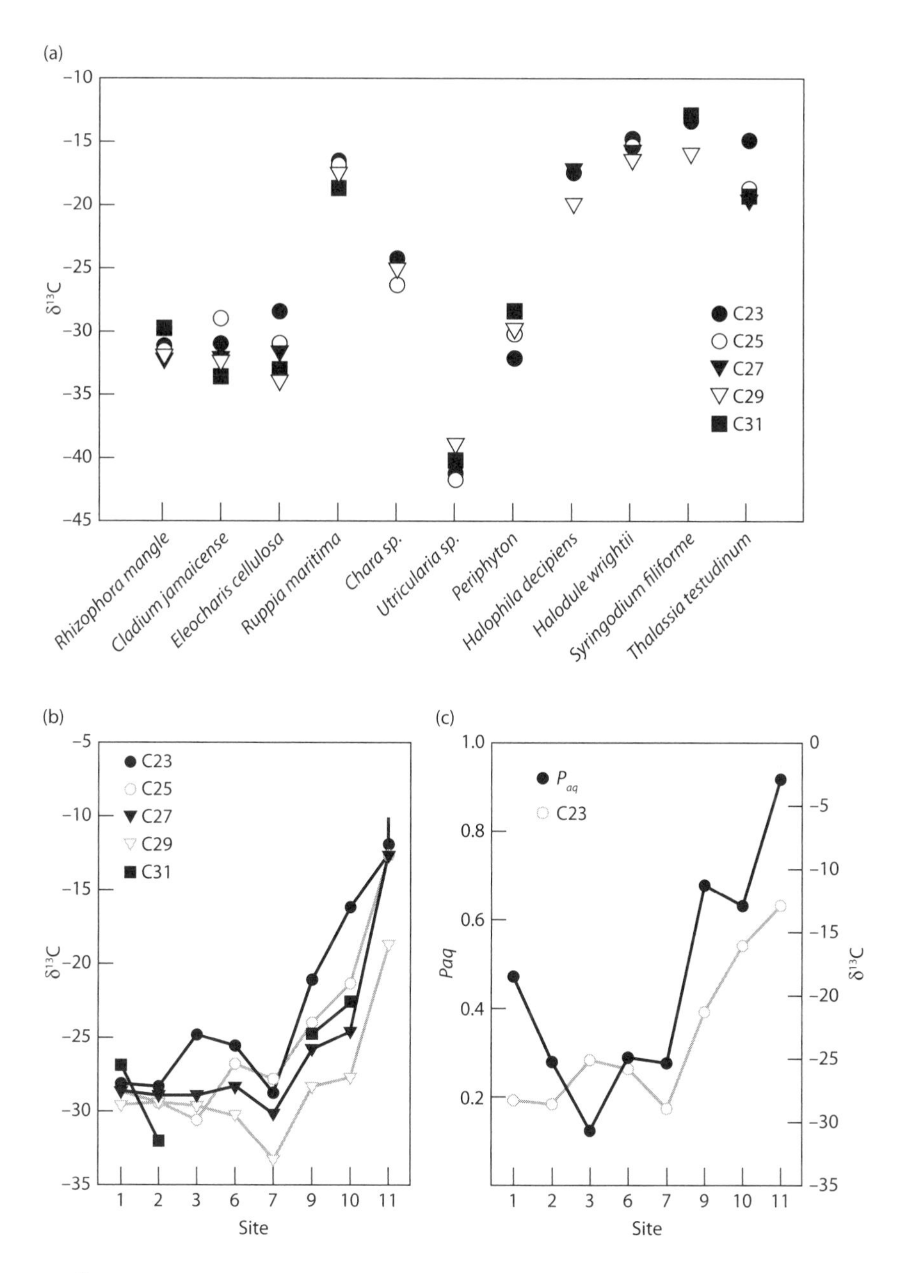

Figure 10.4. $\delta^{13}C$ signatures for *n*-alkanes extracted from plants collected from Everglades National Park, FL (USA) (a) $\delta^{13}C$ signatures of *n*-alkanes (b) and P_{aq} (c) for sites dominated by emergent/submerged freshwater vegetation (sites 1–3), terrestrial dominated estuarine sites (sites 6 and 7), and a region in the coastal zone dominated by submerged marine vegetation (sites 9–11). (Adapted from Mead et al., 2005.)

Since aquatic sources of organic matter such as algae and bacteria are typically dominated by short-chain *n*-alkanes and terrigenous vascular plants are characterized by long-chain *n*-alkanes, the terrestrial-to-aquatic ratio (TAR_{HC}) was developed as a proxy for assessing the relative contributions of autochthonous and allochthonous hydrocarbons in aquatic environments (Meyers, 1997). The ratio of terrigenous-to-aquatic *n*-alkanes is calculated as follows:

$$TAR_{HC} = (C_{27} + C_{29} + C_{31})/(C_{15} + C_{17} + C_{19}) \quad (10.5)$$

When $TAR_{HC} > 1$, contributions from vascular plants derived from the watershed dominate, while $TAR_{HC} > 1$ reflect a dominance of hydrocarbons of aquatic origin.

10.3.2 Alkenes and Isoprenoidal Hydrocarbons as Biomarkers

Unsaturated alkenes with one to four double bonds are typically used as biomarkers for microalgae. Monounsaturated alkenes are likely formed through the decarboxylation of the saturated fatty acids (Weete, 1976). Examples of alkenes with multiple unsaturated bonds include *n*-heneicosa-3,6,9,12,15,18-hexaene (*n*-$C_{21:6}$) and *n*-heneicosa-3,6,9,12,15-pentaene (*n*-$C_{21:5}$). These compounds are thought to originate from the decarboxylation of polyunsaturated fatty acids common to many species of microalgae (e.g., 22:6(*n*-3) fatty acid is abundant in dinoflagelletes). Because these compounds are labile, their presence in sediments may suggest the presence of intact (viable) algal cells (Volkman, 2006). While *n*-$C_{21:5}$ has been observed in water-column samples, it is rarely found in sediments, likely due to its rapid decomposition. These compounds are widely distributed in young aquatic (marine and lacustrine) sediments but disappear in older sediments, likely due to biodegradation and reactions with inorganic sulfur (see below).

A recent paper notes that one C_{25} monounsaturated hydrocarbon (IP25), may be a specific, sensitive, and stable proxy for sea ice in sediments over the Holocene (Belt et al., 2007). Belt et al. (2007) confirmed the identity of IP25 by laboratory synthesis and used the synthesized compound as a reference for quantifying IP25 in a range of sediments along a transect in the Canadian Arctic where sediments were either seasonally covered with ice or had permanent ice cover. Older sediments containing IP25 were dated using radiocarbon methods to at least 9000 yr. Measurements of IP25 in sediment cores, along with other accepted proxies, have the potential to be useful in determining the position of the ice edge throughout the Holocene (at least) in both Arctic and Antarctic regions, which may be important for calibrating climate-prediction models.

Highly unsaturated alkenes are unique to some microalgae, although they are generally not abundant in sediments. Examples include very-long-chain alkenes, such as diunsaturated C_{31} tri- and tetraunsaturated C_{33} and di- and triunsaturated C_{37} and C_{38} compounds (Volkman et al., 1998). While these compounds are generally not found in sediments, C_{37}–C_{39} alkenes, thought to be derived from *Emiliania huxleyi*, were found to be abundant in some Black Sea sediments (Sinninghe Damsté et al., 1995). Alkenes can also reflect organic matter from prokaryotes. C_{19} to C_{29} alkenes were identified in the cynaobacterium, *Anacystis montana*, and $C_{31:3}$ alkene was found in green photosynthetic bacteria (Volkman, 2006).

Another class of alkene biomarkers is the highly branched isoprenoids (HBI). These C_{20}, C_{25}, and C_{30} acyclic alkenes typically contain three to five double bonds (fig. 10.5) and have been found associated with particulate matter and in a variety of sediments (Massè et al., 2004). HBI are often the dominant class of alkenes in sediments with concentrations as high as $40\,\mu g\,g^{-1}$ dry sediment. HBI are thought to reflect contributions from microalgae and have recently been identified in both benthic (Volkman et al., 1994) and pelagic diatoms (Rowland and Robson, 1990; Wraige et al., 1997; Belt et al., 2001a,b). High contents of C_{25} HBI have been found in the diatom *Haslea ostrearia*, and both C_{25} and C_{30} HBI alkenes have been found in diatom strains thought to be *Rhizosolenia setigera* (Volkman et al., 1994). Recent papers have reported C_{25} HBI alkenes isolated from two planktonic marine diatoms belonging to the *Pleurosigma* genus, and a C_{25} HBI triene was identified in *Navicula sclesvicensis*, a freshwater diatom (Belt et al., 2001a,b). While it is unclear what contributes to the

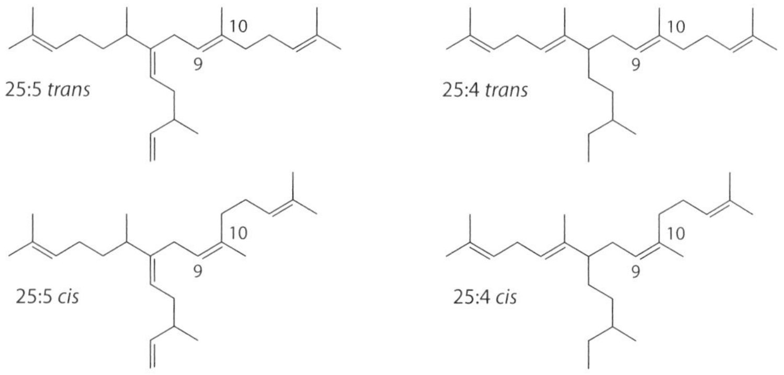

Figure 10.5. Examples of highly branched isoprenoid (HBI) hydrocarbons. (Adapted from Killops and Killops, 2005.)

diverse array of HBI compounds, genetic and environmental factors appear to be important controls on the relative abundances of the various homologues.

Acyclic isoprenoids are present in many organisms but particular compounds provide useful biomarkers for bacteria and Archaea. 2,6,10,15,19-Pentamethyleicosane (fig. 10.2) may be a biomarker for methanogens or Archaea that oxidize methane anaerobically in the water column (Rowland et al., 1982; Freeman et al., 1994; Schouten et al., 2001). A second example of acyclic isoprenoid biomarkers is the biphytanes, which are long-chain isoprenoidal hydrocarbons formed when two phytanyl units join together with a head-to-head linkage (fig. 10.2). These compounds are likely formed from biphytanyl ethers, which are characteristic of some types of archaebacteria, such as methanogens and thermoacidophiles (Killops and Killops, 2005). Some of these compounds may be synthesized directly from bacteria, while others are created during bacterial transformation processes. Lycopane (fig. 10.2), a saturated C_{40} isoprenoidal alkane, may be a product from bacterial reduction of lycopene, while squalane, a saturated C_{30} isoprenoidal alkane, may derive both from archaebacteria as well as diagenetic reduction of squalene (Killops and Killops, 2005). $\delta^{13}C$ analysis of individual hydrocarbons demonstrates a variety of sources for hydrocarbons and suggests that lycopane may also have a photoautotrophic source (Wakeham et al., 1993; Freeman et al., 1994).

10.4 Diagenesis of Hydrocarbons

Ultimately, biogenic organic compounds are defunctionalized during diagenesis, leaving saturated hydrocarbons as the remnants preserved in most ancient sediments and fossil fuels. This occurs through a combination of microbially mediated transformation processes as well as abiotic reactions involving dehydration and rearrangement of organic structures. Understanding product–precursor relationships is important for deciphering the sources of these hydrocarbons preserved in the sediment record. Sterol biomarkers, for example, provide an excellent example of a class of biomarkers where product–precursor relationships have been well established through a combination of laboratory studies, environmental observations of product-precursor relationships, and inverse trends in the abundance of products and precursors with depth in sediment cores (Mackenzie et al., 1982; Wakeham et al., 1984; de Leeuw and Baas, 1986; Wakeham, 1989). Many of these same transformation pathways have been extended to other lipid classes. An excellent review of some of these reaction pathways is provided in Killops and Killops (2005).

Aerobic biodegradation of acyclic alkanes proceeds either by oxidation of the terminal carbon atom (α-oxidation) or oxidation of the carbon atom at the C-2 position (β-oxidation) (fig. 10.6). Although

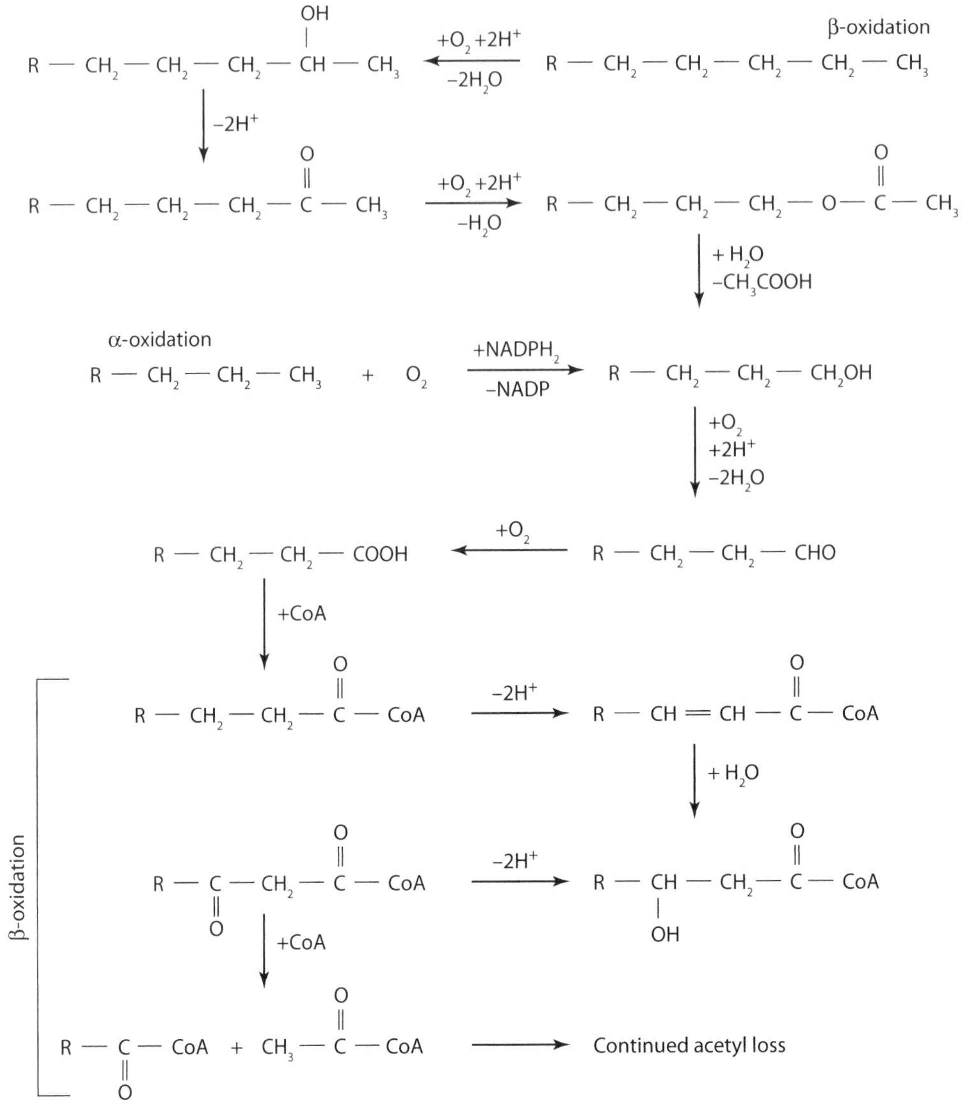

Figure 10.6. Aerobic biodegradation of alkanes by α- and β-oxidation. (Adapted from Killops and Killops, 2005.)

there had been speculation about anaerobic biodegradation of hydrocarbons since the 1940s, it was not until the late 1980s that anaerobic biodegradation of hydrocarbons was demonstrated (see recent review by Grossi et al., 2008). Aeckersberg and colleagues identified a sulfate-reducing bacterium capable of anaerobically oxidizing *n*-alkanes with carbon chain lengths C_{12}–C_{20} (Aeckersberg et al., 1991). Subsequently, anaerobic petroleum oxidation has been demonstrated under both laboratory (Wilkes et al., 2003) and natural conditions (Gieg and Suflita, 2002) using anaerobic metabolites of *n*-alkanes. To date, two mechanisms of anaerobic *n*-alkane degradation have been described, each involving biochemical reactions that differ from those used in aerobic hydrocarbon metabolism. The first pathway involves activation at the subterminal carbon of the alkane and addition to a molecule of fumarate (fig. 10.7a). In a second pathway, the alkane is carboxylated at the C-3 position (fig. 10.7b). Interestingly, both pathways have been observed during the anaerobic degradation of some aromatic hydrocarbons (Spormann and Widdel, 2000). Grossi and colleagues also proposed a third potential

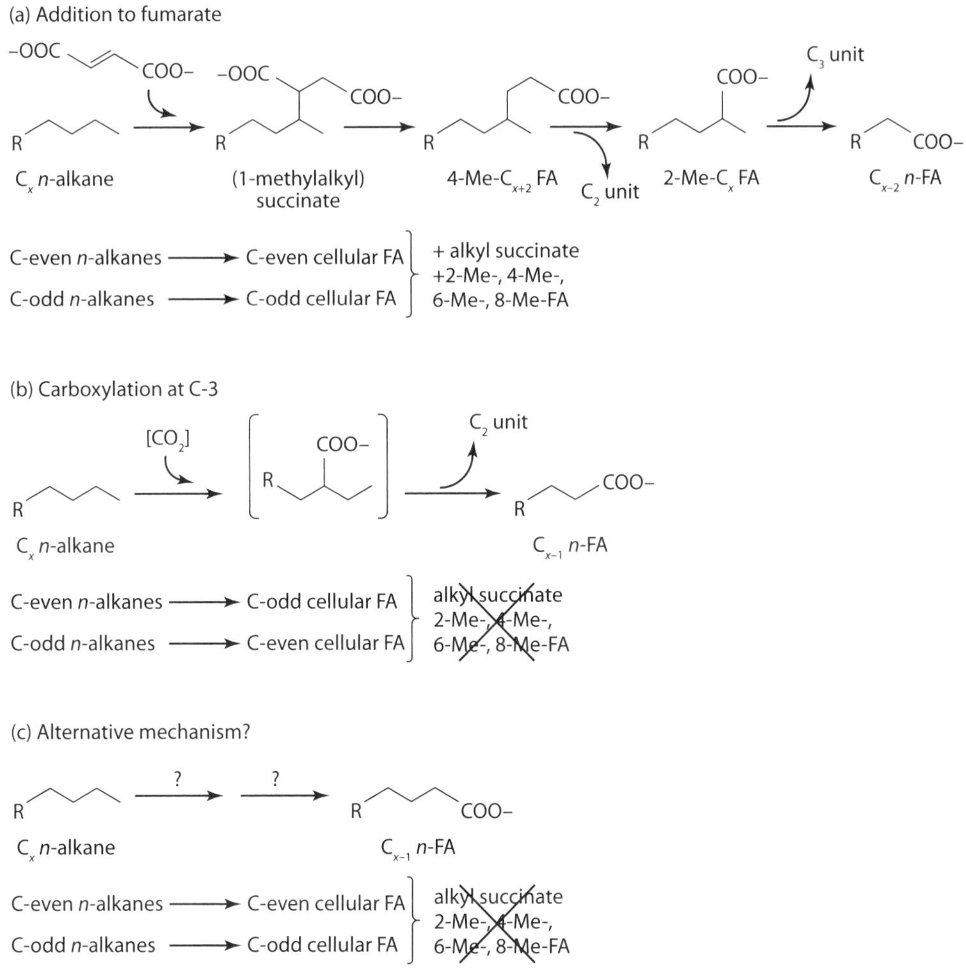

(a) Addition to fumarate

C_x *n*-alkane → (1-methylalkyl) succinate → 4-Me-C_{x+2} FA → (C_2 unit) 2-Me-C_x FA → (C_3 unit) C_{x-2} *n*-FA

C-even *n*-alkanes → C-even cellular FA
C-odd *n*-alkanes → C-odd cellular FA
+ alkyl succinate +2-Me-, 4-Me-, 6-Me-, 8-Me-FA

(b) Carboxylation at C-3

C_x *n*-alkane → [CO_2] → [ ] → (C_2 unit) C_{x-1} *n*-FA

C-even *n*-alkanes → C-odd cellular FA
C-odd *n*-alkanes → C-even cellular FA
alkyl succinate 2-Me-, 4-Me-, 6-Me-, 8-Me-FA (crossed out)

(c) Alternative mechanism?

C_x *n*-alkane → ? → ? → C_{x-1} *n*-FA

C-even *n*-alkanes → C-even cellular FA
C-odd *n*-alkanes → C-odd cellular FA
alkyl succinate 2-Me-, 4-Me-, 6-Me-, 8-Me-FA (crossed out)

Figure 10.7. Pathways for anaerobic degradation of alkanes. (Adapted from Grossi et al., 2008.)

pathway for anaerobic biodegradation of alkanes but the mechanism has not yet been determined (fig. 10.7b) (Grossi et al., 2008). Current areas of research on the topic of anaerobic degradation of alkanes include investigation of the role of co-metabolic reactions involved in the degradation of long-chain ($> C_{20}$) alkanes by pure and mixed cultures and the isolation of anaerobic organisms able to use long-chain ($> C_{20}$) alkanes as growth substrates.

Inorganic sulfur is thought to play an important role in the preservation of hydrocarbons. Organic sulfur compounds may be formed during early diagenesis through the incorporation of inorganic sulfur (sulfide or polysulfides) into functionalized lipids (see reviews by Sinninghe Damsté and De Leeuw, 1990; Werne et al., 2004) (fig. 10.8). The likely mechanism involves the incorporation of elemental S or H_2S, products of sulfate reduction, one of the dominant processes occurring during early diagenesis. Vairavamurthy and colleagues demonstrated that a significant fraction of inorganic polysulfides is bound to the particulate phase of sediments and can react with carbon–carbon unsaturated bonds in organic matter. This mechanism may affect the bioavailability of the reactive organic matter in sediments and play an important role in the preservation of organic matter in anoxic marine sediments (Vairavamurthy et al., 1992). In this way, organic matter may be "protected" from transformation and mineralization through its removal from the low-molecular-weight portion of extractable organic

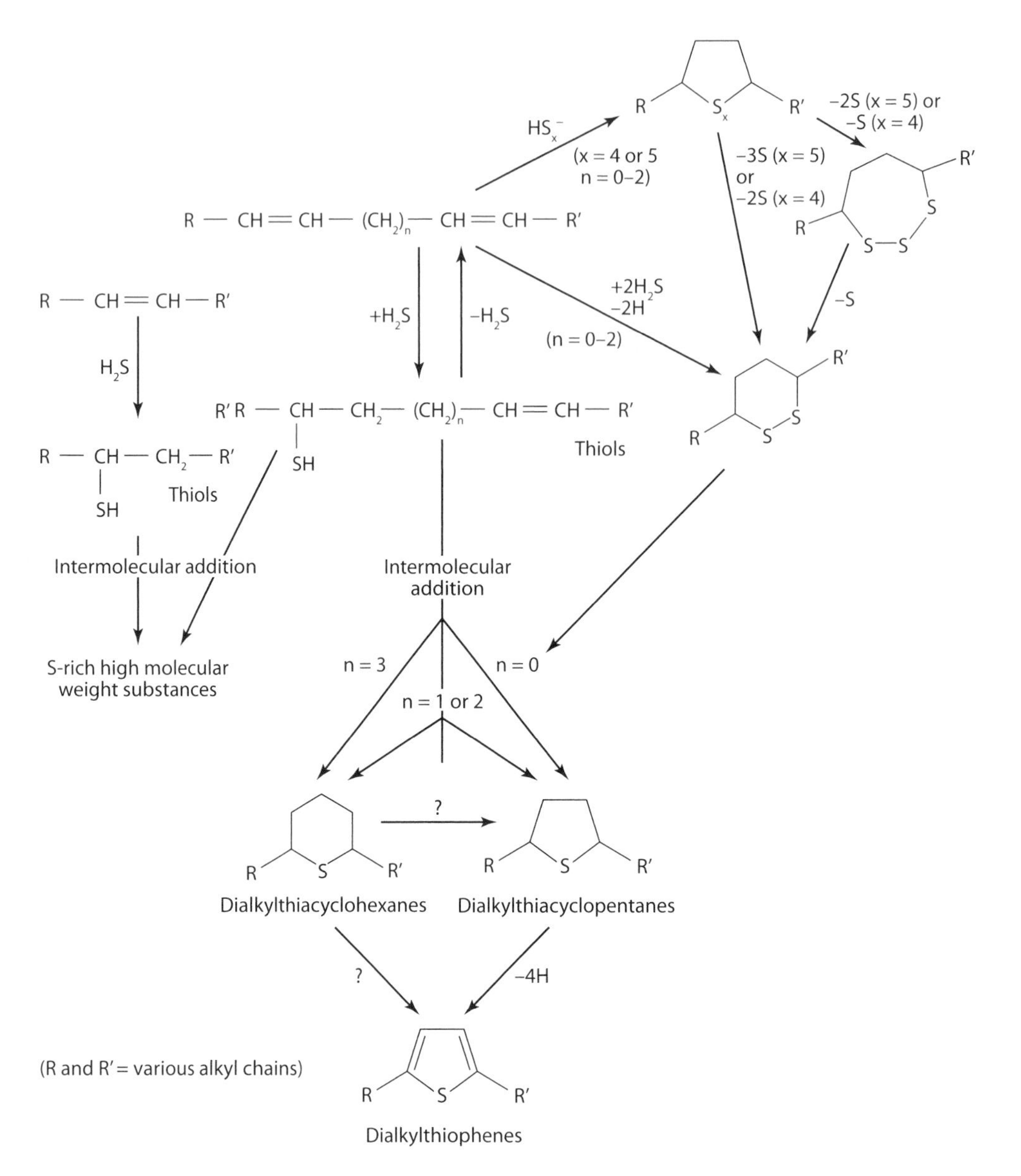

Figure 10.8. Pathways by which inorganic sulfur may be incorporated into alkenes during early diagenesis. (Adapted from Killops and Killops, 2005.)

matter and its incorporation into S-rich, high-molecular-weight substances preserved in the sediment record. As a result, information can be recovered from these biomarkers if they can be released by S-selective *chemolysis*.

Several classes of unsaturated hydrocarbons are capable of incorporating sulfur, including isoprenoidal alkenes from archaebacteria, C_{25} and C_{30} highly branched isoprenoidal alkenes, and phytenes from chlorophyll diagenesis. In the initial step, H_2S reacts across a C=C bond to form a thiol (SH) (fig. 10.8). The thiol group formed from one C=C bond can react with an adjacent double bond in the molecule (intramolecular addition) to form cyclic structures containing S, with the size of the ring dependent on the relative positions of double bonds. In cases where more than one double bond is present in the precursor molecule, the addition and subsequent loss of H_2S can lead to movement of double bonds along the chain (isomerization). The initial products of this reaction are aliphatic but aromatization may occur with increasing maturity.

Several studies have used *Raney-nickel desulfurization* to release biomarker compounds sequestered as organic sulfur compounds (OSC) (Schouten et al., 1995). Macromolecular material from sediment extracts obtained from the Black Sea were treated with Raney nickel, yielding phytane as the dominant hydrocarbon as well as series of C_{25} and C_{30} HBI hydrocarbons, beta-carotane, and isorenieratane (Wakeham et al., 1995). Comparison of concentrations of OSC with intramolecular S-linkages (0.1–0.2 $\mu g\,g^{-1}$ sediment) to the concentrations of OSC with intermolecular S-linkages (up to $50 \mu g\,g^{-1}$ sediment) indicates that most S incorporation in these sediments occurred via intermolecular bonds to macromolecules. Results from this study also suggest that sulfur incorporation into functionalized lipids can occur during the very early stages of sedimentary diagenesis, including at the sediment–water interface.

More recent work has refined our understanding of the timing of organic matter sulfurization reactions during early diagenesis. Using sediments from the Cariaco Basin (Venezuela), Werne et al. (2000) were able to demonstrate conversion of a tricyclic triterpenoid, (17*E*)-13β(H)-malabarica-14(27),17,21-triene to a triterpenoid thiane in the upper 3 m of sediment, representing $\sim$ 10,000 years. Thus, the timing for this reaction in sediments from the Cariaco basin was constrained to a 10-kyr time period. In a subsequent study, Kok et al. (2000) found that steroids in Ace Lake, Antarctica were sulfurized on a similar timescale (1 to 3 kyr). In addition to these observations, other studies have revealed that sulfurization processes can occur over considerably shorter timescales. In the Cariaco basin, some organic sulfur compounds were found in sediments shallower than 3 m, suggesting that sulfurization of these compounds occurred over timescales $<10,000$ years. A thorough review of sulfurization processes and timescales over which they may occur is provided in Werne et al. (2000).

Steroid and triterpenoid hydrocarbons are formed both by microbially mediated transformations occurring during early diagenesis as well as abiotic processes occurring during later stages of diagenesis or catagenesis (fig. 10.3). Dehydration reactions transform sterols to *sterenes*. The dominant pathway involves conversion of Δ^2 (via stera-3,5-dienes), Δ^4 and Δ^5 isomers (see fig. 9.7 in chapter 9). These reactions can proceed via a number of pathways but are generally dominated by reduction reactions, resulting in the ultimate formation of fully saturated *steranes* with the 5α (H) configuration. In other cases, compounds may undergo rearrangement during diagenesis rather than proceeding through reduction reactions. This pathway leads to the formation of diasterenes as the intermediate products. Sterols may also undergo dehydration reactions leading to the formation of diunsaturated hydrocarbons, or steradienes. In the case of 4-methyl sterols, the major products will be 4α-methylsterenes. Interestingly, the behavior of 4-methyl and 4-desmethyl steroids differs during diagenesis, as suggested by the observation that 4-methylsterenes are formed during later stages of diagenesis than the sterenes. This may be due to the fact that dehydration occurs more readily for 3α- than 3β-hydroxy groups. As a result, 3α-stanols may provide a source of sterenes during early diagenesis, while 3β-stanols only undergo dehydration at a later stage of diagenesis (Killops and Killops, 2005). Similarly, pentacyclic triterpenoids undergo defunctionalization during early diagenesis followed by reduction and aromatization during later stages.

10.5 Application of Hydrocarbon Biomarkers

10.5.1 Organic Matter Sources along the River–Estuary Continuum

Hydrocarbon biomarkers can be used to differentiate between algal, higher plant, fossil, and bacterial sources, making their application particularly useful in the coastal zone where organic matter inputs

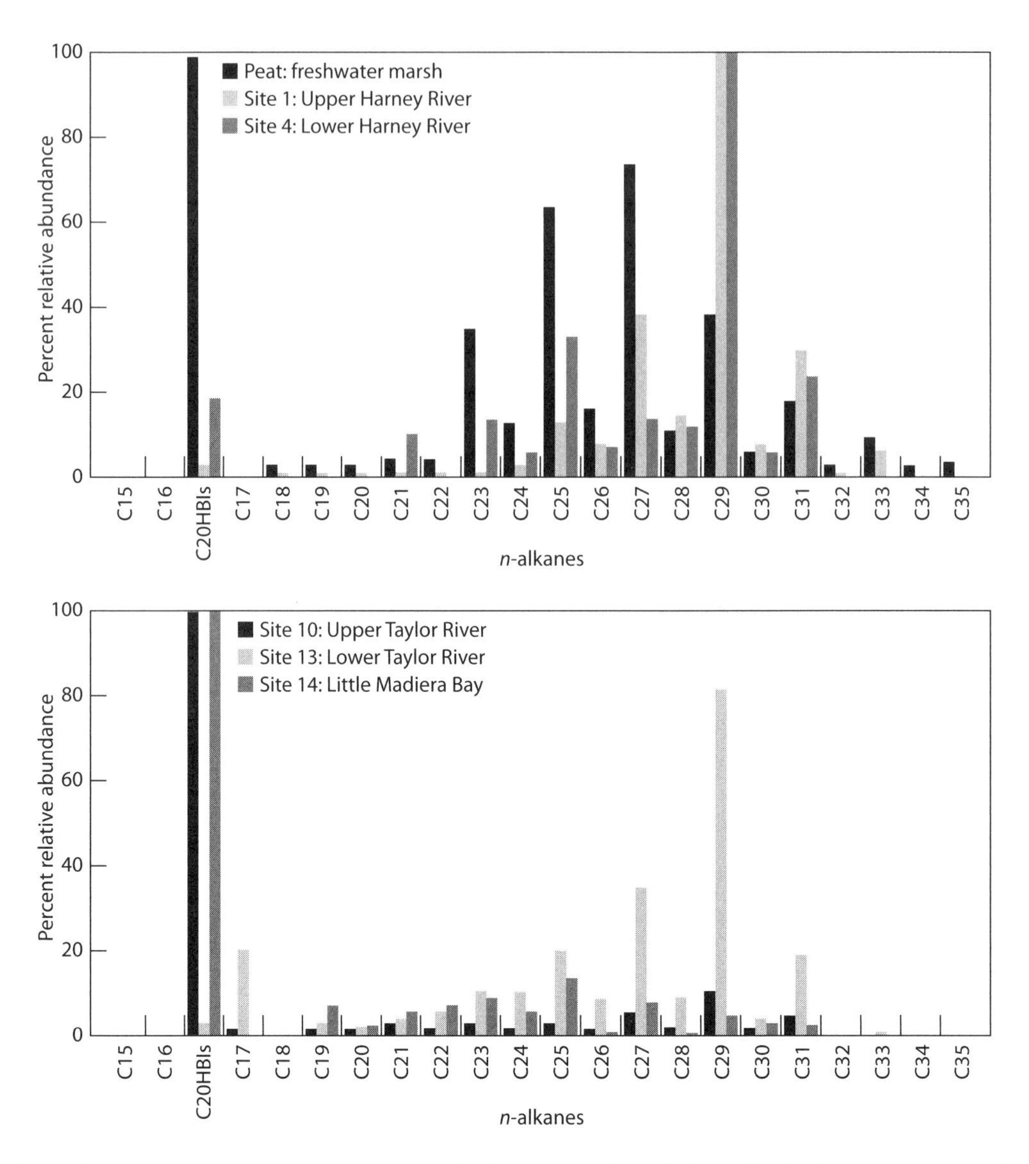

Figure 10.9. Range in *n*-alkanes found along the freshwater–marine salinity gradient of the Harney and Taylor rivers, FL (USA). (Adapted from Jaffe et al., 2001.)

are complex and reflect contributions from several sources. Not surprisingly, several studies have used hydrocarbons as biomarkers for investigating sources of organic matter associated with suspended particulate matter and surface sediments along the riverine–estuarine salinity gradient (Yunker et al., 1994; Jaffè et al., 1995, 2001; Zegouagh et al., 1998; Medeiros and Simoneit, 2008). A good example of this application of hydrocarbon biomarkers is provided by a recent study where Jaffè and colleagues used hydrocarbon biomarkers to determine the origin and transport of organic matter between freshwater and marine end members of two subtropical estuaries located in South Florida (USA) (Jaffè et al., 2001). *n*-Alkanes ranged from C_{15} to C_{35} in the study systems, with an odd-over-even predominance (fig. 10.9). Peat samples collected from a freshwater marsh were characterized by a maximum in the *n*-alkanes at C_{27}, while samples collected from the upper river had maxima at C_{29}. In the upper- and mid-estuary, long-chain *n*-alkanes with a strong odd-over-even predominance dominated, consistent with higher plant sources. Concentrations of *n*-alkanes decreased in the lower estuary, possibly due to dilution by reworked autochthonous OM. Samples collected from Little Madeira Bay were characterized by high concentrations of highly branched isoprenoids (HBI), indicating diatom sources.

Subsequent studies used a combination of *n*-alkane biomarkers, $\delta^{13}C$ analysis of individual *n*-alkanes, and the P_{aq} proxy to investigate contributions from different sources of vegetation along the freshwater–estuarine continuum (Mead et al., 2005). Results from this study showed that vegetation was an important source of sediment organic matter and that the signatures of carbon track changes in vegetation (fig. 10.4b, c). In the freshwater region, P_{aq} values tended to be low (<0.5) and $\delta^{13}C$ values were depleted (-30 to -25 per mil), while coastal sites characterized by submerged marine vegetation had P_{aq} values between 0.6 and 0.9 and enriched $\delta^{13}C$ signatures (around -10 per mil).

There are several examples where hydrocarbon biomarkers have been used with other classes of lipid biomarker compounds to differentiate between autochthonous and allochthonous sources of organic matter. Yunker and colleagues used hydrocarbon biomarkers along with other lipid compounds and multivariate statistical methods to examine sources of dissolved and particulate organic matter in samples collected seasonally from the Mackenzie River, an estuary located in the Arctic (Yunker et al., 1994, 1995). The Mackenzie River was the dominant source of *n*-alkanes derived from vascular plants with freshwater flow and suspended sediment load contributing to the abundance of these compounds. *n*-Alkanes were characterized by an odd–even predominance (OEP) wherever significant higher plant alkanes were present. A suite of alkenes indicated marine production from phytoplankton and zooplankton sources to the Mackenzie River. Alkenes dominated in the water column and sediment trap samples, but tended not to be preserved in surficial sediments because of their reactivity. Overall, this study concluded that river and marine ecosystems each contributed to seasonal variations in the hydrocarbon distribution on the Mackenzie River shelf and that processes such as freshwater flow, particle loading, and marine production controlled hydrocarbon abundance and composition.

10.5.2 Organic Matter Sources in Marine Environments

Hydrocarbons have been widely used for identifying the sources of organic matter delivered to coastal and open marine environments. Given the importance of developing carbon budgets and understanding the exchange of carbon between terrestrial and oceanic reservoirs, considerable efforts have been made to understand the transformations and delivery of terrigenous organic matter to the coastal margins. Hydrocarbon biomarkers have contributed to numerous research projects addressing these questions (Kennicutt Ii and Brooks, 1990; Prahl et al., 1994; Wakeham, 1995; Wakeham et al., 1997, 2002; Zimmerman and Canuel, 2001; Mitra and Bianchi, 2003; Mead and Goni, 2006).

Hydrocarbons have also been used to examine the transformation of suspended and sinking particulate matter in the oceanic water column and in surface sediments (Wakeham, 1995; Wakeham et al., 1997, 2002). In the Black Sea, Wakeham et al. (1991) partitioned hydrocarbons associated with suspended particulate organic matter (POM) into planktonic, terrestrial, bacterial, and fossil sources (fig. 10.10). This study revealed differences in the source signatures of two size classes of suspended particles (<53 and $>53\,\mu m$) as well as changes in the composition of hydrocarbons as POM was transported from surface waters to sediments (Wakeham et al., 1991). In surface waters, hydrocarbons associated with fine-grained suspended POM ($<53\,\mu m$) were dominated by planktonic sources, while hydrocarbons associated with larger suspended POM ($>53\,\mu m$) reflected contributions from planktonic and fossil sources. In the anoxic zone of the Black Sea, fine-grained suspended POM ($>53\,\mu m$) had hydrocarbon signatures dominated by anaerobic bacteria, whereas larger suspended POM ($>53\,\mu m$) had signatures indicative of mixed sources. In contrast, hydrocarbons associated with fecal pellets and sediments did not resemble water column sources but had signatures consistent with

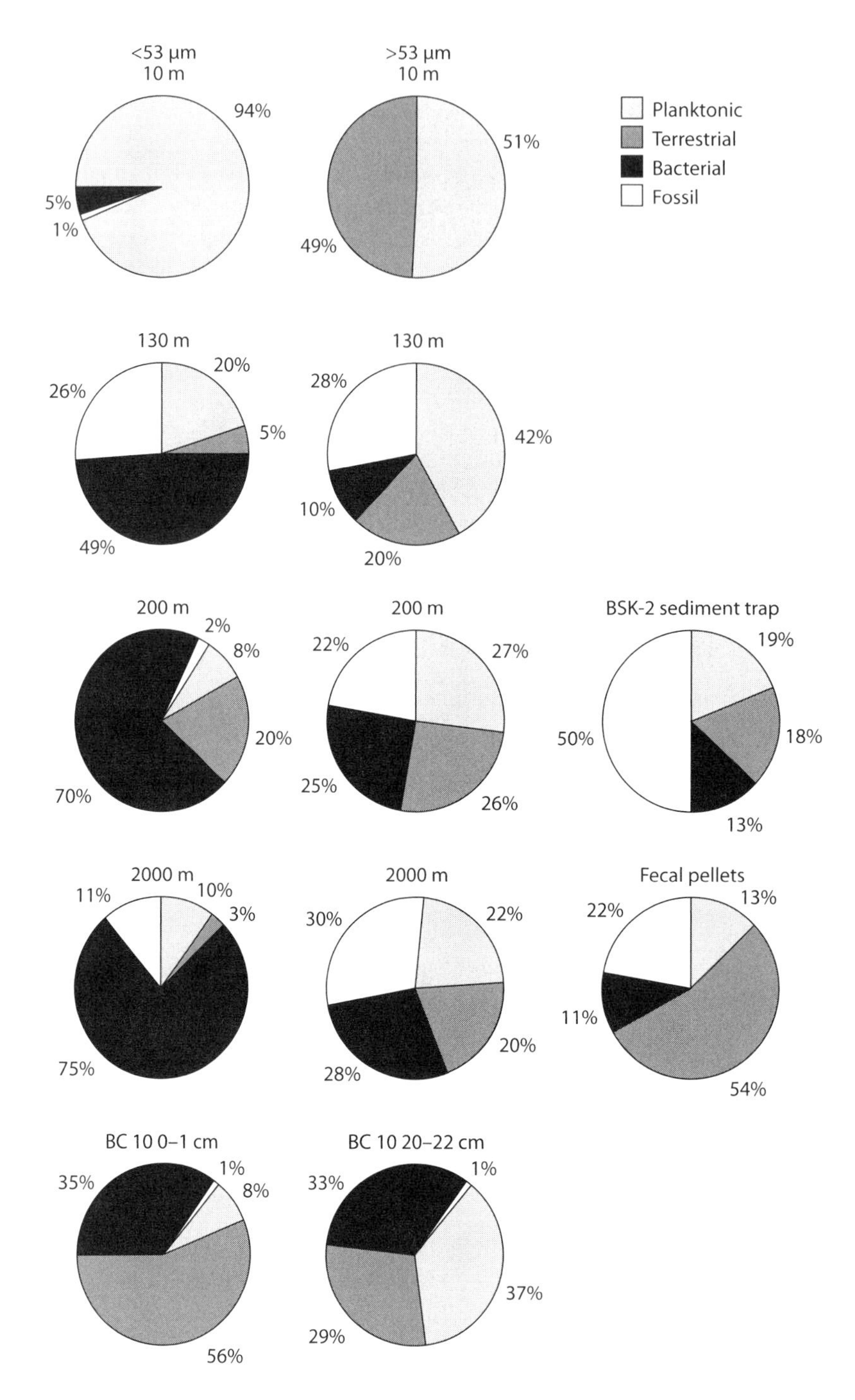

Figure 10.10. Source apportionment of hydrocarbons associated with sinking and suspended particulate organic matter in the Black Sea into planktonic, terrestrial, bacterial, and fossil sources. (Adapted from Wakeham et al., 1991.)

terrestrial sources. Compounds such as sterenes and triterpenes, which are generated by bacterial transformations occurring early diagenesis, increased in abundance as material was transported to depth.

Wakeham (1995) also compared changes in the composition of POM collected in sediment traps deployed in the Black Sea, where the water column is anoxic below about 100 m depth, with that collected from the Pacific Ocean, where the water column is fully oxic. In both systems, POM collected from the euphotic zone contained a mixture of alkanes, alkenes, and isoprenoid compounds, reflecting biogenic and petrogenic sources. Examples of compounds associated with surface water POM included pristase, a biomarker for zooplankton processing of chlorophyll *a*, and HBI, proxies for diatoms. POM collected from depth at the North Pacific differed from surface water samples in several ways, including (1) decreased abundance of algal hydrocarbons, (2) an absence of isoprenoidal hydrocarbons, and (3) the dominance of *n*-alkanes. In contrast, hydrocarbons associated with POM collected from 400-m depth in the anoxic waters of the Black Sea were characterized by signatures from higher plant *n*-alkanes (CPI $\sim$ 2). Pestamethyleicosase (PME), a biomarker compound of anaerobic microbial origin, and hopanoids (e.g., diploptene) characteristic of prokaryotes, such as cyanobacteria, methylotrophs, and chemoheterotrophs, were also observed in the anoxic portion of the water column.

10.5.3 Hydrocarbon Biomarkers in Paleoceanographic and Paleolimnologic Studies

The geochemical stability of hydrocarbons makes this class of biomarkers particularly useful for studying changes in organic matter delivery using sediment core records representing historical and geologic timescales. Likewise, the stability of these compounds makes them useful biomarkers for assessing sources of organic matter in ancient sediments, rocks, and petroleum. We refer readers to several excellent reviews on the application of hydrocarbons as paleo proxies in both lacustrine and marine settings (Meyers, 1997, 2003; Meyers and Ishiwatari, 1993; Eglinton and Eglinton, 2008). One caution in using this class of compounds as biomarkers is that hydrocarbons comprise a small fraction of organic matter in both organisms and sediments (Meyers and Teranes, 2001). Thus, since hydrocarbons are generally preserved preferentially relative to other classes of biomarkers, hydrocarbons may not be representative of the sources of organic matter originally deposited to sediments. As mentioned earlier in this book when describing other classes of biomarkers, we advocate the use of multiple-source proxies.

One of the classic studies where hydrocarbon biomarkers were first utilized to investigate changes in the delivery of organic matter was conducted in Lake Washington (USA) (Wakeham, 1976). Sediments deposited in the 1800s were characterized by low concentrations of hydrocarbons (26 $\mu g\ g^{-1}$) with a dominance of C_{27} *n*-alkane, a compound characteristic of plant waxes. The watershed of the Lake Washington was relatively pristine at this time and predominantly forested. In contrast, sediments deposited in the mid 1970s had considerably higher concentrations of *n*-alkanes (1600 $\mu g\ g^{-1}$) and were dominated by petroleum hydrocarbons. Sediments deposited since the 1950s also contained hydrocarbons diagnostic for algal sources such as C_{17} and C_{19} *n*-alkanes, suggesting eutrophication of Lake Washington. Interestingly, Wakeham and colleagues resampled Lake Washington in 2000 to examine how the lake had changed over the past 25 years (Wakeham et al., 2004). Like many lakes in the United States and Europe, hydrocarbon inputs to Lake Washington have decreased since the 1970s. The delivery mechanisms have also changed with most hydrocarbons (85%) now delivered to the lake by stormwater runoff. Despite the fact that 25 years had elapsed since the initial sampling of Lake Washington, signals from petroleum contamination during the 1970s were still present in the sediment record (fig. 10.11). Results from this study provide a good illustration of the geochemical stability of long-chain hydrocarbons and petroleum hydrocarbons. While the signatures of low-molecular-weight and normal alkanes were absent or low in concentration in the horizon corresponding to $\sim$ 1965, the complex mixture of branched and cyclic hydrocarbons that comprise the UCM, indicating petroleum contamination, persisted in the sedimentary record long after the source of contamination had been eliminated (fig. 10.11a).

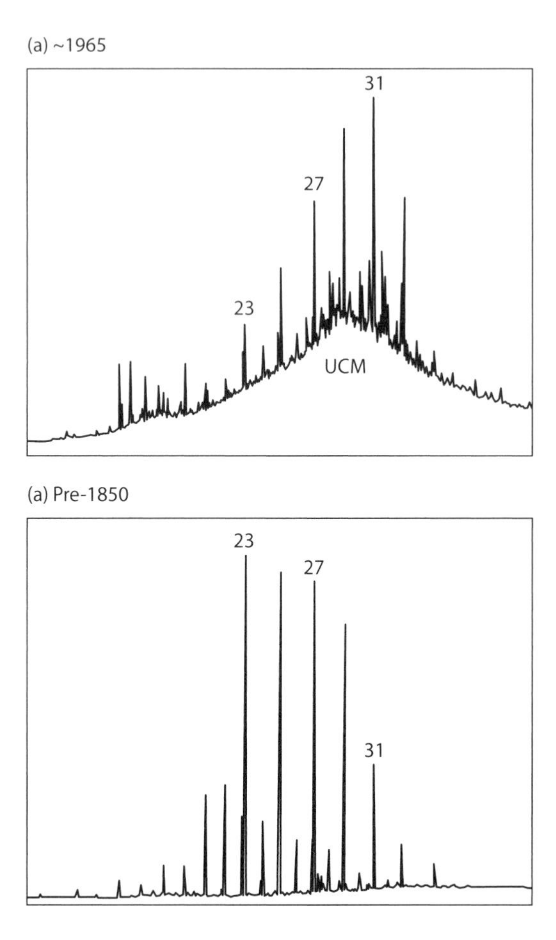

Figure 10.11. Partial gas chromatograms showing aliphatic hydrocarbons in sediment core collected from Lake Washington (USA) in 2000. Aliphatic hydrocarbons included contributions from petroleum and algae in sediments deposited ~ 1850 (a), while hydrocarbons were dominated by plant waxes in the pre-1850 horizon (b). Numbers refer to carbon numbers for *n*-alkanes. 'UCM' the unresolved complex mixture. (Adapted from Wakeham et al., 2004.)

Several studies have used hydrocarbon biomarkers to investigate changes in organic matter delivery resulting from human activities. Meyers and colleagues have used hydrocarbons, along with other biomarkers, in several studies conducted in lakes, with a particular focus on the Laurentian Great Lakes (Meyers and Eadie, 1993; Meyers and Ishiwatari, 1993; Ho and Meyers, 1994; Silliman et al., 1996; Ostrom et al., 1998; Meyers, 2003). Similar studies have been conducted in Lake Apopka, Florida (USA) (Silliman and Schelske, 2003), Lake Planina, a remote mountain lake in northwestern Slovenia (Muri et al., 2004), and Lake Middle Marviken in Sweden (Routh et al., 2007). These studies show a general pattern of a dominance of *n*-alkanes reflecting terrigenous sources corresponding to baseline sediments deposited when the watershed was primarily forested. Between the 1950s and 1970s, hydrocarbon concentrations increase and their signatures reflect increased contributions from petroleum sources as well as increased algal sources from eutrophication.

Sediment lake records of hydrocarbons have also been used at longer timescales to investigate changes in paleoenvironmental conditions in post-Holocene sediments. Along with hydrocarbon biomarkers, these studies often use a combination of complementary proxies, including $\delta^{13}C$, elemental composition, and multiple classes of biomarkers. Recent studies include efforts to determine the effects of changes in paleoenvironmental conditions, including vegetation, hydrology, and climate, on organic matter composition (Filley et al., 2001; Zhou et al., 2005; Luder et al., 2006; Bechtel et al., 2007). A particularly exciting and timely area of research involves the application of novel hydrocarbon biomarkers to determine changes in sea ice extent in polar regions (Belt et al., 2007). This proxy has particular relevance to global climate change research, since polar regions are expected to be regions of rapid change.

10.6 Compound-Specific Isotope Analysis of Hydrocarbons

In addition to the source information provided by hydrocarbon biomarkers alone, $\delta^{13}C$, $\Delta^{14}C$, and δD isotopes of individual hydrocarbons have the potential to contribute novel insights about microbial pathways (Hayes et al., 1990; Hayes, 1993, 2004) as well as paleoenvironmental changes in vegetation and climate (Eglinton and Eglinton, 2008). Initial work examining the $\delta^{13}C$ composition of hydrocarbons preserved in Messel shale revealed that the signatures of individual compounds exceeded those observed within total organic carbon (e.g., $\delta^{13}C$ of hydrocarbons ranged from −73.4 to −20.9 per mil) (Freeman et al., 1990). Differences in the $\delta^{13}C$ composition of hydrocarbons revealed that some compounds originally thought to have the same origin differed in source (e.g., pristane and phytane). This study also revealed that microbial processes modify the signatures of biomarkers preserved in the sediment record. In microbes, isotopic compositions of individual compounds have provided insights about specific processes and environmental conditions. An example is the case where hydrocarbon biomarkers with highly depleted isotopic signatures were used to indicate that bacteria have used methane as a carbon source (Freeman et al., 1990).

The diversity in the $\delta^{13}C$ composition of individual compounds was further examined in studies of the *n*-alkane composition of individual plants (Collister et al., 1994; Canuel et al., 1997). This work revealed the complexities of interpreting $\delta^{13}C$ at the compound-specific level, since individual plants displayed a range of values. Mechanisms for the diverse array of isotopic signatures in primary producers likely arise from reaction kinetics and carbon metabolism. In addition, δD values provide insights about different types of vegetation (C_3 vs. C_4) as well as paleoclimate information such as aridity (Sessions et al., 1999; Sauer et al., 2001; Sessions and Hayes, 2005; Chikaraishi and Naraoka, 2007; Mugler et al., 2008). In a recent study, the *n*-alkane composition and δD values of plant samples from sites along a climatic gradient from northern Finland to southern Italy were examined (Sachse et al., 2006). Results from this study showed that δD values for *n*-alkanes from terrestrial plants record the δD value of precipitation, modified by the amount of isotope enrichment in the leaf water due to the presence or absence of stomata, and meteorological conditions, such as evapotranspiration, relative humidity and soil moisture. Interestingly, while δD values differ for angiosperms vs. gymnosperms and C_3 vs. C_4 plants, recent studies have shown that plant physiology rather than biochemical pathway exerts a more important control on δD values for plant lipids (Krull et al., 2006; Smith and Freeman, 2006).

In addition to stable isotopic analyses of hydrocarbons, $\Delta^{14}C$ analyses of *n*-alkanes have proved useful in differentiating between sources of carbon (Eglinton et al., 1996; Rethemeyer et al., 2004; Uchikawa et al., 2008). Additional applications of $\Delta^{14}C$ analyses of hydrocarbons include source apportionment of PAH (Reddy et al., 2002; Kanke et al., 2004) and studies evaluating bacterial utilization of fossil sources of carbon (Pearson et al., 2005; Slater et al., 2006; Wakeham et al., 2006).

A thorough review of the application of isotopic composition of hydrocarbons to paleoclimate studies is provided in Eglinton and Eglinton (2008). These authors recommend that research be focused on the development of methods that allow for the sensitive and rapid analysis of high-resolution isotope records of biomarkers preserved in sediments, ice cores, tree rings, and corals. A second area of research that is recommended in this review is the ability to analyze multiple isotopes (e.g., ^{13}C, ^{14}C, and D) of individual biomarkers. These approaches are expected to provide new insights about changes in organic carbon composition over Earth's history.

10.7 Summary

Despite the fact that hydrocarbons contain only the elements carbon and hydrogen, this class of biomarkers is remarkably diverse in the structures synthesized by different organisms. As a result, hydrocarbon biomarkers provide a useful tool for identifying organic matter from primary producers, including algae and vascular plants, diverse microorganisms, such as bacteria and archaea, and fossil (petrogenic) inputs to the environment. Future development of methods to analyze multiple isotopes (e.g., ^{13}C, ^{14}C, and D) of individual hydrocarbon biomarkers and combining hydrocarbon biomarkers with other classes of biomarkers show promise for unraveling past changes as recorded in the sediment and rock records. The geochemical stability of this class of biomarkers contributes to the ability to use this class of biomarkers over contemporary to geological timescales.

11. Lipids: Alkenones, Polar Lipids, and Ether Lipids

11.1 Alkenones

Alkenones are long-chain (C_{35} to C_{40}) di-, tri-, and tetra-unsaturated ketones (fig. 11.1). These compounds are produced by a restricted number of species of haptophyte algae (e.g., *Emiliania huxleyi* and *Gephyrocapsa oceanica*) that live over a wide temperature range (2–29°C) in surface waters of the ocean. It is thought that these organisms are able to live under such a large temperature range because they are able to regulate the level to which their lipids are unsaturated, but the function of this class of lipids is largely unknown. Initially, it was thought that alkenones were membrane lipids and the increased degree of unsaturation at colder growth temperatures played a role in maintaining membrane fluidity (Brassell et al., 1986; Prahl and Wakeham, 1987). However, a recent study using the fluorescent lipophilic dye Nile red revealed that these compounds are not membrane lipids but are stored in cytoplasmic vesicles or lipid bodies (Eltgroth et al., 2005). In a laboratory study, these lipid vesicles increased in abundance under nutrient limitation and decreased under prolonged darkness. Eltgroth et al. (2005) showed that this pattern was correlated to polyunsaturated long-chain (C_{37}–C_{39}) alkenones, alkenoates, and alkenes. At this point, it is unknown why nonmembrane lipids should become desaturated at colder temperatures and this is a topic of ongoing study (Gaines et al., 2009).

Several indices for sea surface temperature (SST) have been developed that are based on levels of unsaturation in the C_{37} alkenones. The most common ($U_{37}^{K'}$) is based on the concentrations of tri- to di-unsaturated C_{37} ketones in sediments (Brassell et al., 1986; Prahl and Wakeham, 1987). This proxy

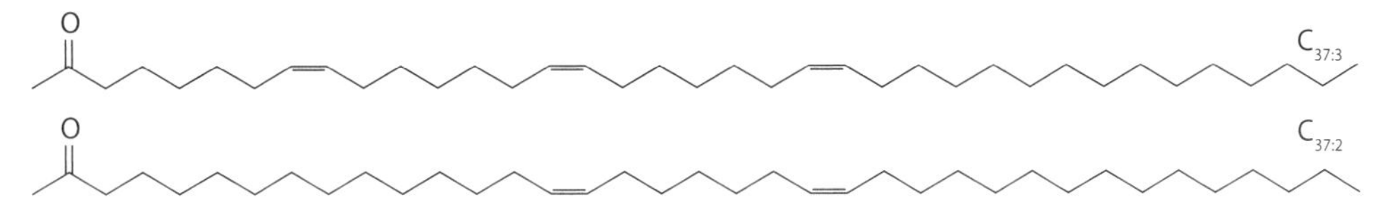

Figure 11.1. Examples of di- and tri-unsaturated C_{37} alkenones.

is calculated as follows:

$$U_{37}^{K'} = \frac{[C_{37:2}]}{[C_{37:2}] + [C_{37:3}]} \tag{11.1}$$

where 37:2 and 37:3 are C_{37} ketones with two and three double bonds, respectively. In addition to this relationship between unsaturation and temperature, long-chain ketones are generally more stable than most unsaturated lipids, resulting in their preservation in the sediment record. Together, these properties (fidelity between sea surface temperature and levels of unsaturation, and long-term preservation) have resulted in the widespread use of alkenones as proxies for paleotemperature (Eglinton and Eglinton, 2008). Although this proxy is often used along with foram microfossils, Mg/Ca ratios and $\delta^{18}O$ (Hoogakker et al., 2009, and references therein), it offers advantages over these proxies since calcium carbonate dissolution influences the preservation of these proxies in oceanic regions lying below the calcium compensation depth. We refer readers to an excellent collection of papers from a symposium held in 1999 highlighting the application of alkenone-based paleoceanographic indicators (Eglinton et al., 2001) as well as a more recent review (Herbert et al., 2003).

Downcore changes in the unsaturation pattern of alkenones corresponding to glacial–interglacial cycles were first observed in a core collected from the Kane Gap region of the northeast tropical Atlantic Ocean (Brassell et al., 1986). Brassell and colleagues (1986) hypothesized that temporal variations in levels of unsaturation in these long-chain ketones were related to SST. Following publication of these results, a temperature calibration for alkenones was developed based on laboratory cultures of *E. huxleyi* grown over a range of temperatures (fig. 11.2). This relationship was further tested using the alkenone composition of particulate matter samples collected from the surface waters of the ocean with known temperatures (Prahl and Wakeham, 1987). Subsequent work established a global calibration using a large number of core top samples collected between 60°S and 60°N in the Atlantic, Indian, and Pacific oceans and representing a considerable range in annual mean SST (0–29°C) (Müller et al., 1998). This study corroborated the positive relationship between $U_{37}^{K'}$ and mixed-layer SST found previously by Prahl and Wakeham (1987):

$$\mathrm{SST}(^{\circ}\mathrm{C}) = (U_{37}^{K'} - 0.044)/0.033 \tag{11.2}$$

Subsequently, a number of global calibrations were evaluated during a community workshop held in Woods Hole, Massachusetts (USA) in 1999 (Herbert, 2001). While these calibrations work well on average, specific calibrations have been developed that work best for polar waters (Sikes and Volkman, 1993) and warm waters (Conte et al., 2001, 2006). Recent research has also focused on factors other than temperature (e.g., nutrients and light) that may affect the $U_{37}^{K'}$ ratio (Prahl et al., 2003). Using laboratory batch cultures of *E. huxleyi*, Prahl et al. (2003) showed that nutrients and light stress had an effect on the alkenone unsaturation index $U_{37}^{K'}$, indicating that physiological factors may contribute to the variability of this index observed in surface sediments .Based on the wide application of this proxy, it has been established that $U_{37}^{K'}$ can be measured to an accuracy of 0.02 units, allowing temperatures to be determined to within 0.5°C (Killops and Killops, 2005).

Since establishing the efficacy of alkenones as a proxy for paleotemperature, this class of biomarkers has become one of the most widely used paleoclimate proxies (Sikes et al., 1991; Conte et al., 1992; Eglinton et al., 2000; Herbert, 2001; Sikes and Sicre, 2002). In many cases, alkenone-based SST has been corroborated with other paleo proxies (see example in fig. 11.3). However, like other biomarkers, further application of this class of compounds has revealed complexities involved in the biosynthesis of these biomolecules, their response to environmental factors, and geochemical stability. *E. huxleyi* is likely the dominant source of alkenones in recent sediments, but its utility is restricted to sediments deposited since the late Pleistocene when this organism first appeared (approximately 250 kyr BP). The ability to use other fossil organisms (e.g., *G. oceanica*) potentially allows the application of alkenones to be extended back to the Eocene (~ 45 Myr BP). However, the relationship between growth temperature

Unsaturation indexes measured in *E. huxleyi* grown in batch culture at different temperatures and harvested at a logarithmic stage of growth

Temperature	U^{K}_{37}*†	U^{K}_{37}†	U^{K}_{37}‡	*n*
5	0.00–0.11	—	—	—
8	—	0.17	0.29	1
10	—	0.31±0.017	0.38±0.027	4
15	0.17, 0.20 0.26, 0.34	0.54±0.016	0.55±0.014	4
20	0.40	0.74±0.017	0.74±0.017	7
25	0.73	0.86±0.017	0.86±0.017	4

* Data extrapolated from Brassel et al. (1986), *Nature* 320, 129–133.
† U^{K}_{37} = [37:2–37:4]/{37:2+37:3+37:4].
‡ U^{K}_{37} = [37:2]/{37:2+37:3].

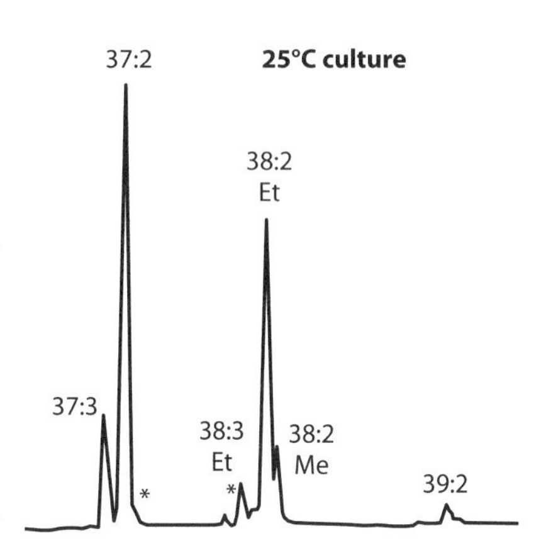

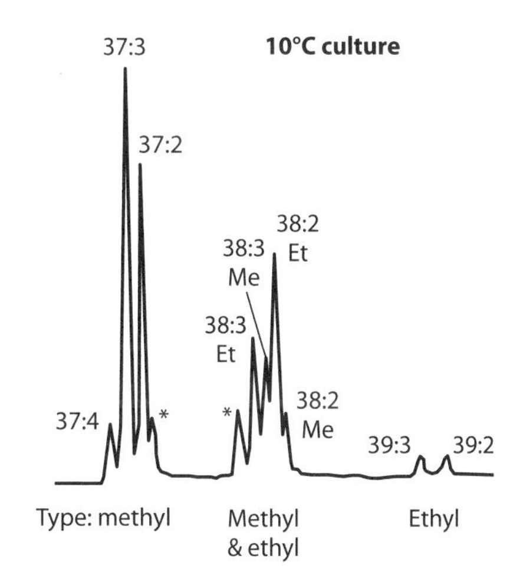

Figure 11.2. Partial gas chromatograms showing differences in levels of unsaturation for cultures of *E. huxleyi* grown at 10 and 25°C. Table 11.1 shows changes in alkenone unsaturation indices with temperature. (Adapted from Prahl and Wakeham, 1987.)

and the ratio of di- to tri-unsaturated alkenones is different for *G. oceanica* than it is for *E. huxleyi* (Volkman et al., 1995), requiring additional calibrations. Additionally, different strains of *E. huxleyi* vary in their response to changes in temperature (Sikes and Volkman, 1993; Conte et al., 1998). Previous studies have also noted that a single calibration of $U^{K'}_{37}$ with SST may not be appropriate for all regions of the ocean (Freeman and Wakeham, 1992; Volkman et al., 1995). Thus, it is important to consider these factors when interpreting the alkenone SST record in different environments and at different timescales.

The geochemical stability of alkenones has also been the topic of considerable research. Several studies have investigated the stability of alkenones under a range of environmental conditions. These studies concluded that while diagenesis influenced the absolute concentrations of alkyl alkenoates, a class of related compounds, there was no evidence for changes in the $C_{37:2}/(C_{37:2}+C_{37:3})$ ratio, leading to preservation of the paleotemperature signal (Teece et al., 1998; Grimalt et al., 2000; Harvey, 2000). While most of these studies have focused on sediment microbial transformations of alkenones, a recent study demonstrated that zooplankton grazing did not influence either the ratio of long-chain alkenones or their stable carbon isotopic compositions. As a result, zooplankton herbivory does not appear to influence the reliability of alkenones as a proxy for SST (Grice et al., 1998).

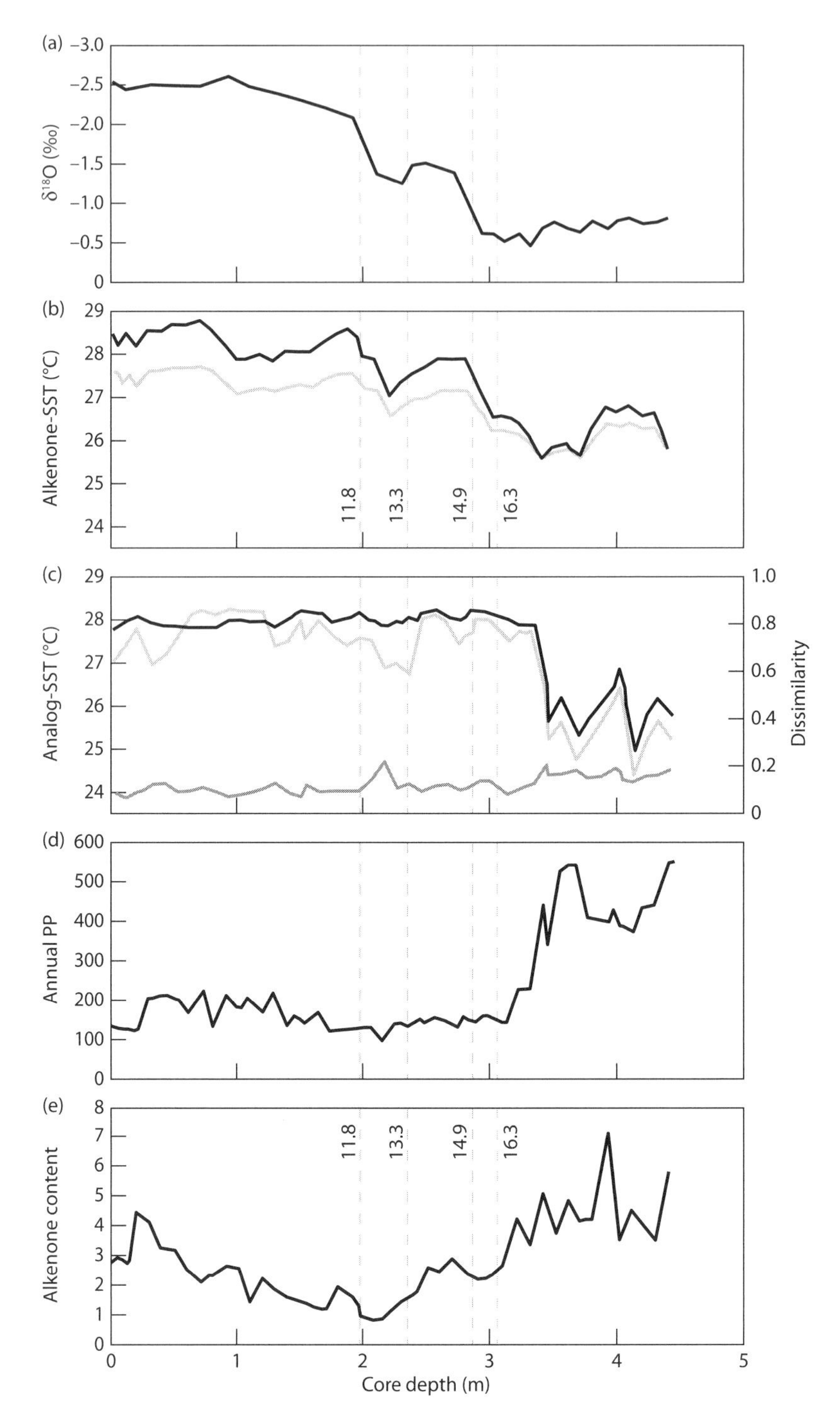

Figure 11.3. Figure showing coherence between different paleo proxies applied to sediments collected from the Arabian Sea. The proxies compared include $\delta^{18}O$ stratigraphy (a), alkenone sea surface temperature (SST) (b), SST determined using modern analog technique (c), estimated primary production from foram transfer coefficient (d), and concentrations of C_{37} alkenones (e; μg per gram dry sediment). (Adapted from Bard, 2001.)

Some uncertainty in the application of alkenones revolves around the tetra-unsaturated compounds. Previous studies have suggested that these tetra-unsaturated compounds may be influenced by salinity, rather than temperature. However, following careful evaluation, no systematic relationship between $C_{37:4}$ alkenone and temperature or salinity was observed across several ocean basins (Sikes and Sicre, 2002). These results suggest that $C_{37:4}$ alkenone levels in the open ocean may respond to other environmental variables; possibilities include changes in growth rate, light, or nutrient supply, as suggested by culture studies. Additional complexities in applying alkenones relate to particle transport times and lateral advection (Schneider, 2001; Ohkouchi et al., 2002). As in other applications of biomarkers, alkenones are best used in combination with other classes of biomarkers or paleotemperature proxies.

Stable carbon, deuterium ($\delta^{13}C$, δD), and radiocarbon isotopic compositions of alkenones have the potential to provide information that is complemetary to paleotemperature. $\delta^{13}C$ analysis of alkenones has been used to provide information about carbon dioxide concentrations, providing a paleo-pCO_2 proxy (Jasper and Hayes, 1990; Jasper et al., 1994; Pagani et al., 2002, 2005; Henderiks and Pagani, 2007). Recent studies have used radiocarbon isotopes to investigate the age of alkenones as a tool for dating foraminifera (Mollenhauer et al., 2003; Ohkouchi et al., 2005). These studies have documented radiocarbon ages for alkenones that are "older" than expected, suggesting complexities in the transport of sediment-associated materials and the potential importance of lateral transport processes. As a result, the use of radiocarbon measurements of foraminifera as a dating tool requires further examination. Methods have also been developed recently for determining the δD composition of alkenones (D'Andrea et al., 2007). Isotopic information combined with alkenone paleotemperature determinations has the potential to provide additional information about changes to the Earth's climate.

11.2 Lipid Membrane Biomarkers

Cellular membranes in bacteria differ in part from those in eukaryotes in that they contain different glycerol diaklyl glycerol tetraether-based lipids (GDGTs). Similarly, archaeal GDGTs may contain 0 to 8 different cyclopentane moieties, with one of particular interest found in Group 1 Crenarcheaota, called crenarcheaol (Hopmans et al., 2000; Sinninghe Damsté et al., 2002; Schouten et al., 2008, 2009, and references therein), which we discuss in further detail in this section. We first begin with some of the new advances in HPLC-MS that allow for analysis of intact polar lipids (IPL) (Sturt et al., 2004), which now provide the full structures of membrane lipids, including information on the headgroups, as well as new applications for GDGTs (e.g., Hopmans et al., 2000).

11.2.1 Intact Polar Membrane Lipids

Components of intact polar membrane lipids (e.g., fatty acids) have traditionally been used as biomarkers for prokaryotes. While this approach has provided many valuable insights (see chapter 8), additional information can be garnered by knowing the associations between fatty acids and the polar molecules from which they derive as well as the composition of the polar molecules themselves. Recent advances in high-performance liquid chromatography–atmospheric chemical ionization–mass spectrometry (HPLC-APCI-MS) now allow for the analysis of IPL, providing new insights about the microbial communities present in sediments and soils (Hopmans et al., 2000; Rutters et al., 2002a; Sturt et al., 2004; Zink and Mangelsdorf, 2004; Pitcher et al., 2009). HPLC-APCI-MS in particular has revealed that glycerol esters and ethers, when examined as intact molecules with their polar head groups, are able to provide taxonomic information about the community of viable prokaryotes. Also,

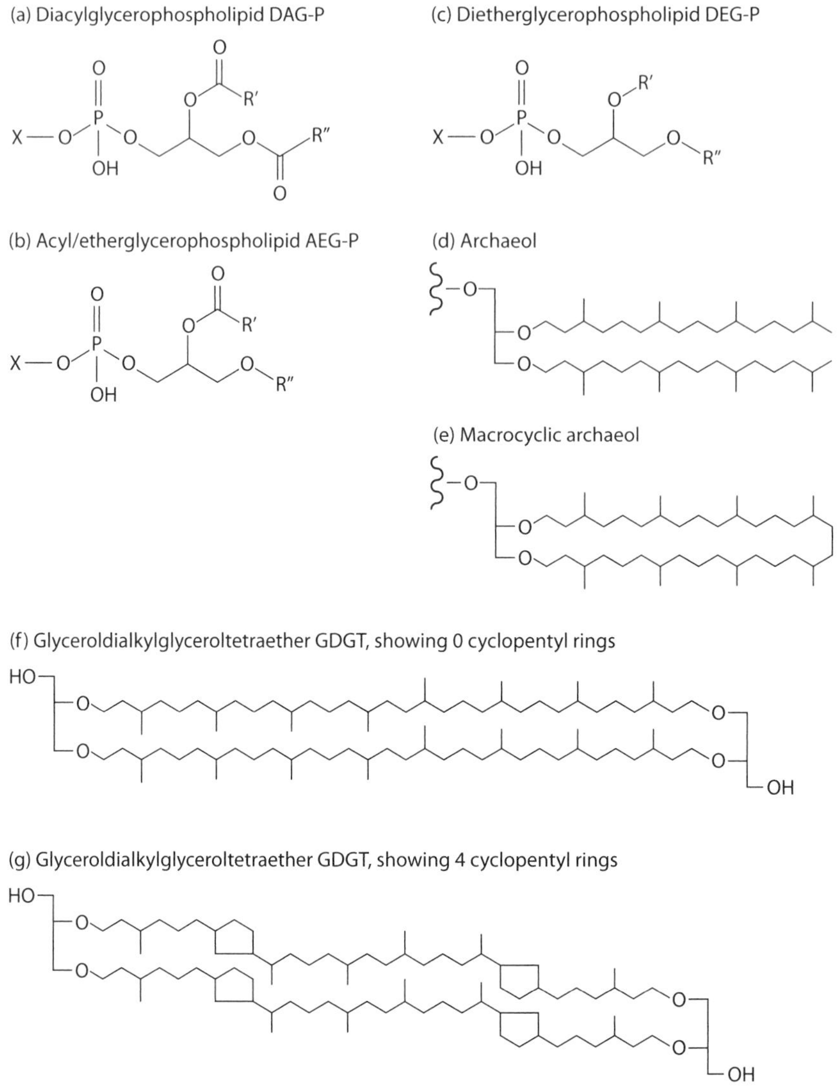

(a) Diacylglycerophospholipid DAG-P

(b) Acyl/etherglycerophospholipid AEG-P

(c) Dietherglycerophospholipid DEG-P

(d) Archaeol

(e) Macrocyclic archaeol

(f) Glyceroldialkylglyceroltetraether GDGT, showing 0 cyclopentyl rings

(g) Glyceroldialkylglyceroltetraether GDGT, showing 4 cyclopentyl rings

Figure 11.4. Example structures showing the diversity of intact polar lipids, including ether- and ester-bound phospholipids (a–c) and glyceryl dialkyl tetraethers (GDGTs) (d–g). Structures include archaeol from methanoogenic archaebacteria (d) and GDGT with 0 and 4 cyclopentyl rings (f and g). (Adapted from Sturt et al., 2004.)

because the polar headgroups are lost soon after a cell dies, these biomarkers allow for the specific characterization of the community of living prokaryotes. Structural features such as the composition of polar headgroups (see list of intact polar lipids and their headgroups in table 11.1; fig. 11.1 in Lipp and Hinrichs, 2009), types of bonding between alkyl groups and the glycerol backbone (see examples provided in fig. 11.4), and chain length and levels of unsaturation in the alkyl side chains contribute to the taxonomic specificity of IPL relative to their nonpolar derivatives. In recent years, IPL have been used to characterize laboratory cultures and environmental samples, providing new insights about the diversity and composition of viable microbial communities, particularly those present in extreme environments such as the deep-sea, deep subsurface sediments and hydrothermal systems.

Table 11.1
Examples of intact polar lipids and their headgroups (headgroups are identified by I through X)

Number	*Name*	*Abbreviation*	*Type*	*Organism*
Polar head groups				
I	Phosphoethanolamine	PE	Phospholipid	Bacteria and archaea
II	Phosphoglycerol	PG	Phospholipid	Bacteria and archaea
III	Phosphocholine	PC	Phospholipid	Bacteria and archaea
IV	Phosphoserine	PS	Phospholipid	Bacteria and archaea
V	Phosphatidic Acid	PA	Phospholipid	Bacteria and archaea
VI	Phosphoaminopentanetetrol	APT	Phospholipid	Bacteria and archaea
VII	Phosphoinositol	PI	Phospholipid	Bacteria and archaea
VIII	Glyco-phosphoethanolamine	GPE	Glycolipid	Bacteria and archaea
IX	Monoglycosidic		Glycolipid	Archaea
X	Diglycosidic		Glycolipid	Archaea
Core lipids				
	Diacylglycerophospholipid	DAG-P	Bacterial core lipids	
	Acyl/etherglycerophospholipid	AEG-P	Bacterial core lipids	
	Dietherglycerophospholipid	DEG-P	Bacterial core lipids	
	Diphosphatidylglycerol	DPG	Bacterial core lipids	
	Archaeol		Archaeal core lipids	
	Macrocyclic archaeol		Archaeal core lipids	
	Glyceroldialkylglyceroltetraether	GDGT	Archaeal core lipids	
	Glyceroldialkylnonitoltetraether	GDNT	Archaeal core lipids	

The analysis of IPL has primarily been conducted using soft ionization methods (e.g., APCI-MS and APCI-TOF-MS), although there are some reports where fast atom bombardment–mass spectrometry (FAB-MS) has been used to analyze IPL in pure cultures (Sprott et al., 1991; Hopmans et al., 2000; Yoshino et al., 2001). The application of FAB-MS has been limited to cases where chromatographic separations were not needed prior to sample analysis. Generally, ESI-MS must be optimized manually to allow separation and fragmentation of the variety of compounds that make up IPL. APCI-MS analyses are often conducted in both positive and negative ion modes. In positive ion mode, fragmentation is restricted to the polar headgroup, providing information about the headgroup and type of bonding. APCI-MS operated in negative ion mode provides complementary information by providing detailed information about the core lipid (i.e., alkyl moieties excluding the polar head groups) (Sturt et al., 2004).

Examples of IPL include the ester- and ether-bound phospholipids (e.g., diacylglycerophospholipids and dietherglycerophospholipids, respectively) (fig. 11.4a, c). Diacylglycerophospholipids (DAG-P) and dietherglycerophospholipids (DEG-P) are the most common membrane lipids in bacteria and eukaryotes (table 11.1). Interestingly, HPLC-APCI-MS analysis has also revealed mixed acyl/etherglycerophospholipids (AEG-P) in some species of sulfate-reducing bacteria (see structure in fig. 11.4b) (Sturt et al., 2004). The intact glyceroldialkylglyceroltetraether glycolipids (GDGT-G) and phospholipids (GDGT-P) are core lipids specific to archaea. As a result, APCI-MS analysis of a suite of IPL, including DAG-P, DEG-P, AEG-P, GDGT-G, and GDGT-P, provides a picture of the microbial community, including bacteria, eukaryotes, and archaea, present in environmental samples (Sturt et al., 2004; Rossel et al., 2008; Lipp and Hinrichs, 2009). In particular, qualitative analysis of the polar headgroup appears to be the most promising application of IPL for characterizing the microbial community. This is because the polar head groups are less susceptible to growth conditions and physical parameters than the alkyl moieties (Hopmans et al., 2000; Sturt et al., 2004). Some examples illustrating the "microbial fingerprints" resulting from analysis of IPL associated with laboratory cultures are provided in fig. 11.5 and in recent papers (Rossel et al., 2008; Lipp and Hinrichs, 2009).

Additionally, recent efforts have investigated the application of intact phospholipids for quantifying microbial biomass living in sediments (Zink et al., 2008). This type of information is useful for studies aimed at quantifying microbial numbers and activity in different environments, including the deep subsurface. Previous attempts to quantify living bacterial biomass have been based on a variety of methods, including direct cell counts as well as analysis of lipid phosphate or phospholipid-linked fatty acids (PLFA). Lipid phosphate and PLFA are usually converted to cell number based on conversion factors developed for model organisms (White et al., 1979; Balkwill et al., 1988). In an alternative approach, Zink and colleagues quantified intact phospholipids in sediments collected from Lake Baikal (Russia) using HPLC-ESI-MS and subsequently converted concentrations of intact phospholipids to cell numbers (Zink et al., 2008). Bacterial cell numbers determined from concentrations of intact phopholipids were higher than bacterial cell counts determined using DAPI, but were within an order of magnitude. One explanation for the discrepancy between the methods may be the conversion factors that were applied to the intact phospholipid concentrations. Thus, future research should investigate the relationship between cell numbers determined by quantification of intact PLs by LC-MS and microbiological (cell counts) approaches.

11.2.2 TEX_{86} as a Paleothermometer

Since the organisms that synthesize alkenones are unique to oceanic environments, paleoclimate studies of continental environments have been held back by the lack of a useful temperature proxy. Recent advances in the analysis of polar lipids by liquid chromatography have led to the development of the TEX_{86} (TetraEther indeX of tetraethers consisting of 86 carbon atoms) temperature proxy (Schouten et al., 2002, 2003b; Powers et al., 2004; Wutcher et al., 2006; Kim et al., 2008; Huguet et al., 2009; Trommer et al., 2009). This index is based on a class of archaeal membrane lipids, GDGTs), which, as discussed previously, occur ubiquitously in marine and lacustrine settings. The TEX_{86} index is based on the number of cyclopentane rings in the GDGTs, providing a useful paleotemperature index for lakes and other sites where alkenones are not produced (Powers et al., 2004; Kim et al., 2008).

Schouten and colleagues first proposed the molecular paleotemperature proxy TEX_{86} for the marine environment based on the distribution of GDGTs, membrane lipids derived from certain Archaea (e.g., Kingdom Crenarchaeota) living in the water column (Schouten et al., 2002, 2003b). With advances in HPLC-APCI-MS, this methodology has recently been improved (Schouten et al., 2007). Microbial ecologists have demonstrated a strong positive relationship between the average number of cyclopentane rings in the GDGTs and growth temperature for laboratory cultures of thermophilic archaea (Gliozzi et al., 1983; Uda et al., 2001). These studies showed that thermophilic archaea were able to adapt to different temperature conditions by incorporating cyclopentane rings into the lipid structure of GDGTs. Schouten and colleagues extended this work to nonthermophilic organisms, demonstrating that the membrane lipids of marine Crenarchaeota also responded to temperature fluctuations by adjusting the number of cyclopentane rings in the GDGT membrane lipids (fig. 11.6). Once the robustness of this relationship was established in modern marine ecosystems, it was used to reconstruct SSTs back into the middle Cretaceous (Schouten et al., 2003a). Since Crenarchaeota were historically known as hyperthermophilic organisms ($> 60°C$), it was very surprising to find them occurring in lakes and oceans, where they are now thought of as representing as much as 20% of the picoplankton in the oceans (Karner et al., 2001). It has also been shown that these "cold"-adapted marine Crenarchaeota actually have an additional GDGT, crenarchaeol, not found in the

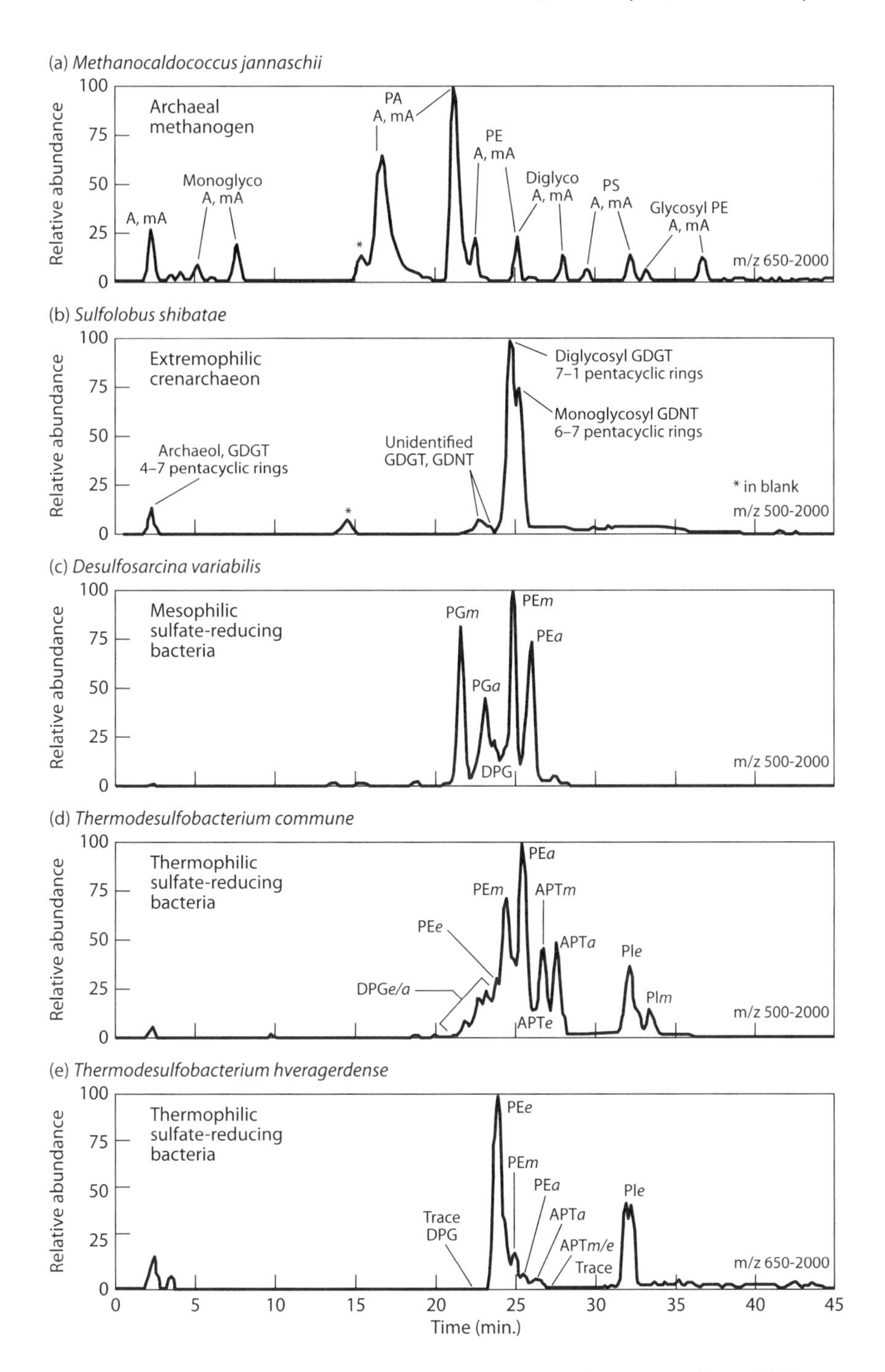

Figure 11.5. Chromatograms showing differences in the intact polar lipid composition of different groups of prokaryotes and archaea. Abbreviations are provided in table 11.1. (Adapted from Sturt et al., 2004.)

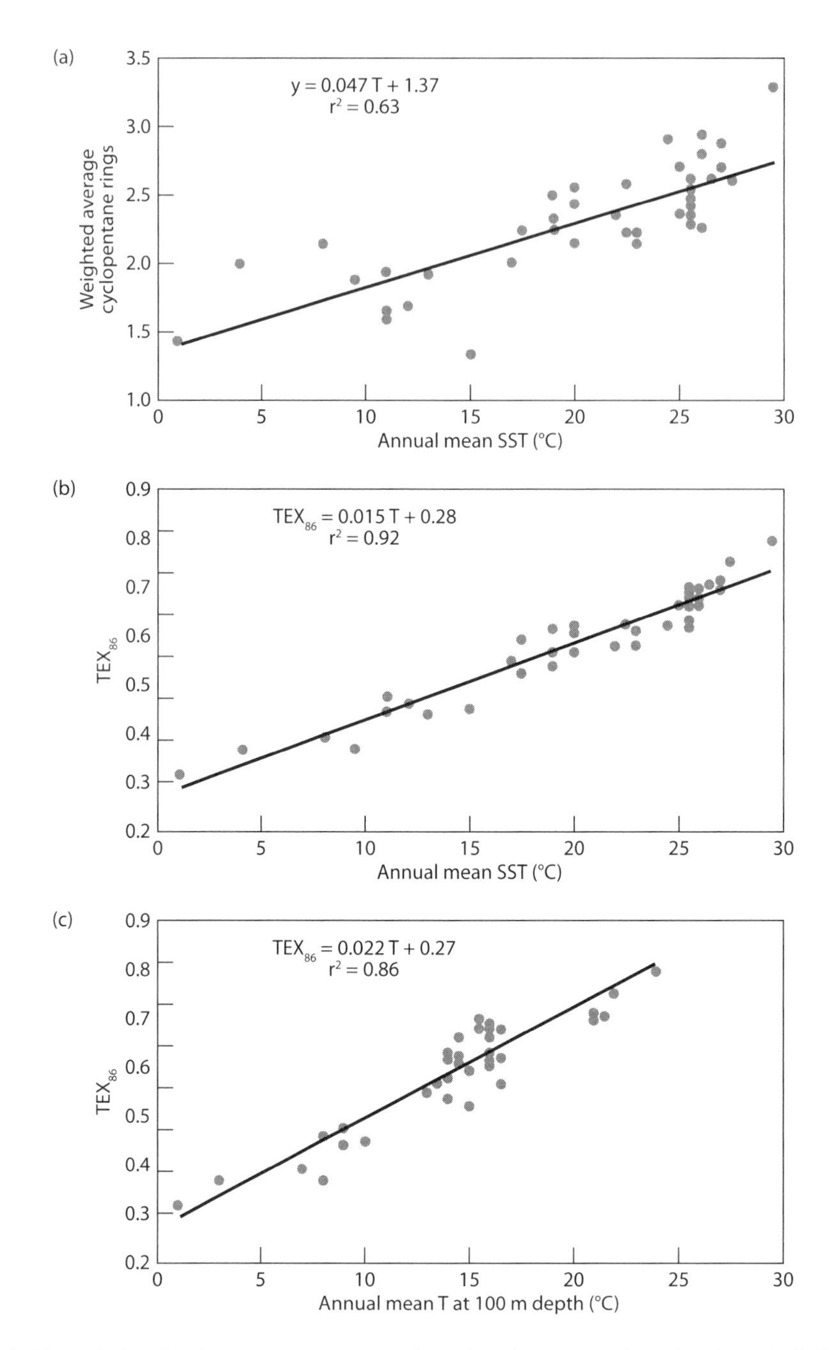

Figure 11.6. The relationship between average number of cyclopentane rings in glycerol dialkyl glycerol tetraethers I–IV (see fig. 11.7 for structures) with annual mean SST ($r^2 = .63$) (a) Relationships between TEX_{86} from surface sediments with annual mean SST ($r^2 = .92$) (b) and TEX_{86} from surface sediments with annual mean temperatures at 100m water depth or bottom water temperatures for deeper samples ($r^2 = .86$) (c) (Adapted from Schouten et al., 2002.)

hyperthermophilic Archaea (Sinninghe Damsté et al., 2002). Based on molecular modeling work, it is also believed that the additional cyclohexane ring found in crenarchaeol is an adaptation to cold temperatures, whereby crenarchaeol prevents dense packing of membrane lipids under these cooler "normal" temperatures (Sinninghe Damsté et al., 2002). There is also a crenarchaeol *regioisomer* that is synthesized in small quantities by Crenarchaeota. In fact, recent work has shown that radiocarbon ages of the crenarchaeol regioisomer of co-occurring GDGTs are different in marine sediments, which may indicate different origins through postdepositional processes (Shah et al., 2008). This study also noted that the other GDGTs were less stable than crenarchaeol regioisomer and were more influenced by postdepositional decay than alkenones. Consequently, GDGTs are likely better indicators than alkenones of historical changes in primary production from local sources, rather than inputs for distant sources contributed via lateral transport, because they will not "survive" long-term postdepositional transport (Shah et al., 2008; Huguet et al., 2009; Kim et al., 2009). However, further work is still needed to resolve any differences in the stability among GDGTs and the effects these may have on GDGT signatures during postdepositional transport. Recent work has also shown that the likely explanation for the typically higher temperature values obtained using TEX_{86} compared to Mg/Ca ratios (e.g., Hoogakker et al., 2009) is, in part, due to artifacts from production of lipids by archaeal populations in sediments that affect the TEX_{86} ratio (Lipp and Hinrichs, 2009). The TEX_{86} is calculated based on the relative proportion of several GDGT compounds with different levels of cyclopentane rings and methyl branches. The index is calculated as follows, where the compound numbers correspond to the structures provided in fig. 11.7b:

$$TEX_{86} = ([IV'] + [VII] + [VIII])/([IV'] + [VI] + [VII] + [VIII]) \tag{11.3}$$

Compound IV′, not shown in fig. 11.7b, is the sterioisomer of crenarcheaol (IV). This index has been extended to continental systems since nonthermophilic crenarchaeota also live in lakes. Powers and colleagues pioneered the application of TEX_{86} to lacustrine environments, demonstrating the presence of nonthermophilic crenarchaeotal biomarkers GDGTs in four large lakes, representing a wide range of climatic conditions (Powers et al., 2004). This study identified a series of isoprenoidal and nonisoprenoidal GDGTs in surface sediments, including the same isoprenoid, archaeal-derived GDGTs (IV–VIII; fig. 11.7b) found in marine sediments, marine particulate matter, and cultures of nonthermophilic Crenarchaeota. The GDGT distribution differed across the lakes, reflecting either changes in the crenarchaeotal community structure or a physiological response to temperature of the Crenarchaeota. Since genetic information suggests a low amount of diversity in freshwater Crenarchaeota (Keough et al., 2003), it is likely that the changes in GDGT composition reflect a response to temperature. For example, the abundance of GDGT V relative to crenarchaeol (IV) was high in the three "cold" lakes, but low in the Lake Malawi (Africa) sediments (fig. 11.7a), similar to what has been observed in marine studies (Schouten et al., 2002, 2003b). Similarly, Powers et al. (2004) found higher relative abundances of the cyclopentane-containing GDGTs (VI–VIII) in the Lake Malawi sediments, consistent with GDGT distributions of populations of marine crenarchaeota adjusting to higher SSTs. Subsequently, Powers and colleagues showed that the relationship between the TEX_{86} and mean annual surface temperature for the lake samples was similar to the relationship found for marine systems, suggesting that the TEX_{86} may be a useful tool for paleotemperature reconstruction in lakes.

11.2.3 BIT Index

GDGTs also provide useful biomarkers for examining contributions of terrigenous carbon to bulk organic matter. The branched and isoprenoid tetraether (BIT) index (the ratio of marine and terrestrial

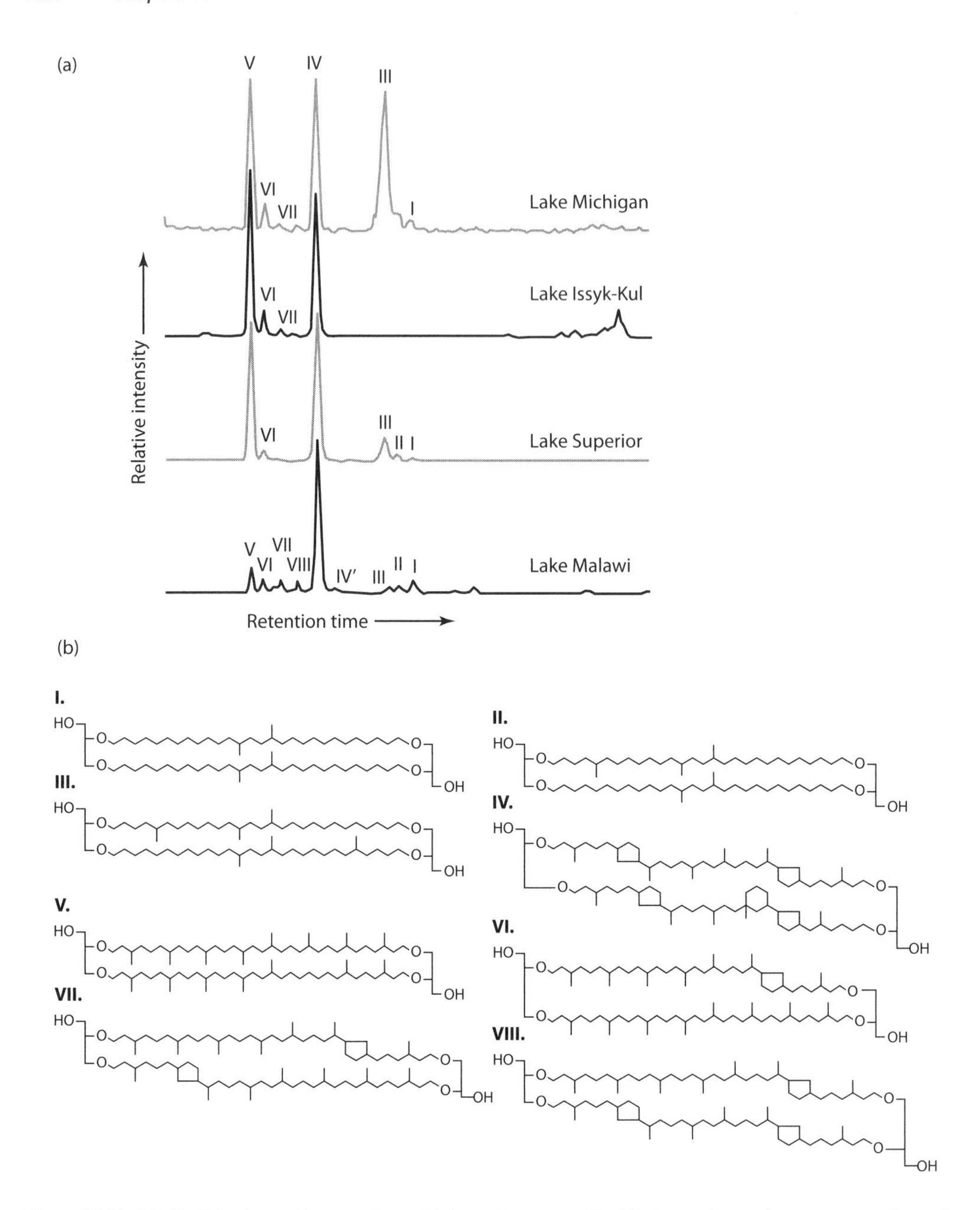

Figure 11.7. (a) Partial chromatogram from high-performance liquid chromatography–mass spectrometry analysis of glycerol dialkyl glycerol tetraether (GDGT) compounds from Lake Michigan (USA), Issyk Kul (Kyrgyzstan), Lake Superior (USA), and Lake Malawi (Africa). Peaks used in the TEX_{86} calculation are shaded. (b) Structures of the GDGT compounds identified (a) and used in the calculation of the TEX_{86} index. GDGTs I–III are nonisoprenoidal and IV–VIII are isoprenoidal compounds commonly found in Crenarchaeota; crenarcheol is compound IV. (Adapted from Powers et al., 2004.)

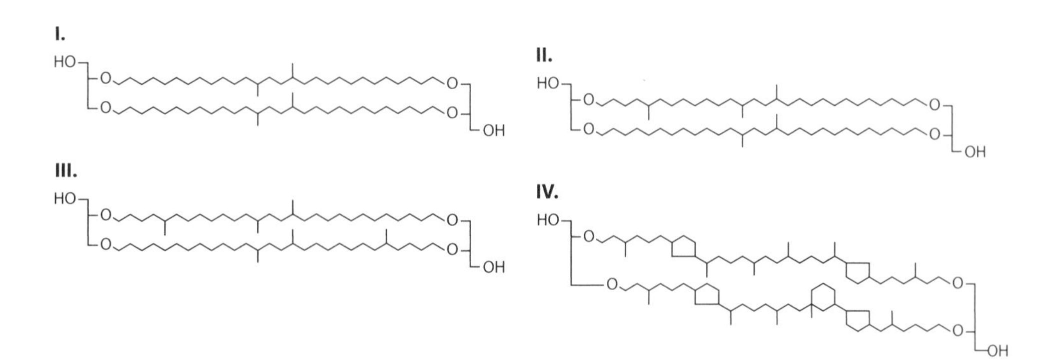

Figure 11.8. Structures corresponding to the four glycerol dialkyl glycerol tetraethers used in the BIT index calculation. Compounds I–III are branched, nonisoprenoidal, bacterial lipids and compound IV is crenarchaeol.

glycerol dialkyl glycerol tetraether [GDGT] membrane lipids) has been shown to correlate with the amount of fluvial organic carbon buried in marine sediments (e.g., Hopmans et al., 2004; Herfort et al., 2006; Weijers et al., 2006). When the BIT index equals zero, it signifies the absence of branched GDGTs (or marine influence). In contrast, a BIT index equal to one indicates the absence of crenarchaeol (or fluvial influence). The BIT index is calculated as follows:

$$\mathrm{BIT} = \frac{[\mathrm{I} + \mathrm{II} + \mathrm{III}]}{[\mathrm{I} + \mathrm{II} + \mathrm{III}] + [\mathrm{IV}]} \tag{11.4}$$

with the structures of the compounds used in the calculation provided in fig. 11.8.

The analysis of terrestrial soil and peat samples shows that branched tetraether lipids are predominant in terrestrial environments. Previous studies have suggested that these compounds likely derive from anaerobic soil bacteria (Hopmans et al., 2004; Weijers et al., 2007). In contrast, crenarchaeol is the characteristic membrane lipid of nonthermophilic Crenarchaeota, which is especially abundant the marine and lacustrine environments. The BIT index was first applied to surface sediments from the Angola Basin, where it correlated to the outflow of the Congo River (Africa). Analyses of particulate organic matter from the North Sea showed relatively higher BIT indices in water column particulate organic matter near large river inputs. Hopmans et al. (2004) conducted a survey of surface sediments collected from marine and lacustrine environments, showing that the BIT index correlated to the relative input of terrestrial organic material from fluvial sources.

Weijers and colleagues subsequently investigated the GDGT composition of 134 soil samples collected from 90 globally distributed locations to study environmental factors controlling GDGT isomers (Weijers et al., 2006, 2007). They concluded that soil bacteria alter their core membrane lipid composition to both pH and temperature. The relative amount of cyclopentyl moieties in the alkyl chains is related to pH, whereas the relative amount of additional methyl branches at the C-5 and C-50 positions may be related to both pH and temperature. As a result, they developed two additional proxies that show promise as biomarkers for reconstructing paleo soil pH and continental air temperatures. The methylation index of branched tetraethers (MBT) was positively correlated with mean air temperature ($r^2 = .62$), while the cyclization ratio of branched tetraethers (CBT) was negatively related to soil pH ($r^2 = .70$). This study suggests that soil-associated GDGTs not only are useful as biomarkers for terrigenous-derived organic matter but can provide paleoenvironmental information. More recently, it has also been shown that crenarcheaol, largely derived from marine Crenachaeota, was significantly more resistant to decay than GDGTs from terrestrially derived soils in sediment from the Madeira Abyssal Plain (Huguet et al., 2009).

In a recent study conducted off the coast of Washington State (USA), the BIT index was compared to other measures of terrigenous organic matter, lignin phenols, and $\delta^{13}C_{TOC}$ (Walsh et al., 2008). Surprisingly, the BIT index did not correlate well with either $\delta^{13}C_{TOC}$ or lignin phenols and suggested much lower contributions from terrigenous sources of organic matter. The authors concluded that the discrepancy may be due to differences in the sources of terrigenous organic matter, since the BIT index likely reflects contributions from soils and peats, while lignin phenols trace inputs from vascular plants. As a result, care must be used in interpreting the BIT index, particularly in environments where soils and peats, are not the dominant sources of terrigenous organic matter. Other more recent studies also supported this idea of BIT being used more ideally as a biomarker of soils (Belicka and Harvey, 2009; Weijers et al., 2009; Smith et al., 2010).

11.3 Summary

In this chapter we discussed several classes of polar lipids, many with applications to the fields of paleoceanography and paleolimnology. Alkenones, for example, have been widely used in paleoceanographic studies, often in concert with other paleotemperature proxies such, as $\delta^{18}O$, Mg/Ca ratios, and fossil assemblages of forams. We also discussed recent developments in high-performance liquid chromatography–mass spectrometry that allow for the analysis of intact polar lipids such as the glycerol dialkyl glycerol tetraethers (GDGTs) and intact phospholipids. These compounds are produced by microorganisms living in both terrestrial and marine systems and have useful application to studies in paleoecology, paleoclimatology, microbial ecology, and organic geochemistry.

The analysis of intact polar molecules has become more widespread since the development of LC-MS techniques. In some cases, these compounds are proving to be more useful for identifying microbial communities associated with soils and sediments (e.g., intact polar lipids) than the more traditional approaches that involve hydrolysis of polar compounds and analysis of their subcomponents (e.g., phospholipid linked fatty acids [PLFA]). This compound class shows great promise for identifying microbes associated with soils and marsh sediments and the delivery of soil-derived organic matter to coastal regions (e.g., BIT index) as well as paleothermometry (TEX_{86}). However, more work is needed to better cross-calibrate among different labs around the world on both indices. The first intercalibration exercise for labs using the BIT and TEX_{86} indices revealed that repeatability (intralaboratory variation) and reproducibility (among laboratory variation) for TEX_{86} measurements across 15 different laboratories analyzing two sediment samples was 0.03 and 0.02 or ± 1 to 2°C – and 0.05 and 0.07 or ± 3 to 4°C, respectively (Schouten et al., 2009). For BIT, repeatability and reproducibility was 0.03 and 0.41, respectively. The large variance in the BIT measurements was attributed to equipment settings and the large difference in molecular weight between GDGTs used in the BIT index. The variability in values obtained in the TEX_{86} was significantly higher than that obtained for similar intercalibration studies for alkenone ($U^{K'}_{37}$) and Mg/Ca paleothermometers, which suggests that improvements are clearly needed to better unify results across different laboratories.

Other improvements are also needed to purify GDGT standards for better quantification of absolute concentrations of GDGTs in the natural environment, but this will not be an easy task. Efforts to provide internal standards, such as the C_{46} GDGT, have already improved the ability, in part, to obtain better absolute abundances of GDGTs (Huguet et al., 2006). Similarly, recent work using HPLC/APCI-ion trap MS has resulted in a rapid screening test for GDGTs (Escala et al., 2007). So, while improvements continue to be made each year in our understating of the applications and limitations of GDGTs as organic proxies, it is clear that further round-robin studies need to continue to provide better the specific details on mass calibrations and tuning setups for LC-MS (Schouten et al., 2009).

12. Photosynthetic Pigments: Chlorophylls, Carotenoids, and Phycobilins

12.1 Background

The primary *photosynthetic pigments* used in absorbing photosynthetically active radiation (PAR) are *chlorophylls*, *carotenoids*, and *phycobilins*—with chlorophyll representing the dominant photosynthetic pigment (Emerson and Arnold, 1932a,b; Clayton, 1971, 1980). Although a greater amount of chlorophyll is found on land, 75% of the global turnover (ca. 10^9 Mg yr^{-1}) occurs in oceans, lakes, and rivers/estuaries (Brown et al., 1991; Jeffrey and Mantoura, 1997). All of the light-harvesting pigments are bound to proteins, making up distinct carotenoid and chlorophyll–protein complexes. These pigment–protein complexes in algae and higher plants are located in the *thylakoid membrane* of chloroplasts (Cohen et al., 1995). The photosynthesizing ability of eukaryotes was made possible by one or more *endosymbiotic* associations between heterotrophic eukaryotes and photosynthetic prokaryotes (or their descendents) (Gray, 1992; Whatley, 1993). Several primary endosymbioses occurred between eukaryotes and cyanobacteria. In one of the lineages, the photosynthetic organism lost much of its genetic independence and became functionally and genetically integrated as plastids—chloroplasts within the host cell (Kuhlbrandt et al., 1994). At least two types of protists—chloroarachniophytes and cryptomonads—acquired *plastids* by forming symbioses with eukaryotic algae.

Photosynthetic organisms have a variety of accessory pigments, on which their classification has been based. Despite this variation, it is generally accepted that all chloroplasts are derived from a single cyanobacterial ancestor. Prochlorophytes are prokaryotes that perform oxygenic photosynthesis using chlorophyll *b*, like land plants and green algae (Chlorophyta), and were proposed to be the ancestors of chlorophyte chloroplasts. However, three known prochlorophytes (*Prochloron didemni, Prochlorothrix hollandica*, and *Prochlorococcus marinus*) have been shown not to be the specific ancestors of chloroplasts, but only diverged members of the cyanobacteria, which contain phycobilins but lack chlorophyll *b* (Olson and Pierson, 1987). Consequently, it has been proposed that the ability to synthesize chlorophyll *b* developed independently several times in prochlorophytes and in the ancestor of chlorophytes. Phylogenetic analyses show that these genes share a common evolutionary origin. This indicates that the progenitors of oxygenic photosynthetic bacteria, including the ancestor of chloroplasts, had both chlorophyll *b* and phycobilins (Hill, 1939; Rutherford et al., 1992; Nugent, 1996).

Chlorophylls are pigments, generally green in color. Chlorophyll itself is actually not a single molecule but a family of related molecules, designated chlorophylls *a*, *b*, *c*, and *d*. Chlorophyll *a* is the molecule found in all plant cells and therefore its concentration is what is most commonly reported in chlorophyll analyses (Jeffrey et al., 1997). The other chlorophylls are more exclusive in their distribution. For example, chlorophyll *d* is generally found in marine red algae, while chlorophylls *b* and *c* are common in green algae and diatoms, respectively (Rowan, 1989; Jeffrey et al., 1997). The relative concentrations of these chlorophylls within cells vary considerably among species of algae. Nevertheless, chlorophyll *a* remains the dominant pigment in virtually all the eukaryotic and prokaryotic cyanobacteria (Cyanobacteria) (Olson and Pierson, 1987), and is the pigment commonly used to estimate algal biomass in aquatic ecosystems. Chlorophylls *a* and *b* contain a propionic acid esterified to a C_{20} phytol; chlorophylls c_1 and c_2 have an acrylic acid that replaces the propionic acid (fig. 12.1). Chlorophylls are composed of cyclic *tetrapyrroles*, which form a *porphyrin* ring that is coordinated around a Mg-chelated complex (Rowan, 1989). The stable ring-shaped structure allows electrons to be 'free' to migrate. Because the electrons move freely, the ring has the potential to gain or lose electrons easily, and thus the potential to provide energized electrons to other molecules. This is the fundamental process of photosynthesis, by which chlorophyll 'captures' the energy of sunlight. There are several kinds of chlorophyll, the most important being chlorophyll *a*. This is the molecule that makes photosynthesis possible, by passing its energized electrons on to molecules that will manufacture sugars (Emerson and Arnold, 1932a,b; Calvin, 1976; Clayton, 1980; Arnon, 1984; Malkin and Niyogi, 2000).

Chloropigments can also be used to resolve contributions between eukaryotic and prokaryotic organisms. For example, eukaryotes and cyanophytes are found in oxygenated waters as opposed to green and purple sulfur bacteria, which are typically found in anoxic or hypoxic environments (Wilson et al., 2004). Bacteriochlorophyll *a* is the dominant light-harvesting pigment in purple sulfur bacteria (e.g., *Chromatiaceae*), and its synthesis is inhibited by the presence of O_2 (Knaff, 1993; Squier et al., 2004). Similarly, bacteriochlorophyll *e* has been shown to be indicative of green sulfur bacteria (e.g., *Chlorobium phaeovibroides* and *C. phaeobacteroides*) (Chen et al., 2001; Repeta et al., 1989), while bacteriochlorophylls *c* and *d* can be found in other *Chlorobium* spp. As a result of their linkages to environmental redox conditions, bacteriochlorophylls have been successfully used as an index for reconstructing paleoredox in aquatic systems (Bianchi et al., 2000a; Chen et al., 2001; Squier et al., 2004; Ruess et al., 2005), as discussed in chapter 2.

The degradation products of chlorophyllous pigments, formed primarily by heterotrophic processes, are the most abundant forms of pigments found in sediments (Daley, 1973; Repeta and Gagosian, 1987; Brown et al., 1991; Bianchi et al., 1993; Chen et al., 2001; Louda et al., 2002). These pigment degradation products may be used to infer the availability of different source materials to consumers. For example, the four dominant tetrapyrrole derivatives of chloropigments (pheopigments) found in marine and freshwater/estuarine systems (chlorophyllide, pheophorbide, pheophytin, pyropheophorbide) are formed during bacterial, autolytic cell lysis, and metazoan grazing activities (Sanger and Gorham, 1970; Jeffrey, 1974; Welschmeyer and Lorenzen, 1985; Bianchi et al., 1988, 1991; Head and Harris, 1996).

More specifically, chlorophyllase-mediated deesterification reactions (loss of the phytol side chain) of chlorophyll yield chlorophyllides when diatoms are physiologically stressed (Holden, 1976; Jeffrey and Hallegraeff, 1987). Pheophytins can be formed from bacterial decay, metazoan grazing, and cell lysis, whereby the Mg is lost from the chlorophyll center (Daley and Brown, 1973). Pheophorbides, which are formed from either removal of the Mg from chlorophyllide or removal of the phytol chain from pheophytin, occur primarily from herbivorous grazing activities (Shuman and Lorenzen, 1975; Welschmeyer and Lorenzen, 1985; Bianchi et al., 2000a). Pyrolized pheopigments, such as

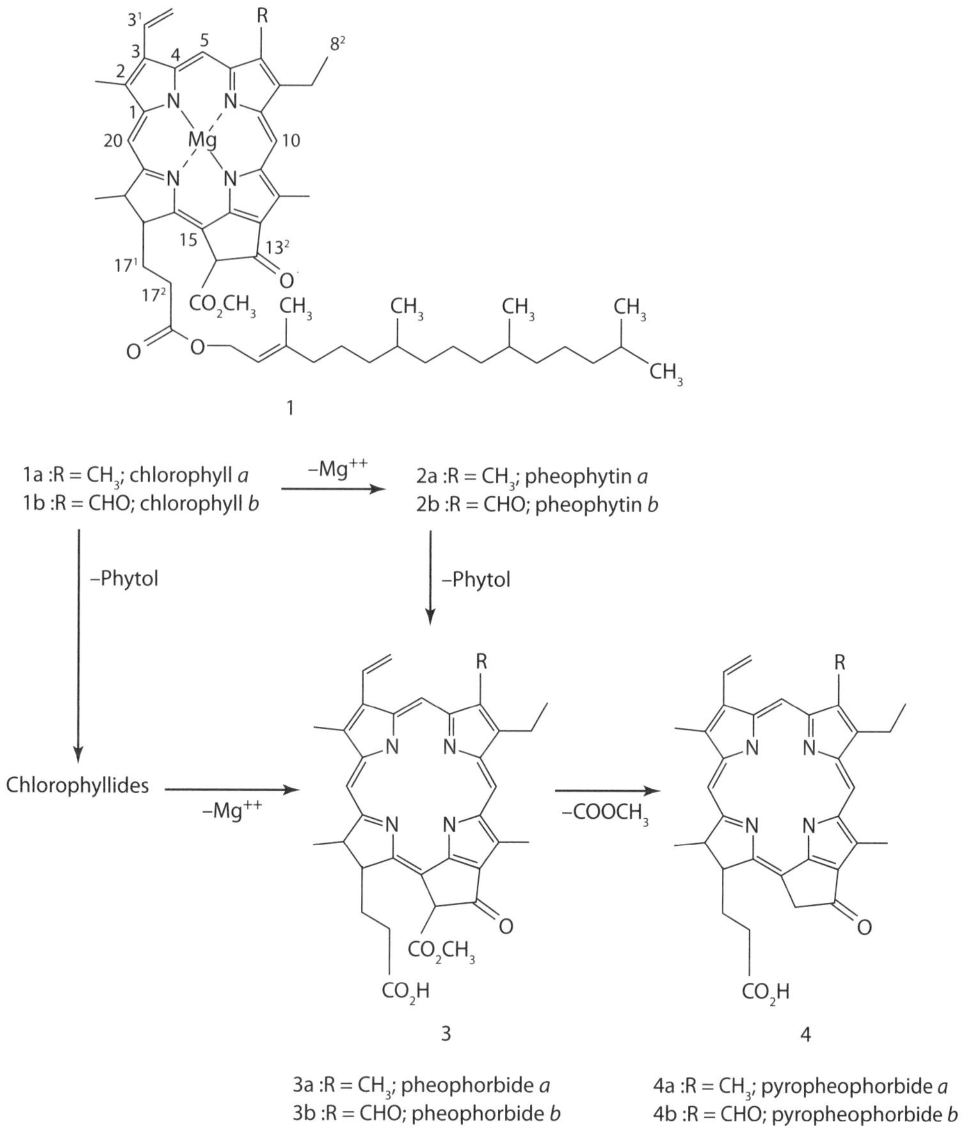

1a :R = CH_3; chlorophyll *a*
1b :R = CHO; chlorophyll *b*

–Mg^{++}

2a :R = CH_3; pheophytin *a*
2b :R = CHO; pheophytin *b*

–Phytol

Chlorophyllides

–$COOCH_3$

3a :R = CH_3; pheophorbide *a*
3b :R = CHO; pheophorbide *b*

4a :R = CH_3; pyropheophorbide *a*
4b :R = CHO; pyropheophorbide *b*

Figure 12.1. Chlorophylls *a* and *b*, which have a propionic acid esterified to a C_{20} phytol, can be degraded to their respective pheopigments: chlorophyllide, pheophorbide, and pyropheophorbide. 'R' representative of changing functional groups. (Adapted from Bianchi, 2007.)

pyropheophorbide and pyropheophytin, are formed by removal of the methylcarboxylate group (–$COOCH_3$) on the isocylic ring from the C_{13} proprionic acid group (Ziegler et al., 1988), and are formed primarily by grazing processes (Hawkins et al., 1986). These compounds can account for the dominant fraction of the breakdown products of pigments in water column particulates (Head and Harris, 1994, 1996) as well as sediments (Keely and Maxwell, 1991; Hayashi et al., 2001; Chen et al., 2003a,b). Cyclic derivatives of chlorophyll *a*, such as cyclopheophorbide *a* enol (Harris et al., 1995; Goericke et al., 2000; Louda et al., 2000) and its oxidized product chlorophyllone *a*, are also abundant in aquatic systems and are believed to be formed in the guts of metazoans (Harradine et al., 1996; Ma and Dolphin, 1996).

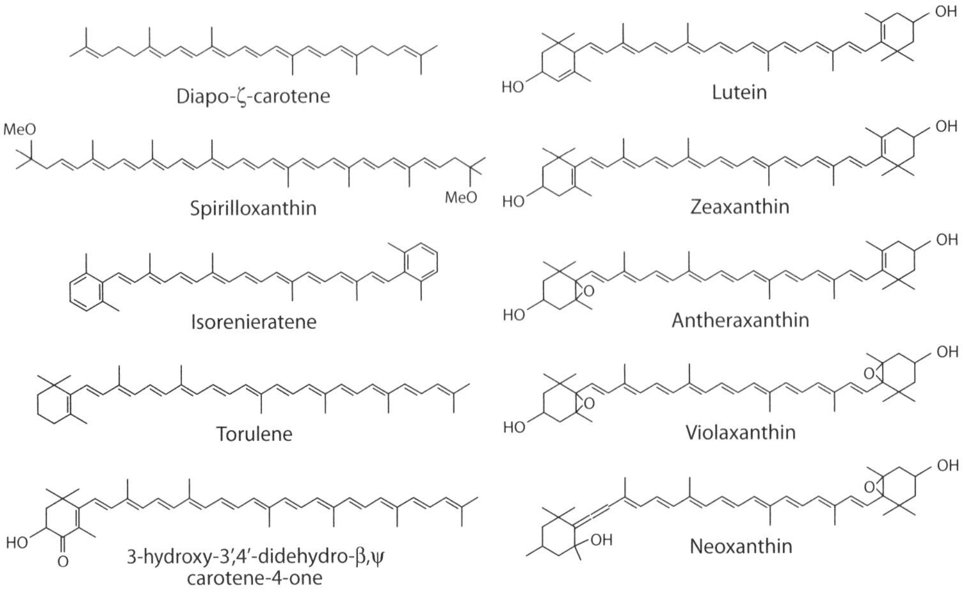

Diapo-ζ-carotene
Lutein
Spirilloxanthin
Zeaxanthin
Isorenieratene
Antheraxanthin
Torulene
Violaxanthin
3-hydroxy-3′,4′-didehydro-β,ψ carotene-4-one
Neoxanthin

Figure 12.2. Carotenoids are found in bacteria, algae, fungi, and higher plants and serve as the principal pigment biomarkers for class-specific taxonomic work in aquatic systems.

Most chlorophyll produced in the euphotic zone is degraded to colorless compounds: the destruction of the chromophore indicates cleavage of the macrocycle (type II reactions) according to (Rutherford, 1989; Brown et al., 1991; Nugent, 1996). The main mechanisms of type II reactions of chlorophylls are photooxidation, involving attack of excited-state singlet oxygen, and enzymatic degradation (Arnon, 1984; Gaossauer and Engel, 1996). Another group of stable nonpolar chlorophyll *a* degradation products is the steryl chlorin esters (SCEs) and carotenol chlorin esters (CCEs) (Furlong and Carpenter, 1988; King and Repeta, 1994; Talbot et al., 1999; Chen et al., 2003a,b; Kowalewska et al., 2004; Kowalewska, 2005). These compounds are formed by esterification reactions between pheophorbide *a* and/or pyropheophorbide *a* and sterols and carotenoids; these compounds are believed to be primarily formed by zooplanktonic and bacterial grazing processes (King and Repeta, 1994; Spooner et al., 1994; Harradine et al., 1996). Recent work has shown that SCEs are formed by anaerobic and facultative aerobic bacteria in sediments and in the guts of invertebrate grazers (Szymczak-Zyla et al., 2008), with many of the SCE precursors formed by benthic organisms (Szymczak-Zyla et al., 2006).

Carotenoids are an important group of tetraterpenoids, consisting of a C_{40} chain with conjugated bonds (Goodwin, 1980; Frank and Cogdell, 1996). These pigments are found in bacteria, algae, fungi, and higher plants and serve as the principal pigment biomarkers for class-specific taxonomic work in aquatic systems (fig. 12.2, table 12.1). Studies on carotenoid biosynthesis have shown lycopene to be an important precursor for the synthesis of most carotenoids (Liaaen-Jensen, 1978; Goodwin, 1980; Schuette, 1983). Carotenoids are found in pigment–protein complexes that are used for light harvesting as well as photoprotection (Frank and Cogdell, 1996; Porra et al., 1997). The carotenoids can be divided into two groups: the carotenes (e.g., *β*-carotene), which are hydrocarbons, and the xanthophylls (e.g., antheraxanthin, violaxanthin, and fucoxanthin), which are molecules that contain at least one oxygen atom (Rowan, 1989). Oxygenated functional groups, such as the 5′6′ epoxide group make certain xanthophylls, such as fucoxanthin, more susceptible to bacterial decay than carotenes (Pfundel and Bilger, 1994; Porra et al., 1997). There are two well-known ‘xanthophyll cycles’ involved in photoprotection of photoautotrophs during excess illumination: (1) green algae and higher plants

Table 12.1
Summary of signature pigments useful as markers of algal groups and processes in the sea

Pigment	*Algal group or process*	*References*
Chlorophylls		
Chl-*a*	All photosynthetic microalgae (except prochlorophytes)	Jeffrey et al. (1997)
Divinyl chl-*a*	Prochlorophytes	Goericke and Repeta (1992)
Chl-*b*	Green algae: clorophytes, prasinophytes, euglenophytes	Jeffrey et al. (1997)
Divinyl chl-*b*	Prochlorophytes	Goericke and Repeta (1992)
Chl-*c* family	Chromophyte algae	Jeffrey (1989)
Chl-c_1	Diatoms, some prymnesiophytes, some freshwater chrysophytes, raphidophytes	Jeffrey (1976b, 1989); Stauber and Jeffrey (1988); Andersen and Mulkey (1983)
Chl-c_2	Most diatoms, dinoflagellates, prymnesiophytes, raphidophytes, cryptophytes	Jeffrey et al. (1975); Stauber and Jeffrey (1988); Andersen and Mulkey (1983)
Chl-c_3	Some prymnesiophytes, one chrysophyte, several diatoms and dinoflagellates	Jeffrey and Wright (1987); Vesk and Jeffrey (1987); Jeffrey (1989); Johnsen and Sakshaug (1993)
Chl-$c_{cs\text{-}170}$	One prasinophyte	Jeffrey (1989)
Phytylated chl-*c*-like[c]	Some prymnesiophytes	Nelson and Wakeham (1989); Jeffrey and Wright (1994); Ricketts (1966); Jeffrey (1989)
Mg3.8 DVP	Some prasinophytes	
Bacteriochlorophylls	Anoxic sediments	Repeta et al. (1989); Repeta and Simpson (1991)
Carotenoids		
Alloxanthin	Cryptophytes	Chapman (1966); Pennington et al. (1985)
19-Butanoyloxyfucoxanthin	Some prymnesiophytes, one chrysophyte, several dinoflagellates	Bjørnland and Liaaen-Jensen (1989); Bjørnland et al. (1989); Jeffrey and Wright (1994)
β, ε-Carotene	Cryptophytes, prochlorophytes, rhodophytes, green algae	Bianchi et al. (1997c); Jeffrey et al. (1997)
β, β-Carotene	All algae except cryptophytes and rhodophytes	Bianchi et al. (1997c); Jeffrey et al. (1997)
Crocoxanthin	Cryptophytes (minor pigment)	Pennington et al. (1985)
Diadinoxanthin	Diatoms, dinoflagellates, prymnesiophytes, chrysophytes, raphidophytes, euglenophytes	Jeffrey et al. (1997)
Dinoxanthin	Dinoflagellates	Johansen et al. (1974); Jeffrey et al. (1975)
Echinenone	Cyanophytes	Foss et al. (1987)
Fucoxanthin	Diatoms, prymnesiophytes, chrysophytes, raphidophytes, several dinoflagellates	Stauber and Jeffrey (1988); Bjørnland and Liaaen-Jensen (1989)

Table 12.1
Continued

Pigment	Algal group or process	References
19-Hexanoyloxyfucoxanthin	Prymnesiophytes, several dinoflagellates	Arpin et al. (1976); Bjørnland and Liaaen-Jensen (1989)
Lutein	Green algae: chlorophytes, prasinophytes, higher plants	Bianchi and Findly (1990); Bianchi et al. (1997c); Jeffrey et al. (1997)
Micromonal	Some prasinophytes	Egeland and Liaaen-Jensen (1992, 1993)
Monadoxanthin	Cryptophytes (minor pigment)	Pennington et al. (1985)
9-*cis*-Neoxanthin	Green algae: chlorophytes, prasinophytes, euglenophytes	Jeffrey et al. (1997)
Peridinin	Dinoflagellates	Johansen et al. (1974); Jeffrey et al. (1975)
Peridininol	Dinoflagellates (minor pigments)	Bjørnland and Liaaen-Jensen (1989)
Prasinoxanthin	Some prasinophytes	Foss et al. (1984)
Pyrrhoxanthin	Dinoflagellates (minor pigment)	Bjørnland and Liaaen-Jensen (1989)
Siphonaxanthin	Several prasinophytes: one euglenophyte	Bjørnland (1989); Fawley and Lee (1990)
Vaucheriaxanthin ester	Eustigmatophytes	Bjørnland and Liaaen-Jensen (1989)
Violaxanthin	Green algae; chlorophytes, prasinophytes: eustigmatophyes	Jeffrey et al. (1997)
Zeaxanthin	Cyanophytes, prochlorophytes, rhodophytes, chlorophytes, eustigmatophyes (minor pigment)	Guillard et al. (1985); Gieskes et al. (1988); Goerieke and Repeta (1992)
Biliproteins		
Allophycocyanins	Cyanophytes, rhodophytes	Rowan (1989)
Phycocyanin	Cyanophytes, cryptophytes rhodophytes (minor pigment)	Rowan (1989)
Phycocrythrin	Cyanophytes, cryptophytes rhodophytes	Rowan (1989)
Chlorophyll degradation products		
Pheophytin-*a*[d]	Zooplankton fecal pellets, sediments	Vernet and Lorenzen (1987); Bianchi et al. (1988, 1991, 2000a)
Pheophytin-*b*[d]	Protozoan fecal pellets	Bianchi et al. (1988, 2000a); Strom (1991, 1993)
Pheophytin-*c*[d]	Protozoan fecal pellets	Strom (1991, 1993)
Pheophorbide-*a*	Protozoan fecal pellets	Strom (1991, 1993); Head et al. (1994); Welschmeyer and Lorenzen (1985)
Pheophorbide-*b*	Protozoan fecal pellets	Strom (1991, 1993)
Chlorophyllide *a*	Senescent diatoms: extraction artifact	Jeffrey and Hallegraeff (1987)
Blue-green chl-*a* derivatives (lactones, 10-hydroxy chlorophylls)	Senescent microalgae	Hallegraeff and Jeffrey (1985)

Table 12.1
Continued

Pigment	*Algal group or process*	*References*
Pyrochlorophyll-*a*	Sediments	Chen et al. (2003a,b)
Pyropheophytin-*a*	Sediments	Chen et al. (2003a,b)
Pyropheophorbide-*a*	Copepod grazing: fecal pellets	Head et al. (1994)
Mesopheophorbide-*a*	Sediments	Chen et al. (2003a,b)

[a]Note that many carotenoids in trace quantities have wider distributions than those shown (Bjørnland and Liaaen-Jensen, 1989).
[b]Note that some "atypical" pigment contents reflect endosymbiotic events (see footnotes to table 2.3, chapter 2).
[c]Two spectrally distinct pigments have been found in these fractions (Garrido et al., 1995).
[d]Multiple zones found in each of these fractions (Strom, 1993).
Source: Modified from Jeffrey et al. (1997).

use the interconversion of zeaxanthin ↔ antherxanthin ↔ violaxathin, where epoxide groups are gained or lost; (2) algae such as chrysophytes and pyrrophytes use the interconversion of zeaxanthin ↔ diatoxanthin ↔ diadinoxanthin (fig. 12.3) (Hager, 1980).

As discussed in chapter 2, plant pigments have been used in aquatic paleoecology to better understand historical changes in bacterial community composition, trophic levels, redox change, acidification of lakes, and past levels of ultraviolet (UV) radiation (Leavitt and Hodgson, 2001; Bianchi, 2007, and references therein). Here we briefly discuss some of the key carotenoids that have been used as fossil pigments. In the case of historical changes in phytoplankton response to UV radiation changes (due to changes in clarity of water column, stratospheric ozone, etc.), sheath pigments like scytonemin have been effective tracers of ecosystem change (Leavitt et al., 1997, 2003, 2009). Scytonemin, a pigment produced in the extracellular sheaths of many cyanaobacteria (Garcia-Pichel and Castenholz, 1991; Garcia-Pichel et al., 1992), has a maximum absorption in the UV-A (315–400 nm) and UV-B (280–315 nm) ranges, conferring some phytoplankton with the advantage of having a natural sunscreen (Vincent et al., 1993; Llewellyn and Mantoura, 1997).

Similarly, other carotenoids have been used to examine past changes in the water clarity index (WCI), such as the isorenieratene:β-carotene ratio (Brown et al., 1984; Hambright et al., 2008). Isorenieratene is a carotenoid also indicative of chemoautotrophic green sulfur bacteria (the brown varieties) Chlorobiaceae (e.g., *Chlorobium phaeovibroides* and *C. phaeobacteroides*), similar to the aforementioned bacteriochlorophyll *e* (Brown et al., 1984; Repeta et al., 1989; Chen et al., 2001). β-Carotene serves as a stable pigment tracer found in virtually all algae (see Bianchi, 2007, and references therein); hence, the isorenieratene:β-carotene ratio allows for estimation of the amounts of Chlorobiaceae relative to all other photosynthetic plankton. Since these bacteria grow in a zone where there is adequate light and sulfides fluxing from bottom sediments to the water column, the ratio serves as an indicator of the WCI (Leavitt and Carpenter, 1989; Hambright et al., 2008). Isorenieratene has also been used to provide further corroboration of microbial sources in paleogenetic studies (e.g., fossil DNA) (e.g., Coolen and Overmann, 1998; Sinninghe Damsté and Coolen, 2006). Another carotenoid, okenone, has been used as a fossil pigment indicative of purple sulfur bacteria (Chromatiaceae) (Overmann et al., 1993; Coolen and Overmann, 1998). Okenone was found to occur as far back as 11,000 yr BP in lake sediments, further supporting the use of pigments as fossil tracers (Overmann et al., 1993).

These more recent examples of fossil pigments are predicated on earlier work that showed carotenoids, such as zeaxanthin, to be preserved in sediments as far back as 56,000 yr BP (Watts and Maxwell, 1977). Along with zeaxanthin, and its isomer lutein, other carotenoids, such as alloxanthin and diatoxanthin, continue to be used as effective tracers of past changes in aquatic ecosystems (e.g., Lami et al., 2000; Bianchi et al., 2000; Leavitt et al., 2003, 2009; Brock et al., 2006; Hambright et al., 2008).

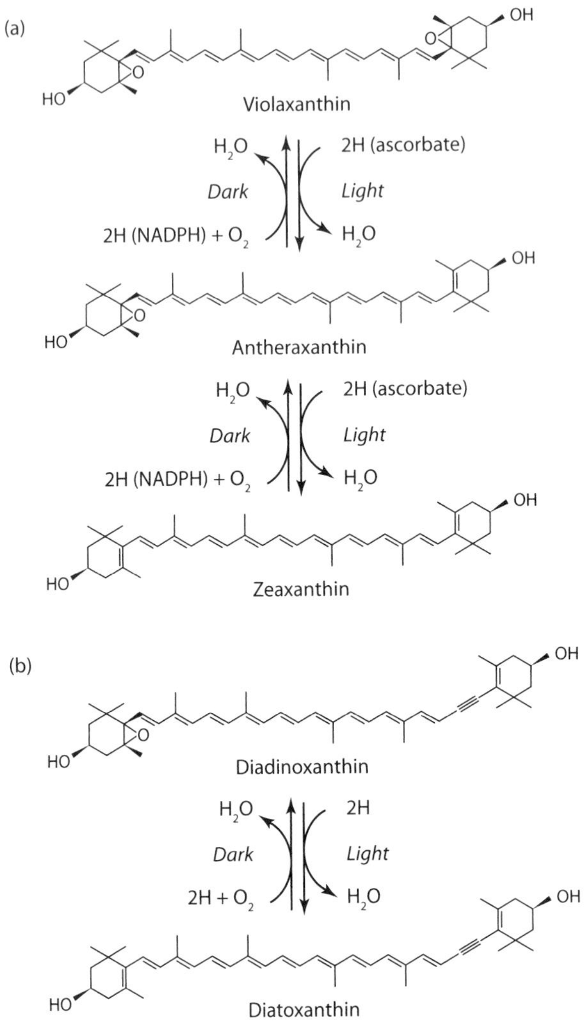

Figure 12.3. Two well-known "xanthophyll cycles" involved in photoprotection of photoautotrophs during excess illumination: (a) green algae and higher plants use the interconversion of zeaxanthin ↔ antherxanthin ↔ violaxathin, where epoxide groups are gained or lost; (b) algae such as chrysophytes and pyrrophytes use the interconversion of zeaxanthin ↔ diatoxanthin ↔ diadinoxanthin. (Adapted from Hager, 1980.)

Phycobilins are water-soluble pigments found in the cytoplasm and/or the *stroma* of the chloroplast. The phycobilins are also more simply referred to as bilins, since their chemical structure is similar to that of the bile pigments *biliverdin* and *bilirubin* (Lemberg and Legge, 1949). These pigments are light-harvesting pigments bound to *apoproteins* in the Cyanophyta, Rhodophyta, and Cryptophyta (Kursar and Alberte, 1983; Colyer et al., 2005). Phycobilins typically absorb in the 540- to 655-nm range (Bermejo et al., 2002), which is between the areas absorbed by carotenoids and chlorophylls. These pigments have not been used as extensively as carotenoids for taxonomic separation due to their more restrictive occurrence across different algal classes and the lack of well-developed high-performance liquid chromatography (HPLC) methods for rapid analyses. Most of the work to date, in Cyanophyta and Rhodophyta, has shown that *phycobiliproteins* are aggregated in a highly ordered protein complex called a *phycobilisome* (PBS), making these phycobilins unique among photosynthetic pigments.

Phycobilisomes are attached to the stromal face of the *thylakoid* and arranged in rows coupled to photosystem II particles. The water-soluble supramolecular protein assemblies of the *phycobilisome* are composed of two types of proteins: the highly colored phycobiliproteins (with their covalently linked tetrapyrrole chormophores) and the generally uncolored, linker proteins (Colyer et al., 2005). Extending into the cytosol, the phycobilisomes consist of a cluster of phycobilin pigments, including *phycocyanin* (blue) and *phycoerythrin* (red), attached by their phycobiliproteins. Phycobiliproteins are covalently linked via thioether bonds to cysteinyl residues; the most common polypeptide-bound phycobilins are shown in fig. 12.4 (Colyer et al., 2005). Phycobilisomes preferentially funnel light energy into photosystem II for the splitting of water and generation of oxygen (Govindjee and Coleman, 1990). While many photosynthetic eubacteria possess photosystem I to oxidize reduced molecules such as H_2S, only Cyanobacteria have photosystem II. The evolution of photosystem II apparently occurred in Cyanobacteria (Michel and Deisenhofer, 1988).

12.2 Chlorophylls, Carotenoids, and Phycobilins Biosynthesis

12.2.1 Chlorophylls

In phase I of chlorophyll biosynthesis, glutamic acid is converted to *5-aminolevulinic acid* (ALA) (fig. 12.5) (Taiz and Zeiger, 2006). This reaction is unusual in that it involves a covalent intermediate whereby glutamic acid is attached to a transfer RNA (tRNA) molecule. This is one of a very small number of examples in biochemistry in which a tRNA is utilized in a process other than protein synthesis. Two molecules of ALA are then condensed to form *porphobilinogen* (PBG), which allows for the formation of the *pyrrole* ring structures in chlorophyll (Clayton, 1980; Armstrong and Apel, 1998). In fact, the four molecules of PBG are specifically used for the assembly of the porphyrin structure in chlorophyll. Phase II, which consists of six enzymatic steps, results in the formation of *protoporphyrin IX*. At this point in the pathway, the fate of the molecule depends on what metal is inserted into the center of the porphyrin. For example, when Mg is inserted, via the enzyme *magnesium chelatase*, further steps will then allow this molecule to be converted into chlorophyll. In contrast, if Fe is inserted it would result in the formation of a *heme*.

Phase III of the chlorophyll biosynthetic pathway involves the formation of the fifth ring (ring E). This occurs through cyclization of one of the propionic acid side chains to form *protochlorophyllide*, where the reduction of a double bond in ring D occurs, using nicotinamide adenine dinucleotide phosphate (NADPH). This process is light dependent in certain higher plants (e.g., angiosperms) and is catalyzed by the enzyme *protochlorophyllide oxidoreductase* (POR). Photosynthetic bacteria, which do not split oxygen during photosynthesis, can also carry-out this reaction in the absence of light, using different enzymes. Cyanobacteria, algae, lower plants, and gymnosperms contain both the light-dependent POR pathway and the light-independent pathway. The final phase (IV) in the chlorophyll biosynthetic pathway is the attachment of the phytol group, which is catalyzed by an enzyme, *chlorophyll synthetase* (Malkin and Niyogi, 2000). The elucidation of the biosynthetic pathways of chlorophylls and related pigments is a difficult task, in part because many of the enzymes are present in low concentrations. However, recent genetic analysis has been used to clarify many aspects of these processes (Suzuki et al., 1997; Armstrong and Apel, 1998).

12.2.2 Carotenoids

Carotenoids are tetraterpenoids, made by bacteria, algae, fungi, and higher plants, generally consisting of a C_{40} chain with conjugated bonds (Britton, 1976; Goodwin, 1980). There are also some carotenoids

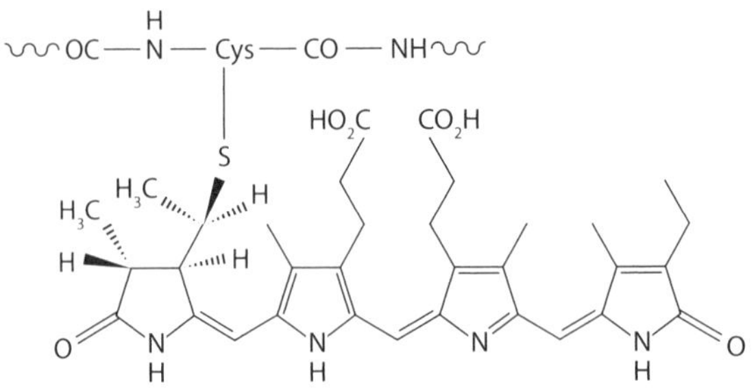

H
OC—N—Cys—CO—NH
S
HO_2C CO_2H
H_3C H_3C H
H H
O N N N N O
H H H

Peptide-linked phycocyanobilin

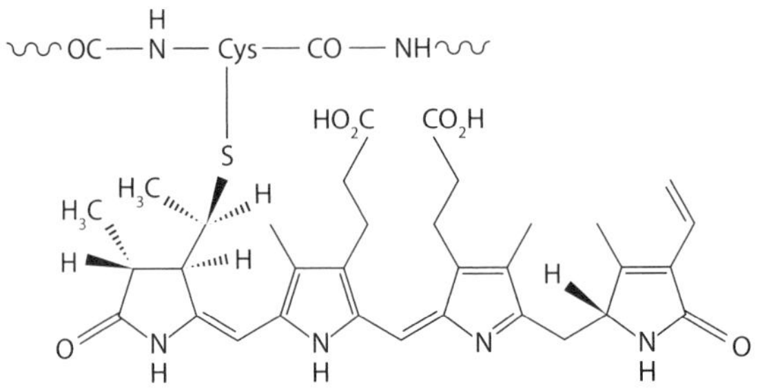

H
OC—N—Cys—CO—NH
S
HO_2C CO_2H
H_3C H_3C H
H H H
O N N N N O
H H H

Peptide-linked phycoerythrobilin

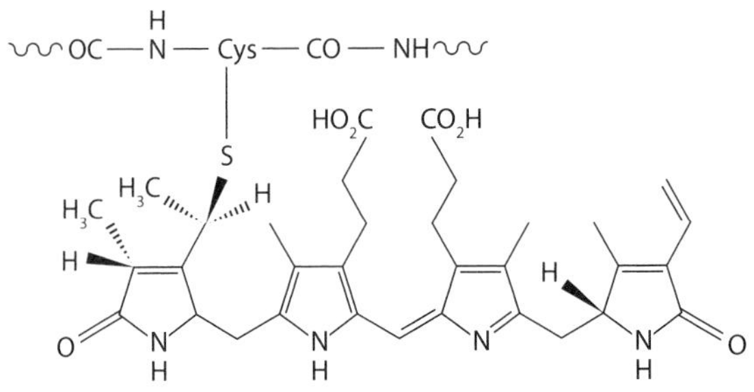

H
OC—N—Cys—CO—NH
S
HO_2C CO_2H
H_3C H_3C H
H H
O N N N N O
H H H

Peptide-linked phycourobilin

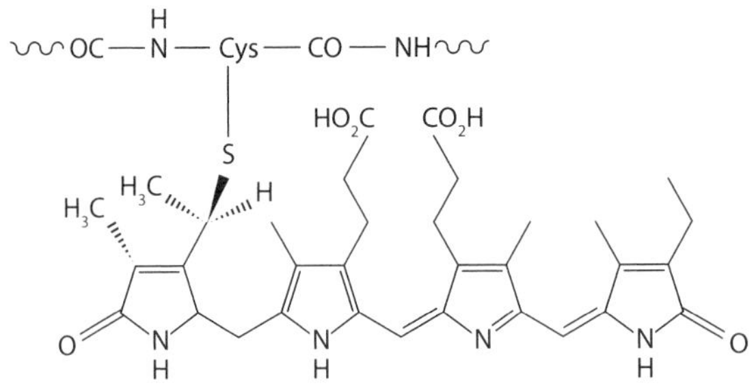

H
OC—N—Cys—CO—NH
S
HO_2C CO_2H
H_3C H_3C H
O N N N N O
H H H

Peptide-linked phycobiliviolin

Figure 12.4. Phycobiliproteins are covalently linked via thioether bonds to cysteinyl residues, the most common polypeptide-bound phycobilins. (Adapted from Colyer et al., 2005.)

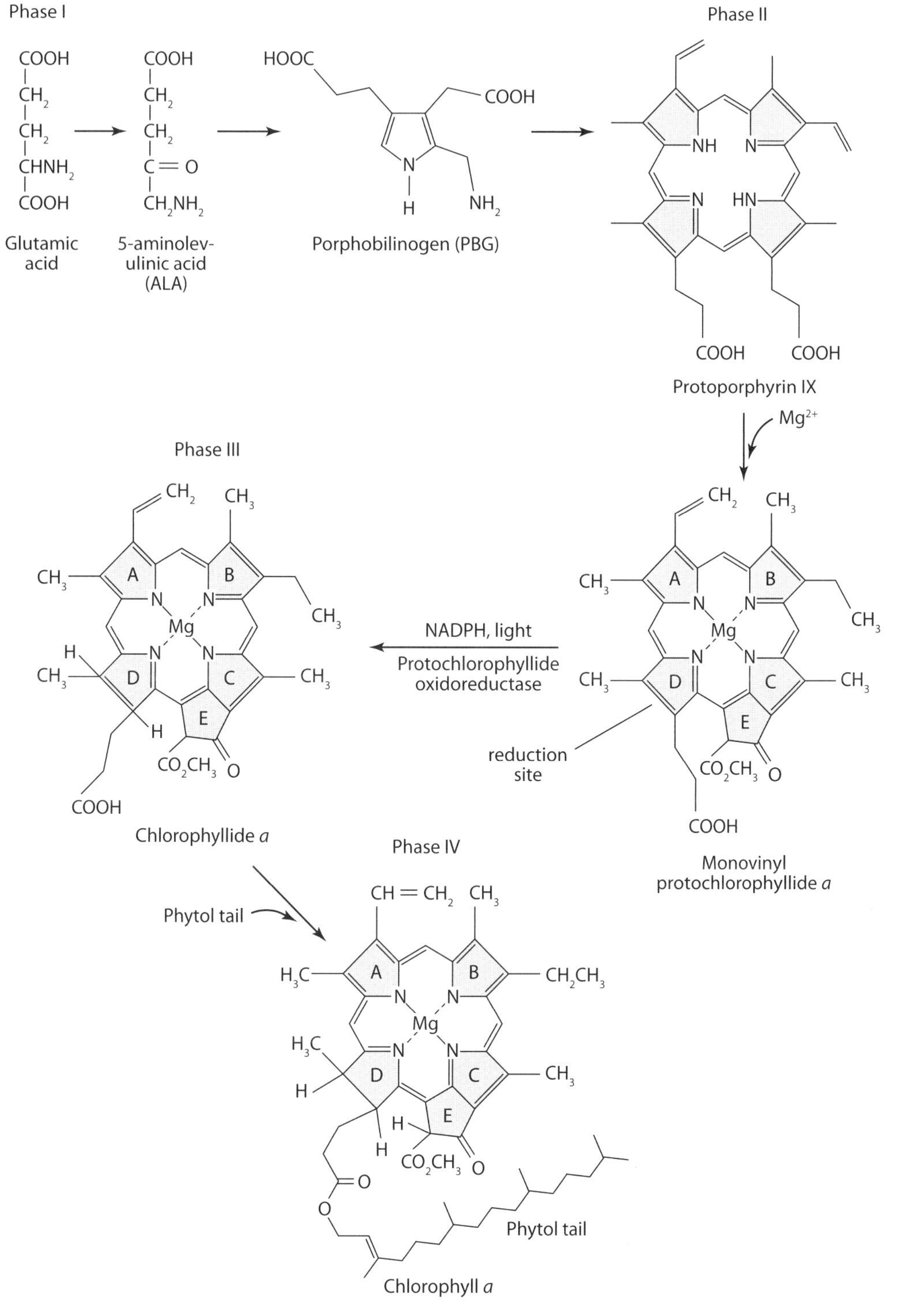

Figure 12.5. Stages of chlorophyll biosynthesis. (Adapted from Taiz and Zeiger, 2006.)

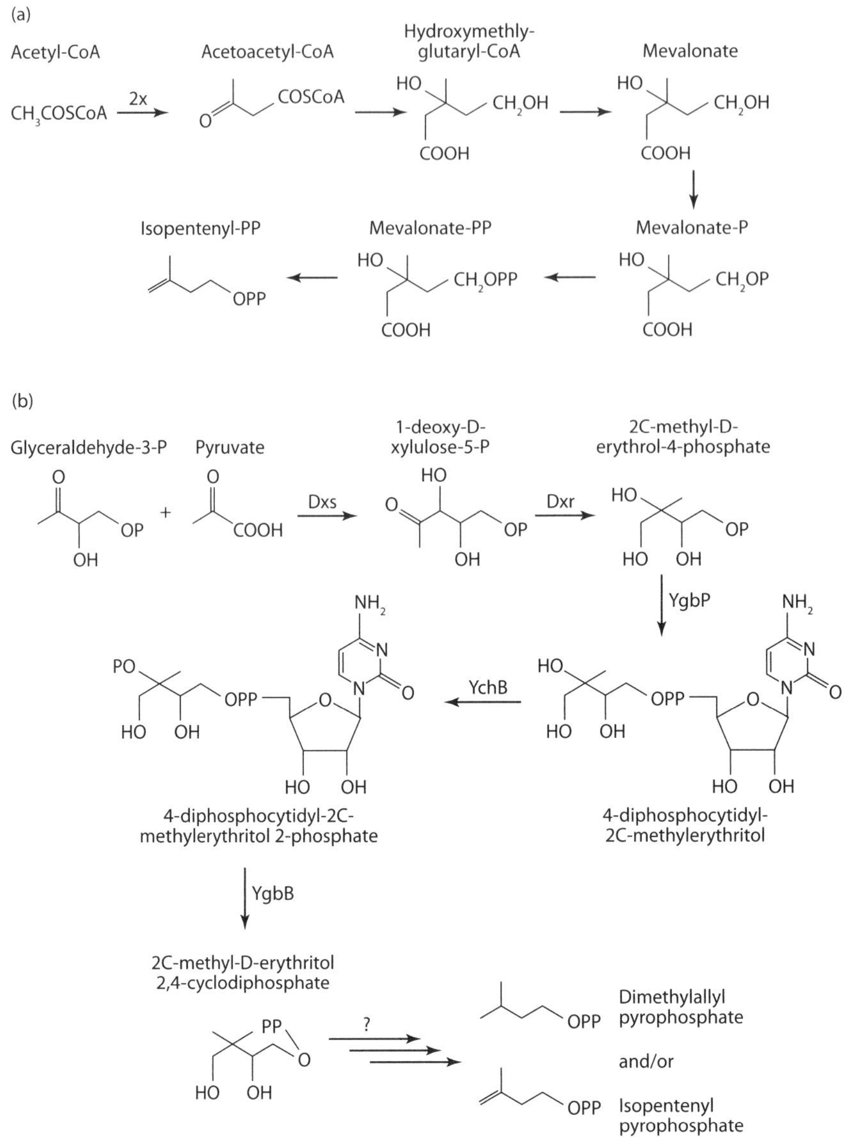

Figure 12.6. In fungi, the mevalonate pathway is used to generate prenyl pyrophosphates, the precursors of carotenoids. (Adapted from Sandmann, 2001.)

made by certain bacteria that are composed of C_{30}, C_{45}, and C_{50} chains, but these represent a small percentage of the estimated 600 carotenoid structures identified. Many algal species have hydroxy groups on the *ionone rings* (α and β) in addition to 5,6-epoxy groups. Many of the enzymes involved in carotenoid synthesis occur within and/or are associated with membranes. In fact, there has been considerable expansion in our knowledge of *carotogenesis* in recent years because of the progress made on the cloning, structure, and functions of the associated genes (Sandmann, 2001). Depending on the organisms involved (e.g., fungi, algae, bacteria) the reaction sequence to the cyclization of carotenoids varies. For example, in fungi, the mevalonate pathway is used to generate prenyl pyrophosphates, the precursors of carotenoids (fig. 12.6) (Sandmann, 2001). Mevalonate is formed via

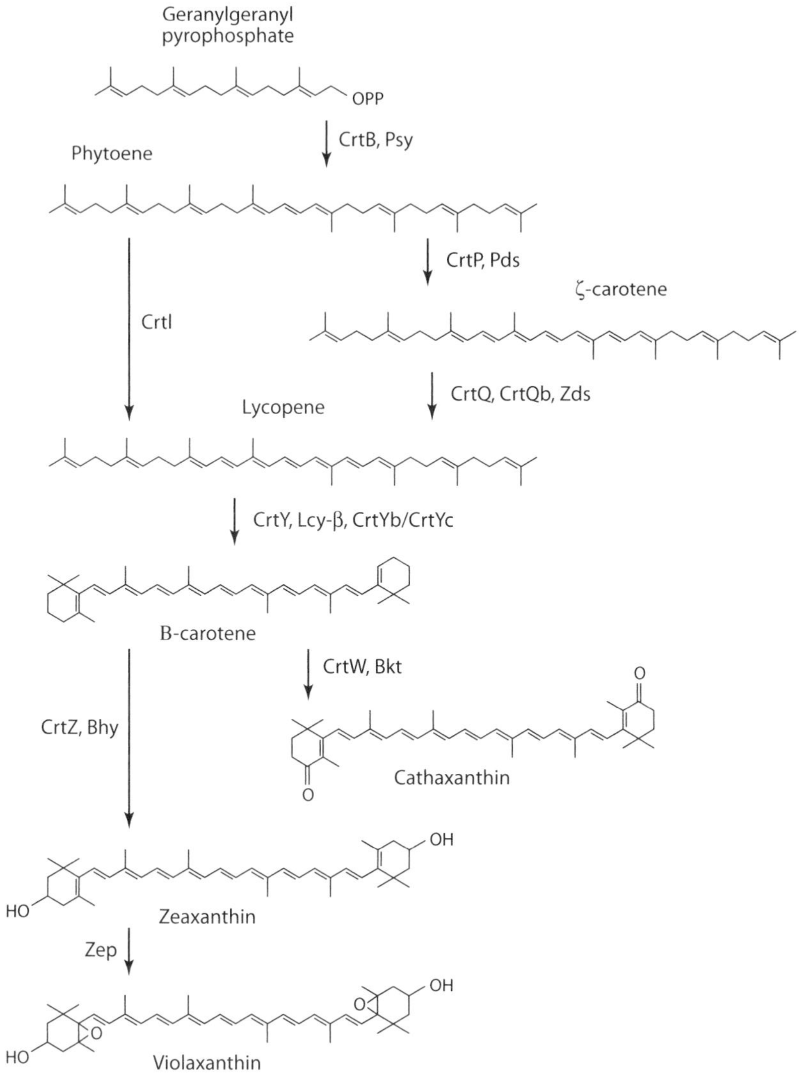

Figure 12.7. Biosynthesis of β-carotene and its oxidation products. (Adapted from Sandmann, 2001.)

acetyl-CoA, with the final product of this pathway resulting in the formation of isopentenyl pyrophosphate (IPP), which is the building block used for the synthesis of terpenoids. In the case of bacteria and plants, the formation of prenyl pyrophosphates, occurs through an alternative pathway called the 1-deoxyxylulose-5-phosphate pathway (fig. 12.6), but not all the reaction steps are currently described in the formation of IPP and/or dimethylallyl pyrophosphate (DMAPP).

Once prenyl pyrophsophates are formed, a number of pathways can occur in the formation of terpenoids. For example, in higher plants geranylgernanyl pyrophosphate (GGPP) is synthesized; in the case of bacteria, farnesyl pyrophosphate (FPP) is formed (not shown) (fig. 12.6) (Sandmann, 2001). The next step in the case of GGPP would be the formation of phytoene, and then a number of dehydrogenase steps that lead to lycopene. The xanthophylls, in this case, canthaxanthin, zeaxanthin, and violaxanthin, are formed as oxidation products of α and β-carotenes (fig. 12.7).

12.2.3 Phycobilins

The general biosynthesis of *phycopbiliproteins* occurs through the conversion of *protoheme* to biliverdin IX_α, followed by a reduction of biliverdin IX_α, which allows for the formation of free phycobilins (Beale and Cornejo, 1983, 1984a,b). The free phycobilins are then covalently *ligated* to *apophycobiliproteins*, which form functional light-harvesting complexes. The currently proposed biosynthetic pathway suggests that 15,16-dihydrobiliverdin IX is the product of biliverdin IX_α reduction (fig. 12.8) (Beale and Cornejo, 1991a,b). This pathway ultimately leads to the formation of 3(*E*)-phycocyanobilin, which is the major phycobilin released from phycocyanin via *methanolysis* (Cole et al., 1967). However, more research is needed to better understand the enzymes that control the oxidation, reduction, and isomerization reactions in construction of the light-harvesting apophycobiliproteins.

12.3 Analysis of Chlorophylls, Carotenoids, and Phycobilins

As mentioned earlier, chlorophyll *a* concentrations have been routinely measured for estimating standing stock and productivity of photoautotrophic cells in freshwater and marine environments. However, it should be noted that although the carbon : chlorophyll *a* (C/Chl-a) ratio of photoautotrophic cells varies considerably as a function of environmental conditions and growth rate (Laws et al., 1983), chlorophyll *a* remains the primary method for determining the biomass of photoautotrophs in aquatic ecosystems. Spectrophotometric analysis of chlorophyll pigments was developed in the 1930s and 1940s (Weber et al., 1986). Richards and Thompson (1952) introduced a *trichromatic* technique that was supposed to measure chlorophylls *a*, *b*, and *c*; these equations attempted to correct for interferences from other chlorophylls at the maximum absorption wavelength for each one. A number of modifications were made to these equations that purportedly produce better estimates of the chlorophylls (Parsons and Strickland, 1963; UNESCO, 1966; Jeffrey and Humphrey, 1975).

Since the 1960s, the *fluorometric* method has generally been used for analyses of field samples. However, many of the early fluorometers could not unambiguously distinguish fluorescence band overlaps when detected with wide bandpass filters (Trees et al., 1985). Moreover, such fluorometers generally underestimated chlorophyll *b* and overestimated chlorophyll *a* when chlorophyll *c* co-occurred in high concentrations (Lorenzen and Downs, 1986; Bianchi et al., 1995; Jeffrey et al., 1997). Finally, pheopigments were generally overestimated and underestimated when large amounts of chlorophyll *b* and *c* were present, respectively (Bianchi et al., 1995; Jeffrey et al., 1997). However, many of the more recently developed fluorometers have improved considerably in their ability to deal with such interference problems. Moreover, in vivo fluorometric analyses over large distances still prove useful for rapid results, over the more time-consuming, albeit accurate, method of HPLC (Jeffrey et al., 1997).

There is general consensus that organic geochemistry as a scientific discipline began in the 1930s with the first modern study of the geochemistry of organic molecules by Treibs (1936), who discovered and described porphyrin pigments in shale, oil, and coal. Thus, it is appropriate that we begin the application of simple chromatographic techniques in separating pigments with the work of Alfred Treibs, the "father of organic geochemistry." However, it was only about 20 years ago that the preferred technique for the analysis of chlorophylls and carotenoids in aquatic ecosystems became HPLC (Gieskes and Kraay, 1983; Mantoura and Llewellyn, 1983; Bidigare et al., 1985; Welschmeyer and Lorenzen, 1985; Jeffrey and Wright, 1987; Repeta and Gagosian, 1987; Bianchi et al., 1988). For example, carotenoids and their degradative products, the xanthophylls, have been shown to be effective biomarkers for different classes of phytoplankton (Jeffrey, 1997), while chlorophyll has been used extensively as a means of estimating phytoplankton biomass (Jeffrey, 1997). Numerous

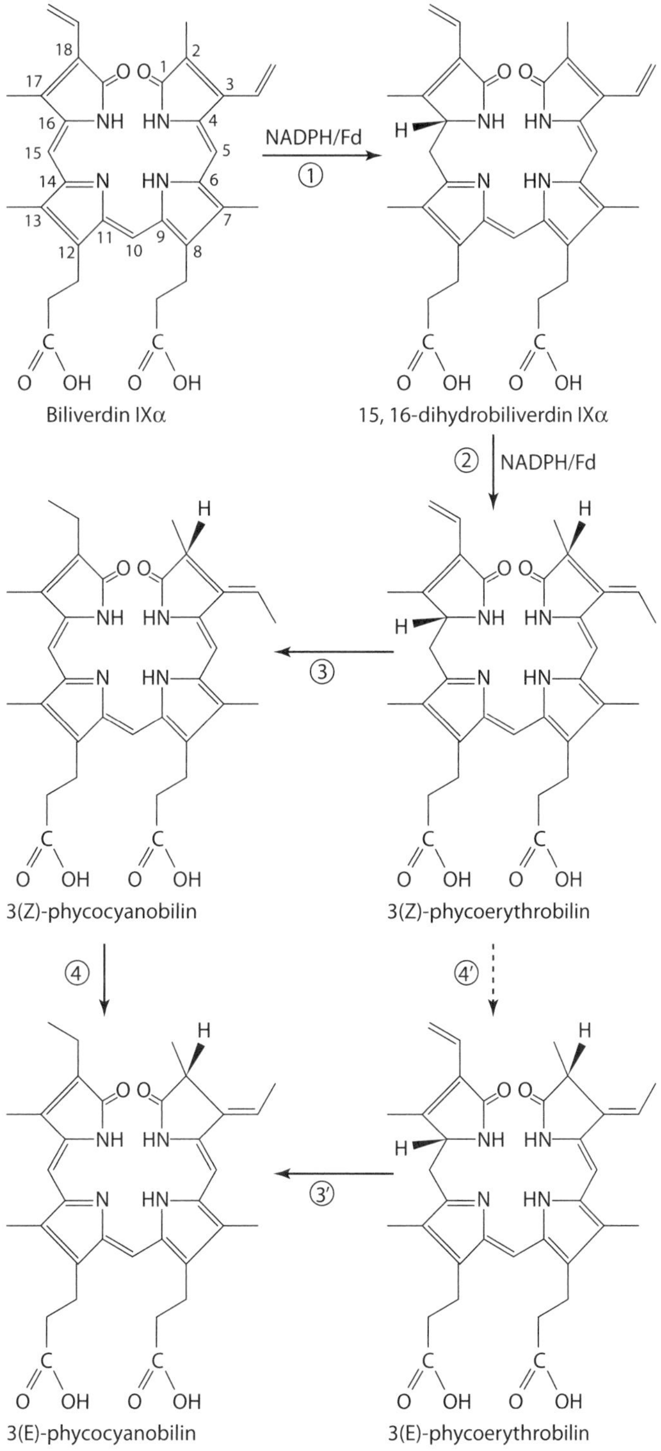

Figure 12.8. The currently proposed biosynthetic pathway for phycobilins from biliverdin suggests that 15,16-dihydrobiliverdin IX is the product of biliverdin IXα reduction. (Adapted from Beale and Cornejo, 1991a,b.)

studies have shown that photopigment concentration correlated well with microscopical counts of different species (Tester et al., 1995; Roy et al., 1996; Meyer-Harms and Von Bodungen, 1997; Schmid et al., 1998), further corroborating that plant pigments are reliable chemosystematic biomarkers. For example, fucoxanthin, peridinin, and alloxanthin are the dominant accessory pigments in diatoms, certain dinoflagellates, and cryptomonads, respectively (Jeffrey, 1997). The application of HPLC to phytoplankton pigment analysis has lowered the uncertainty associated with the measurement of chlorophyll *a* and accessory carotenoids, since compounds are physically separated and individually quantified.

In general, photoautotrophic cells are typically concentrated from aquatic waters by filtration and extracted into an organic solvent (usually acetone), after which, pigments are separated chromatographically and detected by fluorescence and/or absorption spectroscopy (Jeffrey et al., 1997; Bidigare and Trees, 2000). In the case of sediments and/or pure plant materials, extraction times and solvent preference may vary; additional solvents that are commonly used include methanol, dimethylforamide (DMF), dimethyl acetamide (DMA), chloroform, ethanol, and dimethylsulfoxide (DMSO) (Jeffrey et al., 1997). Intercalibration studies using HPLC analysis for algal pigments have also been conducted and provide helpful information on the possibility of developing a universally accepted method (Latasa et al., 1996). Recent work has also shown that once collected, all samples (filters, sediments, etc.) should be frozen immediately at −20°C or colder and that even freeze-dried samples should be stored frozen (Ruess et al., 2005). Finally, another technique that has emerged as an important tool for the identification of phytoplankton is flow cytometry (FCM) (Blanchot and Rodier, 1996; Landry and Kirchman, 2002, and references therein); this method is discussed in more detail in chapter 4, where new analytical methods are examined.

One of the most common HPLC methods was introduced by Wright et al. (1991), allowing detection of 50 chlorophylls, carotenoids, and their degradation products. Briefly, each filter or quantitative amount of sediment is extracted with acetone, and blown to dryness under a stream of N_2 (Chen et al., 2001). Pigment extracts are injected into an HPLC, with a C_{18} column, usually coupled with an in-line photodiode array detector (PDA) and fluorescence detector. The absorbance detector is set at 438 nm and the fluorescence detector at an excitation of 440 nm and an emission of 660 nm. One of the other advantages of this method is that it is capable of separating the lutein and zeaxanthin isomers. Unfortunately, the internal standard used in this method is canthaxanthin, which interferes with zeaxanthin (a biomarker for cyanobacteria); a sample chromatogram is provided in fig. 12.9. Recent work has suggested that vitamin E be used as a replacement for canthaxanthin, since it absorbs below the visible spectrum at 284 nm (Van Heukelem and Thomas, 2001). It should also be noted that while the Wright et al. (1991) method is not capable of separating monovinyl and divinyl chlorophylls, another method was developed that is capable of this separation using a C_8 HPLC technique (Goericke and Repeta, 1993). Pigment standards can be obtained from DHI Water and Environment, Denmark, which are then run individually and as a mixed standard to determine retention times and spectra. Pigment identification is then performed by comparison of retention times and absorption spectra of the peaks in each sample chromatogram to those of the standards. Interlaboratory methods and standard pigment comparisons have also helped to unify the research community in performing pigment analyses (Bidigare et al., 1991; Latasa et al., 1996).

The matrix factorization program CHEMical TAXonomy (CHEMTAX) was introduced to calculate the relative abundance of major algal groups based on concentrations of diagnostic pigments (Mackey et al., 1996; Wright et al., 1996). Recent work has shown that application of the original CHEMTAX reference matrix to southeastern United States estuarine systems produced inaccurate results, as verified by microscopy (Lewitus et al., 2005). However, after modifying the original source matrix from Mackey et al. (1996), general pigment ratios of phytoplankton cultures to the dominant phytoplankton for the region of interest greatly improved the predictive capabilities of CHEMTAX. Some of the limitations of the method include an overestimation of diatom biomass due to the

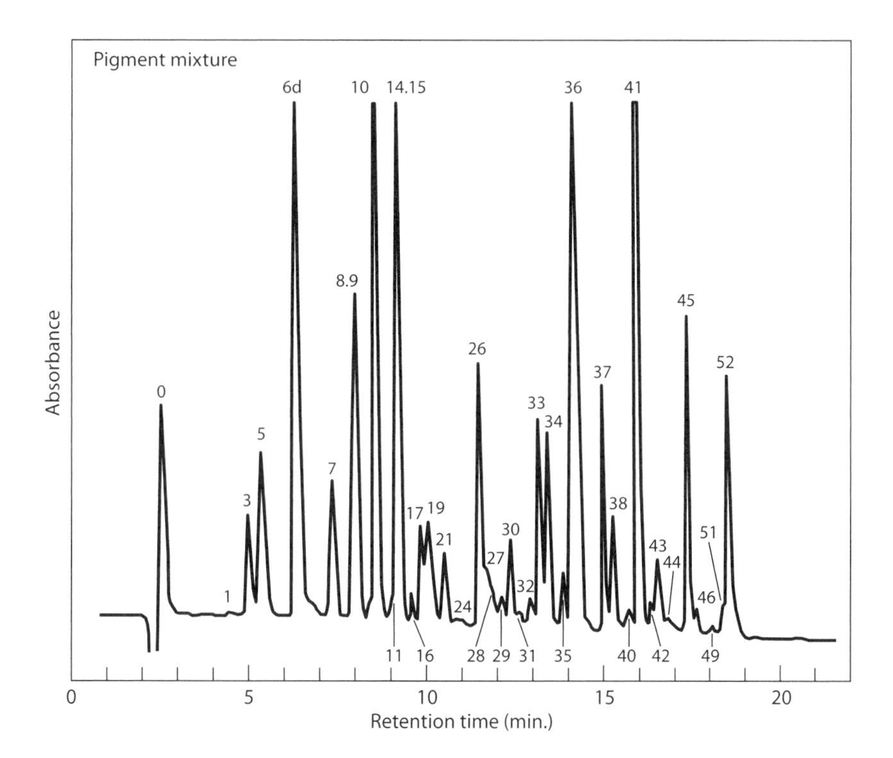

Figure 12.9. HPLC chromatogram of a mixed pigment extract from algal cultures of *Dunaliella tertiolecta* (green alga), *Phaeodactylum tricornutum* (diatom), *Synechococcus* sp. (cyanophyte), *Pycnococcus provasolii* (prasinophyte), *Chroomonas salina* (cryptomonad), *Emiliania huxleyi* (prymnesiophyte), *Pelagococcus subviridis* (chrysophyte), and the authentic carotenoids diadinochrome, siphonein, siphonaxanthin, echinenone, canthaxanthin, lycopene, peridinin, and the internal synthetic standard ethyl 8'-*O*-apocarotenoate. The pigment identifications are as follows: 0, solvent front; 1, chlorophyllide *b*; 3, chlorophyllide *a*; 5, chlorophyll C_3; 6d, chlorophyll $c_1 + c_2$, My2,4D; 7, peridinin; 8, siphonaxanthin; 9, 19'-butanoyloxyfucoxanthin; 10, fucoxanthin; 11, *trans*-neoxanthin; 14, 9'-*cis*-neoxanthin; 15, 19'-butanoyloxyfucoxanthin; 16, *cis*-fucoxanthin neochrome; 17, *cis*-19'-hexanoyloxyfucoxanthin; 19, prasinoxanthin; 21, violaxanthin; 24, *cis*-prasinoxanthin; 26, diadinoxanthin; 27, diadinochrome I; 28, diadinochrome 11; 29, antheraxanthin; 30, alloxanthin; 31, monadoxanthin; 32, diatoxanthin; 33, lutein; 34, zeaxanthin; 35, canthaxanthin; 36, siphonein; 37, chlorophyll *b*; 38, ethyl 8'-β-apocarotenoate; 40, chlorophyll *a* allomer; 41, chlorophyll *a*; 42, chlorophyll *a* epimer; 43, echinenone; 44, unknown carotenoid; 45, lycopene; 46, phaeophytin *b*; 49, β,ψ-carotene; 51, β,ε-carotene; 52, β,β-carotene. (Adapted from Wright et al., 1991.)

inability to differentiate diatoms from taxa with chloroplasts derived from diatom endosymbionts (e.g., dinoflagellates) and a tendency to exclude some raphidophyte species (Llewellyn et al., 2005). More recently, it was shown that if CHEMTAX was altered to include several pigment ratios that are run successively, convergence of valid taxonomic information will be achieved, even in the absence of "seed" ratios for phytoplankton present in the natural samples (Latasa, 2007). Finally, the combined method of selective photopigment HPLC separation coupled with in-line flow scintillation counting using the chlorophyll *a* radiolabeling (^{14}C) technique can also provide information on phytoplankton growth rates under different environmental conditions (Redalje, 1993; Pinckney et al., 1996, 2001).

Ocean color data, remotely sensed from satellites, provides the best practical means of thoroughly monitoring the spatial and temporal variability of near-surface phytoplankton in broad open ocean and coastal environments (Miller et al., 2005). For example, the global distribution of chlorophyll *a* is available daily from satellite-based ocean color sensors, such as *Sea-viewing Wide Field-of-view Sensor* (SeaWiFS) and *Moderate Resolution Imaging Spectroradiometer* (MODIS). These high-resolution (less than or equal to 1 km) chlorophyll *a* images provide invaluable information on processes affecting climate-driven forcing on phytoplankton biomass and productivity (Chavez et al., 1999; Behrenfeld et al., 2001; Seki et al., 2001). Oceanic waters have been differentiated according to their optical characteristics into two classes: case 1 waters, where the presence of chlorophyll and co-varying materials govern the optical properties of seawater, and case 2 waters, where the presence of materials that do not co-vary with chlorophyll govern the optical environment (Miller et al., 2005). In case 1 waters, ocean color analysis has been applied to successfully retrieve pigment concentrations and biomass. Thus, in biologically dominated optical environments (e.g., regions where the inherent optical parameters co-vary with chlorophyll concentration) the biooptical models perform reasonably well, returning pigment concentrations in reasonable agreement with field measurements.

The analysis of phycobiliproteins is more difficult than for chlorophyll and carotenoids because the covalent bilin–protein linkages require more extreme processes for cleavage. To release the tetrapyrroles from their apoproteins, extraction methods may involve hydrolysis with HCl, followed by refluxing in methanol, and final treatment with Hg^{2+} salts and HBr in trifluoroacetic acid (TFA) (Cohen-Bazire and Bryant, 1982). While the free bilins have absorbance properties similar to those of protein-bound bilins, denatured phycobiliproteins degrade to colorless products due to rapid isomerization (Cohen-Bazire and Bryant, 1982; Colyer et al., 2005). Early work on the separation of phycobiliproteins used macro-scale techniques like *electrophoresis, gel filtration*, and *ion exchange columns*, but nothing for micro-scale applications in analyzing phytoplankton (Jeffrey et al., 1997; Colyer et al., 2005). Recently, a two-dimensional electrophoresis method has also been used to separate proteins from the phycobilins in cyanobacteria (Huang et al., 2002).

12.4 Applications

12.4.1 Chlorophylls, Carotenoids, and Phycobilins as Biomarkers in Lakes

A number of studies have examined phytoplankton dynamics in large East African lakes (e.g., Tanganyika, Victoria, and Malawi) (Hecky and Kling, 1981; Talling, 1987; Patterson and Kachinjika, 1995), and have shown consistent seasonal changes. For example, in the dry season (May to September), when there is deep mixing, low light, and higher nutrient availability, diatoms dominate, with some cyanobacterial production occurring at the surface at the end of the dry season. In the wet season (October to April), chlorophyte–chroococcale communities dominate, when there is high light, poor nutrients, and a shallow *epilimnion*. Recent work has also shown that Lake Tanganyika has experienced a significant reduction in phytoplankton abundance and changes in composition that may be related to global climate change (Verburg et al., 2003). A recent study used HPLC analyses for the first time in Lake Tanganyika to better examine the spatial and seasonal changes of photosynthetic pigments over 2 years (Descy et al., 2005). In general, the results showed that pigment concentrations were highest in the upper 60 m of the water column and abruptly decreased below this depth (fig. 12.10). The subsurface chlorophyll *a* maximum typically occurred at about 20 m, with pheopigments becoming more dominant from 40 m and below. This pattern of pheopigment abundance with depth is similar to that found in other lake studies (Yacobi et al., 1996). However, in some cases the observed abundance of pheopigments in Lake Tanganyika was significantly less than observed for some lake systems, where phytoplankton senescence and grazing rates were found to be substantially higher (Carpenter and Bergquist, 1985). Consistent with past studies, diatoms increased during the dry season, but the phytoplankton community had more spatial variability, with chlorophytes and cyanobacteria dominating in the northern and southern regions of the lake, respectively.

Today, the cyanobacteria exhibit remarkable ecophysiological adaptations (Stomp et al., 2004; Huisman et al., 2006), including the ability to thrive in aquatic environments undergoing human- and naturally induced environmental change (Paerl et al., 2001; Paerl and Fulton, 2006). Cyanobacteria

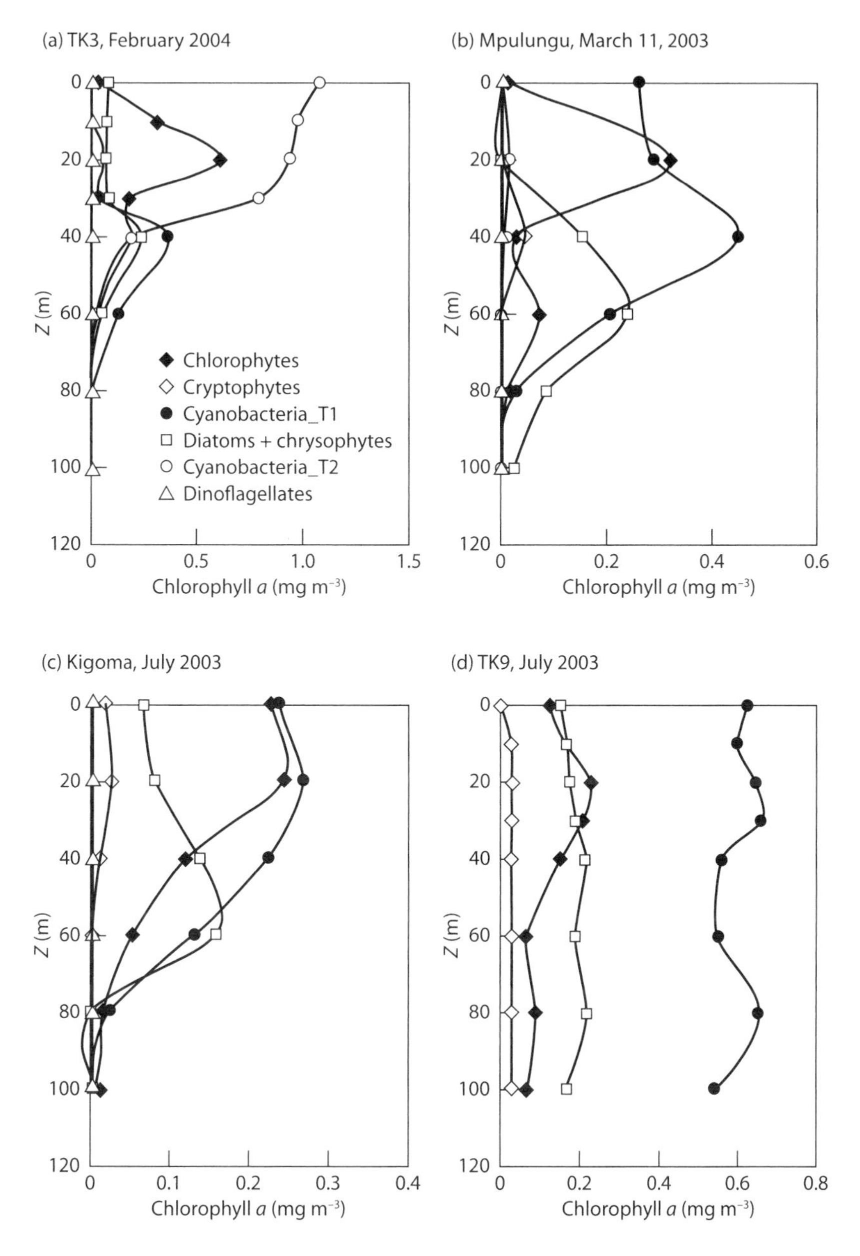

Figure 12.10. HPLC analyses of POC at four locations in Lake Tanganyika (Africa) were used to examine spatial and seasonal changes in phytoplankton composition over two years. T1 and T2 indicate two types of cyanobacteria. (Adapted from Descy et al., 2005.)

dominate in a wide variety of aquatic ecosystems, where they can take advantage of rapidly increasing human activities and impacts, including nutrient overenrichment (eutrophication) and hydrologic modifications (water withdrawal, reservoir construction). Several cyanobacterial genera, including the nitrogen fixers *Anabaena and Cylindrospermopsis* and the non-N_2 fixer *Microcystis*, are notable for being particularly aggressive invaders of these waters, by forming massive surface growths or "blooms" that produce toxins, cause oxygen depletion, and alter food webs, thereby threatening drinking and irrigation water supplies and fishing and recreational use of these waters (Carmichael, 1997; Paerl and Fulton, 2006; Paerl, 2007). In addition, regional and global climate change, specifically warming and alterations of hydrologic cycles, benefit these harmful cyanobacteria (CyanoHABs) by increasing their growth rates, dominance, persistence, and geographic distributions.

Cyanobacterial blooms in eutrophied aquatic ecosystems ranging from small creeks to some of the world's largest lakes have become a problem of global concern (Paerl and Fulton, 2006; Paerl, 2007). In the United States, problems with cyanobacteria in 3 of the 5 Great Lakes, other large lake systems (Lake Okeechobee, Lake Ponchartrain, Great Salt Lake), in midwest, western, and southeast reservoirs, in the wetlands of the southeast, and in estuarine environment of the Pacific Northwest, Florida, and the mid-Atlantic all point to the potential nationwide problems that resource managers will face in coming decades. Cyanobacteria possess phycobilisomes, which, as discussed earlier, have characteristic absorption and fluorescence spectra (Glazer and Bryant, 1975; Bryant et al., 1976). Past work has used the selective fluorescence emission spectra of phycoerythrin for monitoring cyanobacteria in fresh waters (Alberte et al., 1984). This was followed by another study that used phycocyanin to quantitatively determine the relative abundance of cyanobacteria in mixed algal populations (Lee et al., 1995). More recently, the development of a new in situ fluorescence sensing system was described for monitoring cyanobacterial blooms (Asai et al., 2000). Results from this study using a 2-channel fluorescence (excitation of wavelengths set at 620 and 440 nm) showed that a bloom of *Microcystis aeruginosa* was successfully detected in Lake Kasumigaura, Japan. The system was capable of continuous monitoring with time intervals of 25 minutes for each measurement, which included sonification and rinsing.

12.4.2 Chlorophylls, Carotenoids, and Phycobilins as Biomarkers along the River–Estuarine Continuum

Much of the particulate organic carbon (POC) in rivers is derived from both allochthonous (e.g., soil organic matter, algal inputs from streams, and aquatic emergent and submergent wetland vegetation) and autochthonous sources (e.g., phytoplankton, benthic algae) (Onstad et al., 2000; Kendall et al., 2001; Wetzel, 2001). However, dam construction has reduced the suspended load of many rivers around the world, thereby increasing light availability and the potential role of phytoplankton biomass in riverine biogeochemistry (Thorp and Delong, 1994; Humborg et al., 1997, 2000; Ittekkot et al., 2000; Kendall et al., 2001; Sullivan et al., 2001; Duan and Bianchi, 2006). In fact, sediment loads have declined more than 70% in the Mississippi River system (USA) since 1850, which has been attributed to a network of water impoundments and the institution of federal/state/local soil conservation practices since the 1930s (Mossa, 1996).

Recent work has shown that at certain times of the year, phytoplankton represents a significant source of the POC (Duan and Bianchi, 2006) and dissolved organic carbon (DOC) (Bianchi et al., 2004) in the lower Mississippi River. Phytoplankton have also been shown to be an important source of POC in the major upper tributaries of the Mississippi (Kendall et al., 2001). The pigment ratios used for CHEMTAX algorithms in this study were based on ratios for freshwater phytoplankton, collected at depths of 4 to 9 m in northern Wisconsin lakes (USA) (Descy et al., 2000). The relative irradiance at this depth ranged from 3.3 to 10%; these values were chosen to allow for adequate adjustment to low-light conditions in the lower Mississippi River. Diatoms, chlorophytes, cyanobacteria, dinoflagellates, cryptophytes, and euglenoid algae were chosen because they are typically found in many freshwater systems in North America (Wehr and Sheath, 2003), and more specifically in sections of the Ohio River (Wehr and Thorp, 1997). Chlorophyll *a* concentrations in the lower Mississippi River varied from 0.70 to 21.1 $\mu g\ L^{-1}$ ($\bar{X} = 6.34$, CV = 82%) and represented 0.03 to 1.25% of POC (fig. 12.11) (Duan and Bianchi, 2006). Chlorophyll *a* concentrations in the lower Mississippi River were high in

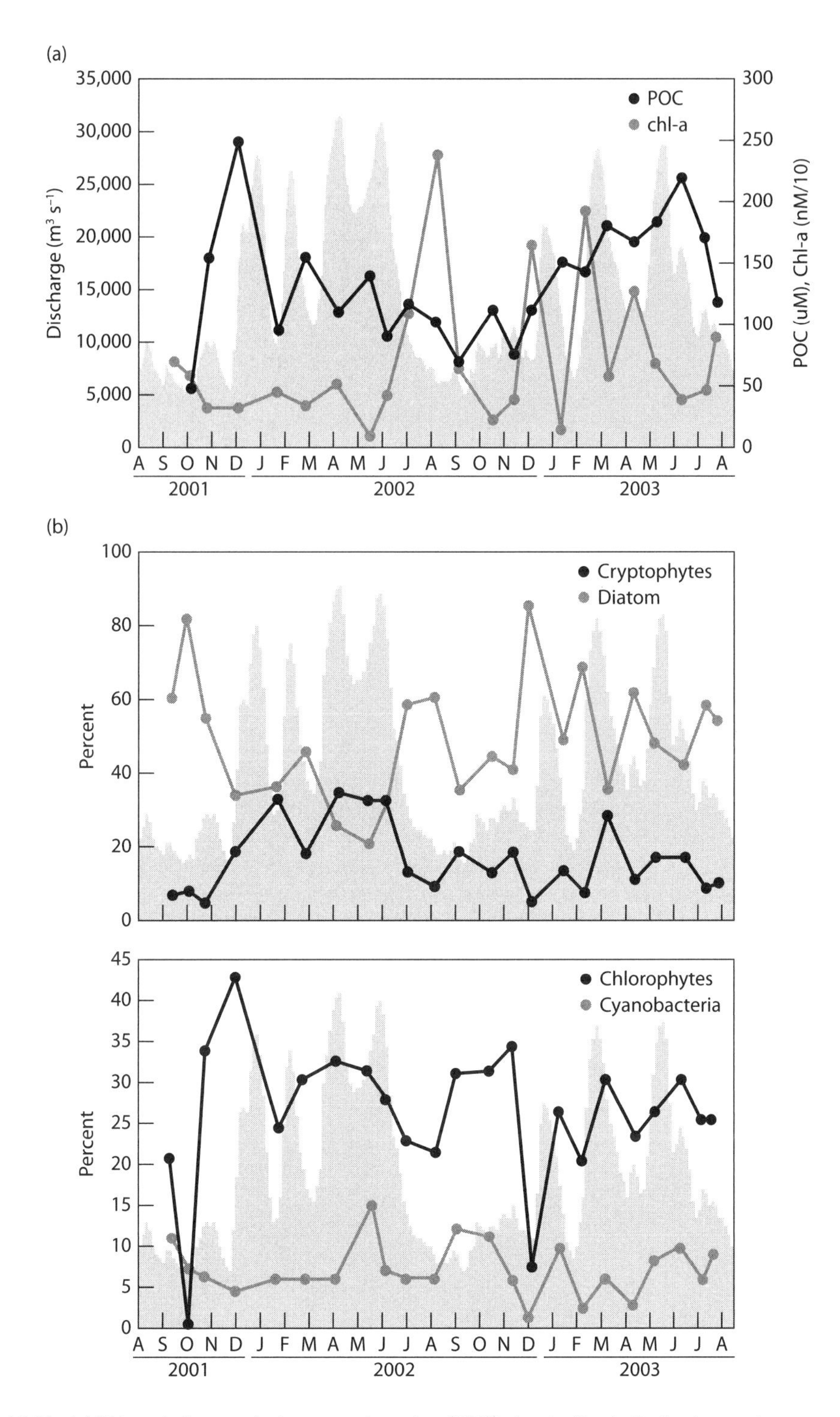

Figure 12.11. (a) Chlorophyll *a*, particulate organic carbon (POC); the shading in the background represents river discharge collected near Baton Rouge, LA (USA) on the left y axis. (b) Percentages of phytoplankton classes in the lower Mississippi River (USA). (Duan and Bianchi, 2006.)

summer during low-flow periods and also during interims in winter and spring, but were not coupled with physical variables and nutrients. This is likely due to a combination of in situ production and inputs from reservoirs, navigation locks, and oxbow lakes in the upper Mississippi and Missouri rivers (USA). Diatoms accounted for more than 60% of total phytoplankton biomass in the lower Mississippi when chlorophyll *a* concentrations were high (fig. 12.11) (Duan and Bianchi, 2006). Chlorophytes and cryptophytes generally showed opposite seasonal trends to diatoms and were more important during the high-flow periods of 2001–2002 and during Fall 2002 in the lower Mississippi. The high, diatom-dominated phytoplankton biomass in the lower Mississippi was likely the result of decreasing total suspended solids (increased damming in the watershed) and increasing nutrients (enhanced agricultural runoff) over the past few decades.

Plant pigments have been shown to be useful biomarkers of organic carbon in estuarine ecosystems (Millie et al., 1993). In general, pigment biomarkers have shown diatoms to be the dominant phytoplankton group in most estuarine systems (Bianchi et al., 1993, 1997, 2002; Lemaire et al., 2002). This is illustrated by a study examining the concentrations of fucoxanthin relative to other diagnostic carotenoids and chlorophyll *a* in European estuaries over different seasons (fig. 12.12) (Lemaire et al., 2002). In another study, CHEMTAX was used to estimate the relative importance of different microalgal groups to total biomass and blooms in the Neuse River estuary (USA) (Pinckney et al., 1998). The estimated overall contribution of cryptomonads, dinoflagellates, diatoms, cyanobacteria, and chlorophytes in the Neuse River estuary from 1994 to 1996 was 23, 22, 20, 18, and 17%, respectively. This study revealed that there was significant variability in the relative abundance of phytoplankton and blooms over a three-year period and that loading of dissolved inorganic nitrogen (DIN) was largely responsible for the blooms of cryptomonads, chlorophytes, and cyanobacteria.

While little recent work has been conducted on phycobilins in estuaries, early work examined the abundance of phycoerythrin and phycocyanin as related to chlorophyll *a* concentrations in the York and Lafayette river estuaries (USA) (Stewart and Farmer, 1984). After an extensive extraction process, first tested on pure cultures of chroococcoid cyanobacteria, four cryptophytes and two rhodophytes, pigment extracts from eight estuarine samples (0.5 to 2 L of filtered water) were analyzed using fluorescence spectroscopy. For phycoerythrin, excitation wavelength was set at 520 nm and scanned from 540 to 600 nm. An excitation wavelength of 580 nm, scanned from 600 to 660 nm, was used for phycocyanin. Results from this study indicated a predominance of cryptophytes (table 12.2), consistent with earlier work in York River (Exton et al., 1983). This is despite the higher phycocyanin concentrations, which would suggest a predominance of coccoid cyanobacteria, because the phycocyanin extinction coefficient (65) is less than that of phycoerythrin in crytophytes (126).

12.4.3 Chlorophylls, Carotenoids, and Phycobilins as Biomarkers in the Ocean

One of the areas of particular interest in the open ocean has been tropical high-nutrient, low-chlorophyll (HNLC) regions (Martin and Fitzwater, 1988; Morel et al., 1991; Falkowski, 1994; Kolber et al., 2002). Understanding the dynamics of these regions has led to some of the most interesting work in the field of oceanography over the last decade as it relates to the Iron Enrichment Experiments (IronEx) (Morel et al., 1991; Landry et al., 2000a,b). Early work has suggested that HNLC regions were characterized by relatively low and constant concentrations of fast-growing phytoplankton (Frost and Franzen, 1992; Landry et al., 2000a,b). As part of the Joint Global Ocean Flux Study (JGOFS), taxon-specific pigment measurements and flow cytometry were used to examine growth and grazing dynamics of picoplankton in HNLC waters of the equatorial Pacific (Landry et al., 2003; Le Borgne and Landry, 2003). This work utilized flow cytometry (Monger and Landry, 1993), HPLC (Wright et al., 1991) and spectrofluorometric analyses (Neveux and Lantoine, 1993) to better understand the dynamics of growth and grazing at the community level (Landry et al., 2003). Pigment

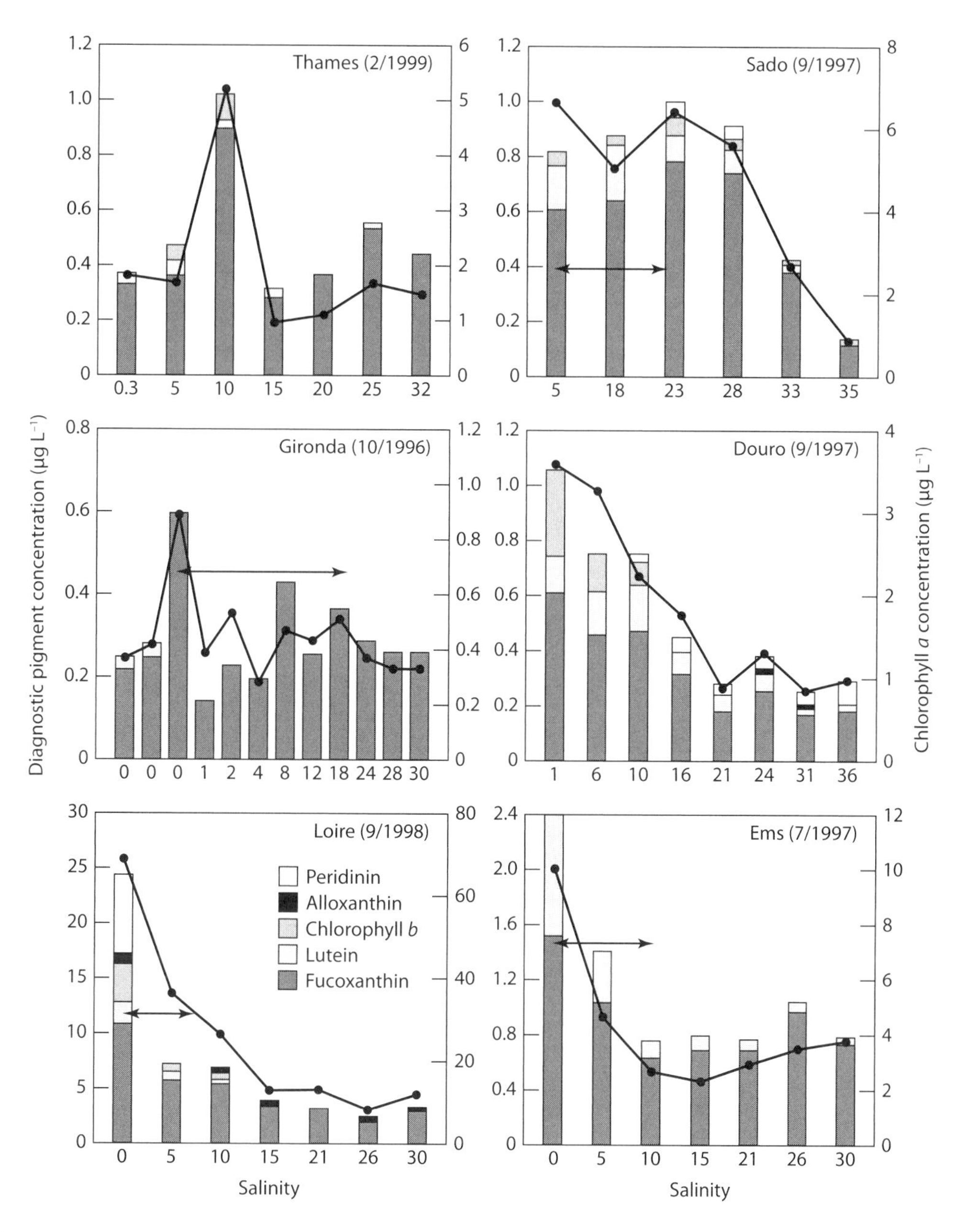

Figure 12.12. Concentrations of diagnostic carotenoids (e.g., peridinin, alloxanthin, lutein, and fucoxanthin) as well as chlorophylls *a* and *b* in European estuaries over different seasons (Adapted from Lemaire et al., 2002.)

concentrations and flow cytometry abundances ($C_{i,o}$) were measured for different dilution treatments, including ambient concentrations in unfiltered seawater and different proportions of unfiltered water in each treatment (Landry et al., 2003). At the end of a 24-hour incubation, final concentrations were determined ($C_{i,t}$). From this information, daily net rates of change (d^{-1}) were calculated using the equation $k_i = \ln(C_{i,t}/C_{i,o})$. These data were then used to estimate microzooplankton grazing rates per day ($m\ d^{-1}$). Finally, daily phytoplankton growth rate ($\mu\ d^{-1}$) was estimated from grazing mortality and

Table 12.2
Phycoerythrin, phycocyanin, and chlorophyll *a* concentrations (μg of pigment per liter of sea-water) in eight estuarine samples

	Phycoerythrin		*Phycocyanin*		
Sample[a]	*F.U.*	*Concn*	*F.U.*	*Concn*	*Chl a concn*
Yl	0.35	0.004	0.30	0.027	1.81
Y2	0.30	0.004	0.35	0.034	1.75
Y3	0.40	0.005	0.38	0.036	2.12
Y4	0.40	0.005	0.35	0.033	1.67
Y5	0.40	0.005	0.49	0.046	2.23
Y6	0.73	0.008	0.70	0.066	3.20
Y7	0.45	0.003	0.49	0.023	—
Ml	0.35	0.008	0.43	0.080	153[b]

[a]Sample volumes varied, thus, final pigment concentrations are not directly proportional to fluorescence units.
[b]This sample contained a dense dinoflagellate bloom.
Note: F.U., fluorescent units.

mean net growth rates in unamended seawater treatments. Taxonomic assignments, based on pigment analyses, were made from known pigment compositions of the different phytoplankton groups (Jeffrey and Vesk, 1997).

Mean dilution rate estimates for phytoplankton, identified from pigments, showed significant differences in growth rates and grazing rates of eukaryotic phytoplankton across groups and at different depths (30 vs. 60 m) (fig. 12.13) (Landry et al., 2003). In particular, phytoplankton growth rates (μ) exceeded grazing rates (m) for all groups at both depths. While the grazing rates of different groups of phytoplankton were similar at the 30 m depth, phytoplankton growth varied considerably. At 60 m, growth and grazing rates were generally lower than at 30 m; the highest grazing rates were on dinoflagellates. Spectrofluorometric and HPLC measurements of monovinyl chlorophyll *a* (MVCHLa), primarily derived from *Prochlorococcus* spp., showed no significant differences at either depth. These results collectively demonstrate that picophytoplankton were largely regulated by grazing activity. If fact, it was estimated that grazing accounted for 69% of the equatorial phytoplankton production in the HNLC region.

It has long been known that small coccoid cyanobacteria comprise a substantial fraction of the global primary production in the open ocean (Johnson and Sieburth, 1979). As discussed earlier, these organisms contain phycobilin pigments that not only interfere with measurements of other plant pigments, but may provide a tool for measuring populations of phycobilin-containing phytoplankton. Early work showed that extraction and fluorometric measurement of phycoerythrin was useful for estimating populations of the cyanobacterium *Trichodesmium* (Moreth and Yentsch, 1970). However, this method did not prove useful for more extraction-resistant cells like *Synechococcus* spp. (Glover et al., 1986). Wyman (1992) developed a method for examining phycoerythrin in *Synechococcus* spp. that allowed for the measurement of phycobilins in open ocean cyanobacteria. This method was tested in the Celtic and Sargasso seas, where *Synechococcus* cell numbers were shown to have an inverse correlation with cell-specific phycoerythrin concentrations, due to light adaptation with water depth (fig. 12.14) (Wyman, 1992). Peak numbers of *Synechococcus* were found at the *nutricline* at both stations.

The Southern Ocean is one of the most productive oceans in the world and has received considerable attention in recent years with respect to global climate issues. The classic pattern of phytoplankton abundance in the Ross Sea continental shelf, believed to be the most productive region in the Southern Ocean, is the occurrence of a large annual phytoplankton bloom (Smith et al., 2006, and references

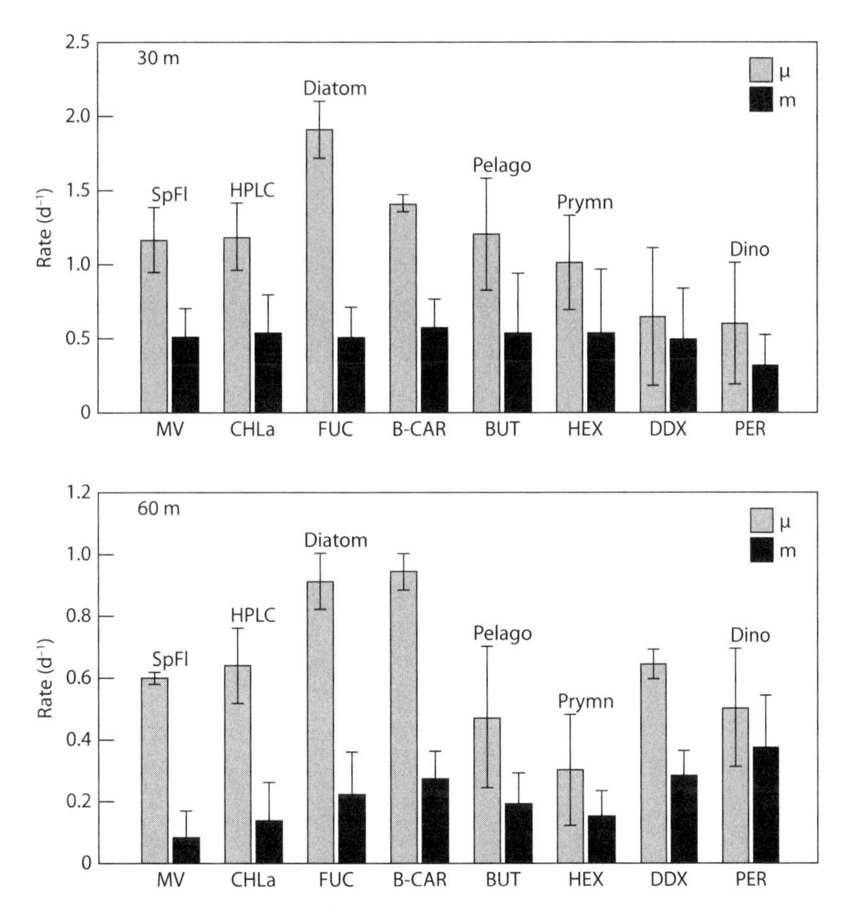

Figure 12.13. Mean phytoplankton growth rates (μ) and grazing rates (m) for phytoplankton, identified from pigments, at 30 and 60 m in the HNCL waters of the equatorial Pacific Ocean. Results from this study showed significant differences in the composition of eukaryotic phytoplankton across groups and at different depths. (Adapted from Landry et al., 2003.)

therein). Using plant pigment biomarkers in the Ross Sea in 2003–2004, it was found that the composition of phytoplankton assemblages were dominated by diatoms and prymnesiophytes (mainly *Phaeocystis antarctica*), with relatively minor contributions by dinoflagellates (fig. 12.15) (Smith et al., 2006). Past work has suggested that if climate change were to increase stratification in the Ross Sea, diatoms would likely replace *Phaeocystis antarctica* (Arrigo et al., 1999). The recent dominant bloom of diatoms may be reflective of changes to come, but the variations observed between years in terms of abundance composition are due to changes in controlling mechanisms on an annual basis, so it is difficult to draw conclusions about long-term changes at this time (Smith et al., 2006).

12.5 Summary

Plant pigments have been effective chemical biomarkers of algal biomass for the past 50 or so years in aquatic sciences. The application of spectrophotometry and fluorometry techniques was involved in their earliest inception as biomarkers. These techniques continue today with more advanced modifications, along with chromatographic techniques that range from HPLC to HPLC-MS.

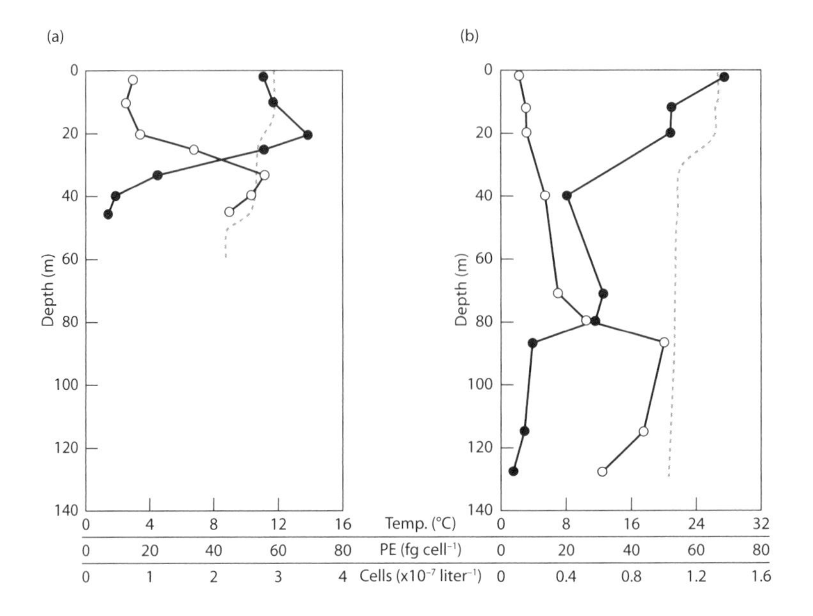

Figure 12.14. *Synechococcus* cell numbers (open circles) and cell-specific phycoerythrin concentrations (closed circles) within the water column of the (a) Celtic (UK) and (b) Sargasso seas. The dotted line shows water column temperature. (Adapted from Wyman, 1992.)

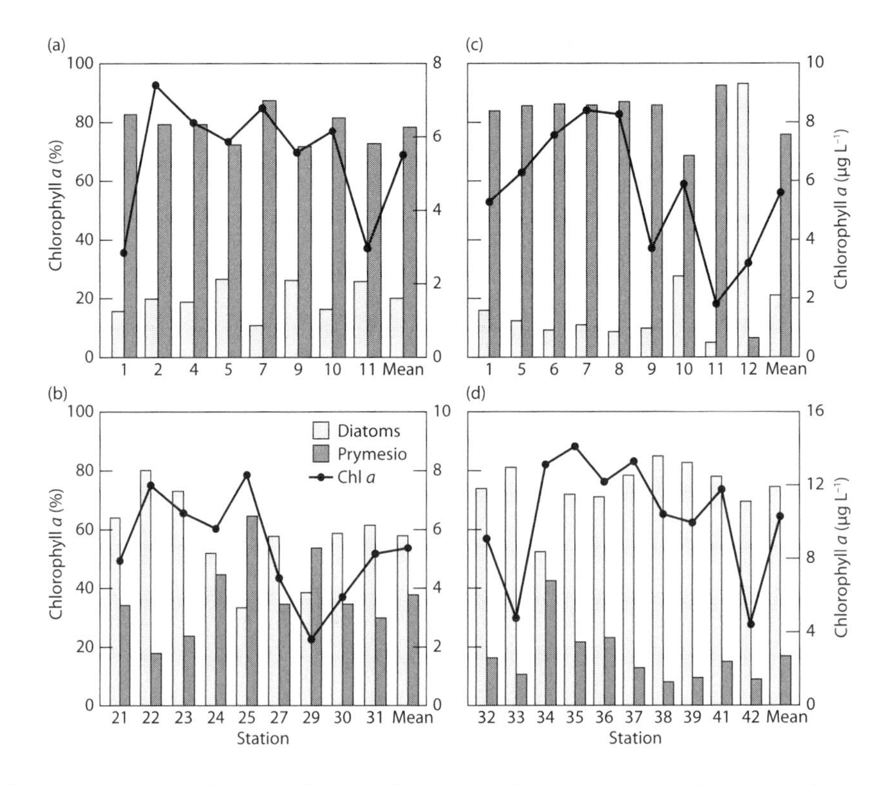

Figure 12.15. Percentages of chlorophyll *a* contributed from diatoms and pymnesiophytes in the Ross Sea (A) December 2001–2002; (B) February 2001–2002; (C) December 2003–2004; and (D) February 2003–2004. These percentages were derived using CHEMTAX. (Adapted from Smith et al., 2006.)

The application of ^{14}C tag experiments with pigments has allowed for a more comprehensive understanding of class-specific responses of phytoplankton to changes in light and nutrient regions in aquatic systems. Satellite imagery has provided a means for a more expansive and rapid assessment of blooms and phytoplankton patterns of distribution around the world. These techniques, with in situ ground-truthing connected with coastal observatories, as well as autonomous underwater vehicles (AUVs), will allow for an even more comprehensive understanding of phytoplankton patterns in deeper waters. The application of carotenoids as indices of UV effects as well as fossil biomarkers of past productivity in aquatic environments will also continue to be useful in our understanding of global climate change.

13. Lignins, Cutins, and Suberins

13.1 Background

Lignin has proven to be a useful chemical biomarker for tracing vascular-plant inputs to aquatic ecosystems (Gardner and Menzel, 1974; Hedges and Parker, 1976; Goñi and Hedges, 1992; Hedges et al., 1997; Bianchi et al., 1999b, 2002, 2007). Cellulose, hemicellulose, and lignin generally make up >75% of the biomass of woody plant materials (Sjöström, 1981). Lignins are a group of macromolecular heteropolymers (600–1000 kDa) found in the cell wall of vascular plants that are made up of hree-dimensional phenylpropanoid units held together by irregular carbon–carbon and diaryl–ether linkages (Sarkanen and Ludwig, 1971; De Leeuw and Largeau, 1993). Perhaps the most common linkage between these units is the arylglycerol-β-aryl ether or β-O-4 link (Adler, 1977). Phenolic compounds in plants, such as lignin, comprise a relatively heterogeneous group of compounds that are primarily synthesized by the *shikimic acid* and *malonic acid pathways*, which is discussed later in this chapter.

Lignin decomposition in soils tends to be relatively slow compared to other plant molecules, such as cellulose. White- and brown-rot fungi are largely responsible for lignin breakdown in aerobic soils (Gleixner et al., 2001), in contrast to bacterial biodegradation (Benner et al., 1986; Bianchi et al., 1999b) and chemical photooxidation (Opsahl and Benner, 1998; Benner and Opsahl, 2001) which are the dominant processes in aquatic systems. Decomposition by white-rot fungi is usually characterized by side-chain oxidation of the propyl side chains of lignin, demethylation of 3- and 5-methoxyl groups, as well as aromatic ring cleavage (Filley et al., 2000). Brown-rot fungi produce 3,4- and 4,5-dihydroxylated phenol groups from demethylation, with only minor side-chain oxidation of propyl groups. Bacteria have also been shown to have the ability to breakdown lignin under aerobic conditions; the aerobic bacteria genera typically associated with wood decay are *Cytopaga* and *Cellvibrio* (Landry et al., 2008). In coastal marine sediments, *Streptomycetes* populations have been shown to play a role in the decay of lignin (Moran et al., 1991). Even under low oxygen conditions, where basidomycetes typically do not thrive, the role of bacteria has been posited to be even more important in lignin decomposition (Landry et al., 2008). These anaerobic bacteria are dominated

Table 13.1
Cutin compositional parameters obtained from conifer needles and bulk sediments of the Dabob Bay (USA) region

Conifer *Sample type*	ω-C_{14}/ ΣC_n^b	ω-C_{16}/ ΣC_{16}	x,ω-C_{16}/ ΣCA	9,10,ω-C_{18}/ ΣCA	8-OH	9-OH
Fir/hemlock mixtures						
Green needle[a]	0.96	0.06	0.78	0.00	0.01	0.94
Ground litter[a]	0.93	0.06	0.77	0.00	0.01	0.93
Cedar needles						
Green needle	1.00	0.08	0.76	0.00	0.01	0.03
Ground litter	0.90	0.07	0.64	0.00	0.00	0.05

[a] Values calculated using a 60/40 weight ratio of fir to hemlock needles, which matches the mean ratio for the fir/hemlock sedimentary isolates.
[b] ΣC_n includes only ω-C_{14} and ω-C_{17} to be consistent with Goñi and Hedges (1990c); ΣCA = total cutin acids; other abbreviations, see text.

by genera such as *Clostridium*, *Bacillus*, *Pseudomonas*, *Arthrobacter*, *Flavobacterium*, and *Spirillum* (Daniel et al., 1987). Based on morphology, the bacteria associated with wood decay have been divided into three groups (e.g., tunneling, erosion, and cavitation) based on their role in the different stages of wood decay (Fors et al., 2008, and references therein). Lignin is generally thought to be highly refractory in anaerobic environments, although early work showed some decay under anaerobic conditions (Benner et al., 1984), and more specifically by methanogenic degradation (Grbic-Galic and Young, 1985). More recently, it has been shown that depolymerization of lignocellulosic material can occur under sulfate-reducing conditions (Pareek et al., 2001; Dittmar and Lara, 2001; Kim et al., 2009). Based on an extensive 4-year experiment of mangrove leaves (*Rhizophora mangle*) decomposing under aerobic conditions, the half-life of lignin was found to be on the order of weeks to months (Opsahl and Benner, 1995). However, other work has shown that long-term *in situ* decomposition rates of lignin from mangrove leaf litter, in sulfate-reducing sediments, was ca. 150 yr (Dittmar and Lara, 2001).

Cutins are lipid polymers in vascular plant tissues that comprise a protective layer called the *cuticle* (Martin and Juniper, 1970; Holloway, 1973). The cuticle, is contiguous with the cell wall and contains a mixture of cutin and aliphatic compounds (mainly C_{24}–C_{34} alkanes, alcohols, and ketones), called *cuticular waxes* (Jetter et al., 2002; Kunst and Samuel, 2003). Due to their close physical association, it is difficult to distinguish between the relative importance of cutins and cuticular waxes in the overall physical properties and biological functions of the cuticle (Jenks et al., 2001). Nevertheless, cutins and cuticular waxes can be analyzed separately, since waxes are easily extracted in organic solvents, while the cutin polymer remains insoluble (Kolattukudy, 1980).

Cutins have also been shown to be effective biomarkers for tracing vascular plant inputs to coastal systems (Kolattukudy, 1980; Cardoso and Eglinton, 1983; Eglinton et al., 1986; Goñi and Hedges, 1990a,b). When cutin is oxidized, using the CuO method commonly used for lignin analyses (Hedges and Ertel, 1982), a series of fatty acids are produced that can be divided into the following three groups: C_{16} hydroxy acids, C_{18} hydroxy acids, and C_n hydroxyl acids (table 13.1)(Goñi and Hedges, 1990a). In general, the cutin matrix is composed mainly of C_{16} and C_{18} mono-, di-, and trihydroxyalkanoic acids, with more dilute concentrations of C_{18} epoxy acids as well as C_{12} to C_{36} *n*-alkanoic acids (Baker et al., 1982; Holloway, 1982). For example, the major cutin acid suite in green firt–hemlock mixtures was dihydroxyhexadecanoic (x, ω-C_{16}) 16-hydroxyhexadecanoic acid (ω-C_{16}), and 14-hydroxytetradecanoic acid (ω-C_{14}) (table 13.1). Minor amounts of aromatic compounds are also present (Kolattukudy, 1980; Nawrath, 2002; Heredia, 2003). It has long been thought that cutin was formed only by interesterification of fatty acids, but it is now clear that glycerol is also esterified to the fatty acid monomers (Graca et al., 2002). The size and exact structure of the polymer (e.g., *dendrimeric* and cross-linked) remain uncertain. In the epidermis of the small flowering plant *Arabidopsis thaliana* (Brasssicacae) (the analog of *Drosophila melanogaster* in plant genetics) (Birnbaum et al., 2003), the major monomers released by polyester depolymerization of delipidated residues are not ω-hydroxy fatty acids, but dicarboxylic acids (Bonaventure et al., 2004; Kurdyukov et al., 2006a,b). Thus, it is not clear whether the rigid classification of cutin as highly enriched in ω-hydroxy fatty acids needs to be changed, or whether the dicarboxylic acids represent a novel polyester domain. Since cutins are not found in wood, they are similar to the *p*-hydroxyl lignin-derived cinnamyl phenols (e.g., trans *p*-coumaric and ferulic acids) and serve as biomarkers of nonwoody vascular plant tissues. However, cutin acids were found to be more reactive than lignins in the degradation of conifer needles (Goñi and Hedges, 1990c), and are not as effective as lignin for tracing vascular plant inputs (Opsahl and Benner, 1995).

Suberin is a waxy substance found in higher plants (e.g., bark and roots) and the primary constituent of cork, as reflected in the species name of the cork oak (*Quercus suber*) (Kolattukudy, 1980; Kolattukudy and Espelie, 1985). Suberin consists of two domains: a polyaromatic (or polyphenolic) and a polyaliphatic domain (Bernards et al., 1995; Bernards, 2002). The polyaromatics are predominantly located within the primary cell wall, and the polyaliphatics are located between the primary cell wall and the *plasmalemma*—both domains are believed to be cross-linked. Suberin is deposited in various inner and outer tissues at specific locations during plant growth (Kolattukudy, 2001; Bernards, 2002; Nawrath, 2002; Kunst et al., 2005; Stark and Tian, 2006). For example, suberin has been chemically identified in seed coat (Espelie et al., 1980; Ryser and Holloway, 1985; Moire et al., 1999), root and stem endodermis (Espelie and Kolattukudy, 1979a; Zeier et al., 1999; Endstone et al., 2003), the bundle sheath of monocots (Espelie and Kolattukudy, 1979b; Griffith et al., 1985), and conifer needles (Wu et al., 2003). In addition to its widespread distribution, suberin is also synthesized in response to stress and wounds (Dean and Kolattukudy, 1976).

Similar to cutin, suberin has been shown to be a major source of hydrolyzable aliphatic lipids in soil organic matter (OM) (Nierop, 1998; Kögel-Knabner, 2000; Naafs and Van Bergen, 2002; Nierop et al., 2003; Otto et al., 2005; Otto and Simpson, 2006). Although suberins have been found in peats and recent sediments (Cardoso and Eglinton, 1983), more work is needed to understand their structure and function for their application as biomarkers (Hedges et al., 1997). Suberin is chemically distinguished from cutin by a highly hydroxycinnamic acid–derived aromatic domain (Bernards et al., 1995; Bernards, 2002), longer fatty alcohol and acid chain lengths (>20 carbons), and dicarboxylic acid monomers (Kolattukudy, 2001). Suberin occurs in the bark and roots of vascular plants as a polyester, primarily composed of C_{16} to C_{30} *n*-alkanoic acids, as well as hydroxyl- and epoxy-substituted *n*-alkanoic acids with phenolic moieties (Kolattukudy and Espelie, 1989). It should also be noted that long-chain homologues (C_{20}–C_{30}), such as *n*-alkanedioic and ω-hydroxyalkanoic acids (suberin acids), have been shown to be exclusively derived from the hydrolyis of suberin. In fact, recent work has shown that ω-hydroxyalkanoic and α, ω-alkanedioic acids were the dominant biomarkers for suberin in grassland soils (Otto et al., 2005). Despite the abundance of the decomposition products of suberin in soils, suberin has been found to be more stable than cutin in soils, because of the higher content of phenolic units in suberin, as well as the nonpolar aliphatic structure of cutin (Nierop et al., 2003).

13.2 Lignin, Cutin, and Suberin Biosynthesis

The *shikimic acid pathway* is involved in the biosynthesis of most plant phenolics (Hermann and Weaver, 1999). The malonic acid pathway, although an important source of phenolic secondary products in fungi and bacteria, is of less significance in higher plants. The shikimic acid pathway, which is common in plants, bacteria, and fungi, uses aromatic amino acids (see chapter 6) to provide the parent compounds for the synthesis of the phenylpropanoid units in lignins (fig. 13.1). Animals have no

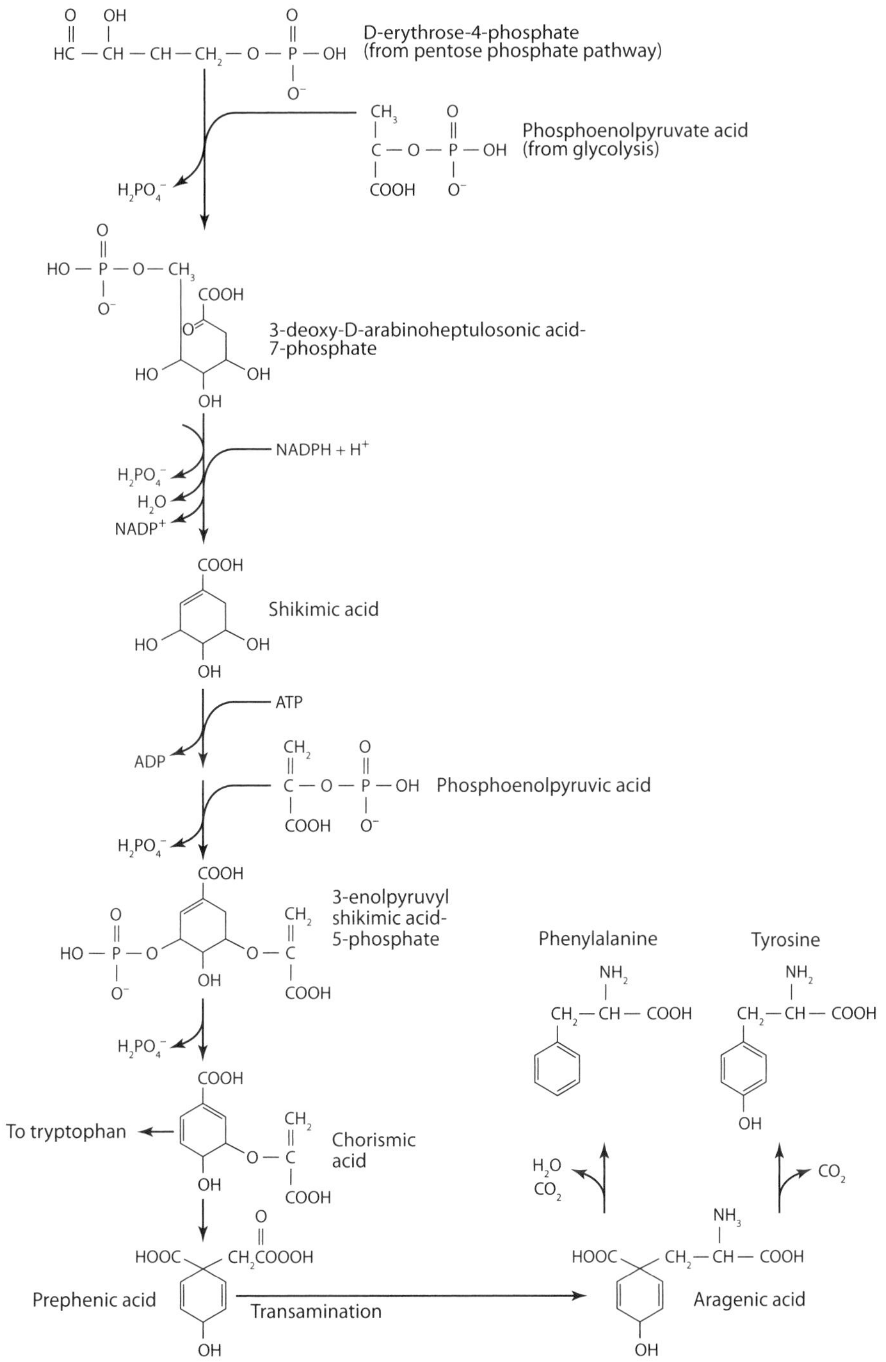

Figure 13.1. The shikimic acid pathway, which is common in plants, bacteria, and fungi, is where aromatic amino acids provide the parent compounds for the synthesis of the phenylpropanoid units in lignins. (Adapted from Goodwin and Mercer, 1972.)

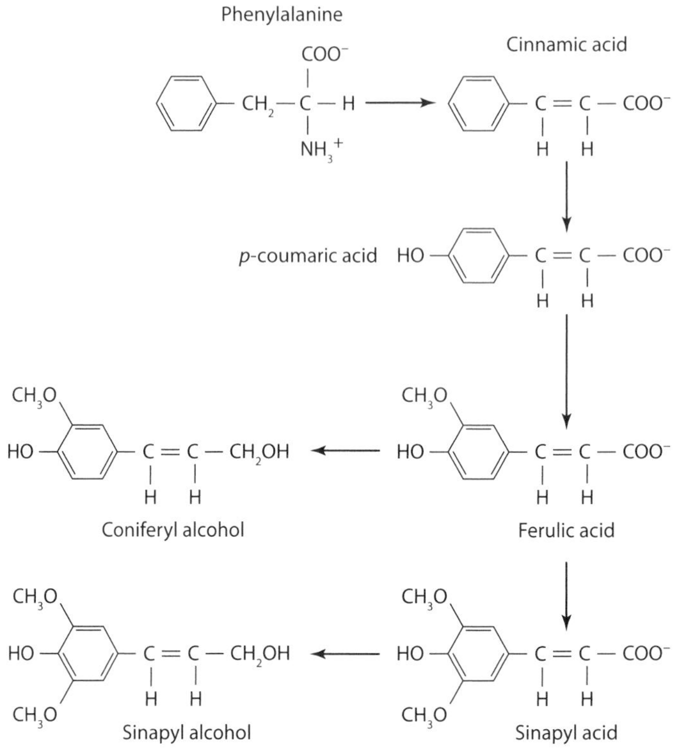

Figure 13.2. Pathways for synthesis of building blocks for lignin. The primary building blocks for lignins are the following monolignols: *p*-coumaryl alcohol, coniferyl alcohol, and sinapyl alcohol. (Adapted from Goodwin and Mercer, 1972.)

way to synthesize the three aromatic amino acids—phenylalanine, tyrosine, and tryptophan—which are therefore essential nutrients in animal diets. The shikimic acid pathway converts simple carbohydrate precursors derived from glycolysis and the pentose phosphate pathway to the aromatic amino acids (Hermann and Weaver, 1999). One of the pathway intermediates is shikimic acid, for which this whole sequence of reactions is named. More specifically, the primary building blocks for lignins are the following monolignols: *p*-coumaryl alcohol, coniferyl alcohol, and sinapyl alcohol (fig. 13.2) (Goodwin and Mercer, 1972). These connecting units are cross-linked by carbon to carbon bonds and the dominant β-O-4 aryl to *aryl ether* bonds, making lignins very stable compounds (fig. 13.3).

Knowledge of the biosynthesis of cutin is based largely on a few early studies, with broad bean (*Vicia faba*) leaves (Croteau and Kolattukudy, 1974; Kolattukudy and Walton, 1972). These studies indicated that cutin biosynthesis involved ω-hydroxylation followed by midchain hydroxylation and incorporation into the polymer and that hydroxy fatty acids can be incorporated into cutin by a reaction requiring CoA and ATP. Specifically, the most common monomers of cutin (e.g., 10,16-dihydroxy C_{16} acid, 18-hydroxy-9,10 epoxy C_{18} acid, and 9,10,18-trihydroxy C_{18} acid), are produced in the epidermal cells by *p*-hydroxylation, in-chain hydroxylation, epoxidation catalyzed by P_{450}-type mixed function oxidase, and epoxide hydration (Blée and Schuber, 1993; Hoffmann-Benning and Kende, 1994). The monomer *acyl* groups are transferred to *hydroxyl* groups in the growing polymer at the extracellular location (Croteau and Kolattukudy, 1974).

More recently, genetic approaches to cutin biosynthesis have been employed. For example, in the *Arabidopsis fatbko* lineage, where a disruption in an acyl-carrier protein (ACP) thioesterase gene results in reduced general availability of *cytosolic* palmitate (Bonaventure et al., 2003), there is an 80% loss

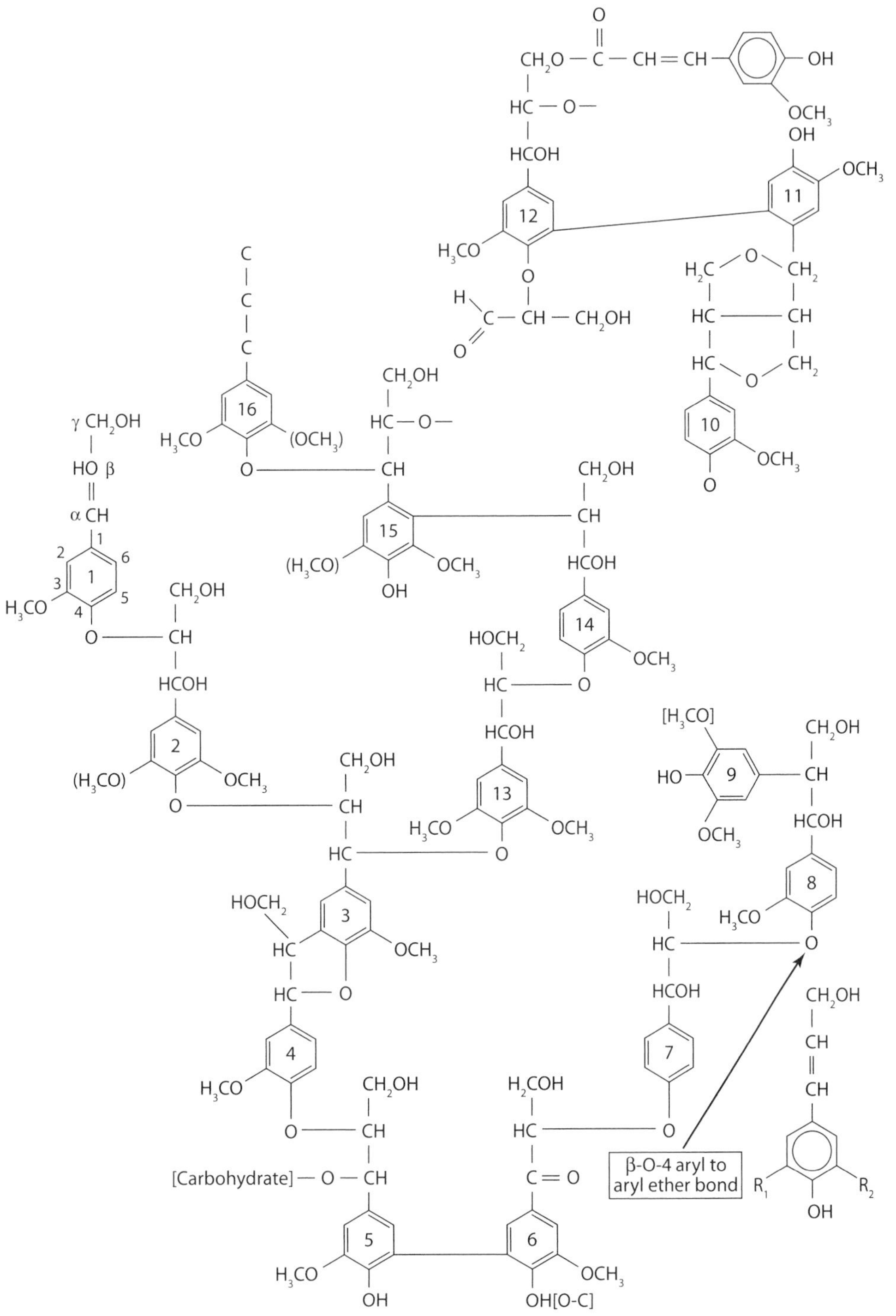

Figure 13.3. Macromolecular structure of lignin. The connecting phenylpropanoid units of lignins are cross-linked by the dominant *β*-O-4 aryl to aryl ether bonds (as shown by arrow), making lignin a very stable macromolecule.

of C_{16} monomers and a compensatory increase in C_{18} monomers in epidermal polyesters (Bonaventure et al., 2004). The first mutant (*ATTL*) demonstrated to be specifically affected in cutin metabolism was reported by Xiao et al. (2004). More specifically, this ethyl methanesulfonate-induced mutant showed enhanced disease severity to a virulent strain of *Pseudomonas syringae*, a 70% decrease in cutin amount, a loose cuticle ultrastructure, and increased permeability to water vapor. *ATTL* (At4g00360 locus, or gene site on chromosome) encodes a P_{450} monooxygenase belonging to the CYP86A family (Xiao et al., 2004) and has been demonstrated to catalyze the ω-hydroxylation of fatty acids (Duan and Schuler, 2005). *ATTL* is thus probably responsible for the synthesis of ω-hydroxy fatty acid and/or dicarboxylic acid monomers found in the epidermal polyesters of *Arabidopsis*. In maize (*Zea mays*) leaves, a peroxygenase has been shown to be involved in the synthesis of C_{18} epoxy monomers of cutin (Lequeu et al., 2003). This enzyme, associated with a *cytochrome* P_{450} monooxygenase (Pinot et al., 1999) and a membrane-bound epoxide hydrolase, can catalyze the formation of these monomers in vitro (Blée and Schuber, 1993).

It remains uncertain whether polymerization reactions of the cutin monomers occur inside or outside the epidermal cells, and how cutin synthesis is coordinated with the deposition of waxes. The characterization of the *Arabidopsis lacs2* mutant (LACS2) indicates that an epidermis-specific, long-chain acyl-CoA synthetase is involved in leaf cuticle formation and barrier function (Schnurr et al., 2004). This further suggests the likely involvement of specific acyl-CoA in cutin biosynthesis. Cuticular waxes of the *Arabidopsis* shoot are found both above the cutin matrix (epicuticular) and embedded in the cutin (intracuticular). The dominant wax components of *Arabidopsis* stems are C_{29} alkane, ketone, and secondary alcohols, as well as less abundant quantities of C_{28} primary alcohol and C_{30} aldehydes (Jenks et al., 2001). Stems typically have much higher wax concentrations than leaves. These saturated aliphatic wax components are synthesized as C_{16} to C_{18} fatty acids in the *plastid*, which is followed by fatty acid elongation (in association with endoplasmic reticulum) to form very-long-chain fatty acids (VLCFA) (Post-Beittenmiller, 1996; Kunst and Samuel, 2003). The function of one of the condensing enzymes of the elongase complex was demonstrated by examining a condensing enzyme (*CER6*) that controls surface wax production in *Arabidopsis*, which has a highly glossy stem phenotype and 7% of wild-type wax (Millar et al., 1999). Subsequent modifications occur to the VLCFAs, but the mechanisms and gene products involved are less well characterized. The mechanism of transport of waxes onto the epidermal surface is better understood as a result of the recent discovery of the condensing enzyme *CER5*, an ATP-binding cassette (ABC) transporter involved in the export of cuticular waxes to the stem surface (Pighin et al., 2004).

The biosynthetic origins and specific chemical structure of suberin are presently a matter of debate. As described earlier, suberin is a heteropolymer consisting of aliphatic and aromatic domains synthesized in a lamellar-like fashion within the cell walls in specific tissues of vascular plants (Bernards et al., 1995; Bernards, 2002). Currently, linkages between the two domains remain unknown. Some common aliphatic monomers include α-hydroxyacids (e.g., 18-hydroxyoctadec-9-enoic acid) and α, ω-diacids (e.g., octadec-9-ene-1,18-dioic acid). The monomers of the polyaromatics are hydroxycinnamic acids and their derivatives, such as feruloyltyramine. A new model for suberin has been developed that suggests a hydroxycinnamic acid–monolignol polyphenolic domain (fig. 13.4) (Graca and Pereira, 2000a,b). This is covalently linked to a glycerol-based polyaliphatic domain and located between the primary cell wall and the plasma membrane (Graca and Pereira, 2000a,b; Graca et al., 2002).

13.3 Analysis of Lignins, Cutins, and Suberins

The ligno-cellulosic complex of the lignin macromolecule is difficult to analyze, but methods using alkaline oxidation with CuO followed by gas chromatography–flame ionization detection (GC-FID)

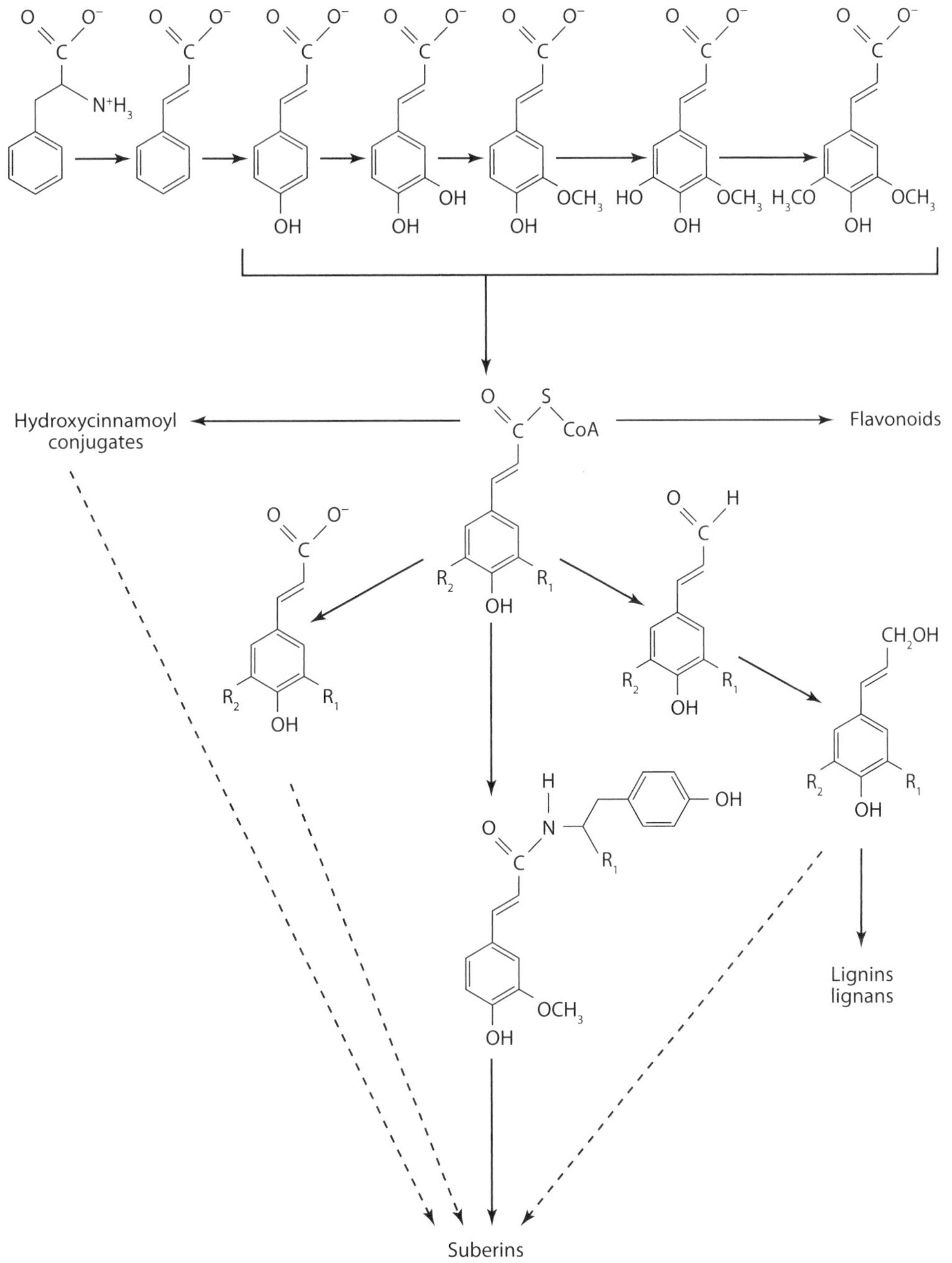

Figure 13.4. Pathway for synthesis of suberin new model for suberin synthesis. This suggests a hydroxycinnamic acid–monolignol polyphenolic domain. (Adapted from Graca and Pereira, 2000a,b; Bernards, 2002.)

and GC–mass spectrometry (GC-MS) have been effective (Hedges and Ertel, 1982; Kögel and Bochter, 1985). This oxidation procedure cleaves aryl ether bonds, which causes phenolic monomers and dimers to be released from the outer reaches of the macromolecule (Johansson et al., 1986). Oxidation of lignin using the CuO oxidation method (Hedges and Parker, 1976; Hedges and Ertel, 1982) yields eleven

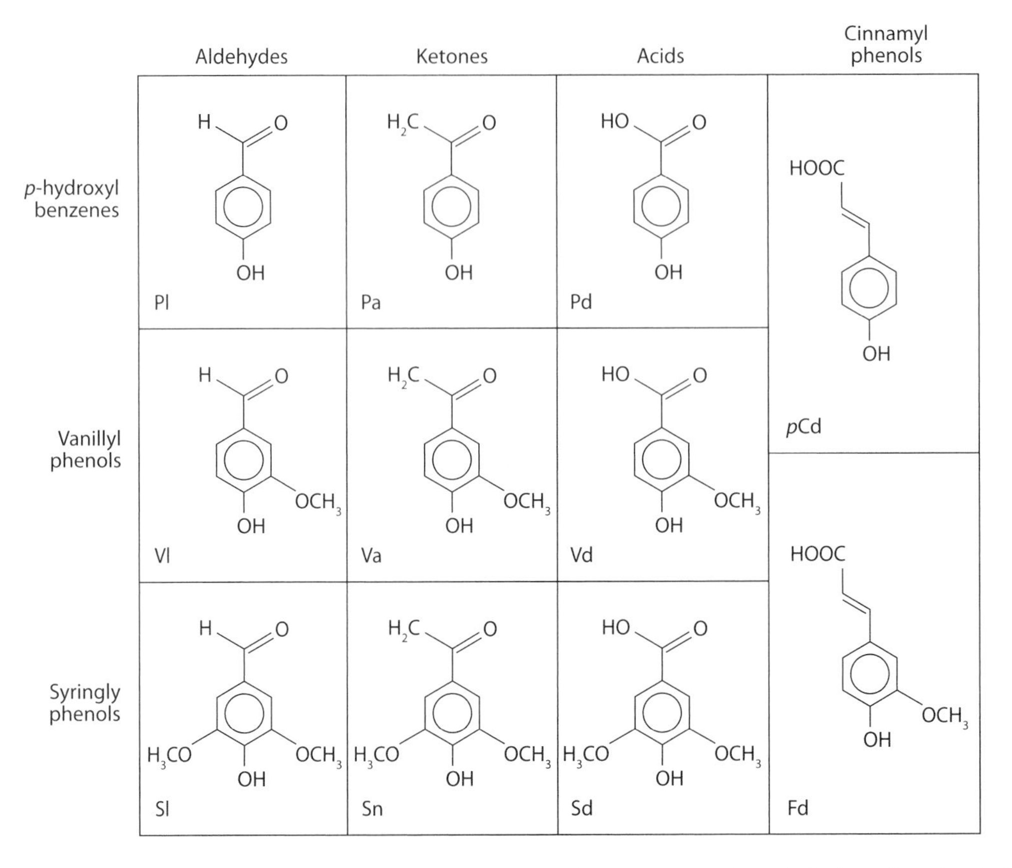

Figure 13.5. Eleven dominant phenolic monomers, yielded from the oxidation of lignin using the CuO oxidation method. These compounds can be separated into four families: *p*-hydroxyl (P), vanillyl (V), syringyl (S), and cinnamyl (C) phenols. The compounds are abbreviated as follows: *p*-hydroxybenzaldehyde (Pl), *p*-hydroxyacetophenone (Pa), *p*-hydroxy benzoic acid (Pd), vanillin (Vl), acetovanillone (Va), vanillic acid (Vd), syringealdehyde (Sl), acetosyringone (Sn), syringic acid (Sd), *p*-coumaric acid (pCd), and ferulic acid (Fd). (Modified from Hedges and Ertel, 1982.)

dominant phenolic monomers, which can be separated into four families: *p*-hydroxyl (P), vanillyl (V), syringyl (S), and cinnamyl (C) phenols (fig. 13.5) (Hedges and Ertel, 1982). Hedges and Parker (1976) used lambda indices to quantify lignin relative to organic carbon. Lambda-6 (Λ_6 or λ) is defined as the total weight in milligrams of the sum of vanillyl (vanillin, acetovanillone, vanillic acid) and syringyl (syringaldehyde, acetosyringone, syringic acid) phenols, normalized to 100 mg of organic carbon (OC), while lambda-8 (Λ_8) includes the cinnamyl (*p*-coumaric and ferulic acid) phenols.

CuO oxidation of lignin has also been shown to produce benzene carboxylic acids (BCAs), which have been proposed to be potential biomarkers for the extent of soil oxidation of OC (Otto et al., 2005; Otto and Simpson, 2006) and tracers of terrestrially and marine-derived OC in aquatic systems (Prahl et al., 1994; Louchouarn et al., 1999; Goñi et al., 2000; Farella et al., 2001; Houel et al., 2006; Dickens et al., 2007). The most prominent BCAs used as tracers are 3,5-dihydroxybenzoic (3,5Bd) and *m*-hydroxybenzoic acids, as well as some tricarboxylic acids. Additionally, the ratio of *p*-hydroxyl phenols (P) to the vanillyl and syringyl phenols (P:[V + S]) has been suggested to be an indicator for the decomposition pathway that results from the demethylation of methoxylated vanillyl and syringyl components, via brown-rot fungi (Dittmar and Lara, 2001). However, since 3,5Bd and *p*-hydroxyl phenols are not exclusively produced by vascular plants (Goñi and Hedges, 1995), this

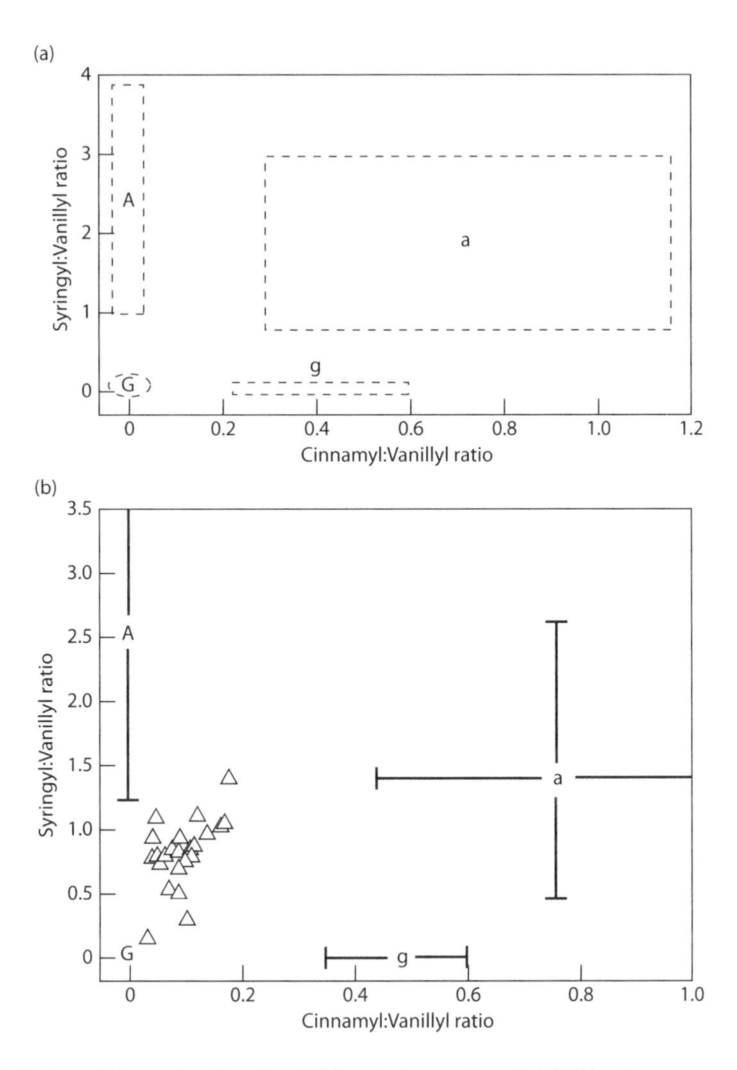

Figure 13.6. (a). Biplots of the syringyl/vanillyl (S/V) and cinnamyl/vanillyl (C/V) ratios can provide information about the relative importance of angiosperm versus gymnosperm sources and woody versus nonwoody source, respectively. A, woody angiosperm; a, nonwoody angiosperm; G, woody gymnosperm; g, nonwoody gymnosperm. (b). Data showing how soil sources (triangles) are difficult to constrain when using the traditional source plot that is based on the ratios for pure plant tissues. (Adapted from Hedges and Mann, 1979.)

ratio has problems when used as an index for diagenetically altered terrestrial OC in marine sediments. For example, the ratio is not very effective when there are significant inputs to sediments from certain brown macroalgae (e.g., kelp), which have 3,5Bd, and plankton-derived amino acids (Houel et al., 2006). Finally, it has been suggested that tannins and other flavanoids may be a source of 3,5Bd (Louchouarn et al., 1999), so further work is still needed to better constrain source inputs of these compounds as biomarkers.

Syringyl derivatives are unique to woody and nonwoody angiosperms, while cinnamyl groups are common to nonwoody angiosperms and gymnosperms (Hedges and Parker, 1976; Hedges and Mann, 1979; Gough et al., 1993). However, all gymnosperm and angiosperm woody and nonwoody tissues yield vanillyl phenols as oxidation products. Hence, the S/V and C/V ratios can provide information about the relative importance of angiosperm versus gymnosperm sources and woody versus nonwoody

Table 13.2
Means and ranges of LPVI

Plant type	LPVI value range		
	Low	*Mean*	*High*
Gymnosperm woods (G)	1	1	1
Nonwoody gymnosperm tissues (g)	12	19	27
Angiosperm woods (A)	67	181	415
Nonwoody angiosperm tissues (a)	378	1090	2782

Source: Lignin phenols data for specific plants were obtained from Hedges and Mann (1979).

tissues, respectively (fig. 13.6a). The *p*-hydroxyl phenols are generally not included when using lignins as indicators of vascular plant inputs to estuarine/coastal systems because these phenols can also be produced by nonlignin constituents (Wilson et al., 1985). However, *p*-hydroxyacetophenone (Pa), primarily derived from lignin (Hedges et al., 1988), can be used as an indicator of lignin when normalized to the other *p*-hydroxyl phenols (e.g., *p*-hydroxylbenzaldehyde and *p*-hydroxylbenzoic acid) that are from lignin-free sources in the ratio Pa/P (Benner et al., 1990). Recent work has suggested that differences in the diagenetic susceptibilities of cinnamyl, syringyl, and vanillyl ($C > S > V$) in highly reactive soils result in pronounced problems when identifying vegetation sources (Tareq et al., 2004). Cinnamyl phenols are more loosely bound to lignin with an ester bond, making them more reactive and less stable in the environment, compared to the syringyl and vanillyl phenols (Dittmar and Lara, 2001). The data in fig. 13.6b demonstrate how soil sources are difficult to constrain when using the traditional source plotbased originally on pure plant ratios (Hedges and Mann, 1979). To help correct for this, the following binary equation, called the lignin phenol vegetation index (LPVI) was developed:

$$\mathrm{LPVI} = [\{S(S+1)/(V+1)+1\} \times \{C(C+1)/(V+1)+1\}] \tag{13.1}$$

where V, S, and C are expressed as percentages of Λ_8 (Tareq et al., 2004). Based on the data shown in table 13.2, there appears to be little overlap in gymnosperm and angiosperm woods, and nonwoody gymnosperms. The application of the LPVI index in natural soils and sedimentary and pelagic systems needs further work but continues to broaden.

During the CuO oxidation procedure most, but not all, of the ether bonds in lignin are broken; however, many of the carbon-to-carbon bonds that link aromatic rings remain intact (Chang and Allen, 1971). Hence, these phenolic dimers have retained the characteristic ring-to-ring and side-chain linkages, providing an additional thirty (or more) CuO products that can be used to identify taxonomic sources of vascular plants beyond the monomeric forms (Goñi and Hedges, 1990a). Moreover, the more numerous dimers may actually appear to be derived exclusively from polymeric lignin, unlike the monomeric forms, which are found in soluble or ether-bound form. Thus, lignin dimers may better document inputs of polymeric lignin in natural systems (Goñi and Hedges, 1990a). However, it was later found that monomers did adequately follow the decomposition of the lignin in a long-term decay experiment of different vascular plant tissues (Opsahl and Benner, 1995).

A newly developed technique involving thermochemolysis in the presence of tetramethylammonium hydroxide (TMAH) has been shown to be successful in analyzing lignin in sediments (Clifford et al., 1995; Hatcher et al., 1995a; Hatcher and Minard, 1996). The TMAH thermochemolysis procedure allows for effective hydrolysis and methylation of esters and ester linkages and results in the cleavage of the major β-O-4 ester bond in lignin (fig. 13.7)(McKinney et al., 1995; Hatcher and Minard, 1996; Filley et al., 1999a). Consequently, this technique has proven to be effective in examining abundances

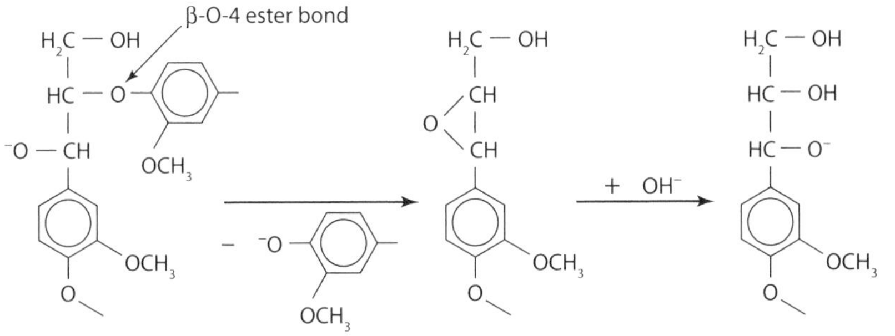

Figure 13.7. The TMAH themochemolysis procedure allows for effective hydrolysis and methylation of esters and ester linkages and results in the cleavage of the major β-O-4 ester bond in lignin. (Adapted from Filley et al., 1999a.)

and sources of vascular plant inputs in both terrestrial (Martin et al., 1994, 1995; Fabbri et al., 1996; Chefetz et al., 2000; Filley, 2003) and aquatic (Pulchan et al., 1997, 2003; Mannino and Harvey, 2000; Galler et al., 2003; Galler, 2004) environments. The TMAH method operates at high temperature (~250–600°C) in a strongly basic salt and effectively methylates the polar hydroxyl and carboxyl functional groups, resulting in a greater diversity of methylated compounds (Hatcher et al., 1995b; Page et al., 2001). Furthermore, the TMAH products tend to retain their side chains and can include tannin-derived or polyphenolic lignin-like structures, which, when analyzed using ^{13}C-labeled TMAH, allow for the determination of microbial demethylation, hydrolysable tannin input, and side-chain oxidation (Filley et al., 1999b, 2006; Nierop and Filley, 2007). This provides greater potential to assess the type and extent of microbial decay. However, compounds that may be produced by nonlignin sources, such as hydrolyzable tannins, which yield the *permethylated* analog of gallic acid, or caffeic acid in the case of the permethylated cinnamyl component, can affect the interpretation of TMAH derived lignin as proxies of terrestrially-derived sources (Filley et al., 2006; Nierop and Filley, 2007).

13.4 Applications

13.4.1 Lignins as Biomarkers in Lakes

To determine the validity of lignin as a biomarker in sediments, a comparison with pollen was performed in Wien Lake (Alaska, USA) sediments (Hu et al., 1999). Wien Lake is an oligotrophic lake in the Alaska range that drains into the Kantishna River, and is dominated by riparian forests comprising balsam poplar (e.g., *Populus glauca* or *P. mariana*). Results from this study revealed that the pollen of *P. glauca* (PIGL) and *P. mariana* (PIMA) had significantly higher concentrations of cinnamyl phenols, particularly *p*-coumaric, than in modern plant tissues (e.g., leaves and needles) (fig 13.8). Conversely, the vanillyl phenols were significantly lower in the pollen than in tissues. This is problematic since the source plots used in the CuO method are based on fresh plants, as shown in fig. 13.6 Thus, this work showed that while there is a strong linkage between lignin biomarkers and fossil pollen, adjustment for pollen effects is needed in lacustrine ecosystems (Ertel and Hedges, 1984; Hu et al., 1999). It should also be noted that while pollen concentrations are generally lower in high- versus low-latitude lakes, the significant differences that exist between pollen and other plant tissues requires adjustments at all latitudes. If not adjusted, lignin data will likely yield overestimates of nonwoody tissues in sediments, which would be further exacerbated by the decay-resistant properties of pollen compared to other plant tissues (Faegri and Iversen, 1992).

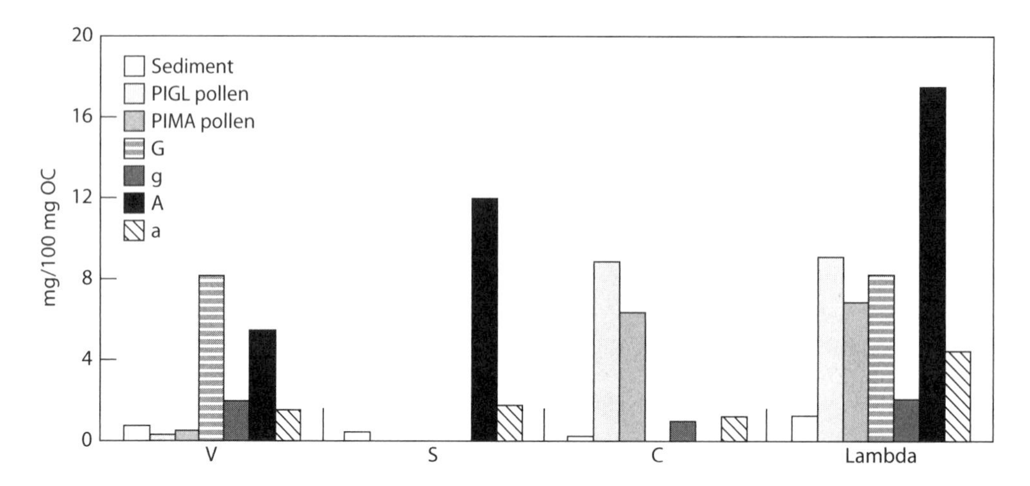

Figure 13.8. The use of lignin as a biomarker was compared with pollen in Wien Lake (AK, USA) sediments, an environment with riparian forests dominated by the balsam poplar species *Populus glauca* (PIGL) and *P. mariana* (PIMA). G, woody gymnosperms; g, nonwoody gymsnoperms; A, woody angiosperms; a, nonwoody angiosperms; V, vanillyl; S, syringyl; C, cinnamyl; lambda is as described in this chapter. (Adapted from Hu et al., 1999.)

13.4.2 Lignins as Biomarkers along the River–Estuarine Continuum

Terrestrial inputs of OC to continental margin waters of the northern Gulf of Mexico are high compared with other coastal margins of the United States because of significant discharge from one of the world's largest river systems—the combined inputs of the Atchafalaya and Mississippi rivers (Hedges and Parker, 1976; Malcolm and Durum, 1976; Eadie et al., 1994; Trefrey et al., 1994; Bianchi et al., 1997a). Bulk OC and carbon isotope measurements made on suspended river particulates and seabed samples within the Mississippi River dispersal system (Eadie et al., 1994; Trefrey et al., 1994) have shown that river particulates are ^{13}C-depleted (relative to marine organic carbon) and that 60 to 80% of the OC buried on the shelf adjacent to the river is marine in origin (as determined by the enriched ^{13}C values of the sediments). Subsequently, it was suggested that much of the terrestrially derived OC delivered to the shelf derives from a mixture of C_3 and C_4 plants and materials from eroded soils in the northwestern grasslands of the Mississippi River drainage basin (Goñi et al., 1997, 1998). These investigators also concluded that C_4 material was transported greater distances offshore because of its association with soils of smaller grain sizes (Goñi et al., 1997, 1998; Onstad et al., 2000; Gordon and Goñi, 2003). C_4 plants generally have isotopic signatures around -14 to -12, which are more enriched than marine sources (-22 to -20). However, the mixture of soils (-28 to -26) with C_4 (-14 to -12) yields a signature similar to marine sources, making use of bulk isotopes alone problematic.

Based on earlier work, Goñi et al. (1997) concluded that the C_3 (^{13}C-depleted) terrestrial material was being deposited on the shelf. However, recent analyses suggest that woody angiosperm material (also ^{13}C-depleted) preferentially settles within the lower Mississippi and in the proximal portion of the dispersal system on the shelf (Bianchi et al., 2002). This suggests that a greater fraction of terrestrially derived carbon in sediments may be represented by OC further offshore, presumably because of selective removal of the more labile, algal fraction of carbon during transport. In fact, recent work has shown that the lignin content of particulate organic matter (POM) in the lower Mississippi River is strongly correlated to river discharge (fig. 13.9) (Bianchi et al., 2007). This further supports the idea that terrestrially derived soil inputs in the upper basin during the spring freshet are linked with inputs of lignin-derived POM to the lower river and Gulf of Mexico.

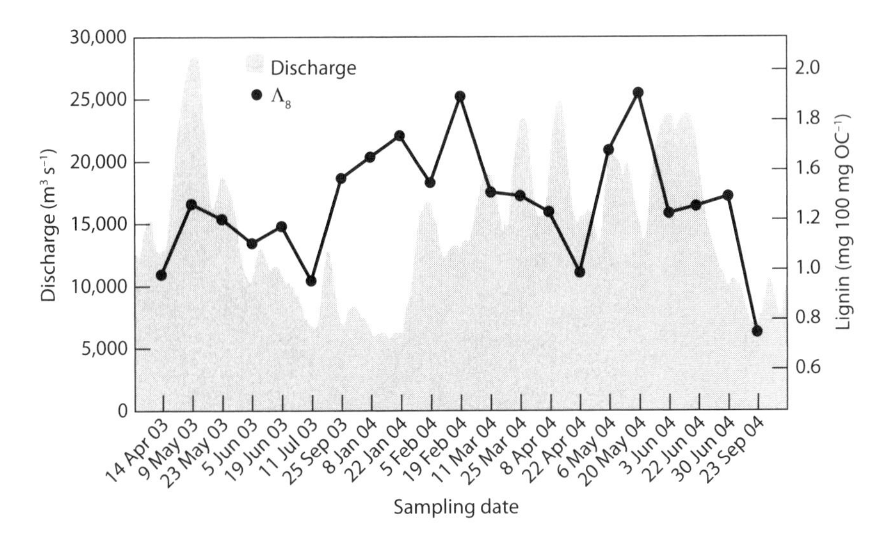

Figure 13.9. Lignin content of POM in the lower Mississippi River (USA) was strongly correlated to river discharge. (Adapted from Bianchi et al., 2007.)

In estuaries it is common to find a gradient of vascular plant inputs of POM with decreasing concentrations moving from the head to the mouth of estuarine systems (Bianchi and Argyrou, 1997; Goñi and Thomas, 2000). In the Baltic Sea, such a gradient is seen in lignin–phenol concentrations in water column particulate OC (POC) (Bianchi et al., 1997a) at stations on the Luleälven River to Baltic Proper, which are situated from the head (north) to the mouth (south) of the estuary. Similarly, soil and sediment samples collected from a forest–brackish marsh-salt gradient in a southeastern U.S. estuary (North Inlet) also indicated that woody and nonwoody gymnosperm and angiosperm sources dominated sedimentary organic matter at the forest site, while nonwoody marsh plants (*Spartina* and *Juncus*) dominated the marsh site (Goñi and Thomas, 2000). Lignin was most degraded at the forest site and least degraded at the salt marsh site where the occurrence of anoxic conditions dominated. Recent work has examined the relative importance of lignin-derived phenols (Λ) versus non-lignin-derived compounds, benzoic acids (B), and *p*-hydroxybenzenes (P), in sediments from rivers of the estuaries in the southeastern United States (fig. 13.10) (Goñi et al., 2003). Benzoic acids and *p*-hydroxybenzenes have commonly been used to identify sources of OM in coastal sediments derived from plankton and soils (Goñi et al., 2000). This work showed that these systems were dominated by lignin-derived compounds from angiosperm and gymnosperm source plants, likely derived from upland coastal forests in this region.

A corresponding gradient in freshwater and terrestrially derived inputs of lignin in DOM has also been documented in riverine/estuarine systems. For example, Benner and Opsahl (2001) found Λ_8 in high-molecular-weight dissolved OM (HMW DOM) in the lower Mississippi River and inner plume region that ranged from 0.10 to 2.30, which is consistent with other HMW DOM measurements in river–coastal margin gradients (Opsahl and Benner, 1997; Opsahl et al., 1999). Controls on the abundance of lignin in DOM in estuarine/coastal systems may involve losses from in situ bacterial breakdown (discussed earlier), flocculation, sorption and desorption processes with resuspended sediments (Guo and Santschi, 2000; Mannino and Harvey, 2000; Mitra et al., 2000), as well as photochemical breakdown (Opsahl and Benner 1998; Benner and Opsahl 2001). In fact, HMW DOM in bottom waters of the Middle Atlantic Bight (MAB), found to be "old" and rich in lignin, was believed to be derived from desorption of sedimentary particles largely derived from Chesapeake Bay and other estuarine systems along the northeastern margin of the United States (Guo and Santschi,

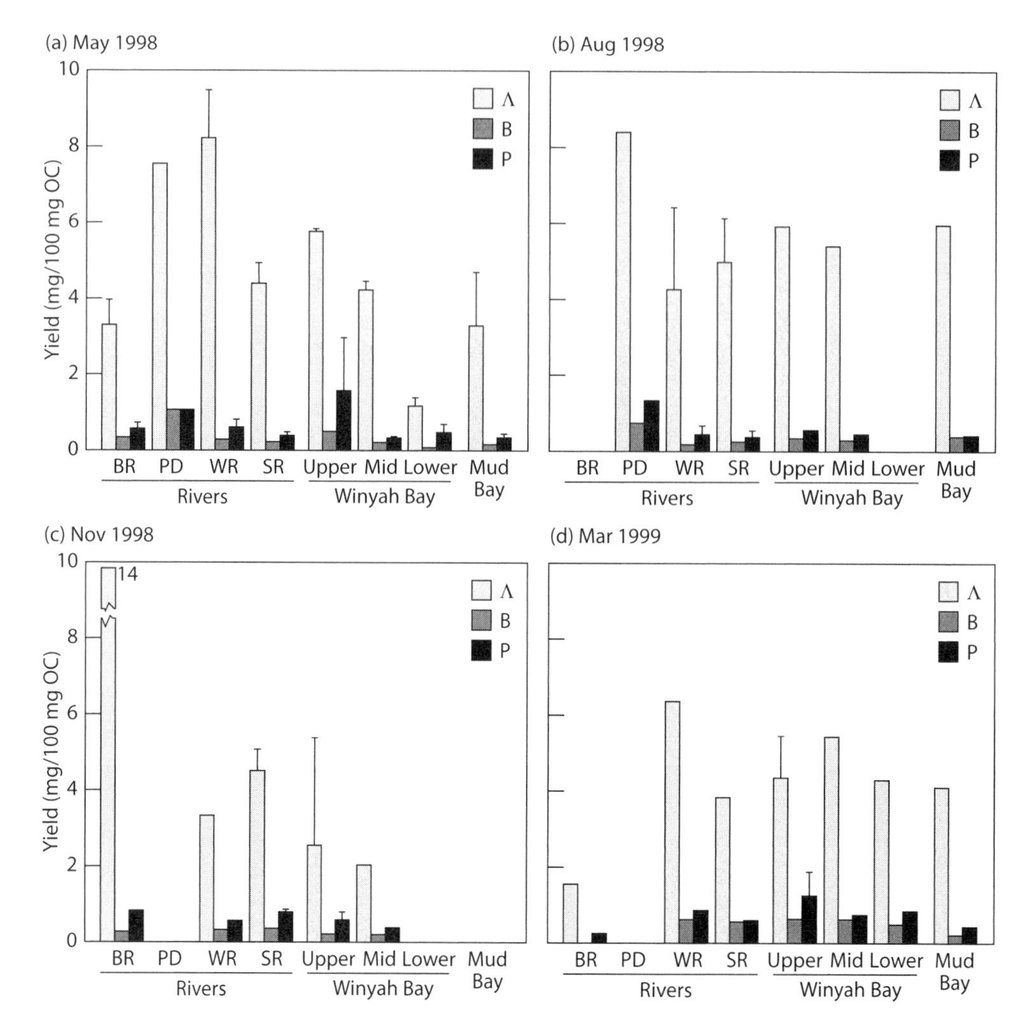

Figure 13.10. Recent work showing the relative importance of lignin-derived phenols (Λ) versus non-lignin-derived compounds, benzoic acids (B), and *p*-hydroxybenzenes (P), in sediments from rivers of the estuaries in the southeastern United States. The station codes describe the location of each collection site and are as follows: BR, Black River; PD, Pee Dee River; WR, Waccamaw River; SR, Sampit River. Upper, Middle, and Lower Winyah Bay stations and Mud Bay stations are also shown. (Adapted from Goñi et al., 2003.)

2000; Mitra et al., 2000). Temporal and spatial variability in the relative importance of such processes can also commonly result in nonconservative behavior of total DOM and concentrations of lignin. The potential importance of these different mechanisms in removing terrestrially derived DOM in estuarine-shelf regions may be responsible for the large deficit in terrigenous DOM in the global ocean (Hedges et al., 1997; Opsahl and Benner, 1997). In another study, chemical constituents (e.g., hydrolyzable neutral sugars and amino acids) in DOM from Russian rivers were not correlated with drainage basin vegetation, while lignin parameters, such as S/V were (fig. 13.11) (Opsahl et al., 1999; Lobbes et al., 2000; Dittmar and Kattner, 2003). These differences were explained by the universal signature of the more labile organic components in these and other soils around the world (Engbrodt, 2001). Nevertheless, it was concluded that since climate change will not alter the dominant components of DOM delivery to the Arctic Ocean from Russian rivers, the lignin signature would be expected to reflect changes in the drainage basin (Dittmar and Kattner, 2003).

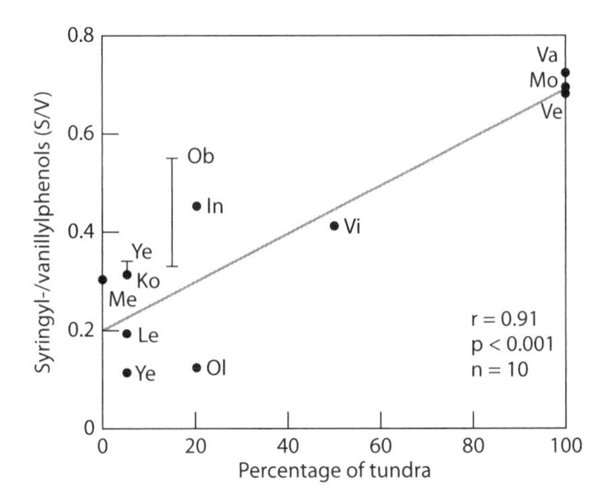

Figure 13.11. Relationship between syringyl/vanillyl phenols vs. percentages of tundra in Russia Rivers. While many other chemical constituents (e.g., hydrolyzable neutral sugars and amino acids) in DOM from Russian rivers were not correlated with drainage basin vegetation, lignin parameters, such as S/V, were. (Adapted from Dittmar and Kattner, 2003.)

13.4.3 Lignins and Cutins as Biomarkers in the Ocean

As mentioned earlier, the CuO method produces a suite of BCAs whose sources have not been well identified. A recent study further examined the sources of BCAs and their potential use as biomarkers in soils and marine sediments (Dickens et al., 2007). It was concluded that BCAs were useful tracers for charcoal in soils and sediments (fig. 13.12). Moreover, it was concluded that while some BCAs were produced by marine biota, 3,5Bd was not (fig. 13.12), and remained a valid biomarker for soil OC, as suggested by earlier studies (e.g., Louchouarn et al., 1999; Farella et al., 2001; Houel et al., 2006). In particular, 3,5Bd concentrations were significantly higher in terrestrial and aromatic organic material than marine sources (fig. 13.12). It was also found that 3,5Bd decreased with increasing distance offshore of the Washington coast (USA), as did Λ—only with significantly less magnitude (fig. 13.12). Conversely, the 3,5Bd to total vanillyl phenols ratio (3,5Bd/V), which has been suggested to be indicative of the extent of OM oxidation in soils (Houel et al., 2006), increased with increasing distance offshore. This ratio was also shown to be significantly correlated with the acid:aldehyde ratio of vanillyl phenols, which corroborated earlier work that demonstrated a more degraded lignin signature with increasing distance offshore the Washington coast (Hedges et al., 1988). Finally, the phenolic OH (OH) and nonphenolic OH (no-OH) ratio (OH/no-OH), which was shown to be high for all terrestrial, aromatic organics, and charcoal samples, also decreased with increasing distance offshore. This may provide another index for separating the relative importance of terrestrial versus marine sources, but further work is needed to better constrain this ratio (Dickens et al., 2007).

A recent study compared dissolved lignin concentrations among the Arctic, Atlantic, and Pacific ocean basins, in surface and deep waters (Hernes and Benner, 2006). In general, the highest and lowest dissolved lignin concentrations were found in the Arctic and Pacific, respectively (fig. 13.13) (Opsahl et al., 1999; Hernes and Benner, 2002, 2006). While most of the differences in total lignin concentrations in surface waters could be explained by a positive correlation with river inputs (fig. 13.13a), the Arctic system also had a shorter residence time and lower annual incident solar radiation, resulting in reduced photooxidation of DOM. Deep waters showed less variability in lignin concentrations, with the Arctic and Atlantic being similar due to their connection in North Atlantic bottom waters. The Pacific likely had the lowest concentrations of total lignin due to "conveyor-belt" deep water circulation between

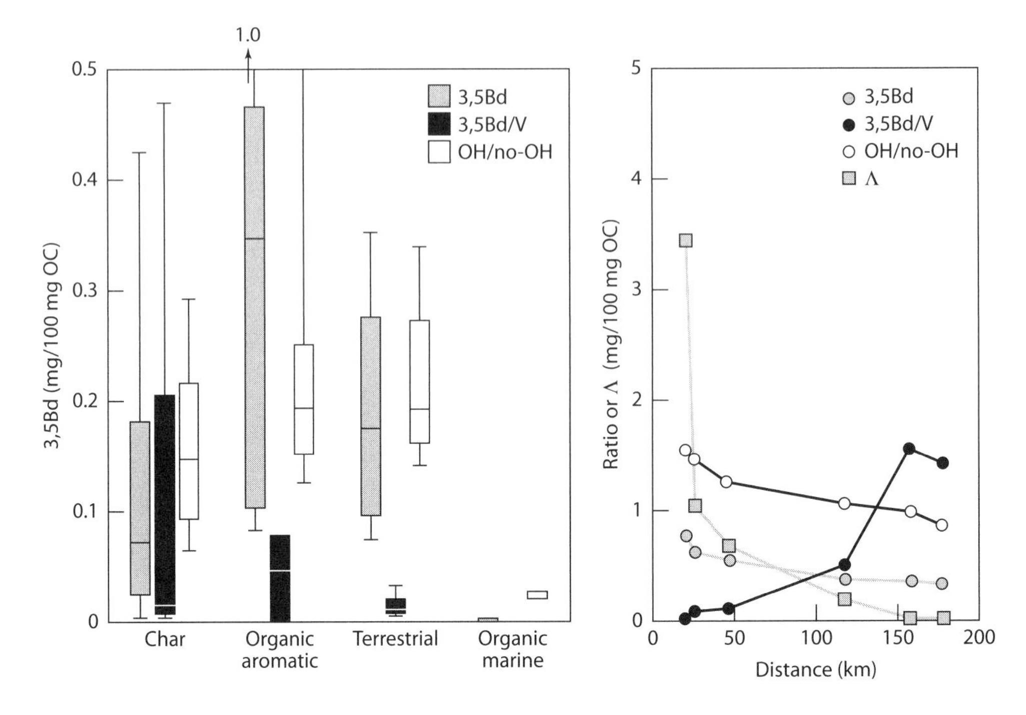

Figure 13.12. Result from a recent study that examined the sources of benzene carboxylic acids (BCAs) and their potential use as biomarkers in soils and marine sediments. Distance = distance offshore Washington coast (USA). (Adapted from Dickens et al., 2007.)

basins (Broecker and Peng, 1982). Pacific deep waters, which are downstream of Atlantic waters, typically had the oldest radiocarbon signature (Bauer et al., 2001) and most degraded signals, resulting in the greatest loss of lignin during transport. Lignin in HMW DOM showed similar patterns to that found in total DOM, despite the HMW fraction typically being more reactive (fig. 13.13b). The lowest S/V ratios in the Arctic were primarily believed to be due to source inputs of gymnosperms from boreal forests (fig. 13.13c) (Opsahl et al., 1999; Lobbes et al., 2000). Although the lignin losses were believed to be greater in the Pacific waters, the Ad/Al_v ratios were not significantly greater (fig. 13.13d), which could not be explained without further investigation.

The application of cutin-derived CuO products as potential biomarkers was first described in a set of papers that examined coastal sediments along the Washington coast (USA) (Goñi and Hedges, 1990a,b). Briefly, cutins are absent in nonvascular plants, in contrast to their presence in vascular nonwoody plant tissues. In particular, the terminally monohydroxylated cutin acids (e.g., 14-hydroxytetradecanoic [ω-C_{14}] and 16-hydroxyhexadecanoic [ω-C_{16}] acids) are found in lower plants (e.g., ferns and clubmosses), while the hydroxylated cutin acids are found in higher plants (e.g., angiosperms and gymnosperms) (Goñi and Hedges, 1990a,b). The dihydroxyhexadecanoic acid predominates in gymnosperms (table 13.1), whereas the 9,10,18-trihydroxyoctadecanoic acid (9,10, ω-C_{18}) dominates in angiosperms. Finally, other structural isomers, such as 8,16-, 9,16-, and 10,16-dihydroxyhexadecanoic acids, allow for further differentiation between monocotyledons and dicotyledons (Goñi and Hedges, 1990a). As discussed with many of the aforementioned biomarkers, the use of ratios in the application of cutin is critical for constraining sources. For example, the ω-$C_{16}/\Sigma C_{16}$ ratio has been used to differentiate between inputs from ferns/clubmosses (0.8 ± 0.1) and higher plants (0.1–0.2), shown for gymnosperms in table 13.1. The ω-$C_{14}\Sigma_{16}$ has also been used to separate lower vascular plants and nonwoody gymnosperms from angiosperms. In some cases elevated

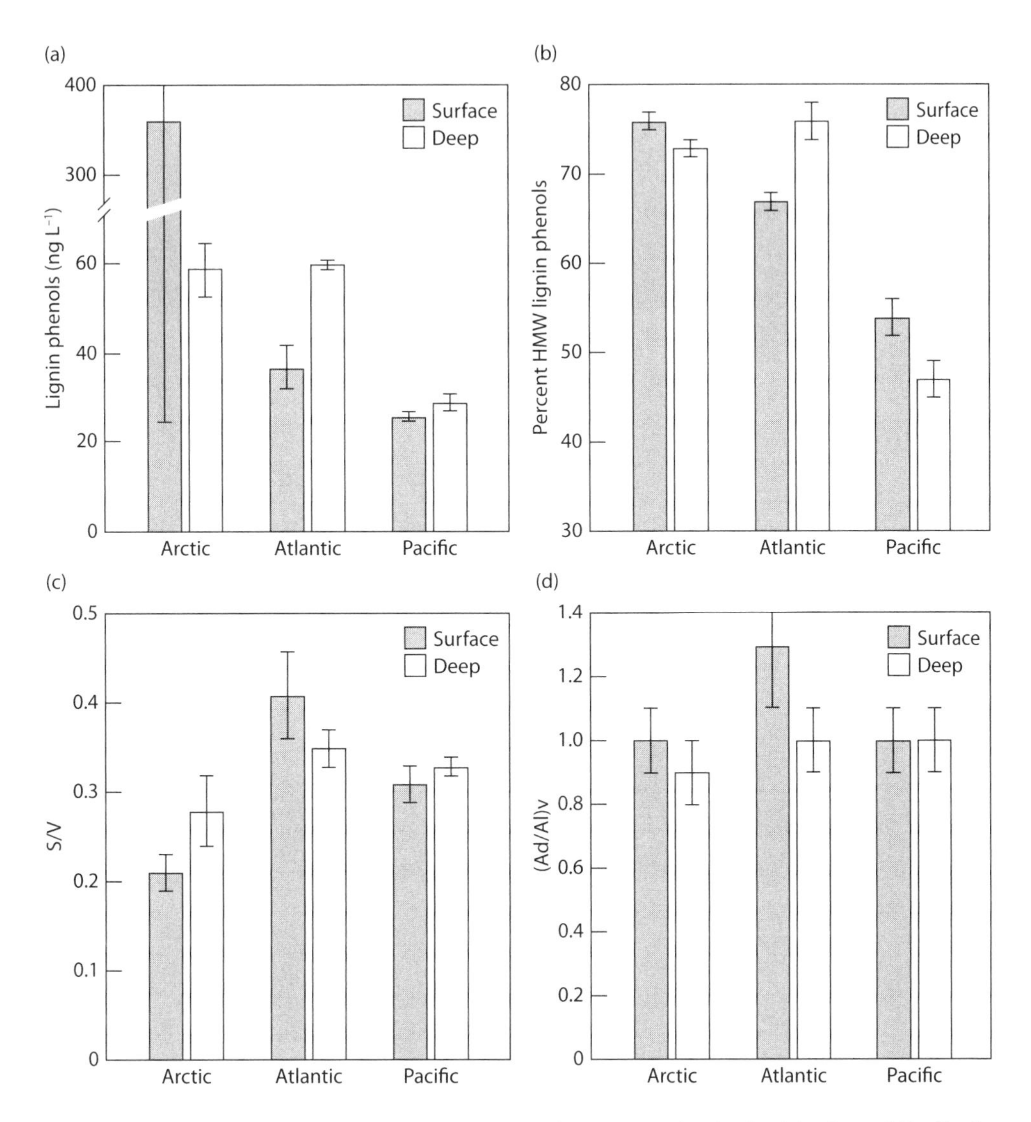

Figure 13.13. A comparison of dissolved lignin concentrations among the Arctic, Atlantic, and Pacific Ocean basins, in surface and deep waters. (Adapted from Hernes and Benner, 2006.)

ω-$C_{16}/\Sigma C_{16}$ ratios may, in part, be due to the presence of suberin, which contains high concentrations of monohydroxylated C_{16} and C_{18} acids (Kolattukudy, 1980). When examining the fate of the major cutin acid suite in fir-hemlock mixtures (ω-C_{16} and ω-C_{14}) (table 13.1, fig. 13.14) from the source tree (green), litter, and surface sediments in Dabob Bay (USA), we find the greatest decrease for C_{16} acids (fig. 13.14).

13.5 Summary

Interest in tracking vascular land plant C inputs to aquatic ecosystems has increased in recent years because of the dramatic effects that land-use change has had on the erosion of soils, vegetation in riparian zones, and loss of wetlands. While the CuO method has remained the most common method

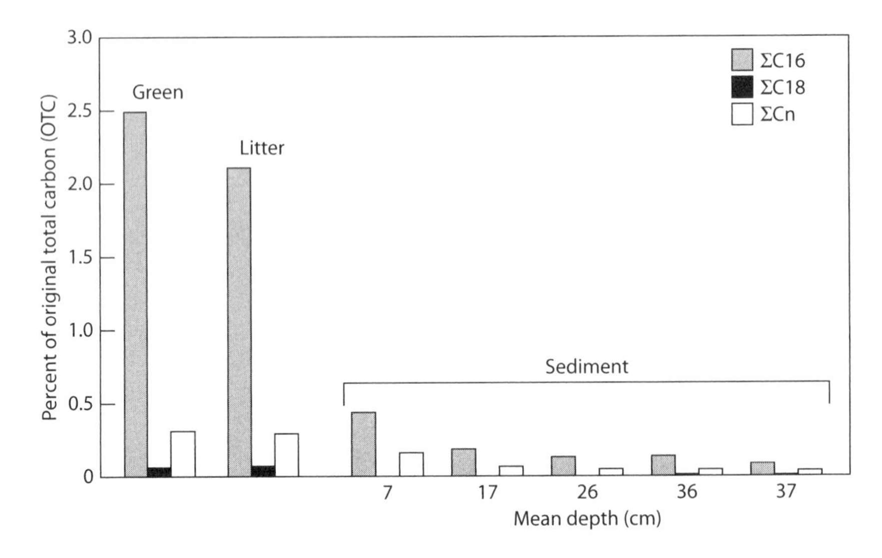

Figure 13.14. The fate of the major cutin acid suite in fir–hemlock mixtures (ω-C16 and ω-C14) from the source tree (green), litter, and surface sediments in Dabob Bay (USA). (Adapted from Goñi and Hedges, 1990.)

used for analyzing cutins and lignin, other new approaches have been introduced in the past 10 years. The TMAH and branched and isopreroid tetraether (BIT) methods (see chapter 11) provide other means for examining different soil components that cannot be examined as well using the CuO method. Lipids also provide another approach for examining terrestrial C inputs to aquatic ecosystems. The role of terrestrially derived C in POC and DOC of coastal ecosystems has gained a renewed interest in recent years, since new and improved analytical methods have documented that these sources are more important than earlier work suggested. For example, compound-specific isotopic analysis (CSIA) has helped in differentiating between sources of C_3 and C_4 plants as well as separating out the ages of specific source inputs. When considering the current rise in sea level along with the aforementioned land-use changes that will continue to result in enhanced inputs of terrestrially-derived materials to aquatic ecosystems, the role of chemical biomarkers in understanding C dynamics will remain vital.

14. Anthropogenic Markers

14.1 Introduction

This chapter will introduce readers to *anthropogenic* compounds representing a spectrum of sources, functional group compositions, physical behaviors, and environmental fates. We will draw from earlier sections of this book to discuss how the processes that control the behavior of naturally occurring organic compounds are similar to those influencing anthropogenic compounds and can aid in developing models to predict the behavior of contaminants in the environment. We will also introduce readers to the use of anthropogenic compounds as biomarkers (*anthropogenic markers*). Despite their often-deleterious consequences to aquatic environments and the organisms living within, contaminants can serve as useful proxies for understanding the sources, transformation, and fate of organic matter in the environment.

Anthropogenic markers are compounds that are produced by humans or human activity and are introduced either intentionally or unintentionally to the atmosphere or aquatic or sedimentary environments. These compounds may be natural or synthetic and include both harmless and toxic molecules. Generally speaking, anthropogenic compounds are classified as pollutants because they are mutagens or carcinogens and often have deleterious effects on human health or ecology. Some anthropogenic compounds are described as *xenobiotic*, referring to the fact that they do not have a natural source. There has been a dramatic increase in the number, amount, and applications of xenobiotic substances since World War II and the presence of these compounds in the environment provides unambiguous evidence of pollution. Generally, these compounds are present in the environment at trace concentrations, but their impact to humans and aquatic organisms is often disproportionate to their concentration due to their toxic, *teratogenic* or *carcinogenic* nature. Many anthropogenic compounds are also geochemically stable, resulting in a compounding of their environmental impacts due to their persistence. In this chapter, we introduce readers to current research focusing on traditional (i.e., introduced since World War II) classes of anthropogenic compounds, such as polychlorinated biphenyls (PCBs), chlorinated pesticides and herbicides, surfactants, polycyclic aromatic hydrocarbons (PAHs), and petroleum hydrocarbons. Additionally, with recent advances in mass spectrometry, new classes of anthropogenic compounds are now detectable in the environment. These *emerging contaminants* include estrogens, endocrine-disrupting compounds, antimicrobial agents, caffeine, phthalates, and polybrominated diphenyl ethers (PBDEs). Use of these compounds has increased dramatically in recent decades and modern methods of measurement suggest they are globally distributed (Schwarzenbach et al., 2006). PBDEs, for example, have been widely used as flame retardants since the 1970s and are found in both remote and highly urbanized regions (Hites, 2004).

Upon entering the environment, a variety of processes control contaminant transport, including exchange from one medium to another (e.g., volatilization, particle sorption–desorption), dilution in air or water, abiotic and biotic alteration to compounds that are more or less toxic, and their absorption and concentration by organism ingestion. Most of these processes and the rates at which they occur are influenced by environmental conditions, such as temperature, pH, salinity, *redox* conditions, wind and water velocity, and light availability. Additionally, the availability of electron acceptors, microbial community composition, and microbial activity influence the fate of contaminants (Braddock et al., 1995; Slater et al., 2005; Castle et al., 2006). The physicochemical properties of the compounds are also important, since the behavior and fate of anthropogenic compounds are determined by properties such as vapor pressure, aqueous solubility, partition coefficients, and their tendency to evaporate (e.g., Henry's law constant) (Schwarzenbach et al., 2003). Anthropogenic compounds exhibit a wide range in physical behaviors and bioaccumulation (or bioconcentration) factors (Mackay et al., 1992a, b; Cousins and Mackay, 2000; Cetin and Odabasi, 2005; Cetin et al., 2006). Bioconcentration refers to the increased concentration of a contaminant within the tissues of an organism relative to the concentration of the contaminant in the surrounding medium. To illustrate the spectrum in environmental behaviors for anthropogenic compounds, the physicochemical constants for some of the compounds discussed in this chapter are presented in table 14.1.

14.2 Petroleum Hydrocarbons

Since natural sources of hydrocarbons were covered in chapter 10, this chapter focuses exclusively on anthropogenic sources of this compound class. Petroleum hydrocarbons enter the environment in a variety of forms and through a range of processes (table 14.2). In some cases, the environmental consequences of petroleum hydrocarbons are immediate and apparent (e.g., spills), while in other cases they may be chronic and less obvious (e.g. non-point-source pollution). Generally speaking, chronic inputs are the dominant source of petroleum hydrocarbons, but spills and other catastrophic events often receive more attention. There are several excellent reviews of petroleum hydrocarbons in the environment and we refer readers to the recent book by Peters et al. (2005), as well as recently published review articles (Boehm and Page, 2007) and book chapters (Bayona and Albaiges, 2006).

Once in the environment, petroleum hydrocarbons exhibit a range of environmental effects due to their physical properties and behavior, bioavailability, and toxicity (table 14.3). While many hydrocarbons are persistent, others undergo *weathering*, the process by which compounds break down at the Earth's surface due to mechanical action and reaction with air, water, and organisms. Weathering is a selective process. As a result, the process is often described for a specific compound relative to a stable compound(s). For example, research shows that 17α, 21β(H)-hopane is conserved during weathering of petroleum. As a result, changes in the concentration of specific compounds due to weathering are often expressed relative to this compound using the following equation:

$$\% \text{ compound depleted} = [1 - (C_w/C_0)/H_0/H_w] \times 100 \quad (14.1)$$

Table 14.1
Physical constants for representative classes of anthropogenic organic contaminants at 25 °C

Compound or class	*Origin*	*Henry's law constant Pa m³mol⁻¹*	*Log K_{ow} (mol L^{-1} octanol)/(mol L^{-1} water)*	*Vapor pressure Pa*	*Bioconcentration factor log BCF*
Linear alkylbenzenes	Wastewater		6.9 to 9.3[1]		
Polycyclic aromatic hydrocarbons[2]	Combustion products	0.9 to 232.8	3.3 to 67.5	0.0885 to 197	1.5 to 5.2
Phthalate esters[2]	Plastics		3.57 to 8.18	1.3×10^{-5} to 0.22	
Bisphenol A[6]	Plastics		2.2 to 3.8		5.1 to 13.8
Polychlorinated biphenyls[2]	Capacitors and insulating fluids	1.7 to 151.4	3.9 to 8.26	3×10^{-5} to 0.6	2.1 to 7.1
Aroclor mixtures		5 to 900	4.1 to 7.5	2×10^{-4} to 2.0	0.1 to 5.5
Polybrominated diphenyl ethers	Flame retardants	0.04 to 4.83[3]	5.7 to 8.3[4]		
Chlorinated dibenzo-*p*-dioxins	Combustion products	0.7 to 11.7	4.3 to 8.2	9.53×10^{-7} to 0.51	1 to 6.3
DDT[7]	Pesticide	1.3	6.4[2]	4.7×10^{-4} to 1.6×10^{-3}	
Hexachloro-cyclohexane (HCH)	Agrochemical		3.8[1]	3×10^{-3} to 1.07×10^{-1}	
17β-Estradiol	Wastewater		4.01[5]		
Caffeine	Wastewater		−0.07[5]		

Note: Values are approximate since the constants can be derived by a variety of methods.
[1]Bayona and albaiges (2006).
[2]Date from Schwarzenbach et al. (1994), Mackay et al. (1992), or Cousins and Mackay (2000).
[3]Date from Cetin and Odabasi (2005).
[4]Braekevelt et al. (2003).
[5]Vallack et al. (1998).
[6]Staples et al. (1998).
[7]Bamford et al. (1998).

where the concentration of the compound in a weathered sample (C_w) is expressed relative to its initial concentration (C_0), normalized to the initial and weathered concentration of this hopane (H_0 and H_w). Generally, compound ratios that exhibit a large range provide useful tools for quantifying the extent of weathering while compound ratios that remain stable in the environment make good proxies for tracing hydrocarbon sources.

The broad field of *environmental forensics* has developed in response to the need to trace a wide variety of contaminants in the environment (Slater, 2003; Wang et al., 2006; Philp, 2007). Just as biomarkers can be used to trace the sources of natural organic matter in the environment, hydrocarbon "fingerprints" and their isotopic signatures ($\delta^{13}C$ and $\Delta^{14}C$) can be used to trace the origin of petroleum (Bence et al., 1996; Reddy et al., 2002b). Like other biomarkers, this information is most useful when the signatures are both source specific and environmentally stable. For example, $\delta^{13}C$ signatures and biomarkers provided powerful tools for distinguishing *Exxon Valdez* residues on the shoreline of Prince William Sound from tars that were linked to oil produced in Southern California (USA) (Kvenvolden et al., 1993; Bence et al., 1996). Residues from the *Exxon Valdez* had $\delta^{13}C$ signatures of -29.4 ± 0.1, while residues from California oils had $\delta^{13}C$ signatures of -23.7 ± 0.2, similar to those found at some sites on the shoreline of Prince William Sound (Alaska, USA). Additionally, triterpane biomarkers provided evidence that tar residues were from crude oil generated from the Monterey Formation (California, USA) rather than from *Exxon Valdez* oil (fig. 14.1). In this study, the presence of oleanane

Table 14.2
Natural and anthropogenic sources of oil pollution

			Environment		*Scale of impact*		
	Type of input	*Source of input*	*Hydrosphere*	*Atmosphere*	*Local*	*Regional*	*Global*
	Natural	Seeps, erosion of sediments or rocks	+	–	+	?	–
		Biosynthesis by marine organisms	+	–	+	+	+
Anthropogenic	Marine	Marine oil transportation (e.g. accidents, operational discharges from tankers)	+	–	+	+	?
		Marine nontanker shipping (operational, accidental, and illegal discharges)	+	–	+	?	–
		Offshore oil production (drilling discharges, accidents)	+	+	+	?	–
	Onshore	Sewage waters	+	–	+	+	?
		Oil terminals, pipelines, trucks	+	–	+	–	–
		Rivers, land runoff	+	–	+	+	?
	Combustion	Incomplete fuel combustion	–	+	+	+	?

Source: Adapted from Peters et al. (2005).

Table 14.3
Environmental effects of petroleum hydrocarbons

Group type	*Abundance in oil*	*Bioavailability*	*Persistence in tissues*	*Toxicity*
Aliphatic hydrocarbons	High	High	Low	Low
Cyclic hydrocarbons (e.g., cycloalkanes, cycloalkenes)	High	High	Low	Low
Aromatics, alkylaromatics, N- and S-heterocycles	High	High	Species-dependent	Species-dependent
Polars (e.g., acids, phenols, thiols, thiophenols)	Low (?)	Low (?)	Low	Low
Elements (e.g., sulfur, vanadium, nickel, iron)	Low	Low	Low	Low
Insoluble components (e.g., asphaltenes, resins, tar)	Low	Low	Low	Low

Source: From Peters et al. (2005).

and C_{28}- bisnorhopane indicated that the beach residues derived from a mature petroleum source similar to local oils from Southern Alaska as well as those from the Monterey Formation. Oils from the Monterey Formation contained both oleanane and C_{28}-bisnorhopane, while North Slope (Alaska) oils contained only oleanane. However, these compounds were entirely absent from the *Exxon Valdez* crude (fig. 14.1a). As a result, this study concluded that the tars were probably released during an event that occurred 25 years before the *Exxon Valdez* spill.

Other features, such as the unresolved complex mixture (UCM), can be used to characterize oil contamination in the environment. In a recent study, Reddy and coworkers used gas chromatography

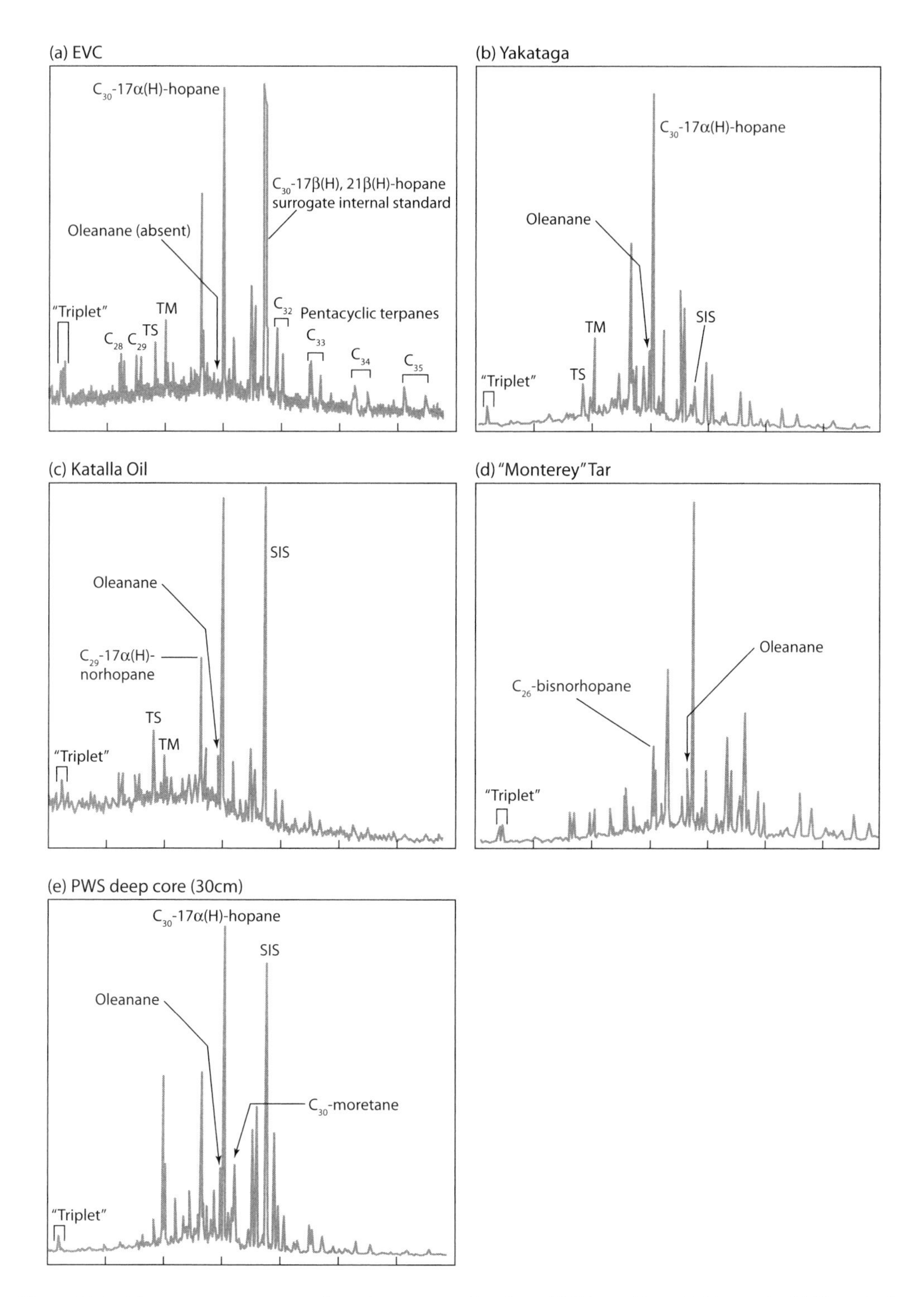

Figure 14.1. Chromatograms comparing triterpane profiles (m/z 191) from *Exxon Valdez* crude (EVC), with oils from Gulf of Alaska seeps (Yakataga and Katalla) and Monterey tars with hydrocarbons extracted from Prince William Sound sediments. (Adapted from Kvenvolden et al.,1993 and Bence et al., 1996.)

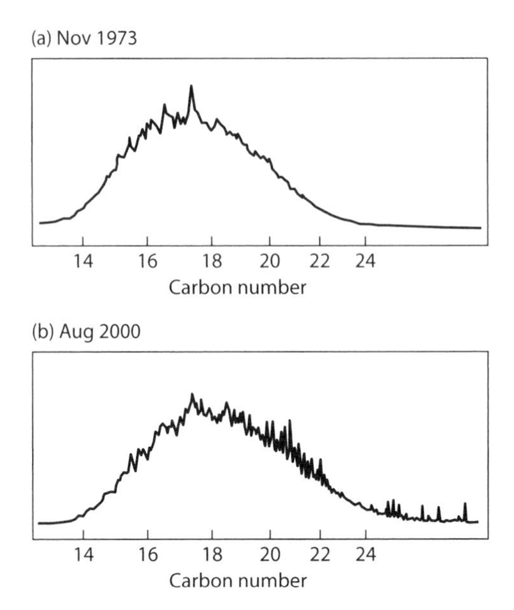

Figure 14.2. Chromatograms comparing size of the unresolved complex mixture (UCM) in marsh sediments collected following the Falmouth oil spill in 1973 and nearly thirty years later. (Adapted from Reddy et al., 2001.)

and two-dimensional gas chromatography to examine whether compositional changes in the UCM had occurred over a thirty-year time period since an oil spill took place in Falmouth, Massachusetts (USA) (Reddy et al., 2002b). This study revealed that concentrations of the UCM were similar to those observed following the spill (fig. 14.2). In addition, using two-dimensional gas chromatography, the investigators showed that while *n*-alkanes had degraded over the time since the spill, pristane and phytane, as well as other branched hydrocarbons, had persisted in the sediments. Consistent with these findings, a recent study in Lake Washington (USA) also documented the presence of hydrocarbons deposited approximately 25 years before the study was conducted (Wakeham et al., 2004), highlighting the persistence of petroleum hydrocarbons in the environment.

Another tool that has proved useful for characterizing the fate of petroleum in the environment is the use of radiocarbon isotopes. Since radiocarbon is absent in petroleum, this isotope can be used as an *inverse tracer* to examine the utilization of petrogenic carbon by microbial communities (Slater et al., 2005; Wakeham et al., 2006). Slater et al. (2005) examined $\Delta^{14}C$ signatures of phospholipid linked fatty acids (PLFA) to investigate whether sediment microbes incorporated petroleum residues into their biomass. They found that $\Delta^{14}C$ signatures of PLFA matched those of sediment organic matter, suggesting that microbes were assimilating labile sources of natural organic matter rather than petroleum. In contrast, Wakeham and colleagues (2006) found that PLFA derived from eubacteria were depleted in radiocarbon at a marsh site in Georgia (USA) contaminated with petroleum relative to an uncontaminated site. These authors estimated that 6 to 10% of carbon in bacterial PLFAs at the oiled site could derive from petroleum residues. These results demonstrate that weathered petroleum may contain components of sufficient lability to be a carbon source for sediment microorganisms. However, given the contradictory findings from these studies, factors such as the composition of the microbial community and availability of labile sources of carbon must be considered when trying to predict whether sediment microbes will use petroleum components.

Radiocarbon was also used in a recent study to investigate the associations between fossil fuel-derived hydrophobic organic contaminants (HOCs) and different pools of natural organic matter (White et al., 2008). Using approaches that combined isotope mass balance with gas

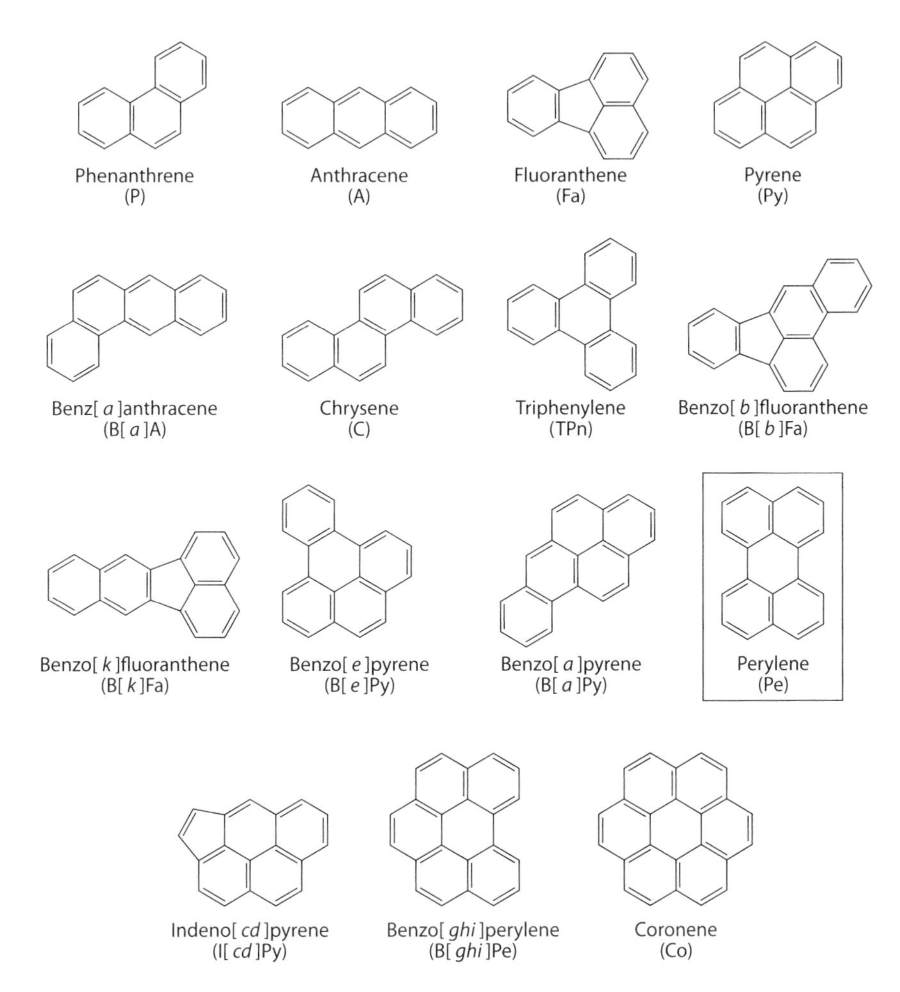

Figure 14.3. Structures of polycyclic aromatic hydrocarbons showing fused benzene rings. Anthropogenic sources contribute to the high concentrations of these compounds in the environment. However, most compounds are also produced at low concentrations by natural combustion processes, such as forest fires and biomass burning. Perylene, identified in the box, is derived from diagenetic processes. (Adapted from Killops and Killops, 2005.)

chromatography–mass spectrometry, this study used radiocarbon to trace HOCs (radiocarbon-free) within chemically defined fractions of contemporary OM in marine and freshwater sediments as well as in one dust sample and one soil sample. Samples were sequentially treated by solvent extraction followed by saponification. This study showed that most of the HOCs were associated with the solvent-extractable pool, suggesting they may be mobile and bioavailable.

14.3 Polycyclic Aromatic Hydrocarbons

Polycyclic aromatic hydrocarbons (PAHs) are a class of aromatic hydrocarbons that contain multiple fused benzene rings (fig. 14.3). There are three general categories of PAHs: (1) *petrogenic* compounds that are derived from slow maturation of organic matter under geothermal conditions, (2) *pyrogenic* compounds, which are derived from incomplete combustion of recent (e.g., biomass burning) and fossil (e.g., coal) organic matter, and (3) *diagenetic* products derived from biogenic precursors (Killops and Killops, 2005; Walker et al., 2005). While natural sources exist, anthropogenic activities, including combustion of fossil fuels and release of noncombusted petroleum through leaks and spills, constitute the primary source of PAHs to the environment. Primary sources of PAHs include fireplaces, wood-burning stoves, heating furnaces, automobiles, oils spills, and asphalt and other road-building materials. Secondary sources include aeolian transport, river runoff, and atmospheric deposition (wet and dry) (Dickhut and Gustafson, 1995; Dickhut et al., 2000). Once in the environment, PAHs may be transformed by abiotic processes such as *photolysis* (Plata et al., 2008), decomposed by microbial processes (Coates et al., 1996, 1997; Hayes et al., 1999; Rothermich et al., 2002), or ingested by organisms and passed up the food chain (MacDonald et al., 2000). Ultimately, because of the hydrophobic nature of these contaminants, the dominant process influencing most PAHs is sediment deposition and burial (Chiou et al., 1998).

Many PAHs (or ratios of PAHs) can be used as proxies for both the source and weathering of petroleum in the environment. Weathering parameters are based on compounds that have different ratios of evaporation, solubility, adsorption onto clay and mineral surface, photooxidation, and/or biodegradation. Generally speaking, the number of rings and degree of *alkylation* are the major parameters that influence the weathering of PAHs. As compounds increase in size and angularity, the hydrophobility and electrochemical stability of PAHs increase (Zander, 1983; Harvey, 1997; Kanaly and Harayama, 2000). Ultimately, the stability and hydrophobicity of PAH molecules contribute to their persistence in the environment. For example, naphthalenes, which consist of two fused benzene rings, are easily weathered due to their evaporation and solubility, while high molecular PAHs with four or more benzene rings tend to be persistent in the environment. In addition, alkylation provides increased stability, as evidenced by preferential preservation of isomers with C_{3+} alkyl groups over time.

In addition to structural features, associations between PAHs and organic and inorganic phases influence their partitioning and fate. PAHs may be associated with dissolved, *colloidal*, particulate, and sedimentary phases and the organic content and composition of these phases influences PAH distributions. For example, recent studies have noted changes in the colloid–water partition coeffficients (K_{coc}) for pyrene dependent on the dominant sources of organic matter in the water column (Gustafsson et al., 2001). In this study, the pyrene K_{coc} decreased approximately fourfold from values of $12.9 \pm 0.9 \times 10^3$ L_w/kg_{oc}^{-1} in the terrestrial runoff dominated regime to values around $2.9 \pm 0.7 \times 10^3$ L_w/kg_{oc}^{-1}, when phytoplankton production dominated the sources of organic matter in surface waters. Other studies have suggested that PAHs that enter the water column through air–sea exchange may associate with phytoplankton in surface waters, providing a vector for transferring PAHs to the aquatic food web rather than deposition to sediments (Arzayus et al., 2001; Countway et al., 2003). This is consistent with a study conducted in Chesapeake Bay (USA) showing differences in the sources of PAH associated with water column phases vs. those associated with sediment phases (Dickhut et al., 2000). Automotive sources of PAHs dominated particulate matter in the water column, while coal was the dominant source in sediments. In sediments and soils, a number of studies have demonstrated that carbonaceous materials, such as soot, *black carbon,* and kerogen, influence the partitioning of PAHs (Ribes et al., 2003; Cornelissen et al., 2005). The influence of soot phase on PAHs behavior in the environment has implications for the bioavailability and this class of contaminants and should be incorporated into sediment quality criteria.

Several tools have been developed to apportion PAHs between their potential sources, including use of PAH isomer ratios, as well as stable carbon and radiocarbon isotope compositions of PAHs. Petrogenic sources usually have higher proportions of alkylated homologues relative to their parent PAH (e.g., methylphenanthrenes [MP] > phenanthrene [P]), while combustion sources are usually depleted in alkyl groups since higher temperature combustion processes form PAHs through condensation reactions and/or ring fusion (Page et al., 1996; Dickhut et al., 2000; Walker et al., 2005). As a result, MP/P ratios > 1 are indicative of petrogenic PAHs, while ratios < 1 indicate pyrogenic

Table 14.4
PAH isomer ratios for major emission sources

Source	*BbA/chrysene*	*BbF/BkF*	*BaP/BeP*	*IP/BghiP*
Automobiles	0.53 (0.06)	1.26 (0.19)	0.88 (0.13)	0.33 (0.06)
Coal/coke	1.11 (0.06)	3.70 (0.17)	1.48 (0.03)	1.09 (0.03)
Wood	0.79 (0.13)	0.92 (0.16)	1.52 (0.19)	0.28 (0.05)
Smelters	0.60 (0.06)	2.69 (0.20)	0.81 (0.04)	1.03 (0.15)

Note: BaA, benz[*a*]anthracene; BbF, benzo[*b*]fluoranthene; BkF, benzo[*k*]fluoranthene; BaP, benzo[*a*]pyrene; BeP, benzo[*e*]pyrene; IP, indeno(l23cd) pyrene; and BghiP, benzo(*ghi*)perylene.
Source: Dickhut et al. (2000), *Environmental Science & Technology* 34:4635–4640.

sources (Bouloubassi and Saliot, 1993). Specific compounds (Budzinski et al., 1997; Boehm et al., 2001), as well as ratios of PAH isomers (Dickhut et al., 2000; Walker and Dickhut, 2001), can be used to differentiate between different combustion sources (table 14.4). In addition, statistical methods such as linear regression analysis and principal components analysis have been used widely to differentiate between different sources of PAH to the environment (Yunker et al., 1995; Burns et al., 1997; Arzayus et al., 2001). These tools have demonstrated that PAHs often derive from multiple sources, often with different modes of delivery, which determine the behavior and fate of PAH in the environment.

Compound-specific isotope analysis (CSIA) of PAH provides an additional tool for identifying different sources of PAH (O'Malley et al., 1994; Ballentine et al., 1996; Lichtfouse et al., 1997; Walker et al., 2005). O'Malley and colleagues (1994) pioneered this approach by demonstrating a high precision in isotopic values as well as the fidelity of isotopic signature after vaporization, photolytic decomposition, and microbial degradation. O'Malley et al. (1994) also pointed out that co-elution and contributions from the UCM can add uncertainty to isotopic signatures. More recently, investigators have adopted multiple cleanup steps to increase the accuracy and precision of CSIA analyses so that co-elution of the UCM with targeted peaks is minimized. For example, Walker et al. (2005) examined $\delta^{13}C$ isotopic compositions of PAH associated with sediment samples collected from the highly contaminated Elizabeth River Estuary, Virginia (USA). This study yielded precise $\delta^{13}C$ compositions of PAH, revealing that a former wood-treatment facility and coal use (either historical or current) were the predominant sources of PAHs (fig. 14.4).

More recently, radiocarbon analyses have been used as a tool for apportioning sources of PAH in environmental samples (Reddy et al., 2002c; Kanke et al., 2004; Mandalakis et al., 2004; Zencak et al., 2007). Differences between in the radiocarbon age of combustion products derived from fossil fuels (C-14-free) versus those from modern biomass (contemporary C-14) provide a useful tool for differentiating between sources of PAH. Radiocarbon provides a tool for differentiating between sources of PAH in cases of overlapping $\delta^{13}C$ signatures (Mandalakis et al., 2004). Radiocarbon can also be used in combination with source information derived from CSIA and isomer ratios. In a recent study, radiocarbon revealed that nonfossil material contributed significantly (35–65%) to atmospheric PAH pollution, even in urban and industrialized areas (Zencak et al., 2007). In another study, radiocarbon analysis of PAH extracted from a sediment core showed that the fraction of modern carbon values (Pies et al., 2008) of individual PAHs (phenanthrene, alkylphenanthrenes, fluoranthene, pyrene, and benz[*a*]anthracene) ranged from 0.06 to 0.21, suggesting that these compounds were derived mostly from fossil fuel combustion (Kanke et al., 2004). This study also revealed differences in radiocarbon compositions of individual PAH throughout the sediment core, leading the authors to conclude that phenanthrene received a greater contribution from biomass burning than alkylphenanthrenes throughout the core. Overall, these studies indicate that radiocarbon provides a useful tool for identifying the sources and origin of PAHs in the environment.

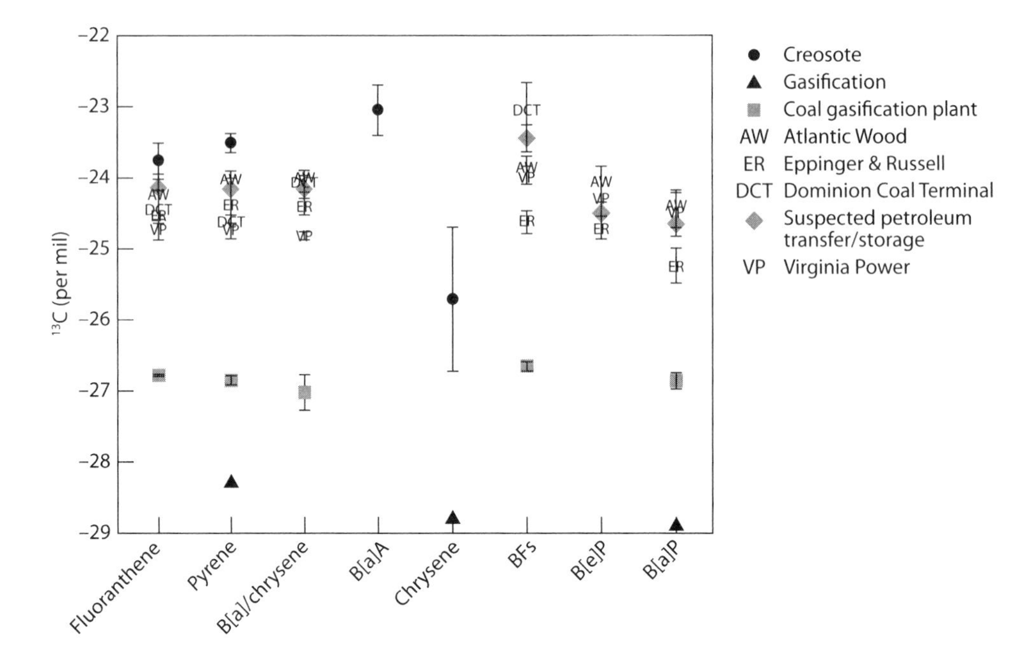

Figure 14.4. $\delta^{13}C$ values for individual PAHs from literature sources and from samples collected from Elizabeth River, VA (USA). (Adapted from Walker et al., 2005.)

14.4 Organochlorine Pesticides

Organochlorine pesticides include a broad spectrum of compounds with a wide range of behaviors. For this reason, we will focus on a few examples of organochlorine pesticides, with an emphasis on compounds that are widely distributed and persistent in the environment. Specific properties of these compounds, including their hydrophobic nature, scavenging by particulate matter, and accumulation in sediments, have contributed to their persistence in the environment as "legacy" contaminants. While the Stockholm Convention (ratified in 2001) banned the use of several organochlorine pesticides, large stores of organochlorine pesticides are still present in the environment in sediments. As a result, these contaminants continue to be available to organisms living in sediments and can be released to the environment when sediments are disturbed by erosion, resuspension, and dredging. It is worth noting here that many of the pesticides that have been developed more recently, break down more easily and are less damaging to the environment.

14.4.1 DDT and Related Compounds

DDT (1,1,1-trichloro-2,2-bis(*p*-chlorophenyl)ethane) is an organochlorine pesticide used for the control of insects, with no natural sources in the environment (WHO, 1979). The effectiveness of DDT for insect control was recognized during World War II, and after 1945 DDT was used extensively as an insecticide in agriculture (Brooks 1979). DDE (1,1-dichloro-2,2-bis(*p*-chlorophenyl)ethylene) and DDD (1,1-dichloro-2,2-bis(*p*-chlorophenyl) ethane) were minor contaminants introduced by the manufacturing of DDT, but most of the DDE and DDD present in the environment today derives from the breakdown of DDT. Environmental problems associated with DDT and its persistence in the environment resulted in DDT being banned in the United States in 1972. However, DDT was used in the United Kingdom until the 1980s and is still used in many countries in tropical regions today.

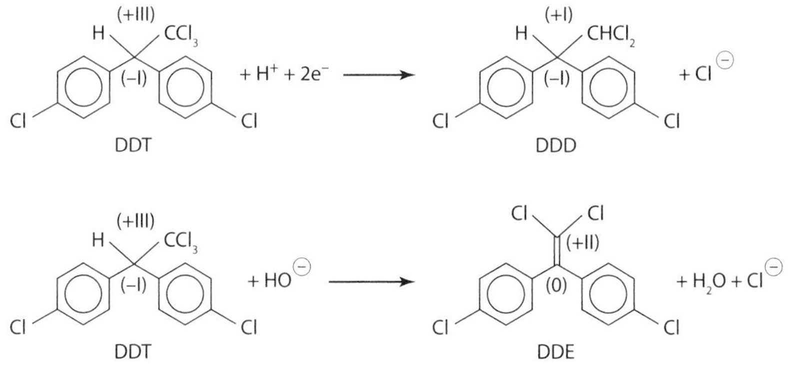

Figure 14.5. Pathways showing DDT degradation to DDD and DDE. (From Schwarzenbach et al., 1993.)

DDT is thought to degrade to DDD and DDE by microbial processes, primarily under anaerobic conditions, with the dominant pathway being reductive dechlorination (fig. 14.5) (Schwarzenbach et al., 1993; Killops and Killops, 2005). Following reductive dechlorination, the CCl_3 group on DDT can be converted to a COOH group, yielding 2,2-di(*p*-chlorophenyl)acetic acid (DAA). Since DAA is polar and water soluble, its behavior and fate in the environment differs from that of DDD and DDE. DDD and DDE, the nonpolar breakdown products of DDT, have similar chemical and physical properties to DDT, causing them to be highly persistent in the environment (WHO, 1979). As a result, concentrations of this class of contaminants are often reported as DDTr, or the sum of the DDT and its decay product residues (sum of DDT, DDD, and DDE contents). DDTr have extremely low solubility in water but are relatively soluble in organic solvents. Thus, DDTr in soils are associated with the organic matter common in fine-grained sediment. Removal of DDTr from agricultural lands occurs primarily when soil particles from DDT-sprayed fields are mobilized from agricultural lands during floods and washed into the drainage ditches and ultimately into the local rivers. Riverine transport is the primary vector by which DDTr enters the ocean.

While the persistence of DDTr has many negative consequences for the environment, the well-established use patterns for this class of compounds and their geochemical stability have made them useful tracers for sediment accumulation and transport (Bopp et al., 1982; Paull et al., 2002; Canuel et al., 2009). For example, Paull et al. (2002) used DDT and its degradation products as a tracer for the transport of fine sediment from coastal agricultural lands into the deep sea. In a recent study in the Sacramento–San Joaquin River Delta in California (USA), Canuel et al. (2009) used a variety of anthropogenic tracers, including total DDE, to quantify sediment and organic carbon accumulation rates. This study documented changes in the amount of sediment and carbon delivered to the delta over the time period from the 1940s to present and noted a marked reduction in sediment and carbon accumulation since the 1970s, coincident with changes in land use and completion of several large reservoir projects on California.

14.4.2 Lindane (*γ-Hexachlorocyclohexane*)

Lindane is a chlorinated pesticide that is similar to DDT in its behavior and persistence in the environment as well as its toxicological properties. There are two main forms of hexachlorocyclohexane HCH: technical HCH is a mixture of three isomers of HCH (55–80% α-HCH, 5–14% β-HCH and 8–15% γ-HCH), while lindane is essentially pure γ-HCH. Lindane has been used widely as an insecticide, since its insecticidal properties were discovered in the 1940s (Bayona and Albaiges, 2006). Like DDT, there has been a gradual trend toward the use of less persistent and less bioaccumulative pesticides, such as lindane. However, in some developing countries, especially in tropical regions, organochlorine pesticides are still used in agriculture and for other purposes. Since HCH are water soluble, they are less likely to bioaccumulate than DDT and its products (Vallack et al., 1998). However, the high vapor pressure of HCH makes it amenable to atmospheric transport over long ranges, resulting in the ubiquitous distribution of HCH in the environment as well as transport to regions that are highly sensitive, such as polar regions (Vallack et al., 1998 and references therein).

An area of active research is the study of enantiomer selectivity of chiral pollutants and its impact on pollutant exposure and fate. Several studies have investigated the chirality of chlorinated pesticides such as *o, p*'-DDT, *o, p*'-DDD, α-HCH (hexachlorocyclohexane), and *cis*- and *trans*-chlordane (Garrison et al., 2006 and references therein). Chirality can aid in elucidating the pathways influencing the fate of persistent organic pollutants, since abiotic reactions are not enantiospecific while biotic reactions (metabolism, microbial transformations, etc.) generally prefer a specific enantiomer. Interestingly, Padma et al. (2003) observed that enantiomer ratios (ERs) of α-HCH in surface water in the York River estuary, Virginia (USA), were inversely related to α-HCH concentrations, with equal abundances of both isomers (racemic mixtures) at the head of the estuary, where bacterioplankton activity was high and α-HCH concentrations were low, relative to at the mouth of the estuary where the microbial activity was lower. Enantiomer ratios have also been informative in elucidating the processes influencing the sources and distribution of pollutants in transport studies (Garrison et al., 2006). Padma and Dickhut (2002), for example, observed that seasonal changes in the relative importance of precipitation and runoff, volatilization, and microbial degradation controlled the fate and transport of HCHs in a temperate estuary. Garrison and colleagues (2006) provide several suggestions for research on the environmental occurrence, fate, and exposure of chiral compounds that should be considered in risk assessments. They recommend that research is needed in two areas: (1) investigations of the potential for target-inactive enantiomers to produce unintended effects on nontarget species, and (2) studies that determine how environmental conditions affect the relative persistence of enantiomers.

14.5 Polychlorinated Biphenyls

Properties of polychlorinated biphenyls (PCBs) such as their high chemical and thermal stabilities, electrical resistance, and low volatility contributed to their widespread use as dielectric fluids in capacitors and transformers, lubricants, and hydraulic fluids. Commercial production of PCBs began in the United States in 1929, with 1.2 Mt of PCBs produced between and 1929 and 1986. Most of this production occurred in the northern hemisphere. Environmental restrictions were imposed in the 1970s and production of PCBs in most parts of the world ended in 1986 (Breivik et al., 2002). Given the persistent nature of this class of compounds, 65% of this production is still in use or can be found in landfills. Like other classes of contaminants, PCBs enter the environment by direct and indirect means. These compounds are distributed globally through atmospheric transport processes, similar to other legacy persistent organic pollutants (POPs) (Dickhut and Gustafson, 1995; Dachs et al., 2002; Lohmann et al., 2007). As a result, PCBs are found in sediments in coastal regions (Jonsson et al., 2003) as well as remote regions, including the Arctic and Antarctic (Risebrough et al., 1976; Macdonald et al., 2000).

PCBs vary in the number and positions of chlorine atoms (table 14.5; fig. 14.6a), and it is the substitution of chlorine atoms that controls the persistence of this class of contaminants in the environment. Although a total of 209 possible congeners exist in theory, approximately 130 are likely to occur in commercial formulations. The PCB congeners with chlorine substitutions in 2,4,5-, 2,3,5-, and 2,3,6-positions have the longest half-lifes, which range from 3 weeks to 2 years in air to

Table 14.5
PCB formulae and the number of congeners associated with different levels of chlorine substitution

PCB formula	*Number of congeners*
$C_{12}H_9Cl$	3
$C_{12}H_8Cl_2$	12
$C_{12}H_7Cl_3$	24
$C_{12}H_6Cl_4$	42
$C_{12}H_5Cl_5$	46
$C_{12}H_4Cl_6$	42
$C_{12}H_3Cl_7$	24
$C_{12}H_2Cl_8$	12
$C_{12}HCl_9$	3
$C_{12}Cl_{10}$	1
Total	209

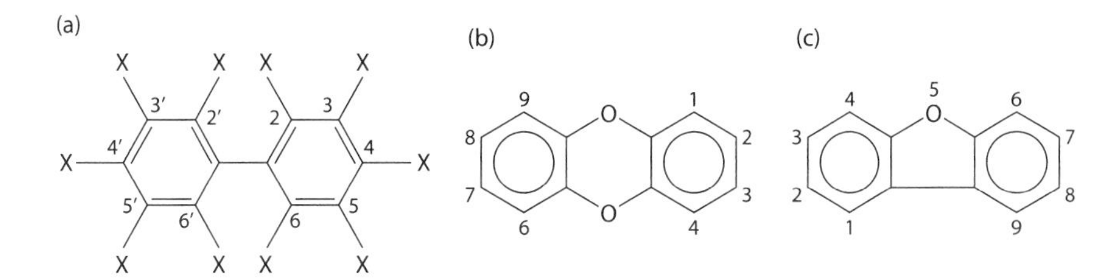

Figure 14.6. General structure for polychlorinated biphenyls (a), dibenzo-*p*-dioxin (b), and dibenzofuran (c), showing numbering scheme for carbon atoms (numbers) and positions where chlorine atoms can be substituted. See table 14.5 for number of congeners associated with different levels of chlorination of PCBs.

more than 6 years in aerobic sediments (Bayona and Albaiges, 2006). PCBs degrade by two distinct biological processes in the environment: aerobic oxidative processes and anaerobic reductive processes (Abramowicz, 1995). These processes occur alone or may be coupled to one another. Anaerobic dechlorination of PCBs, for example, results in the conversion of highly chlorinated PCBs to lightly chlorinated ortho-enriched congeners that can subsequently be degraded by a wide range of aerobic bacteria.

State-of-the-art methods for identifying the sources of PCBs and their fate in the environment include the use of compound-specific isotope analysis (Yanik et al., 2003; Horii et al., 2005a,b). Horii et al. (2005) examined carbon isotopic ratios ($\delta^{13}C$) of chlorobiphenyl (CB) congeners in several technical preparations of PCBs using two-dimensional gas chromatography–combustion furnace–isotope ratio mass spectrometry (2 DGC-IRMS). This study showed that PCB preparations with similar chlorine content, but different geographical origin, exhibited a range in $\delta^{13}C$ values. PCB preparations from Eastern European countries had $\delta^{13}C$ values ranging from -34.4 (Delors) to -22.0 parts per thousand (Sovol). PCB mixtures also showed increased depletion in ^{13}C with increasing chlorine content. Results from this study suggest that $\delta^{13}C$ values of technical mixtures of PCBs may be useful for identifying the sources of these compounds in the environment.

Chlorine isotopes may provide an additional tool for tracing the sources and fate of anthropogenic compounds. This area of study is still in its infancy since fractionation factors (Hofstetter et al., 2007) and the effects of enzyme-catalyzed chlorination on isotopic signatures have only recently been

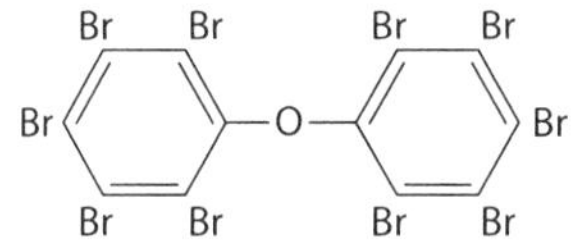

Figure 14.7. Example structure of a polybrominated diphenyl ether (PBDE).

elucidated (Reddy et al., 2002a). Radiocarbon isotopes provide another evolving tool for resolving the sources of chlorinated contaminants and their products in the environment (Reddy et al., 2004). The premise for this approach is that industrial compounds are manufactured from petrochemicals, which are ^{14}C-free, while natural compounds should have "modern" ^{14}C signatures. Reddy and colleagues suggest that the use of C and Cl isotopes may prove more valuable when these isotopes are used in combination with one another rather than as single isotope tools.

14.6 Polybrominated Diphenyl Ethers

Polybrominated diphenyl ethers (PBDEs) have become one of the most ubiquitous classes of contaminants in the environment (e.g., see recent reviews by DeWit (2002) and Hites (2004). These compounds are used extensively as flame retardants in many types of consumer products and their use has increased dramatically in recent decades (Hale et al., 2002, 2006). Concentrations of PBDEs in human blood, milk, and tissues have increased exponentially by a factor of ~100 during the last 30 yr, with a higher rate of increase in the United States (~35 ng g lipid^{-1}) than in Europe (~2 ng g lipid^{-1}).

PBDEs are commercially available as three products: the penta-, octa-, and deca-products (see representative structure in fig. 14.7). The penta-product contains 2, 2′, 4, 4′-tetrabromodiphenyl ether (BDE-47), 2, 2′, 4, 4′, 5-pentabromodiphenyl ether (BDE-99), 2, 2′, 4, 4′, 6-pentabromodiphenyl ether (BDE-100), 2, 2′, 4, 4′, 5, 5′-hexabromodiphenyl ether (BDE-153), and 2, 2′, 4, 4′, 5, 6'-tetrabromodiphenyl ether (BDE-154), in a ratio of about 9:12: 2:1:1 (Hites, 2004). The octa-product contains several hexa- to nonabrominated congeners and the deca-product is almost entirely composed of decabromodiphenyl ether (BDE-209). In general, environmental concentrations of BDE-209 appear to be increasing, while penta-BDE burdens in Europe may have peaked.

PBDEs have been found in air (Hoh and Hites, 2005; Venier and Hites, 2008), water (Oros et al., 2005), fish (Boon et al., 2002; Jacobs et al., 2002), birds, marine mammals (Stapleton et al., 2006), and people. The general trend is one of increasing concentrations of these compounds over time (fig. 14.8). Concentrations of PBDEs are now higher than PCBs in many environmental samples (Hale et al., 2006). These contaminants exhibit a range of physical–chemical behaviors, which influence their transport and fate in the environment (table 14.1). Generally, the more brominated compounds are thought to accumulate in sediments near emission sites, while compounds that are less brominated are more volatile and water soluble, contributing to their *bioaccumulation*. However, the transport and fate of these contaminants is still poorly understood and is an active area of research (Dachs et al., 2002; Lohmann et al., 2007). Volatile PBDEs are generally thought to dominate in the vapor phase, and may be carried to remote locations by long-range atmospheric transport (Chiuchiolo et al., 2004). In contrast, other PBDEs (e.g., BDE-209) predominate on particulates. A wide variety of compounds used as flame retardants, including PBDEs, have been detected in sewage sludges, which may contribute to contamination of soils (Hale et al., 2006). Sediments integrate environmental burdens, resulting in higher concentrations of PBDEs near point sources.

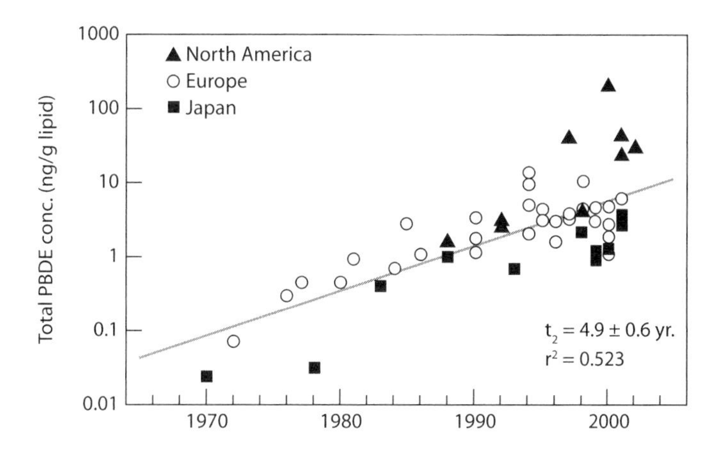

Figure 14.8. Total PBDE concentrations in human blood, milk, and tissue (in ng g^{-1} lipid). Results show a trend of increasing concentrations over time. Open circles refer to samples collected from Europe, closed circles represent samples from Japan, and triangles refer to samples from North America. (Adapted from Hites, 2004.)

The decomposition of these compounds is still not well understood, although some laboratory and field observations suggest debromination may be possible to a limited extent (Hale et al., 2006). The behavior and fate of PBDEs is still poorly understood and tools such as stable carbon isotopes may provide useful tools for tracing the sources of these contaminants in the environment. A recent study utilizing $\delta^{13}C$ analysis of individual congeners in two technical polybrominated diphenyl ether products showed that values were usually more negative (depleted) as the degree of bromination increased (Vetter et al., 2008). Clearly, additional studies focused on the distribution and behavior of this class of contaminants are needed since use is on the rise and potential adverse affects to wildlife and humans are still poorly understood (De Wit, 2002).

14.7 Polychlorinated Dibenzodioxins and Dibenzofurans

Additional classes of halogenated contaminants include the polychlorinated dibenzodioxins (PCDDs) and furans (PCDFs), which are produced during combustion of organic matter and fuels at temperatures below 800 °C as well as during some industrial processes (Bayona and Albaiges, 2006). The physical properties of the PCDDs and PCDFs are similar to those for the PCBs (e.g., *lipophilic*, low volatility, long half-life, and tendency to bioaccumulate) (table 14.1), resulting in similar patterns for their distribution and fate in the environment. Like PCBs, the likelihood of toxic effects and persistence in the environment associated with these classes of contaminants is driven by the degree of chlorination and position of the chlorine substitutions. Generally, higher toxicity is found for congeners with chlorine substitutions in the 2-, 3-, 7-, and 8-positions (fig. 14.6b, c), with the tetrachloro (2,3,7,8-TCDD) isomer being most toxic (Bayona and Albaiges, 2006).

A number of recent studies have examined concentration and distribution patterns for PCDDs and PCDFs in the environment, including characterization of air concentrations and atmospheric deposition processes (Lohmann and Jones, 1998; Ogura et al., 2001; Lohmann et al., 2006; Castro-Jimenez et al., 2008) as well as studies quantifying sediment concentrations (Fattore et al., 1997; Isosaari et al., 2002; Eljarrat et al., 2005). Together, these studies show higher concentrations of PCDDs and PCDFs in coastal environments and near point sources than in remote regions. A second area of research has been investigations of the concentrations of these compounds in aquatic organisms (Zhang and Jiang, 2005) and their toxic effects (Van Den Berg et al., 1998). Additional studies have focused on identifying the sources and fate of these contaminants. Baker and colleagues (2000) found that while combustion is the predominant source of PCDDs and PCDFs, photochemical synthesis of octachlorinated dibenzo-*p*-dioxin (OCDD) from pentachlorophenol (PCP) may occur during reactions with hydroxyl radical in the atmosphere resulting in a significant source of OCDD to the environment from atmospheric condensed water (Baker and Hites, 2000). Additional research is needed to understand the various biogeochemical and geophysical processes influencing the distribution and fate of these contaminants in order to predict the global fate of PCDDs and PCDFs accurately (Lohmann et al., 2006, 2007).

14.8 Emerging Contaminants

The number of organic contaminants has multiplied over the past two decades as new chemicals have been introduced to the environment and analytical detection capabilities have improved. Collectively, these compounds are referred to as *emerging contaminants*. These compounds include new chemicals produced in the development of industry, agriculture, medicine, and personal care products. These contaminants enter the environment through a number of pathways, thereby influencing their effects on aquatic organisms as well as their environmental fates (fig. 14.9). The identification of many of these compounds in the environment is largely due to new methods of analysis that have enabled the analysis of more polar compounds and increased the sensitivity of existing methods (e.g., high-performance liquid chromatography with positive electrospray ionization and tandem mass spectrometry; HPLC/ESI-MS/MS) (Richardson and Ternes, 2005; Richardson, 2006). For example, methods such as liquid chromatography tandem mass spectrometry (LC/MS/MS) and gas chromatography tandem mass spectrometry (GC/MS/MS) provide the ability to detect compounds at concentrations ranging from 0.08 to 0.33 $ng\,L^{-1}$ and 0.05 to 2.4 $ng\,L^{-1}$, respectively (Ternes, 2001). Other methods, such as solid phase micro extraction combined with high-performance liquid chromatography (SPME-HPLC), have detection limits in the range 0.064 to 1.2 $ng\,L^{-1}$. Far less is known about the environmental and toxicological properties of emerging contaminants than about those of the more traditional contaminants, such as PAHs and PCBs.

14.8.1 Pharmaceutical and Personal Care Products

One class of emerging contaminants of particular environmental concern is the pharmaceutical and personal care products (PPCPs). These include pharmaceuticals used by humans and animals for clinical diagnosis, treatment, or prevention of diseases. Examples include estrogens, which are used in oral contraceptives, to increase athletic performance, and to enhance growth in animal production. Many of these compounds are *endocrine-disrupting compounds* (EDCs) and some have been linked to alterations in the expression of sexual characteristics in aquatic organisms (Kelly and Di Giulio, 2000; Ankley et al., 2003; Nash et al., 2004; Jensen et al., 2006; Brian et al., 2007). While initial research focused on the feminization of males resulting from exposure to estrogenic compounds, more recent studies have shown that androgenic contaminants can also impose male characteristics on females (Durhan et al., 2006). Much of this research has been conducted in laboratory studies testing the responses of aquatic organisms to individual EDCs. However, aquatic organisms are exposed to complex mixtures of contaminants in the environment, often resulting in interactive and additive effects due to the presence of various mixtures of EDCs. Additional concerns arise since many aquatic organisms are exposed throughout their life-cycle and across multiple generations. This increases the possibility of continual but undetectable or unnoticed effects on aquatic organisms, which means that effects could accumulate so slowly that major change goes undetected until the cumulative level of these effects reaches a level of irreversible change (Daughton and Ternes, 1999).

In addition to EDCs, a wide variety of pharmaceuticals enter aquatic systems through wastewater and septic systems. These include a variety of antibiotics used in the treatment of humans and

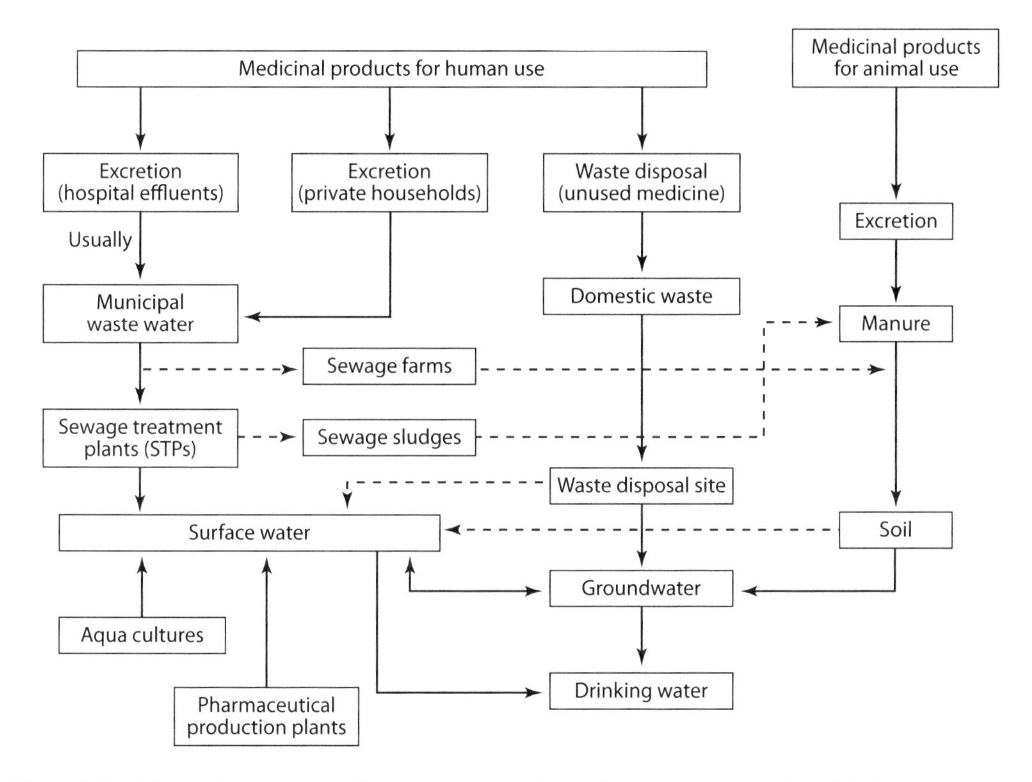

Figure 14.9. Schematic showing possible sources and pathways for the occurrence of medicinal/pharmaceutical compounds in the aquatic environment. (Adapted from Herberer, 2002.)

animals. It is thought that the presence of these compounds in the environment may accelerate or spread resistance of bacterial pathogens, which may alter microbial communities and organisms at higher trophic levels. Other pharmaceuticals such as antidepressants are known to influence serotonin metabolism, and sensitivities to these compounds have been reported in bivalves, crustaceans, and other aquatic communities. For example, the antidepressant fluoxetine has been linked to changes in the development and reproduction of *Daphnia magna* (Flaherty and Dodson, 2005). This study also showed that the aquatic toxicity of pharmaceutical mixtures can be unpredictable and complex compared to individual pharmaceutical effects and that the timing and duration of pharmaceutical exposure influence aquatic toxicity.

Personal care compounds include bath additives, shampoos, soaps, skin care products, hair sprays, oral hygiene products, cosmetics, perfumes, aftershave, and food supplements. Additional classes of PPCPs include synthetic musks, parabens, triclosan, and various agents used in sunscreen. Synthetic musks are refractory to biodegradation and ubiquitous in the environment. Antimicrobial agents such as parabens, which are widely used as preservatives in cosmetics, and triclosan, an antiseptic agent in toothpaste, may have biocide properties. The bacteriocide triclosan and methyl triclosan (a transformation product of triclosan) were measured in lakes and in a river in Switzerland at concentrations of up to 74 and 2 ng L^{-1}, respectively (Lindstrom et al., 2002). Both compounds likely derive from wastewater treatment plants, with methyl triclosan probably being formed by biological methylation. However, their environmental fates differed. Laboratory experiments showed that triclosan (in its dissociated form) was rapidly decomposed in lake water when exposed to sunlight (half-life less than 1 h in August), while methyl triclosan and nondissociated triclosan were not influenced by photodegradation (Lindstrom et al., 2002).

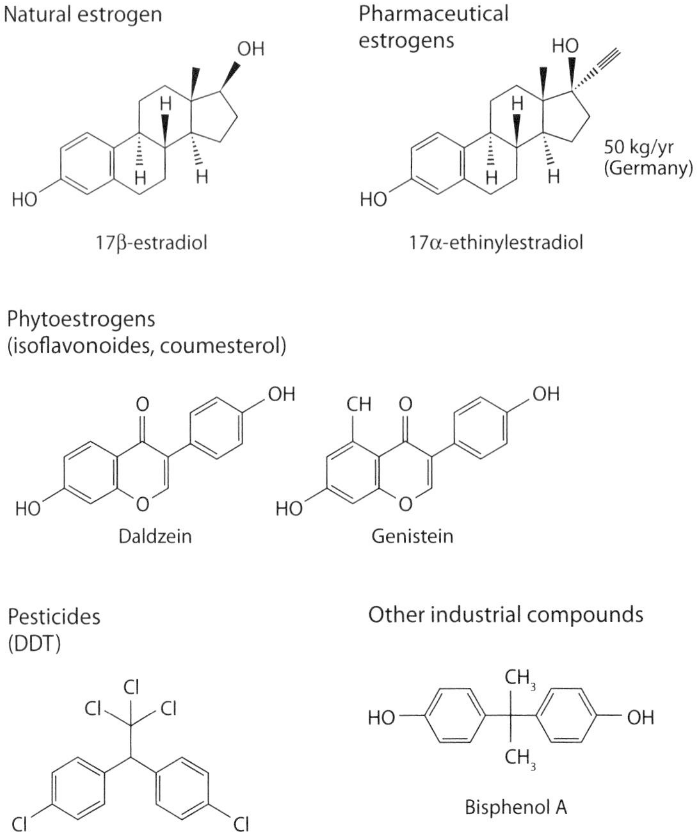

Figure 14.10. Examples of a natural estrogen and anthropogenic compounds that mimic or induce estrogen-like response in organisms.

PPCPs enter the environment in a number of ways, including sewage treatment plants, septic systems, and direct discharge into streams (fig. 14.9). They may also be delivered by runoff or groundwater from farms and fish farms. Exposure routes differ for drugs used by humans and animals and their behavior depends on factors such as susceptibility to microbial degradation and binding capacity to solids. Like other xenobiotics, these compounds may have biological effects at the cellular level as well as to specific organs, organisms, and populations. For many of the antibiotics and endocrine disruptors, the effects may be irreversible even at low concentrations. Estrogenic EDCs (e-EDCs) are natural hormonal estrogens or chemicals that mimic or induce estrogen-like response in organisms (fig. 14.10). These chemicals have the potential to affect the endocrine system of humans and other organisms. They include natural hormones and geogenic compounds as well as synthetic compounds such as pharmaceuticals, pesticides, industrial chemicals, and heavy metals. Estrogenic-disrupting compounds are found in wastewater, surface water, sediments, and drinking water and may cause responses even at low concentrations.

14.8.2 Anthropogenic Compounds as Molecular Markers

A number of wastewater components have been used as proxies for anthropogenic inputs to freshwater and coastal waters. Compounds used for this application of anthropogenic markers include surfactants

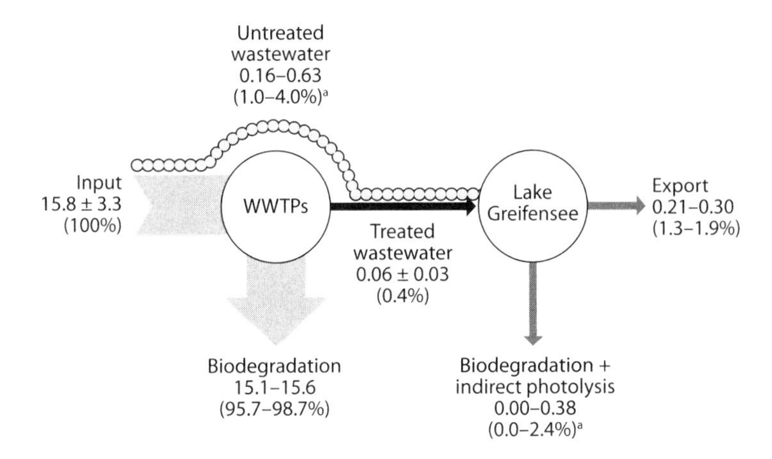

Figure 14.11. Mass balance model for caffeine in a Swiss lake during winter. (Adapted from Buerge et al., 2003.)

(Eganhouse et al., 1983; Bruno et al., 2002; Sinclair and Kannan, 2006), caffeine (Standley et al., 2000; Lindstrom et al., 2002; Buerge et al., 2003, 2006; Peeler et al., 2006), and fecal steroids (see chapter 9). Initial work using surfactants as a tool for tracing point-source pollution to coastal waters involved the application of linear alkylbenzenes (Eganhouse et al., 1983). In recent years, a broader range of surfactants, including branched alkylbenzenes, linear alkylsolfonates, trialkylamines, and fluorescent whitening agents, have been used to trace wastewater inputs to offshore regions as well septic system inputs to groundwaters (Swartz et al., 2006). These compounds, and their products, differ in their susceptibility to transformation in the environment, providing tracers for near-source tracing and long-distance transport of wastewater inputs as well as geochronometers. Ferguson and colleagues recently demonstrated that nonylphenol polyethoxylate (NPEO) surfactants and their metabolites were good markers for tracing anthropogenic waste inputs into Jamaica Bay, New York (USA) (Ferguson et al., 2001, 2003). Sediment and surface water concentrations coincided with trends in two traditional tracers of anthropogenic waste, concentrations of sediment-bound silver and waterborne fecal coliform. However, the use of NPEO surfactants requires further consideration since Standley and colleagues (2000) subsequently identified nonylphenol surfactants in animal waste, suggesting this compound class may be a general marker for domestic, industrial, and agricultural waste.

In recent years, caffeine has been used as an anthropogenic marker for wastewater contamination (Gardinali and Zhao, 2002; Buerge et al., 2003, 2006; Peeler et al., 2006). Caffeine is present in a broad array of beverages and food products and although its use varies throughout the world, the global average consumption is about 70 mg per person per day (Buerge et al., 2003). It is also used in many pharmaceutical products. Buerge and colleagues investigated the concentration of caffeine in influent and effluent waters of wastewater treatment plants (WWTP) and surface waters in lakes and rivers in Switzerland (Buerge et al., 2003). They found efficient removal of caffeine from WWTP (average influent concentrations of 7–73 $\mu g\,L^{-1}$ and effluent concentrations of 0.03 to 9.5 $\mu g\,L^{-1}$). With the exception of remote mountain lakes, surface water concentrations of caffeine ranged from 6 to 250 $ng\,L^{-1}$, and correlated with anthropogenic burdens ($r^2 = 0.67$). This study also investigated the fate of caffeine in surface water and showed that most of the caffeine was removed by biodegradation (95.7 to 98.7%), with smaller losses attributed to biodegradation + indirect photolysis (0.0 to 2.4%) and flushing (1.3 to 1.9%) (fig. 14.11).

In a more recent study, Peeler and colleagues examined the use of caffeine as a tracer for anthropogenic sources of organic matter by examining relationships between caffeine and water-quality indicators, such as nitrate and fecal coliform bacteria in surface waters and groundwater representing rural and urban systems (Peeler et al., 2006). These authors found that caffeine and nitrate were correlated in rural areas and linked to population centers and their WWTP. However, bacteria were not related to these variables, since the highest bacterial abundance was found in an isolated wetland. In contrast, in surface waters of Sarasota Bay (USA), an urban system, caffeine was generally correlated with fecal coliform abundance and nitrate. However, these authors recommend that multiple variables should be used when tracing anthropogenic inputs, since the processes controlling the sources and fate of these variables will differ. For example, because caffeine may be susceptible to photochemical degradation, it may be a better tracer for anthropogenic inputs in groundwater than surface waters. In contrast, microbial decomposition processes primarily influence nitrate removal.

14.9 Summary

While this chapter provides examples of a variety of anthropogenic compounds, it is not exhaustive. We refer readers to review articles cited throughout the text for further reading. Overall, new research efforts in the field of contaminants will likely focus on identifying the sources, distribution, and fate of anthropogenic compounds in the environment, utilizing new analytical tools that will enhance our detection of contaminants in a variety of environmental matrices, including air, water, organism tissues, and sediments. Another area of research will likely involve the development of more sophisticated models that incorporate a variety of abiotic and biotic processes influencing the distribution and fate of anthropogenic compounds. Future studies should also consider the role of global environmental change and its potential influence on the input, distribution, and fate of anthropogenic compounds.

I Atomic Weights of Elements

Atomic number	*Name*	*Symbol*	*Atomic weight*
1	Hydrogen	H	1.0079
2	Helium	He	4.0026
3	Lithium	Li	6.9410
4	Beryllium	Be	9.0122
5	Boron	B	10.811
6	Carbon	C	12.011
7	Nitrogen	N	14.007
8	Oxygen	O	15.999
9	Fluorine	F	18.998
10	Neon	Ne	20.180
11	Sodium	Na	22.990
12	Magnesium	Mg	24.305
13	Aluminum	Al	26.982
14	Silicon	Si	28.086
15	Phosphorus	P	30.974
16	Sulfur	S	32.066
17	Chlorine	Cl	35.453
18	Argon	Ar	39.948
19	Potassium	K	39.098
20	Calcium	Ca	40.078
21	Scandium	Sc	44.956
22	Titanium	Ti	47.956
23	Vanadium	V	50.942
24	Chromium	Cr	51.996
25	Manganese	Mn	54.938
26	Iron	Fe	55.845
27	Cobalt	Co	58.933
28	Nickel	Ni	58.693
29	Copper	Cu	63.546
30	Zinc	Zn	65.392
31	Gallium	Ga	69.723

Atomic number	*Name*	*Symbol*	*Atomic weight*
32	Germanium	Ge	72.612
33	Arsenic	As	74.922
34	Selenium	Se	78.963
35	Bromine	Br	79.904
36	Krypton	Kr	83.800
37	Rubidium	Rb	85.468
38	Strontium	Sr	87.520
39	Yttrium	Y	88.906
40	Zirconium	Zr	91.224
41	Niobium	Nb	92.906
42	Molybdenum	Mo	95.940
43	Technetium	Te	98.906
44	Ruthenium	Ru	101.07
45	Rhodium	Rh	102.91
46	Palladium	Pd	106.42
47	Silver	Ag	107.87
48	Cadmium	Cd	112.41
49	Indium	In	114.82
50	Tin	Sn	118.71
51	Antimony	Sb	121.76
52	Tellurium	Te	127.60
53	Iodine	I	126.90
54	Xenon	Xe	131.29
55	Cesium	Cs	132.91
56	Barium	Ba	137.33
57	Lanthanum	La	138.91
58	Cerium	Ce	140.12
59	Praseodymium	Pr	140.91
60	Neodymium	Nd	144.24
61	Promethium	Pm	146.92
62	Samarium	Sm	150.36
63	Europium	Eu	151.96
64	Gadolinium	Gd	157.25
65	Terbium	Tb	158.93
66	Dysprosium	Dy	162.50
67	Holmium	Ho	164.93
68	Erbium	Er	167.26
69	Thulium	Tm	168.93
70	Ytterbium	Yb	173.04
71	Lutetium	Lu	174.97
72	Hafnium	Hf	178.49
73	Tantalum	Ta	180.95
74	Tungsten	W	183.84
75	Rhenium	Re	186.21
76	Osmium	Os	190.23
77	Iridium	Ir	192.22
78	Platinum	Pt	195.08
79	Gold	Au	196.97
80	Mercury	Hg	200.59
81	Thallium	Tl	204.38
82	Lead	Pb	207.20

Atomic number	*Name*	*Symbol*	*Atomic weight*
83	Bismuth	Bi	208.98
84	Polonium	Po	209.98
85	Astatine	At	209.99
86	Radon	Rn	222.02
87	Francium	Fr	223.02
88	Radium	Ra	226.03
89	Actinium	Ac	227.03
90	Thorium	Th	232.04
91	Protactinium	Pa	231.04
92	Uranium	U	238.03
93	Neptunium	Np	237.05
94	Plutonium	Pu	239.05
95	Americium	Am	241.06
96	Curium	Cm	244.06
97	Berkelium	Bk	249.08
98	Californium	Cf	252.08
99	Einsteinium	Es	252.08
100	Fermium	Fm	257.10
101	Mendelevium	Md	258.10
102	Nobelium	No	259.10
103	Lawrencium	Lr	262.11

II Useful SI Units and Conversion Factors

SI Prefix Units	
atto (a)	$= 10^{-18}$
femto (f)	$= 10^{-15}$
pico (p)	$= 10^{-12}$
nano (n)	$= 10^{-9}$
micro (μ)	$= 10^{-6}$
milli (m)	$= 10^{-3}$
centi (c)	$= 10^{-2}$
deci (d)	$= 10^{-1}$
deca (da)	$= 10^{1}$
hecto (h)	$= 10^{2}$
kilo (k)	$= 10^{3}$
mega (M)	$= 10^{6}$
giga (G)	$= 10^{9}$
tera (T)	$= 10^{12}$
peta (P)	$= 10^{15}$
exa (E)	$= 10^{18}$

Conversion Factors	
Force	
1 newton (N)	$= kg\,m\,s^{-2}$
1 dyne (dyn)	$= 10^{-5}$ N
Pressure	
1 pascal (Pa)	$= kg\,m^{-1}\,s^{-2}$
1 torr (torr)	= 133.32 Pa
1 atmosphere (atm)	= 760 torr
	= 12.5 psi
Temperature	
$^{\circ}C = 5/9(^{\circ}F\text{-}32)$	
$K = 273.15 + ^{\circ}C$	

Energy

1 kcal = 1000 cal
1 cal = 4.184 Joule

Speed

1 knot = 1 nautical mile h^{-1}
= 1.15 statute miles h^{-1}
= 1.85 kilometers h^{-1}
Velocity of sound in water (salinity = 35) = 1, 507 m s^{-1}

Volume

1 cubic kilometer (km^3) = 10^9 m^3
= 10^{15} cm^3
1 cubic meter (m^3) = 1000 liters
1 liter (L) = 1000 cm^3
= 1.06 liquid quarts
1 milliliter (mL) = 0.001 liter
= 1 cm^3

Length

1 kilometer (km) = 10^3 m
1 centimeter (cm) = 10^{-2} m
1 millimeter (mm) = 10^{-3} m
1 micrometer (μm) = 10^{-6} m
1 nanometer (nm) = 10^{-9} m
Angstrom (Å) = 10^{-10} m

1 fathom = 6 feet
= 1.83 m
1 statute mile = 5280 feet
= 1.6 km
= 0.87 nautical mile
1 nautical mile = 6076 feet
= 1.85 km
= 1.15 statute miles

Mass

1 kilogram (kg) = 1000 g
1 milligram (mg) = 0.001 g
1 ton (tonne) (t) = 1 metric ton
= 10^6 g

Area

1 square centimeter (cm^2) = 0.155 square inch
= 100 square millimeters
1 square meter (m^2) = 10.8 square feet
1 square kilometer (km^2) = 0.386 square statute miles
= 0.292 square nautical miles
= 10^6 square meters
= 247.1 acres
1 hectare = 10, 000 square miles

III Physical and Chemical Constants

Avogadro's number (N) $= 6.022137 \times 10^{23}$ mol^{-1}
Boltzmann's constant (k) $= 1.380658 \times 10^{-23}$ J K^{-1}
Faraday's constant (F) $= 9.6485309 \times 10^{4}$ C mol^{-1}
Gas constant (R) $= 8.3145$ J mol^{-1} K^{-1}
Planck's constant (h) $= 6.6260755 \times 10^{-34}$ J s^{-1}

Glossary

Accuracy—The degree to which measurements of a quantity reflect its actual (true) value.

Acidic sugar—Phosphorylated derivatives of carbohydrates that contain a carboxylate group. When the acidic sugar is at a neutral pH, it has a negative charge.

Acyl—A functional group derived from a carboxylic acid, also known as an alkanoyl.

Acyl phosphate—An organic compound containing a phosphate group linked by one of the oxygen atoms to an acyl (RC=O) group. They are named by citing the acyl group first, then naming any substituents attached to the phosphate oxygen atoms before the word *phosphate*. For example, a common acyl phosphate is acetyl adenosyl phosphate.

Acylheteropolysaccharides—Polymers formed from two or more different kinds of monosaccharides, which also contain an acyl functional group (which can be derived from carboxylic acid functional groups). The generic acyl group contains compounds also described as esters, amides, ketones, and aldehydes. These compounds are commonly produced by phytoplankton and constitute a large part of high-molecular-weight dissolved organic matter (HMWDOM) in aquatic systems, possibly contributing to the recalcitrance of this material.

S-Adenosyl-L-methionine—A substrate involved with the methyl group transfer. It is used in several pathways, including transmethylation, transulfuration, and aminopropylation.

Alcohol (alkanol)—A class of organic compounds containing at least one hydroxyl (–OH) group.

Aldehyde—A class of organic compounds containing the functional group –CHO.

Alditols—Linear molecules that do not form cyclic structures as aldoses do. Also known as sugar alcohols.

Aldolase reactions—Chemical reactions catalyzed by the enzymes that are considered aldolases. These enzymes catalyze aldol cleavage.

Aldose—A monosaccharide containing one aldehyde group per molecule.

Algaenan—The nonhydrolyzable biopolymer found in the cell walls of marine and freshwater algae.

Aliphatic—Refers to a class of hydrocarbons that contain saturated and/or single unsaturated carbon bonds (i.e., contain no benzene (aromatic) rings).

Alkaloids—A large and diverse group of nitrogen-containing compounds usually with an alkaline pH. Many have pharmacological effects and are used as medications.

Alkane—An organic compound that contains only carbon and hydrogen. This class of hydrocarbons contains only saturated carbon bonds, which can be straight, cyclic, or branched. Alkanes are also called paraffins.

Alkenes (or olefins)—Hydrocarbons that contain a carbon–carbon double bond. They tend to be more reactive than alkanes because of the higher energy bonds. Aromatic compounds are drawn as cyclic alkenes but their properties are different; thus, they are not considered true alkenes.

Alkenoates—α, β-Unsaturated esters. They generally have methyl and ethyl substitutions and are related to alkenones.

Alkenones—A class of organic compounds with a carbonyl group produced by phytoplankton. These are useful biomarkers because they resist degradation and can be used to reconstruct sea surface temperatures.

Alkoxy—An alcohol derivative where a carbon and hydrogen chain (alkyl group) is linked to oxygen.

Alkylation—The transfer of an alkyl group (usually a hydrocarbon) from one molecule to another.
Alkynes—Hydrocarbons that contain a carbon–carbon triple bond. They are much rarer in nature than alkenes and alkanes, and are also more reactive.
Allelopathy—Biochemicals produced by an organism that may effect growth or development of another organism.
Allochthonous—Refers to carbon or organic matter inputs from outside an environment. For example, terrigenous inputs to the ocean are referred to as allochthonous.
Alpha decay—The loss of an alpha particle (nucleus of a helium-4 atom) from the nucleus, which results in a decrease in the atomic number by two units (two protons) and the mass number by four units (two protons and two neutrons).
Amide—An organic compound containing the acyl functional group (R–C=O) linked to a nitrogen.
Amine (aliphatic amines and polyamines)—An organic derivative of ammonia where one or more of the H atoms are substituted with an alkyl moiety. Aliphatic amines contain only nonaromatic alkyl substituents, and may be cyclic or acyclic, saturated or unsaturated. A polyamine is a compound that contains two or more amino groups. Like ammonia, amines contain a lone pair of electrons, making them basic and nucleophilic.
Amino acids—Organic compounds that contain amino ($-NH_2$) and carboxylic acid (–COOH) groups. L-Amino acids are the building blocks of protein in living organisms. These compounds are present in a variety of pools in the environment, including dissolved free amino acids (amino acids analyzed without hydrolysis), dissolved combined amino acids (amino acids analyzed following hydrolysis), etc.
Amino sugars—An organic compound that contains an amine group that replaces the hydroxyl group.
5-Aminolevulinic acid—The first compound in the porphyrin synthesis pathway that leads to the production of heme and eventually chlorophyll.
Amphipathic—Describes a chemical compound possessing both hydrophilic (water-loving) and hydrophobic (water-avoiding) properties. These compounds are also called amphiphilic.
Amphoteric—Refers to a substance that can act as either an acid or a base. Amino acids, which contain an amino group and a carboxylic acid group, can donate or accept a proton, depending on whether the pH of the medium is high or low, respectively.
Anaerobic—Refers to metabolism in the absence of oxygen (i.e., under anoxic conditions).
Anomers—Stereoisomers of cyclic saccharides that differ only in the configuration at the hemiacetal or anomeric carbon. Anomers are identified as α or β, depending on the relationship of the oxygen atom outside of the saccharide ring and the oxygen atom attached to the configurational carbon.
Anoxic—Refers to an area of sediment or water where oxygen is not present.
Anoxygenic photosynthesis—An autotrophic process by which carbon is fixed into cellular biomass but oxygen is not produced. It is typically carried out by green and purple bacteria.
Anthesis—The time in which a flower is in full bloom.
Anthocyanes—Pigments found in the vacuole that could be red, purple, or blue, depending on the pH.
Anthropogenic—Refers to the effects, processes, or materials that are derived from human activities or influences.
Apophycobiliproteins—Hydrophobic linker proteins found in some phycobilisomes.
Apoproteins—Protein moieties of a molecule or complex.
Arenes (or aromatic compounds)—Cyclic hydrocarbons, with conjugated double-bond systems, that differ in geometry from and are more stable than their hypothetical localized counterparts. Many natural compounds, such as hormones, have aromatic moieties. Aromaticity gets its name from the fragrances (e.g., cherries, almonds, peaches) observed by early chemists working with such compounds.
Aromatic—Refers to a type of hydrocarbon that contains one or more benzene rings. Examples include polycyclic aromatic hydrocarbons, monoaromatic steroids, and some sulfur compounds, such as benzothiophenes and porphyrins.
Aryl ether (bonds)—The covalent bonds that hold the phenolic structures of lignin together.
Atomic number—The number of protons in an element. The atomic number is identified as Z. The mass number refers to the sum of the number of neutrons (N) and protons (mass number = $N + Z$).
Autochthonous—Refers to sources of carbon (or organic matter) produced within the environment where it is found (e.g., marine phytoplankton is an autochthonous source in the ocean).
Autotrophs—Organisms that fix carbon dioxide to produce cellular biomass.
Beer-Lambert law—Relates the absorption of light to several properties of the material or substance through which the light is traveling.

Benthic—Occurring at the ocean bottom.

Beta decay (or negatron decay)—Occurs when a neutron changes to a proton and a negatron (negatively charged electron) is emitted, thereby increasing the atomic number by one unit.

Bilirubin—A yellow tetrapyrrole pigment that is produced by heme catabolism.

Biliverdin—A green tetrapyrrole pigment that is produced by heme catabolism.

Bioaccumulation—The amassing of a substance in an organism because more of the substance is being absorbed within the organism than lost. Bioaccumulation occurs most frequently for lipid-soluble compounds, such as pesticides, methylated mercury, or tetraethyl lead.

Biomarkers—Organic compounds in water, soils, sediments, rocks, coal, or petroleum that can be linked to molecules synthesized by living organisms. Biomarkers are also referred to as biological markers, molecular fossils, fossil molecules, or geochemical fossils.

Biosphere—All of the Earth's ecosystems as one.

Bitumen—The fraction of organic matter that can be extracted from rocks using organic solvents. Unlike oil, which migrates, bitumen originates in the rock in which it is found.

Black carbon—A group of thermogenic carbon-containing compounds formed through incomplete combustion of fossil fuels and biomass.

Calvin cycle (or Calvin-Benson cycle)—The biochemical process by which carbon dioxide is fixed into cellular biomass by autotrophic organisms.

Carbamyl phosphate—A compound containing a phosphate group linked by one of the oxygen atoms to a carboxamide ($-C(=O)NH_2$) functionality. Carbamyl phosphate is a key intermediate in the biochemical conversion of ammonia to urea.

Carbohydrates—Ketones or aldehydes with two or more hydroxyl groups. They are the most abundant biomolecules and have many different functions. Carbohydrates are classified by the number of simple sugars (i.e., monosaccharide [simple sugar], disaccharide [two monosaccharides bonded together], and oligosaccharides [three or more monosaccharide chains]).

Carbon preference index (CPI)—A measure of the relative amounts of compounds containing odd and even numbers of carbon atoms. The CPI proxy can provide insight into whether hydrocarbon biomarkers reflect contributions from vascular plants (high CPI) or fossil fuel sources (low CPI).

Carbonic anhydrase—An enzyme that catalyzes the conversion of carbon dioxide to bicarbonate and protons during photosynthesis.

Carboxylic acids—Compounds that contain a carboxyl (COOH) functional group. These compounds are generally composed of straight or methyl-branched carbon chains and may be saturated (no double bonds) or unsaturated (one or more double bonds). *Iso*-branched fatty acids have a methyl group at the n-1 position, while anteiso-branched fatty acids have a methyl group at the n-2 position.

Carcinogenic—Refers to any substance or agent that leads to the production of a cancer.

β-Carotene (β-carotane)—An orange pigment typically used as an accessory pigment in photosynthesis. This compound is a saturated tetraterpenoid biomarker ($C_{40}H_{78}$) comprising eight isoprene units.

Carotenoid—A class of yellow to orange pigments used as accessory pigments in photosynthesis to absorb green light. They include hydrocarbons (carotenes) and oxygenated derivatives (xanthophylls). Carotenoids are tetraterpenoids (C_{40}) and are composed of eight isoprene units.

Catagenesis—A process by which buried organic matter is altered over a time period greater than 1000 years at temperatures of 50–150 °C.

Cellulose—A polysaccharide composed exclusively of glucose monomers. It is the most abundant neutral sugar in vascular plants.

Chemical ionization (CI)—A type of gas chromatography—mass spectrometry analysis where by a reagent gas (typically methane or ammonia) is introduced into the ion source where it is ionized to yield primary and secondary ions. The compound to be analyzed is introduced from the GC column and reacts with the secondary ions in the source. Ions are formed by proton transfer or hydride abstraction, yielding principal ions (either $(M + 1)^+$ or $(M - 1)^+$). These ions are useful for determining the molecular weight of the compound of interest.

Chemical shift—The dependence of nuclear magnetic energy levels in nuclear magnetic resonance (NMR) on the electronic environment in a molecule. Differences in field strengths are caused by the shielding or deshielding of electrons associated with adjacent atoms; this electron interference is referred to as a chemical shift. The chemical shift of a nucleus will be proportional to the external magnetic field. Since many spectrometers have different magnetic field strengths, it is common to report the shift independent of the external field strength.

Chemocline—The depth in the water column or sediments at which oxygen disappears and hydrogen sulfide first appears. This zone marks the transition from aerobic respiration to anaerobic respiration.

Chemosynthesis—the process by which bacteria use energy from chemicals, such as hydrogen sulfide, to convert water and carbon dioxide to carbohydrates.

Chiral—Refers to a molecule that has a non-super-imposable mirror image. Chiral molecules are characterized by having an asymmetric carbon atom. A distinguishing characteristic of chiral compounds is their ability to rotate polarized light. Molecules may rotate polarized light clockwise (dextrorotation) or counterclockwise (levorotation). Compounds with these properties are said to have optical activity and consist of chiral molecules. If a chiral molecule is dextrorotary, its enantiomer will be levorotary, and vice versa. In fact, the enantiomers will rotate polarized light the same number of degrees, but in opposite directions.

Chlorophyll—A green pigment that is functional in photosynthesis and, therefore, found only in plants. There are several forms of chlorophyll. The two main types are chlorophyll *a*, which absorbs purple and orange light, and chlorophyll *b*, which absorbs blue and orange light.

Chloropigments—The total chlorophylls and chlorophyll degradation products found in a specific sample.

Chlorophyll synthetase—A transferase enzyme that alkylates chlorophyllide (giving chlorophyllide its phytyl tail), which converts it into chlorophyll *a*.

Choline—An ammonium salt containing nitrogen typically found attached to phospholipids in the lipid bilayer.

Citric acid cycle (Krebs cycle)—A metabolic pathway that occurs in the mitochondria of organisms that use oxygen in cellular respiration. This cycle produces carbon dioxide and water from carbohydrates, protein, and fats to ultimately produce energy.

Colloidal—Refers to a chemical mixture in which one substance is evenly dispersed in another. Colloidal particles in the environment are usually smaller than 0.45 μm but are not considered truly dissolved.

Conjugated bonds—Multiple bonds that alternate with single bonds such that double-bond electron density can be delocalized over multiple nuclei. This lowers overall energy and increases stability of the molecule, and will often cause changes in molecular geometry, such as planarization, to align molecular orbitals and accommodate electron delocalization.

Coulomb repulsion—Electrostatic repulsion.

Cuticle—The waxy coating on the surface of stems and leaves that helps prevent desiccation in terrestrial plants.

Cuticular waxes—These waxes fill up the cuticular membrane, whereas *epicuticular* waxes cover the surface of the cuticular membrane.

Cutins—The waxy part of the cuticle that contains ester-linked fatty acids that have been hydroxylated.

Cyanobacteria—Prokaryotes that are photosynthetic and contain chlorophyll. Also known as blue-green algae.

Cyclopentane—An aliphatic, cyclic molecule with the chemical formula C_5H_{10}. It is a colorless liquid with a petrol-like odor at room temperature, and is highly flammable.

Cytochrome—A protein containing a heme prosthetic group, which functions by carrying electrons. Cytochromes are usually one-electron transfer agents and the heme iron atom is converted between Fe^{2+} and Fe^{3+}.

Cytosolic—Refers to anything within the cytosol or intracellular fluid/cytoplasmic matrix of cells.

Decarboxylation—An elimination reaction that results in the loss of CO_2. This reaction is unique to compounds that have a carbonyl group two carbon atoms away from (β position to) a carboxylic acid group. The reaction occurs by a cyclic mechanism resulting in an enol intermediate.

Dehydration—An elimination reaction that results in the loss of H_2O from an alcohol, generally giving an alkene product. The reaction can occur by treatment of an alcohol with a strong acid. In biological pathways, dehydration reactions generally occur with substrates in which the –OH is positioned two carbon atoms away from (β position to) a carbonyl group.

Dendrimeric—Refers to the structure of a polymer in which branches of subunits are strung off of a central "spine."

Detritus—Particulate matter that enters into a marine or aquatic system. Organic detritus refers to materials from decaying organic matter.

de Vries effect—A 100-year oscillation of radiocarbon from the atmosphere, resulting in radiocarbon aging of samples.

Diagenesis—Chemical, physical, or biological process by which buried organic matter is altered over a time period less than 1000 years at temperatures of 50–150 °C. Diagenesis occurs before oil generation but includes the formation of methane from microbial processes.
Diagenetic—Refers to the products resulting from chemical or physical change that occurs in sediments between the times of deposition and solidification, under low temperatures and pressures.
Diastereomers—Isomers that differ in configuration at only one chiral center and are non-mirror images of each other.
Diazotrophs—Microorganisms that are able to fix atmospheric N_2 into more usable forms of nitrogen (like ammonia) and are thus able to grow without external sources of fixed nitrogen. They are characterized by the possession of nitrogenase systems that catalyze nitrogen fixation.
Dicotyledon—A flowering plant whose seed has a pair of embryonic leaves.
Dioxygenases—An enzyme that transfers both oxygen atoms from O_2 onto a substrate.
Dipeptides—A molecule consisting of two amino acids joined by a peptide bond (addition of the amino group of one amino acid to the carbonyl carbon of the other followed by the elimination of H_2O).
Direct temperature-resolved mass spectrometry (DT-MS)—A rapid measurement that provides fingerprints of molecular-level characteristics across a wide range of compound classes, including both desorbable and pyrolyzable components. DT-MS is very sensitive (requires only microgram quantities of sample), allowing it to be used with other methods, including particle-sorting methods such as flow cytometry.
Dissolved combined amino acids—Amino acids linked or bound to proteins and/or sugars. These compounds must be hydrolyzed using acid before their constituent amino acids can be analyzed.
Dissolved combined neutral sugars—Usually refers to the neutral sugars (monomers) present in a sample that represent any free monomers as well as the monomers liberated after an acid hydrolysis step.
Dissolved free amino acids—Amino acids that exist in uncombined forms. They can be analyzed directly, rather than requiring acid hydrolysis.
Dysaerobic—Refers to environments characterized by low concentrations of dissolved oxygen (0.1 to 1.0 $mL\,L^{-1}$).
Ebullition—The liberation of a gas (e.g., by bubbling or boiling).
Electromagnetic spectrum—The range of frequencies of electromagnetic radiation, from below frequencies of what we use for radio transmissions to above the frequencies of gamma radiation.
Electron capture—A type of radioactive decay where a proton is changed to a neutron after combining with the captured extranuclear electron (from the K shell). This form of radioactive decay results in the atomic number decreasing by one unit.
Electron impact (EI)—A mode of gas chromatography–mass spectrometry analysis where effluent leaving the GC column is bombarded with electrons (usually at 70 eV). This causes fragmentation of the molecules, resulting in a molecular ion ($M + e^- \rightarrow M^{+\cdot} + 2e^-$) and additional fragments. The *molecular ion* ($M^{+\cdot}$) provides information about the molecular weight of the molecule of interest, while the fragmentation pattern (*mass spectrum*) provides a unique fingerprint based on the structure of a given molecule.
Electrophoresis—A separation technique that disperses particles in a fluid under the influence of an electric field.
Electrospray ionization mass spectrometry (ESI-MS)—An ionization technique used in mass analysis of biological macromolecules. ESI-MS ionizes molecules without fragmentation by using electricity to disperse the analyte into a fine aerosol, from which characteristic mass–charge signals can be analyzed.
Emerging contaminants—New classes of pollutants that have increased in production over the past few decades and have recently been found in the environment due to an increase in analytical sensitivity. These compounds have been found to have deleterious effects on the ecosystem.
Endocrine-disrupting compounds—Exogenous substances that behave similarly to hormones and disrupt the physiologic function of endogenous hormones.
Endosymbionts—Organisms that live within the cells or body of a host and are biologically linked.
Endosymbiotic—Refers to the relationship between two organisms in which one organism resides/lives within the body or cells of the other organism.
Enzyme—A large globular protein molecule that catalyzes biological reactions by altering the mechanism such that the reaction activation energy is lower than it would be without the presence of the enzyme. An *ecto-enzyme* is excreted by and works outside the biological cell by which it is produced, while an *endo-enzyme* works within the cell.
Epidermis—The outermost layering of cells on an organism.

Epilimnion—The top layer in a thermally stratified lake.

Epimerase reactions—A reaction in which enzymes catalyze the stereochemical inversion of the configuration of substituents at an asymmetric carbon, interconverting the α and β epimers.

Ester—A compound formed by the reaction of a carboxylic acid and an alcohol resulting in the elimination of a molecule of water. Ester bonds (R–CO–O–R') are readily cleaved (hydrolyzed) by the addition of water in the presence of an acid or a base. *Saponification* refers to base hydrolysis of esters. The ester functional group is composed of a carbon–oxygen double bond joined via carbon to another oxygen atom. *Ester linkage* refers to the bond between fatty acids and glycerol that characterizes true fats.

Estuary—A body of water in which there is input of freshwater from rivers and streams and saltwater from the ocean. Salinity changes along the estuary due to the mixing of fresh water with seawater.

Ethers—A class of organic compounds in which two hydrocarbon groups are linked by an oxygen atom. Membrane lipids bound by an ether linkage are one of the distinguishing characteristics of Archaea.

Eukaryote—An organism that contains membrane-bound organelles and a nucleus.

Eutrophication—A process that occurs when water becomes enriched with excess inorganic nutrients. These excess nutrients often lead to an enrichment in organic matter beyond what the system is able to process. Respiration of this excess organic matter often leads to oxygen depletion.

Extremophiles—Organisms that grow under extreme environmental conditions such as high (thermophiles) or low temperature (psychrophiles) or under very acidic (*acidophiles* thrive at low pH) or basic conditions (*alkaliphiles*). *Halophiles* live in very salty environments.

Fast atom bombardment (FAB)—An ionization technique used in mass spectrometry that allows for the bombardment of the analyte/matrix mixture with a fast particle beam. The particle beam is typically a neutral inert gas (e.g., argon) at 4 to 10 keV, making the process a relatively soft ionization. The particle beam interacts with the analyte mixture, causing collisions and disruptions that result in the ejection (sputtering) of analyte ions. These ions are then captured and focused before entering the mass analyzer.

Fats—See Lipid.

Fatty acids—See Carboxylic acids.

Ferrodoxin—An acidic, low-molecular-weight iron–sulfur protein found in various organisms that serves as an electron carrier in many redox reactions, such as photosynthesis and nitrogen fixation, but lacks classical enzyme function.

First order—Refers to radioactive decay where the number of atoms decomposing per unit of time is proportional to the number present. First order can also be used to describe rate constants for organic matter diagenesis. In this case, the rate of organic matter diagenesis is proportional to the amount of labile organic matter.

Fischer projection—A two-dimensional representation of the stereochemistry of an organic molecule, depicting bonds as horizontal or vertical lines. Carbon atoms are usually depicted as the center of crossed lines, and the projection is oriented with the C1 carbon at the top.

Flavanoids (flavonones)—Compounds that do not contain ketones but are polyhydroxyl and polyphenol and are found in plants. They tend to have antioxidant qualities.

Flow cytometry—A method that separates particles on the basis of morphological and chemical criteria including size, shape, and autofluorescence characteristics.

Fluorescamine—A reagent used for the detection of primary amines and peptides in picomolar quantities. The reaction occurs almost instantaneously in aqueous media at room temperature to yield highly fluorescent, readily detectable products.

Fluorometric—Refers to the measurement of the emission of visible light by a substance that has absorbed light of a different wavelength.

Fourier transform cyclotron resonance (FT-ICR) mass spectrometry—An analytical technique that determines the mass-to-charge ratio of ions based on the cyclotron frequency of the ions in a fixed magnetic field. The signal or mass spectrum is extracted from the data by performing a Fourier transform.

Furanose—Refers to a group of carbohydrates that form a five-membered ring consisting of four carbon atoms and one oxygen atom. Furanose rings are formed from the reaction of the ketone group and the alcohol group on the carbohydrate.

Gamma (γ)-glutamyl kinase—An enzyme that is used in the proline biosynthetic pathway. It causes feedback inhibition.

Gamma (γ) radiation—Belongs to a class of electromagnetic radiation. It consists of quanta or packets of energy transmitted in the form of wave motion. Gamma radiation travels very large distances and loses its energy by interacting with atomic electrons. It is difficult to absorb completely. It can travel through the body.

Gel filtration—A chromatographic method that separates molecules in aqueous solution based on their size, usually through a column.

Geolipid—A combination of biological lipid material and diagenetically derived lipid material.

Glucosinolate—An organic compound whose sulfur and nitrogen originate from an amino acid.

Glutamate dehydrogenase—The enzyme responsible for catalyzing the oxidative deamination reaction of glutamate, which is a product of the first step of excess amino acid degradation, to form ammonium ion. Ammonium ion is usually then converted to urea and excreted.

Glycerides—Esters of the alcohol glycerol. The glycerol molecule has three hydroxyl (OH) groups, allowing it to react with one, two, or three carboxylic acids to form monoglycerides, diglycerides, and triglycerides (also called *triacylglycerols*).

Glycerol dialkyl glycerol tetraethers (GDGTs)—A class of membrane lipids found in archaebacteria containing long-chained hydrocarbons linked to glycerols in the R configuration at each end through ether bonds. The hydrocarbons are almost twice as long as those found in eukaryotes.

Glycolysis—The process by which glucose is broken down into two molecules of pyruvate, creating cellular energy.

Glycoproteins—Molecules in living organisms that are composed of a carbohydrate linked through its anomeric center to a protein. These molecules are components of cell walls and are crucial in the process of cellular recognition.

$1,4'$-β-Glycoside linkages—Linkages that connect sugars to other biomolecules, and such as other sugars. The $1,4'$-β linkage is between the 4'-OH on one monosaccharide and the C-1 (beta) of another monosaccharide. Cellulose is a polysaccharide made from $1,4'$-β linkages of glucose molecules.

Glycosides (cardiac and cyanogenic)—A sugar bound to a noncarbohydrate with a glycosidic bond.

Half-life ($t_{1/2}$)—Refers to the time required for half of the initial number of atoms to decay.

Haptophytes (also called prymnesiophytes)—Unicellular marine phytoplankton, typically covered in tiny scales or plates composed of carbohydrates and calcium deposits. Coccolithophores are probably the best-known group of haptophytes and have an exoskeleton of calcareous plates called coccoliths. Coccolithophores are one of the most abundant marine phytoplankton, especially in the open ocean, and are widely used as microfossils in geologic and geochemical studies.

Hatch-Slack pathway—Also known as the C_4 carbon-fixation pathway. A pathway found in some plants that is used to fix carbon dioxide. The first metabolite produced is a four-carbon compound vs. the three-carbon compound found in other plant species. This pathway is believed to have evolved much more recently than the C_3 pathway and reduces the possible loss of energy to photorespiration that occurs during the C_3 pathway.

Haworth projections—Used to represent the cyclic structure of monosaccharides, showing the three-dimensional structure. Carbon and hydrogen atoms are usually implicit in Haworth projections, but carbon atoms are numbered, with C1 being the anomeric carbon.

Heme—A heterocyclic organic compound. An example is a *porphyrin*, which contains an iron atom in the center of a large heterocyclic ring.

Hemicellulose—A term used to describe several matrix heteropolysaccharides that have amorphous structures providing little strength to plant cell walls.

Heterotroph—An organism that receives its energy and carbon from organic matter.

Hexosamines—Derivatives of hexoses formed by replacing a hydroxyl group with an amino group.

Homologous series—A class of organic molecules with similar general formulas and chemical properties due to common functional groups, which shows a gradation of physical properties (i.e., boiling point) with incremental increases in size, molecular weight, extent of substitution, etc. For example, butane, pentane, and hexane comprise a homologous series where each successive addition of a CH_2 results in an increase in boiling point of approximately 30 °C.

Hopanoids—A class of pentacyclic lipids that are similar to sterols and function as bacterial membranes. Hopanes are the dominant triterpanes in petroleum.

Hydrocarbon—Also known as petroleum hydrocarbon. An organic compound consisting of carbon and hydrogen atoms. Hydrocarbons may be aliphatic or aromatic.

Hydrogenation—The chemical process in which there is an addition of hydrogen. This process typically reduces an unsaturated compound to a saturated compound.

Hydrolyzable amino acids—Amino acids that are cleaved using acid hydrolysis.

Hydrophobic—Refers to nonpolar compounds that are not generally soluble in water. In contrast, *hydrophilic* refers to polar compounds that "like" water. *Amphiphilic* compounds are partly hydrophobic and partly hydrophilic.

Hydrothermal—Refers to systems that are usually located at ocean ridge spreading centers, and are environments where reactions between seawater and newly formed oceanic crust occur.

Hydroxyl—A functional group containing an oxygen atom covalently bonded to a hydrogen atom (–OH).

Infrared radiation (IR)—The midrange of the electromagnetic spectrum, which covers 2500 to 25,000 nm.

Ion exchange columns—Columns used to separate ions or charged particles, like proteins and amino acids.

Ion trap mass spectrometer—A quadrupole ion trap that uses a direct current and radio frequency oscillating alternating current electric fields to trap ions. These traps can be used as selective mass filters.

Ionone rings—Commonly found in aromatic compounds, but also found in the carotenoid pigments that can be converted into retinol and retinal.

Isomerase reactions—Reactions in which enzymes catalyze the structural rearrangement of isomers, or compounds with the same molecular formula but different structure.

Isoprenoids—A class of hydrocarbons composed of isoprene (2-methylbuta-1,3-diene) units linked together in head-to-tail or tail-to-tail conformations. Isoprenoids can be cyclic or acyclic.

Isostatic rebound—The upward movement of the Earth's crust following the melting of ice sheets formed during the last glacial period. Regions of northern Europe (especially Scotland, Fennoscandia, and northern Denmark), Siberia, Canada, and the Great Lakes of Canada and the United States are influenced by this process.

Isotopes—Atoms whose nuclei contain the same number of protons but different numbers of neutrons. Isotopes may be *radioactive* (i.e., the process by which an atom is transformed from one element to another by emission of a subatomic particle) or stable (isotopes that do not decay radioactively). Carbon has two stable isotopes (^{12}C and ^{13}C) and one radioactive isotope (^{14}C).

Kendrick mass—A technique used for hydrocarbon systems, where IUPAC masses are converted to Kendrick masses by multiplying each mass by 14.00000/14.01565. This technique simplifies the display of complex data by rectilinearizing the peak patterns.

Kerogen—Insoluble organic matter within sedimentary rock that is formed through the processes of catagenesis or diagenesis.

α-Ketoglutarate—An intermediate anion in the Kreb cycle. It is a keto acid that is formed when glutamate is deaminated.

Ketones—Compounds that contain at least one carbonyl group (a carbon double bonded to an oxygen atom) that is bonded to two other carbon atoms within the molecule.

Ketose—A saccharide or sugar that contains at least one ketone group per molecule.

Lacustrine—Refers to a lake environment.

Ligated—Refers to the process of producing peptides from amino acids or joining two pieces of DNA.

Lignin—An organic polymer that provides structure in vascular plants. Degradation of lignin yields phenolic acid. The phenolic consitutuents of lignin are used as biomarkers for vascular plants.

Lipid—Operationally defined as a class of organic compounds that are insoluble in water but soluble in organic solvents. Lipids comprise one of four classes of biochemicals, with the others being carbohydrate, protein, and nucleic acids. Lipid often refers specifically to fats, waxes, steroids, and phospholipids. Lipids are the precursors of petroleum.

Lipophilic—Refers to substances that have a strong affinity for lipids, or that may dissolve in or absorb lipids.

Lipoproteins—Any of a group of conjugated proteins in which at least one of the components is a lipid.

Littoral zone—The part of the beach that occurs between the high-tide and low-tide water level. In a lake, it is the area in which light will penetrate to the bottom.

Macromolecules—Large molecules made up of diverse structural subunits with a variety of elements and functional groups. These are opposite to *polymers*, which are composed of repeating units.

Magnesium chelatase—is an enzyme (a ligase) that forms nitrogen–D-metal bonds in coordination complexes. Magnesium chelatase is an important enzyme in porphyrin and chlorophyll metabolism. An alternate name for this class of enzymes is *Mg-Protoporphyrin IX*.

Magnetic resonance—A property in which magnetic nuclei absorb energy and then radiate back out any applied electromagnetic pulse. Any stable nuclide that contains an odd number of protons and/or neutrons has an intrinsic magnetic moment, and therefore a positive spin, whereas nuclides with even numbers of protons and/or neutrons have a spin = 0. ^1H and ^{13}C are most commonly studied using magnetic resonance.

Malonic acid pathway—A biosynthetic pathway that forms phenolic compounds (or precursors of lignin) in lower plants, fungi, and bacteria.

Malonyl-CoA decarboxylase deficiency—A state that results when the enzyme that catalyzes the process of converting malonyl-CoA to acetyl-CoA and carbon dioxide is not present. Malonic acid becomes the product and blocks the Krebs cycle. Lactic acid also builds up, decreasing the pH of the blood and causing detrimental symptoms in humans.

Marine snow—A macroscopic aggregate in seawater that consists of detritus, living organisms (bacteria, copepods, etc.), and some inorganic matter (like clay minerals and particles). Marine snow is an important component of the biological pump, which transports products and wastes from the surface waters to the deep ocean, but much of it is reorganized or remineralized in the upper 1000 m of the water column.

Mass number—See Atomic number.

Mass spectrometry—An analytical method where by atoms are bombarded with electrons, resulting in positively charged particles, which are then passed through an electromagnetic field. The ratio of mass to charge is then measured based on the motion through the field. Mass spectrometry is used to identify compounds either through the formation of a molecular ion or through characteristic fragementation patterns that can be used as a "fingerprint."

Mass spectrum—See Electron impact.

Matrix-assisted laser desorption ionization (MALDI)—An ionization technique used in mass spectrometry that can analyze molecules with high molecular masses (e.g., >10, 000 Da). It is a soft ionization technique that uses a matrix species that mixes with the analyte; the role of the matrix species is to create a mixture that strongly absorbs UV radiation.

Mean life (τ)—The average life expectancy of a radioactive atom. It is expressed as the inverse of the decay constant ($\tau = 1/\lambda$).

Mesophyll—The thick area between epidermal layers where photosynthesis takes place and which consists of parenchyma tissue.

Mesotrophic—Refers to waters or environments with intermediate levels of productivity (not considered oligotrophic, but not as nutrient-rich as eutrophic systems).

Methanolysis—The reverse reaction of transesterification. It exchanges the organic group of an ester with the organic group of an alcohol.

Methoxy—A methyl group ($-CH_3$) linked to an oxygen.

O-Methyl sugars—Aldoses or uronic acids with one or more hydroxyl groups replaced by an etherified methyl group.

Methylotrophic bacteria—A group of heterotrophic microorganisms that use reduced one-carbon species (e.g., methane, methanol, methylated amines, and methylated sulfur compounds) as their sole source of carbon and energy.

Methyltransferase—An enzyme that transfers a sulfur-bound methyl group to a substrate.

Micelles—Small, spherical structures composed of molecules that attract one another to reduce surface tension. The head of the molecule is *hydrophilic* (water-loving), while the interior portion is *hydrophobic* (water-avoiding).

Microfibrils—Fine fiber-like strands of structural support material consisting of glycoproteins and cellulose.

Moderate resolution imaging spectroradiometer (MODIS)—A scientific instrument launched into space on a satellite that captures data in different spectral bands at various resolutions to provide measurements in large-scale global dynamics, such as cloud cover and Earth's radiation budget.

Multiwavelength UV detector—An analytical instrument that allows responses to be monitored and collected from multiple wavelengths. It is often used in liquid chromatographic applications.

Multiwavelength fluorescence detector—A type of detector used in HPLC that contains two monochromators. One is used to select the excitation wavelength, and the second is used to select the fluorescence wavelength or produce a fluorescence spectrum.

Mycosporine-like amino acids—Natural products derived from dihydropyrogallols and amino acids. Marine organisms and plants are known to use these molecules for protection from harmful UV radiation.

N_2 fixation—The biological process of reducing atmospheric N_2 into ammonium, which can then be incorporated into organic matter. This process is restricted to certain species of bacteria that contain the enzyme nitrogenase.

Ninhydrin reaction—A reaction used to detect ammonium or primary or secondary amines in amino acids. Ninhydrin (2,2-dihydroxyindane-1,3-dione) reacts with amino acids to give a deep blue or yellow color, depending on the type of amine present. The amount of amino acid can be quantified by the optical absorbance of the solution.

Nitrate reductase—An NADH/NADPH-dependant enzyme that catalyzes the biological conversation of nitrate to nitrite.

Nitrite reductase—A ferrodoxin-dependant enzyme that catalyzes the biological conversion of nitrite to ammonia.

Nonprotein amino acids—Amino acids that are not constituents of proteins. Examples include D-alanine, an amino acid in peptidoglycan, a bacterial cell wall constituent.

Normal-phase HPLC—A type of high-performance liquid chromatograph (HPLC) where the column is polar (e.g., silica-based) and the mobile phase(s) are nonpolar.

Nucleic acids—A class of biomacromolecules composed of polymers of the nucleotides ribonucleic acid (RNA) and deoxyribonucleic acid (DNA). Nucleic acids are one of four biochemical classes whose functions include protein synthesis, regulation of cellular processes, and conveying genetic information.

Nucleotide—A unit of nucleic acids that is made up of a pentose sugar, a phosphate, and a nitrogenous base. Ribonucleic acid (RNA) is an example of a nucleotide that is found in all organisms. Deoxyribonucleic acid (DNA) is a nucleotide found in most organisms.

Nutricline—A sharp gradient in nutrients within a column of water.

Oddo-Harkins rule—States that nuclides with even atomic numbers are more abundant than nuclides with odd numbers.

Olefinic-containing alkenes—Hydrocarbons containing unsaturated straight or branched chains with at least one double bond.

Oligotrophic—Refers to waters or environments with low biological productivity. This usually is a consequence of low nutrient concentrations or low concentrations of a specific productivity limiting nutrient (e.g., iron).

Oxic—Containing oxygen.

Parenchyma tissue—The cells of nonwoody plants that make up the ground tissue. These cells are used for storage of food. In animals, parenchyma relates to the functional part of an organ.

Pectin—A structural heteropolysaccharide contained in the primary cell walls of terrestrial plants. Pectins are complex polysaccharides that contain 1,4-α-D-galactosyluronic acid residues. Examples include homogalacturonans, substituted galacturonans, and thamnogalacturonans.

Pelagic—Refers to the deep parts of the ocean that are removed from land.

Pentose phosphate pathway—A process that reduces $NADP^+$ to NADPH and produces other sugars as an alternative to glycolysis. The enzymes involved in this pathway are located in the cytosol of cells and operate mainly in liver and adipose cells.

Peptidoglycan—A macromolecule consisting of linear polysaccharide chains cross-linked by short peptides, which forms a layer outside the plasma membrane of bacteria, forming the cell wall and providing structural support. Also called *murein*.

Permethylated—Refers to a compound where all carbon atoms have a methyl group bonded to them.

Petrogenic—Refers to something that is the product of rock formation, or that is produced when rocks form.

Phenols—Compounds that contain an aromatic ring attached to the hydroxyl group. This attachment makes the compounds acidic.

Phenylpropanoids—Organic compounds found in plants derived from phenylalanine. They contain a 6-carbon ring linked to oxygen with a 3-carbon side chain. They are important in the structure and protection of the plant.

Pheopigments—Nonphotosynthetic pigments that are the degradation products of chlorophyll.

Phospholipids—A class of lipids that function as the major component of all cell membranes. Most phospholipids contain a diglyceride, a phosphate group, and a simple organic molecule such as choline. An exception is *sphingomyelin*, which has sphingosine, an amino alcohol formed from palmitate and serine, as its backbone. The *phospholipid bilayer* is a membrane consisting of two layers of phospholipids.

Photodiode array detector (PDA)—A series of hundreds or thousands of photodetectors that are capable of converting light energy into current or voltage, used in analytical instrumentation involving transmission, absorption, or emission of light.

Photolysis—The chemical decomposition of materials due to light.

Photorespiration—Oxygen consumption due to the oxidation of phosphoglycolate. This process occurs when the amount of oxygen is high but the amount of carbon dioxide is low. No net ATP is produced during this process.

Photosynthate—The general term for any chemical products of photosynthesis; usually refers to carbohydrates or simple sugars.

Photosynthesis—The use of light energy to convert carbon dioxide into carbohydrates through reduction with water. Oxygen is produced as a waste product.

Phycobilins—Water-soluble accessory pigments found in red algae and cyanobacteria.

Phycobiliproteins—Water-soluble proteins found in cyanobacteria and red algae that transfer light energy to chlorophylls during photosynthesis.

Phycobilisome—A complex formed by a phycobilin covalently bound to a phycobiliprotein that acts as a chromophore to capture light.

Phycocyanin—A blue photosynthetic pigment in cyanobacteria and red algae.

Phycoerythrin—A red photosynthetic pigment (phycobilin) found in cyanobacteria and red algae.

Phylogenetic—Refers to a classification system for evaluating how similar organisms are based on evolution. Molecular sequencing is one tool used to identify how an organism should be classified relative to related organisums.

Phylogeny—The classification of species into evolutionary trees based on their evolutionary development or history. Phylogenetic relationships provide higher taxonomic grouping of organisms.

Phytoalexins—An antimicrobial toxin that attacks an intruding pathogen.

Phytol—An alcohol with 20 carbons that includes four methyl groups. Phytols belong to the class of compounds isoprenoids, which are made up of isoprene units.

Phytoplankton—Unicellular plants and algae that are photosynthetic and occur in the photic zone of water bodies.

Pigments—A class of compounds that provide coloration and enhance absorption of light over specific wavelengths during photosynthesis. The dominant pigment in most photosynthetic organisms is chlorophyll *a*. Pigments are used as biomarkers for understanding algal community structure and (paleo)-productivity. In addition, photosynthetic organisms have *accessory pigments*, which allow them to absorb additional light from shorter wavelengths. Accessory pigments capture light energy in photosynthesis and pass along the adsorbed light energy compounds, such as carotenoids and a variety of chlorophylls and related tetrapyrroles.

Plasmalemma—Another name for plasma membrane or cell membrane.

Plastid—One of a family of plant organelles such as a chloroplast, chromoplast, or amyloplast/leucoplast.

Polysaccharides—Two or more monosaccharides (carbohydrate containing a sugar unit) attached by glycosidic bonds used for energy storage. Examples of polysaccharides are cellulose, hemicellulose, and pectin.

Porphobilinogen—A pyrrole produced by aminolevulinate dehydratase that is involved in porphyrin metabolism.

Porphyrins—A biomarker that originates from a source such as chlorophyll and is characterized by a tetrapyrrole ring containing vanadium or nickel.

Porphyrin ring—An organic ring structure, that contains an iron atom in the center of the large heterocyclic ring.

Positron—A positively charged electron. During radioactive decay, release of a positron results in a proton becoming a neutron and a decrease in the atomic number by one unit.

Precision—The degree to which repeated measurements under unchanged conditions show the same results. Precision is also called reproducibility or repeatability.

Principal components analysis—A mathematical procedure that reduces the dimensionality of data by transforming a number of correlated variables into a smaller number of uncorrelated variables that express as much of the correlation in the data as possible. Used for exploratory data analysis and for the creation of predictive models.

Prokaryote—A single-celled organism that does not have membrane-bound organelles or a nucleus.

Proteins—A macromolecule consisting of chains of amino acids linked by peptide bonds.

Proteolytic enzymes—Enzymes that cause proteolysis, the breakdown of proteins, by hydrolyzing the peptide bonds that link the amino acid monomers together.

Proteomics—The field of study related to the structure and function of proteins, an organism's entire expressed set of which is referred to as the *proteome*.

Protochlorophyllide—The immediate precursor of chlorophyll *a*. It is highly fluorescent and lacks the phytol side chain of chlorophyll.

Protochlorophyllide oxidoreductase—The enzyme that is part of porphyrin and chlorophyll metabolism and that produces protochlorophyllide.

Protoheme—A complex organic pigment containing iron, which is the precursor to heme.

Protoporphyrin IX—A part of porphyrin metabolism, which is a direct precursor of heme and a carrier for divalent cations.

Psychrophiles—Organisms that live at extremely cold temperatures, such as those found in Antarctica or at the bottom of the ocean.

Pulsed amperometric detection (PAD)—A detector commonly used in anion-exchange chromatography to provide a sensitive and selective means of determining carbohydrates and other oxidizable species, such as amines and sulfur compounds. Pulsed amperometric detection works by applying a repeating potential vs. time wave form to the working electrode in a flowthrough detector cell.

Purines—One of the two classes of bases in DNA and RNA. Guanine (G) and adenine (A) are both purine bases.

Pyran—A six-membered heterocyclic ring that contains two double bonds. The ring is composed of five carbon atoms and one oxygen atom.

Pyranose—Refers to a group of carbohydrates that form a six-membered ring consisting of five carbon atoms and one oxygen atom. Pyranose rings are formed by the dehydration reaction of the hydroxyl group on C-5 of a sugar with the aldehyde group on C-1.

Pyrimidines—A class of nitrogen-containing, aromatic compounds present in nucleic acids. Cytosine (C) and thymine (T) are the two pyrimidines found in DNA molecules.

Pyrogenic—Refers to something that is the product of heat.

Pyrrole—A heterocyclic aromatic organic compound that is an important component of hemes, chlorins and other macrocycles.

Quadrupole—A mass filter used in mass spectrometry. A quadrupole has four voltage-carrying rods that serve to give ions, which travel between them, oscillations. Only ions with the right mass/charge ratio (m/z) can undergo these oscillations without hitting one of these rods.

Racemization—The process of the interconversion between two enantiomers of a chiral molecule, resulting in an overall conversion of a mixture in which one enantiomer exists in excess to an equal mixture of the two enantiomers.

Radioactivity—The spontaneous adjustment of nuclei of unstable nuclides to a more stable state.

Radionuclide—See Isotope.

Raney-nickel desulfurization—A desulfurization reaction (reducing thioacetals to saturated hydrocarbons) carried out with the solid, fine-grained, Ni–Al catalyst developed by Murray Raney in 1926. In organic syntheses, it is often used as an intermediate step in reduction of aldehydes and ketones to hydrocarbons.

Redox—Short for reduction–oxidation reaction, which describes any chemical reaction in which the atoms within the reaction show a change in oxidation state. *Oxidation* is the part of the reaction where a loss of electrons or an increase in oxidation state occurs. *Reduction* is the part of the reaction where a gain of electrons or decrease in oxidation state occurs.

Retention time—The elapsed time between an analyte's injection onto a column (stationary phase) and the analyte's elution/detection. Compounds with different retention times are separated because of differing intermolecular interactions with the stationary phase.

Reverse osmosis–electrodialysis (RO-ED)—The combined process that transports salt ions from one solution to another through both reverse-osmotic membrane flow and ion-exchange membranes under the influence of an electric potential. This process has recently been adapted for use in isolating dissolved organic matter from aquatic samples.

Reversed phase–HPLC—A form of column chromatography used to separate, identify, and quantify compounds based on interactions between the polar or aqueous mobile phase and the nonpolar column (stationary phase).

Ribosomes—Small organelles in the cytoplasm of living organisms consisting of rRNA and enzymes. Ribosomes are involved in the production of proteins by translating messenger RNA.

Saponins—A metabolite in which a sugar is bound to a compound that is not a carbohydrate. When in solution, saponins create the foam that occurs when the solution is shaken.

Sapropel—A deposit formed from decaying organic matter resulting from an anoxic environment.

Sea-viewing Wide Field-of-view Sensor (SeaWiFS)—A NASA program that provides quantitative data on global ocean biooptical properties from ocean color data. Characteristics of the ocean, such as chlorophyll *a* and water clarity, are measured from the satellite orbiting Earth with the SeaWiFS sensor on it.

Secondary ion mass spectrometry—An analytical technique used to analyze the composition of solid surfaces and thin films by focusing an ion beam on the material and collecting and analyzing the ejected "secondary" ions.

Seeps—Places where liquid or gaseous hydrocarbons collect as they escape through fissures and fractures in rocks where they were formed. Also known as hydrocarbon seeps.

Shielding—In an applied magnetic field (B_0), such as that used in NMR spectroscopy, electrons circulate and produce an induced field (B_i) which opposes the applied field. The effective field at the nucleus will be $B = B_0 - B_i$. As a result, the nucleus is described as experiencing a diamagnetic shielding.

Shikimic acid pathway (shikimate)—Participates in the biosynthesis of most plant phenolics (see chapter 13 for more details). The shikimic acid pathway converts simple carbohydrate precursors derived from glycolysis and the pentose phosphate pathway to aromatic amino acids.

Solid-phase extraction—A process that separates dissolved or suspended compounds from other compounds in the mixture based on certain physical or chemical properties. This process is usually used to concentrate and purify samples for analysis.

Spallation—A type of nuclear reaction in which a photon or particle hits a nucleus and causes it to emit many other particles or photons.

Specific activity (*A*)—A term used to describe the observed counting rate for a radionuclide sample. The specific activity of a sample is usually described by the following equation: $A = \lambda N$, where λ = the rate of decay and N = number of atoms.

Stable isotopes—Atoms that contain the same number of protons and electrons, but varying numbers of neutrons. A different neutron count changes the mass of the atom but not the chemical identity. Isotopes are stable if they are not radioactive.

Steranes—Diagenetic and catagenetic products of steroids and sterols. The sterone structure makes up the core of steroids and sterols and is all that remains after the loss of functional groups during diagenesis and catagenesis. Sterones are often used as biomarkers for eukaryote-derived material.

Stereoisomers—Compounds that have the same molecular formula and sequence of bonded atoms, but differ in the three-dimensional orientation of their atoms in space. *Structural isomers* share the same molecular formula, but the bond connections and/or their order between different atoms/groups differ.

Steroids—A class of polycyclic compounds comprising three six-membered rings and one five-membered ring with a wide range of side chains and functional groups. Steroids are synthesized both by plants (*phytosterols*) and animals.

Sterols—A class of tetracyclic organic compounds consisting of four rings; three of the rings have six carbon atoms and one ring has five carbon atoms and a hydroxyl (–OH) group. These compounds are important membrane and hormone components in eukaryotic organisms and are widely used as biomarkers. *Steranes* are hydrocarbons that originate from sterols.

Stomata—Open pores on the stem and leaf epidermis that allow exchange of carbon dioxide and oxygen. Parenchyma surround the pore opening and regulate its size.

Stroma—The matrix between the grana in chloroplasts where the "dark" or biochemical reactions of photosynthesis occur.

Suberins—A class of compounds similar to cutin. Suberins contain ester-linked fatty acids that have been hydroxylated.

Submarine vents—Fissures in the oceanic crust from which geothermally heated water is expelled. Dense communities of organisms dependant on chemosynthetic bacteria and archaea form around the mineral-rich hot water that erupts from these fissures. Also known as hydrothermal vents.

Symmetry rule—States that a stable nuclide with a low atomic number (Z) has approximately the same number of neutrons and protons, whereby the N/Z ratio is near unity.

Tannins—Polyphenols found in plant material that will bind to and precipitate proteins, or shrink proteins, amino acids, and alekloids. The breakdown of tannins is what causes the ripening of fruit and the aging of wine.

Teratogenic—Refers to a chemical or substance that is capable of interfering with the development of an embryo or fetus, usually causing defects.

Terpenoids (terpenes)—A class of lipids containing two or more isoprenes.

Terrestrial—Refers to something that originated from land.

Tetrapyrroles—Compounds containing four pyrrole rings. Bilirubin, phycobilins, porphyrins, and chlorophylls are all examples of tetrapyrroles.

Thermokarst—A land surface that forms as ice-rich permafrost thaws. It occurs extensively in Arctic areas, and on a smaller scale in mountainous areas such as the Himalayas and the Swiss Alps.

Thermophiles—Organisms able to grow at high temperatures, such as in hot springs. The highest temperature at which a prokaryotic organism has been recorded to grow is 121 °C. The highest temperature at which eukaryotes have been known to survive is 60 °C.

Thylakoid—A sac-like membranous structure of the grana of chloroplasts, which houses chlorophyll and therefore where the "light" reactions of photosynthesis occur.

Thylakoid membrane—The membrane that surrounds the thylakoid lumen and contains many intramembranous proteins that are essential to the photosynthetic process.

Time-of-flight (TOF)-mass spectrometry—A mass separation technique in which ions are accelerated by an electric field. The velocity of the ion depends on the mass-to-charge ratio. The time it takes a particular ion to travel a known distance helps determine the mass to charge ratio of the particular ion.

Transketolase reactions—Reactions that link the pentose phosphate pathway and glycolysis: converting 5-carbon sugars to 6-carbon sugars.

Triacylglycerols (or triglycerides)—A class of lipids composed of three fatty acids esterified to glycerol. These compounds function as short-term energy stores in marine organisms, such as zooplankton.

Trichromatic—Refers to three colors, or three pigments, such as chlorophyll *a*, *b*, and *c*.

Triterpenoids—A class of lipids, similar to sterols, widely used as biomarkers. These compounds are composed of six isoprene units ($\sim C_{30}$).

Trophic level—Refers to organisms with identical feeding habits. *Trophic relationships* explain the interactions between organisms feeding at a specific level or across levels in a food chain or food web. Trophic level 1 consists of *primary producers* (plants), trophic level 2 refers to herbivores, trophic level 3 to small carnivores, etc.

Ultrafiltration—A method for separating samples into different sizes based on molecular size. It typically uses membranes with different cutoffs (e.g., 1, 5, and 10 kDa) to isolate fractions of dissolved organic matter in aquatic systems.

Ultrafiltered dissolved organic matter—The dissolved fraction of organic matter that passes through the ultrafiltration process, usually within the size range of 10^{-7} to 10^{-9} m.

Ultraviolet (UV) radiation—The portion of the electromagnetic spectrum that has wavelengths shorter than visible light and longer than x-rays, a range of 10 to 400 nm.

Uronic acids (glucuronic and galacturonic acids)—Minor sugars that are differentiated from other aldoses by the replacement of the methanolic carbon opposite the carbonyl carbon with a carboxylic acid group.

Vacuole—An organelle that is membrane bound and stores water, pigments, and various other substances.

van der Waals interactions—Intermolecular attractions between one molecule and a neighboring molecule. All molecules experience these relatively weak forces (compared to chemical bonds). These interactions play important roles in the organization and interaction of molecules in all states of matter.

van Krevelen diagrams—Two-dimensional scatter plots of H : C ratios on the *y*-axis versus O : C ratios on the *x*-axis. These plots have been used to identify the dominant compound classes (e.g., lipid, carbohydrate, protein) and maturity of petroleum and coal. They have also been used to aid in the interpretation of data obtained using FT-ICR MS.

Vascular tissues—Tissues that transport fluids and nutrients in plants and provide support. Xylem and phloem are the two main components of vascular tissue.

Visible (Vis) radiation—The portion of the electromagnetic spectrum that can be detected by the human eye, a range of wavelengths from 380 to 750 nm.

Water use efficiency (WUE)—The efficiency with which plants use water while fixing CO_2. WUE is the ratio of the energy capture by photosynthesis per unit of water lost to transpiration.

Waxes (or wax esters)—Serve as energy stores in plants, animals, and microbes. They also act as a protective outer coating on plant tissues. Waxes are composed of a long-chain alcohol linked to a long-chain fatty acid by an ester bond.

Weathering—Describes a variety of mechanical and chemical processes that cause exposed rocks to break down.

Xenobiotic—Refers to chemicals found in an organism that are not typically expected to be present in that particular organism.

Zooplankton—Microscopic animals and protozoans that feed on phytoplankton.

Zwitterions—Dipolar ions, ionic molecules that contain a positive and negative charge, such as an amino acid in a neutral solution with a protonated amino group ($-NH_3^+$) and a deprotonated carboxylic acid group ($-COO^-$).

Bibliography

Preface

DeCaprio, A. P. 2006. *Toxicologic biomarkers.* Culinary and Hospitality Industry Publications Services, Weimar, TX.

Eglinton, G., and M. Calvin. 1967. Chemical fossils. *Scientific American* **216:** 32–43.

Eglinton, G., P. M. Scott, T. Belsky, A. L. Burlingame, W. Richter, and M. Calvin. 1964. Occurrence of isoprenoid alkanes in a Precambrian sediment. Pp. 41–74 *in* G. D. Hobson and M. C. Louis. eds., *Advances in organic geochemistry.* Pergamon Press, Oxford, UK.

Gaines, S. M., G. Eglinton, and J. Rüllkolter. 2009. *Echoes of life—What fossil molecules reveal about earth history.* Oxford University Press, New York.

Hunt, J. H. 1996. *Petroleum geochemistry and geology*, 2nd ed. Freeman, New York.

Killops, S., and V. Killops. 2005. *Introduction to organic geochemistry*, 2nd ed. Backwell, Malden, MA.

Marsh, A., and K. R. Tenore. 1990. The role of nutrition in regulating the population dynamics of opportunistic, surface deposit-feeders in a mesohaline community. *Limnology and Oceanography* **35:** 710–724.

Meyers, P. A. 1997. Organic geochemical proxies of paleoceanographic, paleolimnologic, and paieoclimatic processes. *Organic Geochemistry* **27:** 213–250.

Meyers, P. A. 2003. Application of organic geochemistry to paleolimnological reconstructions: a summary of examples from the Laurentian Great Lakes. *Organic Geochemistry* **34:** 261–290.

Peters, K. E., C. C., Walters, and J. M. Moldowan. 2005. *The biomarker guide.* 2nd ed.: *Biomarkers and isotopes in petroleum systems and earth history.* Cambridge University Press, Cambridge, UK.

Tenore, K. R., and E. Chesney, Jr. 1985. The effects of interaction of rate of food supply and population density on the bioenergetics of the opportunistic polychaete, *Capitella capitala* (type 1). *Limnology and Oceanography* **30:** 1188–1195.

Timbrell, J. A. 1998. Biomarkers in toxicology (review). *Toxicology* **129:** 1–12.

Treibs, A., 1934. Organic mineral substances, II: occurrence of chlorophyll derivatives in an oil shale of the upper triassic. *Annals.* **509:** 103–14.

Wilson, S. H. and W. A. Suk. 2002. *Biomarkers of environmentally associated disease.* Lewis, Boca Raton, FL.

Chapter 1

Arnold, G. L., A. D. Anbar, J. Barling, and T. W. Lyons. 2004. Molybdenum isotope evidence for widespread anoxia in mid-proterozoic oceans. *Science* **304:** 87–90.

Atsatt, P. R., and D. J. O'Dowd. 1976. Plant defense guilds. *Science* **193:** 24–29.

Bacon, M. A. 2004. Water use efficiency in plant biology. Pp. 1–26 *in* M. A. Bacon, ed., *Water use efficiency in plant biology.* Blackwell, Oxford, UK.

Badger, M. R., and G. D. Price. 1994. The role of carbonic anhydrase in photosynthesis. *Annual Review of Plant Physiology and Plant Molecular Biology* **45:** 369–392.

Banthorpe, D. V., and B. V. Charlwood. 1980. The terpenoids. Pp. 185–220 *in* E. A. Bell and B. V. Charlwood, eds., *Encyclopedia of plant physiology*, n.S. Springer, Berlin.

Bassham, J., A. Benson, and M. Calvin. 1950. The path of carbon in photosynthesis. *Journal of Biological Chemistry* **185:** 781–787.

Bekker, A., and others. 2004. Dating the rise of atmospheric oxygen. *Nature* **427:** 117–120.

Bell, E. A., and B. V. Charlwood. 1980. Secondary plant products. *In Encyclopedia of plant physiology*. Springer, Berlin.

Bowers, W. S., T. Ohta, J. S. Cleere, and P. A. Marsella. 1986. Discovery of insect antijuvenile hormones in plants. *Science* **193:** 542–547.

Brock, T. D., M. T. Madigan, J. M. Martinko, and J. Parker. 1994. *Biology of microorganisms*, 7th ed. Prentice Hall, Upper Saddle River, NS.

Brocks, J. J., G. A. Logan, R. Buick, and R. E. Summons. 1999. Archean molecular fossils and the early rise of eukaryotes. *Science* **285:** 1033–1036.

Dakora, F. F. 1995. Plant flavonoids: Biological molecules for useful exploitation. *Australian Journal of Plant Physiology* **22:** 87–99.

Delong, E. F. 2006. Archaeal mysteries of the deep revealed. *Proceedings of the National Academy of Sciences of the USA* **103:** 6417–6418.

Delong, E. F., and N. R. Pace. 2001. Environmental diversity of bacteria and archaea. *Systematic Biology* **50:** 1–9.

Edwards, R., and J. A. Gatehouse. 1999. Secondary metabolism. Pp. 193–218. *in Plant biochemistry and molecular biology*. Wiley, New York.

Ehleringer, J. R., and R. K. Monson. 1993. Evolutionary and ecological aspects of photosynthetic pathway variation. *Annual Review of Ecology and Systematics* **24:** 411–439.

Ehleringer, J. R., T. E. Cerling, and M. D. Dearing. 2005. *A history of atmospheric* CO_2 *and its effect on plants, animals, and ecosystems*. Springer, New York.

Engel, M. H., and S. A. Macko. 1986. Stable isotope evaluation of the origins of amino acids in fossils. *Nature* **323:** 615–625.

Facchini, P. J., and V. De Luca. 1995. Phloem-specific expression of tyrosin/dopa decarboxylase genes and the biosynthesis of isoquinoline alkaloids in opium poppy. *Plant Cell* **7:** 1811–1821.

Facchini, P. J., C. Penzes-Yos, N. Samanani, and B. Kowalchuk. 1998. Expression patterns conferred by tyrosine/dihydroxyphenylaline decarboxylase promoters from opium poppy are conserved in transgenic tobacco. *Plant Physiology* **118:** 69–81.

Facchini, P. J., L. Kara, L. Huber-Allanach, and W. Tari. 2000. Plant aromatic L-amino acid decarboxylases: Evolution, biochemistry, regulation, and metabolic engineering applications. *Phytochemistry* **54:** 121–138.

Farquhar, J., H. Bao, and M. Thiemens. 2000. Atmospheric influence of earth's earliest sulfur cycle. *Science* **289:** 756–758.

Galimov, E. M. 2001. *Phenomenon of life: Between equilibrium and nonlinearity—Origin and principles of evolution*. Urss, Moscow.

———. 2004. Phenomenon of life: Between equilibrium and nonlinearity. *Origins of Life and Evolution of the Biosphere* **34:** 599–613.

———. 2005. Redox evolution of the earth caused by a multi-stage formation of its core. *Earth and Planetary Science Letters* **233:** 263–276.

———. 2006. Isotope organic geochemistry. *Organic Geochemistry* **37:** 1200–1262.

Giovannoni S. J., and U. Stingl. 2005. Molecular diversity and ecology of microbial plankton. *Nature* **427:** 343–348.

Han, T. M., and B. Runnegar. 1992. Megascopic eukaryotic algae from the 2.1 billion year old Negaunne iron formation, Michigan. *Science* **257:** 232–235.

Harborne, J. B. 1977. *Introduction to ecological biochemistry*. Academic Press, London.

Harborne, J. B., T. J. Marby, and H. Marby. 1975. *The flavonoids*. Chapman Hall, London.

Haslam, E. 1989. *Plant polyphenols: Vegetable tannins revisited*. Cambridge University Press, Cambridge, UK.

Herrmann, K. M., and L. M. Weaver. 1999. The shikimate pathway. *Annual Review of Plant Physiology and Plant Molecular Biology* **50:** 473–503.

Ingalls, A. E., and others. 2006. Quantifying archaeal community autotrophy in the mesopelagic ocean using natural radiocarbon. *Proceedings of the National Academy of Sciences of the USA* **103:** 6442–6447.

Kashefi, K., and D. R. Lovley. 2003. Extending the upper temperature limit for life. *Science* **301:** 934.

Kates, M., D. J. Kushner, and A. T. Matheson. 1993. *The biochemistry of archaea (archaebacteria)*. Elsevier Science, Amsterdam.

Killops, S., and V. Killops. 2005. *Introduction to organic geochemistry*. Blackwell, Malden, MA.

Kossel, A. 1891. Ueber die chemische zusammensetzung der zelle. *DuBois-Reymond's Archives* **181:** 181–186.

Kosslak, R. M., R. Bookland, J. Barkei, H. E. Paaren, and E. R. Appelbaum. 1987. Induction of *Bradyrhizobium japonicum* common nod genes by isoflavones isolated from glycine max. *Proceeding of the National Academy of Sciences of the USA* **84:** 7428–7432.

Kump, L. R., J. F. Kasting, and R. G. Crane. 2010. *Earth system*, 3rd ed. Prentice Hall, Upper Saddle River, NJ.

Lehninger, A., D. Nelson, and M. Cox. 1993. *Principles of biochemistry*, 2nd ed. Worth, New York.

Luckner, M. 1984. *Secondary metabolism in microorganisms, plants, and animals*. Springer, Berlin.

Mabry, T. J., K. R. Markham, and M. B. Thomas. 1970. *The systematic identification of flavonoids*. Springer, Berlin.

Miller, S. L. 1953. Production of amino acids under possible primitive earth conditions. *Science* **117:** 245.

Miller, S. L., and H. C. Urey. 1959. Organic compound synthesis on the primitive earth. *Science* **130:** 245.

Monson, R. K. 1989. On the evolutionary pathways resulting in C_4 photosynthesis and crassulacean acid metabolism (CAM). *Advances in Ecological Research* **19:** 57–110.

Palenik, B. 2002. The genomics of symbiosis: Hosts keep the baby and the bath water. *Proceedings of the National Academy of Sciences of the USA* **99:** 11996–11997.

Pant, P., and R. P. Ragostri. 1979. The triterpenoids. *Phytochemistry* **18:** 1095–1108.

Peters, K. E., C. C. Walters, and J. M. Moldowan. 2005. Biomarkers and isotopes in the environment and human history. Pp. 471 *in The biomarker guide*. Cambridge University Press, Cambridge, UK.

Ransom, S. L., and M. Thomas. 1960. Crassulacean acid metabolism. *Annual Review of Plant Physiology* **11:** 81–110.

Reysenbach, A. L., A. B. Banta, D. R. Boone, S. C. Cary, and G. W. Luther. 2000. Microbial essentials at hydrothermal vents. *Nature* **404:** 835.

Sagan, C., and C. Chyba. 1997. The early faint sun paradox: Organic shielding of ultraviolet-labile greenhouse gases. *Science* **276:** 1217–1220.

Schenk, J.E.A., R. G. Herrmann, K. W. Jeon, N. E. Muller, and W. Schwemmler. 1997. *Eukaryotism and symbiosis*. Springer, New York.

Schopf, J. W., and B. M. Packer. 1987. Early archean (3.3 to 3.5 ga-old) fossil microorganisms from the Warrawoona Group, western Austrailia. *Science* **237:** 70–73.

Seigler, D. S. 1975. Isolation and characterization of naturally occurring cyanogenic compounds. *Phytochemistry* **14:** 9–29.

Sogin, M. L. 2000. Evolution of eukaryotic ribosomal RNA genes. Pp. 260–262 *in* S. Parker, ed., *McGraw-Hill yearbook of science and technology*. McGraw-Hill, New York.

Swain, T. 1966. *Comparative phytochemistry*. Academic Press, London.

Thorington, G., and L. Margulis. 1981. *Hydra viridis*: Transfer of metabolites between hydra and symbiotic algae. *Biological Bulletin* **160:** 175–188.

Voet, D., and J. G. Voet. 2004. *Biochemistry*. Wiley, New York.

Waterman, P. G., and S. Mole. 1994. *Analysis of phenolic plant metabolites*. Blackwell Scientific.

Woese, C. R. 1981. Archaebacteria. *Scientific American* **244:** 98–122.

Woese, C. R., O. Kandler, and M. L. Wheelis. 1990. Towards a natural system of organisms: Proposal for the domains archaea, bacteria, and eucarya. *Proceedings of the National Academy of Sciences of the USA* **87:** 4576–4579.

Chapter 2

Aller, R. C. 1998. Mobile deltaic and continental shelf muds as suboxic, fluidized bed reactors. *Marine Chemistry* **61:** 143–155.

Almendros, G., R. Frund, F. J. Gonzalez-Vila, K. M. Haider, H. Knicker, and H. D. Ludemann. 1991. Analysis of C-13 and N-15 CPMAS NMR spectra of soil organic matter and composts. *FEBS Letters* **282:** 119–121.

Aluwihare, L. L., D. J. Repeta, S. Pantoja, and C. G. Johnson. 2005. Two chemically distinct pools of organic nitrogen accumulate in the ocean. *Science* **308:** 1007–1010.

Arzayus, K. M., and E. A. Canuel. 2005. Organic matter degradation in sediments of the York River estuary: Effects of biological vs. physical mixing. *Geochimica et Cosmochimica Acta* **69:** 455–463.

Aspinall, G. O. 1970. Pectins, plant gums, and other plant polysaccharides. Pp. 515–536 *in* W. Pigman and D. Horton, eds., *The carbohydrates. Chemsitry and biochemistry*, 2nd ed. Academic Press, New York.

Aufdenkampe, A. K., J. I. Hedges, J. E. Richey, A. V. Krusche, and C. A. Llerena. 2001. Sorptive fractionation of dissolved organic nitrogen and amino acids onto fine sediments within the amazon basin. *Limnology and Oceanography* **46:** 1921–1935.

Benner, R., P. G. Hatcher, and J. I. Hedges. 1990. Early diagenesis of mangrove leaves in a tropical estuary: Bulk chemical characterization using solid-state ^{13}C NMR and elemental analyses. *Geochemica et Cosmochimica acta* **54:** 2003–2013.

Benner, R., J. D. Pakulski, M. Mccarthy, J. I. Hedges, and P. G. Hatcher. 1992. Bulk chemical characteristics of dissolved organic matter in the ocean. *Science* **255:** 1561–1564.

Bianchi, T. S. 2007. *Biogeochemistry of estuaries.* Oxford University Press, New York.

Bianchi, T. S., and E. A. Canuel. 2001. Organic geochemical tracers in estuaries. *Organic Geochemistry* **32:** 451.

Bianchi, T. S., and S. Findlay. 1991. Decomposition of Hudson Estuary macrophytes: Photosynthetic pigment transformations and decay constants. *Estuaries* **14:** 65–73.

Bianchi, T. S., B. Johansson, and R. Elmgren. 2000. Breakdown of phytoplankton pigments in baltic sediments: Effects of anoxia and loss of deposit-feeding macrofauna. *Journal of Experimental Marine Biology and Ecology* **251:** 161–183.

Bianchi, T. S., and others. 2002. Do sediments from coastal sites accurately reflect time trends in water column phytoplankton? A test from Himmerfjarden Bay (Baltic Sea proper). *Limnology and Oceanography* **47:** 1537–1544.

Clark, L. L., E. D. Ingall, and R. Benner. 1998. Marine phosphorus is selectively remineralized. *Nature* **393:** 426.

Cloern, J. E., E. A. Canuel, and D. Harris. 2002. Stable carbon and nitrogen isotope composition of aquatic and terrestrial plants of the San Francisco Bay estuarine system. *Limnology and Oceanography* **47:** 713–729.

Conley, D. J., C. L. Schelske, and E. F. Stoermer. 1993. Modification of the biogeochemical cycle of silica with eutrophication. *Marine Ecology-Progress Series* **101:** 179–192.

Cooper, S. R., and G. S. Brush. 1993. A 2,500-year history of anoxia and eutrophication in Chesapeake Bay. *Estuaries* **16:** 617–626.

Day, J. W., C.A.S. Hall, W. M. Kemp, and A. Yanez-Aranciba. 1989. *Estuarine ecology.* Wiley, New York.

Duan, S., and T. S. Bianchi. 2006. Seasonal changes in the abundance and composition of plant pigments in particulate organic carbon in the lower Mississippi and Pearl rivers (USA). *Estuaries* **29:** 427–442.

Eadie, B. J., B. A. Mckee, M. B. Lansing, J. A. Robbins, S. Metz, and J. H. Trefrey. 1994. Records of nutrient-enhanced coastal ocean productivity in sediments from the Louisiana continental shelf. *Estuaries* **17:** 754–765.

Elmgren, R., and U. Larsson. 2001. Eutrophication in the Baltic Sea area: Integrated coastal managemant issues. Pp. 15–35 *in* B. von Bodungen and R. K. Turner, eds., *Science and integrated coastal management.* Dahlem University Press, Berlin.

Emerson, S., and J. Hedges. 2008. *Chemical oceanography and the marine carbon cycle.* Cambridge University Press, Cambridge, UK.

Emerson, S., and J. Hedges. 2009. *Chemical oceanography and the marine carbon cycle.* Cambridge University Press, New York.

Filley, T. R., K. H. Freeman, T. S. Bianchi, M. Baskaran, L. A. Colarusso, and P. G. Hatcher. 2001. An isotopic biogeochemical assessment of shifts in organic matter input to holocene sediments from Mud Lake, Florida. *Organic Geochemistry* **32:** 1153–1167.

Fry, B. 2006. Stable isotope ecology. Springer Science+Business Media, New York.

Gong, C., and D. J. Hollander. 1997. Differential contribution of bacteria to sedimentary organic matter in oxic and anoxic environments, Santa Monica Basin, California. *Organic Geochemistry* **26:** 545–563.

Goñi M. A., and J. I. Hedges. 1995. Sources and reactivities of marine-derived organic matter in coastal sediments as determined by alkaline CuO oxidation. *Geochimica et Cosmochimica Acta* **59:** 2965–2981.

Gordon, E. S., and M. A. Goni. 2003. Sources and distribution of terrigenous organic matter delivered by the Atchafalaya River to sediments in the northern Gulf of Mexico. *Geochimica et Cosmochimica Acta* **67:** 2359–2375.

Hartnett, H. E., R. G. Keil, J. I. Hedges, and A. H. Devol. 1998. Influence of oxygen exposure time on organic carbon preservation in continental margin sediments. *Nature* **391:** 572–574.

Harvey, H. R., and S. A. Macko. 1997. Kinetics of phytoplankton decay during simulated sedimentation: Changes in lipids under oxic and anoxic conditions. *Organic Geochemistry* **27:** 129–140.

Harvey, H. R., J. H. Turttle, and J. T. Bell. 1995. Kinetics of phytoplankton decay during simulated sedimentation: Changes in biochemical composition and microbial activity under oxic and anoxic conditions. *Geochimica et Cosmochimica Acta* **59:** 3367–3377.

Hatcher, P. G. 1987. Chemical structural studies of natural lingin by dipolar dephasing solid-state C-13 nuclear magnetic resonance. *Organic Geochemistry* **11:** 31–39.

Hebting, Y., and others. 2006. Early diagenetic reduction of organic matter: Implications for paleoenvironmental reconstruction. *Geochimica et Cosmochimica Acta* **70:** A239–A239.

Hedges, J. I. 1992. Global biogeochemical cycles: Progress and problems. *Marine Chemistry* **39:** 67–93.

Hedges, J. I., and R. G. Keil. 1995. Sedimentary organic matter preservation: an assessment and speculative synthesis. *Marine Chemistry* **49:** 81–115.

Hedges, J. I., and F. G. Prahl. 1993. Pp. 237–253 *in* S. M. a. M. Engel ed., *Organic geochemistry principles and applications.* Plenum Press, New York.

Hedges, J. I., P. G. Hatcher, J. R. Ertel, and K. J. Meyersschulte. 1992. A comparison of dissolved humic substances from seawater with Amazon River counterparts by C-13-NMR spectrometry. *Geochimica et Cosmochimica Acta* **56:** 1753–1757.

Hedges, J. I., J. A. Baldock, Y. Gelinas, C. Lee, M. L. Peterson, and S. G. Wakeham. 2002. The biochemical and elemental compositions of marine plankton: A NMR perspective. *Marine Chemistry* **78:** 47–63.

Hupfer, M., R. Gachter, and R. Giovanoli. 1995. Transformation of phosphorus species in settling seston and during early sediment diagenesis. *Aquatic Sciences* **57:** 305–324.

Hurd, D. C., and D. W. Spencer. 1991. Marine particles: Analysis and characterization. *Geophysical Monograph* **63:** 472.

Ingall, E. D., P. A. Schroeder, and R. A. Berner. 1990. Characterization of organic phosphorus in marine sediments by ^{31}P NMR. *Chemical Geology* **84:** 220–223.

Knicker, H. 2000. Solid-state 2-D double cross polarization magic angle spinning ^{15}N ^{13}C NMR spectroscopy on degraded algal residues. *Organic Geochemistry* **31:** 337–340.

———. 2001. Incorporation of inorganic nitrogen into humified organic matter as revealed by solid-state 2-D ^{15}N, ^{13}C NMR spectroscopy. *Abstracts of Papers of the American Chemical Society* **221:** U517–U517.

———. 2002. The feasibility of using DCPMAS ^{15}N ^{13}C NMR spectroscopy for a better characterization of immobilized ^{15}N during incubation of ^{13}C- and ^{15}N-enriched plant material. *Organic Geochemistry* **33:** 237–246.

Knicker, H., and H. D. Ludemann. 1995. ^{15}N and ^{13}C CPMAS and solution NMR studies of ^{15}N enriched plant material during 600 days of microbial degradation. *Organic Geochemistry* **23:** 329–341.

Kohnen, M.E.L., S. Schouten, J.S.S. Damste, J. W. Deleeuw, D. A. Merritt, and J. M. Hayes. 1992. Recognition of paleobiochemicals by a combined molecular sulfur and isotope geochemical approach. *Science* **256:** 358–362.

Leavitt, P. R. 1993. A review of factors that regulate carotenoids and chlorophyll deposition and fossil pigment abundance. *Journal of Paleolimnology* **1:** 201–214.

Leavitt, P. R., and S. R. Carpenter. 1990. Aphotic pigment degradation in the hypolimion: Implications for sedimentation studies and paleoliminology. *Limnology and Oceanography* **35:** 520–534.

Leavitt, P. R., and D. A. Hodgson. 2001. Sedimentary pigments. Pp. 2–21 *in* J. P. Smol, H.J.B. Birks and W. M. Last, eds., *Tracking enviromental changes using lake sediments.* Kluwer, New York.

Louchouarn, P., M. Lucotte, R. Canuel, J.-P. Gagne, and L.-F. Richard. 1997. Sources and early diagenesis of lignin and bulk organic matter in the sediments of the lower St. Lawrence estuary and the Saguenay fjord. *Marine Chemistry* **58:** 3–26.

McCallister, S. L., J. E. Bauer, J. E. Cherrier, and H. W. Ducklow. 2004. Assessing the sources and ages of organic matter supporting river and estuarine bacterial production: A multiple-isotope (δ^{14}C, δ^{13}C, and δ^{15}N) approach. *Limnology and Oceanography* **49:** 1687–1702.

Meyers, P. A. 1997. Organic geochemical proxies of paleoceanographic, paleolimnologic, and paleoclimatic processes. *Organic Geochemistry* **27:** 213–250.

———. 2003. Applications of organic geochemistry to paleolimnological reconstructions: A summary of examples from the Laurentian Great Lakes. *Organic Geochemistry* **34:** 261–289.

Meyers, P. A., and R. Ishiwatari. 1993. Lacustrine organic geochemistry:-An overview of indicators of organic matter sources and diagenesis in lake sediments. *Organic Geochemistry* **20:** 867–900.

Mopper, K., A. Stubbins, J. D. Ritchie, H. M. Bialk, and P. G. Hatcher. 2007. Advanced instrumental approaches for characterization of marine dissolved organic matter: Extraction techniques, mass spectrometry, and nuclear magnetic resonance spectroscopy. *Chemical Reviews* **107:** 419–442.

Nanny, M. A., and R. A. Minear. 1997. Characterization of soluble unreactive phosphorus using ^{31}P nuclear magnetic resonance spectroscopy. *Marine Geology* **139:** 77–94.

Nixon, S. W. 1988. Physical energy inputs and the comparative ecology of lake and marine ecosystems. *Limnology and Oceanography* **33:** 1005–1025.

Olcott, A. N. 2007. The utility of lipid biomarkers as paleoenvironmental indicators. *Palaios* **22:** 111–113.

Orem, W. H., and P. G. Hatcher. 1987. Early diagenesis of organic matter in a sawgrass peat from the Everglades, Florida. *International Journal of Coal Geology* **8:** 33–54.

Raymond, P., and J. E. Bauer. 2001a. DOC cycling in a temperate estuary: A mass balance approach using natural ^{14}C and ^{13}C isotopes. *Limnology and Oceanography* **46:** 655–667.

———. 2001b. Use of ^{14}C and ^{13}C natural abundances for evaluating riverine, estuarine and coastal DOC and POC sources and cycling: A review and synthesis. *Organic Geochemistry* **23:** 469–485.

Reuss, N., D. J. Conley, and T. S. Bianchi. 2005. Preservation conditions and the use of sediment pigments as a tool for recent ecological reconstruction in four northern european estuaries. *Marine Chemistry* **95:** 283–302.

Rice, D. L., and R. B. Hanson. 1984. A kinetic model for detritus nitrogen: Role of the associated bacteria in nitrogen accumulation. *Bulletin of Marine Science* **35:** 326–340.

Russell, M., J. O. Grimalt, W. A. Hartgers, C. Taberner, and J. M. Rouchy. 1997. Bacterial and algal markers in sedimentary organic matter deposited under natural sulphurization conditions (Lorca Basin, Murcia, Spain). *Organic Geochemistry* **26:** 605–625.

Schnitzer, M., and C. M. Preston. 1986. Analysis of humic acids by solution and solid-state ^{13}C nuclear magnetic resonance. *Soil Science Society of America Journal* **50:** 326–331.

Sterner, R. W., and J. J. Elser. 2002. *Ecological stoichiometry—the biology of elements from molecules to the biosphere.* Princeton University Press, Princeton, NJ.

Sun, M. Y., and S. G. Wakeham. 1998. A study of oxic/anoxic effects on degradation of sterols at the simulated sediment–water interface of coastal sediments. *Organic Geochemistry* **28:** 773–784.

Sun, M. Y., R. C. Aller, and C. Lee. 1994. Spatial and temporal distributions of sedimentary chloropigments as indicators of benthic processes in Long Island Sound. *Journal of Marine Research* **52:** 149–176.

Sun, M. Y., S. G. Wakeham, and C. Lee. 1997. Rates and mechanisms of fatty acid degradation in oxic and anoxic coastal marine sediments of Long Island Sound, New York, USA. *Geochimica et Cosmochimica Acta* **61:** 341–355.

Sun, M. Y., R. C. Aller, C. Lee, and S. G. Wakeham. 2002a. Effects of oxygen and redox oscillation on degradation of cell-associated lipids in surficial marine sediments. *Geochimica et Cosmochimica Acta* **66:** 2003–2012.

Sun, M. Y., W. J. Cai, S. B. Joye, H. Ding, J. Dai, and J. T. Hollibaugh. 2002b. Degradation of algal lipids in microcosm sediments with different mixing regimes. *Organic Geochemistry* **33:** 445–459.

Sun, M.-Y., L. Zou, J. H. Dai, H. B. Ding, R. A. Culp, and M. I. Scranton. 2004. Molecular carbon isotopic fractionation of algal lipids during decomposition in natural oxic and anoxic seawaters. *Organic Geochemistry* **35:** 895–908.

Tani, Y., and others. 2002. Temporal changes in the phytoplankton community of the southern basin of Lake Baikal over the last 24,000 years recorded by photosynthetic pigments in a sediment core. *Organic Geochemistry* **33:** 1621–1634.

Tenore, K. R., L. Cammen, S.E.G. Findlay, and N. Phillips. 1982. Perspectives of research on detritus: Do factors controlling the availability of detritus to macro-consumers depend on its source? *Journal of Marine Research* **40:** 473–490.

Tunnicliffe, V. 2000. A fine-scale record of 130 years of organic carbon deposition in an anoxic fjord, Saanich Inlet, British Columbia. *Limnology and Oceanography* **45:** 1380–1387.

Wakeham, S. G., and E. A. Canuel. 2006. Degradation and preservation of organic matter in marine sediments. *in* J. K. Volkman, ed., *Handbook of environmental Chemistry*, Springer, Vol. 2. Berlin.

Watts, C. D., and J. R. Maxwell. 1977. Carotenoid diagenesis in a marine sediment. *Geochimica et Cosmochimica Acta* **41:** 493–497.

Webster, J. R., and E. F. Benfield. 1986. Vascular plant breakdown in fresh-water ecosystems. *Annual Review of Ecology and Systematics* **17:** 567–594.

Zang, X., and P. G. Hatcher. 2002. A PY-GC-MS and NMR spectroscopy study of organic nitrogen in Mangrove Lake sediments. *Organic Geochemistry* **33:** 201–211.

Zang, X., R. T. Nguyen, H. R. Harvey, H. Knicker, and P. G. Hatcher. 2001. Preservation of proteinaceous material during the degradation of the green alga *Botryococcus braunii*: A solid-state 2D ^{15}N ^{13}C NMR spectroscopy study. *Geochimica et Cosmochimica Acta* **65:** 3299–3305.

Zimmerman, A. R., and E. A. Canuel. 2000. A geochemical record of eutrophication and anoxia in Chesapeake Bay sediments: Anthropogenic influence on organic matter composition. *Marine Chemistry* **69:** 117–137.

———. 2002. Sediment geochemical records of eutrophication in the mesohaline Chesapeake Bay. *Limnology and Oceanography* **47:** 1084–1093.

Chapter 3

Anderson, E. C., and W. F. Libby. 1951. World-wide distribution of natural radiocarbon. *Physical Review* **81:** 64–69.

Anderson, E. C., W. F. Libby, S. Weinhouse, A. F. Reid, A. D. Kirshenbaum, and A. V. Grosse. 1947. Radiocarbon from cosmic radiation. *Science* **105:** 576–576.

Anderson T. F., and M. A. Arthur. 1983. Stable isotopes of oxygen and corbon and their application to sedimentologic and paleoenvironmental problems. Pp. 1–151 *in* M. A. Arthur, T. F. Anderson, I. R. Kaplan, J. Veizer, and L. S. Land, eds., Society of Economic Paleontology and Mineralogy.

Arnold, J. R., and W. F. Libby. 1949. Age determinations by radiocarbon content: Checks with samples of known age. *Science* **110:** 678–680.

Bauer, J. E. 2002. Biogeochemistry and cycling of carbon in the northwest Atlantic continental margin: Findings of the Ocean Margins Program. *Deep-Sea Research Part II: Topical Studies in Oceanography* **49:** 4271–4272.

Bauer, J. E., and E.R.M. Druffel. 1998. Ocean margins as a significant source of organic matter to the deep open ocean. *Nature* **392:** 482–485.

Bauer, J. E., E.R.M. Druffel, D. M. Wolgast, S. Griffin, and C. A. Masiello. 1998. Distributions of dissolved organic and inorganic carbon and radiocarbon in the eastern North Pacific continental margin. *Deep-Sea Research Part II: Topical Studies in Oceanography* **45:** 689–713.

Benner, R., M. L. Fogel, E. K. Sprague, and R. E. Hodson. 1987. Depletion of ^{13}C in lignin and its implications for stable carbon isotope studies. *Nature* **329:** 708–710.

Bennett, C. L., R. P. Beukens, M. R. Clover, H. E. Gove, R. B. Liebert, A. E. Litherland, K. H. Purser, and W. E. Sondheim. 1977. Radiocarbon dating using electrostatic accelerators: negative ions provide the key. *Science* **198:** 508–510.

Bianchi, T. S., S. Mitra, and M. Mckee. 2002. Sources of terrestrially-derived carbon in the Lower Mississippi River and Louisiana shelf: Implications for differential sedimentation and transport at the coastal margin. *Marine Chemistry* **77:** 211–223.

Bousquet, P., and others. 2006. Contribution of anthropogenic and natural sources to atmospheric methane variability. *Nature* **443:** 439–443.

Boutton, T. W. 1991. Stable carbon isotope ratios of natural materials, II: Atmospheric terrestrial, marine, and freshwater enviroments. Pp. 173–185 *in* D. C. Coleman and B. Fry, eds., *Carbon isotope techniques*. Academic Press, New York.

Broecker, W. S., and T. H. Peng. 1982. *Tracers in the sea*. Eldigio Press, Palisades NY.

Caraco, N. F., G. Lampman, J. J. Cole, K. E. Limburg, M. L. Pace, and D. Fischer. 1998. Microbial assimilation of DIN in a nitrogen rich estuary: implications for food quality and isotope studies. *Marine Ecology—Progress Series* **167:** 59–71.

Casper, P., S. C. Maberly, G. H. Hall, and B. J. Finlay. 2000. Fluxes of methane and carbon dioxide from a small productive lake to the atmosphere. *Biogeochemistry* **49:** 1–19.

Cherrier, J., J. E. Bauer, E.R.M. Druffel, R. B. Coffin, and J. P. Chanton. 1999. Radiocarbon in marine bacteria: Evidence for the ages of assimilated carbon. *Limnology and Oceanography* **44:** 730–736.

Cifuentes, L. A., J. H. Sharp, and M. L. Fogel. 1988. Stable carbon and nitrogen isotope biogeochemistry in the Delaware estuary. *Limnology and Oceanography* **33:** 1102–1115.

Clayton, R. N. 2003. Isotopic self-shielding of oxygen and nitrogen in the solar nebula. *Geochimica et Cosmochimica Acta* **67:** A68–A68.

Cloern, J. E., E. A. Canuel, and D. Harris. 2002. Stable carbon and nitrogen isotope composition of aquatic and terrestrial plants of the San Francisco Bay estuarine system. *Limnology and Oceanography* **47:** 713–729.

Craig, H. 1954. ^{13}C variations in sequoia rings and the atmosphere. *Science* **119:** 141–143.

———. 1961. Standards for reporting concentations of deuterium and oxygen-18 in natural waters. *Science* **133:** 1833ff.

Currin, C. A., S. Y. Newell, and H. W. Paerl. 1995. The role of standing dead *Spartina alterniflora* and benthic microalge in salt-marsh food webs: Considerations based on multiple stable-isotope analysis. *Marine Ecology—Progress Series* **121:** 99–116.

Deegan, C. E., and R. H. Garritt. 1997. Evidence for spatial variability in estuarine food webs. *Marine Ecology Progress Series* **147:** 31–47.

Degens, E. T., M. Behrendt, B. Gotthard, and E. Reppmann. 1968. Metabolic fractionation of carbon isotopes in marine plankton, 2: Data on samples collected off the coasts of Peru and Equador. *Deep-Sea Research* **15:** 11–ff.20.

Deines, P. 1980. The isotopic composition of reduced carbon. Pp. 329–406 *in* P. Fitz and J. C. Fontes, eds., *Handbook of enviromental isotope geochemistry,* Vol. I: The terrestrial enviroment. Elsevier, Amsterdam.

DeNiro, M. J., and S. Epstein. 1978. Carbon isotopic evidence for different feeding patterns in 2 *Hyrax* species occupying the same habitat. *Science* **201:** 906–908.

———. 1981a. Influence of diet on the distribution of nitrogen isotopes in animals. *Geochimica et Cosmochimica Acta* **45:** 341–351.

———. 1981b. Isotopic composition of cellulose from aquatic organisms. *Geochimica et Cosmochimica Acta* **45:** 1885–1894.

Druffel, E.R.M., J. E. Bauer, S. Griffin, S. R. Beaupre, and J. Hwang. 2008. Dissolved inorganic radiocarbon in the North Pacific Ocean and Sargasso Sea. *Deep-Sea Research, Part I: Oceanographic Research Papers* **55:** 451–459.

Druffel, E.R.M., P. M. Williams, J. E. Bauer, and J. R. Ertel. 1992. Cycling of dissolved and particulate organic matter in the open ocean. *Journal of Geophysical Research—Oceans* **97:** 15639–15659.

Eglinton, T. I., and A. Pearson. 2001. Ocean process tracers: Single compound radiocarbon measurements. Pp. 2786–2795 *in Encyclopedia of ocean sciences.* Academic Press, San Diego, CA.

Ehleringer, J. R., R. F. Sage, L. B. Flanagan, and R. W. Pearcy. 1991. Climate change and the evolution of C_4 photosynthesis. *Trends in Ecology & Evolution* **6:** 95–99.

Elmore, D., and F. M. Phillips. 1987. Accelerator mass spectrometry for measurement of long-lived radioisotopes. *Science* **236:** 543–550.

Farquhar, G. D., M. C. Ball, S. Voncaemmerer, and Z. Roksandic. 1982. Effect of salinity and humidity on $\delta^{13}C$ value of halophytes: Evidence for diffusional isotope fractionation determined by the ratio of inter-cellular atmospheric partial pressure of CO_2 under different environmental conditions. *Oecologia* **52:** 121–124.

Farquhar, G. D., J. R. Ehleringer, and K. T. Hubick. 1989. Carbon isotope discrimination and photosynthesis. *Annual Review of Plant Physiology and Plant Molecular Biology* **40:** 503–537.

Faure, G. 1986. *Principles of isotope geology.* Wiley, Hoboken, NJ.

Fogel, M. L., and L. A. Cifuentes. 1993. Isotope fractionation during primary production, Pp. 73–98 *in* M. H. Engel and S. A. Macko, eds., *Organic geochemistry: Principles and applications.* Plenum Press, New York.

Fogel, M. L., L. A. Cifuentes, D. J. Velinsky, and J. H. Sharp. 1992. Relationship of carbon availability in estuarine phytoplankton to isotopic composition. *Marine Ecology-Progress Series* **82:** 291–300.

Freeman, K. H., J. M. Hayes, J. M. Trendel, and P. Albrecht. 1990. Evidence from carbon isotope measurements for diverse origins of sedimentary hydrocarbons. *Nature* **343:** 254–256.

Fry, B. 2006. *Stable isotope ecology.* Springer, New York.

Fry, B., and P. L. Parker. 1979. Animal diet in texas seagrass meadows: $\delta^{13}C$ evidence for the importance of benthic plants. *Estuarine and Coastal Marine Science* **8:** 499–509.

Fry, B., and E. B. Sherr. 1984. $\delta^{13}C$ measurements as indicators of carbon flow in marine and fresh-water ecosystems. *Contributions in Marine Science* **27:** 13–47.

Goering, J., V. Alexander, and N. Haubenstock. 1990. Seasonal variability of stable carbon and nitrogen isotope ratios of organisms in a North Pacific bay. *Estuarine Coastal and Shelf Science* **30:** 239–260.

Goni, M. A., K. C. Ruttenberg, and T. I. Eglinton. 1998. A reassessment of the sources and importance of land-derived organic matter in surface sediments from the Gulf of Mexico. *Geochimica et Cosmochimica Acta* **62:** 3055–3075.

Guo, L., and P. H. Santschi. 1997. Isotopic and elemental characterization of colloidal organic matter from the Chesapeake Bay and Galveston Bay. *Marine Chemistry* **59:** 1–15.

Guo, L. D., P. H. Santschi, L. A. Cifuentes, S. Trumbore, and J. Southon. 1996. Cycling of high-molecular-weight dissolved organic matter in the Middle Atlantic Bight as revealed by carbon isotopic (^{13}C and ^{14}C) signatures. *Limnology and Oceanography* **41:** 1242–1252.

Guy, R. D., D. M. Reid, and H. R. Krouse. 1986. Factors affecting $^{13}C/^{12}C$ ratios of inland halophytes. I: Controlled studies on growth and isotopic composition of *Puccinella nuttalliana*. Canadian Journal of Botany. **64:** 2693–2699.

Guy, R. D., M. L. Fogel, J. A. Berry, and T. C. Hoering. 1987. Isotope fractionation during oxygen production and consumption by plants. Pp. 597–600 *in* J. Biggens, ed., *Progress in Photosynthetic Research III*. Martinus Nijhoff, Dordrecht, The Netherlands.

Hayes, J. M. 1983. Geochemical evidence bearing on the origin of aerobiosis, an speculative interpretation. Pp. 291–301 *in* J. W. Schopf, ed., *The Earth's earliest biosphere: Its origin and evolution*. Princeton University Press, Princeton, NJ.

Hayes, J. M. 1993. Factors controlling ^{13}C contents of sedimentary organic compounds: Principles and evidence. *Marine Geology* **113:** 111–125.

Hayes, J. M. 2001. Fractionation of carbon and hydrogen isotopes in biosynthetic processes. *Reviews in Mineralogy and Geochemistry* **43:** 225–277.

Hayes, J. M., R. Takigiku, R. Ocampo, H. J. Callot, and P. Albrecht. 1987. Isotopic compositions and probable origins of organic molecules in the Eocene Messel Shale. *Nature* **329:** 48–51.

Hayes, J. M., K. H. Freeman, B. N. Popp, and C. H. Hoham. 1990. Compound specific isotopic analyses: A novel tool for reconstruction of ancient biogeochemical processes. *Organic Geochemistry* **16:** 1115–1128.

Hedges, J. I., W. A. Clark, P. D. Quay, J. E. Richey, A. H. Devol, and U. D. Santos. 1986. Compositions and fluxes of particulate organic material in the Amazon River. *Limnology and Oceanography* **31:** 717–738.

Hein, R., P. J. Crutzen, and M. Heimann. 1997. An inverse modeling approach to investigate the global atmospheric methane cycle. *Global Biogeochemical Cycles* **11:** 43–76.

Hoefs, J. 1980. *Stable isotope geochemistry*. Springer, Heidelberg.

———. 2004. *Stable isotope geochemistry*, 5th ed. Springer, Berlin.

Holmes, R. M., and others. 2000. Flux of nutrients from Russian rivers to the Arctic Ocean: Can we establish a baseline against which to judge future changes? *Water Resources Research* **36:** 2309–2320.

Horrigan, S. G., J. P. Montoya, J. L. Nevins, and J. J. Mccarthy. 1990. Natural isotopic composition of dissolved inorganic nitrogen in the Chesapeake Bay. *Estuarine Coastal and Shelf Science* **30:** 393–410.

Hughes, E. H., and E. B. Sherr. 1983. Subtidal food webs in a Georgia estuary: delta-C-13 analysis. *Journal of Experimental Marine Biology and Ecology* **67:** 227–242.

Hughes, J. E., L. A. Deegan, B. J. Peterson, R. M. Holmes, and B. Fry. 2000. Nitrogen flow through the food web in the oliohaline zone of a New England estuary. *Ecology* **81:** 433–452.

Incze, L. S., L. M. Mayer, E. B. Sherr, and S. A. Macko. 1982. Carbon inputs to bivalve mollusks: A comparison of two estuaries. *Canadian Journal of Fisheries and Aquatic Sciences* **39:** 1348–1352.

Jasper, J. P., and J. M. Hayes. 1990. A carbon isotope record of CO_2 levels during the late Quaternary. *Nature* **347:** 462–464.

Kamen, M. D. 1963. History of ^{14}C. *Science* **141:** 861–862.

Kurie, F.N.D. 1934. A new mode of disintegration induced by neutrons. *Physical Review* **45:** 904–905.

Laws, E. A., B. N. Popp, R. R. Bidigare, M. C. Kennicutt, and S. A. Macko. 1995. Dependence of phytoplankton carbon isotopic composition on growth rate and [CO_2aq]: Theoretical considerations and experimental results. *Geochimica Cosmochimica Acta* **59:** 1131–1138.

Libby, W. F. 1952. Chicago radiocarbon dates, 3. *Science* **116:** 673–681.

———. 1982. Nuclear dating: An historical perspective. Pp. 1–4 *in Nuclear and Chemical Dating Techniques*, American Chemical Society Symposium Series, Vol. 176, Washington, DC.

Lucas, W. J., and J. A. Berry. 1985. Inorganic carbon transport in aquatic photosynthetic organisms. *Physiologia Plantarum* **65:** 539–543.

Macko, S. A., M. F. Estep, M. H. Engel, and P. E. Hare. 1986. Kinetic fractionation of stable nitrogen isotopes during amino acid transamination. *Geochimica et Cosmochimica Acta* **50:** 2143–2146.

Macko, S. A., R. Helleur, G. Hartley, and P. Jackman. 1990. Diagenesis of organic matter: A study using stable isotopes of individual carbohydrates. *Organic Geochemistry* 16: 1129–1137.

Mariotti, A., and others. 1981. Experimental determination of nitrogen kinetic isotope fractionation—some principles: Illustration for the denitrification and nitrification processes. *Plant and Soil* **62:** 413–430.

Martens, C. S., N. E. Blair, C. D. Green, and D. J. Desmarais. 1986. Seasonal variations in the stable carbon isotopic signature of biogenic methane in a coastal sediment. *Science* **233:** 1300–1303.

McCallister, S. L., J. E. Bauer, J. Cherrier, and H. W. Ducklow. 2004. Assessing sources and ages of organic matter supporting river and estuarine bacterial production: A multiple-isotope ($\Delta^{14}C$, $\delta^{13}C$, and $\delta^{15}N$) approach. *Limnology and Oceanography* **49:** 1687–1702.

McNichol, A. P., and L. I. Aluiwhare. 2007. The power of radiocarbon in biogeochemical studies of the marine carbon cycle: Insights from studies of dissolved and particulate organic carbon (DOC and POC). *Chemical Reviews* **107:** 443–466.

Mitra, S., T. S. Bianchi, L. Guo, and P. H. Santschi. 2000. Terrestrially-derived dissolved organic matter in Chesapeake Bay and the Middle Atlantic Bight. *Geochimica et Cosmochimica Acta* **64:** 3547–3557.

Montoya, J. P. 1994. Nitrogen isotope fractionation in the modern ocean: implications for the sedimentary record. Pp. 259–280 *in* R. Zahn, T. F. Pedersen, M. A. Kaminski and L. Labeyrie eds., *Carbon cycling in the glacial ocean: Constraints on the ocean's role in global change*. Springer, Berlin.

Nier, A. O. 1947. A mass spectrometer for isotope and gas analysis analysis. *Review of Scientific Instruments* **18:** 398–411.

O'Brien, B. J. 1986. The use of natural and anthropogenic C-14 to investigate the dynamics of soil organic carbon. *Radiocarbon* **28:** 358–362.

O'Leary, M. H. 1981. Carbon isotope fractionation in plants. *Phytochemistry* **20:** 553–567.

———. 1988. Carbon isotopes in photosynthesis. *Bioscience* **38:** 328–336.

Ostlund, H. G., and C.G.H. Rooth. 1990. The North Atlantic tritium and radiocarbon transients 1972–1983. *Journal of Geophysical Research—Oceans* **95:** 20147–20165.

Park, R., and S. Epstein, 1960. Carbon isotope fractionation during photosynthesis. *Geochimica Cosmochimica Acta* **21:** 110–126.

Parsons, T. R., and Y. L. Lee Chen. 1995. The comparative ecology of a subartic and tropical estuarine ecosystem as measured with carbon and nitrogen isotopes. *Estuarine Coastal Shelf Science* **41:** 215–224.

Peterson, B. J., and B. Fry. 1987. Stable isotopes in ecosystem studies. *Annual Review of Ecology and Systematics* **18:** 293–320.

———. 1989. Stable isotopes in ecosystems studies. *Annual Review of Ecology and Systematics* **18:** 293–320.

Peterson, B. J., and R. W. Howarth. 1987. Sulfur, carbon, and nitrogen isotopes used to trace organic-matter flow in the salt-marsh estuaries of Sapelo Island, Georgia. *Limnology and Oceanography* **32:** 1195–1213.

Peterson, B. J., R. W. Howarth, and R. H. Garritt. 1985. Multiple stable isotopes used to trace the flow of organic matter in estuarine food webs. *Science* **227:** 1361–1363.

———. 1986. Sulfur and carbon isotopes as tracers of salt-marsh organic-matter flow. *Ecology* **67:** 865–874.

Ralph, E. K. 1971. Carbon-14 dating, Pp. 1–48 *in* H. N. Michael and E. K. Ralph, eds., *Dating techniques for the archeologist*. MIT Press, Cambridge, MA.

Rau, G. H., T. Takahashi, and D.J.D. Marais. 1989. Latitudinal variations in plankton $\delta^{13}C$: Implications for CO_2 and productivity in past oceans. *Nature* **341:** 516–518.

Rau, G. H., T. Takahashi, D. J. Desmarais, D. J. Repeta, and J. H. Martin. 1992. The relationship between $\delta^{13}C$ of organic matter and [CO_2(aq)] in ocean surface water: Data from a JGOFS site in the northeast Atlantic Ocean and a model. *Geochimica Cosmochimica Acta* **56:** 1413–1419.

Rau, G. H., U. Riebesell, and D. Wolfgladrow. 1997. CO_2aq-dependent photosynthetic $\delta^{13}C$ fractionation in the ocean: A model versus measurements. *Global Biogeochemical Cycles* **11:** 267–278.

Raymond, P. A., and J. E. Bauer. 2001a. DOC cycling in a temperate estuary: A mass balance approach using natural ^{14}C and ^{13}C isotopes. *Limnology and Oceanography* **46:** 655–667.

———. 2001b. Riverine export of aged terrestrial organic matter to the North Atlantic Ocean *Nature* **409:** 497–500.

———. 2001c. Use of ^{14}C and ^{13}C natural abundances for evaluating riverine, estuarine, and coastal DOC and POC sources and cycling: A review and synthesis. *Organic Geochemistry* **32:** 469–485.

Rishter, D. D., D. Markewitz, S. E., Trumbore, and C. G., Wells. 1999. Rapid accumulation and turnover of soil carbon in a re-establishing forest. *Nature* **400:** 56–58.

Sachs, J. P. 1997. *Nitrogen isotopes in chlorophyll and the origin of eastern Mediterranean sapropels.* Massachusetts Institute of Technology/Woods Hole Oceanographic Institution, Woods Hole, MA.

Santschi, P. H., and others. 1995. Isotopic evidence for the contemporary origin of high-molecular-weight organic matter in oceanic environments. *Geochimica et Cosmochimica Acta* **59:** 625–631.

Schiff, S. L., R. Aravena, S. Trumbore, and P. J. Dillon. 1990. Dissolved organic-carbon cycling in forested watersheds: A carbon isotope approach. *Water Resources Research* **26:** 2949–2957.

Schouten, S., and others. 1998. Biosynthetic effects on the stable carbon isotopic compositions of algal lipids: Implications for deciphering the carbon isotopic biomarker record. *Geochimica et Cosmochimica Acta* **62:** 1397–1406.

Sharp, Z. 2007. *Principles of stable isotope geochemistry.* Pearson Prentice Hall, Upper Saddle River, NJ.

Sigleo, A. C., and S. A. Macko. 1985. Stable isotope and amino acid composition of estuarine dissolved colloidal material. Pp. 29–46 *in* A. C. Sigleo and A. Hattori, eds., *Marine and estuarine geochemistry.* Lewis, Boca Raton, FL.

Simenstad, C. A., D. O. Duggins, and P. D. Quay. 1993. High turnover of inorganic carbon in kelp habitats as a cause of $\delta^{13}C$ variability in marine food webs. *Marine Biology* **116:** 147–160.

Smethie, W. M., H. G. Ostlund, and H. H. Loosli. 1986. Ventilation of the deep Greenland and Norwegian seas: Evidence from ^{85}Kr, tritium, ^{14}C and ^{39}Ar. *Deep-Sea Research, Part A: Oceanographic Research Papers* **33:** 675–703.

Smith, B. N., and S. Epstein. 1971. Two categories of $^{13}C/^{12}C$ ratios for higher plants. *Plant Physiology* **47:** 380–384.

Spiker, E. C., and M. Rubin. 1975. Petroleum pollutants in surface and groundwater as indicated by C-14 activity of dissolved organic carbon. *Science* **187:** 61–64.

Stuiver, M. 1978. Atmospheric carbon dioxide and carbon reservoir changes. *Science* **199:** 253–258.

Stuiver, M., and H. A. Polach. 1977. Reporting of ^{14}C data: Discussion. *Radiocarbon* **19:** 355–363.

Stuiver, M., and P. D. Quay. 1981. A 1600-year-long record of solar change derived from atmospheric ^{14}C levels. *Solar Physics* **74:** 479–481.

Suess, H. E. 1958. Radioactivity of the atmosphere and hydrosphere. *Annual Review of Nuclear Science* **8:** 243–256.

———. 1968. Climatic changes, solar activity and the cosmic ray production rate of radiocarbon. *Meterological Monographs* **8:** 146–150.

Sullivan, M. J., and C. A. Moncreiff. 1990. Edaphic algae are an important component of salt marsh food-webs: Evidence from multiple stable isotope analyses. *Marine Ecology Progress Series* **62:** 149–159.

Trumbore, S. 2000. Age of soil organic matter and soil respiration: Radiocarbon constraints on belowground ^{14}C dynamics. *Ecological Applications* **10:** 399–411.

Trumbore, S., J. S. Vogel, and J. Southon. 1989. AMS ^{14}C measurements of fractionated soil organic-matter: An approach to deciphering the soil carbon cycle. *Radiocarbon* **31:** 644–654.

Valley, J. W., and C. M. Graham. 1993. Cryptic grain-scale heterogeneity of oxygen isotope ratios in metamorphic magnetite. *Science* **259:** 1729–1733.

Voss, M., and U. Struck. 1997. Stable nitrogen and carbon isotopes as indicators of eutrophication of the Oder River (Baltic Sea). *Marine Chemistry* **59:** 35–49.

Wahlen, M., and others. 1989. ^{14}C in methane sources and in atmospheric methane: The contribution from fossil carbon. *Science* **245:** 286–290.

Wainright, S. C., M. P. Weinstein, K. W. Able, and C. A. Currin. 2000. Relative importance of benthic microalgae, phytoplankton and the detritus of smooth cordgrass *Spartina alterniflora* and the common reed *Phragmites australis* to brackish-marsh food webs. *Marine Ecology—Progress Series* **200:** 77–91.

Walter, K. M., S. A. Zimov, J. P. Chanton, D. Verbyla, and F. S. Chapin. 2006. Methane bubbling from Siberian thaw lakes as a positive feedback to climate warming. *Nature* **443:** 71–75.

Walter, K. M., M. Engram, C. R. Duguay, M. O. Jefferies, and F. S. Chapin. 2008. The potential use of synthetic aperture radar for estimating methane ebullition from Arctic lakes. *Journal of the American Water Resources Association* **44:** 305–315.

Westerhausen, L., J. Poynter, G. Eglinton, H. Erlenkeuser, and M. Sarnthein. 1993. Marine and terrigenous origin of organic-matter in modern sediments of the Equatorial East Atlantic: The $\delta^{13}C$ and molecular record. *Deep-Sea Research, Part I—Oceanographic Research Papers* **40:** 1087–1121.

Whiticar, M. J., E. Faber, and M. Schoell. 1986. Biogenic methane formation in marine and fresh-water environments: CO_2 reduction vs acetate fermentation isotope evidence. *Geochimica et Cosmochimica Acta* **50:** 693–709.

Williams, P. M., and E.R.M. Druffel. 1987. Radiocarbon in dissolved organic matter in the central North Pacific Ocean. *Nature* **330:** 246–248.

Williams, P. M., and L. I. Gordon. 1970. ^{13}C:^{12}C ratios in dissolved and particulate organic matter in sea. *Deep-Sea Research* **17:** 19–27.

Chapter 4

Abelson, P. H., and T. C. Hoering. 1961. Carbon isotope fractionation in formation of amino acids by photosynthetic organisms. *Proceedings of the National Academy of Sciences of the USA.* **47:** 623–632.

Almendros, G., R. Frund, F. J. Ganzalez-Vila, K. M. Haider, H. Knicker, and H. D. Ludemann. 1991. Analysis of ^{13}C and ^{15}N CPMAS NMR-spectra of soil organic matter and composts. *Federation of European Biochemistry Society* **282:** 119–121.

Aluwihare, L. I., D. J. Repeta, and R. F. Chen. 1997. A major biopolymeric component to dissolved organic carbon in surface sea water. *Nature* **387:** 166–169.

———. 2002. Chemical composition and cycling of dissolved organic matter in the Mid-Atlantic Bight. *Deep-Sea Research, Part II: Topical Studies in Oceanography* **49:** 4421–4437.

Amon, R.M.W., and R. Benner. 1996. Bacterial utilization of different size classes of dissolved organic matter. *Limnology and Oceanography* **41:** 41–51.

Arnarson, T. S., and R. G. Keil. 2001. Organic-mineral interactions in marine sediments studied using density fractionation and X-ray photoelectron spectroscopy. *Organic Geochemistry* **32:** 1401–1415.

Asamoto, E., 1991. *FT-ICR/MS: Analytical applications of Fourier transform ion cyclotron resonance mass spectrometry.* VCH, New York.

Bauer, J. E., P. M. Williams, and E.R.M. Druffel. 1992. C-14 activity of dissolved organic-carbon fractions in the north-central Pacific and Sargasso Sea. *Nature* **357:** 667–670.

Beckett, R., and B. T. Hart. 1993. Use of field-flow fractionation techniques to characterize aquatic particles, colloids and macromolecules. *in* J. V. Buffle and H. P. Leeuwen, eds., *Environmental Particles.* Lewis, Chelsea, MI.

Benner, R., J. D. Pakulski, M. McCarthy, J. I. Hedges, and P. G. Hatcher. 1992. Bulk chemical characteristics of dissolved organic matter in the ocean. *Science* **255:** 1561–1564.

Bennett, C. L., R. P. Beukens, M. R. Clover, H. E. Gove, R. B. Liebert, A. E. Litherland, K. H. Purser, W. E. Sondheim, and others. 1977. Radiocarbon dating using electrostatic accelerators: Negative ions provide the key. *Science* **198:** 508–510.

Bianchi, T. S., C. Lambert, and D. Biggs. 1995. Distribution of chlorophyll-*a* and phaeopigments in the northwestern Gulf of Mexico: A comparison between fluorimetric and high-performance liquid chromatography measurements. *Bulletin of Marine Science* **56:** 25–32.

Blomberg, J., P. J. Schoenmakers, J. Beens, and R. Tijssen. 1997. Compehensive two-dimensional gas chromatography (GC-GC) and its applicability to the characterization of complex (petrochemical) mixtures. *Journal of High Resolution Chromatography* **20:** 539–544.

Brandes, J. A., and others. 2004. Examining marine particulate organic matter at sub-micron scales using scanning transmission X-ray microscopy and carbon X-ray absorption near edge structure spectroscopy. *Marine Chemistry* **92:** 107–121.

Brandes, J. A., E. Ingall, and D. Paterson. 2007. Characterization of minerals and organic phosphorus species in marine sediments using soft X-ray fluorescence spectromicroscopy. *Marine Chemistry* **103:** 250–265.

Buesseler, K. O., and others. 2000. A comparison of the quantity and composition of material caught in a neutrally buoyant versus surface-tethered sediment trap. *Deep-Sea Research, Part I: Oceanographic Research Papers* **47:** 277–294.

———. 2007. An assessment of the use of sediment traps for estimating upper ocean particle fluxes. *Journal of Marine Research* **65:** 345–416.

Chen, N., T. S. Bianchi, and J. M. Bland. 2003. Novel decomposion products of chlorophyll-*a* in continental shelf (Louisiana shelf) sediments: Formation and transformation of carotenol chlorine esters. *Geochimica Cosmochimica Acta* **67:** 2027–2042.

Clark, L. L., E. D. Ingall, and R. Benner. 1998. Marine phosphorus is selectively remineralized. *Nature* **393:** 426.

Coble, P. G., 1996. Characterization of marine and terrestrial DOM in seawater using excitation-emission matrix spectroscopy. *Marine Chemistry* **51:** 323–346.

Coble, P.G., 2007. Marine optical biogeochemistry: The chemistry of ocean color. *Chemical Reviews.* **107:** 402–418.

Coppola, L., O. Gustafsson, P. Andersson, T. I. Eglinton, M. Uchida, and A. F. Dickens. 2007. The importance of ultrafine particles as a control on the distribution of organic carbon in Washington margin and Cascadia basin sediments. *Chemical Geology* **243:** 142–156.

Cortes, H. J., B. Winniford, J. Luong, and M. Pursch. 2009. Comprehensive two dimensional gas chromatography review. *Journal of Separation Science* **32:** 883–904.

Cory, R. M., and D. M. McKnight. 2005. Fluorescence spectroscopy reveals ubiquitous presence of oxidized and reduced quinones in DOM. *Environmental Science and Technology* **39:** 8142–8149.

Crowe, D. E., J. W. Valley, and K. L. Baker. 1990. Microanalysis of sulfur-isotope ratios and zonation by laser microprobe. *Geochimica et Cosmochimica Acta* **54:** 2075–2092.

Dandonneau, Y., and A. Niang. 2007. Assemblages of phytoplankton pigments along a shipping line through the North Atlantic and tropical Pacific. *Progress in Oceanography* **73:** 127–144.

Dorsey, J., C. M. Yentsch, C. McKenna, and S. Mayo. 1989. Rapid analytical technique for assessment of metabolic activity. *Cytometry* **10:** 622–628.

Dria, K. J., J. R. Sachleben, and P. G. Hatcher. 2002. Solid-state carbon-13 nuclear magnetic resonance of humic acids at high magnetic field strengths. *Journal of Environmental Quality* **31:** 393–401.

Dunbar, R. C., and T. B. McMahan, 1998. Unimolecular reactions by ambient blackbody radiation activation. *Science* **147:** 194–197.

Eglinton, T. I., and A. Pearson. 2001. Ocean process tracers: Single compound radiocarbon measurements, Pp. 2786–2795, *in Encyclopedia of ocean sciences.* Academic Press, San Diego, CA (USA).

Eglinton, T. I., L. I. Aluwihare, J. E. Bauer, E.R.M. Druffel, and A. P. McNichol. 1996. Gas chromatographic isolation of individual compounds from complex matrices for radiocarbon dating. *Analytical Chemistry* **68:** 904–912.

Eglinton, T. I., B. C. BenitezNelson, A. Pearson, A. P. McNichol, J. E. Bauer, and E.R.M. Druffel. 1997. Variability in radiocarbon ages of individual organic compounds from marine sediments. *Science* **227:** 769–799.

Engel, M. H., S. A. Macko, Y. Qian, and J. A. Silfer. 1994. Stable-isotope analysis at the molecular-level: A new approach for determining the origins of amino acids in the Murchison meteorite. *Advances in Space Research* **15:** 99–106.

Engelhaupt, E., and T. S. Bianchi. 2001. Sources and composition of high-molecular-weight dessolved organic carbon in a southern Louisiana tidal stream (Bayou Trepagnier). *Limnology and Oceanography* **46:** 917–926.

Fellman, J. B., M. P. Miller, R. M. Cory, D. V. D'Amore, and D. White. 2009. Characterizing dissolved organic matter using PARAFAC modeling of fluorescence spectroscopy: A comparison of two models. *Environmental Science and Technology* **43:** 6228–6234.

Floge, S. A., and M. L. Wells. 2007. Variation in colloidal chromophoric dissolved organic matter in the Damariscotta estuary, Maine. *Limnology and Oceanography* **52:** 32–45.

Freeman, K. H., J. M. Hayes, J.-M. Trendel, and P. Albrecht. 1989. Evidence from GC-MS carbon-isotopic measurements for multiple origins of sedimentary hydrocarbons. *Nature* **353:** 254–256.

Freeman, K. H., J. M. Hayes, J. M. Trendel, and P. Albrecht. 1990. Evidence from carbon isotope measurements for diverse orgins of sedimentary hydrocarbons. *Nature* **343:** 254–256.

Fray, B. 2006. *Stable isotope ecology.* Springer, New York.

Frysinger, G. S., R. B. Gaines, and C. M. Reddy. 2002. GC × GC: A new analytical tool for environmental forensics. *Environmental Forensics* **3:** 27–34.

Frysinger, G. S., R. B. Gaines, L. Xu, and C. M. Reddy. 2003. Resolving the unresolved complex mixture in petroleum-contaminated sediments. *Environmental Science and Technology* **37:** 1653–1662.

Giddings, J. C. 1993. Field-flow fractionation: Analysis of macromolecular, colloidal and particulate materials. *Science* **260:** 1456–1465.

Goni, M. A., and T. I. Eglinton. 1996. Stable carbon isotopic analyses of lignin-derived CuO oxidation products by isotope ratio monitoring gas chromatography mass spectrometry (IRM-GC-MS). *Organic Geochemistry* **24:** 601–615.

Goutx, M., and others. 2007. Composition and degradation of marine particles with different settling velocities in the northwestern Mediterranean Sea. *Limnology and Oceanography* **52:** 1645–1664.

Guo, L. D., and P. H. Santschi. 1997. Composition and cycling of colloids in marine environments. *Reviews of Geophysics* **35:** 17–40.

Guo, L. D., N. Tanaka, D. M. Schell, and P. H. Santschi. 2003. Nitrogen and carbon isotopic composition of high-molecular-weight dissolved organic matter in marine environments. *Marine Ecology-Progress Series* **252:** 51–60.

Hansell, D. A., and C. A. Carlson. 2002. *Biogeochemistry of marine dissolved organic matter*. Academic Press, Amsterdam.

Hasselhov, M., B. Lyven, C. Haraldsson, and D. Turner. 1996. A flow field-flow fractionation system for size fractionation of dissolved organic matter in seawater and freshwater using on-channel preconcentration. *6th International Symposium of FFFF.*

Hatcher, P. H. 1987. Chemical structural studies of natural lignin by dipolar dephasing solid state ^{13}C nuclear magnetic resonance. *Organic Geochemistry* **11:** 31–39.

Hayes, J. M., K. H. Freeman, B. N. Popp, and C. H. Hoham. 1990. Compound-specific isotopic analyses: A novel tool for reconstruction of ancient biogeochemical processes. *Organic Geochemistry* **16:** 1115–1128.

Hedges, J. I., and R. G. Keil. 1995. Sedimentary organic-matter preservation: An assessment and speculative synthesis. *Marine Chemistry* **49:** 81–115.

Hedges, J. I., P. H. Hatcher, J. R. Ertel, and K. Meyers-Schulte. 1992. A comparison of dissolved humic substances from seawater with Amazon River counterparts by ^{13}C-NMR spectrometry. *Geochimica et Cosmochimica Acta* **56:** 1753–1757.

Hedges, J. I., B. A. Bergamaschi, and R. Benner. 1993a. Comparative analyses of DOC and DON in natural waters. *Marine Chemistry* **41:** 121–134.

Hedges, J. I., C. Lee, S. G. Wakeham, P. J. Hernes, and M. L. Peterson. 1993b. Effects of poisons and preservatives on the fluxes and elemental compositions of sediment trap materials. *Journal of Marine Research* **51:** 651–668.

Hedges, J. I., J. A. Baldock, Y. Gelinas, C. Lee, M. L. Peterson, and S. G. Wakeham. 2002. The biochemical and elemental compositions of marine plankton: A NMR perspective. *Marine Chemistry* **78:** 47–63.

Hertkorn, N., R. Benner, M. Frommberger, P. Schmitt-Kopplin, M. Witt, K. Kaiser, A. Kettrup, and J. I. Hedges. 2006. Characterization of a major refractory component of marine dissolved organic matter. *Geochimica et Cosmochimica Acta* **70:** 2990–3010.

Hoefs, J. 2004. *Stable isotope geochemistry*, 5th ed. Springer, Berlin.

Hughey, A. C., C. L. Hendrickson, R. P. Rodgers, and A. G. Marshall. 2001. Kendrick mass defect spectrum: A compact visual analysis ultra-resolution broadband mass spectra. *Analytical Chemistry* **73:** 4676–4681.

Hupfer, M., R. Gachter, and H. Ruegger. 1995. Polyphosphate in lake sediments: ^{31}P NMR spectroscopy as a tool for its identification. *Limnology and Oceanography* **40:** 610–617.

Ingall, E. D., P. A.Schroeder, and R. A. Berner. 1990. The nature of organic phosphorus in marine sediments: new insights from ^{31}P NMR. *Geochimica et Cosmochimica Acta* **54:** 2617–2620.

Ingalls, A. E., and A. Pearson. 2005. Ten years of compound-specific radiocarbon analysis. *Oceanography* **18:** 18–31.

Ingalls, A. E., R. F. Anderson, and A. Pearson. 2004. Radiocarbon dating of diatom-bound organic compounds. *Marine Chemistry* **92:** 91–105.

Jeffrey, S. W., R.F.C. Mantoura, and S. W. Wright. 1997. *Phytoplankton pigments in oceanography: Guidelines to modern methods.* UNESCO, Paris.

Keil, R. G., D. B. Montlucon, F. G. Prahl, and J. I. Hedges. 1994. Sorptive preservation of labile organic matter in marine sediments. *Nature* **370:** 549–552.

Keil, R. G., L. M. Mayer, P. D. Quay, J. E. Richey, and J. I. Hedges. 1997. Loss of organic matter from riverine particles in deltas. *Geochimica et Cosmochimica Acta* **61:** 1507–1511.

Kendrick, E. 1963. A mass scale based on high resolution mass spectrometry of organic compounds. *Analytical Chemistry* **35:** 2146–2154.

Knicker, H. 2000. Solid-state 2D double cross polarization magic angle spinning ^{15}N ^{13}C NMR spectroscopy on degraded algal residues. *Organic Geochemistry* **31:** 337–340.

Knicker, H., and H. D. Ludemann. 1995. N-15 and C-13 CPMAS and solution NMR studies of N-15 enriched plant material during 600 days of microbial degradation. *Organic Geochemistry* **23:** 329–341.

Koprivnjak, J. F., and others. 2009. Chemical and spectroscopic characterization of marine dissolved organic matter isolated using coupled reverse osmosis–electrodialysis. *Geochimica et Cosmochimica Acta* **73:** 4215–4231.

Kovac, N., O. Bajt, J. Faganneli, B. Sket, and B. Orel. 2002. Study of macroaggragate composition using FT-IR and ^{1}H-NMR spectroscopy. *Marine Chemistry* **78:** 205–215.

Kowalczuk, P., J. Ston-Egiert, W. J. Cooper, R. F. Whitehead, and M. J. Durako. 2005. Characterization of the chromophoric dissolved organic matter (CDOM) in the Baltic Sea by excitation emission matrix fluorescence spectroscopy. *Marine Chemistry* **96:** 273–292.

Kujawinski, E. B., M. A. Freitas, X. Zang, P. G. Hatcher, K. B. Green-Church, and R. B. Jones. 2002. The application of electrospray ionization mass spectrometry (ESI MS) to the structural characterization of natural organic matter. *Organic Geochemistry* **33:** 171–180.

Kujawinski, E. B., R. Del Vecchio, N. V. Blough, G. C. Klein, and A. G. Marshall. 2004. Probing molecular-level transformations of dissolved organic matter: Insights on photochemical degradation and protozoan modification of DOM from electrospray ionization Fourier transform ion cyclotron resonance mass spectrometry. *Marine Chemistry* **92:** 23–37.

Lamborg, C. H., and others. 2008. The flux of bio- and lithogenic material associated with sinking particles in the mesopelagic "twilight zone" of the northwest and north central Pacific Ocean. *Deep-Sea Research, Part II: Topical Studies in Oceanography* **55:** 1540–1563.

Lee, C., J. I. Hedges, S. G. Wakeham, and N. Zhu. 1992. Effectiveness of various treatments in retarding microbial activity in sediment trap material and their effects on the collection of swimmers. *Limnology and Oceanography* **37:** 117–130.

Legendre, L., and J. Le Fevre. 1989. Hydrodynamical singularities as controls of recycled versus export production in oceans. Pp. 49–63 *in* W. H. Berger, V. S. Smetacek, and G. Wefer, eds., *Productivity of the ocean: Present and past.* Wiley, Chichester, UK.

Legendre, L., and C. M. Yentsch. 1989. Overview of flow cytometry and image analysis in biological oceanography and limnology. *Cytometry* **10:** 501–510.

Levitt, M. H. 2001. *Spin dynamics: Basics of nuclear magnetic resonance. Wiley*, New York.

Loh, A. N., J. E. Bauer, and E.R.M. Druffel. 2004. Variable ageing and storage of dissolved organic components in the open ocean. *Nature* **430:** 877–881.

Macko, S. A., M. E. Uhle, M. H. Engel, and V. Andrusevich. 1997. Stable nitrogen isotope analysis of amino acid enantiomers by gas chromatography combustion/isotope ratio mass spectrometry. *Analytical Chemistry* **69:** 926–929.

Marriott, P. J., P. Haglund, and R.C.Y. Ong. 2003. A review of environmental toxicant analysis by using multidimensional gas chromatography and comprehensive GC. *Clinica Chimica Acta* **328:** 1–19.

Mayer, L. M. 1994. Surface-area control of organic-carbon accumulation in continental-shelf sediments. *Geochimica et Cosmochimica Acta* **58:** 1271–1284.

McCarthy, J. F., J. Ilavsky, J. D. Jastrow, L. M. Mayer, E. Perfect, and J. Zhuang. 2008. Protection of organic carbon in soil microaggregates via restructuring of aggregate porosity and filling of pores with accumulating organic matter. *Geochimica et Cosmochimica Acta* **72:** 4725–4744.

McMahon, G. 2007. *Analytical instrumentation: a guide to laboratory, portable and miniaturized instruments.* Wiley, Chichester, UK.

McNichol, A. P., and L. I. Aluiwhare. 2007. The power of radiocarbon in biogeochemical studies of the marine carbon cycle: Insights from studies of dissolved and particulate organic carbon (DOC and POC). *Chemical Reviews* **107:** 443–466.

McNichol, A. P., J. R. Ertel, and T. I. Eglinton. 2000. The radiocarbon content of individual lignin-derived phenols: Technique and initial results. *Radiocarbon* **42:** 219–227.

Messaud, F. A., and others. 2009. An overview on field-flow fractionation techniques and their applications in the separation and characterization of polymers. *Progress in Polymer Science* **34:** 351–368.

Minor, E. C., T. I. Eglinton, J. J. Boon, and R. Olson. 1999. Protocol for the characterization of oceanic particles via flow cytometric sorting and direct temperature-resolved mass spectrometry. *Analytical Chemistry* **71:** 2003–2013.

Minor, E. C., J. J. Boon, H. R. Harvey, and A. Mannino. 2001. Estuarine organic matter composition as probed by direct temperature-resolved mass spectrometry and traditional geochemical techniques. *Geochimica et Cosmochimica Acta* **65:** 2819–2834.

Minor, E. C., J. P. Simjouw, J. J. Boon, A. E. Kerkhoff, and J. van der Horst. 2002. Estuarine/marine UDOM as characterized by size-exclusion chromatography and organic mass spectrometry. *Marine Chemistry* **78:** 75–102.

Mondello, L., P. Q. Tranchida, P. Dugo, and G. Dugo. 2008. Comprehensive two-dimensional gas chromatography–mass spectrometry: A review. *Mass Spectrometry Reviews* **27:** 101–124.

Mopper, K. A., and C. A. Schultz, 1993. Fluorescence as a possible tool for studying the nature and water column distribution of DOC components. *Marine Chemistry* **41:** 229–238.

Mopper, K., A. Stubbins, J. D. Ritchie, H. M. Bialk, and P. G. Hatcher. 2007. Advanced instrumental approaches for characterization of marine dissolved organic matter: extraction techniques, mass spectrometry, and nuclear magnetic resonance spectroscopy. *Chemical Reviews* **107:** 419–442.

Muller, R. A. 1977. Radioisotope dating with a cyclotron. *Science* **196:** 489.

Murphy, K. R., C. A. Stedmon, D. Waite, and G. M. Ruiz. 2008. Distinguishing between terrestrial and autochthonous organic matter sources in marine environments using fluorescence spectroscopy. *Marine Chemistry* **108:** 40–58.

Nanny, M. A., and R. A. Minear. 1997. Characterization of soluble unreactive phosphorus using ^{31}P nuclear magnetic resonance spectroscopy. *Marine Geology* **139:** 77–94.

Nelson, D. E., R. G. Korteling, and W. R. Stott. 1977. Carbon-14: Direct detection at natural concentrations. *Science* **198:** 507–508.

Nelson, R. K., and others. 2006. Tracking the weathering of an oil spill with comprehensive two-dimensional gas chromatography. *Environmental Forensics* **7:** 33–44.

Nier, A. O. 1947. A mass spectrometer for isotope and gas analysis analysis. *Review of Scientific Instruments* **18:** 398–411.

Nunn, B. L., and R. G. Keil. 2005. Size distribution and chemistry of proteins in Washington coast sediments. *Biogeochemistry* **75(2):** 177–200.

Nunn, B. L., and R. G. Keil. 2006. A comparison of non-hydrolyte methods for extracting amino acids and proteins from coastal marine sediments. *Marine Chemistry* **89:** 31–42.

Olson, R. J., A. Shalapyonok, and H. M. Sosik. 2003. An automated submersible flow cytometer for analyzing pico- and nanophytoplankton: FlowCytobot. *Deep Sea Research (Part I)* **50:** 301–315.

Orem, W. H., and P. G. Hatcher. 1987. Solid-state ^{13}C NMR studies of dissolved organic matter in pore waters from different depositional environments. *Organic Geochemistry* **11:** 73–82.

Parlianti, E., K. Worz, L. Geoffrey, and M. Lamotte. 2000. Dissolved organic matter fluorescence stectroscopy as a tool to estimate biological activity in a coastal zone submitted to anthropogenic inputs. *Organic Geochemistry* **31:** 1765–1781.

Pearson, A., A. P. McNichol, R. J. Schneider, K. F. Von Reden, and Y. Zheng. 1998. Microscale AMS C-14 measurement at NOSAMS. *Radiocarbon* **40:** 61–75.

Peters, K. H., and J. M. Moldowan. 1993. *The biomarker guide: Interpreting molecular fossils in petroleum and ancient sediments*. Prentice-Hall, Englewood Cliffs, NJ.

Peters, K. H., C. C. Walters, and J. M. Moldowan. 2005. *The biomarker guide*. Pp. 198–251. Cambridge University Press, Cambridge, UK.

Peterson, M. L., S. G. Wakeham, C. Lee, M. A. Askea, and J. C. Miquel. 2005. Novel techniques for collection of sinking particles in the ocean and determining their settling rates. *Limnology and Oceanography-Methods* **3:** 520–532.

Prahl, F. G., J.-F. Rontani, J. K. Volkman, M. A. Sparrow, and I. M. Royer. 2006. Unusual C_{35} and C_{36} alkenones in a paleoceanographic benchmark strain of *Emiliania huxleyi*. *Geochimica et Cosmochimica Acta* **70:** 2856–2867.

Quan, T., and D. J. Repeta. 2005. Radiocarbon analysis of individual amino acids from marine high molecular weight dissolved organic matter. Abstract in ASLO Aquatic Sciences Meeting, Salt Lake City, UT.

Ransom, B., D. Kim, M. Kastner, and S. Wainwright. 1998. Organic matter preservation on continental slopes: Importance of mineralogy and surface area. *Geochimica et Cosmochimica Acta* **62:** 1329–1345.

Reddy, C. M., and others. 2002. The West Falmouth oil spill after thirty years: The persistence of petroleum hydrocarbons in marsh sediments. *Environmental Science and Technology* **36:** 4754–4760.

Repeta, D. J., and L. I. Aluwihare. 2006. Radiocarbon analysis of neutral sugars in high-molecular-weight dissolved organic carbon: Implications for organic carbon cycling. *Limnology and Oceanography* **51:** 1045–1053.

Richards, F. A., and T. F. Thompson. 1952. The estimation and characterization of plankton populations by pigment analyses, II: A spectrophotometric method for the estimation of plankton pigments. *Journal of Marine Research* **11:** 156–172.

Rowland, S. J., and J. N. Robson. 1990. The widespread occurrence of highly branched acyclic C_{20}, C_{25} and C_{30} hydrocarbons in recent sediments and biota: A review. *Marine Environmental Research* **30:** 191–216.

Schimpf, M. E., K. Caldwell, and J. C. Giddings. 2000. *Field-flow fractionation handbook.* Wiley-Interscience, New York.

Schnitzer, M., and C. M. Preston. 1986. Analysis of humic acids by solution and solid-state carbon-13 nuclear magnetic resonance. *American Journal Soil Science Society* **50:** 326–331.

Schure, M. R., M. E. Schimpf, and P. D. Schettler. 2000. Retention-normal mode. Pp. 31–48 *in* M. E. Schimpf, K. Caldwell, and J. C. Giddings, eds., *Field-flow-fractionation handbook.* Wiley-Interscience, New York.

Sharp, Z. 2006. *Principles of stable isotope geochemistry.* Prentice Hall, New York.

Simjouw, J. P., E. C. Minor, and K. Mopper. 2005. Isolation and characterization of estuarine dissolved organic matter: Comparison of ultrafiltration and C-18 solid-phase extraction techniques. *Marine Chemistry* **96:** 219–235.

Sinninghe Damsté, J. S., S. Schouten, J. W. de Leeuw, A.C.T. van Duin, and J.A.J. Geenevasen. 1999. Identification of novel sulfur-containing steroids in sediments and petroleum: Probable incorporation of sulfur into [delta]5,7-sterols during early diagenesis. *Geochimica et Cosmochimica Acta* **63:** 31–38.

Skoczynska, E., P. Korytar, and J. De Boer. 2008. Maximizing chromatographic information from environmental extracts by GC×GC-ToF-MS. *Environmental Science and Technology* **42:** 6611–6618.

Solomons, T.W.G. 1980. *Organic chemistry.* Wiley, New York.

Stedmon, C. A., and R. Bro. 2008. Characterizing DOM fluorescence with PARAFAC: A tutorial. *Limnology and Oceanography Methods* **6:** 572–579.

Stedmon, C. A., and S. Markager 2005. Resolving the variability in DOM fluorescence in a temperate estuary and its catchment using PARAFAC. *Limnology and Oceanography* **50:** 686–697.

Stedmon, C. A., S. Markager, and R. Bro. 2003. Tracing DOM in aquatic environments using a new approach to fluorescence spectroscopy. *Marine Chemistry* **82:** 239–254.

Steen, A. D., C. Arnosti, L. Ness, and N. V. Blough. 2006. Electron paramagnetic resonance spectroscopy as a novel approach to measure macromolecule-surface interactions and activities of extracellular enzymes. *Marine Chemistry* **101:** 266–276.

Stenson, A. C., A. G. Marshall, and W. T. Cooper. 2003. Exact masses and chemical formulas of individual Suwannee River fulvic acids from ultrahigh resolution electrospray ionization Fourier transform ion cyclotron resonance mass spectra. *Analytical Chemistry* **75:** 1275–1284.

Trees, C. C., M. C. Kennicut, and J. M. Brooks. 1985. Errors associated with the standard fluorimetric determination of chlorophylls and phaeopigments. *Marine Chemistry* **17:** 1–12.

Valley, J. W., and C. M. Graham. 1993. Cryptic grain-scale heterogenity of oxygen isotope ratios in metamorphic magnetite. *Science* **259:** 1729–1733.

Vetter, T. A., E. M. Perdue, E. Ingall, J. F. Koprivnjak, and P. H. Pfromm. 2007. Combining reverse osmosis and electrodialysis for more complete recovery of dissolved organic matter from seawater. *Separation and Purification Technology* **56:** 383–387.

Viskari, P. J., C. S. Kinkada, and C. L. Colyer. 2001. Determination of phycobiliproteins by capillary electrophoresis with laser-induced fluorescence detection. *Electrophoresis* **22:** 2327–2335.

von Muhlen, C., C. A. Zini, E. B. Caramao, and P. J. Marriott. 2006. Characterization of petrochemical samples and their derivatives by comprehensive two-dimensional gas chromatography. *Quimica Nova* **29:** 765–775.

Wakeham, S. G., J. I. Hedges, C. Lee, and T. K. Pease. 1993. Effects of poisons and preservatives on the composition of organic matter in a sediment trap experiment. *Journal of Marine Research* **51:** 669–696.

Wakeham, S. G., and others. 2007. Microbial ecology of the stratified water column of the Black Sea as revealed by a comprehensive biomarker study. *Organic Geochemistry* **38:** 2070–2097.

Wardlaw, G. D., J. S. Arey, C. M. Reddy, R. K. Nelson, G. T. Ventura, and D. L. Valentine. 2008. Disentangling oil weathering at a marine seep using GC×GC: Broad metabolic specificity accompanies subsurface petroleum biodegradation. *Environmental Science and Technology* **42:** 7166–7173.

Weber, C. I., L. I. Fay, G. B. Collins, D. E. Rathke, and J. Tobin. 1986. A review of methods for the analysis of chlorophyll in periphyton and plankton of marine and freshwater systems. *Ohio State University Sea Grant Program Tech. Bull.*

Wietzorrek, J., M. Stadler, and V. Kachel. 1994. Flow cytometric imaging: A novel tool for identification of marine organisms OCEANS. *Oceans Engineering for Today Technology and Tomorrow's Preservation Proceedings* **1:** I/688–I/693.

Yoon, T. H. 2009. Applications of soft X-ray spectromicroscopy in material and environmental sciences. *Applied Spectroscopy Reviews* **44:** 91–122.

Zang, X., R. T. Nguyen, H. R. Harvey, H. Knicker, and P. G. Hatcher. 2001. Preservation of proteinaceous material during the degradation of the green alga *Botryococcus braunii*: A solid-state 2D ^{15}N ^{13}C NMR spectroscopy study. *Geochimica Cosmochimica Acta* **65:** 3299–3305.

Zimmerman, A. R., J. Chorover, K. W. Goyne, and S. L. Brantley. 2004a. Protection of mesopore-adsorbed organic matter from enzymatic degradation. *Environmental Science & Technology* **38:** 4542–4548.

Zimmerman, A. R., K. W. Goyne, J. Chorover, S. Komarneni, and S. L. Brantley. 2004b. Mineral mesopore effects on nitrogenous organic matter adsorption. *Organic Geochemistry* **35:** 355–375.

Chapter 5

Alldredge, A., U. Passow, and B. Logan. 1993. The abundance and significance of a class of large, transparent organic particles in the ocean. *Deep-Sea Research* **40:** 1131–1140.

Aluwihare, L. I., D. J. Repeta, and R. F. Chen. 1997. A major biopolymeric component to dissolved organic carbon in surface sea water. *Nature* **387:** 166–169.

Amend, J. P., A. C. Amend, and M. Valenza. 1998. Determination of volatile fatty acids in the hot springs of Vulcano, Aeolian Islands, Italy. *Organic Geochemistry* **28:** 699–705.

Amon, R.M.W., and R. Benner. 2003. Combined neutral sugars as indicators of the diagenetic state of dissolved organic matter in the Arctic Ocean. *Deep-Sea Research, Part I: Oceanographic Research Papers* **50:** 151–169.

Amon, R.M.W., H. P. Fitznar, and R. Benner. 2001. Linkages among the bioreactivity, chemical composition, and diagenetic state of marine dissolved organic matter. *Limnology and Oceanography* **46:** 287–297.

Arnosti, C., and M. Holmer. 1999. Carbohydrate dynamics and contributions to the carbon budget of an organic-rich coastal sediment. *Geochimica et Cosmochimica Acta* **63:** 393–403.

Arnosti, C., and D. J. Repeta. 1994. Oligosaccharide degradation by an aerobic marine bacteria: Characterization of an experimental system to study polymer degradation in sediments. *Limnology and Oceanography* **39:** 1865–1877.

Arnosti, C., D. J. Repeta, and N. V. Blough. 1994. Rapid bacterial degradation of polysaccharides in anoxic marine systems. *Geochimica et Cosmochimica Acta* **58:** 2639–2652.

Aspinall, G. O. 1970. Pectins, plant gums, and other plant polysaccharides. Pp. 515–536 *in* W. Pigman and D. Horton, eds., The carbohydrates; chemistry and biochemistry, vol. IIB. Academic Press, New York.

———. 1983. CRC Handbook of chromatography—Carbohydrates, vol. 1: Churms, S. C. *Journal of the American Chemical Society* **105:** 5963–5964.

Benner, R., and K. Kaiser. 2003. Abundance of amino sugars and peptidoglycan in marine particulate and dissolved organic matter. *Limnology and Oceanography* **48:** 118–128.

Bergamaschi, B. A., J. S. Walters, and J. I. Hedges. 1999. Distributions of uronic acids and O-methyl sugars in sinking and sedimentary particles in two coastal marine environments. *Geochimica et Cosmochimica Acta* **63:** 413–425.

Bianchi, T. S., T. Filley, K. Dria, and P. G. Hatcher. 2004. Temporal variability in sources of dissolved organic carbon in the lower Mississippi River. *Geochimica et Cosmochimica Acta* **68:** 959–967.

Biddanda, B. A., and R. Benner. 1997. Carbon, nitrogen, and carbohydrate fluxes during the production of particulate and dissolved organic matter by marine phytoplankton. *Limnology and Oceanography* **42:** 506–518.

Bishop, C. T., and H. J. Jennings. 1982. Immunology of polysaccharides. Pp. 292–325 *in* G. O. Aspinall, ed., *The polysaccharides*. Academic Press, New York.

Borch, N. H., and D. L. Kirchmann. 1997. Concentration and composition of dissolved combined neutral sugars (polysaccharides) in seawater determined by HPLC-PAD. *Marine Chemistry* **57:** 85–95.

Boschker, H.T.S., E.M.J. Dekkers, R. Pele, and T. E. Cappenberg. 1995. Sources of organic carbon in the littoral of Lake Gooimeer as indicated by stable carbon-isotope and carbohydrate compositions. *Biogeochemistry* **29:** 89–105.

Buffle, J. 1990. The analytical challenge posed by fulvic and humic compounds. *Analytica Chimica Acta* **232:** 1–2.

Buffle, J., and G. G. Leppard. 1995. Characterization of aquatic colloids and macromolecules, 1: Structure and behavior of colloidal material. *Environmental Science and Technology* **29:** 2169–2175.

Burdige, D. J., A. Skoog, and K. Gardner. 2000. Dissolved and particulate carbohydrates in contrasting marine sediments. *Geochimica et Cosmochimica Acta* **64:** 1029–1041.

Burney, C. M., and J. M. Sieburth. 1977. Dissolved carbohydrate in seawater, 2: A spectrophotometric procedure for total carbohydrate analysis and polysaccharide estimation. *Marine Chemistry* **5:** 15–28.

Cauwet, G. 2002. DOM in the coastal zone. Pp. 579–602 *in* D. A. Hansell and C. A. Carlson, eds., *Biogeochemistry of marine dissolved organic matter*. Academic Press, San Diego, CA.

Compiano, A. M., J. C. Romano, F. Garabetian, P. Laborde, and I. Delagiraudiere. 1993. Monosaccharide composition of particulate hydrolyzable sugar fraction in surface microlayers from brackish and marine waters. *Marine Chemistry* **42:** 237–251.

Cowie, G. L., and J. I. Hedges. 1984a. Carbohydrate sources in a coastal marine-enviroment. *Geochimica et Cosmochimica Acta* **48:** 2075–2087.

———. 1984b. Determination of neutral sugars in plankton, sediments, and wood by capillary gas chromatography of equilibrated isomeric mixtures. *Analytical Chemistry* **56:** 497–504.

———. 1994. Biochemical indicators of diagenetic alteration in natural organic-matter mixtures. *Nature* **369:** 304–307.

Cowie, G. L., J. I. Hedges, and S. E. Calvert. 1992. Sources and relative reactivities of amino acids, neutral sugars, and lignin in an intermittently anoxic marine environment. *Geochimica et Cosmochimica Acta* **56:** 1963–1978.

Cowie, G. L., J. I. Hedges, F. G. Prahl, and G. L. De Lange. 1995. Elemental and biochemical changes across an oxidation front in a relict turbidite: An oxygen effect. *Geochimica et Cosmochimica Acta* **59:** 33–46.

Danishefsky, I., and G. Abraham. 1970. Purification and composition of chondroitin 6-sulfate-protein. *Federation Proceedings* **29:** A869–A902.

Decho, A. W., and G. J. Herndl. 1995. Microbial activities and the transformation of organic matter within mucilaginous material. *Science of the Total Environment* **165:** 33–42.

Duursma, E. K., and R. Dawson. 1981. *Marine organic chemistry*. Elsevier, Amsterdam.

Fazio, S. A., D. J. Uhlinger, J. H. Parker, and D. C. White. 1982. Estimations of uronic acids as quantitative measures of extracellular and cell-wall polysaccharide polymers from environmental samples. *Applied and Environmental Microbiology* **42:** 1151–1159.

Gremm, T. J., and L. A. Kaplan. 1997. Dissolved carbohydrates in streamwater determined by HPLC and pulsed amperometric detection. *Limnology and Oceanography* **42:** 385–393.

Gueguen, C., R. Gibin, M. Pardos, and J. Dominik. 2004. Water toxicity and metal contamination assessment of a polluted river: The Upper Vistula River (Poland). *Applied Geochemistry* **19:** 153–162.

Hamilton, S. E., and J. I. Hedges. 1988. The comparative geochemistries of lignins and carbohydrates in an anoxic fjord. *Geochimica et Cosmochimica Acta* **52:** 129–142.

Handa, N. 1970. Dissolved and particulate carbohydrates. Pp. 129–152 *in* D. W. Hood, ed., *Organic matter in natural waters*. Inst. Marine Science, University of Alaska, Fairbanks, AK.

Hayes, J. M. 2001. Fractionation of carbon and hydrogen isotopes in biosynthetic processes. Pp. 225–277 *in* J. S. Valley and D. R. Cole, eds., *Organic geochemistry of contemporaneous and ancient sediments*. Mineralogical Society of America, Chantilly, YA.

Hedges, J. I., W. A. Clark, and G. L. Cowie. 1988. Organic-matter sources to the water column and surficial sediments of a marine bay. *Limnology and Oceanography* **33:** 1116–1136.

Hernes, P. J., J. I. Hedges, M. L. Peterson, S. G. Wakeham, and C. Lee. 1996. Neutral carbohydrate geochemistry of particulate material in the central Equatorial Pacific. *Deep-Sea Research, Part II: Topical Studies in Oceanography* **43:** 1181–1204.

Horrigan, S. G., A. Hagstrom, K. Koike, and F. Azam. 1988. Inorganic nitrogen utilization by assemblages of marine bacteria in seawater culture. *Marine Ecology Progress Series* **50**: 147–150.

Hung, C. C., and P. H. Santschi. 2000. Spectrophotometric determination of total uronic acids in seawater using cation-exchange separation and pre-concentration lyophilization. *Analytica Chimica Acta* **427**: 111–117.

Hung, C. C., and P. H. Santschi. 2001. Spectrophotometric determination of total uronic acids in seawater using cation-exchange separation and pre-concentration by lyophilization. *Analytica Chimica Acta* **427**: 111–117.

Hung, C. C., D. G. Tang, K. W. Warnken, and P. H. Santschi. 2001. Distributions of carbohydrates, including uronic acids, in estuarine waters of Galveston Bay, Texas. *Marine Chemistry* **73**: 305–318.

Hung, C. C., L. D. Guo, P. H. Santschi, N. Alvarado-Quiroz, and J. M. Haye. 2003a. Distributions of carbohydrate species in the Gulf of Mexico. *Marine Chemistry* **81**: 119–135.

Hung, C. C., L. D. Guo, G. E. Schultz, J. L. Pinckney, and P. H. Santschi. 2003b. Production and flux of carbohydrate species in the Gulf of Mexico. *Global Biogeochemical Cycles* **17**.

Ittekkot, V., U. Brockmann, W. Michaelis, and E. T. Degens. 1981. Dissolved free and combined carbohydrates during a phytoplankton bloom in the northern North Sea. *Marine Ecology-Progress Series* **4**: 299–305.

Ittekkot, V., E. T. Degens, and U. Brockmann. 1982. Monosaccharide composition of acid-hydrolyzable carbohydrates in particulate matter during a plankton bloom. *Limnology and Oceanography* **27**: 770–776.

Johnson, D. C., and W. R. Lacourse. 1990. Liquid chromatography with pulsed electrochemical detection at gold and platinum electrodes. *Analytical Chemistry* **62**: A589–A597.

Johnson, K. M., and J. M. Sieburth. 1977. Dissolved carbohydrates in seawater, I: A precise spectrophotometric analysis for monosaccharides. *Marine Chemistry* **5**: 1–13.

Jorgensen, N.O.G., and R. E. Jensen. 1994. Microbial fluxes of free monosaccharides and total carbohydrates in fresh-water determined by HPLC-PAD. *FEMS Microbiology Ecology* **14**: 79–93.

Kaiser, K., and R. Benner. 2000. Determination of amino sugars in environmental samples with high salt content by high performance anion exchange chromatography and pulsed amperometric detection. *Analytical Chemistry* **72**: 2566–2572.

Keith, S. C., and C. Arnosti. 2001. Extracellular enzyme activity in a river–bay–shelf transect: Variations in polysaccharide hydrolysis rates with substrate and size class. *Aquatic Microbial Ecology* **24**: 243–253.

Kenne, L., and B. Lindberg. 1983. Bacterial polysaccharides. Pp. 287–365 *in* G. O. Aspinall, ed., *The polysaccharides*. Academic Press, New York.

Kirchman, D. L. 2003. The contribution of monomers and other low-molecular weight compounds to the flux of dissolved organic material in aquatic ecosystems. Pp. 218–237 *in* S.E.G. Findlay and R. L. Sinsabaugh, eds., *Aquatic ecosystems: Interactivity of dissolved organic matter*. Academic Press, San Diego, CA.

Kirchmann, D. L., and N. H. Borch. 2003. Fluxes of dissolved combined neutral sugars (polysaccharides) in the Delaware estuary. *Estuaries* **26**: 894–904.

Klok, J., H. C. Cox, M. Baas, J. W. De Leeuw, and P. A. Schenck. 1984a. Carbohydrates in recent marine sediments, II: . Occurrence and fate of carbohydrates in a recent stromatolitic deposit: Solar Lake, Sinai. *Organic Geochemistry* **7**: 101–109.

Klok, J., H. C. Cox, M. Baas, P.J.W. Schuyl, J. W. De Leeuw, and P. A. Schenck. 1984b. Carbohydrates in recent marine sediments, I: Origin and significance of deoxy- and *O*-methyl-monosaccharides. *Organic Geochemistry* **7**: 73–84.

Lee, R. F. 1980. *Phycology*. Cambridge University Press, Cambridge, UK.

Leppard, G. G. 1993. *Particulate matter and aquatic contaminants*. Lewis, Boca Raton, FL.

Liebezeit, G., M. Bolter, I. F. Brown, and R. Dawson. 1980. Dissolved free amino acids and carbohydrates at pycnocline boundaries in the Sargasso Sea and related microbial activity. *Oceanologica Acta* **3**: 357–362.

Lucas, C. H., J. Widdows, and L. Wall. 2003. Relating spatial and temporal variability in sediment chlorophyll *a* and carbohydrate distribution with erodibility of a tidal flat. *Estuaries* **26**: 885–893.

Lyons, W. B., H. E. Gaudette, and A. D. Hewitt. 1979. Dissolved organic matter in pore waters of carbonate sediments from Bermuda. *Geochimica et Cosmochimica Acta* **43**: 433–437.

Macko, S. A., R. Helleur, G. Hartley, and P. Jackman. 1989. Diagenesis of organic matter: A study using stable isotopes of individual carbohydrates. *Advanced Organic Geochemistry* **16**: 1129–1137.

Martens, C. S., C. A. Kelley, J. P. Chanton, and W. J. Showers. 1992. Carbon and hydrogen isotopic characterization of methane from wetlands and lakes of the Yukon-Kuskokwim delta, western Alaska. *Journal of Geophysical Research-Atmospheres* **97**: 16689–16701.

Medeiros, P. M., and B.R.T. Simoneit. 2007. Analysis of sugars in environmental samples by gas chromatography–mass spectrometry. *Journal of Chromatography* A **1141:** 271–278.

Minor, E. C., J. P. Simjouw, J. J. Boon, A. E. Kerkhoff, and J. Van Der Horst. 2002. Estuarine/marine UDOM as characterized by size-exclusion chromatography and organic mass spectrometry. *Marine Chemistry* **78:** 75–102.

Minor, E. C., J. J. Boon, H. R. Harvey, and A. Mannino. 2001. Estuarine organic matter composition as probed by direct temperature-resolved mass spectrometry and traditional geochemical techniques. *Geochimica et Cosmochimica Acta* **65:** 2819–2834.

Modzeles, J. E., W. A. Laurie, and B. Nagy. 1971. Carbohydrates from Santa-Barbara Basin sediments: gas chromatographic–mass spectrometric analysis of trimethylsilyl derivatives. *Geochimica et Cosmochimica Acta* **35:** 825–832.

Moers, M.E.C., and S. R. Larter. 1993. Neutral monosaccharides from a hypersaline tropical environment: Applications to the characterization of modern and ancient ecosystems. *Geochimica et Cosmochimica Acta* **57:** 3063–3071.

Mopper, K. 1977. Sugars and uronic acids in sediment and water from Black Sea and North Sea with emphasis on analytical techniques. *Marine Chemistry* **5:** 585–603.

Mopper, K., and K. Larsson. 1978. Uronic and other organic acids in Baltic Sea and Black Sea sediments. *Geochimica et Cosmochimica Acta* **42:** 153–163.

Mopper, K., C. A. Schultz, L. Chevolot, C. Germain, R. Revuelta, and R. Dawson. 1992. Determination of sugars in unconcentrated seawater and other natural waters by liquid chromatography and pulsed amperometric detection. *Environmental Science and Technology* **26:** 133–138.

Murrell, M. C., and J. T. Hollibaugh. 2000. Distribution and composition of dissolved and particulate organic carbon in northern San Francisco Bay during low flow conditions. *Estuarine Coastal and Shelf Science* **51:** 75–90.

Myklestad, S. M. 1997. A sensitive and rapid method for analysis of dissolved mono- and polysaccharides in seawater. *Marine Chemistry* **56:** 279–286.

Nissenbaum, A., and I. R. Kaplan. 1972. Chemical and isotopic evidence for in-situ origin of marine humic substances. *Limnology and Oceanography* **17:** 570ff.

Ochiai, M., and T. Nakajima. 1988. Distribution of neutral sugar and sugar decomposing bacteria in small stream water. *Archiv für Hydrobiologie* **113:** 179–187.

Ogner, G. 1980. The complexity of forest soil carbohydrates as demonstrated by 27 different O-methyl monosaccharides, 10 previously unknown in nature. *Soil Science* **129:** 1–4.

Opsahl, S., and R. Benner. 1999. Characterization of carbohydrates during early diagenesis of five vascular plant tissues. *Organic Geochemistry* **30:** 83–94.

Paez-Osuna, F., H. Bojorquez-Leyva, and C. Green-Ruiz. 1998. Total carbohydrates: Organic carbon in lagoon sediments as an indicator of organic effluents from agriculture and sugar-cane industry. *Environmental Pollution* **102:** 321–326.

Painter, T. J. 1983. Algal polysaccharides. Pp. 195–285 *in* G. O. Aspinall, ed., *The polysaccharides*. Academic Press, New York.

Pakulski, J. D., and R. Benner. 1992. An improved method for the hydrolysis and MBTH analysis of dissolved and particulate carbohydrates in seawater. *Marine Chemistry* **40:** 143–160.

———. 1994. Abundance and distribution of carbohydrates in the ocean. *Limnology and Oceanography* **39:** 930–940.

Parsons, T. R., M. Takahashi, and B. Hargrave. 1984. *Biological oceanographic processes*. Pergamon Press, Oxford, UK.

Passow, U. 2002. Production of transparent exopolymer particles (TEP) by phyto- and bacterioplankton. *Marine Ecology Progress Series* **236:** 1–12.

Percival, E. 1970. Algal carbohydrates. Pp. 537–568 *in* W. Pigman and D. Horton, eds., *The carbohydrates: Chemistry and biochemistry*, 2nd ed. Academic Press, New York.

Quayle, W. C., and P. Convey. 2006. Concentration, molecular weight distribution and neutral sugar composition of DOC in maritime Antarctic lakes of differing trophic status. *Aquatic Geochemistry* **12:** 161–178.

Raymond, P. A., and J. E. Bauer. 2000. Bacterial consumption of DOC during transport through a temperate estuary. *Aquatic Microbial Ecology* **22:** 1–12.

———. 2001. DOC cycling in a temperate estuary: A mass balance approach using natural ^{14}C and ^{13}C isotopes. *Limnology and Oceanography* **46:** 655–667.

Repeta, D. J., T. M. Quan, L. I. Aluwihare, and A. M. Accardi. 2002. Chemical characterization of high molecular weight dissolved organic matter in fresh and marine waters. *Geochimica et Cosmochimica Acta* **66:** 955–962.

Rich, J. H., H. W. Ducklow, and D. L. Kirchman. 1996. Concentrations and uptake of neutral monosaccharides along 140° W in the Equatorial Pacific: Contribution of glucose to heterotrophic bacterial activity and the dom flux. *Limnology and Oceanography* **41:** 595–604.

Rich, J., M. Gosselin, E. Sherr, B. Sherr, and D. L. Kirchman. 1997. High bacterial production, uptake and concentrations of dissolved organic matter in the central Arctic Ocean. *Deep-Sea Research, Part II: Topical Studies in Oceanography* **44:** 1645–1663.

Rocklin, R. D., and C. A. Pohl. 1983. Determination of carbohydrates by anion-exchange chromatography with pulsed amperometric detection. *Journal of Liquid Chromatography* **6:** 1577–1590.

Santschi, P. H., E. Balnois, K. J. Wilkinson, J. W. Zhang, J. Buffle, and L. D. Guo. 1998. Fibrillar polysaccharides in marine macromolecular organic matter as imaged by atomic force microscopy and transmission electron microscopy. *Limnology and Oceanography* **43:** 896–908.

Senior, W., and L. Chevolot. 1991. Studies of dissolved carbohydrates (or carbohydrate-like substances) in an estuarine environment. *Marine Chemistry* **32:** 19–35.

Simoneit, B.R.T., R. N. Leif, and R. Ishiwatari. 1996. Phenols in hydrothermal petroleums and sediment bitumen from Guaymas basin, Gulf of California. *Organic Geochemistry* **24:** 377–388.

Skoog, A., and R. Benner. 1997. Aldoses in various size fractions of marine organic matter: Implications for carbon cycling. *Limnology and Oceanography* **42:** 1803–1813.

Skoog, A., B. Biddanda, and R. Benner. 1999. Bacterial utilization of dissolved glucose in the upper water column of the Gulf of Mexico. *Limnology and Oceanography* **44:** 1625–1633.

Skoog, A., K. Whitehead, F. Sperling, and K. Junge. 2002. Microbial glucose uptake and growth along a horizontal nutrient gradient in the North Pacific. *Limnology and Oceanography* **47:** 1676–1683.

Skoog, A., P. Vlahos, K. L. Rogers, and J. P. Amend. 2007. Concentrations, distributions, and energy yields of dissolved neutral aldoses in a shallow hydrothermal vent system of Vulcano, Italy. *Organic Geochemistry* **38:** 1416–1430.

Steen, A. D., L. J. Hamdan, and C. Arnosti. 2008. Dynamics of dissolved carbohydrates in the Chesapeake Bay: Insights from enzyme activities, concentrations, and microbial metabolism. *Limnology and Oceanography* **53:** 936–947.

Stephen, A. M. 1983. Other plant polysaccharides. Pp. 97–193 *in* G. O. Aspinall, ed., *The polysaccarides*. Academic Press, New York.

Svensson, E., A. Skoog, and J. P. Amend. 2004. Concentration and distribution of dissolved amino acids in a shallow hydrothermal system, Vulcano Island (Italy). *Organic Geochemistry* **35:** 1001–1014.

Tolhurst, T., M. Consalvey, and D. Paterson. 2008. Changes in cohesive sediment properties associated with the growth of a diatom biofilm. *Hydrobiologia* **596:** 225–239.

Tranvik, L. J., and N.O.G. Jorgensen. 1995. Colloidal and dissolved organic matter in lake water: Carbohydrate and amino-acid composition, and ability to support bacterial growth. *Biogeochemistry* **30:** 77–97.

Uhlinger, D. J., and D. C. White. 1983. Relationship between physiological status and formation of extracellular polysaccharide glycocalyx in *Pseudomonas atlantica*. *Applied and Environmental Microbiology* **45:** 64–70.

Voet, D., and J. G. Voet. 2004. *Biochemistry*. Wiley, New York.

Walters, J. S., and J. I. Hedges. 1988. Simultaneous determination of uronic acids and aldoses in plankton, plant tissues, and sediment by capillary gas chromatography of *N*-hexylaldonamide and alditol acetates. *Analytical Chemistry* **60:** 988–994.

Wheeler, P. A., and D. L. Kirchman. 1986. Utilization of inorganic and organic nitrogen by bacteria in marine systems. *Limnology and Oceanography* **31:** 998–1009.

Whistler, R. L., and E. L. Richards. 1970. Hemicellulose. Pp. 447–468 *in* W. Pigman and D. Horton, eds., *The carbohydrates: Chemistry and biochemistry*, 2nd ed. Academic Press, New York.

Wicks, R. J., M. A. Moran, L. J. Pittman, and R. E. Hodson. 1991. Carbohydrate signatures of aquatic macrophytes and their dissolved degradation products as determined by a sensitive high-performance ion chromatography method. *Applied and Environmental Microbiology* **57:** 3135–3143.

Wilkinson, K. J., A. Joz-Roland, and J. Buffle. 1997. Different roles of pedogenic fulvic acids and aquagenic biopolymers on colloid aggregation and stability in freshwaters. *Limnology and Oceanography* **42:** 1714–1724.

Yamaoka, Y. 1983. Carbohydrates in humic and fulvic acids from Hiroshima Bay sediments. *Marine Chemistry* **13:** 227–237.

Zhou, J., K. Mopper, and U. Passow. 1998. The role of surface-active carbohydrates in the formation of transparent exopolymer particles by bubble adsorption of seawater. *Limnology and Oceanography* **43:** 1860–1871.

Zou, L., and others. 2004. Bacterial roles in the formation of high-molecular-weight dissolved organic matter in estuarine and coastal waters: Evidence from lipids and the compound-specific isotopic ratios. *Limnology and Oceanography* **49:** 297–302.

Zucker, W. V. 1983. Tannins: Does structure determine function—An ecological perspective? *American Naturalist* **121:** 335–365.

Chapter 6

Alberts, J. J., Z. Filip, M. T. Price, J. I. Hedges, and T. R. Jacobsen. 1992. CuO-oxidation products, acid, hydrolyzable monosaccharides and amino acids of humic substances occurring in a salt-marsh estuary. *Organic Geochemistry* **18:** 171–180.

Aminot, A., and R. Kerouel. 2006. The determination of total dissolved free primary amines in seawater: Critical factors, optimized procedure and artifact correction. *Marine Chemistry* **98:** 223–240.

Amon, R.M.W., and R. Benner. 1996. Bacterial utilization of different size classes of dissolved organic matter. *Limnology and Oceanography* **41:** 41–51.

Aufdenkampe, A. K., J. I. Hedges, J. E. Richey, A. V. Krusche, and C. A. Llerena. 2001. Sorptive fractionation of dissolved organic nitrogen and amino acids onto fine sediments within the Amazon basin. *Limnology and Oceanography* **46:** 1921–1935.

Azevedo, R. A., P. Arruda, W. L. Turner, and P. J. Lea. 1997. The biosynthesis and metabolism of the aspartate derived amino acids in higher plants. *Phytochemistry* **46:** 395–419.

Benson, J. R., and P. E. Hare. 1975. Ortho-phthalaldehyde-fluorogenic detection of primary amines in picomole range: Comparison with fluorescamine and ninhydrin. *Proceedings of the National Academy of Sciences of the USA* **72:** 619–622.

Bhushan, R., and S. Joshi. 1993. Resolution of enantiomers of amino acids by HPLC. *Biomedical Chromatography* **7:** 235–250.

Billen, G. 1984. Heterotrophic utilization and regeneration of nitrogen. Pp. 313–355 *in* J. E. Hobbie and P. J. Williams, eds., *Heterotrophic utilization and regeneration of nitrogen*. Plenum Press, New York.

Bjellqvist, B., K. Ek, Righetti, E. Gianazza, A. Görg, R. Westermeir, and W. Postel. 1982. Isoelectric focusing in immobilized pH gradients: Principle, methodology and some applications. *Journal of Biochemical and Biophysical Methods* **6:** 317–339.

Boetius, A., and K. Lochte. 1994. Regulation of microbial enzymatic degradation of organic-matter in deep-sea sediments. *Marine Ecology-Progress Series* **104:** 299–307.

Bradford, M. M. 1976. Rapid and sensitive method for quantitation of microgram quantities of protein utilizing principle of protein-dye binding. *Analytical Biochemistry* **72:** 248–254.

Burdige, D. J. 1989. The effects of sediment slurrying on microbial processes, and the role of amino acids as substrates for sulfate reduction in anoxic marine sediments. *Biogeochemistry* **8:** 1–23.

———. 2002. Sediment pore waters. Pp. 612–653 *in* D. A. Hansell and C. A. Carlson, eds., *Biogeochemistry of marine dissolved organic matter*. Academic Press, San Diego, CA.

Burdige, D. J., and C. S. Martens. 1988. Biogeochemical cycling in an organic-rich coastal marine basin, 10: The role of amino acids in sedimentary carbon and nitrogen cycling. *Geochimica et Cosmochimica Acta* **52:** 1571–1584.

———. 1990. Biogeochemical cycling in an organic-rich coastal marine basin, 11: The sedimentary cycling of dissolved, free amino acids. *Geochimica et Cosmochimica Acta* **54:** 3033–3052.

Campbell, M. K., and F. O. Shawn. 2007. *Biochemistry*. Thomson Learning, Belmont, CA.

Caughey, M. E. 1982. *A study of dissolved organic matter in pore waters of carbonate-rich sediment cores from the Florida Bay*. Thesis, University of Texas at Dallas.

Christensen, D., and T. H. Blackburn. 1980. Turnover of tracer (^{14}C, ^{3}H labeled) alanine in inshore marine sediments. *Marine Biology* **58:** 97–103.

Coffin, R. B. 1989. Bacterial uptake of dissolved free and combined amino acids in estuarine waters. *Limnology and Oceanography* **34:** 531–542.

Colombo, J. C., N. Silverberg, and J. N. Gearing. 1998. Amino acid biogeochemistry in the Laurentian Trough: vertical fluxes and individual reactivity during early diagenesis. *Organic Geochemistry* **29:** 933–945.

Constantz, B., and S. Weiner. 1988. Acidic macromolecules associated with the mineral phase of scleractinian coral skeletons. *Journal of Experimental Zoology* **248:** 253–258.

Cowie, G. L., and J. I. Hedges. 1994. Biochemical indicators of diagenetic alteration in natural organic-matter mixtures. *Nature* **369:** 304–307.

Cowie, G. L., J. I. Hedges, and S. E. Calvert. 1992. Sources and relative reactivities of amino acids, neutral sugars, and lignin in an intermittently anoxic marine environment. *Geochimica et Cosmochimica Acta* **56:** 1963–1978.

Crawford, C. C., J. E. Hobbie, and K. L. Webb. 1974. Utilization of dissolved free amino acids by estuarine microorganisms. *Ecology* **55:** 551–563.

Dauwe, B., and J. J. Middelburg. 1998. Amino acids and hexosamines as indicators of organic matter degradation state in North Sea sediments. *Limnology and Oceanography* **43:** 782–798.

Dauwe, B., J. J. Middleburg, V. Rijswijk, J. Sinke, P.M.J. Herman, and C.H.R. Heip. 1999. Enzymatically hydrolysable amino acids in the North Sea sediments and their possible implication for sediment nutritional values. *Journal of Marine Research* **57:** 109–134.

Dawson, R., and K. Gocke. 1978. Heterotrophic activity in comparison to free amino-acid concentrations in Baltic Sea water samples. *Oceanologica Acta* **1:** 45–54.

Dawson, R., and G. Liebezeit. 1981. *The analytical methods for the characterisation of organics in seawater.* Elsevier, Amsterdam.

———. 1983. Determination of amino acids. Pp. 319–330. *in* K. Grasshoff, M. Ehrhardt, and K. Kremling, eds., *Methods of seawater analysis*, 2nd ed. Verlag Chemie, Weinheim.

Dawson, R., and R. G. Pritchard. 1978. Determination of alpha-amino acids in seawater using a fluorimetric analyzer. *Marine Chemistry* **6:** 27–40.

De Stefano, C., C. Foti, A. Gianguzza, and S. Sammartano. 2000. The interaction of amino acids with major constituents of natural waters at different ionic strengths. *Marine Chemistry* **72:** 61–76.

Degens, E. T. 1977. Man's impact on nature. *Nature* **265:** 14–14.

Degens, E. T., J. M. Hunt, J. H. Reuter, and W. E. Reed. 1964. Data on the distribution of amino acids and oxygen isotopes in petroleum brine waters of various geologic ages. *Sedimentology* **3:** 199–225.

Ding, X., and S. M. Henrichs. 2002. Adsorption and desorption of proteins and polyamino acids by clay minerlas and marine sediments. *Marine Chemistry* **77:** 225–237.

Dittmar, T. 2004. Evidence for terrigenous dissolved organic nitrogen in the Arctic deep sea. *Limnology and Oceanography* **49:** 148–156.

Dittmar, T., H. P. Fitznar, and G. Kattner. 2001. Origin and biogeochemical cycling of organic nitrogen in the eastern Arctic Ocean as evident from D- and L-amino acids. *Geochimica et Cosmochimica Acta* **65:** 4103–4114.

Duan, S., and T. S. Bianchi. 2007. Particulate and dissolved amino acids in the lower Mississippi and Pearl rivers. *Marine Chemistry* **107:** 214–229.

Fuhrman, J. 1990. Dissolved free amino-acid cycling in an estuarine outflow plume. *Marine Ecology-Progress Series* **66:** 197–203.

Gardner, W. S., and R. B. Hanson. 1979. Dissolved free amino acids in interstitial waters of georgia salt-marsh soils. *Estuaries* **2:** 113–118.

Gardner, W. S., and P. A. St. John. 1991. High-performance liquid-chromatographic method to determine ammonium ion and primary amines in seawater. *Analytical Chemistry* **63:** 537–540.

Garrasi, C., E. T. Degens, and K. Mopper. 1979. Free amino-acid composition of seawater obtained without desalting and preconcentration. *Marine Chemistry* **8:** 71–85.

Gibbs, R. J. 1967. Geochemistry of Amazon River system, I: Factors that control salinity and composition and concentration of suspended solids. *Geological Society of America Bulletin* **78:** 1203–1209.

Gocke, K. 1970. Investigations on release and uptake of amino acids and polypeptides by plankton organisms. *Archiv fur Hydrobiologie* **67:** 285–367.

Goodfriend, G. A., and H. B. Rollins. 1998. Recent barrier beach retreat in Georgia: Dating exhumed salt marshes by aspartic acid racemization and post-bomb radiocarbon. *Journal of Coastal Research* **14:** 960–969.

Guo, L., P. H. Santschi, and T. S. Bianchi. 1999. Dissolved organic matter in estuaries of the Gulf of Mexico. Pp. 269–299 *in* T. S. Bianchi, J. Pennock and R. R. Twilley, eds., *Biogeochemistry of Gulf of Mexico Estuaries.* Wiley, New York.

Gupta, L. P., and H. Kawahata. 2000. Amino acid and hexosamine composition and flux of sinking particulate matter in the Equatorial Pacific at 175° E longitude. *Deep-Sea Research, Part I: Oceanographic Research Papers* **47:** 1937–1960.

Hanson, R. B., and W. S. Gardner. 1978. Uptake and metabolism of two amino acids by anaerobic microorganisms in four diverse salt-marsh soils. *Marine Biology* **46:** 101–107.

Haugen, J. E., and R. Lichtentaler. 1991. Amino-acid diagenesis, organic-carbon and nitrogen mineralization in surface sediments from the inner Oslo Fjord, Norway. *Geochimica et Cosmochimica Acta* **55:** 1649–1661.

Hedges, J. I. 1978. Formation and clay mineral reactions of melanoidins. *Geochimica et Cosmochimica Acta* **42:** 69–76.

Hedges, J. I., and P. E. Hare. 1987. Amino-acid adsorption by clay minerals in distilled water. *Geochimica et Cosmochimica Acta* **51:** 255–259.

Hedges, J. I., and others. 1994. Origins and processing of organic matter in the Amazon River as indicated by carbohydrates and amino acids. *Limnology and Oceanography* **39:** 743–761.

Hedges, J. I., and others. 2000. Organic matter in Bolivian tributaries of the Amazon River: A comparison to the lower mainstream. *Limnology and Oceanography* **45:** 1449–1466.

Hellebust, J. A. 1965. Excretion of some organic compounds by marine phytoplankton. *Limnology and Oceanography* **10:** 192–206.

Henrichs, S. M., and J. W. Farrington. 1979. Amino acids in interstitial waters of marine sediments. *Nature* **279:** 319–322.

———. 1987. Early diagenesis of amino acids and organic matter in two coastal marine sediments. *Geochimica et Cosmochimica Acta* **51:** 1–15.

Henrichs, S. M., and S. F. Sugai. 1993. Adsorption of amino acids and glucose by sediments of Resurrection Bay, Alaska, USA: Functional-group effects. *Geochimica et Cosmochimica Acta* **57:** 823–835.

Henrichs, S. M., J. W. Farrington, and C. Lee. 1984. Peru upwelling region sediments near 15° S, 2: Dissolved free and total hydrolyzable amino-acids. *Limnology and Oceanography* **29:** 20–34.

Hollibaugh, J. T., and F. Azam. 1983. Microbial degradation of dissolved proteins in seawater. *Limnology and Oceanography* **28:** 1104–1116.

Honjo, S., R. Francois, S. Manganini, J. Dymond, and R. Collier. 2000. Particle fluxes to the interior of the Southern Ocean in the western Pacific sector along 170° W. *Deep-Sea Research, Part II: Topical Studies in Oceanography* **47:** 3521–3548.

Hoppe, H. G. 1983. Significance of exoenzymatic activities in the ecology of brackish water: Measurements by means of methylumbelliferyl substrates. *Marine Ecology-Progress Series* **11:** 299–308.

———. 1991. Microbial extracellular enzyme activity: A new key parameter in aquatic ecology. Pp. 60–83 *in* R. J. Chrost, ed., *Microbial enzymes in aquatic environments.* Springer, New York.

Howarth, R. W., R. Marino, J. Lane, and J. J. Cole. 1988a. Nitrogen fixation in fresh-water, estuarine, and marine ecosystems 1: Rates and importance. *Limnology and Oceanography* **33:** 669–687.

Howarth, R. W., R. Marino, and J. J. Cole. 1988b. Nitrogen fixation in fresh water, estuarine, and marine ecosystems, 2: Biogeochemical controls. *Limnology and Oceanography* **33:** 688–701.

Huheey, J. E. 1983. *Inorganic chemistry*, 3rd ed. Harper & Row, New York.

Ingalls, A. E., C. Lee, S. G. Wakeham, and J. I. Hedges. 2003. The role of biominerals in the sinking flux and preservation of amino acids in the Southern Ocean along 170° W. *Deep-Sea Research, Part II: Topical Studies in Oceanography* **50:** 713–738.

Ittekkot, V., and R. Arain. 1986. Nature of particulate organic-matter in the river Indus, Pakistan. *Geochimica et Cosmochimica Acta* **50:** 1643–1653.

Ittekkot, V., and S. Zhang. 1989. Pattern of particulate nitrogen transport in world rivers. *Global Biogeochemical Cycle* **3:** 383–391.

Ittekkot, V., E. T. Degens, and S. Honjo. 1984. Seasonality in the fluxes of sugars, amino acids, and amino sugars to the deep ocean—Panama Basin. *Deep-Sea Research, Part A: Oceanographic Research Papers* **31:** 1071–1083.

Jaffe, D. A. 2000. The nitrogen cycle. Pp. 322–342 *in* M. C. Jacobson, R. J. Charlson, H. Rodhe, and G. H. Orians, eds., *Earth system science: From biogeochemical cycles to global change*. Academic Press, San Diego, CA.

Jones, A. D., C.H.S. Hitchcock, and G. H. Jones. 1981. Determination of tryptophan in feeds and feed ingredients by high-performance liquid chromatography. *Analyst* **106:** 968–973.

Jorgensen, N.O.G. 1979. A theoretical model of the stable sulfur isotope distribution in marine sediments. *Geochimica et Cosmochimica Acta* **43:** 363–374.

———. 1984. Microbial activity in the water–sediment interface: Assimilation and production of dissolved free amino acids. *Oceanus* **10:** 347–365.

———. 1987. Free amino acids in lakes: Concentrations and assimilation rates in relation to phytoplankton and bacterial production. *Limnology and Oceanography* **32:** 97–111.

Jorgensen, N.O.G., K. Mopper, and P. Lindroth. 1980. Occurrence, origin, and assimilation of free amino acids in an estuarine environment. *Ophelia* **1:** 179–192.

Jorgensen, N.O.G., P. Lindroth, and K. Mopper. 1981. Extraction and distribution of free amino acids and ammonium in sediment interstitial waters from the Limfjord, Denmark. *Oceanologica Acta* **4:** 465–474.

Josefsson, B., P. Lindroth, and G. Ostling. 1977. Automated fluorescence method for determination of total amino acids in natural waters. *Analytica Chimica Acta* **89:** 21–28.

Kanaoka, Y., T. Takahashi, H. Nakayama, K. Takada, T. Kimura, and S. Sakakibara. 1977. Organic fluorescence reagent, 4: Synthesis of a key fluorogenic amide, "L-arginine-4-methylcoumaryl-7-amide (L-arg-mca) and its derivatives: Fluorescence assays for trypsin and papain. *Chemical & Pharmaceutical Bulletin* **25:** 3126–3128.

Keil, R. G., and D. L. Kirchman. 1991a. Contribution of dissolved free amino acids and ammonium to the nitrogen requirements of heterotrophic bacterioplankton. *Marine Ecology-Progress Series* **73:** 1–10.

———. 1991b. Dissolved combined amino acids in marine waters as determined by a vapor-phase hydrolysis method. *Marine Chemistry* **33:** 243–259.

———. 1993. Dissolved combined amino acids: Chemical form and utilization by marine bacteria. *Limnology and Oceanography* **38:** 1256–1270.

Keil, R. G., E. Tsamakis, J. C. Giddings, and J. I. Hedges. 1998. Biochemical distributions (amino acids, neutral sugars, and lignin phenols) among size-classes of modern marine sediments from the Washington coast. *Geochimica et Cosmochimica Acta* **62:** 1347–1364.

Keil, R. G., E. Tsamakis, and J. I. Hedges. 2000. Early diagenesis of particulate amino acids in marine systems. Pp. 69–82 *in* G. A. Goodfriend, M. J. Cpllins, M. L. Fogel, S. A. Macko, and J. F. Wehmiller, eds., *Perspectives in amino acid and protein geochemistry*. Oxford University Press, New York.

King, K., and P. E. Hare. 1972. Amino-acid composition of planktonic foraminifera: Paleobiochemical approach to evolution. *Science* **175:** 1461–1463.

Kroger, N., R. Deutzmann, and M. Sumper. 1999. Polycationic peptides from diatom biosilica that direct silica nanosphere formation. *Science* **286:** 1129–1132.

Laemmli, U. K. 1970. Cleavage of structural proteins during assembly of head of bacteriophage-T4. *Nature* **227:** 680–685.

Landen, A., and P.O.J. Hall. 2000. Benthic fluxes and pore water distributions of dissolved free amino acids in the open Skagerrak. *Marine Chemistry* **71:** 53–68.

Lee, C. C. Cronin. 1982. The vertical flux of particulate organic nitrogen in the sea: Decomposition of amino acids in the Peru upwelling area and the Equatorial Atlantic. *Journal of Marine Research* **40:** 227–251.

Lee, C., and J. L. Bada. 1977. Dissolved amino acids in Equatorial Pacific, Sargasso Sea, and Biscayne Bay. *Limnology and Oceanography* **22:** 502–510.

Lee, C., S. G. Wakeham, and J. I. Hedges. 2000. Composition and flux of particulate amino acids and chloropigments in Equatorial Pacific seawater and sediments. *Deep-Sea Research, Part I: Oceanographic Research Papers* **47:** 1535–1568.

Liebezeit, G., and B. Behrends. 1999. Determination of amino acids. Pp. 541–546 *in* K. Grasshoff, K. Kremling, and M. Ehrhardt, eds., *Methods of seawater analysis*, 3rd ed. Wiley-VCH, Weinheim.

Lindroth, P., and K. Mopper. 1979. High-performance liquid-chromatographic determination of subpicomole amounts of amino acids by precolumn fluorescence derivatization with *ortho*-phthaldialdehyde. *Analytical Chemistry* **51:** 1667–1674.

Lomstein, B. A., and others. 1998. Budgets of sediment nitrogen and carbon cycling in the shallow water of Knebel Vig, Denmark. *Aquatic Microbial Ecology* **14:** 69–80.

Long, R. A., and F. Azam. 1996. Abundant protein-containing particles in the sea. *Aquatic Microbial Ecology* **10:** 213–221.

Matsudaira, P. 1987. Sequence from picomole quantities of proteins electroblotted onto polyvinylidene difluoride membranes. *Journal of Biological Chemistry* **262:** 10035–10038.

Mayer, L. M. 1986. 1st occurrence of polysolenoxylon Krausel and Dolianiti from the Rio Bonto Formation, Permian, Parana basin. *Anais Da Academia Brasileira De Ciencias* **58:** 509–510.

Mayer, L. M., L. L. Schick, and F. W. Setchell. 1986. Measurement of protein in nearshore marine sediments. *Marine Ecology-Progress Series* **30:** 159–165.

Mayer, L. M., L. L. Schick, T. Sawyer, C. J. Plante, P. A. Jumars, and R. L. Self. 1995. Bioavailable amino acids in sediments: A biomimetic, kinetics-based approach. *Limnology and Oceanography* **40:** 511–520.

Middelboe, M., and D. L. Kirchman. 1995. Bacterial utilization of dissolved free amino acids, dissolved combined amino acids and ammonium in the Delaware Bay estuary: Effects of carbon and nitrogen limitation. *Marine Ecology-Progress Series* **128:** 109–120.

Mitchell, J. G., A. Okubo, and J. Fuhrman. 1985. Microzones surrounding phytoplankton form the basis for a stratified marine microbial ecosystem. *Nature* **316:** 58–59.

Mopper, K., and R. Dawson. 1986. Determination of amino acids in sea water—Recent chromatographic developments and future directions. *Science of the Total Environment* **49:** 115–131.

Mopper, K., and P. Lindroth. 1982. Diel and depth variations in dissolved free amino acids and ammonium in the Baltic Sea determined by shipboard HPLC analysis. *Limnology and Oceanography* **27:** 336–347.

Mulholland, M. R., P. M. Gilbert, G. M. Berg, L. Van Heukelem, S. Pantoja, and C. Lee. 1998. Extracellular amino acid oxidation by microplankton: A cross-ecosystem comparison. *Aquatic Microbial Ecology* **15:** 141–152.

Mulholland, M. R., C. J. Gobler, and C. Lee. 2002. Peptide hydrolysis, amino acid oxidation, and nitrogen uptake in communities seasonally dominated by *Aureococcus anophagefferens*. *Limnology and Oceanography* **47:** 1094–1108.

Nguyen, R. T., and H. R. Harvey. 1994. A rapid microscale method for the extraction and analysis of protein in marine samples. *Marine Chemistry* **45:** 1–14.

———. 1997. Protein and amino acid cycling during phytoplankton decomposition in oxic and anoxic waters. *Organic Geochemistry* **27:** 115–128.

———. 1998. Protein and amino acid cycling during phytoplankton decomposition in oxic and anoxic water (vol. 27, p. 115, 1997). *Organic Geochemistry* **29:** 1019–1022.

Nikaido, H., and M. Vaara. 1985. Molecular basis of bacterial outer-membrane permeability. *Microbiological Reviews* **49:** 1–32.

North, B. B. 1975. Primary amines in California coastal waters: Utilization by phytoplankton. *Limnology and Oceanography* **20:** 20–27.

Nunn, B. L., and R. G. Keil. 2005. Size distribution and chemistry of proteins in Washington coast sediments. *Biogeochemistry* 75(**2**): 177–200.

Nunn, B. L., and R. G. Keil. 2006. A comparison of non-hydrolytic methods for extracting amino acids and proteins from coastal marine sediments. *Marine Chemistry* **89:** 31–42.

Oakley, B. R., D. R. Kirsch, and N. R. Morris. 1980. A simplified ultrasensitive silver stain for detecting proteins in polyacrylamide gels. *Analytical Biochemistry* **105:** 361–363.

O'Farrell, P. H. 1975. High resolution two-dimensional electrophoresis of proteins. *Journal Biological Chemistry* **250:** 4007–4021.

Palenik, B., and F.M.M. Morel. 1991. Amine oxidases of marine phytoplankton. *Applied and Environmental Microbiology* **57:** 2440–2443.

Palmork, K. H., M.E.U. Taylor, and R. Coates. 1963. The crystal structure of aberrant otoliths. *Acta Chemica Scandinavica* **17:** 1457–1458.

Pantoja, S., and C. Lee. 1994. Cell-surface oxidation of amino acids in seawater. *Limnology and Oceanography* **39:** 1718–1726.

———. 1999. Peptide decomposition by extracellular hydrolysis in coastal seawater and salt marsh sediment. *Marine Chemistry* **63:** 273–291.

Pantoja, S., C. Lee, and J. F. Marecek. 1997. Hydrolysis of peptides in seawater and sediment. *Marine Chemistry* **57:** 25–40.

Payne, J. W. 1980. Transport and utilization of peptides by bacteria. Pp. 211–256 *in* J. W. Payne, ed., *Microorganisms and nitrogen sources:Transport and utilization of amino acids, peptides, proteins, and related substrates*. Wiley, Chichester, UK.

Perez, M. T., C. Pausz, and G. J. Herndl. 2003. Major shift in bacterioplankton utilization of enantiomeric amino acids between surface waters and the ocean's interior. *Limnology and Oceanography* **48:** 755–763.

Preston, R. L. 1987. Occurrence of D-amino acids in higher organisms: A survey of the distribution of D-amino acids in marine invertebrates. *Comparative Biochemistry and Physiology B: Biochemistry and Molecular Biology* **87:** 55–62.

Preston, R. L., H. Mcquade, O. Oladokun, and J. Sharp. 1997. Racemization of amino acids by invertebrates. *Bulletin Mount Desert Island Biology Lab* **36:** 86.

Ramagli, L. S., and L. V. Rodriguez. 1985. Quantitation of microgram amounts of protein in two-dimensional polyacrylamide-gel electrophoresis sample buffer. *Electrophoresis* **6:** 559–563.

Rheinheimer, G., K. Gocke, and H. G. Hoppe. 1989. Vertical distribution of microbiological and hydrographic-chemical parameters in different areas of the Baltic Sea. *Marine Ecology-Progress Series* **52:** 55–70.

Righetti, P. G., and F. Chillemi. 1978. Isoelectric focusing of peptides. *Journal of Chromatography* **157:** 243–251.

Riley, J. P., and D. A. Segar. 1970. Seasonal variation of free and combined dissolved amino acids in the Irish Sea. *Journal of the Marine Biological Association of the United Kingdom* **50:** 713–720.

Rittenberg, S. C., K. O. Emery, J. Hulsemann, E. T. Degens, and R. C. Fayand. 1963. Biogeochemistry of sediments in experimental Mohole. *Journal of Sedimentary Research* **33:** 140–172.

Robbins, L. L., and K. Brew. 1990. Proteins from the organic matrix of core-top and fossil planktonic foraminifera. *Geochimica et Cosmochimica Acta* **54:** 2285–2292.

Romankevich, E. A. 1984. *Geochemistry of organic matter in the ocean*. Springer, Heidelberg.

Rosenfeld, J. K. 1979. Amino acid diagensis and adsorption in nearshore anoxic sediments. *Limnology and Oceanography* **24:** 1014–1021.

Roth, M. 1971. Fluorescence reaction for amino acids. *Analytical Chemistry* **43**(7): 880–882.

Saijo, S., and E. Tanoue. 2004. Characterization of particulate proteins in Pacific surface waters. *Limnology and Oceanography* **49:** 953–963.

———. 2005. Chemical forms and dynamics of amino acid-containing particulate organic matter in Pacific surface waters. *Deep-Sea Research, Part I: Oceanographic Research Papers* **52:** 1865–1884.

Schleifier, K. H., and O. Kandler. 1972. Peptidoglycan types of bacterial cell walls and their taxonomic implications. *Bacteriological Reviews* **36:** 407–477.

Sedmak, J. J., and S. E. Grossberg. 1977. Rapid, sensitive, and versatile assay for protein using coomassie brilliant blue g250. *Analytical Biochemistry* **79:** 544–552.

Seiler, N. 1977. Chromatography of biogenic amines, 1: Generally applicable separation and detection methods. *Journal of Chromatography* **143:** 221–246.

Sellner, K. G., and E. W. Nealley. 1997. Diel fluctuations in dissolved free amino acids and monosaccharides in Chesapeake Bay dinoflagellate blooms. *Marine Chemistry* **56:** 193–200.

Setchell, F. W. 1981. Particulate protein measurement in oceanographic samples by dye binding. *Marine Chemistry* **10:** 301–313.

Sharp, J. H. 1983. Estuarine organic chemistry as a clue to mysteries of the deep oceans. *Marine Chemistry* **12:** 232–232.

Shick, J. M., and W. C. Dunlop. 2002. Mycosporine-like amino acids and related gadusols: Biosynthesis, accumulation, and UV-protective functions in aquatic organisms. *Annual Review of Physiology* **64:** 223–262.

Siegel, A., and E. T. Degens. 1966. Concentration of dissolved amino acids from saline waters by ligand-exchange chromatography. *Science* **151:** 1098–1101.

Smith, D. C., M. Simon, A. L. Alldredge, and F. Azam. 1992. Intense hydrolytic enzyme activity on marine aggregates and implications for rapid particle dissolution. *Nature* **359:** 139–142.

Sommaruga, R., and F. Garcia-Pichel. 1999. UV-absorbing mycosporine-like compounds in planktonic and benthic organisms from a high-mountain lake. *Archiv fur Hydrobiologie* **144:** 255–269.

Somville, M., and G. Billen. 1983. A method for determining exoproteolytic activity in natural waters. *Limnology and Oceanography* **28:** 190–193.

Spitzy, A., and V. Ittekkot. 1991. Dissolved and particulate organic matter in rivers. Pp. 5–17 *in* R. F. C. Mantoura, J. M. Martin, and R. Wollast, eds., *Ocean margin processes in global change*. Wiley, Chichester, UK.

Starikov, N. D., and R. I. Korzhiko. 1969. Amino acids in the Black Sea. *Oceanology–USSR* **9:** 509–512.

Sugai, S. F., and S. M. Henrichs. 1992. Rates of amino-acid uptake and mineralization in Resurrection Bay (Alaska) sediments. *Marine Ecology-Progress Series* **88:** 129–141.

Summons, R. E., P. Albrecht, G. Mcdonald, and J. M. Moldowan. 2008. Molecular biosignatures. *Space Science Reviews* **135:** 133–159.

Tanoue, E. 1992. Vertical distribution of dissolved organic carbon in the North Pacific as determined by the high-temperature catalytic oxidation method. *Earth and Planetary Science Letters* **111:** 201–216.

———. 1996. Characterization of the particulate protein in Pacific surface waters. *Journal of Marine Research* **54:** 967–990.

Tanoue, E., S. Nishiyama, M. Kamo, and A. Tsugita. 1995. Bacterial membranes: Possible source of a major dissolved protein in seawater. *Geochimica et Cosmochimica Acta* **59:** 2643–2648.

Tanoue, E., M. Ishii, and T. Midorikawa. 1996. Discrete dissolved and particulate proteins in oceanic waters. *Limnology and Oceanography* **41:** 1334–1343.

Tartarotti, B., G. Baffico, P. Temporetti, and H. E. Zagarese. 2004. Mycosporine-like amino acids in planktonic organisms living under different UV exposure conditions in Patagonian lakes. *Journal of Plankton Research* **26:** 753–762.

Trefry, J. H., S. Metz, T. A. Nelsen, R. P. Trocine, and B. J. Eadie. 1994. Transport of particulate organic carbon by the Mississippi River and its fate in the Gulf of Mexico. *Estuaries* **17:** 839–849.

Tsugita, A., T. Uchida, H. W. Mewes, and T. Ataka. 1987. A rapid vapor-phase acid (hydrochloric acid and trifluoroacetic acid) hydrolysis of peptide and protein. *Journal of Biochemistry* **102:** 1593–1597.

Van Mooy, B.A.S., R. G. Keil, and A. H. Devol. 2002. Impact of suboxia on sinking particulate organic carbon: Enhanced carbon flux and preferential degradation of amino acids via denitrification. *Geochimica et Cosmochimica Acta* **66:** 457–465.

Volkman, J. K., and E. Tanoue. 2002. Chemical and biological studies of particulate organic matter in the ocean. *Journal of Oceanography* **58:** 265–279.

Wakeham, S. G. 1999. Monocarboxylic, dicarboxylic and hydroxy acids released by sequential treatments of suspended particles and sediments of the Black Sea. *Organic Geochemistry* **30:** 1059–1074.

Wang, X. C., and C. Lee. 1993. Adsorption and desorption of aliphatic amines, amino acids and acetate by clay minerals and marine sediments. *Marine Chemistry* **44:** 1–23.

———. 1995. Decomposition of aliphatic amines and amino acids in anoxic salt-marsh sediment. *Geochimica et Cosmochimica Acta* **59:** 1787–1797.

Whelan, J. K., and K. Emeis. 1992. Sedimentation and preservation of amino acid compounds and carbohydrates in marine sediments. Pp. 167–200 *in* J. K. Whelan and J. W. Farrington, eds., *Productivity, accumulation, and preservation of organice matter: recent and ancient sediments.* Columbia University Press, New York.

Whitehead, K. and J. I. Hedges. 2005. Photodegradation and photosensitization of mycosporine-like amino acids. *Journal of Photochemistry and Photobiology B: Biology* **80:** 115–121.

Winterbum, P. J., and C. F. Phelps. 1972. Significance of glycosylated proteins. *Nature* **236:** 147–151.

Wittenberg, J. B. 1960. The source of carbon monoxide in the float of the Portugese man-of-war, *Physalia physalis l. Journal of Experimental Biology* **37:** 698–704.

Chapter 7

Allen, A. E., M. G. Booth, M. E. Frischer, P. G. Verity, J. P. Zehr, and S. Zani. 2001. Diversity and detection of nitrate assimilation genes in marine bacteria. *Applied and Environmental Microbiology* **67:** 5343–5348.

Alzerreca, J. J., J. M. Norton, and M. G. Klotz. 1999. The amo operon in marine, ammonia-oxidizing gamma-proteobacteria. *FEMS Microbiology Letters* **180:** 21–29.

Armbrust, E. V., and others. 2004. The genome of the diatom *Thalassiosira pseudonana*: Ecology, evolution, and metabolism. *Science* **306:** 79–86.

Arp, D. J., P.S.G. Chain, and M. G. Klotz. 2007. The impact of genome analyses on our understanding of ammonia-oxidizing bacteria. *Annual Review of Microbiology* **61:** 503–528.

Behrens, S., and others. 2008. Linking microbial phylogeny to metabolic activity at the single-cell level by using enhanced element labeling-catalyzed reporter deposition fluorescence in situ hybridization (EL-FISH) and nanosims. *Applied and Environmental Microbiology* **74:** 3143–3150.

Bosak, T., R. M. Losick, and A. Pearson. 2008. A polycyclic terpenoid that alleviates oxidative stress. *Proceedings of the National Academy of Sciences of the USA* **105:** 6725–6729.

Braker, G., J. Z. Zhou, L. Y. Wu, A. H. Devol, and J. M. Tiedje. 2000. Nitrite reductase genes (nirK and nirS) as functional markers to investigate diversity of denitrifying bacteria in Pacific Northwest marine sediment communities. *Applied and Environmental Microbiology* **66:** 2096–2104.

Brocks, J. J., and A. Pearson, eds. 2005. *Building the biomarker tree of life.* Mineralogical Society of America, Chantilly, VA.

Brocks, J. J., G. A. Logan, R. Buick, and R. E. Summons. 1999. Archean molecular fossils and the early rise of eukaryotes. *Science* **285:** 1033–1036.

Brocks, J. J., R. Buick, R. E. Summons, and G. A. Logan. 2003. A reconstruction of archean biological diversity based on molecular fossils from the 2.78 to 2.45 billion-year-old mount bruce supergroup, Hamersley Basin, Western Australia. *Geochimica et Cosmochimica Acta* **67:** 4321–4335.

Brum, J. R. 2005. Concentration, production and turnover of viruses and dissolved DNA pools at stn Aloha, North Pacific subtropical gyre. *Aquatic Microbial Ecology* **41:** 103–113.

Brum, J. R., G. F. Steward, and D. M. Karl. 2004. A novel method for the measurement of dissolved deoxyribonucleic acid in seawater. *Limnology and Oceanography Methods* **2:** 248–255.

Bulow, S. E., C. A. Francis, G. A. Jackson, and B. B. Ward. 2008. Sediment denitrifier community composition and nirS gene expression investigated with functional gene microarrays. *Environmental Microbiology* **10:** 3057–3069.

Casciotti, K. L., and B. B. Ward. 2001. Dissimilatory nitrite reductase genes from autotrophic ammonia-oxidizing bacteria. *Applied and Environmental Microbiology* **67:** 2213–2221.

Cherrier, J., J. E. Bauer, E.R.M. Druffel, R. B. Coffin, and J. P. Chanton. 1999. Radiocarbon in marine bacteria: Evidence for the ages of assimilated carbon. *Limnology and Oceanography* **44:** 730–736.

Coffin, R. B., and L. A. Cifuentes. 1999. Stable isotope analysis of carbon cycling in the Perdido estuary, Florida. *Estuaries* **22:** 917–926.

Coffin, R. B., D. J. Velinsky, R. Devereux, W. A. Price, and L. A. Cifuentes. 1990. Stable carbon isotope analysis of nucleic acids to trace sources of dissolved substrates used by estuarine bacteria. *Applied and Environmental Microbiology* **56:** 2012–2020.

Collier, J. L., B. Brahamsha, and B. Palenik. 1999. The marine cyanobacterium *Synechococcus* sp. WH7805 requires urease (urea amidohydrolase, EC 3.5.1.5) to utilize urea as a nitrogen source: Molecular-genetic and biochemical analysis of the enzyme. *Microbiology* **145:** 447–459.

Coolen, M.J.L., and J. Overmann. 1998. Analysis of subfossil molecular remains of purple sulfur bacteria in a lake sediment. *Applied and Environmental Microbiology* **64:** 4513–4521.

———. 2007. 217 000-year-old DNA sequences of green sulfur bacteria in mediterranean sapropels and their implications for the reconstruction of the paleoenvironment. *Environmental Microbiology* **9:** 238–249.

Coolen, M.J.L., and others. 2004a. Evolution of the methane cycle in Ace Lake (Antarctica) during the Holocene: Response of methanogens and methanotrophs to environmental change. *Organic Geochemistry* **35:** 1151–1167.

Coolen, M.J.L., G. Muyzer, W.I.C. Rijpstra, S. Schouten, J. K. Volkman, and J.S.S. Damste. 2004b. Combined DNA and lipid analyses of sediments reveal changes in Holocene haptophyte and diatom populations in an Antarctic lake. *Earth and Planetary Science Letters* **223:** 225–239.

Coolen, M.J.L., A. Boere, B. Abbas, M. Baas, S. G. Wakeham, and J.S.S. Damste. 2006. Ancient DNA derived from alkenone-biosynthesizing haptophytes and other algae in holocene sediments from the Black Sea. *Paleoceanography* **21**, PA1005, doi 10.1029/2005PA001188.

Corinaldesi, C., A. Dell'anno, and R. Danovaro. 2007a. Early diagenesis and trophic role of extracellular DNA in different benthic ecosystems. *Limnology and Oceanography* **52:** 1710–1717.

———. 2007b. Viral infection plays a key role in extracellular DNA dynamics in marine anoxic systems. *Limnology and Oceanography* **52:** 508–516.

Creach, V., F. Lucas, C. Deleu, G. Bertru, and A. Mariotti. 1999. Combination of biomolecular and stable isotope techniques to determine the origin of organic matter used by bacterial communities: Application to sediment. *Journal of Microbiological Methods* **38:** 43–52.

Culley, A. I., A. S. Lang, and C. A. Suttle. 2006. Metagenomic analysis of coastal RNA virus communities. *Science* **312:** 1795–1798.

Danovaro, R., C. Corinaldesi, G. M. Luna, and A. Dell'anno. 2006. Molecular tools for the analysis of DNA in marine environments, Pp. 105–126 *in* J. K. Volkman ed., *Marine organic matter biomarkers, isotopes and DNA: The handbook of environmental chemistry*. Springer, Berlin.

Danovaro, R., and others. 2008. Major viral impact on the functioning of benthic deep-sea ecosystems. *Nature* **454:** 1084–1027.

Deflaun, M. F., J. H. Paul, and W. H. Jeffrey. 1987. Distribution and molecular weight of dissolved DNA in subtropical estuarine and oceanic environments. *Marine Ecology-Progress Series* **38:** 65–73.

Dell'Anno, A., S. Bompadre, and R. Danovaro. 2002. Quantification, base composition and fate of extracelluar DNA in marine sediments. *Limnology and Oceanography* **47:** 899–905.

Dell'Anno, A., and R. Danovaro. 2005. Extracellular DNA plays a key role in deep-sea ecosystem functioning. *Science* **309:** 2179–2179.

Dell'Anno, A., C. Corinaldesi, S. Stavrakakis, V. Lykousis, and R. Danovaro. 2005. Pelagic–benthic coupling and diagenesis of nucleic acids in a deep-sea continental margin and an open-slope system of the eastern Mediterranean. *Applied and Environmental Microbiology* **71:** 6070–6076.

Delong, E. F. 1992. Archaea in coastal marine environments. *Proceedings of the National Academy of Sciences of the USA* **89:** 5685–5689.

Delong, E. F., and D. M. Karl. 2005. Genomic perspectives in microbial oceanography. *Nature* **437:** 336–342.

Dupont, S., K. Wilson, M. Obst, H. Skold, H. Nakano, and M. C. Thorndyke. 2007. Marine ecological genomics: When genomics meets marine ecology. *Marine Ecology-Progress Series* **332:** 257–273.

Ertefai, T. F., and others. 2008. Vertical distribution of microbial lipids and functional genes in chemically distinct layers of a highly polluted meromictic lake. *Organic Geochemistry* **39:** 1572–1588.

Fischer, W. W., and A. Pearson. 2007. Hypotheses for the origin and early evolution of triterpenoid cyclases. *Geobiology* **5:** 19–34.

Francis, C. A., K. J. Roberts, J. M. Beman, A. E. Santoro, and B. B. Oakley. 2005. Ubiquity and diversity of ammonia-oxidizing archaea in water columns and sediments of the ocean. *Proceedings of the National Academy of Sciences of the USA* **102:** 14683–14688.

Francis, C. A., J. M. Beman, and M.M.M. Kuypers. 2007. New processes and players in the nitrogen cycle: The microbial ecology of anaerobic and archaeal ammonia oxidation. *Isme Journal* **1:** 19–27.

Fuhrman, J. A., D. E. Comeau, A. Hagstrom, and A. M. Chan. 1988. Extraction from natural planktonic microorganisms of DNA suitable for molecular biological studies. *Applied and Environmental Microbiology* **54:** 1426–1429.

Fuhrman, J. A., K. McCallum, and A. A. Davis. 1992. Novel major archaebacterial group from marine plankton. *Nature* **356:** 148–149.

Giovannoni, S. J., T. B. Britschgi, C. L. Moyer, and K. G. Field. 1990. Genetic diversity in Sargasso Sea bacterioplankton. *Nature* **345:** 60–63.

Glockner, F. O., and others. 2003. Complete genome sequence of the marine planctomycete *Pirellula* sp. strain 1. *Proceedings of the National Academy of Sciences of the USA* **100:** 8298–8303.

Hinrichs, K. U., J. M. Hayes, S. P. Sylva, P. G. Brewer, and E. F. Delong. 1999. Methane-consuming archaebacteria in marine sediments. *Nature* **398:** 802–805.

Hoehler, T. M., M. J. Alperin, D. B. Albert, and C. S. Martens. 1994. Field and laboratory studies of methane oxidation in an anoxic marine sediment: Evidence for a methanogen-sulfate reducer consortium. *Global Biogeochemical Cycles* **8:** 451–463.

Huang, W. E., and others. 2007. Raman fish: Combining stable-isotope raman spectroscopy and fluorescence in situ hybridization for the single cell analysis of identity and function. *Environmental Microbiology* **9:** 1878–1889.

Inagaki, F., H. Okada, A. I. Tsapin, and K. H. Nealson. 2005. The paleome: A sedimentary genetic record of past microbial communities. *Astrobiology* **5:** 141–153.

Jahren, A. H., G. Petersen, and O. Seberg. 2004. Plant DNA: A new substrate for carbon stable isotope analysis and a potential paleoenvironmental indicator. *Geology* **32:** 241–244.

Jetten, M.S.M. 2008. The microbial nitrogen cycle. *Environmental Microbiology* **10:** 2903–2909.

Jorgensen, N.O.G., and C. S. Jacobsen. 1996. Bacterial uptake and utilization of dissolved DNA. *Aquatic Microbial Ecology* **11:** 263–270.

Kaneko, T., and S. Tabata. 1997. Complete genome structure of the unicellular cyanobacterium *Synechocystis* sp. PCC6803. *Plant Cell Physiology* **38:** 1171–1176.

Kaneko, T., and others. 1996. Sequence analysis of the genome of the unicellular cyanobacterium *Synechocystis* sp. strain PCC6803, II: Sequence determination of the entire genome and assignment of potential protein-coding regions. *DNA Research* **3:** 109–136.

Karner, M. B., E. F. Delong, and D. M. Karl. 2001. Archaeal dominance in the mesopelagic zone of the Pacific Ocean. *Nature* **409:** 507–510.

Kelley, C. A., R. B. Coffin, and L. A. Cifuentes. 1998. Stable isotope evidence for alternative bacterial carbon sources in the Gulf of Mexico. *Limnology and Oceanography* **43:** 1962–1969.

Kirchman, D. L. 2008. *Microbial ecology of the oceans.* Wiley, Hoboken, NJ.

Kirchman, D. L., L. Yu, B. M. Fuchs, and R. Amann. 2001. Structure of bacterial communties in aquatic systems as revealed by filter PCR. *Aquatic Microbial Ecology* **26:** 13–22.

Koenneke, M., A. E. Bernhard, J. R. De La Torre, C. B. Walker, J. B. Waterbury, and D. A. Stahl. 2005. Isolation of an autotrophic ammonia-oxidizing marine archaeon. *Nature* **437:** 543–546.

Kramer, J. G., M. Wyman, J. P. Zehr, D. G. Capone. 1996. Diel variability in transcription of the structural gene for glutamine synthetase (*glnA*) in natural populations of the marine diazotrophic cyanobacterium *Trochodesmium theiebautii. FEMS Microbiology Ecology* **21:** 187–196.

Kreuzer-Martin, H. W. 2007. Stable isotope probing: Linking functional activity to specific members of microbial communities. *Soil Science Society of America Journal* **71:** 611–619.

Lam, P., P. Lama, G. Lavika, M. M. Jensena, J. van de Vossenbergb, M. Schmidb, D. Woebkena, D. Gutie'rrezc, R. Amanna, M.S.M. Jettenb, and M.M.M. Kuypersathers. 2007. Linking crenarchaeal and bacterial nitrification to anammox in the Black Sea. *Proceedings of the National Academy of Sciences of the USA* **104:** 7104–7109.

Lange, B. M., T. Rujan, W. Martin, and R. Croteau. 2000. Isoprenoid biosynthesis: The evolution of two ancient and distinct pathways across genomes. *Proceedings of the National Academy of Sciences of the USA* **97:** 13172–13177.

Lindell, D., E. Padan, and A. F. Post 1998. Regulation of *ntcA* expression and nitrite uptake in the marine *Synechococcus* sp. strain WH 7803. *Journal of Bacteriology* **180:** 1878–1886.

Lipp, J. S., Y. Morono, F. Inagaki, and K. U. Hinrichs. 2008. Significant contribution of archaea to extant biomass in marine subsurface sediments. *Nature* **454:** 991–994.

McCallister, S. L., J. E. Bauer, J. E. Cherrier, and H. W. Ducklow. 2004. Assessing sources and ages of organic matter supporting river and estuarine bacterial production: A multiple-isotope ($\delta^{14}C$, $\delta^{13}C$, and $\delta^{15}N$) approach. *Limnology and Oceanography* **49:** 1687–1702.

Moldowan, J. M., and N. M. Talyzina. 2009. Biogeochemical evidence for dinoflagellate ancestors in the early Cambrian. *Science* **281:** 1168–1170.

Munn, C. B. 2004. *Marine microbiology: Ecology & applications.* Garland Science/BIOS Scientific, London.

Nauhaus, K., T. Treude, A. Boetius, and M. Kruger. 2005. Environmental regulation of the anaerobic oxidation of methane: A comparison of ANME-I and ANME-II communities. *Environmental Microbiology* **7:** 98–106.

Orphan, V. J., C. H. House, K. U. Hinrichs, K. D. Mckeegan, and E. F. Delong. 2002. Multiple archaeal groups mediate methane oxidation in anoxic cold seep sediments. *Proceedings of the National Academy of Sciences of the USA* **99:** 7663–7668.

Palenik, B., B. Brahamsha, F. W. Larimer, M. Land, L. Hauser, P. Chain, J. Lamerdin J., W. Regala, E. E. Allen, J. McCarren, I. Paulsen, A. Dufresne, F. Partensky, E. A. Webb, and J. Waterbury. 2003. The genome of a motile marine *Synechococcus. Nature* **424:** 1037–1042.

Paul, J. H., W. H. Jeffrey, A. W. David, M. F. DeFlaun, and L. H. Cazaercs. 1989. Turnover of extracelluar DNA in eutrophic and oligotrophic freshwater environments of southwest Florida. *Applied Environmental Microbiology* **55(7):** 1623–1828.

Paul, J. H., L. H. Cazares, A. W. David, M. F. Deflaun, and W. H. Jeffrey. 1991a. The distribution of dissolved DNA in an oligotrophic and a eutrophic river of southwest Florida. *Hydrobiologia* **218:** 53–63.

Paul, J. H., S. C. Jiang, and J. B. Rose. 1991b. Concentration of viruses and dissolved DNA from aquatic environments by vortex flow filtration. *Applied and Environmental Microbiology* **57:** 2197–2204.

Pearson, A., M. Budin, and J. J. Brocks. 2004a. Phylogenetic and biochemical evidence for sterol synthesis in the bacterium *Gemmata obscuriglobus* (vol. 100, p. 15352, 2003). *Proceedings of the National Academy of Sciences of the USA* **101:** 3991–3991.

Pearson, A., A. L. Sessions, K. J. Edwards, and J. M. Hayes. 2004b. Phylogenetically specific separation of rRNA from prokaryotes for isotopic analysis. *Marine Chemistry* **92:** 295–306.

Pearson, A., S.R.F. Page, T. L. Jorgenson, W. W. Fischer, and M. B. Higgins. 2007. Novel hopanoid cyclases from the environment. *Environmental Microbiology* **9:** 2175–2188.

Pinchuk, G. E., and others. 2008. Utilization of DNA as a sole source of phosphorus, carbon, and energy by *Shewanella* spp.: Ecological and physiological implications for dissimilatory metal reduction. *Applied and Environmental Microbiology* **74:** 1198–1208.

Prosser, J. I., and G. W. Nicol. 2008. Relative contributions of archaea and bacteria to aerobic ammonia oxidation in the environment. *Environmental Microbiology* **10:** 2931–2941.

Raghoebarsing, A. A., and others. 2006. A microbial consortium couples anaerobic methane oxidation to denitrification. *Nature* **440:** 918–921.

Reitzel, K., J. Ahlgren, A. Gogoll, H. S. Jensen, and E. Rydin. 2006. Characterization of phosphorus in sequential extracts from lake sediments using P-31 nuclear magnetic resonance spectroscopy. *Canadian Journal of Fisheries and Aquatic Sciences* **63:** 1686–1699.

Romanowski, G., M. G. Lorenz, and W. Wackernagel. 1991. Adsorption of plasmid DNA to mineral surfaces and protection against DNAase-I. *Applied and Environmental Microbiology* **57:** 1057–1061.

Rotthauwe, J. H., K. P. Witzel, and W. Liesack. 1997. The ammonia monooxygenase structural gene amoA as a functional marker: Molecular fine-scale analysis of natural ammonia-oxidizing populations. *Applied and Environmental Microbiology* **63:** 4704–4712.

Scala, D. J., and L. J. Kerkhof. 1998. Nitrous oxide reductase (nosZ) gene-specific PCR primers for detection of denitrifiers and three nosZ genes from marine sediments. *FEMS Microbiology Letters* **162:** 61–68.

Schippers, A., and L. N. Neretin. 2006. Quantification of microbial communities in near-surface and deeply buried marine sediments on the Peru continental margin using real-time PCR. *Environmental Microbiology* **8:** 1251–1260.

Schwartz, E., S. Blazewicz, R. Doucett, B. A. Hungate, S. C. Hart, and P. Dijkstra. 2007. Natural abundance δ^{15}N and δ^{13}C of DNA extracted from soil. *Soil Biology and Biochemistry* **39:** 3101–3107.

Sessions, A. L., S. P. Sylva, and J. M. Hayes. 2005. Moving-wire device for carbon isotopic analyses of nanogram quantities of nonvolatille organic carbon. *Analytical Chemistry* **77:** 6519–6527.

Sinninghe Damsté, J. S., and M.J.L. Coolen. 2006. Fossil DNA in Cretaceous black shales: Myth or reality? *Astrobiology* **6:** 299–302.

Sorensen, K. B., and A. Teske. 2006. Stratified communities of active archaea in deep marine subsurface sediments. *Applied and Environmental Microbiology* **72:** 4596–4603.

Stepanauskas, R., and M. E. Sieracki. 2007. Matching phylogeny and metabolism in the uncultured marine bacteria, one cell at a time. *Proceedings of the National Academy of Sciences* **104:** 9052–9057.

Sterner, R. W., and J. J. Elser. 2002. Ecological stoichiometry the biology of elements from molecules to the biosphere. Princeton University Press, Princeton, NJ.

Strous, M., and others. 1999. Missing lithotroph identified as new planctomycete. *Nature* **400:** 446–449.

———. 2006. Deciphering the evolution and metabolism of an anammox bacterium from a community genome. *Nature* **440:** 790–794.

Teske, A. P. 2006. Microbial communities of deep marine subsurface sediments: Molecular and cultivation surveys. *Geomicrobiology Journal* **23:** 357–368.

Valentine, D. L., and W. S. Reeburgh. 2000. New perspectives on anaerobic methane oxidation. *Environmental Microbiology* **2:** 477–484.

Vieira, R. P., and others. 2007. Archaeal communities in a tropical estuarine ecosystem: Guanabara Bay, Brazil. *Microbial Ecology* **54:** 460–468.

Volkman, J. K. 2005. Sterols and other triterpenoids: Source specificity and evolution of biosynthetic pathways. *Organic Geochemistry* **36:** 139–159.

Wang, Q. F., H. Li, and A. F. Post. 2000. Nitrate assimilation genes of the marine diazotrophic, filamentous cyanobacterium *Trichodesmium* sp. strain wh9601. *Journal of Bacteriology* **182:** 1764–1767.

Ward, B. B. 2008. Phytoplankton community composition and gene expression of functional genes involved in carbon and nitrogen assimilation. *Journal of Phycology* **44:** 1490–1503.

Ward, B. B., D. Eveillard, J. D. Kirshtein, J. D. Nelson, M. A. Voytek, and G. A. Jackson. 2007. Ammonia-oxidizing bacterial community composition in estuarine and oceanic environments assessed using a functional gene microarray. *Environmental Microbiology* **9:** 2522–2538.

Wilms, R., H. Sass, B. Kopke, H. Koster, H. Cypionka, and B. Engelen. 2006. Specific bacterial, archaeal, and eukaryotic communities in tidal-flat sediments along a vertical profile of several meters. *Applied and Environmental Microbiology* **72:** 2756–2764.

Wuchter, C. and others. 2006. Archaeal nitrification in the ocean. *Proceedings of the National Academy of Sciences of the USA* **103:** 12317–12322.

Zehr, J. P., and L. A. McReynolds. 1989. Use of degenerate oligonucleotides for amplification of the nifH gene from the marine cyanobacterium *Trichodesmium thiebautii*. *Applied and Environmental Microbiology* **55:** 2522–2526.

Zehr, J. P., and B. B. Ward. 2002. Nitrogen cycling in the ocean: New perspectives on processes and paradigms. *Applied Environmental Microbiology* **68:** 1015–1024.

Zehr, J. P., S. R. Bench, E. A. Mondragon, J. McCarren, and E. F. Delong. 2007. Low genomic diversity in tropical oceanic N_2-fixing cyanobacteria. *Proceedings of the National Academy of Sciences of the USA* **104:** 17807–17812.

Chapter 8

Ahlgren, G., L. Lundstedt, M. Brett, and C. Forsberg. 1990. Lipid composition and food quality of some freshwater phytoplankton for cladoceran zooplankters. *Journal of Plankton Research* **12:** 809–818.

Aries, E., P. Doumenq, J. Artaud, M. Acquaviva, and J. C. Bertrand. 2001. Effects of petroleum hydrocarbons on the phospholipid fatty acid composition of a consortium composed of marine hydrocarbon-degrading bacteria. *Organic Geochemistry* **32:** 891–903.

Arzayus, K. M., and E. A. Canuel. 2005. Organic matter degradation in sediments of the York River estuary: Effects of biological vs. physical mixing. *Geochimica et Cosmochimica Acta* **69:** 455–464.

Bec, A., D. Desvilettes, A. Vera, C. Lemarchand, D. Fontvieille, and G. Bourdier. 2003. Nutritional quality of freshwater heterotrophic flagellate: Trophic upgrading of its microalgal diet for *Daphnia hyalina*. *Aquatic Microbial Ecology* **32:** 203–207.

Bianchi, T. S. 2007. *Biogeochemistry of estuaries*. Oxford University Press, New York.

Bigot, M., A. Saliot, X. Cui, and J. Li. 1989. Organic geochemistry of surface sediments from the Huanghe estuary and adjacent Bohai Sea (China). *Chemical Geology* **75:** 339–350.

Bligh, E. G., and W. J. Dyer. 1959. A rapid method of total lipid extraction and purification. *Canadian Journal of Biochemistry and Physiology* **37:** 911–917.

Bodineau, L., G. Thoumelin, V. Beghin, and M. Wartel. 1998. Tidal time-scale changes in the composition of particulate organic matter within the estuarine turbidity maximum zone in the macrotidal Seine estuary, France: The use of fatty acid and sterol biomarkers. *Estuarine, Coastal and Shelf Science* **47:** 37–49.

Boschker, H.T.S., and J. J. Middelburg. 2002. Stable isotopes and biomarkers in microbial ecology. *FEMS Microbiology Ecology* **40:** 85–95.

Boschker, H.T.S., S. C., Nold, P. Wellsbury, D. Bos, W. de Graaf, R. Pel, R. J. Parkes, and T. E. Cappenberg. 1998. Direct linking of microbial populations to specific biogeochemical processes by ^{13}C-labelling of biomarkers. *Nature* **392:** 801–805.

Boschker, H.T.S., A. Wielemaker, B.E.M. Schaub, and M. Holmer. 2000. Limited coupling of macrophyte production and bacterial carbon cycling in the sediments of *Zostera* spp. Meadows. *Marine Ecology Progress Series* **203:** 181–189.

Bossio, D. A., J. A. Fleck, K. M. Scow, and R. Fujii. 2006. Alteration of soil microbial communities and water quality in restored wetlands. *Soil Biology and Biochemistry* **38:** 1223–1233.

Bouillon, S., and H.T.S. Boschker. 2005. Bacterial carbon sources in coastal sediments: A review based on stable isotope data of biomarkers. *Biogeosciences Discussions* **2:** 1617–1644.

Bouillon, S., T. Moens, N. Koedam, F. Dahdouh-Guebas, W. Baeyens, and F. Dehairs. 2004. Variability in the origin of carbon substrates for bacterial communities in mangrove sediments. *FEMS Microbiology Ecology* **49:** 171–179.

Brett, M. T., and D. C. Müller-Navarra. 1997. The role of essential fatty acids in aquatic food web processes. *Freshwater Biology* **38:** 483–499.

Brinis, A., A. Méjanelle, G. Momzikoff, J. Gondry, V. Fillaux, V. Point, and A. Saliot 2004. Phospholipid ester-linked fatty acids composition of size-fractionated particles at the top ocean surface. *Organic Geochemistry* **35:** 1275–1287.

Budge, S. M., S. J. Iverson, and H. N. Koopman. 2006. Studying trophic ecology in marine ecosystems using fatty acids: A primer on analysis and interpretation. *Marine Mammal Science* **22:** 759–801.

Bull, I. D., N. R. Parekh, G. H. Hall, P. Ineson, and R. P. Evershed. 2000. Detection and classification of atmosphereric methane oxidizing bacteria in soil. *Nature* **405:** 175–178.

Canuel, E. A. 2001. Relations between river flow, primary production and fatty acid composition of particulate organic matter in San Francisco and Chesapeake bays: A multivariate approach. *Organic Geochemistry* **32:** 563–583.

Canuel, E. A., and C. S. Martens. 1996. Reactivity of recently deposited organic matter: Degradation of lipid compounds near the sediment–water interface. *Geochimica et Cosmochimica Acta* **60:** 1793–1806.

Canuel, E. A., J. E. Cloern, D. B. Ringelburg, J. B. Guckert, and G. H. Rau. 1995. Molecular and isotropic tracers used to examine sources of organic matter and its incorporation into the food webs of San Francisco Bay. *Limnology and Oceanography* **40:** 67–81.

Canuel, E. A., K. H. Freeman, and S. G. Wakeham. 1997. Isotopic compositions of lipid biomarker compounds in estuarine plants and surface sediments. *Limnology and Oceanography* **42:** 1570–1583.

Canuel, E. A., A. C. Spivak, E. J. Waterson, and J. E. Duffy. 2007. Biodiversity and food web structure influence short-term accumulation of sediment organic matter in an experimental seagrass system. *Limnology and Oceanography* **52:** 590–602.

Cavaletto, J. F., and W. S. Gardner. 1998. Seasonal dynamics of lipids in freshwater benthic invertebrates. Pp. 109–131 *in* M. T. Arts and B. C. Wainman, eds., *Lipids in freshwater ecosystems*. Springer, New York.

Chikaraishi, Y., and H. Naraoka. 2005. δ^{13}C and δD identification of sources of lipid biomarkers in sediments of Lake Haruna (Japan). *Geochimica et Cosmochimica Acta* **69:** 3285–3297.

Christie, W. W. 1998. Gas chromatography–mass spectrometry methods for structural analysis of fatty acids. *Lipids* **33:** 343–354.

Cifuentes, L. A., and G. G. Salata. 2001. Significance of carbon isotope discrimination between bulk carbon and extracted phospholipid fatty acids in selected terrestrial and marine environments. *Organic Geochemistry* **32:** 613–621.

Countway, R. E., R. M. Dickhut, and E. A. Canuel. 2003. Polycyclic aromatic hydrocarbon (PAH) distributions and associations with organic matter in surface waters of the York River, VA estuary. *Organic Geochemistry* **34:** 209–224.

Cranwell, P. A. 1982. Lipids of aquatic sediments and sedimenting particulates. *Progress Lipid Research* **21:** 271–308.

Crossman, Z. M., P. Ineson, and R. P. Evershed. 2005. The use of ^{13}C labelling of bacterial lipids in the characterisation of ambient methane-oxidising bacteria in soils. *Organic Geochemistry* **36:** 769–778.

Dalsgaard, J., M. St. John, G. Kattner, D. Müller-Navarra, and W. Hagen. 2003. Fatty acid trophic markers in the pelagic marine environment. *Advances in Marine Biology* **46:** 225–340.

David, V., B. Sautour, R. Galois, and P. Chardy. 2006. The paradox high zooplankton biomass-low vegetal particulate organic matter in high turbidity zones: What way for energy transfer? *Journal of Experimental Marine Biology and Ecology* **333:** 202–218.

De Leeuw, J. W., W.I.C. Riipstra, and L. R. Mur. 1992. The absence of long-chain alkyl diols and alkyl keto-1-ols in cultures of the cyanobacterium aphanizomenon flos-aquae. *Organic Geochemistry* **18:** 575–578.

Derieux, S., J. Fillaux, and A. Saliot. 1998. Lipid class and fatty acid distributions in particulate and dissolved fractions in the north Adriatic Sea. *Organic Geochemistry* **29:** 1609–1621.

Dijkman, N. A., and J. C. Kromkamp. 2006. Phospholipid-derived fatty acids as chemotaxonomic markers for phytoplankton: Application for inferring phytoplankton composition. *Marine Ecology-Progress Series* **324:** 113–125.

Erwin, J. 1973. Comparative biochemistry of fatty acids in eukaryotic microorganisms. Pp. 41–143 *in* J. A. Erwin, ed., *Lipids and biomembranes of eukaryotic microorganisms*. Academic Press, New York.

Goutx, M., S. G. Wakeham, C. Lee, M. Duflos, C. Gui, Z. Liugue, B. Moriceau, R. Sempéré, M. Tedetti, and J. Xue. 2007. Composition and degradation of marine particles with different settling velocities in the northwestern mediterranean sea. *Limnology and Oceanography* **52:** 1645–1664.

Grob, R. L. 1977. *Modern practice of gas chromatography*. Wiley, Hoboken, NJ.

Grossi, V., S. Caradec, and F. Gilbert. 2003. Burial and reactivity of sedimentary microalgal lipids in bioturbated mediterranean coastal sediments. *Marine Chemistry* **81:** 57–69.

Guckert, J. B., C. P. Antworth, P. D. Nichols, and D. C. White. 1985. Phospholipd, ester-linked fatty acid profiles as reproducible assays for changes in prokaryotic community structure of estuarine sediments. *FEMS Microbiology Ecology* **31:** 147–158.

Haddad, R. I., C. S. Martens, and J. W. Farrington. 1992. Quantifying early diagenesis of fatty acids in a rapidly-accumulating coastal marine sediment. *Organic Geochemistry* **19:** 205–216.

Ho, E. S., and P. A. Meyers. 1994. Variability of early diagenesis in lake sediments: Evidence from the sedimentary geolipid record in an isolated tarn. *Chemical Geology* **112:** 309–324.

Hu, J., H. Zhang, and P. A. Peng. 2006. Fatty acid composition of surface sediments in the subtropical Pearl River estuary and adjacent shelf, southern China. *Estuarine, Coastal and Shelf Science* **66:** 346–356.

Iverson, S. J., C. Field, W. D. Bowen, and W. Blanchard. 2004. Quantitative fatty acid signature analysis: A new method of estimating predator diet. *Ecological Monographs* **74:** 211–235.

Jonasdottir, S. H. 1994. Effects of food quality on the reproductive success of *Acartia tonsa* and *Acartia hudonica*: Laboratory observations. *Marine Biology* **121:** 67–81.

Jonasdottir, S. H., D. Fields, and S. Pantoja. 1995. Copepod egg production in Long Island Sound, USA, as a function of the chemical composition of seston. *Marine Ecology-Progress Series* **119:** 87–98.

Jones, R. H. and K. J. Flynn. 2005. Nutritional status and diet composition affect the value of diatoms as copepod prey. *Science* **307:** 1457–1459.

Kaneda, T. 1991. Iso- and anteiso-fatty acids in bacteria: Biosynthesis, function and taxonomic significance. *Microbiology Reviews* **55:** 288–302.

Kawamura, K., R. Ishiwatari, and K. Ogura. 1987. Early diagenesis of organic matter in the water column and sediments: Microbial degradation and resynthesis of lipids in Lake Haruna. *Organic Geochemistry* **11:** 251–264.

Khan, M., and J. Scullion. 2000. Effect of soil on microbial responses to metal contamination. *Environmental Pollution* **110:** 115–125.

Klein Bretelar, W.C.M., N. Schogt, M. Baas, S. Schouten, and K. W. Kraay. 1999. Trophic upgrading of food quality by protozoans enhancing copepod growth: Role of essential lipids. *Marine Biology* **135:** 191–198.

Klein Bretelar, W.C.M., N. Schogt, and S. Rampen. 2005. Effect of diatom nutrient limitation on cepopod development: Role of essential lipids. *Marine Ecology-Progress Series* **291:** 125–133.

Kolattukudy, P. E. 1980. Biopolyester membranes of plants: Cutin and suberin. *Science* **208:** 990–1000.

Leblond, J. D., and P. J. Chapman. 2000. Lipid class distribution of highly unsaturated long chain fatty acids in marine dinoflagellates. *Journal of Phycology* **36:** 1103–1008.

Lee, C., and S. G. Wakeham, eds. 1988. *Organic matter in seawater:* J. P. Rilly, ed., biochemical processes. Pp. 1–51 *in Chemical oceanography*, vol. 9. Academic Press, San Diego, CA.

Lee, C., S. Wakeham, and C. Arnosti. 2004. Particulate organic matter in the sea: The composition conundrum. *Ambio* **33:** 565–575.

Loh, A. N., J. E. Bauer, and E. A. Canuel. 2006. Dissolved and particulate organic matter source–age characterization in the upper and lower Chesapeake Bay: A combined isotope and biomarker approach. *Limnology and Oceanography* **51:** 1421–1431.

MaCnaughton, S. J., T. L. Jenkins, M. H. Wimpee, M. R. Cormier, and D. C. White. 1997. Rapid extraction of lipid biomarkers from pure culture and environmental samples using pressurized accelerated hot solvent extraction. *Journal of Microbiological Methods* **31:** 19–27.

Mannino, A., and H. R. Harvey. 1999. Lipid composition in particulate and dissolved organic matter in the Delaware estuary: Sources and diagenetic patterns. *Geochimica et Cosmochimica Acta* **63:** 2219–2235.

Matsuda, H., and T. Koyama. 1977. Early diagenesis of fatty acids in lacustrine sediments, II: A statistical approach to changes in fatty acid composition from recent sediments and some source materials. *Geochimica et Cosmochimica Acta* **41:** 1825–1834.

Mayzaud, P., J. P. Chanut, and R. G. Ackman. 1989. Seasonal changes in the biochemical composition of marine particulate matter with special reference to fatty acids and sterols. *Marine Ecology Progress Series* **56:** 189–204.

McCallister, S. L., J. E. Bauer, H. W. Ducklow, and E. A. Canuel. 2006. Sources of estuarine dissolved and particulate organic matter: A multi-tracer approach. *Organic Geochemistry* **37:** 454–468.

McKinley, V. L., A. D. Peacock and D. C. White. 2005. Microbial community PLFA and PHB responses to ecosystem restoration in tallgrass prairie soils. *Soil Biology and Biochemistry* **37**: 1946–1958.

Metcalfe, L. D., A. A. Schmitz, and J. R. Pelka. 1966. Rapid preparation of fatty acid esters from lipids for gas chromatographic analysis. *Analytical Chemistry* **38**: 514–515.

Meyers, P. A. 1997. Organic geochemical proxies of paleoceanographic, paleolimnologic, and paleoclimatic processes. *Organic Geochemistry* **27**: 213–250.

———. 2003. Applications of organic geochemistry to paleolimnological reconstructions: A summary of examples from the Laurentian Great Lakes. *Organic Geochemistry* **34**: 261–289.

Meyers, P. A., and B. J. Eadie. 1993. Sources, degradation and recycling of organic matter associated with sinking particles in Lake Michigan. *Organic Geochemistry* **20**: 47–56.

Meyers, P. A., and R. Ishiwatari. 1993. Lacustrine organic geochemistry: An overview of indicators of organic matter sources and diagenesis in lake sediments. *Organic Geochemistry* **20**: 867–900.

Meyers, P. A., R. A. Bourbonniere, and N. Takeuchi. 1980. Hydrocarbons and fatty acids in two cores of Lake Huron sediments. *Geochimica et Cosmochimica Acta* **44**: 1215–1221.

Meyers, P. A., M. J. Leenheer, B. J. Eaoie, and S. J. Maule. 1984. Organic geochemistry of suspended and settling particulate matter in Lake Michigan. *Geochimica et Cosmochimica Acta* **48**: 443–452.

Michener, R. H. and D. M. Schell. 1994. Stable isotope ratios as tracers in marine aquatic food webs. Pp. 138–157 *in* K. Lajtha and R. H. Michener, eds., *Stable isotopes in ecology and environmental science*. Blackwell Scientific, oxford, UK.

Middelburg, J. J., C. Barranguet, H.T.S. Boschker, P.M.J. Herman, T. Moens, and C.H.R. Heip. 2000. The fate of intertidal microphytobenthos carbon: An in situ ^{13}C-labeling study. *Limnology and Oceanography* **45**: 1224–1234.

Mollenhauer, G., and T. I. Eglinton. 2007. Diagenetic and sedimentological controls on the composition of organic matter preserved in California borderland basin sediments. *Limnology and Oceanography* **52**: 558–576.

Müller-Navarra, D. C. 1995a. Biochemical versus mineral limitation in *Daphnia*. *Limnology and Oceanography* **40**: 1209–1214.

———. 1995b. Evidence that a highly unsaturated fatty acid limits *Daphnia* growth in nature. *Archiv fur Hydrobiolgie* **132**: 297–307.

Müller-Navarra, D. C., M. T. Brett, A. M. Liston, and C. R. Goldman. 2000. A highly unsaturated fatty acid predicts carbon transfer between primary producers and consumers. *Nature* **403**: 74–77.

Müller-Navarra, D. C., Brett, M.T., Park, S., Chandra, S, Ballntyne, A. P., Zorita, E. and Goldman, C. R. 2004. Unsaturated fatty acid content in seston and tropho-dynamic coupling in lakes. *Nature* **427**: 69–72.

Muri, G., and S. G. Wakeham. 2006. Organic matter and lipids in sediments of Lake Bled (NW Slovenia): Source and effect of anoxic and oxic depositional regimes. *Organic Geochemistry* **37**: 1664–1679.

Muri, G., S. G. Wakeham, T. K. Pease, and J. Faganeli. 2004. Evaluation of lipid biomarkers as indicators of changes in organic matter delivery to sediments from Lake Planina, a remote mountain lake in NW Slovenia. *Organic Geochemistry* **35**: 1083–1093.

Napolitano, G. E. 1999. Fatty acids as trophic and chemical markers in freshwater ecosystems. Pp. 21–44 *in* M. T. Arts and B. C. Wainmann, eds., *Lipids in freshwater ecosystems*. Springer, New York.

Nichols, P., J. Guckert, and D. White. 1986. Determination of monounsaturated fatty acid double-bond position and geometry for microbial monocultures and complex consortia by capillary GC-MS of their dimethyl disulphide adducts. *Journal of Microbiological Methods* **5**: 49–55.

Oren, A., and P. Gurevich. 1993. Characterization of the dominant halophilic archaea in a bacterial bloom in the Dead Sea. *FEMS Microbial Ecology* **12**: 249–256.

Park, S., M. T. Brett, D. C. Müller-Navarra, and C. R. Goldman. 2002. Essential fatty acid content and the phosphorus to carbon ration in cultured algae as indicators of food quality for *Daphnia*. *Freshwater Biology* **47**: 1377–1390.

Park, S., M. T. Brett, D. C. Müller-Navarra, S.-C. Shin, and C. R. Goldman. 2003. Heterotrophic nanoflagellates and increased essential fatty acids during *Microcystis* decay. *Aquatic Microbial Ecology* **33**: 201–205.

Parkes, R. J., N.J.E. Dowling, D. C. White, R. A. Herbert, and G. R. Gibson. 1993. Characterization of sulphate-reducing bacterial populations within marine and estuarine sediments with different rates of sulphate reduction. *FEMS Microbiology Letters* **102**: 235–250.

Pearson, A., A. P. McNichol, B. C. Benitez-Nelson, J. M. Hayes, and T. I. Eglinton. 2001. Origins of lipid biomarkers in Santa Monica basin surface sediment: A case study using compound-specific $\delta^{14}C$ analysis. *Geochimica et Cosmochimica Acta* **65**: 3123–3137.

Peterson, B. J., and B. Fry. 1987. Stable isotopes in ecosystems studies. *Annual Review of Ecology and Systematics* **18**: 293–320.

Petsch, S. T., T. I. Eglinton, and K. J. Edwards. 2001. ^{14}C-dead living biomass: Evidence for microbial assimilation of ancient organic carbon during shale weathering. *Science* **292**: 1127–1131.

Pinturier-Geiss, L., J. Laureillard, C. Riaux-Gobin, J. Fillaux, and A. Saliot. 2001. Lipids and pigments in deep-sea surface sediments and interfacial particles from the western Crozet basin. *Marine Chemistry* **75**: 249–266.

Poerschmann, J., and R. Carlson. 2006. New fractionation scheme for lipid classes based on "in-cell fractionation" using sequential pressurized liquid extraction. *Journal of Chromatography A* **1127**: 18–25.

Polymenakou, P. N., A. Tselepides, E. G. Stephanou, and S. Bertilsson. 2006. Carbon speciation and composition of natural microbial communities in polluted and pristine sediments of the eastern Mediterranean Sea. *Marine Pollution Bulletin* **52**: 1396–1405.

Quemeneur, M., and Y. Marty. 1992. Sewage influence in a macrotidal estuary: Fatty acid and sterol distributions. *Estuarine, Coastal and Shelf Science* **34**: 347–363.

Rajendran, N., O., Matsuda, N. Inamura, and Y. Urushigawa, 1992. Determination of microbial biomass and its community structure from the distribution of phospholipid ester-linked fatty acids in sediments of Hiroshima Bay. *Estuarine, Coastal and Shelf Science* **34**: 501–514.

Rajendran, N., Y. Suwa, and Y. Urushigawa. 1993. Distribution of phospholipid ester-linked fatty acid biomarkers for bacteria in the sediment of Ise Bay, Japan. *Marine Chemistry* **42**: 39–56.

Rajendran, N., O. Matsuda, R. Rajendran, and Y. Urushigawa. 1997. Comparative description of microbial community structure in surface sediments of eutrophic bays. *Marine Pollution Bulletin* **34**: 26–33.

Ramos, C. S., C. C. Parrish, T.A.O. Quibuyen, and T. A. Abrajano. 2003. Molecular and carbon isotopic variations in lipids in rapidly settling particles during a spring phytoplankton bloom. *Organic Geochemistry* **34**: 195–207.

Saliot, A., J. Tronczynski, P. Scribe, and R. Letolle. 1988. The application of isotopic and biogeochemical markers to the study of the biochemistry of organic matter in a macrotidal estuary, the Loire, France. *Estuarine, Coastal and Shelf Science* **27**: 645–669.

Saliot, A., C. C. Parrish, N. Sadouni, I. Bouloubassi, J. Fillaux, and G. Cauwet. 2002. Transport and fate of Danube delta terrestrial organic matter in the northwest Black Sea mixing zone. *Marine Chemistry* **79**: 243–259.

Schmidt, K., and S. H. Jonasdottir. 1997. Nutritional quality of two cyanobacteria: How rich is "poor" food? *Marine Ecology Progress Series* **151**: 1–10.

Scribe, P., J. Fillaux, J. Laureillard, V. Denant, and A. Saliot. 1991. Fatty acids as biomarkers of planktonic inputs in the stratified estuary of the Krka River, Adriatic Sea: Relationship with pigments. *Marine Chemistry* **32**: 299–312.

Shi, W., M.-Y. Sun, M. Molina, and R. E. Hodson. 2001. Variability in the distribution of lipid biomarkers and their molecular isotopic composition in Altamaha estuarine sediments: Implications for the relative contribution of organic matter from various sources. *Organic Geochemistry* **32**: 453–467.

Sinninghe Damsté, J. S., B. E. Van Dongen, W.I.C. Rijpstra, S. Schouten, J. K. Volkman, and J.A.J. Geenevasen. 2001. Novel intact glycolipids in sediments from an Antarctic lake (Ace Lake). *Organic Geochemistry* **32**: 321–332.

Skerratt, J. H., P. D. Nichols, J. P. Bowman, and L. I. Sly. 1992. Occurrence and significance of long-chain (ω-1)-hydroxy fatty acids in methane-utilizing bacteria. *Organic Geochemistry* **18**: 189–194.

Slater, G. F., R. K. Nelson, B. M. Kile, and C. M. Reddy. 2006. Intrinsic bacterial biodegradation of petroleum contamination demonstrated in situ using natural abundance, molecular-level ^{14}C analysis. *Organic Geochemistry* **37**: 981–989.

Smedes, F., and T. K. Askland. 1999. Revisiting the development of the Bligh and Dyer total lipid determination method. *Marine Pollution Bulletin* **38**: 193–201.

Spivak, A. C, E. A. Canuel, J. E. Duffy, and J. P. Richardson. 2007. Top-down and bottom-up controls on sediment organic matter composition in an experimental seagrass ecosystem. *Limnology and Oceanography* **52**: 2595–2607.

Stowasser, G., G. J. Pierce, C. F. Moffat, M. A. Collins, and J. W. Forsythe. 2006. Experimental study on the effect of diet on fatty acid and stable isotope profiles of the squid *Lolliguncula brevis*. *Journal of Experimental Marine Biology and Ecology* **333:** 97–114.

Sun, M.-Y., and J. Dai. 2005. Relative influences of bioturbation and physical mixing on degradation of bloom-derived particulate organic matter: Clue from microcosm experiments. *Marine Chemistry* **96:** 201–218.

Sun, M. Y., and S. G. Wakeham. 1994. Molecular evidence for degradation and preservation of organic matter in the anoxic Black Sea Basin. *Geochemica Cosmochimica Acta* **58:** 3395–3406.

Sun, M.-Y., and S. G. Wakeham. 1999. Diagenesis of planktonic fatty acids and sterols in Long Island Sound sediments: Influences of a phytoplankton bloom and bottom water oxygen content. *Journal of Marine Research* **57:** 357–385.

Sun, M.-Y., S. G. Wakeham, and C. Lee. 1997. Rates and mechanisms of fatty acid degradation in oxic and anoxic coastal marine sediments of Long Island Sound, New York, USA. *Geochimica et Cosmochimica Acta* **61:** 341–355.

Sun, M.-Y. I., R. C. Aller, C. Lee, and S. G. Wakeham. 2002. Effects of oxygen and redox oscillation on degradation of cell-associated lipids in surficial marine sediments. *Geochimica et Cosmochimica Acta* **66:** 2003–2012.

Tenzer, G. E., P. A. Meyers, J. A. Robbins, B. J. Eadie, N. R. Morehead, and M. B. Lansing. 1999. Sedimentary organic matter record of recent environmental changes in the St. Mary's River ecosystem, Michigan–Ontario border. *Organic Geochemistry* **30:** 133–146.

Van Den Meersche, K., J. J. Middelburg, K. Soetart, P. Van Rijswijk, H.T.S. Boschker, and C.H.R. Heip. 2004. Carbon–nitrogen coupling and algal–bacterial interactions during an experimental bloom: Modeling a ^{13}C tracer experiment. *Limnology and Oceanography* **49:** 862–878.

Veloza, A. J., F.-L. E. Chu, and K. W. Tang. 2006. Trophic modification of essential fatty acids by hetertrophic protists and its effects on the fatty acid composition of the copepod *Acartia tonsa*. *Marine Biology* **148:** 779–788.

Volkman, J. K., ed., 2006. *Lipid markers for marine organic matter*. Pp. 27–70 *in* J. K. Volkman, ed., *Marine organic matter: Biomarkers, isotopes and DNA*. Springer, Heidelberg.

Volkman, J. K., S. M. Barrett, S. I. Blackburn, M. P. Mansour, E. L. Sikes, and F. Gelin. 1998. Microalgal biomarkers: A review of recent research developments. *Organic Geochemistry* **29:** 1163–1179.

Wacker, A., P. Becher, and E. Von Elert. 2002. Food quality effects of unsaturated fatty acids on larvae of the zebra mussel *Dreissena polymorpha*. *Limnology and Oceanography* **47:** 1242–1248.

Wakeham, S., and J. A. Beier. 1991. Fatty accid and sterol biomarkers as indicators of particulate matter source and alteration processes in the Black Sea. *Deep Sea Research* **38:** S943–S968.

Wakeham, S. G. 1985. Wax esters and triacylglycerols in sinking particulate matter in the Peru upwelling area (15°, 75° W). *Marine Chemistry* **17:** 213–235.

———. 1995. Lipid biomarkers for heterotrophic alteration of suspended particulate organic matter in oxygenated and anoxic water columns of the ocean. *Deep Sea Research, Part I: Oceanographic Research Papers* **42:** 1749–1771.

Wakeham, S. G., and C. Lee, eds. 1993. Production, transport and alteration of particulate organic matter in the marine water column. Pp. 145–169 *in* M. H. Engel and S. A. Macko, eds., *Organic geochemistry principles and applications*. Plenum Press, New York.

Wakeham, S. G., C. Lee, J. I. Hedges, P. J. Hernes, and M. L. Peterson. 1997. Molecular indicators of diagenetic status in marine organic matter. *Geochimica et Cosmochimica Acta* **61:** 5363–5369.

Wakeham, S. G., M. L. Peterson, J. I. Hedges, and C. Lee. 2002. Lipid biomarker fluxes in the Arabian Sea, with a comparison to the Equatorial Pacific Ocean. *Deep Sea Research, Part II: Topical Studies in Oceanography* **49:** 2265–2301.

Wakeham, S. G., A. P. McNichol, J. E. Kostka, and T. K. Pease. 2006. Natural-abundance radiocarbon as a tracer of assimilation of petroleum carbon by bacteria in salt marsh sediments. *Geochimica et Cosmochimica Acta* **70:** 1761–1771.

Wakeham, S. G. and N. M. Frew. 1982. Glass capillary gas chromatography–mass spectrometry of wax esters, steryl esters and triacylglycerols. *Lipids* **17:** 831–843.

Weers, P.M.M. and R. D. Gulati. 1997. Effect of the addition of polyunsaturated fatty acids to the diet on the growth and fecundity of *Daphnia galeata*. *Freshwater Biology* **38:** 721–729.

White, D. C. 1994. Is there anything else you need to understand about the microbiota that cannot be derived from analysis of nucleic acids? *Microbial Ecology* **28:** 163–166.

White, D. C., W. M. Davis, J. S. Nickels, J. D. King, and R. J. Bobbie. 1979. Determination of the sedimentary microbial biomass by extractible lipid phosphate. *Oecologia* **40:** 51–62.

Whyte, J.N.C. 1988. Fatty acid profiles from direct methanolysis of lipids in tissue of cultured species. *Aquaculture* **75:** 193–203.

Yano, Y., A. Nakayama, and K. Yoshida. 1997. Distribution of polyunsaturated fatty acids in bacteria present in intestines of deep-sea fish and shallow-sea poikilothermic animals. *Applied and Environmental Microbiology* **63:** 2572–2577.

Zak, D. R., W. E. Holmes, D. C. White, A. D. Peacock, and D. Tilman. 2003. Plant diversity, soil microbial communities, and ecosystem function: Are there any links? *Ecology* **84:** 2042–2050.

Zimmerman, A. R., and E. A. Canuel. 2000. A geochemical record of eutrophication and anoxia in Chesapeake Bay sediments: Anthropogenic influence on organic matter composition. *Marine Chemistry* **69:** 117–137.

———. 2001. Bulk organic matter and lipid biomarker composition of Chesapeake Bay surficial sediments as indicators of environmental processes. *Estuarine, Coastal and Shelf Science* **53:** 319–341.

———. 2002. Sediment geochemical records of eutrophication in the mesohaline Chesapeake Bay. *Limnology and Oceanography* **47:** 1084–1093.

Zou, L., X. Wang, J. Callahan, R. A. Culp, R. F Chen, M. Altabet, and M. Sun. 2004. Bacterial roles in the formation of high-molecular-weight dissolved organic matter in estuarine and coastal waters: Evidence from lipids and the compound-specific isotopic ratios. *Limnology and Oceanography* **49:** 297–302.

Chapter 9

Bachtiar, T., J. P. Coakley and M. J. Risk. 1996. Tracing sewage contaminated sediments in Hamilton Harbor using selected geochemical indicators. *The Science of the Total Environment* **179:** 3–16.

Barrett, S. M., J. K. Volkman, G. A. Dunstan, and J. M. Leroi. 1995. Sterols of 14 species of marine diatoms (Bacillariophyta). *Journal of Phycology* **31:** 360–369.

Bayona, J. M., A. Farran, and J. Albaiges. 1989. Steroid alcohols and ketones in coastal waters of the western Mediterranean: Sources and seasonal variability. *Marine Chemistry* **27:** 79–104.

Beier, J. A., S. G. Wakeham, C. H. Pilskaln, and S. Honjo. 1991. Enrichment in saturated compounds of Black Sea interfacial sediment. *Nature* **351:** 642–644.

Belicka, L. L., R. W. Macdonald, M. B. Yunker, and H. R. Harvey. 2004. The role of depositional regime on carbon transport and preservation in Arctic Ocean sediments. *Marine Chemistry* **86:** 65–88.

Bosak, T., R. M. Losick, and A. Pearson. 2008. A polycyclic terpenoid that alleviates oxidative stress. *Proceedings of the National Academy of Sciences of the USA* **105:** 6725–6729.

Boucher Y., and W. F. Doolittle. 2000. The role of lateral gene transfer in the evolution of isoprenoid biosynthesis pathways. *Molecular Microbiology* **37:** 703–716.

Brocks, J. J., R. Buick, R. E. Summons, and G. A. Logan. 2003. A reconstruction of Archean biological diversity based on molecular fossils from the 2.78 to 2.45 billion-year-old Mount Bruce Supergroup, Hamersley basin, Western Australia. *Geochimica et Cosmochimica Acta* **67:** 4321–4335.

Brooks, C., E. Horning, and J. Young. 1968. Characterization of sterols by gas chromatography–mass spectrometry of the trimethylsilyl ethers. *Lipids* **3:** 391–402.

Bull, I. D., M. M. Elhmmali, D. J. Roberts, and R. P. Evershed. 2003. The application of steroidal biomarkers to track the abandonment of a Roman wastewater course at the Agora (Athens, Greece). *Archaeometry* **45:** 149–161.

Canuel, E. A., K. H. Freeman, and S. G. Wakeham. 1997. Isotopic compositions of lipid biomarker compounds in estuarine plants and surface sediments. *Limnology and Oceanography* **42:** 1570–1583.

Carreira, R. S., A.L.R. Wagener, and J. W. Readman. 2004. Sterols as markers of sewage contamination in a tropical urban estuary (Guanabara Bay, Brazil): Space–time variations. *Estuarine Coastal Shelf Science* **60:** 587–598.

Chalaux, N., H. Takada, and J. M. Bayona. 1995. Molecular markers in Tokyo Bay sediments: Sources and distribution. *Marine Environmental Research* **40:** 77–92.

Chen, N., T. S. Bianchi, and J. M. Bland. 2003a. Novel decomposition products of chlorophyll-*a* in continental shelf (Louisiana shelf) sediments: Formation and transformation of carotenol chlorin esters. *Geochimica et Cosmochimica Acta* **67:** 2027–2042.

———. 2003b. Implications for the role of pre versus post-depositional transformation of chlorophyll-*a* in the Lower Mississippi River and Louisiana shelf. *Marine Chemistry* **81:** 37–55.

Chikaraishi, Y. 2006. Carbon and hydrogen isotopic composition of sterols in natural marine brown and red macroalgae and associated shellfish. *Organic Geochemistry* **37:** 428–436.

Chikaraishi, Y., and H. Naraoka. 2005. $\delta^{13}C$ and δD identification of sources of lipid biomarkers in sediments of Lake Haruna (Japan). *Geochimica et Cosmochimica Acta* **69:** 3285–3297.

Conte, M. H., L.A.S. Madureira, G. Eglinton, D. Keen, and C. Rendall. 1994. Millimeter-scale profiling of abyssal marine sediments: Role of bioturbation in early sterol diagenesis. *Organic Geochemistry* **22:** 979–990.

Dachs, J., J. M. Bayona, S. W. Fowler, J.-C. Miquel, and J. Albaigès. 1998. Evidence for cyanobacterial inputs and heterotrophic alteration of lipids in sinking particles in the Alboran Sea (SW Mediterranean). *Marine Chemistry* **60:** 189–201.

Disch, A., J. Schwender, C. Müller, K. H. Lichtenthaler, and M. Rohmer. 1998. Distribution of the mevalonate and glyceraldehyde phosphate/pyruvate pathways for isoprenoid biosynthesis in unicellular algae and the cyanobacterium *Synechocystis* PCC 6714. *Journal of Biochemistry* **333:** 381–388.

Eganhouse, R. P. and P. M. Sherblom. 2001. Anthropogenic organic contaminants in the effluent of a combined sewer overflow: Impact on Boston Harbor. *Marine Environmental Research* **51:** 51–74.

Farrimond, P., I. M. Head, and H. E. Innes. 2000. Environmental influence on the biohopanoid composition of recent sediments. *Geochimica et Cosmochimica Acta* **64:** 2985–2992.

Fernandes, M. B., M.-A. Sicre, J. N. Cardoso, and S. J. Macedo. 1999. Sedimentary 4-desmethyl sterols and *n*-alkanols in an eutrophic urban estuary, Capiberibe River, Brazil. *The Science of the Total Environment* **231:** 1–16.

Gagosian, R. B., S. O. Smith, C. Lee, J. W. Farrington, and N. M. Frew. eds. 1980. Steroid transformations in recent marine sediments. Pp. 407–419 *in* Advances in A. G. Douglas and J. R. Maxwell, eds. *organic geochemistry*, vol. 12. Pergamon Press, Oxford, UK.

Goericke, R., A. Shankle, and D. J. Repeta. 1999. Novel carotenol chlorin esters in marine sediments and water column particulate matter. *Geochimica et Cosmochimica Acta* **63:** 2825–2834.

González-Oreja, J. A., and J. Saiz-Salinas. 1998. Short-term spatiotemporal changes in urban pollution by means of faecal sterols analysis. *Marine Pollution Bulletin* **36:** 868–875.

Grimalt, J. O. and J. Albaiges. 1990. Characterization of the depositional environments of the Ebro delta (western Mediterranean) by the study of sedimentary lipid markers. *Marine Geology* **95:** 207–224.

Grimalt, J. O., P. Fernandez, J. M. Bayona, and J. Albaiges. 1990. Assessment of faecal sterols and ketones as indicators of urban sewage inputs to coastal waters. *Environmental Science and Technology* **24:** 357–363.

Harvey, H. R., and S. A. Macko. 1997. Kinetics of phytoplankton decay during simulated sedimentation: Changes in lipids under oxic and anoxic conditions. *Organic Geochemistry* **27:** 129–140.

Harvey, H. R., S. A. Bradshaw, S.C.M. O'Hara, G. Eglinton, and E.D.S. Corner. 1988. Lipid composition of the marine dinoflagellate *Scrippsiella trochoidea. Phytochemistry* **27:** 1723–1729.

Hayes, J. M. 2001. Fractionation of carbon and hydrogen isotopes in biosynthetic processes. *Reviews in Mineral Geochemistry* **43:** 225–277.

Hinrichs, K. U., R. R. Schneider, P.J.M. ¸Ller, and J. Rullkötter. 1999. A biomarker perspective on paleoproductivity variations in two late quaternary sediment sections from the southeast Atlantic Ocean. *Organic Geochemistry* **30:** 341–366.

Itoh, N., Y. Tani, Y. Soma, and M. Soma. 2007. Accumulation of sedimentary photosynthetic pigments characterized by pyropheophorbide *a* and steryl chlorin esters (SCEs) in a shallow eutrophic coastal lake (Lake Hamana, Japan). *Estuarine, Coastal and Shelf Science* **71:** 287–300.

Jeng, W.-L., and B. C. Han (1996) Coprostanol in a sediment core from the anoxic Tan-Shui estuary, Taiwan. *Estuarine, Coastal and Shelf Science* **42:** 727–735.

Jones, G., P. D., Nichols, P. M., Shaw, M., Goodfellow, and A. G., O'Donnel, eds. 1994. Analysis of microbial sterols and hopanoids. Pp. 163–165 *in Chemical methods in prokaryotic systematics.* Wiley, New York.

Kannenberg, E. L., and K. Poralla. 1999. Hopanoid biosynthesis and function in bacteria. *Naturwissenschaften* **86:** 168–176.

Kennedy, J. A., and S. C. Brassell. 1992. Molecular stratigraphy of the Santa Barbara basin: Comparison with historical records of annual climate change. *Organic Geochemistry* **19:** 235–244.

Killops, S., and V. Killops. 2005. *Introduction to organic geochemistry*, 2nd ed. Blackwell, Oxford, UK.

King, L. L., and D. J. Repeta. 1994. Phorbin steryl esters in black sea sediment traps and sediments: A preliminary evaluation of their paleooceanographic potential. *Geochimica et Cosmochimica Acta* **58:** 4389–4399.

King, L. L., and S. G. Wakeham. 1996. Phorbin steryl ester formation by macrozooplankton in the Sargasso Sea. *Organic Geochemistry* **24:** 581–585.

Koch, B. P., J. Rullkötter, and R. J. Lara. 2003. Evaluation of triterpenols and sterols as organic matter biomarkers in a mangrove ecosystem in northern Brazil. *Wetlands Ecology and Management* **11:** 257–263.

Kohl, W., A. Gloe, and H. Reichenbach. 1983. Steroids from the myxobacterium *Nannocystis exedens*. *Journal of General Microbiology* **129:** 1629–1635.

Lange B. M., T. Rujan, W. Martin, and R. Croteau. 2000. Isoprenoid biosynthesis: The evolution of two ancient and distinct pathways across genomes. *Proceedings of the National Academy of Sciences of the USA* **97:** 13172–13177.

Laureillard, J., and A. Saliot. 1993. Biomarkers in organic matter produced in estuaries: A case study of the Krka estuary (Adriatic Sea) using the sterol marker series. *Marine Chemistry* **43:** 247–261.

LeBlanc, L. A., J. S. Latimer, J. T. Ellis, and J. G. Quinn. 1992. The geochemistry of coprostanol in waters and surface sediments from Narragansett Bay. *Estuarine, Coastal and Shelf Science* **34:** 439–458.

Leeming, R., V. Latham, M. Rayner, and P. Nichols, eds. 1997. Detecting and distinguishing sources of sewage pollution in Australian inland and coastal waters and sediments. Pp. 306–319, *in Molecular markers in environmental biochemistry*. American Chemical Society, Symposium Series, vol. 671. Washington, DC.

Lichtenthaler, H. K. 1999. The 1-deoxy-D-xylulose-5-phosphate pathway of isoprenoid biosynthesis in plants. *Annual Review of Plant Physiology and Plant Molecular Biology* **50:** 47–65.

Lichtenthaler, H. K., J. Schwender, A. Disch, and M. Rohmer. 1997. Biosynthesis of isoprenoids in higher plant chloroplasts proceeds via a mevalonate-independent pathway. *FEBS Letters* **400:** 271–274.

Loh, A. N., J. E. Bauer, and E. A. Canuel. 2006. Dissolved and particulate organic matter source-age characterization in the upper and lower Chesapeake Bay: A combined isotope and biochemical approach. *Limnology and Oceanography* **51:** 1421–1431.

Mannino, A., and H. R. Harvey. 1999. Lipid composition in particulate and dissolved organic matter in the Delaware estuary: Sources and diagenetic patterns. *Geochimica et Cosmochimica Acta* **63:** 2219–2235.

Matsumoto, K., K. Yamada, and R. Ishiwatari. 2001. Sources of 24-ethylcholest-5-en-3[beta]-ol in Japan Sea sediments over the past 30,000 years inferred from its carbon isotopic composition. *Organic Geochemistry* **32:** 259–269.

McCaffrey, M. A., and others. 1994. Paleoenvironmental implications of novel C_{30} steranes in Precambrian to Cenozoic age petroleum and bitumen. *Geochimica et Cosmochimica Acta* **58:** 529–532.

McCallister, S. L., J. E. Bauer, H. W. Ducklow, and E. A. Canuel. 2006. Sources of estuarine dissolved and particulate organic matter: A multi-tracer approach. *Organic Geochemistry* **37:** 454–468.

Mèjanelle, L., and J. Laureillard. 2008. Lipid biomarker record in surface sediments at three sites of contrasting productivity in the tropical North Eastern Atlantic. *Marine Chemistry* **108:** 59–76.

Meyers, P. A. 1997. Organic geochemical proxies of paleoceanographic, paleolimnologic, and paleoclimatic processes. *Organic Geochemistry* **27:** 213–250.

Meyers, P. A., and R. Ishiwatari. 1993. Lacustrine organic geochemistry: An overview of indicators of organic matter sources and diagenesis in lake sediments. *Organic Geochemistry* **20:** 867–900.

Mudge, S. M. and M. J. Bebianno. 1997. Sewage contamination following an accidental spillage in the Ria Formosa, Portugal. *Marine Pollution Bulletin* **34:** 163–170.

Mudge, S. M., and C. E. Norris. 1997. Lipid biomarkers in the Conwy estuary (North Wales, UK): A comparison between fatty alcohols and sterols. *Marine Chemistry* **57:** 61–84.

Ourisson, G., and P. Albrecht. 1992. Hopanoids, 1: Geohopanoids—The most abundant natural products on earth? *Accounts of Chemical Research* **25:** 398–402.

Ourisson, G., P. Albrecht, and M. Rohmer. 1979. The hopanoids: Paleochemistry and biochemistry of a group of natural products. *Pure and Applied Chemistry* **51:** 709–729.

Pancost, R. D., K. H. Freeman, and S. G. Wakeham. 1999. Controls on the carbon–isotope compositions of compounds in Peru surface waters. *Organic Geochemistry* **30:** 319–340.

Patterson, G., and W. Nes. 1991. *Physiology and biochemistry of sterols*. American Oil Chemists' Society, Champaign, IL.

Pearson, A. 2007. Factors controlling C-14 contents of organic compounds in oceans and sediments. *Geochimica et Cosmochimica Acta* **71:** A768–A768.

Pearson, A., T. I. Eglinton, and A. P. McNichol. 2000. An organic tracer for surface ocean radiocarbon. *Paleoceanography* **15:** 541–550.

Pearson, A., A. P. McNichol, B. C. Benitez-Nelson, J. M. Hayes, and T. I. Eglinton. 2001. Origins of lipid biomarkers in Santa Monica basin surface sediment: A case study using compound-specific $\delta^{14}C$ analysis. *Geochimica et Cosmochimica Acta* **65:** 3123–3137.

Pearson, E. J., P. Farrimond, and S. Juggins. 2007. Lipid geochemistry of lake sediments from semi-arid Spain: Relationships with source inputs and environmental factors. *Organic Geochemistry* **38:** 1169–1195.

Peters, K. E., C. C. Walters, and J. M. Moldowan. 2005a. *The biomarker guide* vol. 1: *Biomarkers and isotopes in the environment and human history*, 2nd ed, Cambridge University Press, Cambridge, UK.

———. 2005b. *The biomarker guide: Biomarkers and isotopes in petroleum exploration and earth history*. Cambridge University Press, Cambridge, UK.

Pratt, L. M., R. E. Summons, and G. B. Hieshima. 1991. Sterane and triterpane biomarkers in the Precambrian Nonesuch Formation, North American midcontinent rift. *Geochimica et Cosmochimica Acta* **55:** 911–916.

Quéméneur, M., and Y. Marty. 1992. Sewage influence in a macrotidal estuary: fatty acids and sterol distributions. *Estuarine, Coastal and Shelf Science* **34:** 347–363.

Readman, J. W., R.F.C. Mantoura, C. A. Llewellyn, M. R. Preston, and A. D. Reeves. 1986. The use of pollutant and biogenic markers as source discriminants of organic inputs to estuarine sediments. *International Journal of Environmental Analytical Chemistry* **27:** 29–54.

Rieder C., G. Strauβ, G. Fuchs, D. Arigoni, A. Bacher, and W. Eisenreich. 1998. Biosynthesis of the diterpene verrucosan-2β-ol in the phototrophic eubacterium *Chlorflexus aurantiacus*. *Journal of Biological Chemistry* **273:** 18099–18108.

Riffè-Chalard, C., L. Verzegnassi, and F.O.G. ¸Laáar. 2000. A new series of steryl chlorin esters: Pheophorbide a steryl esters in an oxic surface sediment. *Organic Geochemistry* **31:** 1703–1712.

Sacchettini, J. C., and C. D. Poulter. 1997. Creating isoprenoid diversity. *Science* **277:** 1788–1789.

Sauer, P. E., T. I. Eglinton, J. M. Hayes, A. Schimmelmann, and A. L. Sessions. 2001. Compound-specific D/H ratios of lipid biomarkers from sediments as a proxy for environmental and climatic conditions. *Geochimica et Cosmochimica Acta* **65:** 213–222.

Schouten, S., M.J.L. Hoefs, and J. S. Sinninghe Damsté. 2000. A molecular and stable carbon isotopic study of lipids in late quaternary sediments from the Arabian Sea. *Organic Geochemistry* **31:** 509–521.

Schwender J., C. Gemünden, and K. H. Lichtenthaler. 2001. Chlorophyta exclusively use the 1-deoxyxylulose 5-phosphate/2-C-methylerythritol 4-phosphate pathway for the biosynthesis of isoprenoids. *Planta* **212:** 416–423.

Seguel, C. G., S. M. Mudge, C. Salgado, and M. Toledo. 2001. Tracing sewage in the marine environment: Altered signatures in Concepciûn Bay, Chile. *Water Research* **35:** 4166–4174.

Seto, H., H. Watanabe, and K. Furihata. 1996. Simultaneous operation of the mevalonate and non-mevalonate pathways in the biosynthesis of isopentenyl diphosphate in *Streptomyces aeriouvifer*. *Tetrahedron Letters* **37:** 7979–7982.

Sherblom, P. M., M. S. Henry, and D. Kelly, eds. 1997. Questions remain in the use of coprostanol and epicoprostanol as domestic waste markers: Examples from coastal Florida. Pp. 320–331 *in* R. P. Eganhouse, ed., *Molecular markers in environmental geochemistry*. American Chemical Society, Washington, DC.

Sherwin, M. R., E. S. Van Vleet, V. U. Fossato, and F. Dolci. 1993. Coprostanol (5β-cholestan-3β-ol) in lagoonal sediments and mussels of Venice, Italy. *Marine Pollution Bulletin* **26:** 501–507.

Shi, W., M.-Y. Sun, M. Molina, and R. E. Hodson. 2001. Variability in the distribution of lipid biomarkers and their molecular isotopic composition in Altamaha estuarine sediments: Implications for the relative contribution of organic matter from various sources. *Organic Geochemistry* **32:** 453–467.

Squier, A. H., D. A. Hodgson, and B. J. Keely. 2002. Sedimentary pigments as markers for environmental change in an Antarctic lake. *Organic Geochemistry* **33:** 1655–1665.

Summons, R. E., J. K. Volkman, and C. J. Boreham. 1987. Dinosterane and other steroidal hydrocarbons of dinoflagellate origin in sediments and petroleum. *Geochimica et Cosmochimica Acta* **51:** 3075–3082.

Summons, R. E., L. L. Jahnke, J. M. Hope, and G. A. Logan. 1999. 2-Methylhop-anoids as biomarkers for cyanobacterial oxygenic photosynthesis. *Nature* **400:** 554–557.

Summons, R. E., A. S. Bradley, L. L. Jahnke, and J. R. Waldbauer. 2006. Steroids, triterpenoids and molecular oxygen. *Philosophical Transactions of the Royal Society B* **361:** 951–968.

Sun, M. Y., and S. G. Wakeham. 1998. A study of oxic/anoxic effects on degradation of sterols at the simulated sediment–water interface of coastal sediments. *Organic Geochemistry* **28:** 773–784.

Talbot, H. M., R. N. Head, R. P. Harris, and J. R. Maxwell. 1999. Steryl esters of pyrophaeophorbide *b*: A sedimentary sink for chlorophyll *b*. *Organic Geochemistry* **30:** 1403–1410.

———. 2000. Discrimination against 4-methyl sterol uptake during steryl chlorin ester production by copepods. *Organic Geochemistry* **31:** 871–880.

Talbot, H. M., R. E. Summons, L. L. Jahnke, C. S. Cockell, M. Rohmer, and P. Farrimond. 2008. Cyanobacterial bacteriohopanepolyol signatures from cultures and natural environmental settings. *Organic Geochemistry* **39:** 232–263.

ten Haven, H. L., T. M. Peakman, and J. Rullkötter. 1992a. [delta]2-Triterpenes: Early intermediates in the diagenesis of terrigenous triterpenoids. *Geochimica et Cosmochimica Acta* **56:** 1993–2000.

———. 1992b. Early diagenetic transformation of higher-plant triterpenoids in deep-sea sediments from Baffin Bay. *Geochimica et Cosmochimica Acta* **56:** 2001–2024.

Tyagi, P., D. Edwards, and M. Coyne. 2008. Use of sterol and bile acid biomarkers to identify domesticated animal sources of fecal pollution. *Water, Air, and Soil Pollution* **187:** 263–274.

Venkatesan, M. I. and I. R. Kaplan. 1990. Sedimentary coprostanol as an index of sewage addition in Santa Monica basin, southern California. *Environmental Science and Technology* **24:** 208–214.

Venkatesan, M. I., and C. A. Santiago. 1989. Sterols in ocean sediments: Novel tracers to examine habitats of cetaceans, pinnipeds, penguins and humans. *Marine Biology* **102:** 431–437.

Volkman, J. K. 1986. A review of sterol markers for marine and terrigenous organic matter. *Organic Geochemistry* **9:** 83–99.

———. 2003. Sterols in microorganisms. *Applied Microbiologcial Biotechnology* **60:** 495–506.

———. 2005. Sterols and other triterpenoids: Source specificity and evolution of biosynthetic pathways. *Organic Geochemistry* **36:** 139–159.

———, ed. 2006. *Lipid markers for marine organic matter*. Springer, Berlin.

Volkman, J. K., S. M. Barrett, S. I. Blackburn, M. P. Mansour, E. L. Sikes, and F. Gelin. 1998. Microalgal biomarkers: A review of recent research developments. *Organic Geochemistry* **29:** 1163–1179.

Volkman, J. K., A. T. Revill, P. I. Bonham, and L. A. Clementson. 2007. Sources of organic matter in sediments from the Ord River in tropical Northern Australia. *Organic Geochemistry* **38:** 1039–1060.

Wakeham, S. G. 1982. Organic matter from a sediment trap experiment in the Equatorial North Atlantic: Wax esters, steryl esters, triacylglycerols and alkyldiacylglycerols. *Geochimica et Cosmochimica Acta* **46:** 2239–2257.

———. 1995. Lipid biomarkers for heterotrophic alteration of suspended particulate organic matter in oxygenated and anoxic water columns of the ocean. *Deep Sea Research, Part I: Oceanographic Research Papers* **42:** 1749–1771.

Wakeham, S. G., and E. A. Canuel. 1990. Fatty acids and sterols of particulate matter in a brackish and seasonally anoxic coastal salt pond. *Organic Geochemistry* **16:** 703–713.

Wakeham, S. G., and J. R. Ertel. 1988. Diagenesis of organic matter in suspended particles and sediments in the Cariaco Trench. *Organic Geochemistry* **13:** 815–822.

Wakeham, S. G., and N. M. Frew. 1982. Glass capillary gas chromatography-mass spectrometry of wax esters, steryl esters and triacylglycerols. *Lipids* **17:** 831–843.

Wakeham, S. G., R. B. Gagosian, J. W. Farrington, and E. A. Canuel. 1984. Sterenes in suspended particulate matter in the eastern tropical North Pacific. *Nature* **308:** 840–843.

Wardroper, A.M.K. 1979. *Aspects of the geochemistry of polycyclic isoprenoids*. Thesis, University of Bristol, Bristol, UK.

Waterson, E. J., and E. A. Canuel. 2008. Sources of sedimentary organic matter in the Mississippi River and adjacent Gulf of Mexico as revealed by lipid biomarker and $\delta^{13}C_{TOC}$ analyses. *Organic Geochemistry* **39:** 422–439.

Xu, Y., and R. Jaffè. 2007. Lipid biomarkers in suspended particles from a subtropical estuary: Assessment of seasonal changes in sources and transport of organic matter. *Marine Environmental Research* **64:** 666–678.

Yunker, M. B., R. W. MacDonald, D. J. Veltkamp, and W. J. Cretney. 1995. Terrestrial and marine biomarkers in a seasonally ice-covered arctic estuary: Integration of multivariate and biomarker approaches. *Marine Chemistry* **49:** 1–50.

Zimmerman, A. R., and E. A. Canuel. 2000. A geochemical record of eutrophication and anoxia in Chesapeake Bay sediments: Anthropogenic influence on organic matter composition. *Marine Chemistry* **69:** 117–137.

———. 2001. Bulk organic matter and lipid biomarker composition of Chesapeake Bay surficial sediments as indicators of environmental processes. *Estuarine, Coastal and Shelf Science* **53:** 319–341.

———. 2002. Sediment geochemical records of eutrophication in the mesohaline Chesapeake Bay. *Limnology and Oceanography* **47:** 1084–1093.

Chapter 10

Aeckersberg, F., F. Bak, and F. Widdel. 1991. Anaerobic oxidation of saturated hydrocarbons to CO_2 by a new type of sulfate-reducing bacterium. *Archive Microbiology* **156:** 5–14.

Ageta, H., and Y. Arai. 1984. Fern constituents: Cycloartane triterpenoids and allied compounds from *Polypodium formosanum* and *P. niponicum*. *Phytochemistry* **23:** 2875–2884.

Bechtel, A., M. Widera, R. F. Sachsenhofer, R. Gratzer, A. Luecke, and M. Woszczyk. 2007. Biomarkers and geochemical indicators of Holocene environmental changes in coastal Lake Sarbsko (Poland). *Organic Geochemistry* **38:** 1112–1131.

Belt, S. T., G. Massè, W. G. Allard, J.-M. Robert, and S. J. Rowland. 2001a. C_{25} highly branched isoprenoid alkenes in planktonic diatoms of the *Pleurosigma genus*. *Organic Geochemistry* **32:** 1271–1275.

———. 2001b. Identification of a C_{25} highly branched isoprenoid triene in the freshwater diatom *Navicula sclesvicensis*. *Organic Geochemistry* **32:** 1169–1172.

Belt, S. T., G. Massè, S. J. Rowland, M. Poulin, C. Michel, and B. Leblanc. 2007. A novel chemical fossil of palaeo sea ice: IP25. *Organic Geochemistry* **38:** 16–27.

Blumer, M., M. M. Mullin and D. W. Thomas. 1964. Pristane in the marine environment. *Helgoland Marine Research* **10:** 187–201.

Brassell, S. C. and G. Eglinton. 1982. Molecular geochemical indicators in sediments. *In* M. L. Sohn, ed., *Organic marine geochemistry*, American Chemical Society, Washington, DC.

Brassell, S. C., A.M.K. Wardroper, I. D. Thompson, J. R. Maxwell, and G. Eglinton. 1981. Specific acyclic isoprenoids as biological markers of methanogenic bacteria in marine sediments. *Nature* **290:** 693–696.

Bray, E. E., and E. D. Evans. 1961. Distribution of *n*-paraffins as a clue to recognition of source beds. *Geochimica et Cosmochimica Acta* **22:** 2–15.

Canuel, E. A., K. H. Freeman, and S. G. Wakeham. 1997. Isotopic compositions of lipid biomarkers compounds in estuarine plants and surface sediments. *Limnology and Oceanography* **42:** 1570–1583.

Chikaraishi, Y., and H. Naraoka. 2007. δ^{13}Cand δD relationships among three *n*-alkyl compound classes (*n*-alkanoic acid, *n*-alkane and *n*-alkanol) of terrestrial higher plants. *Organic Geochemistry* **38:** 198–215.

Collister, J. W., G. Rieley, B. Stern, G. Eglinton, and B. Fry. 1994. Compound-specific δ^{13}C analyses of leaf lipids from plants with differing carbon dioxide metabolisms. *Organic Geochemistry* **21:** 619–627.

Cranwell, P. A. 1982. Lipids of aquatic sediments and sedimenting particulates. *Progress in Lipid Research* **21:** 271–308.

———. 1984. Lipid geochemistry of sediments from Upton Broad, a small productive lake. *Organic Geochemistry* **7:** 25–37.

de Leeuw, J. W., and M. Baas. 1986. Early stage diagenesis of steroids. Pp. 103–123 *in* R. B. Johns, ed., *Biological Markers in the sedimentary record*. Elsevier, Amsterdam.

Eglinton, G., and R. J. Hamilton. 1967. Leaf epicuticular waxes. *Science* **156:** 1322–1335.

Eglinton, T. I., and G. Eglinton. 2008. Molecular proxies for paleoclimatology. *Earth and Planetary Science Letters* **275:** 1–16.

Eglinton, T. I., L. I. Aluwihare, J. E. Bauer, E.R.M. Druffel, and A. P. McNichol. 1996. Gas chromatographic isolation of individual compounds from complex matrices for radiocarbon dating. *Analytical Chemistry* **68:** 904–912.

Ficken, K. J., B. Li, D. L. Swain, and G. Eglinton. 2000. An *n*-alkane proxy for the sedimentary input of submerged/floating freshwater aquatic macrophytes. *Organic Geochemistry* **31:** 745–749.

Filley, T. R., K. H. Freeman, T. S. Bianchi, M. Baskaran, L. A. Colarusso, and P. G. Hatcher. 2001. An isotopic biogeochemical assessment of shifts in organic matter input to Holocene sediments from Mud Lake, Florida. *Organic Geochemistry* **32:** 1153–1167.

Freeman, K. H., J. M. Hayes, J.-M. Trendel, and P. Albrecht. 1990. Evidence from carbon isotope measurements for diverse origins of sedimentary hydrocarbons. *Nature* **343:** 254–256.

Freeman, K. H., S. G. Wakeham, and J. M. Hayes. 1994. Predictive isotopic biogeochemistry: Hydrocarbons from anoxic marine basins. *Organic Geochemistry* **21:** 629–644.

Gieg, L. M., and J. M. Suflita. 2002. Detection of anaerobic metabolites of saturated and aromatic hydrocarbons in petroleum-contaminated aquifers. *Environmental Science and Technology* **36:** 3755–3762.

Grimalt, J., and J. Albaiges. 1987. Sources and occurrence of C_{12}-C_{22} *n*-alkane distributions with even carbon-number preference in sedimentary environments. *Geochimica et Cosmochimica Acta* **51:** 1379–1384.

Grossi, V., C. Cravo-Laureau, R. Guyoneaud, A. Ranchou-Peyruse, and A. Hirschler-Rèa. 2008. Metabolism of *n*-alkanes and *n*-alkenes by anaerobic bacteria: A summary. *Organic Geochemistry* **39:** 1197–1203.

Hayes, J. M. 1993. Factors controlling ^{13}C contents of sedimentary organic compounds: Principles and evidence. *Marine Geology* **113:** 111–125.

———. 2004. Isotopic order, biogeochemical processes, and earth history: Goldschmidt lecture, Davos, Switzerland, August 2002. *Geochimica et Cosmochimica Acta* **68:** 1691–1700.

Hayes, J. M., K. H. Freeman, B. N. Popp, and C. H. Hoham. 1990. Compound-specific isotopic analyses: A novel tool for reconstruction of ancient biogeochemical processes. *Organic Geochemistry* **16:** 1115–1128.

Ho, E. S., and P. A. Meyers. 1994. Variability of early diagenesis in lake sediments: Evidence from the sedimentary geolipid record in an isolated tarn. *Chemical Geology* **112:** 309–324.

Howard, D. L. 1980. *Polycyclic triterpenes of anaerobic photosynthetic bacterium, Rhodomicrobium vannielii.* Thesis, University of California at Los Angeles, Los Angeles, CA.

Jaffè, R., G. A. Wolff, A. Cabrera, and H. Carvajal Chitty. 1995. The biogeochemistry of lipids in rivers of the Orinoco basin. *Geochimica et Cosmochimica Acta* **59:** 4507–4522.

Jaffè, R., R. Mead, M. E. Hernandez, M. C. Peralba, and O. A. Diguida. 2001. Origin and transport of sedimentary organic matter in two subtropical estuaries: A comparative, biomarker-based study. *Organic Geochemistry* **32:** 507–526.

Kanke, H., M. Uchida, T. Okuda, M. Yoneda, H. Takada, Y. Shibata, and M. Morita. 2004. Compound-specific radiocarbon analysis of polycyclic aromatic hydrocarbons (PAHs) in sediments from an urban reservoir. *Nuclear Instruments and Methods in Physics Research Section B*: Beam Interactions with Materials and Atoms **223–24:** 545–554.

Kennicutt II, M. C., and J. M. Brooks. 1990. Unusually normal alkane distributions in offshore New Zealand sediments. *Organic Geochemistry* **15:** 193–197.

Killops, S., and V. Killops. 2005. Introduction to organic geochemistry, 2nd ed. Blackwell, Oxford, UK.

Kok, M. D., W.I.C. Rijpstra, L. Robertson, J. Volkman, and J. S. Sinninghe Damsté. 2000. Early steroid sulfurisation in surface sediments of a permanently stratified lake (Ace Lake, Antarctica). *Geochimica et Cosmochimica Acta* **64:** 1425–1436.

Krull, E., D. Sachse, I. Mugler, A. Thiele, and G. Gleixner. 2006. Compound-specific $\delta^{13}C$ and $\delta^{2}H$ analyses of plant and soil organic matter: A preliminary assessment of the effects of vegetation change on ecosystem hydrology. *Soil Biology and Biochemistry* **38:** 3211–3221.

Lee, R. F. and A. R. Loeblich III. 1971. Distribution of 21:6 hydrocarbon and its relationship to 22:6 fatty acid in algae. *Phytochemistry* **10:** 593–602.

Lüder, B., G. Kirchner, A. Lucke, and B. Zolitschka. 2006. Palaeoenvironmental reconstructions based on geochemical parameters from annually laminated sediments of Sacrower See (Northeastern Germany) since the 17th century. *Journal of Paleolimnology* **35:** 897–912.

Mackenzie, A. S., S. C. Brassell, G. Eglinton and J. R. Maxwell. 1982. Chemical fossils: The geological fate of steroids. *Science* **217:** 491–504.

Marzi, R., B. E. Torkelson, and R. K. Olson. 1993. A revised carbon preference index. *Organic Geochemistry* **20:** 1303–1306.

Massè, G., S. T. Belt, W. Guy Allard, C. Anthony Lewis, S. G. Wakeham, and S. J. Rowland. 2004. Occurrence of novel monocyclic alkenes from diatoms in marine particulate matter and sediments. *Organic Geochemistry* 35: 813–822.

Mead, R., and M. A. Goni. 2006. A lipid molecular marker assessment of sediments from the northern Gulf of Mexico before and after the passage of Hurricane Lili. *Organic Geochemistry* **37:** 1115–1129.

Mead, R., Y. Xu, J. Chong, and R. Jaffè. 2005. Sediment and soil organic matter source assessment as revealed by the molecular distribution and carbon isotopic composition of *n*-alkanes. *Organic Geochemistry* **36:** 363–370.

Medeiros, P. M., and B.R.T. Simoneit. 2008. Multi-biomarker characterization of sedimentary organic carbon in small rivers draining the northwestern United States. *Organic Geochemistry* **39:** 52–74.

Meyers, P. A. 1997. Organic geochemical proxies for paleoceanographic, paleolimnologic and paleoclimatic processes. *Organic Geochemistry* **27:** 213–250.

———. 2003. Applications of organic geochemistry to paleolimnological reconstructions: A summary of examples from the Laurentian Great Lakes. *Organic Geochemistry* **34:** 261–289.

Meyers, P. A., and B. J. Eadie. 1993. Sources, degradation and recycling of organic matter associated with sinking particles in Lake Michigan. *Organic Geochemistry* **20:** 47–56.

Meyers, P. A., and R. Ishiwatari. 1993. Lacustrine organic geochemistry: An overview of indicators of organic matter sources and diagenesis in lake sediments. *Organic Geochemistry* **20:** 867–900.

Meyers, P. A., and J. L. Teranes, eds. 2001. Sediment organic matter. Pp. 239–270 *in Tracking Environmental change using lake sediment, Vol. 2: Physical and geochemical methods*. Kluwer Academic, Dordrect, The Netherlands.

Mitra, S., and T. S. Bianchi. 2003. A preliminary assessment of polycyclic aromatic hydrocarbon distributions in the lower Mississippi River and Gulf of Mexico. *Marine Chemistry* **82:** 273–288.

Moldowan, J. M., J. Dahl, B. J. Huizinga, F. J. Fago, L. J. Hickey, T. M. Peakman, and D. W. Taylor. 1994. The molecular fossil record of oleanane and its relation to angiosperms. *Science* **265:** 768–771.

Mugler, I., D. Sachse, M. Werner, B. Xu, G. Wu, T. Yaoand, and G. Gleixner. 2008. Effect of lake evaporation on d values of lacustrine *n*-alkanes: A comparison of Nam Co (Tibetan plateau) and Holzmaar (Germany). *Organic Geochemistry* **39:** 711–729.

Muri, G., S. G. Wakeham, T. K. Pease, and J. Faganeli. 2004. Evaluation of lipid biomarkers as indicators of changes in organic matter delivery to sediments from Lake Planina, a remote mountain lake in NW Slovenia. *Organic Geochemistry* **35:** 1083–1093.

Nishimura, M., and E. W. Baker. 1986. Possible origin of *n*-alkanes with a remarkable even-to-odd predominance in recent sediments. *Geochimica et Cosmochimica Acta* **50:** 299–305.

Ostrom, P. H., N. E. Ostrom, J. Henry, B. J. Eadie, P. A. Meyers, and J. A. Robbins. 1998. Changes in the trophic state of Lake Erie: Discordance between molecular and bulk sedimentary records. *Chemical Geology* **152:** 163–179.

Pearson, A., J. S. Seewald, and T. I. Eglinton. 2005. Bacterial incorporation of relict carbon in the hydrothermal environment of Guaymas basin. *Geochimica et Cosmochimica Acta* **69:** 5477–5486.

Peters, K. E., C. C. Walters, and J. M. Moldowan. 2005. *The biomarker guide, I: Biomarkers and isotopes in the environment and human history*, 2nd ed. Cambridge University Press. Cambridge, UK.

Philippi, G. T. 1965. On the depth, time and mechanism of petroleum generation. *Geochimica et Cosmochimica Acta* **29:** 1021–1049.

Prahl, F. G., J. R. Ertel, M. A. Goni, M. A. Sparrow, and B. Eversmeyer. 1994. Terrestrial organic carbon contributions to sediments on the Washington margin. *Geochimica et Cosmochimica Acta* **58:** 3035–3048.

Reddy, C. M., L. Xu, T. I. Eglinton, J. P. Boon, and D. J. Faulkner. 2002. Radiocarbon as a tool to apportion the sources of polycyclic aromatic hydrocarbons and black carbon in environmental samples. *Environmental Science and Technology* **36:** 1774–1782.

Rethemeyer, J., C. Kramer, G. Gleixner, B. John, T. Yamashita, H. Flessa, N. Andersen, M. J. Nadeau, and P. M. Grootes. 2004. Complexity of soil organic matter: AMS ^{14}C analysis of soil lipid fractions and individual compounds. *Radiocarbon* **46:** 465–473.

Risatti, J. B., S. J. Rowland, D. A. Yon, and J. R. Maxwell. 1984. Stereochemical, studies of acyclic isoprenoids, XII: Lipids of methanogenic bacteria and possible contributions to sediments. *Organic Geochemistry*. **6:** 93–103.

Rogge, W. F., P. M. Medeiros, and B.R.T. Simoneit. 2007. Organic marker compounds in surface soils of crop fields from the San Joaquin Valley fugitive dust characterization study. *Atmospheric Environment* **41:** 8183–8204.

Routh, J., P. A. Meyers, T. Hjorth, M. Baskaran, and R. Hallberg. 2007. Sedimentary geochemical record of recent environmental changes around Lake Middle Marviken, Sweden. *Journal of Paleolimnology* **37:** 529–545.

Rowland, S. J., and J. N. Robson. 1990. The widespread occurrence of highly branched acyclic C_{20}, C_{25} and C_{30} hydrocarbons in recent sediments and biota. *Marine Environmental Research* **30:** 191–216.

Rowland, S. J., N. A. Lamb, C. F. Wilkinson, and J. R. Maxwell. 1982. Confirmation of 2,6,10,15,19-pentamethyleicosane in methanogenic bacteria and sediments. *Tetrahedron Letters* **23:** 101–104.

Sachse, D., J. Radke, and G. Gleixner. 2006. δ values of individual *n*-alkanes from terrestrial plants along a climatic gradient: Implications for the sedimentary biomarker record. *Organic Geochemistry* **37:** 469–483.

Sauer, P. E., T. I. Eglinton, J. M. Hayes, A. Schimmelmann, and A. L. Sessions. 2001. Compound-specific D/H ratios of lipid biomarkers from sediments as a proxy for environmental and climatic conditions. *Geochimica et Cosmochimica Acta* **65:** 213–222.

Scanlan, R. S., and J. E. Smith. 1970. An improved measure of the odd-to-even predominace in the normal alkanes of sediment extracts and petroleum. *Geochimica et Cosmochimica Acta* **34:** 611–620.

Schouten, S., J. S. Sinninghe Damsté, M. Baas, A. C. Kock-Van Dalen, M.E.L. Kohnen and J. W. De Leeuw. 1995. Quantitative assessment of mono- and polysulfide-linked carbon skeletons in sulfur-rich macromolecular aggregates present in bitumen and oils. *Organic Geochemistry* **23:** 765–775.

Schouten, S., M.J.L. Hoefs, M. P. Koopmans, H. J. Bosch, and J. S. Sinninghe Damsté. 1998. Structural characterization, occurrence and fate of archaeal ether-bound acyclic and cyclic biphytanes and corresponding diols in sediments. *Organic Geochemistry* **29:** 1305–1319.

Schouten, S., S. G. Wakeham, and J. S. Sinninghe Damsté. 2001. Evidence for anaerobic methane oxidation by archaea in euxinic waters of the Black Sea. *Organic Geochemistry* **32:** 1277–1281.

Sessions, A. L., and J. M. Hayes. 2005. Calculation of hydrogen isotopic fractionations in biogeochemical systems. *Geochimica et Cosmochimica Acta* **69:** 593–597.

Sessions, A. L., T. W. Burgoyne, A. Schimmelmann, and J. M. Hayes. 1999. Fractionation of hydrogen isotopes in lipid biosynthesis. *Organic Geochemistry* **30:** 1193–1200.

Shiea, J., S. C. Brassell, and D. M. Ward. 1990. Mid-chain branched mono- and dimethyl alkanes in hot spring cyanobacterial mats: A direct biogenic source for branched alkanes in ancient sediments? *Organic Geochemistry* **15:** 223–231.

Silliman, J. E., and C. L. Schelske. 2003. Saturated hydrocarbons in the sediments of Lake Apopka, Florida. *Organic Geochemistry* **34:** 253–260.

Silliman, J. E., P. A. Meyers, and R. A. Bourbonniere. 1996. Record of postglacial organic matter delivery and burial in sediments of Lake Ontario. *Organic Geochemistry* **24:** 463–472.

Silliman, J. E., P. A. Meyers, and B. J. Eadie. 1998. Perylene: An indicator of alteration processes or precursor materials? *Organic Geochemistry* **29:** 1737–1744.

Silliman, J. E., P. A. Meyers, P. H. Ostrom, N. E. Ostrom, and B. J. Eadie. 2000. Insights into the origin of perylene from isotopic analyses of sediments from Saanich Inlet, British Columbia. *Organic Geochemistry* **31:** 1133–1142.

Silliman, J. E., P. A. Meyers, B. J. Eadie, and J. Val Klump. 2001. A hypothesis for the origin of perylene based on its low abundance in sediments of Green Bay, Wisconsin. *Chemical Geology* **177:** 309–322.

Sinninghe Damsté, J. S., and J. W. De Leeuw. 1990. Analysis, structure and geochemical significance of organically-bound sulphur in the geosphere: State of the art and future research. *Organic Geochemistry* **16:** 1077–1101.

Sinninghe Damsté, J. S., A.-M. W.E.P. Erkes, W. Irene, C. Rijpstra, J. W. De Leeuw, and S. G. Wakeham. 1995. C_{32}–C_{36} polymethyl alkenes in Black Sea sediments. *Geochimica et Cosmochimica Acta* **59:** 347–353.

Sinninghe Damsté, J. S., W.I.C. Rijpstra, S. Schouten, H. Peletier, M.J.E.C. van der Maarel, W. C. Gieskes. 1999. A C_{25} highly branched isoprenoid alkene and C_{25} and C_{27} *n*-polyenes in the marine diatom *Rhizosolenia setigera*. *Organic Geochemistry* **30:** 95–100.

Sinninghe Damsté, J. S., M.M.M. Kuypers, S. Schouten, S. Schulte, J. Rullkötter. 2003. The lycopane/C_{31} *n*-alkane ratio as a proxy to assess paleooxocity during sediment deposition. *Earth and Planetary Science Letters* **209:** 215–226.

Slater, G. F., R. K. Nelson, B. M. Kile, and C. M. Reddy. 2006. Intrinsic bacterial biodegradation of petroleum contamination demonstrated in situ using natural abundance, molecular-level ^{14}C analysis. *Organic Geochemistry* **37:** 981–989.

Smith, F. A., and K. H. Freeman. 2006. Influence of physiology and climate on δ^2D of leaf wax *n*-alkanes from C_3 and C_4 grasses. *Geochimica et Cosmochimica Acta* **70:** 1172–1187.

Spormann, A. M., and F. Widdel. 2000. Metabolism of alkylbenzenes, alkanes, and other hydrocarbons in anaerobic bacteria. *Biodegradation* **11:** 85–105.

Tornabene, T. G. and T. A. Langworthy. 1979. Biphytanyl and diphytanyl glycerol ether lipids of methanogenic Archaebacteria. *Science* **203:** 51–53.

Tornabene, T. G., T. A. Langworthy, G. Holzer, and J. Oró. 1979. Squalenes, phytanes and other isoprenoids as major neutral lipids of methanogenic and thermoacidophilic "archaebacteria." *Journal of Molecular Evolution* **13:** 73–83.

Uchikawa, J., B. N. Popp, J. E. Schoonmaker, and L. Xu. 2008. Direct application of compound-specific radiocarbon analysis of leaf waxes to establish lacustrine sediment chronology. *Journal of Paleolimnology* **39:** 43–60.

Vairavamurthy, A., K. Mopper, and B. F. Taylor. 1992. Occurrence of particle-bound polysulfides and significance of their reaction with organic matters in marine sediments. *Geophysical Research Letters* **19:** 2043–2046.

Viso, A.-C., D. Pesando, P. Bernard, and J.-C. Marty. 1993. Lipid components of the Mediterranean seagrass *Posidonia oceanica*. *Phytochemistry* **34:** 381–387.

Volkman, J. K. 2005. Sterols and other triterpenoids: Source specificity and evolution of biosynthetic pathways. *Organic Geochemistry* **36:** 139–159.

Volkman, J. K., ed. 2006. *Lipid markers for marine organic matter*, Pp. 27–70 *in*: J. K. Volkman, ed. *Marine organic matter: Biomarkers, isotopes and DNA*. Springer, Berlin.

Volkman, J. K., and J. R. Maxwell. 1986. Acyclic isoprenoids as biological markers. Pp. 1–42 *in* R. B. Johns, ed., *Biological markers in the sedimentary record*, Elsevier, New York.

Volkman, J. K., S. M. Barrett, and G. A. Dunstan. 1994. C_{25} and C_{30} highly branched isoprenoid alkenes in laboratory cultures of two marine diatoms. *Organic Geochemistry* **21:** 407–414.

Volkman, J. K., S. M. Barrett, S. I. Blackburn, M. P. Mansour, E. L. Sikes, and F. Gelin. 1998. Microalgal biomarkers: A review of recent research developments. *Organic Geochemistry* **29:** 1163–1179.

Wakeham, S. G. 1976. A comparative survey of petroleum hydrocarbons in lake sediments. *Marine Pollution Bulletin* **7:** 206–211.

———. 1989. Reduction of stenols to stanols in particulate matter at oxic–anoxic boundaries in sea water. *Nature* **342:** 787–790.

———. 1995. Lipid biomarkers for heterotrophic alternation of suspended particulate organic matter in oxygenated and anoxic water columns of the ocean. *Deep-Sea Research II* **42:** 1749–1771.

Wakeham, S. G., R. B. Gagosian, J. W. Farrington, and E. A. Canuel. 1984. Sterenes in suspended particulate matter in the eastern tropical North Pacific. *Nature* **308:** 840–843.

Wakeham, S. G., J. A. Beier, and C. H. Clifford, eds. 1991. *Organic matter sources in the Black Sea as inferred from hydrocarbon distributions*. Kluwer Academic, Dordrecht, The Netherlands.

Wakeham, S. G., K. H. Freeman, T. K. Pease, and J. M. Hayes. 1993. A photoautotrophic source for lycopane in marine water columns. *Geochimica et Cosmochimica Acta* **57:** 159–165.

Wakeham, S. G., J.S.S. Damste, M.E.L. Kohnen, and J. W. Deleeuw. 1995. Organic sulfur-compounds formed during early diagenesis in Black Sea sediments. *Geochimica et Cosmochimica Acta* **59:** 521–533.

Wakeham, S. G., C. Lee, J. I. Hedges, P. J. Hernes, and M. L. Peterson. 1997. Molecular indicators of diagenetic status in marine organic matter. *Geochimica et Cosmochimica Acta* **61:** 5363–5369.

Wakeham, S. G., M. L. Peterson, J. I. Hedges, and C. Lee. 2002. Lipid biomarker fluxes in the Arabian Sea, with a comparison to the Equatorial Pacific Ocean. *Deep-Sea Research II* **49:** 2265–2301.

Wakeham, S. G., J. Forrest, C. A. Masiello, Y. Gelinas, C. R. Alexander, and P. R. Leavitt. 2004. Hydrocarbons in Lake Washington sediments: A 25-year retrospective in an urban lake. *Environmental Science and Technology* **38:** 431–439.

Wakeham, S. G., A. P. McNichol, J. E. Kostka, and T. K. Pease. 2006. Natural-abundance radiocarbon as a tracer of assimilation of petroleum carbon by bacteria in salt marsh sediments. *Geochimica et Cosmochimica Acta* **70:** 1761–1771.

Weete, J. D., ed. 1976. *Algal and fungal waxes*. Elsevier, Amsterdam.

Werne, J. P., D. J. Hollander, A. Behrens, P. Schaeffer, P. Albrecht, and J.S.S. Sinninghe Damsté. 2000. Timing of early diagenetic sulfurization of organic matter: A precursor–product relationship in Holocene sediments of the anoxic Cariaco basin, Venezuela. *Geochimica et Cosmochimica Acta* **64:** 1741–1751.

Werne, J. P., D. J. Hollander, T. W. Lyons, and J. S. Sinnghe-Damsté. 2004. Organic sulfur biogeochemistry: Recent advances and future research directions. *in: Sulfur* Pp. 135–150 *in* J. Amend, K. Edwards, and T. Lyons, eds., *Biogeochemistry: Past and present*, Geological Society of America Special Paper 379.

Wilkes, H., S. Kühner, C. Bolm, T. Fischer, A. Classen, F. Widdel, and R. Rabus. 2003. Formation of *n*-alkane- and cycloalkane-derived organic acids during anaerobic growth of a denitrifying bacterium with crude oil. *Organic Geochemistry* **34:** 1313–1323.

Wraige, E. J., S. T. Belt, C. A. Lewis, D. A. Cooke, J. M. Robert, G. Massé, and S. J. Rowland. 1997. Variations in structures and distributions of C_{25} highly branched isoprenoid (HBI) alkenes in cultures of the diatom, *Haslea ostrearia (simonsen). Organic Geochemistry* **27:** 497–505.

Wraige, E. J., L. Johns, S. T. Belt, G. Massé, J.-M. Robert, and S. Rowland. 1999. Highly branched C_{25} isoprenoids in axenic cultures of *Haslea ostrearia. Phytochemistry* **51:** 69–73.

Yunker, M. B., R. W. MacDonald, D. J. Veltkamp, and W. J. Cretney. 1995. Terrestrial and marine biomarkers in a seasonally ice-covered arctic estuary: Integration of multivariate and biomarker approaches. *Marine Chemistry* **49:** 1–50.

Yunker, M. B., R. W. MacDonald, and B. G. Whitehouse. 1994. Phase associations and lipid distributions in the seasonally ice-covered Arctic estuary of the Mackenzie Shelf. *Organic Geochemistry* **22:** 651–669.

Zegouagh, Y., S. Derenne, C. Largeau, G. Bardoux, and A. Mariotti. 1998. Organic matter sources and early diagenetic alterations in Arctic surface sediments (Lena River delta and Laptev Sea, Eastern Siberia), II: Molecular and isotopic studies of hydrocarbons. *Organic Geochemistry* **28:** 571–583.

Zhou, W., S. Xie, P. A. Meyers, and Y. Zheng. 2005. Reconstruction of late glacial and holocene climate evolution in southern China from geolipids and pollen in the Dingnan Peat sequence. *Organic Geochemistry* **36:** 1272–1284.

Zimmerman, A. R., and E. A. Canuel. 2001. Bulk organic matter and lipid biomarker composition of Chesapeake Bay surficial sediments as indicators of environmental processes. *Estuarine and Coastal Shelf Science* **53:** 319–341.

Chapter 11

Balkwill, D. L., F. R. Leach, J. T. Wilson, J. F. McNabb, and D. C. White. 1988. Equivalence of microbial biomass measures based on membrane lipid and cell-wall components, adenosine-triphosphate, and direct counts in subsurface aquifer sediments. *Microbial Ecology* **16:** 73–84.

Belicka, L. L., and H. R. Harvey. 2009. The sequestration of terrestrial organic carbon in Arctic Ocean sediments: a comparison of methods and implications for regional carbon budgets. *Geochimica et Cosmochimica Acta* **73:** 6231–6248.

Brassell, S. C., G. Eglinton, I. T. Marlowe, U. Pflaumann, and M. Sarnthein. 1986. Molecular stratigraphy: A new tool for climatic assessment. *Nature* **320:** 129–133.

Conte, M. H., G. Eglinton, and L.A.S. Madureira. 1992. Long-chain alkenones and alkyl alkenoates as paleotemperature indicators: Their production, flux and early sedimentary diagenesis in the eastern North Atlantic. *Organic Geochemistry* **19:** 287–298.

Conte, M. H., A. Thompson, D. Lesley, and R. P. Harris. 1998. Genetic and physiological influences on the alkenone/alkenoate versus growth temperature relationship in *Emiliania huxleyi* and *Gephyrocapsa oceanica. Geochimica et Cosmochimica Acta* **62:** 51–68.

Conte, M. H., J. C. Weber, L. L. King, and S. G. Wakeham. 2001. The alkenone temperature signal in western North Atlantic surface waters. *Geochimica et Cosmochimica Acta* **65:** 4275–4287.

Conte, M. H., M. A. Sicre, C. Rühlemann, J. C. Weber, S. Schulte, D. Schulz-Bull, and T. Blanz. 2006. Global temperature calibration of the alkenone unsaturation index ($U_{37}^{K'}$) in surface waters and comparison with surface sediments. *Geochemistry Geophysics and Geosystems* **7:** 1–22.

D'Andrea, W. J., Z. H. Liu, M. D. Alexandre, S. Wattley, T. D. Herbert, and Y. S. Huang. 2007. An efficient method for isolating individual long-chain alkenones for compound-specific hydrogen isotope analysis. *Analytical Chemistry* **79:** 3430–3435.

Eglinton, T. I., and G. Eglinton. 2008. Molecular proxies for paleoclimatology. *Earth and Planetary Science Letters* **275:** 1–16.

Eglinton, T., M. Conte, G. Eglinton, and J. Hayes. 2000. Alkenone biomarkers gain recognition as molecular paleoceanographic proxies. *Eos Transactions AGU* **81:** 253–253.

Eglinton, T. I., M. H. Conte, G. Eglinton, and J. M. Hayes. 2001. Proceedings of a workshop on alkenone-based paleoceanographic indicators. *Geochemistry, Geophysics, and Geosystems* **2:** Paper number 2000GC000122.

Eltgroth, M. L., R. L. Watwood, and G. V. Wolfe. 2005. Production and cellular localization of neutral long-chain lipids in the haptophyte algae *Isochrysis galbana* and *Emiliani huxleyi*. *Journal of Phycology* **41:** 1000–1009.

Escala, M., A. Rosell-Melé, and P. Masqué. 2007. Rapid screening of glycerol dialkyl glycerol tetraethers in continental Eurasia samples using HPLC/APCI-ion trap mass spectrometry. *Organic Geochemistry* **38:** 161–164.

Freeman, K. H., and S. G. Wakeham. 1992. Variations in the distributions and isotopic compositions of alkenones in Black Sea particles and sediments. *Organic Geochemistry* **19:** 277–285.

Gaines, S. M., G. Eglinton, and J. Rullkötter. 2009. Deep sea mud. Pp. 101–152, *in Echoes of life—What fossil molecules reveal of Earth history*. Oxford University Press, New York.

Gliozzi, A., G. Paoli, M. Derosa, and A. Gambacorta. 1983. Effect of isoprenoid cyclization on the transition temperature of lipids in thermophilic archaebacteria. *Biochimica et Biophysica Acta* **735:** 234–242.

Grice, K., W. Breteler, S. Schouten, V. Grossi, J. W. De Leeuw, and J.S.S. Damsté. 1998. Effects of zooplankton herbivory on biomarker proxy records. *Paleoceanography* **13:** 686–693.

Grimalt, J., J. Rullkotter, M. Sicre, R. Summons, J. Farrington, H. Harvey, M. Goñi, and K. Sawada. 2000. Modifications of the C_{37} alkenone and alkenoate composition in the water column and sediment: Possible implications for sea surface temperature estimates in paleoceanography. *Geochemistry, Geophysics, and Geosystems* **1:** Paper number 2000GC000053.

Harvey, H. R. 2000. Alteration processes of alkenones and related lipids in water columns and sediments. *Geochemistry, Geophysics, and Geosystems* **1:** Paper number 2000GC000054.

Henderiks, J., and M. Pagani. 2007. Refining ancient carbon dioxide estimates: Significance of coccolithophore cell size for alkenone-based pCO_2 records. *Paleoceanography* **22:** 1–12.

Herbert, T. 2001. Review of alkenone calibrations (culture, water column, and sediments). *Geochemistry, Geophysics, and Geosystems* **2:** Paper number 2000GC000055.

Herbert, T. D., D. H. Heinrich, and K. T. Karl. 2003. Alkenone paleotemperature determinations. Pp. 391–432 *in Treatise on Geochemistry*. H. D. Halland and K. K. Turekian, eds., Pergamon, Oxford, UK.

Herfort, L., S. Schouten, J. P. Boon, M. Woltering, M. Baas, J.W.H. Weijers, and J. S. Sinninghe Damsté. 2006. Characterization of transport and deposition of terrestrial organic matter in the southern North Sea using the BIT index. *Limnology and Oceanography* **51:** 2196–2205.

Hoogakker, B., G. P. Klinkhammer, H. Elderfield, E. Rohling, and C. Hayward. 2009. Mg/Ca paleothermometry in high salinity environments. *Earth and Planetary Science Letters* **284:** 583–589.

Hopmans, E. C., S. Schouten, R. D. Pancost, M.T.J. van der Meer, and J. S. Sinninghe Damsté. 2000. Analysis of intact tetraether lipids in archael cell material and sediments by high performance liquid chromatography/atmospheric pressure chemical ionization mass spectrometry. *Rapid Communication in Mass Spectrometry* **14:** 585–589.

Hopmans, E. C., J.W.H. Weijers, E. Schefuss, L. Herfort, J.S.S. Damsté, and S. Schouten. 2004. A novel proxy for terrestrial organic matter in sediments based on branched and isoprenoid tetraether lipids. *Earth and Planetary Science Letters* **224:** 107–116.

Huguet, C., E. C. Hopmans, W. Febo-Ayala, D. H. Thomopson, J. S. Shinninghe Damsté, and S. Schouten. 2006. An improved method to determine the absolute abundance of glycerol dibiphytanyl glycerol tetraether lipids. *Organic Geochemistry* **37:** 1036–1041.

Huguet, C., J. H. Kim, G. J. de Lange, J. S. Sinninghe Damsté, and S. Schouten. 2009. Effects of long term oxic degradation on the$U_{37}^{K'}$; TEX_{86} and BIT organic proxies. *Organic Geochemistry* **40:** 1188–1194.

Jasper, J. P., and J. M. Hayes. 1990. A carbon isotope record of CO_2 levels during the late Quaternary. *Nature* **347:** 462–464.

Jasper, J. P., J. M. Hayes, A. C. Mix, and F. G. Prahl. 1994. Photosynthetic fractionation of C-13 and concentrations of dissolved CO_2 in the central Equatorial Pacific during the last 255,000 years. *Paleoceanography* **9:** 781–798.

Karner, M., E. F. Delong, and D. M. Karl. 2001. Archaeal dominance in the 606 mesopelagic zone of the Pacific Ocean. *Nature* **409:** 507–510.

Keough, B. P., T. M. Schmidt, and R. E. Hicks. 2003. Archaeal nucleic acids in picoplankton from great lakes on three continents. *Microbial Ecology* **46:** 238–248.

Killops, S., and V. Killops. 2005. *Introduction to organic geochemistry*, 2nd ed., Blackwell, Oxford, UK.

Kim, J. H., S. Schouten, E. C. Hopmans, B. Donner, and J.S.S. Damsté. 2008. Global sediment core-top calibration of the TEX_{86} paleothermometer in the ocean. *Geochimica et Cosmochimica Acta* **72:** 1154–1173.

Kim, J.-H., X. Crosta, E. Michel, S. Schouten, J. Duprat, and J. S. Sinnighe Damsté 2009. Impact of lateral transport on organic proxies in the Southern Ocean. *Quaternary Research* **71:** 246–250.

Lipp, J. S., and K.-U. Hinrichs. 2009. Structural diversity and fate of intact polar lipids in marine sediments. *Geochimica et Cosmochimica Acta* **73:** 6816–6833.

Mollenhauer, G., T. I. Eglinton, N. Ohkouchi, R. R. Schneider, P. J. Müller, P. M. Grootes, and J. Rullkötter. 2003. Asynchronous alkenone and foraminifera records from the Benguela upwelling system. *Geochemica et Cosmochimica Acta* **67:** 1157–1171.

Müller, P. J., G. Kirst, G. Ruhland, I. Von Storch, and A. Rosell-Melé. 1998. Calibration of the alkenone paleotemperature index $U_{37}^{K'}$ based on core-tops from the eastern South Atlantic and the global ocean (60° N-60° S). *Geochimica et Cosmochimica Acta* **62:** 1757–1772.

Ohkouchi, N., T. I. Eglinton, L. D. Keigwin, and J. M. Hayes. 2002. Spatical and temporal offsets between proxy records in a sediment drift. *Science* **298:** 1224–1227.

Ohkouchi, N., L. Xu, C. M. Reddy, D. Montlucon, and T. I. Eglinton. 2005. Radiocarbon dating of alkenones from marine sediments; I: Isolation protocol. *Radiocarbon* **47:** 401–412.

Pagani, M., K. H. Freeman, N. Ohkouchi, and K. Caldeira. 2002. Comparison of water column [CO_2aq] with sedimentary alkenone-based estimates: A test of the alkenone–CO_2 proxy. *Paleoceanography* **17:** Paper number doi:10.1029/2002PA000756,2002.

Pagani, M., J. Zachos, K. H. Freeman, B. Tipple, and S. Boharty. 2005. Marked decline in atmospheric carbon dioxide concentrations during the paleogene. *Science* **309:** 600–603.

Pitcher, A., E. C. Hopmans, S. Schouten, and J. S. Sinninghe Damsté. 2009. Separation of core and intact polar archael tetrether lipids using silica columns: Insights into living and fossil biomass contributions. *Organic Geochemistry* **40:** 12–19.

Powers, L. A., J. P. Werne, T. C. Johnson, E. C. Hopmans, J.S.S. Damsté, and S. Schouten. 2004. Crenarchaeotal membrane lipids in lake sediments: A new paleotemperature proxy for continental paleoclimate reconstruction? *Geology* **32:** 613–616.

Prahl, F. G., and S. G. Wakeham. 1987. Calibration of unsaturation patterns in long-chain ketone compositions for paleotemperature assessment. *Nature* **330:** 367–369.

Prahl, F. G., G. V. Wolfe, and M. A. Sparrow. 2003. Physiological impacts on alkenone paleothermometry. *Paleoceanography* **18:** 1025–1031.

Rossel, P. A., J. S. Lipp, H. F. Fredricks, J. Arnds, A. Boetius, M. Elvert, K.-U. Hinrichs. 2008. Intact polar lipids of anaerobic methanotrophic archaea and associated bacteria. *Organic Geochemistry* **39:** 992–999.

Rutters, H., H. Sass, H. Cypionka, and J. Rullkotter. 2002a. Microbial communities in a Wadden Sea sediment core: Clues from analyses of intact glyceride lipids, and released fatty acids. *Organic Geochemistry* **33:** 803–816.

———. 2002b. Phospholipid analysis as a tool to study complex microbial communities in marine sediments. *Journal of Microbiological Methods* **48:** 149–160.

Schneider, R. 2001. Alkenone temperature and carbon isotope records: Temporal resolution, offsets, and regionality. *Geochemistry, Geophysics, and Geosystematics* **2:** Paper number 2000GC000060.

Schouten, S., E. C. Hopmans, E. Schefuss, and J.S.S. Damsté. 2002. Distributional variations in marine crenarchaeotal membrane lipids: A new tool for reconstructing ancient sea water temperatures? *Earth and Planetary Science Letters* **211:** 205–206.

Schouten, S., E. C. Hopmans, A. Forster, Y. Van Breugel, M.M.M. Kuypers, and J.S.S. Damsté. 2003a. Extremely high sea-surface temperatures at low latitudes during the middle cretaceous as revealed by archaeal membrane lipids. *Geology* **31:** 1069–1072.

———. 2003b. Distributional variations in marine crenarchaeotal membrane lipids: A new tool for reconstructing ancient sea water temperatures? *Earth and Planetary Science Letters* **211:** 205–206.

Schouten, S., C. Huguet, E. C. Hopmans, M.V.M. Kienhuis, and J.S.S. Damsté. 2007. Analytical methodology for TEX_{86} paleothermometry by high-performance liquid chromatography/atmospheric pressure chemical ionization–mass spectrometry. *Analytical Chemistry* **79:** 2940–2944.

Schouten, S., E. C. Hopmans, M. Bass, H. Boumann, S. Standfest, M. Konneke, D. A. Stahl, and J. S. Sinninghe Damsté. 2008. Intact membrane lipids of *Candidatus Nitrosopumilus maritimus,* a cultivated representative of the cosmopolitan mesophilic group I Crenarcheaota. *Applied Environmental Microbiology* **74:** 2433–2440.

Schouten, S., E. C. Hopmans, J. van der Meer, A. Mets, E. Bard, T. S. Bianchi, A. Diefendorf, M. Escala, K. H. Freeman, Y. Furukawa, C. Huguet, A. Ingalls, G. Menot-Combes, A. J. Nederbragt, M. Oba, A. Pearson, E. J. Pearson, A. Rosell-Mele, P. Schaeffer, S. R. Shah, T. Shanahan, R. W. Smith, R. Smittenberg, H. M. Talbot, M. Uchida, B.A.S. Van Mooy, M. Yamamoto, Z. Zhang, and J. Sinninghe Damsté. 2009. An interlaboratory study of TEX_{86} and BIT analysis using high-performance liquid chromatography–mass spectrometry. *Geochemistry, Geophysics, and Geosystems:* **10:** Paper Number 1029/2008GC002221.

Shah, S. R., G. Mollenhauer, N. Ohkouchi, T. I. Eglinton, and A. Pearson. 2008. Origins of archaeal tetraether lipids in sediments: insights from radiocarbon analysis. *Geochimica Cosmochimica Acta* **72:** 4577–4594.

Sikes, E. L., and M. A. Sicre. 2002. Relationship of the tetra-unsaturated C_{37} alkenone to salinity and temperature: Implications for paleoproxy applications. *Geochemistry, Geophysics, and Geosystems* **3:** 1–11.

Sikes, E. L., and J. K. Volkman. 1993. Calibration of alkenone unsaturation ratios ($U_{37}^{k'}$) for paleotemperature estimation in cold polar waters. *Geochimica et Cosmochimica Acta* **57:** 1883–1889.

Sikes, E. L., J. W. Farrington, and L. D. Keigwin. 1991. Use of the alkenone unsaturation ratio $U_{37}^{k'}$ to determine past sea-surface temperatures: Core-top SST calibrations and methodology considerations. *Earth and Planetary Science Letters* **104:** 36–47.

Sinninghe Damsté, J. S., S. Schouten, E. C. Hopmans, A.C.T. van Duin, and J.A.J. Geenevasen. 2002. Crenarchaeol: The characteristic core glycerol dibiphytanyl glycerol tetraether membrane lipid of cosmopolitan pelagic Crenarchaeota. *Journal of Lipid Research* **43:** 1641–1651.

Smith, R. W., T. S. Bianchi, and C. Savage. 2010. Comparison of lignin-phenols and branched/isoprenoid tetraethers (BIT index) as indices of terrestrial organic matter in Doubtful Sound, Fiordland, New Zealand. *Organic Geochemistry.* **41:** 281–290.

Sprott, G. D., M. Meloche, and J. C. Richards. 1991. Proportions of diether, macrocyclic diether, and tetraether lipids in *Methanococcus jannaschii* grown at different temperatures. *Journal of Bacteriology* **173:** 3907–3910.

Sturt, H. F., R. E. Summons, K. Smith, M. Elvert, and K.-U. Hinrichs. 2004. Intact polar membrane lipids in prokaryotes and sediments deciphered by high-performance liquid chromatography/electrospray ionization multistage mass spectrometry: New biomarkers for biogeochemistry and microbial ecology. *Rapid Communications in Mass Spectrometry* **18:** 617–628.

Teece, M. A., J. M. Getliff, J. W. Leftley, R. J. Parkes, and J. R. Maxwell. 1998. Microbial degradation of the marine prymnesiophyte *Emiliania huxleyi* under oxic and anoxic conditions as a model for early diagenesis: Long chain alkadienes, alkenones and alkyl alkenoates. *Organic Geochemistry* **29:** 863–880.

Trommer, G., M. Siccha, M.T.J. van der Meer, S. Schouten, J. S. Sinninghe Damsté, H. Schulz, C. Hemelben, and M. Kucera. 2009. Distribution of Crenarchaeota tetraether membrane lipids in surface sediments from the Red Sea. *Organic Chemistry* **40:** 724–731.

Uda, I., A. Sugai, Y. H. Itoh, and T. Itoh. 2001. Variation in molecular species of polar lipids from *Thermoplasma acidophilum* depends on growth temperature. *Lipids* **36:** 103–105.

Volkman, J. K., S. M. Barrett, S. I. Blackburn, and E. L. Sikes. 1995. Alkenones in *Gephyrocapsa-oceanica:* Implications for studies of paleoclimate. *Geochimica et Cosmochimica Acta* **59:** 513–520.

Walsh, E. M., A. E. Ingalls, and R. G. Keil. 2008. Sources and transport of terrestrial organic matter in Vancouver Island fjords and the Vancouver–Washington margin: A multiproxy approach using $\delta^{13}C$ (org), lignin phenols, and the ether lipid BIT index. *Limnology and Oceanography* **53:** 1054–1063.

Weijers, J.W.H., S. Schouten, O. C. Spaargaren, and J.S.S. Damsté. 2006. Occurrence and distribution of tetraether membrane lipids in soils: Implications for the use of the TEX_{86} proxy and the bit index. *Organic Geochemistry* **37:** 1680–1693.

Weijers, J.W.H., S. Schouten, J. C. Van Den Donker, E. C. Hopmans, and J.S.S. Damsté. 2007. Environmental controls on bacterial tetraether membrane lipid distribution in soils. *Geochimica et Cosmochimica Acta* **71:** 703–713.

Weijers, J.W.H., S. Schouten, E. Schefuß, R. R. Schneider, and J. S. Sinninghe Damsté. 2009. Disentangling marine, soil and plant organic carbon contributions to continental margin sediments: A multi-proxy approach in a 20,000 year sediment record from the Congo deep-sea fan. *Geochimica et Cosmochimica Acta* **73:** 119–132.

White, D. C., W. M. Davis, J. S. Nickels, J. D. King, and R. J. Bobbie. 1979. Determination of the sedimentary microbial biomass by extractable lipid phosphate. *Oecologia* **40:** 51–62.

Wuchter, C., S. Schouten, S. Wakeham, and J. S. Sinninghe Damsté. 2006. Archael tetraether membrane lipid fluxes in the northeastern Pacific and the Arabian Sea: Implications for TEX_{86} paleothemometry. *Paleooceanography* **21:** doi:10:1029/2006pa001279.

Yoshino, J. I., Y. Sugiyama, S. Sakuda, T. Kodama, H. Nagasawa, M. Ishii, and Y. Igarashi. 2001. Chemical structure of a novel aminophospholipid from *Hydrogenobacter thermophilus* strain tk-6. *Journal of Bacteriology* **183:** 6302–6304.

Zink, K. G., and K. Mangelsdorf. 2004. Efficient and rapid method for extraction of intact phospholipids from sediments combined with molecular structure elucidation using LC-ESI-MS-MS analysis. *Analytical and Bioanalytical Chemistry* **380:** 798–812.

Zink, K. G., K. Mangelsdorf, L. Granina, and B. Horsfield. 2008. Estimation of bacterial biomass in subsurface sediments by quantifying intact membrane phospholipids. *Analytical and Bioanalytical Chemistry* **390:** 885–896.

Chapter 12

Alberte, R. S., A. M. Wood, T. A. Kursar, and R.R.L. Guillard. 1984. Novel phycoerythrins in marine *Synechococcus* spp.: Characterization and evolutionary and ecological implications. *Plant Physiology* **75:** 732–739.

Andersen, R. A., and T. J. Mulkey. 1983. The occurrence of chlorophylls c_1 and c_2 in the Chrysophyceae. *Journal of Phycology* **19:** 289–294.

Armstrong, G., and K. Apel. 1998. Molecular and genetic analysis of light-dependent chlorophyll biosynthesis. *Methods in Enzymology*. **297:** 237–244.

Arnon, D. I. 1984. The discovery of photosynthetic phosphorylation. *Trends in Biochemical Sciences* **9:** 258–262.

Arpin, N., W. A. Svec, and S. Liaaen-Jensen. 1976. A new fucoxanthin-related carotenoid from *Coccolithus huxleyi*. *Phytochemistry* **15:** 529–532.

Arrigo, K. R., D. H. Robinson, D. L. Worthen, R. B. Dunbar, G. R. DiTullio, M. van Woert, and M. P. Lizotte. 1999. Phytoplankton community structure and the drawdown of nutrients and CO_2 in the Southern Ocean. *Science* **283:** 365–367.

Asai, R., S. McNiven, K. Ikebukuro, I. Karube, Y. Horiguchi, S. Uchiyama, A. Yoshida, and Y. Masuda. 2000. Development of a fluorometric sensor for the measurement of phycobilin pigment and application to freshwater phytoplankton. *Field Analytical Chemistry and Technology* **4:** 53–61.

Beale, S. I., and J. Cornejo. 1983. Biosynthesis of phycocyanobilin from exogenous labeled biliverdin in *Cyanidium caldarium*. *Archives of Biochemistry and Biophysics* **227:** 279–286.

———. 1984a. Enzymatic heme oxygenase activity in soluble extracts of the unicellular red alga, *Cyanidium caldarium*. *Archives of Biochemistry and Biophysics* **235:** 371–384.

———. 1984b. Enzymic transformation of biliverdin to phycocyanobilin by extracts of the unicellular red alga *Cyanidium caldarium*. *Plant Physiology* **76:** 7–15.

———. 1991a. Biosynthesis of phycobilins: 15,16-Dihydrobiliverdin-ix-alpha is a partially reduced intermediate in the formation of phycobilins from biliverdin-ix-alpha. *Journal of Biological Chemistry* **266:** 22341–22345.

———. 1991b. Biosynthesis of phycobilins: 3(z)-Phycoerythrobilin and 3(z)-phycocyanobilin are intermediates in the formation of 3(e)-phycocyanobilin from biliverdin-ix-alpha. *Journal of Biological Chemistry* **266:** 22333–22340.

Behrenfeld, M. J., J. T. Randerson, C. R. McClain, G. C. Feldman, S. O. Los, C. J. Tucker, P. G. Falkowski, C. B. Field, R. Frouin, W. E. Esaias, D. D. Kolber, and N. H. Pollack. 2001. Biospheric primary production during an ENSO transition. *Science* **291:** 2594–2597.

Bermejo, R., E. Fernandez, J. M. Alvarez-Pez, and E. M. Talavera. 2002. Labeling of cytosine residues with biliproteins for use as fluorescent DNA probes. *Journal of Luminescence* **99:** 113–124.

Bianchi, T. S. 2007. *Biogeochemistry of estuaries*. Oxford University Press, New York.

Bianchi, T. S., and S. Findlay. 1990. Plant pigments as tracers of emergent and submergent macrophytes from the Hudson River. *Canadian Journal of Fisheries and Aquatic Sciences*. **47:** 92–494.

Bianchi, T. S., R. Dawson, and P. Sawangwong. 1988. The effects of macrobenthic deposit-feeding on the degradation of chloropigments in sandy sediments. *Journal of Experimental Marine Biology and Ecology* **122:** 243–255.

Bianchi, T. S., S. Findlay, and D. Fontvielle. 1991. Experimental degradation of plant materials in Hudson River sediments,1: Heterotrophic transformations of plant pigments. *Biogeochemistry* **12:** 171–187.

Bianchi, T. S., S. Findlay, and R. Dawson. 1993. Organic matter sources in the water column and sediments of the Hudson River estuary: The use of plant pigments as tracers. *Estuarine, Coastal and Shelf Science* **36:** 359–376.

Bianchi, T. S., C. Lambert, and D. C. Biggs. 1995. Distribution of chlorophyll-a and pheopigments in the northwestern Gulf of Mexico: A comparison between fluorometric and high-performance liquid-chromatography measurements. *Bulletin of Marine Science* **56:** 25–32.

Bianchi, T. S., M. Baskaran, J. Delord, and M. Ravichandran. 1997a. Carbon cycling in a shallow turbid estuary of southeast Texas: The use of plant pigments as biomarkers. *Estuaries* **20:** 404–415.

Bianchi, T. S., C. D. Lambert, P. H. Santschi, and L. Guo. 1997b. Sources and transport of land-derived particulate and dissolved organic matter in the Gulf of Mexico (Texas shelf/slope): The use of lignin-phenols and loliolides as biomarkers. *Organic Geochemistry* **27:** 65–78.

Bianchi, T. S., K. Kautsky, and M. Argyrou. 1997c. Dominant chlorophylls and carotenoids in macroalgae of the Baltic Sea (Baltic proper): Their use as potential biomarkers. *Sarsia* **82:** 55–62.

Bianchi, T. S., B. Johansson, and R. Elmgren. 2000a. Breakdown of phytoplankton pigments in Baltic sediments: Effects of anoxia and loss of deposit-feeding macrofauna. *Journal of Experimental Marine Biology and Ecology* **251:** 161–183.

Bianchi, T. S., P. Westman, C. Rolff, E. Engelhaupt, T. Andren, and R. Elmgren, 2000b. Cyanobacterial blooms in the Baltic Sea: Natural or human-induced? *Limnology and Oceanography* **45:** 716–726.

Bianchi, T. S., E. Engelhaupt, B. A. McKee, S. Miles, R. Elmgren, S. Hajdu, C. Savage, C., and M. Baskaran. 2002. Do sediments from coastal sites accurately reflect time trends in water column phytoplankton? A test from Himmerfjarden Bay (Baltic Sea proper). *Limnology and Oceanography* **47:** 1537–1544.

Bianchi, T. S., T. Filley, K. Dria, and P. G. Hatcher. 2004. Temporal variability in sources of dissolved organic carbon in the lower Mississippi River. *Geochimica et Cosmochimica Acta* **68:** 959–967.

Bidigare, R. R., and C. C. Trees. 2000. HPLC phytoplankton pigments: Sampling, laboratory methods, and quality assurance procedures. Pp. 154–161 *in* J. Mueller and G. Fargion, eds., *Ocean optics protocols for satellite ocean color sensor validation*, revision 2. (NASA technical memorandum 2000-209966). NASA, Greenbelt, MD.

Bidigare, R. R., M. C. Kennicutt, and J. M. Brooks. 1985. Rapid determination of chlorophylls and their degradation products by high-performance liquid chromatography. *Limnology and Oceanography* **30:** 432–435.

Bidigare, R. R., M. C. Kennicutt, W. L. Keeney-Kennicutt, and S. A. Macko. 1991. Isolation and purification of chlorophyll-*a* and chlorophyll-*b* for the determination of stable carbon and nitrogen isotope compositions. *Analytical Chemistry* **63:** 130–133.

Bjørnland, T., and S. Liaaen-Jensen. 1989. Distribution patterns of carotenoids in relation to chromophyte phylogeny and systematics. Pp. 37–60 *in* J.C. Green and B.S.C. Diver, eds., *The chromophyte algae: Problems and perspectives*. Clarendon Press, Oxford, UK.

Blanchot, J., and M. Rodier. 1996. Picophytoplankton abundance and biomass in the western tropical Pacific Ocean during the 1992 El Nino year: Results from flow cytometry. *Deep-Sea Research, Part I: Oceanographic Research Papers* **43:** 877–895.

Britton, G. 1976. Later reactions of carotenoids biosynthesis. *Pure and Applied Chemistry* **47:** 223–236.

Brock, C. S., P. R. Leavitt, D. E. Schindler, S. P. Johnson, and J. W. Moore. 2006. Spatial variability of stable isotopes and fossil pigments in surface sediments of Alaskan coastal lakes: Constraints on quantitative estimates of past salmon abundance. *Limnology and Oceanography* **51:** 1637–1647.

Brown, S. B., J. D. Houghton, and G.A.F. Hendry. 1991. Chlorophyll breakdown, Pp. 465–489 *in* H. Scheer, ed., *Chlorophylls*. CRC Press, Boca Raton, FL.

Brown, S. R., H. J. McIntosh, and J. P. Smol. 1984. Recent paleolimnology of a meromictic lake: Fossil pigments of photosynthetic bacteria. *Verhandlungen der Internationalen Ver Limnologie* **22:** 1357–1360.

Bryant, D. A., A. N. Glazer, and F. A. Eiserling. 1976. Characterization and structural properties of major biliproteins of *Anabaena* sp. *Archives of Microbiology* **110:** 61–75.

Calvin, M. 1976. Photosynthesis as a resource for energy and materials. *Photochemistry and Photobiology* **23:** 425–444.

Carmichael, W. W. 1997. The cyanotoxins. Advances in botanical research **27:** 211–256.

Carpenter, S. R., and A. M. Bergquist. 1985. Experimental tests of grazing indicators based on chlorophyll a degredation products. *Archiv fur Hydrobiologie* **102:** 303–317.

Chapman, V. J. 1966. Three new carotenoids isolated from algae. *Phytochemistry* **5:** 331–1333.

Chavez, F. P., P. G. Strutton, G. E. Friederich, R. A. Feely, G. C. Feldman, D. G. Foley, and M. J. McPhaden. 1999. Biological and chemical response of the Equatorial Pacific Ocean to the 1997–98 El Nino. *Science* **286:** 2126–2131.

Chen, N. H., T. S. Bianchi, B. A. McKee, and J. M. Bland. 2001. Historical trends of hypoxia on the Louisiana shelf: Application of pigments as biomarkers. *Organic Geochemistry* **32:** 543–561.

Chen, N. H., T. S. Bianchi, and J. M. Bland. 2003a. Novel decomposition products of chlorophyll-*a* in continental shelf (Louisiana shelf) sediments: Formation and transformation of carotenol ester. *Geochimica et Cosmochimica Acta* **67:** 2027–2042.

———. 2003b. Fate of chlorophyll-*a* in the lower Mississippi River and Lousiana shelf: Implications for pre- versus post-depositional decay. *Marine Chemistry* **83:** 37–55.

Clayton, R. K. 1971. *Light and living matter: A guide to the study of photobiology*. McGraw-Hill, New York.

———. 1980. Photosynthesis: Physical mechanisms and chemical patterns. Cambridge University Press, Cambridge, MA.

Cohen, Y., S. Yalovsky, and R. Nechushtai. 1995. Integration and assembly of photosynthetic protein complexes in chloroplast thylakoid membranes *Biochemica et Biophysica Acta* **1241:** 1–30.

Cohen-Bazire, G., and D. A. Bryant. 1982. Phycobilisomes: Composition and structure. Pp. 143–190 *in* N. G. Carr and B. A. Whitton, eds., *The biology of cyanobacteria*. Blackwell Scientific, Cambridge, MA.

Cole, W. J., C. Oheocha, A. Moscowit, and W. R. Krueger. 1967. Optical activity of urobilins derived from phycoerythrobilin. *European Journal of Biochemistry* **3:** 202–203.

Colyer, C. L., C. S. Kinkade, P. J. Viskari, and J. P. Landers. 2005. Analysis of cyanobacterial pigments and proteins by electrophoretic and chromatographic methods. *Analytical and Bioanalytical Chemistry* **382:** 559–569.

Coolen, M.J.L., and J. Overmann. 1998. Analysis of subfossil molecular remains of purple sulfur bacteria in a lake sediment. *Applied and Environmental Microbiology* **64:** 4513–4521.

Daley, R. J. 1973. Experimental characterization of lacustrine chlorophyll diagenesis, 2: Bacterial, viral and herbivore grazing effects. *Archiv fur Hydrobiologie* **72:** 409–439.

Daley, R. J., and S. R. Brown. 1973. Experimental characterization of lacustrine chlorophyll diagenesis, 1: Physiological and enviromental effects. *Archiv fur Hydrobiologie* **72:** 277–304.

Descy, J. P., H. W. Higgins, D. J. Mackey, J. P. Hurley, and T. M. Frost. 2000. Pigment ratios and phytoplankton assessment in northern Wisconsin lakes. *Journal of Phycology* **36:** 274–286.

Descy, J. P., M. A. Hardy, S. Stenuite, S. Pirlot, B. Leporcq, I. Kimirei, B. Sekadende, S. R. Mwaitega, and D. Sinyenza. 2005. Phytoplankton pigments and community composition in Lake Tanganyika. *Freshwater Biology* **50:** 668–684.

Duan, S. W., and T. S. Bianchi. 2006. Seasonal changes in the abundance and composition of plant pigments in particulate organic carbon in the lower Mississippi and Pearl rivers. *Estuaries and Coasts* **29:** 427–442.

Egeland, E. S., and S. Liaaen-Jensen. 1992. Eight new carotenoids from a chemosystematic evaluation of Prasinophyceae. Proceedings, 7th International Symposium on Marine Natural Products, Capri.

Egeland, E. S., and S. Liaaen-Jensen. 1993. New carotenoids and chemosystematics in the Prasinophyceae. Proc. 10th Symposium on Carotenoids, Norway.

Emerson, R., and W. Arnold. 1932a. A separation of the reactions in photosynthesis by means of intermittent light. *Journal of General Physiology* **15:** 391–420.

———. 1932b. The photochemical reaction in photosynthesis. *Journal of General Physiology* **16:** 191–205.

Exton, R. J., W. M. Houghton, W. E. Esaias, L. W. Haas, and D. Hayward. 1983. Spectral differences and temporal stability of phycoerythrin fluorescence in estuarine and coastal waters due to the domination of labile cryptophytes and stabile cyanobacteria. *Limnology and Oceanography* **28:** 1225–1230.

Falkowski, P. G. 1994. The role of phytoplankton photosynthesis in global biogeochemical cycles. *Photosynthesis Research* **39:** 235–258.

Foss, P., R. R. Guillard, and, S. Liaanen-Jensen. 1984. Prasinoxanthin: A chemosystematic marker. *Phytochemistry* **23:** 1629–1633.

Foss, P., R. A. Levin, and S. Liaaen-Jensen. 1987. Carotenoids of *Prochloron* sp. (Prochlorophyta). *Phycologia* **26:** 142–144.

Frank, H. A., and R. J. Cogdell. 1996. Carotenoids in photosynthesis. *Photochemistry and Photobiology* **63:** 257–264.

Frost, B. W., and N. C. Franzen. 1992. Grazing and iron limitation in the control of phytoplankton stock and nutrient concentration: A chemostat analog of the Pacific Equatorial upwelling zone. *Marine Ecology Progress Series* **83:** 291–303.

Furlong, E. T., and R. Carpenter. 1988. Pigment preservation and remineralization in oxic coastal marine sediments *Geochimica et Cosmochimica Acta* **52:** 87–99.

Gaossauer, A., and N. Engel. 1996. Chlorophyll catabolism: Structures, mechanisms, conversions. *Journal of Photochemistry and Photobiology B: Biology* **32:** 141–151.

Garcia-Pichel, F., and R. W. Catenholz. 1991. Characterization and biological implications of scytonemin, a cyanobacterial sheath pigment. *Journal of Phycology* **27:** 395–409.

Garcia-Pichel, F., N. D. Sherry, and R. W. Castenholz. 1992. Evidence for an ultraviolet sunscreen role of the extracellular pigment scytonemin in the terrestrial cynaobacterium *Chlorogloeopsis* sp. *Photochemistry and Photobiology* **56:** 17–23.

Garrido, J., M. Zapata. And S. Muñoz. 1995. Spectral characterization of new chlorophyll *c* pigments isolated from *Emiliana huxleyi (Prymnesiophyceae)* by high-performance liquid chromatography. *Journal of Phychology* **31:** 761–768.

Gieskes, W. W., and G. W. Kraay. 1983. Unknown chlorophyll a derivatives in the North Sea and the tropical Atlantic Ocean revealed by HPLC analysis. *Limnology and Oceanography* **28:** 757–766.

Gieskes, W.W.C., G. W. Kraay, W. Nontji, A. Setiapermana, and D. Sutomo. 1988. Monsoonal alternation of a mixed and layered structure in the phytoplankton of the euphotic zone of the Banda Sea (Indonesia): A mathematical analysis of algal pigment fingerprints. *Netherlands Journal of Sea Research* **22:** 123–137.

Glazer, A. N., and D. A. Bryant. 1975. Allophycocyanin *b* (lambda max 671, 618 nm): New cyanobacterial phycobiliprotein. *Archives of Microbiology* **104:** 15–22.

Glover, H., L. Campbell, and B. B. Prezelin. 1986. Contribution of *Synechococcus* spp. to size fractionated primary productivity in three water masses in the northwest Atlantic Ocean. *Marine Biology* **91:** 193–203.

Goericke, R., and D. J. Repeta. 1992. The pigments of *Prochlorococcus marinus*: The presence of divinyl chlorophyll *a* and *b* in a marine prokaryote. *Limnology and Oceanography* **37:** 425–433.

Goericke, R., and D. J. Repeta. 1993. Chlorophyll-*a* and chlorophyll-*b* and divinyl chlorophyll-*a* and chlorophyll-*b* in the open subtropical North Atlantic Ocean. *Marine Ecology-Progress Series* **101:** 307–313.

Goericke, R., S. L. Strom, and R. A. Bell. 2000. Distribution and sources of cyclic pheophorbides in the marine environment. *Limnology and Oceanography* **45:** 200–211.

Goodwin, T. W. 1980. *The biochemistry of the carotenoids*, 2nd ed. Chapman & Hall, London.

Govindjee, F., and W. J. Coleman. 1990. How plants make oxygen. *Scientific American* **262:** 50–58.

Gray, M. W. 1992. The endosymbiont hypothesis revisited. *International Review of Cytology: A Survey of Cell Biology* **141:** 233–357.

Guillard, R.R.L., L. S. Murphy, P. Foss, and S. Liaaen-Jensen. 1985. *Synechcoccus* spp. as a likely zeaxanthin-dominant ultraphytoplankton in the North Atlantic. *Limnology and Oceanography* **30:** 412–414.

Hager, A. 1980. The reversible, light-induced conversions of xanthophylls in the chloroplast. Pp. 57–79 *in* F. C. Czygan, ed., *Pigments in plants*. Fischer, Stuttgart.

Hallegraeff, G. M., and S. W. Jeffrey 1985. Description of new chlorophyll *a* alteration products in marine phytoplankton. *Deep-Sea Research* **32:** 697–705.

Hambright, K. D., T. Zohary, W. Eckert, S. S. Schwartz, C. L. Schelske, K. R. Laird, and P. R. Leavitt. 2008. Exploitation and destabilization of a warm, freshwater ecosystem through engineered hydrological change. *Ecological Applications* **18:** 1591–1603.

Harradine, P. J., P. G. Harris, R. N. Head, R. P. Harris, and J. R. Maxwell. 1996. Steryl chlorine estrers are formed by zooplankton herbivory. *Geochimica et Cosmochimica Acta* **60:** 2265–2270.

Harris, P. G., J. F. Carter, R. N. Head, R. P. Harris, G. Eglinton, and J. R. Maxwell. 1995. Identification of chlorophyll transformation products in zooplankton faecal pellets and marine sediments extracts by liquid chromatography/mass spectrometry atmospheric pressure chemical ionization. *Rapid Communications in Mass Spectrometry* **9**: 1177–1183.

Hawkins, A.J.S., B. L. Bayne, R.F.C. Mantoura, and C. A. Llewellyn. 1986. Chlorophyll degradation and absorption through the digestive system of the blue mussel *Mytilus edulis*. *Journal of Experimental Marine Biology and Ecology* **96**: 213–223.

Hayashi, M., K. Furuya, and H. Hattori. 2001. Spatial hetogeneity in distributions of chlorophyll a derivatives in the subarctic North Pacific during summer. *Journal of Oceanography* **57**: 323–331.

Head, E.J.H., and L. R. Harris. 1994. Feeding selectivity by copepods grazing on natural mixtures of phytoplankton determined by HPLC analysis of pigments. *Marine Ecology-Progress Series* **110**: 75–83.

Head, E.J.H., and L. R. Harris. 1996. Chlorophyll destruction by *Calanus* spp. Grazing on phytoplankton: Kinetics, effects of ingection rate and feeding history, and a mechanistic interpretation. *Marine Ecology-Progress Series* **135**: 223–235.

Head, E.J.H., B. T. Hargrave, and D. V. Subba Rao. 1994. Accumulation of a phaeophorbide a-like pigment in sediment traps during late stages of a spring bloom: A product of dying algae? *Limnology and Oceanography* **39**: 176–181.

———. 1996. Chlorophyll destruction by *Calanus* spp. grazing on phytoplankton: Kinetics, effects of ingestion rate and feeding history, and a mechanistic interpretation. *Marine Ecology-Progress Series* **135**: 223–235.

Hecky, R. E., and H. J. Kling. 1981. The phytoplankton and protozooplankton of the euphotic zone of Lake Tanganyika: Species composition, biomass, chlorophyll content, and spatio-temporal distribution. *Limnology and Oceanography* **26**: 548–564.

Hill, R. 1939. Oxygen produced by isolated chloroplasts. *Proceedings of the Royal Society of London Series B: Biological Sciences* **127**: 192–210.

Holden, M. 1976. Chlorophylls. Pp. 2–37 *in* T. W. Goodwin, ed., *Chemistry and biochemistry of plant pigments*, 2nd ed. Academic Press, London.

Huang, F., I. Parmryd, F. Nilsson, A. L. Persson, H. B. Pakrasi, B. Andersson, and B. Norling. 2002. Proteomics of *Synechocystis* sp. strain pcc 6803: Identification of plasma membrane proteins. *Molecular and Cellular Proteomics* **1**: 956–966.

Huisman, J., N.N.P. Thi, D. M. Karl, and B. Sommeijer. 2006. Reduced mixing generates oscillations and chaos in the oceanic deep chlorophyll maximum. *Nature* **439**: 322–325.

Humborg, C., V. Ittekkot, A. Cociasu, and B. Vonbodungen. 1997. Effect of Danube River dam on Black Sea biogeochemistry and ecosystem structure. *Nature* **386**: 385–388.

Humborg, C., D. J. Conley, L. Rahm, F. Wulff, A. Cociasu, and V. Ittekkot. 2000. Silica retention in river basins: Far-reaching effects on biogeochemistry and aquatic food webs in coastal marine enviroments. *Ambio* **29**: 45–50.

Ittekkot, V., C. Humborg, and P. Schafer. 2000. Hydrological alterations and marine biogeochemistry: A silicate issue? *Bioscience* **50**: 776–782.

Jeffrey, S. W. 1974. Profiles of photosynthetic pigments in ocean using thin-layer chromatography. *Marine Biology* **26**: 101–110.

Jeffrey, S. W. 1976a. The occurrence of chlorophyll c_1 and c_2 in algae. *Journal of Phycology* **12**: 349–354.

Jeffrey, S. W. 1976b. A report on green algal pigments in the central North Pacific Ocean. *Marine Biology* **37**: 33–37.

Jeffrey, S. W. 1989. Chlorophyll *c* pigments and their distribution in the chromophyte algae. Pp. 13–36 *in* J. C. Green, B.S.C. Leadbeater, and W. L. Diver, eds. *The Chromophyte algae: Problems and perspectives*. Clarendon Press, Oxford, UK.

———. 1997. Application of pigment methods to oceanography. Pp. 127–178 *in* S. W. Jeffrey, R.F.C. Mantoura, and S. W. Wright eds., *Phytoplankton pigments in oceanography*. UNESCO, Paris.

Jeffrey, S. W., and G. M. Hallegraeff. 1987. Phytoplankton pigments, species and photosynthetic pigments in a warm-core eddy of the East Australian current,. I: Summer populations. *Marine Ecology-Progress Series* **3**: 285–294.

Jeffrey, S. W., and G. F. Humphrey. 1975. New spectrophotometric equations for determining chlorophylls *a*, *b*, c_1 and c_2 in higher-plants, algae and natural phytoplankton. *Biochemie und Physiologie der Pflanzen* **167**: 191–194.

Jeffrey, S. W., and R.F.C. Mantoura. 1997. Development of pigment methods for oceanography: SCOR-supported working groups and objectives. Pp. 19–31 *in* S. W. Jeffrey, R.F.C. Mantoura, and S. W. Wright, eds., *Phytoplankton pigments in oceanography*. UNESCO, Paris.

Jeffrey, S. W., and M. Vesk. 1997. Introduction to marine phytoplankton and their pigment signatures. Pp. 37–84 *in* S. W. Jeffrey, R.F.C. Mantoura, and S. W. Wright, eds., *Phytoplankton pigments in oceanography: A guide to advanced methods*. SCOR-UNESCO, Paris.

Jeffrey, S. W., and S. W. Wright. 1987. A new spectrally distinct component in preparations of chlorophyll-*c* from the micro-alga *Emiliania huxleyi* (prymnesiophyceae). *Biochimica et Biophysica Acta* **894:** 180–188.

Jeffrey, S. W., and S. W. Wright. 1994. Photosynthetic pigments in the Haptophyta. Pp. 111–132 *in* J. C. Green and B.S.C. Leadbeater, eds., *The haptophyte algae*, Clarendon Press, Oxford, UK.

Jeffrey, S. W., M. Sielicki, and F. T. Haxo. 1975. Chloroplast pigment patterns in dinoflagellates. *Journal of Phycology* **11:** 374–384.

Jeffrey, S. W., R.F.C. Mantoura, and S. W. Wright, eds. 1997. *Phytoplankton pigments in oceanography*. UNESCO, Paris.

Johansen, J. E., W. A. Svec, S. Liaaen-Jensen, and F. T. Haxo. 1974. Carotenoids of the Dinophyceae. *Phytochemistry* **13:** 2261–2271.

Johnsen, G., and E. Sakshaug. 1993. Bio-optical characteristics and photoadaptive responses in the toxic and bloom-forming dinoflagellates *Gymnodinium aureolum*, *G. galatheanum*, and two strains of *Prorocentrum minimum*. *Journal of Phycology* **29:** 627–642.

Johnson, P. W., and J. M. Sieburth. 1979. Chroococcoid cyanobacteria in the sea: Ubiquitous and diverse phototropic biomass. *Limnology and Oceanography* **24:** 928–935.

Keely, B. J., and J. R. Maxwell. 1991. Structural characterization of the major chlorins in a recent sediment. *Organic Geochemistry* **17:** 663–669.

Kendall, C., S. R. Silva, and V. J. Kelly. 2001. Carbon and nitrogen isotopic compositions of particulate organic matter in four large river systems across the United States. *Hydrological Processes* **15:** 1301–1346.

King, L. L., and D. J. Repeta. 1994. Novel pyropheophorbide steryl esters in Black Sea sediments. *Geochimica et Cosmochimica Acta* **55:** 2067–2074.

Knaff, D. B. 1993. The cytochrome-bc1 complexes of photosynthetic purple bacteria. *Photosynthesis Research* **35:** 117–133.

Kolber, Z. S., R. T. Barber, K. H. Coale, S. E. Fitzwateri, R. M. Greene, K. S. Johnson, S. Lindley, and P. G. Falkowski 2002. Iron limitation of phytoplankton photosynthesis in the Equatorial Pacific Ocean. *Nature* **371:** 145–149.

Kowalewska, G. 2005. Algal pigments in sediments as a measure of eutrophication in the Baltic environment. *Quaternary International* **130:** 141–151.

Kowalewska, G., B. Wawrzyniak-Wydrowska, and M. Szymczak-Zyla. 2004. Chlorophyll *a* and its derivatives in sediments of the Odra estuary as a measure of its eutrophication. *Marine Pollution Bulletin* **49:** 148–153.

Kuhlbrandt, W., D. N. Wang, and Y. Fujiyoshi. 1994. Atomic model of plant light-harvesting complex by electron crystallography. *Nature* **367:** 614–621.

Kursar, T. A., and R. S. Alberte. 1983. Photosynthetic unit organization in a red alga: Relationships between light-harvesting pigments and reaction centers. *Plant Physiology* **72:** 409–414.

Lami, A., P. Guilizzoni, and A. Marchetto. 2000. High resolution analysis of fossil pigments, carbon, nitrogen and sulfur in the sediment of eight European Alpine lakes: The MOLAR project. *Journal of Limnology* **59:** 15–28.

Landry, M. R., J. Constantinou, M. Latasa, S. L. Brown, R. R. Bidigare, and M. E. Ondrusek. 2000a. Biological response to iron fertilization in the eastern Equatorial Pacific (IronEx II), II: Dynamics of phytoplankton growth and microzooplankton grazing. *Marine Ecology-Progress Series* **201:** 57–72.

Landry, M. R., J. Constantinou, M. Latasa, S. L. Brown, R. R. Bidigare, and M. E. Ondrusek. 2000b. Biological response to iron fertilization in the eastern Equatorial Pacific (IronEx II), III: Microplankton community abundances and biomass. *Marine Ecology-Progress Series* **201:** 27–42.

Landry, M. R., and D. L. Kirchman. 2002. Microbial community structure and variability in the tropical Pacific. *Deep-Sea Research, Part II: Topical Studies in Oceanography* **49:** 2669–2693.

Landry, M. R., S. L. Brown, J. Neveux, C. Dupouy, J. Blanchot, S. Christensen, and R. R. Bidigare, 2003. Phytoplankton growth and microzooplankton grazing in high-nutrient, low-chlorophyll waters of the

Equatorial Pacific: Community and taxon-specific rate assessments from pigment and flow cytometric analyses. *Journal of Geophysical Research-Oceans* **108:** 8142, doi:10.1029/2000JC000744.

Latasa, M. 2007. Improving estimations of phytoplankton class abundances using CHEMTAX. *Marine Ecology-Progress Series* **329:** 13–21.

Latasa, M., R. R. Bidigare, M. E. Ondrusek, and M. C. Kennicutt. 1996. HPLC analysis of algal pigments: A comparison exercise among laboratories and recommendations for improved analytical performance. *Marine Chemistry* **51:** 315–324.

Laws, E. A., D. M. Karl, D. G. Redalje, R. S. Jurick, and C. D. Winn. 1983. Variability in ratios of phytoplankton carbon and RNA to ATP and chlorophyll *a* in batch and continuous cultures. *Journal of Phycology* **19:** 439–445.

Le Borgne, R., and M. R. Landry. 2003. EBENE: A JGOFS investigation of plankton variability and trophic interactions in the Equatorial Pacific (18°). *Journal of Geophysical Research-Oceans* **108:** 8136, doi:10.1029/2001JC001252.

Leavitt, P. R., and S. R. Carpenter. 1989. Effects of sediment mixing and benthic algal production on fossil pigment startigraphies. *Journal of Paleolimniology* **2:** 147–158.

Leavitt, P. R., and Hodgson, D. A. 2001. Sedimentary pigments. Pp. 2–21 *in* J. P Smol, H.J.B. Birks, and W. M. Last, eds., *Tracking environmental changes using lake sediments*. Kluwer, New York.

Leavitt, P. R., R. D. Vinebrooke, D. B. Donald, J. P. Smol, and D. W. Schindler. 1997. Past ultraviolet environments in lakes derived from fossil pigments. *Limnology and Oceanography* **44:** 757–773.

Leavitt, P. R., B. F. Cumming, J. P. Smol, M. Reasoner, R. Pienitz, and D. A. Hodgson. 2003. Climatic control of ultraviolet radiation effects on lakes. *Limnology and Oceanography* **48:** 2062–2069.

Leavitt, P. R., S. C. Fritz, N. J. Anderson, P. A. Baker, T. Blenckner, L. Bunting, J. Catalan, D. J. Conley, W. O. Hobbs, E. Jeppesen, A. Korhola, S. McGowan, K. Ruhland, J. A. Rusak, G. L. Simpson, N. Solovieva, and J. Werne. 2009. Paleolimnological evidence of the effects on lakes of energy and mass transfer from climate and humans. *Limnology and Oceanography* **54:** 2230–2348.

Lee, T., M. Tsuzuki, T. Takeuchi, K. Yokoyama, and I. Karube. 1995. Quantitative determination of cyanobacteria in mixed phytoplankton assemblages by an *in vitro* fluorimetric method. *Analytica Chimica Acta* **302:** 81–87.

Lemaire, E., G. Abril, R. De Wit, and H. Etcheber. 2002. Distribution of phytoplankton pigments in nine European estuaries and implications for an estuarine typology. *Biogeochemistry* **59:** 5–23.

Lemberg, R., and J. W. Legge. 1949. *Hematin compounds and bile pigments*. Interscience, New York.

Lewitus, A. J., D. L. White, R. G. Tymowski, M. E. Geesey, S. N. Hymel, and P. A. Nobel. 2005. Adapting the CHEMTAX method for assessing phytoplankton taxonomic composition in southeastern U.S. estuaries. *Estuaries* **28:** 160–172.

Liaaen-Jensen, S. 1978. Marine carotenoids. Pp. 1–73 *in* P. J. Scheuer, ed., *Marine natural products: Chemical and biological perspectives*. Academic Press, New York.

Llewellyn, C. A., and R.F.C. Mantoura. 1997. A UV absorbing compound in HPLC pigment chromatograms obtained from Icelandic basin phytoplankton. *Marine Ecology Progress Series* **158:** 283–287.

Llewellyn, C. A., J. R. Fishwick, and J. C. Blackford. 2005. Phytoplankton community assemblage in the English Channel: A comparison using chlorophyll *a* derived from HPLC-CHEMTAX and carbon derived from microscopy cell counts. *Journal of Plankton Research* **27:** 103–119.

Lorenzen, C. J., and J. N. Downs. 1986. The specific absorption coefficients of chlorophyllide-*a* and pheophorbide-*a* in 90-percent acetone, and comments on the fluorometric determination of chlorophyll and pheopigments. *Limnology and Oceanography* **31:** 449–452.

Louda, J. W., J. W. Loitz, D. T. Rudnick, and E. W. Baker. 2000. Early diagenetic alteration of chlorophyll-*a* and bacteriochlorophyll-*a* in a contemporaneous marl ecosystem, Florida Bay. *Organic Geochemistry* **31:** 1561–1580.

Louda, J. W., L. Liu, and E. W. Baker. 2002. Senescence- and death-related alteration of chlorophylls and carotenoids in marine phytoplankton. *Organic Geochemistry* **33:** 1635–1653.

Ma, L. F., and D. Dolphin. 1996. Stereoselective synthesis of new chlorophyll *a* related antioxidants isolated from marine organisms. *Journal of Organic Chemistry* **61:** 2501–2510.

Mackey, M. D., D. J. Mackey, H. W. Higgins, and S. W. Wright. 1996. CHEMTAX—a program for estimating class abundances from chemical markers: Application to HPLC measurements of phytoplankton. *Marine Ecology Progress Series* **144:** 265–283.

Malkin, R., and K. Niyogi. 2000. Photosynthesis. Pp. 568–628 *in* B. Buchanan, W. Gruissem, and R. Jones, eds., *Biochemistry and molecular biology of plants*. American Society of Plant Physiologists, Rockville, MD.

Mantoura, R.F.C., and C. A. Llewellyn. 1983. The rapid-determination of algal chlorophyll and carotenoid pigments and their breakdown products in natural waters by reverse-phase high-performance liquid chromatography *Analytica Chimica Acta* **151**: 297–314.

Martin, J. H., and S. E. Fitzwater. 1988. Iron-deficiency limits phytoplankton growth in the northeast Pacific subarctic. *Nature* **331**: 341–343.

Meyer-Harms, B., and B. Von Bodungen. 1997. Taxon-specific ingestion rates of natural phytoplankton by calanoid copepods in a estuarine enviroment (Pomeranian Bight, Baltic Sea) determined by cell counts and HPLC analyses of marker pigments. *Marine Ecology-Progress Series* **153**: 181–190.

Michel, H., and J. Deisenhofer. 1988. Relevance of the photosynthetic reaction center from purple bacteria to the structure of photosystem-II. *Biochemistry* **27**: 1–7.

Miller, R. L., C. E. Del Castillo, and B. A. McKee, eds. 2005. *Remote sensing of coastal aquatic environments.* Springer, Dordrecht, The Netherlands.

Millie, D. F., H. W. Paerl, and J. P. Hurley. 1993. Microalgal pigment assessments using high-performance liquid chromatography: A synopsis of organismal and ecological applications. *Canadian Journal of Fisheries and Aquatic Sciences* **50**: 2513–2527.

Monger, B. C., and M. R. Landry. 1993. Flow cytometric analysis of marine–bacteria with Hoechst 33342. *Applied and Enivromental Microbiology* **59**: 905–911.

Morel, F.M.M., D. A. Dzombak, and N. M. Price. 1991. Heterogeneous reactions in coastal waters. Pp. 165–180 *in* R.F.C. Mantoura, J. M. Martin, and R. Wollast, eds., *Ocean margin processes in global change*, Wiley, New York.

Moreth, C. M., and C. S. Yentsch. 1970. A sensitive method for determination of open ocean phytoplankton phycoerythrin pigments by fluorescence. *Limnology and Oceanography* **15**: 313–317.

Mossa, J. 1996. Sediment dynamics in the lowermost Mississippi River. *Engineering Geology* **45**: 457–479.

Nelson, J. R., and S. G. Wakeham. 1989. A phytol-substituted chlorophyll *c* from *Emiliana huxleyi* (Prymnesiophyceae). *Journal of Phycology* **25**: 761–766.

Neveux, J., and F. Lantoine. 1993. Spectrofluorometric assay of chlorophylls and pheopigments using the least-squares approximation technique. *Deep-Sea Research, Part I: Oceanographic Research Papers* **40**: 1747–1765.

Nugent, J.H.A. 1996. Oxygenic photosynthesis–electron transfer in photosystem I and photosystem II. *European Journal of Biochemistry* **237**: 519–531.

Olson, J. M., and B. K. Pierson. 1987. Origin and evolution of photosynthetic reaction centers. *Origins of Life and Evolution of the Biospheres* **17**: 419–430.

Onstad, G. D., D. E. Canfield, P. D. Quay, and J. I. Hedges. 2000. Sources of particulate organic matter in rivers from the continental USA: Lignin phenol and stable carbon isotope compositions. *Geochimica et Cosmochimica Acta* **64**: 3539–3546.

Overmann, J., G. Sandmann, K. J. Hall, and T. G. Northcote. 1993. Fossil carotenoids and paleolimnology of meromictic Mahoney Lake, British Columbia, Canada. *Aquatic Sciences* **55**: 1015–1621.

Paerl, H. W. 2007. Nutrient and other environmental controls of harmful cyanobacterial blooms along the freshwater–marine continuum. *In* H. K. Hudnell, ed., Proceedings of the interagency, international symposium on cyanobacterial harmful algal blooms. *Advances in Experimental Medicine and Biology.*

Paerl, H. W., and R. S. Fulton. 2006. Ecology of harmful cyanobacteria. Pp. 65–107 *in* E. Graneli and J. Turner, eds., *Ecology of harmful marine algae*. Springer, Berlin.

Paerl, H. W., R. S. Fulton, P. H. Moisander, and J. Dyble. 2001. Harmful freshwater algal blooms, with an emphasis on cyanobacteria. *The Scientific World* **1**: 76–113.

Parsons, T. R., and J.D.H. Strickland. 1963. Discussion of spectrophotometric determination of marine-plant pigments, with revised equations for ascertaining chlorophylls and carotenoids. *Journal of Marine Research* **21**: 155–163.

Patterson, G., and O. Kachinjika. 1995. Limnology and phytoplankton ecology. Pp. 1–67 *in* A. Menz, ed., *The fishery potential and productivity of the pelagic zone of Lake Malawi/Niassa*. Natural Resource Institute. Chatham, UK.

Pennington, F. C., F. T. Haxo, G. Borch, G., and S. Liaaen-Jensen. 1985. Carotenoids of Cryptophyceae. *Biochemical Systematic Ecology* **13**: 215–219.

Pfundel, E., and W. Bilger. 1994. Regulation and possible function of the violaxanthin cycle. *Photosynthesis Research* **42:** 89–109.

Pinckney, J. L., D. F. Millie, K. E. Howe, H. W. Paerl, and J. P. Hurley. 1996. Flow scintillation counting of C-14-labeled microalgal photosynthetic pigments. *Journal of Plankton Research* **18:** 1867–1880.

Pinckney, J. L., H. W. Paerl, M. B. Harrington, and K. E. Howe. 1998. Annual cycles of phytoplankton community-structure and bloom dynamics in the Neuse River estuary, North Carolina. *Marine Biology* **131:** 371–381.

Pinckney, J. L., T. L. Richardson, D. F. Millie, and H. W. Paerl. 2001. Application of photopigment biomarkers for quantifying microalgal community composition and *in situ* growth rates. *Organic Geochemistry* **32:** 585–595.

Porra, R. J., E. E. Pfundel, and N. Engel. 1997. Metabolism and function of photosynthetic pigments. Pp. 85–126 *in* S. W. Jeffrey, R.F.C. Mantoura, and S. W. Wright, eds., *Phytoplankton pigments in oceanography*. UNESCO, Paris.

Redalje, S. J. 1993. ACS directory of graduate research on disc. *Journal of Chemical Information and Computer Sciences* **33:** 796–796.

Repeta, D. J., and R. B. Gagosian. 1987. Carotenoid diagenesis in recent marine sediments, 1: The Peru continental-shelf (15°S, 75°W). *Geochimica et Cosmochimica Acta* **51:** 1001–1009.

Repeta, D. J., and D. J. Simpson. 1991. The distribution and cycling of chlorophyll, bacteriochlorophyll and carotenoids in the Black Sea. *Deep-Sea Research* **38:** 969–984.

Repeta, D. J., D. J. Simpson, B. B. Jorgensen, and H. W. Jannasch. 1989. Evidence for anoxygenic photosynthesis from the distribution of bacteriochlorophylls in the Black Sea. *Nature* **342:** 69–72.

Richards, T. A., and T. G. Thompson. 1952. The estimation and characterization of plankton populations by pigment analyses, II: A spectrophotometric method for the estimation of plankton pigments. *Journal of Marine Research* **11:** 156–172.

Ricketts, T. R, 1966. Magnesium 2,4-divinylphaeoporphyrin a, monomethyl ester, a photochlorophyll-like pigment presence in some unicelluar flagellates. *Phytochemistry* **5:** 223–229.

Rowan, K. S. 1989. *Photosynthetic pigments of algae*. Cambridge University Press, Cambridge, UK.

Roy, S., J. P. Chanut, M. Gosselin, and T. Sime-Ngando. 1996. Characterization of phytoplankton communities in the lower St. Lawerence estuary using HPLC-detected pigments and cell microscopy. *Marine Ecology-Progress Series* **142:** 55–73.

Ruess, N., D. J. Conley, and T. S. Bianchi. 2005. Preservation conditions and the use of sediment pigments as a tool for recent ecological reconstruction in four northern European estuaries. *Marine Chemistry* **95:** 283–302.

Rutherford, A. W. 1989. Photosystem-II, the water-splitting enzyme. *Trends in Biochemical Sciences* **14:** 227–232.

Rutherford, D. W., C. T. Chiou, and D. Kile. 1992. Influence of soil organic matter composition on the partitioning of organic compounds. *Environmental Science and Technology* **26:** 336–340.

Sandmann, G. 2001. Carotenoid biosynthesis and biotechnological application. *Archives of Biochemistry and Biophysics* **385:** 4–12.

Sanger, J. E., and E. Gorham. 1970. Diversity of pigments in lake sediments and its ecological significance. *Limnology and Oceanography* **15:** 59–69.

Schmid, H., F. Bauer, and H. B. Stich. 1998. Determination of algal biomass with HPLC pigment analysis from lakes of different trophic state in comparison to microscopically measured biomass. *Journal of Plankton Research* **20:** 1651–1661.

Schuette, H. R. 1983. Secondary plant substances: Aspects of carotenoid biosynthesis. *Progress in Botany* **45:** 120–135.

Seki, M. P., J. J. Polovina, R. E. Brainard, R. R. Bidigare, C. L. Leonard, and D. G. Foley. 2001. Biological enhancement at cyclonic eddies tracked with goes thermal imagery in Hawaiian waters. *Geophysical Research Letters* **28:** 1583–1586.

Shuman, F. R., and C. J. Lorenzen. 1975. Quantitative degradation of chlorophyll by a marine herbivore. *Limnology and Oceanography* **20:** 580–586.

Smith, W. O., Jr., A. R., Shields, J. A. Peloquin, G. Catalano, S. Tozzi, M. S. Dinniman, and V. A. Asper. 2006. Interannual variations in nutrients, net community production, and biogeochemical cycles in the Ross Sea. *Deep-Sea Research* **53:** 815–833.

Sinninghe-Damsté, J. S., and M.J.L. Coolen. 2006. Fossil DNA in Cretaceous black shales: myth or reality? *Astrobiology* **6:** 299–302.

Spooner, N., H. R. Harvey, G.E.S. Pearce, C. B. Eckardt, and J. R. Maxwell. 1994. Biological defunctionalisation of chlorophyll in the aquatic environment, 2: Action of endogenous algal enzymes and aerobic bacteria. *Organic Geochemistry* **22:** 773–780.

Squier, A. H., D. A. Hodgson, and B. J. Keely. 2004. Identification of bacteriophaeophytin a esterified with geranylgeraniol in an Antarctic lake sediment. *Organic Geochemistry* **35:** 203–207.

Stauber, J. L., and S. W. Jeffrey. 1988. Photosynthetic pigments in fifty-one species of marine diatoms. *Journal of Phycology* **24:** 158–172.

Stewart, D. E., and F. H. Farmer. 1984. Extraction, identification, and quantitation of phycobiliprotein pigments from phototrophic plankton. *Limnology and Oceanography* **29:** 392–397.

Stomp, M., J. Huisman, F. de Jongh, A. J. Veraart, D. Gerla, M. Rijkeboer, B. W. Ibelings, U.I.A. Wollenzien, and L. J. Stal. 2004. Adaptive divergence in pigment composition promotes phytoplankton biodiversity. *Nature* **432:** 104–107.

Strom, S. L., 1991. Growth and grazing rates of an herbivorous dinoflagellate (*Gymnodinium* sp.) from the open subarctic Pacific Ocean. *Marine Ecology-Progress Series* **78:** 103–113.

Strom, S. L. 1993. Production of phaeopigments by marine protozoa: Results of laboratory experiments analyzed by HPLC. *Deep-Sea Research* **40:** 57–80.

Sullivan, B. E., F. G. Prahl, L. F. Small, and P. A. Covert. 2001. Seasonality of phytoplankton production in the Columbia River: A natural or anthropogenic pattern? *Geochimica et Cosmochimica Acta* **65:** 1125–1139.

Suzuki, J. Y., D. W. Bollivar, and C. E. Bauer. 1997. Genetic analysis of chlorophyll biosynthesis. *Annual Review of Genetics* **31:** 61–89.

Szymczak-Zyla, M., B. Wawrzyniak-Wydrowska, and G. Kowalewska. 2006. Products of chlorophyll *a* transformation by selected benthic organisms in the Odra estuary (southern Baltic Sea). *Hydrobiologia* **554:** 155–164.

Szymczak-Zyla, M., G. Kowalewska, and J. Louda. 2008. The influence of microorganisms on chlorophyll *a* degradation in the marine environment. *Limnology and Oceanography* **53:** 851–862.

Taiz, L., and E. Zeiger. 2006. *Plant physiology,* 4th ed. Sinauer Associates, Sunderland, MA.

Talbot, H. M., R. Head, R. P. Harris, and J. R. Maxwell. 1999. Distribution and stability of steryl chlorin esters in copepod faecal pellets from diatom grazing. *Organic Geochemistry* **30:** 1163–1174.

Talling, J. F. 1987. The phytoplankton of Lake Victoria (East Africa). *Archiv fur Hydrobiologie* **25:** 229–256.

Tester, P. A., M. E. Geesey, C. Z. Guo, H. W. Paerl, and D. F. Millie. 1995. Evaluating phytoplankton dynamics in the Newport River estuary (North Carolina, USA) by HPLC-derived pigment profiles. *Marine Ecology-Progress Series* **124:** 237–245.

Thorp, J. H., and M. D. Delong. 1994. The riverine productivity model: An heuristic view of carbon sources and organic processing in large river ecosystems. *OIKOS* **70:** 305–308.

Trees, C. C., M. C. Kennicutt, and J. M. Brooks. 1985. Errors associated with the standard fluorimetric determination of chlorophylls and phaeopigments. *Marine Chemistry* **17:** 1–12.

Treibs, A. 1936. Chlorophyll- und Häminderivate in organischen Mineralstoffen. *Angewandte Chemie* **49:** 682–688.

UNESCO. 1966. Monographs on oceanographic methodology, 1: *Determination of photosynthetic pigments in sea water*. United Nations Educational, Scientific and Cultural Organization, Paris.

Van Heukelem, L., and C. S. Thomas. 2001. Computer-assisted high-performance liquid chromatography method development with applications to the isolation and analysis of phytoplankton pigments. *Journal of Chromatography A* **910:** 31–49.

Verburg, P., R. E. Hecky, and H. J. Kling. 2003. Ecological consequences of a century of warming in Lake Tanganyika. *Science* **301:** 505–507.

Vernet, M., and C. J. Lorenzen. 1987. The presence of chlorophyll *b* and the estimation of phaeopigments in marine phytoplankton. *Journal of Plankton Research* **9:** 255–265.

Vesk, M., and S.W. Jeffrey. 1987. Ultrastructure and pigments of two strains of the picoplanktonic alga *Pelagococcus subviridis* (Chrysophyceae). *Journal of Phycology* **23:** 322–336.

Vincent, W. F., M. T. Downes, R. W. Castenholz, and C. Howard-Williams. 1993. community structure and pigment organisation of cyanbacteria-dominated microbial mats in Antarctica. *European Journal of Phycology* **28:** 213–221.

Watts, C. D., and Maxwell, J. R. 1977. Carotenoid diagenesis in a marine sediment. *Geochimica et Cosmochimica Acta* **41:** 493–497.

Weber, C. I., L. A. Fay, G. B. Collins, D. E. Rathke, and J. Tobin. 1986. A review of methods for the analysis of chlorophyll in periphyton and plankton of marine and freshwater systems. Ohio State University Sea Grant Program Tech Bull OHSU-TB-15.

Wehr, J. D., and R. G. Sheath. 2003. Freshwater habitats of algae. Pp. 757–776 *in Freshwater algae of North America: Ecology and classification*. Academic Press, San Diego.

Wehr, J. D., and J. H. Thorp. 1997. Effects of navigation dams, tributaries, and littoral zones on phytoplankton communities in the Ohio River. *Canadian Journal of Fisheries and Aquatic Sciences* **54:** 378–395.

Welschmeyer, N., and C. J. Lorenzen. 1985. Chlorophyll budgets: Zooplankton grazing and phytoplankton growth in a temperate fjord and the central Pacific gyres. *Limnology and Oceanography* **30:** 1–21.

Wetzel, R. G. 2001. Fundamental processes within natural and constructed wetland ecosystems: Short-term versus long-term objectives. *Water Science and Technology* **44:** 1–8.

Whatley, J. M. 1993. The endosymbiotic origin of chloroplasts. *International Review of Cytology: A Survey of Cell Biology* **144:** 259–299.

Wilson, M. A., R. L. Airs, J. E. Atkinson, and B. J. Keely. 2004. Bacteriovirdins: Novel sedimentary chlorines providing evidence for oxidative processes affecting paleobacterial communities. *Organic Geochemistry* **35:** 199–202.

Wright, S. W., S. W. Jeffrey, R.F.C. Mantoura, C. A. Llewellyn, T. Bjørland, D. Repta, and N. Welschmeyer. 1991. Improved HPLC method for the analysis of chlorophylls and carotenoids from marine phytoplankton. *Marine Ecology Progress Series* **77:** 183–196.

Wright, S. W., D. P. Thomas, H. J. Marchant, H. W. Higgins, M. D. Mackey, and D. J. Mackey. 1996. Analysis of phytoplankton of the Australian sector of the southern ocean: Comparisons of microscopy and size frequency data with interpretations of pigment HPLC data using the 'CHEMTAX' matrix factorisation program. *Marine Ecology-Progress Series* **144:** 285–298.

Wyman, M. 1992. An in-vivo method for the estimation of phycoerythrin concentrations in marine cyanobacteria (*Synechococcus* spp). *Limnology and Oceanography* **37:** 1300–1306.

Yacobi, Y. Z., U. Pollingher, Y. Gonen, V. Gerhardt, and A. Sukenik. 1996. HPLC analysis of phytoplankton pigments from Lake Kinneret with special reference to the bloom-forming dinoflagellate *Peridinium gatunense* (Dinophyceae) and chlorophyll degradation products. *Journal of Plankton Research* **18:** 1781–1796.

Ziegler, R., A. Blaheta, N. Guha, and B. Schonegge. 1988. Enzymatic formation of pheophorbide and pyrophephorbide during chlorophyll degradation in a mutant of chlorella *Fusca shihira. Journal of Plant Physiology* **132:** 327–332.

Chapter 13

Adler, E. 1977. Lignin chemistry—past, present and future. *Wood Science Technology* **11:** 169–218.

Baker, C. J., S. L. McCormick, and D. F. Bateman. 1982. Effects of purified cutin esterase upon the permeability and mechanical strength of cutin membranes. *Phytopathology* **72:** 420–423.

Bauer, J. E., E.R.M. Druffel, D. M. Wolgast, and S. Griffin. 2001. Sources and cycling of dissolved and particulate organic radiocarbon in the northwest Atlantic continental margin. *Global Biogeochemical Cycles* **15:** 615–636.

Benner, R., and S. Opsahl. 2001. Molecular indicators of the sources and transformations of dissolved organic matter in the Mississippi River plume. *Organic Geochemistry* **32:** 597–611.

Benner, R., A. E. Maccubin, and R. E. Hodson. 1984. Anaerobic degradation of the lignin and polysaccharide components of lignocellulose and synthetic lignin by sediment microflora. *Applied Environmental Microbiology* **47:** 988–1004.

Benner, R., M. A. Moran, and R. E. Hodson. 1986. Biogeochemical cycling of lignocellulosic carbon in marine and fresh- water ecosystems: Relative contributions of procaryotes and eucaryotes. *Limnology and Oceanography* **31:** 89–100.

Benner, R., K. Weliky, and F. I. Hedges. 1990. Early diagenesis of mangrove leaves in a tropical estuary: molecular-level analyses of neutral sugars and lignin-phenols. *Geochimica et Cosmochimica Acta* **54:** 1991–2001.

Bernards, M. A. 2002. Demystifying suberin. *Canadian Journal of Botany* **80:** 227–240.

Bernards, M. A., M. L. Lopez, and J. Zajicek. 1995. Hydroxycinnamic acid-derived polymers constitute the polyaromatic domain of suberin. *Journal of Biological Chemistry* **270:** 7382–7386.

Bianchi, T. S., and M. E. Argyrou. 1997. Temporal and spatial dynamics of particulate organic carbon in the Lake Pontchartrain estuary, southeast Louisiana, U.S.A. *Estuarine, Coastal and Shelf Science* **45:** 287–297.

Bianchi, T. S., C. Rolff, and C. Lambert. 1997. Sources and composition of particulate organic carbon in the Baltic Sea: The use of plant pigments and lignin-phenols as biomarkers. *Marine Ecology-Progress Series* **156:** 25–31.

Bianchi, T. S., M. Argyrou, and H. F. Chipett. 1999. Contribution of vascular-plant carbon to surface sediments across the coastal margin of Cyprus (eastern Mediterranean). *Organic Geochemistry* **30:** 287–297.

Bianchi, T. S., S. Mitra, and M. McKee. 2002. Sources of terrestrially-derived carbon in the Lower Mississippi River and Louisiana shelf: Implications for differential sedimentation and transport at the coastal margin. *Marine Chemistry* **77:** 211–223.

Bianchi, T. S., L. A. Wysocki, M. Stewart, T. R. Filley, and B. A. McKee. 2007. Temporal variability in terrestrially-derived sources of particulate organic carbon in the lower Mississippi River. *Geochimica et Cosmochimica Acta* **71:** 4425–4437.

Birnbaum, K., D. E. Shasha, J. Y. Wang, J. W. Jung, G. M. Lambert, D. W. Galbraith, and P. N. Benfey. 2003. A gene expression map of the *Arabidopsis* root. *Science* **302:** 1956–1960.

Blee, E., and F. Schuber. 1993. Biosynthesis of cutin monomers: Involvement of a lipoxygenase/peroxygenase pathway. *Journal of Plants* **4:** 113–123.

Bonaventure, G., J. J. Salas, M. R. Pollard, and J. B. Ohlrogge. 2003. Disruption of the FATB gene in *Arabidopsis* demonstrates an essential role of saturated fatty acids in plant growth. *Plant Cell* **15:** 1020–1033.

Bonaventure, G., F. Beisson, J. Ohlrogge, and M. Pollard. 2004. Analysis of the aliphatic monomer composition of polyesters associated with *Arabidopsis epidermis*: Occurrence of octadeca-*cis*-6, *cis*-9-diene-1,18-dioate as the major component. *Plant Journal* **40:** 920–930.

Broecker, W. S., and T. H. Peng. 1982. *Tracers in the sea.* LDGEO Press, Palissades, NY.

Cardoso, J. N., and G. Eglinton. 1983. The use of hydroxy acids as geochemical indicators. *Geochimica et Cosmochimica Acta* **47:** 723–730.

Chang, H. M., and G. G. Allen. 1971. Oxidation. Pp. 433–485 *in* K. V. Sarkanen and C. H. Ludwig, eds., *Lignins.* Wiley Interscience, New York.

Chefetz, B., Y. Chen, C. E. Clapp, and P. G. Hatcher. 2000. Characterization of organic matter in soils by thermochemolysis using tetramethylammonium hydroxide (TMAH). *Soil Science Society of America* **64:** 583–589.

Clifford, D. J., D. M. Carson, D. E. McKinney, J. M. Bortiatynski, and P. G. Hatcher. 1995. A new rapid technique for the characterization of lignin in vascular plants: Thermochemolysis with tetramethylammonium hydroxide (TMAH). *Organic Geochemistry* **23:** 169–175.

Croteau, R., and P. E. Kolattukudy. 1974. Biosynthesis of hydroxy fatty acid polymers: Enzymatic synthesis of cutin from monomer acids by cell-free preparations from epidermis of *Vicia faba* leaves. *Biochemistry* **13:** 3193–3202.

Daniel, G. F., T. Nilsson, and A. P. Singh. 1987. Degradation of lignocellulosics by unique tunnel-forming bacteria. *Canadian Journal of Microbiology* **33:** 943–948.

De Leeuw, D., and C. Largeau. 1993. A review of macromolecular organic compounds that comprise living organisms and their role in kerogen, coal, and petroleum formation. Pp. 23–72 *in* M. H. Engel and S. A. Macko, eds., *Organic geochemistry: Principle and applications.* Plenum Press, New York.

Dean, B. B., and P. E. Kolattukudy. 1976. Synthesis of suberin during wound-healing in jade leaves, tomato fruit, and bean pods. *Plant Physiology* **58:** 411–416.

Dickens, A. F., J. A. Gudeman, Y, Gélinas, J. A. Baldock, W. Tinner, F. S. Hu, and J. I. Hedges. 2007. Sources and distribution of CuO-derived benzene carboxylic acids in soils and sediments. *Organic Geochemistry* **38:** 1256–1276.

Dittmar, T., and G. Kattner. 2003. The biogeochemistry of the river and shelf ecosystem of the Arctic Ocean: A review. *Marine Chemistry* **83:** 103–120.

Dittmar, T., and R. J. Lara. 2001. Molecular evidence for lignin degradation in sulfate-reducing mangrove sediments (Amazônia, Brazil). *Geochimica et Cosmochimica Acta* **65:** 1417–1428.

Duan, H., and M. A. Schuler. 2005. Differential expression and evolution of the Arabidopsis CYP86A subfamily. *Plant Physiology* **137:** 1067–1081.

Eadie, B. J., B. A. McKee, M. B. Lansing, J. A. Robbins, S. Metz, and J. H. Trefrey. 1994. Records of nutrient-enhanced coastal ocean productivity in sediments from the Louisiana continental shelf. *Estuaries* **17:** 754–765.

Eglinton, G., D. H. Hunneman, and K. Douraghi-Zadeh. 1986. Gas chromatographic–mass spectrometric studies of long chain hydroxy acids, II: The hydroxy acids and fatty acids of a 5000-year-old lacustrine sediment. *Tetrahedron* **24:** 5929–5941.

Endstone, D. E., C. A. Peterson, and F. Ma. 2003. Root endodermis and exodermis: Structure, function, and responses to the environment. *Journal of Plant Growth Regulation.* **21:** 335–351.

Engbrodt, R. 2001. Biogeochemistry of dissolved carbohydrates in the Arctic. *Report of Polar Research* **396:** 106.

Ertel, J. R., and J. I. Hedges. 1984. Sources of sedimentary humic substances: Vascular plant debris. *Geochimica et Cosmochimica Acta* **48:** 2065–2074.

Espelie, K. E., and P. E. Kolattukudy. 1979a. Composition of the aliphatic components of suberin of the endodermal fraction from the first internode of etiolated *Sorghum* seedlings. *Plant Physiology* **63:** 433–435.

———. 1979b. Composition of the aliphatic components of suberin from the bundle sheaths of *Zea mays* leaves. *Plant Science Letters* **15:** 225–230.

Espelie, K. E., N. Z. Sadek, and P. E. Kolattukudy. 1980. Composition of suberin-associated waxes from the subterranean storage organs of seven plants, parsnip, carrot, rutabaga, turnip, red beet, sweet potato and potato. *Planta* **148:** 468–476.

Fabbri, D., G. Chiavari, and G. C. Galletti. 1996. Characterization of soil humin by pyrolysis (methylation)–gas chromatography/mass spectrometry: Structural relationships with humic acids. *Journal Analytical Applied Pyrolysis* **37:** 161–172.

Faegri, K., and J. Iverson. 1992. *Textbook of pollen analysis*, 4th ed. Wiley, New York.

Farella, N., M. Lucotte, P. Louchouarn, and M. Roulet. 2001. Deforestation modifying terrestrial organic transport in the Rio Tapajós, Brazilian Amazon. *Organic Geochemistry* **32:** 1143–1458.

Filley, T. R. 2003. Assessment of fungal wood decay by lignin analysis using tetramethylammonium hydroxide (TMAH) and ^{13}C-labeled TMAH thermochemolysis. Pp. 119–139 *in* B. Goodell, D. D. Nicholas, and T. P. Schultz, eds., *Wood deterioration and preservation: advances in our changing world.* ACS Symposium Series, Washington, DC.

———. 1999b. Tetramethylammonium hydroxide (TMAH) thermochemolysis: Proposed mechanisms based upon the application of C-13-labeled TMAH to a synthetic model lignin dimer. *Organic Geochemistry* **30:** 607–621.

Filley, T. R., R. D. Minard, and P. G. Hatcher. 1999a. Tetramethylammonium hydroxide (TMAH) thermochemolysis: Proposed mechanisms based upon the application of ^{13}C-labeled TMAH to a synthetic model lignin dimer. *Organic Geochemistry* **30.**

Filley, T. R., P. G. Hatcher, W. C. Shortle, and R. T. Praseuth. 2000. The application of ^{13}C-labeled tetramethylammonium hydroxide (^{13}C-TMAH) thermochemolysis to the study of fungal degradation of wood. *Organic Geochemistry* **31:** 181–198.

Filley, T. R., K. G. J. Nierop, and Y. Wang. 2006. The contribution of polyhydroxyl aromatic compounds to tetramethylammonium hydroxide lignin-based proxies. *Organic Geochemistry* **37:** 711–727.

Fors, Y., T. Nilsson, E. Risberg, M. Sandstrom, and P. Torssander. 2008. Sulfur accumulation in pinewood (*Pinus sylvestris*) induced by bacteria in a simulated seabed environment: Implications for marine archaeological wood and fossil fuels. *International Biodeterioration and Biodegradation* **62:** 336–347.

Galler, J. J. 2004. *Estuarine and bathymetric controls on seasonal storage in the lower Mississippi and Atchafalaya rivers.* Tulane University, New Orleand, LA.

Galler, J. J., T. S. Bianchi, M. A. Allison, R. Campanella, and L. A. Wysocki. 2003. Biogeochemical implications of levee confinement in the lower-most Mississippi River. *EOS* **84:** 469–476.

Gardner, W. S., and D. W. Menzel. 1974. Phenolic aldehydes as indicators of terrestrially derived organic matter in the sea. *Geochimica et Cosmochimica Acta* **38:** 813–822.

Gleixner, G., C. Czimczi, K. Kramer, B. M. Luhker, and M.W.I. Schmidt. 2001. Plant compounds and their turnover and stabilization as soil organic matter. Pp. 201–215 *in* M. Schulze et al., eds., *Global biogeochemical cycles in the climate system.* Academic Press, San Diego.

Goñi, M. A., and J. I. Hedges. 1990a. Potential applications of cutin-derived CuO reaction products for discriminating vascular plant sources in natural environments. *Geochimica et Cosmochimica Acta* **54:** 3073–3083.

———. 1990b. Cutin derived CuO reaction products from purified cuticles and tree leaves. *Geochimica Cosmochimica Acta* **54:** 3065–3072.

———. 1990c. The diagenetic behavior of cutin acids in buried conifer needles and sediments from a coastal marine environment. *Geochimica et Cosmochimica Acta* **54:** 3083–3093.

———. 1992. Lignin dimers: Structures, distribution, and potential geochemical applications. *Geochimica et Cosmochimica Acta* **56:** 4025–4043.

———. 1995. Sources and reactivities of marine-derived organic matter in coastal sediments as determined by alkaline CuO oxidation. *Geochimica et Cosmochimica Acta* **59**.

Goñi, M. A., and K. A. Thomas. 2000. Sources and transformations of organic matter in surface soils and sediments from a tidal estuary (North Inlet, South Carolina, U.S.A). *Estuaries* **23:** 548–564.

Goñi, M. A., K. C. Ruttenberg, and T. I. Eglinton. 1997. Sources and contribution of terrigenous organic carbon to surface sediments in the Gulf of Mexico. *Nature* **389:** 275–278.

———. 1998. A reassessment of the sources and importance of land-derived organic matter in surface sediments from the Gulf of Mexico. *Geochimica et Cosmochimica Acta* **62:** 3055-3075.

Goñi, M. A., M. B. Yunker, R. W. MacDonald, and T. I. Eglinton. 2000. Distribution and sources of organic biomarkers in arctic sediments from the Mackenzie River and Beaufort shelf. *Marine Chemistry* **71:** 23–51.

Goñi, M. A., M. J. Teixeira, and D. W. Perkey. 2003. Sources and distribution of organic matter in a river-dominated estuary (Winyah Bay, SC, USA). *Estuarine Coastal and Shelf Science* **57:** 1023–1048.

Goodwin, T. W., and E. I. Mercer. 1972. *Introduction to plant biochemistry*. Pergamon Press, New York.

Gordon, E. S., and M. A. Goñi. 2003. Sources and distribution of terrigenous organic matter delivered by the Atchafalaya River to sediments in the northern Gulf of Mexico. *Geochimica et Cosmochimica Acta* **67:** 2359–2375.

Gough, M. A., R.F.C. Mantoura, and M. Preston. 1993. Terrestrial plant biopolymers in marine sediments. *Geochimica et Cosmochimica Acta* **57:** 945–964.

Graca, J., and H. Pereira. 2000a. Methanolysis of bark suberins: Analysis of glycerol and acid monomers. *Phytochemical Analytica* **11:** 45–51.

———. 2000b. Suberin in potato periderm: Glycerol, long chain monomers, and glyceryl and feruloyl dimers. *Journal of Agricultural Food Chemistry* **48:** 5476–5483.

Graca, J., L. Schreiber, J. Rodrigues, and H. Pereira. 2002. Glycerol and glyceryl esters of omega-hydroxyacids in cutins. *Phytochemistry* **61:** 205–215.

Grbic-Galic, D., and L. Y. Young. 1985. Methane fermentation of ferrulate and benzoate: Anaerobic degradation pathways. *Applied Environmental Microbiology* **50:** 292–297.

Griffith, M., N.P.A. Huner, K. E. Espeilie, and P. E. Kolattukudy. 1985. Lipid polymers accumulate in the epidermis and mestome sheath cell walls during low temperature development of winter rye leaves. *Protoplasma* **125:** 53–64.

Guo, L., and P. H. Santschi. 2000. Sedimentary sources of old high molecular weight dissolved organic carbon from the ocean margin benthic nepheloid layer. *Geochimica et Cosmochimica Acta* **64:** 651–660.

Hatcher, P. G., and R. D. Minard. 1996. Comparison of dehydrogenase polymer (DHP) lignin with native lignin from gymnosperm wood by thermochemolysis using tetramethylammonium hydroxide (TMAH). *Organic Geochemistry* **24:** 593–600.

Hatcher, P. G., M. A. Nanny, R. D. Minard, S. C. Dible, and D. M. Carson. 1995. Comparison of two thermochemolytic methods for the analysis of lignin in decomposing gymnosperm wood: The CuO oxidation method and the method of thermochemolysis with tetramethylammonium hydroxide (TMAH). *Organic Geochemistry* **23:** 881–888.

Hedges, J. I., and J. R. Ertel. 1982. Characterization of lignin by gas capillary chromatography of cupric oxide oxidation products. *Analytical Chemistry* **54:** 174–178.

Hedges, J. I., and D. C. Mann. 1979. The characterization of plant tissues by their lignin oxidation products. *Geochimica et Cosmochimica Acta* **43:** 1809–1818.

Hedges, J. I., and P. L. Parker. 1976. Land-derived organic matter in the surface sediments from the Gulf of Mexico. *Geochimica et Cosmochimica Acta* **40:** 1019–1029.

Hedges, J. I., W. A. Clark, and G. L. Cowie. 1988. Organic matter sources to the water column and surficial sediments of a marine bay. *Limnology and Oceanography* **33:** 1116–1136.

Hedges, J. I., R. Keil, and R. Benner. 1997. What happens to terrestrially-derived organic matter in the ocean? *Organic Geochemistry* **27:** 195–212.

Heredia, A. 2003. Biophysical and biochemical characteristics of cutin, a plant barrier biopolymer. *Biochimica Biophysica Acta* **1620:** 1–7.

Hermann, K. M., and L. M. Weaver. 1999. The shikimate pathway. *Annual Review of Plant Physiology and Plant Molecular Biology* **50:** 473–503.

Hernes, P. J., and R. Benner. 2002. Transport and diagenesis of dissolved and particulate terrigenous organic matter in the North Pacific Ocean. *Deep Sea Research, Part I* **49:** 2119–2132.

———. 2006. Terrigenous organic matter sources and reactivity in the North Atlantic Ocean and a comparison to the Arctic and Pacific oceans. *Marine Chemistry* **100:** 66–79.

Hoffmann-Benning, S., and H. Kende. 1994. Cuticle biosynthesis in rapidly growing internodes of deepwater rice. *Plant Physiology* **104:** 719–723.

Holloway, P. J. 1973. Cutins of *Malus pumila* fruits and leaves. *Phytochemistry* **12:** 2913–2920.

———. 1982. Structure and histochemistry of plant cuticular membranes: An overview. Pp. 1–32 *in The plant cuticle*. Academic Press, New York.

Houel, S., P. Louchouarn, M. Lucotte, R. Canuel, and B. Ghaleb. 2006. Translocation of soil organic matter following reservoir impoundment in boreal systems: Implications for in situ production. *Limnology and Oceanography* **51:** 1497–1513.

Hu, F. S. and others. 1999. Abrupt changes in North American climates during early Holocene times. *Nature* **400:** 437–440.

Jenks, M., L. Anderson, R. S. Tuesinik, and M. H. Williams. 2001. Leaf cuticular waxes of potted rose cultivars as affected by plant development, drought and paclobutrazol treatments. *Physiology of Plants* **112:** 62–70.

Jetter, R., A. Klinger, and S. Schaffer. 2002. Very long-chain phenylpropyl and phenylbutyl esters from *Taxus baccata* needle cuticular waxes. *Phytochemistry* **61:** 579–587.

Johansson, M. B., I. Kogel, and W. Zech. 1986. Changes in the lignin fraction of spruce and pine needle litter during decomposition as studied by some chemical methods. *Soil Biology and Biochemistry* **18:** 611–619.

Kim, J., M. Kim, and W. Bae. 2009. Effect of oxidized leacheate on degradation of lignin by sulfate-reducing bacteria. *Water Management and Research*. **27:** 520–526.

Kogel-Knabner, I. 2000. Analytical approaches for characterizing soil organic matter. *Organic Geochemistry* **31:** 609–625.

Kogel, I., and R. Bochter. 1985. Amino sugar determination in organic soils by capillary gas chromatography using a nitrogen-selective detector. *Zeitschrift für Pflanzenernährung und Bodenkunde* **148:** 261–267.

Kolattukudy, P. E. 1980. Biopolyester membranes of plants: Cutin and suberin. *Science* **208:** 990–1000.

———. 2001. Polyesters in higher plants. *Advances in Biochemical Engineering Biotechnology* **71:** 1–49.

Kolattukudy, P. E., and K. Espelie. 1985. Biosynthesis of cutin, suberin and associated waxes. *in* H. Higuchi, ed., *Biosynthesis and biodegradation of wood components*. Academic Press, New York.

———.1989. Chemistry, biochemistry, and function of suberin and associated waxes. Pp. 304–367 *in* J. W. Rowe, ed., *Natural products of woody plants, I*. Springer, Heidelberg.

Kolattukudy, P. E., and T. J. Walton. 1972. Structure and biosynthesis of hydroxy fatty acids of cutin in *Vicia faba* leaves. *Biochemistry* **11:** 1897–1907.

Kunst, L., and A. L. Samuels. 2003. Biosynthesis and secretion of plant cuticular wax. *Progress in Lipid Research* **42:** 51–80.

Kunst, L., A. L. Samuels, and R. Jetter. 2005. The plant cuticle: Formation and structure of epidermal surfaces. Pp. 270–302 *in* D. Murphy, ed., *Plant lipids: Biology, utilisation and manipulation*. Blackwell, Oxford, UK.

Kurdyukov, S., A. Faust, C. Nawrath, S. Bär, D. Voisin, N. Efremova, R. Franke, L. Schreiber, H. Saedler, J. Métraux, and A. Yephremov. 2006a. The epidermis-specific extracellular BODYGUARD controls cuticle development and morphogenesis of *Arabidopsis*. *Plant Cell* **18:** 321–339.

———. 2006b. Genetic and biochemical evidence for involvement of HOTHEAD in the biosynthesis of long-chain α,ω-dicarboxylic fatty acids and formation of extracellular matrix. *Planta* **224:** 315–329.

Landry, E. T., J. I. Mitchell, S. Hotchkiss, and R. A. Eaton. 2008. Bacterial diversity associated with archeological waterlogged wood: ribosomal RNA clone libraries and denaturing gradient gel electrophoresis (DGGE). *International Biodeterioration and Biodegradation* **61:** 106–116.

Lequeu, J., M. L., Fauconnier, A. Chammai, R. Bronner, E. Blee. 2003. Formation of plant cuticle: Evidence for the occurrence of the peroxygenase pathway. *Plant Journal* **36:** 155–164.

Lobbes, J., H. P. Fitznar, and G. Kattner. 2000. Biogeochemical characteristics of dissolved and particulate organic matter in Russian rivers entering the Arctic Ocean. *Geochimica et Cosmochimica Acta* **64:** 2973–2983.

Louchouarn, P., M. Lucotte, and N. Farella. 1999. Historical and geographical variations of sources and transport of terrigenous organic matter within a large-scale coastal environment. *Organic Geochemistry* **30:** 675–699.

Malcolm, R. I., and W. H. Durum. 1976. Organic carbon and nitrogen concentrations and annual organic carbon load of six selected rivers of the U.S. Geological Survey Water-Supply Paper.

Mannino, A., and H. R. Harvey. 2000. Biochemical composition of particles and dissolved organic matter along an estuarine gradient: Sources and implications for DOM reactivity. *Limnology and Oceanography* **45:** 775–788.

Martin, F., F. J. Gonzalez-Vila, J. C. Del Rio, and T. Verdejo. 1994. Pyrolysis derivatization of humic substances, I: Pyrolysis of fulvic acids in the presence of tetramethylammonium hydroxide. *Journal of Analytical Applied Pyrolysis* **28:** 71–80.

———. 1995. Pyrolysis derivatization of humic substances, II: Pyrolysis of soil humic acids in the presence of tetramethylammonium hydroxide. *Journal of Analytical Applied Pyrolysis* **31:** 75–83.

Martin, J. T., and B. E. Juniper. 1970. *The cuticles of plants*. Edward Arnold, Edinburgh, UK.

McKinney, D. E., D. M. Carson, D. J. Clifford, R. D. Minard, and P. G. Hatcher. 1995. Off-line thermochemolysis versus flash pyrolysis for the in situ methylation of lignin: Is pyrolysis necessary? *Journal of Analytical and Applied Pyrolysis* **34:** 41–46.

Millar, A. A., S. Clemens, S. Zachgo, E. M. Giblin, D. C. Taylor, and L. Kunst. 1999. CUT1, an *Arabidopsis* gene required for cuticular wax biosynthesis and pollen fertility, encodes a very-long-chain fatty acid condensing enzyme. *Plant Cell* **11:** 825–832.

Mitra, S., T. S. Bianchi, L. Guo, and P. H. Santschi. 2000. Terrestrially-derived dissolved organic matter in Chesapeake Bay and the Middle-Atlantic Bight. *Geochimica et Cosmochimica Acta* **64:** 3547–3557.

Moire, L., A. Schmutz, A. Buchala, B. Yan, R. E. Stark, and U. Ryser. 1999. Glycerol is a suberin monomer: New experimental evidence for an old hypothesis. *Plant Physiology* **119:** 1137–1146.

Naafs, D.F.W., and P. F. van Bergen. 2002. Effects of pH adjustments after base hydrolysis: Implications for understanding organic matter in soils. *Geoderma* **106:** 191–217.

Nawrath, C. 2002. The biopolymers cutin and suberin. *In* C. R. Somerville and M. Meyerowitz, eds., *The Arabidopsis book*. American Society of Plant Biologists, Rockville, MD.

Nierop, K.G.J. 1998. Origin of aliphatic compounds in a forest soil. *Organic Geochemistry* **29:** 1009–1016.

Nierop, K.G.J., and T. R. Filley. 2007. Assessment of lignin and (poly-)phenol transformations in oak (*Quercus robur*) dominated soils by ^{13}C-TMAH thermochemolysis. *Organic Geochemistry* **13:** 551–565.

Nierop, K.G.J., F. Dennis, D.F.W. Naafs, and J. M. Verstraten. 2003. Occurrence and distribution of ester-bound lipids in Dutch coastal dune soils along a pH gradient. *Organic Geochemistry* **34:** 719–729.

Onstad, G. D., D. E. Canfield, P. D. Quay, and J. I. Hedges. 2000. Sources of particulate organic matter in rivers from the continental USA: Lignin phenol and stable carbon isotope compositions. *Geochimica et Cosmochimica Acta* **64:** 3539–3546.

Opsahl, S., and R. Benner. 1995. Early diagenesis of vascular plant tissues: Lignin and cutin decomposition and biogeochemical implications. *Geochimica et Cosmochimica Acta* **59.**

———. 1997. Distribution and cycling of terrigenous dissolved organic matter in the ocean. *Nature* **386:** 480–482.

———. 1998. Photochemical reactivity of dissolved lignin in river and ocean waters. *Limnology and Oceanography* **43:** 1297–1304.

Opsahl, S., R. Benner, and R.M.W. Amon. 1999. Major flux of terrigenous dissolved organic matter through the Arctic Ocean. *Limnology and Oceanography* **44:** 2017–2023.

Otto, A., and M. J. Simpson. 2006. Evaluation of CuO oxidation parameters for determining the source and stage of lignin degradation in soil. *Biogeochemistry* **80:** 121–142.

Otto, A., C. Shunthirasingham, and M. J. Simpson. 2005. A comparison of plant and microbial biomarkers in grassland soils from the Prairie Ecozone of Canada. *Organic Geochemistry* **36:** 425–448.

Page, D. W., J. A. van Leeuen, K. M. Spark, and D. E. Mulcahy. 2001. Tracing terrestrial compounds leaching from two reservoir catchments as input to dissolved organic matter. *Marine and Freshwater Research* **52:** 223–233.

Pareek, S., J. I. Azuma, S. Matsui, and Y. Shimizu. 2001. Degradation of lignin and lignin model compound under sulfate reducing condition. *Water Science and Technology* **44:** 351–358.

Pighin, J. A., H. Zheng, L. J. Balakshin, I. P. Goodman, T. L. Western, R. Jetter, L. Kunst, A. L. Samuels. 2004. Plant cuticular lipid export requires an ABC transporter. *Science* **306:** 702–704.

Pinot, F., I. Benveniste, J. Salaun, O. Loreau, J. Noel, L. Schreiber, and F. Durst. 1999. Production in vitro by the cytochrome P450 CYP94A1 of major C18 cutin monomers and potential messengers in plant–pathogen interactions: enantioselectivity studies. *Biochemistry Journal* **342:** 27–32.

Post-Beittenmiller, D. 1996. Biochemistry and molecular biology of wax production in plants. *Annual Review Plant Physiology Plant Molecular Biology* **47:** 405–430.

Prahl, F. G., J. R. Ertel, M. A. Goni, M. A. Sparrow, and B. Eversmeyer. 1994. Terrestrial organic carbon contributions in sediments on the Washington margin. *Geochimica et Cosmochimica Acta* **58:** 3035–3048.

Pulchan, K., T. A. Abrajano, and R. J. Helleur. 1997. Characterization of tetramethylammonium hydroxide thermochemolysis products of near-shore marine sediments using gas chromatography/mass spectrometry and gas chromatography/combustion/isotope ratio mass spectrometry. *Journal Analytical Applied Pyrolysis* **42:** 135–150.

Pulchan, K. J., R. J. Helleur, and T. A. Abrajano. 2003. TMAH thermochemolysis characterization of marine sedimentary organic matter in a Newfoundland fjord. *Organic Geochemistry* **34:** 305–317.

Ryser, U., and P. J. Holloway. 1985. Ultrastructure and chemistry of soluble and polymeric lipids in cell walls from seed coats and fibres of *Gossypium* species. *Planta* **163:** 151–163.

Sarkanen, K. V., and C. H. Ludwig. 1971. *Lignins: Occurrence, formation, structure, and reactions.* Wiley-Interscience, New York.

Schnurr, J., J. Schockey, and J. Browse. 2004. The acyl-CoA synthetase encoded by LACS2 is essential for normal cuticle development in *Arabidopsis. Plant Cell* **16:** 629–642.

Sjostrom, E. 1981. *Wood chemistry, fundamentals and applications.* Academic Press, London.

Stark, R. E., and S. Tian. 2006. The cutin bioploymer matrix. Pp. 126–144 *in* M. Rieder, ed., *Biology of the plant cuticle.* Blackwell, Cambridge, MA.

Tareq, S. M., N. Tanaka, and O. Keiichi. 2004. Biomarker signature in tropical wetland: Lignin phenol vegetation index (LPVI) and its implications for reconstructing the paleoenvironmen. *Science in the Total Environment* **324:** 91–103.

Trefrey, J. H., S. Metz, T. A. Nelsen, T. P. Trocine, and B. A. Eadie. 1994. Transport and fate of particulate organic carbon by the Mississippi River and its fate in the Gulf of Mexico. *Estuaries* **17:** 839–849.

Wilson, J. O., I. Valiela, and T. Swain. 1985. Sources and concentrations of vascular plant material in sediments of Buzzards Bay, Massachusetts, USA. *Marine Chemistry* **90:** 129–137.

Wu, X. Q., J. X. Lin, J. M. Zhu, Y. X. Hu, K. Hartmann, and L. Schreiber. 2003. Casparian strips in needles of *Pinus bungeana*: Isolation and chemical characterization. *Plant Physiology* **117:** 421–424.

Xiao, F., M. Goodwin, Y. Xiao, Z. Sun, D. Baker, X. Tang, M. A. Jenks, and J. Zhou, 2004. *Arabidopsis* CYP86A2 represses *Pseudomonas syringae* type III genes and is required for cuticle development. *EMBO Journal* **23:** 2903–2913.

Zeier, J., K. Ruel, U. Ryser, and L. Schreiber. 1999. Chemical analysis and immunolocalisation of lignin and suberin in endodermal and hypodermal/rhizodermal cell walls of developing maize (*Zea mays*) primary roots. *Planta* **209:** 1–12.

Chapter 14

Abramowicz, D. A. 1995. Aerobic and anaerobic PCB biodegradation in the environment. *Environmental Health Perspectives* **103:** 97–99.

Ankley, G. T., and others. 2003. Effects of the androgenic growth promoter 17-*b*-trenbolone on fecundity and reproductive endocrinology of the fathead minnow. *Environmental Toxicological Chemistry* **22:** 1350–1360.

Arzayus, K. M., R. M. Dickhut, and E. A. Canuel. 2001. Fate of atmospherically deposited polycyclic aromatic hydrocarbons (PAHs) in Chesapeake Bay. *Environmental Science and Technology* **35:** 2178–2183.

Baker, J. I., and R. A. Hites. 2000. Is combustion the major source of polychlorinated dibenzo-*p*-dioxins and dibenzofurans to the environment? A mass balance investigation. *Environmental Science and Technology* **34:** 2879–2886.

Ballentine, D. C., S. A. Macko, V. C. Turekian, W. P. Gilhooly, and B. Martincigh. 1996. Compound specific isotope analysis of fatty acids and polycyclic aromatic hydrocarbons in aerosols: Implications for biomass burning. *Organic Geochemistry* **25:** 97–104.

Bamford, H. A., J. E. Baker, and D. L. Poster. 1998. Review of methods and measurements of selected hydrophobic organic contaminant aqueous solubilities, vapor pressures and air-water partition coefficients. NIST

Special Publication 928, National Institute of Standards and Technology, U.S. Government Printing Office, Washington, DC.

Bayona, J. M., and J. Albaiges. 2006. Sources and fate of organic contaminants in the marine environment. Pp. 323–370 *in* J. K. Volkman, ed., *Handbook of environmental chemistry*, Vol. 2, Part N: *Marine organic matter—Biomarkers, isotopes and DNA*. Springer, Berlin.

Bence, A. E., K. A. Kvenvolden, and M. C. Kennicutt. 1996. Organic geochemistry applied to environmental assessments of Prince William Sound, Alaska, after the *Exxon Valdez* oil spill: A review. *Organic Geochemistry* **24:** 7–42.

Boehm, P. D., and D. S. Page. 2007. Exposure elements in oil spill risk and natural resource damage assessments: A review. *Human and Ecological Risk Assessment* **13:** 418–448.

Boehm, P. D., D. S. Page, W. A. Burns, A. E. Bence, P. J. Mankiewicz, and J. S. Brown. 2001. Resolving the origin of the petrogenic hydrocarbon background in Prince William Sound, Alaska. *Environmental Science and Technology* **35:** 471–479.

Boon, J. P., and others. 2002. Levels of polybrominated diphenyl ether (PBDE) flame retardants in animals representing different trophic levels of the North Sea food web. *Environmental Science and Technology* **36:** 4025–4032.

Bopp, R. F., H. J. Simpson, C. R. Olsen, R. M. Trier, and N. Kotysk. 1982. Chlorinated hydrocarbons and radionuclide chronologies in sediments of the Hudson River and estuary, New York. *Environmental Science and Technology* **16:** 676–681.

Bouloubassi, I., and A. Saliot. 1993. Investigation of anthropogenic and natural organic inputs in estuarine sediments using hydrocarbon markers (NAH, LAB, PAH). *Oceanologica Acta* **16:** 145–161.

Braddock, J. F., J. E. Lindstrom, and E. J. Brown. 1995. Distribution of hydrocarbon-degrading microorganisms in sediments from Prince William Sound, Alaska, following the *Exxon Valdez* oil spill. *Marine Pollution Bulletin* **30:** 125–132.

Braekevelt, E., S. A. Tittlemier and G. T. Tomy. 2003. Direct measurement of octanol–water partition coefficients of some environmentally relevant brominated diphenyl ether congeners. *Chemosphere* **51:** 563–567.

Breivik, K., A. Sweetman, J. M. Pacyna, and K. C. Jones. 2002. Towards a global historical emission inventory for selected PCB congeners—A mass balance approach, 1: Global production and consumption. *Science of the Total Environment* **290:** 181–198.

Brian, J. V., C. A. Harris, M. Scholze, A. Kortenkamp, P. Booy, M. Lamoree, G. Pojana, N. Jonkers, A. Marcomini, and J. P. Sumpter. 2007. Evidence of estrogenic mixture effects on the reproductive performance of fish. *Environmental Science and Technology* **41:** 337–344.

Brooks G. T. 1979. Pp. 1–46 *in Chlorinated insecticides*, vol. I CRC Press, Boca Raton, FL.

Bruno, F., R. Curini, A. Di Corcia, I. Fochi, M. Nazzari, and R. Samperi. 2002. Determination of surfactants and some of their metabolites in untreated and anaerobically digested sewage sludge by subcritical water extraction followed by liquid chromatography–mass spectrometry. *Environmental Science and Technology* **36:** 4156–4161.

Budzinski, H., I. Jones, J. Bellocq, C. Pierard, and P. Garrigues. 1997. Evaluation of sediment contamination by polycyclic aromatic hydrocarbons in the Gironde estuary. *Marine Chemistry* **58:** 85–97.

Buerge, I. J., T. Poiger, M. D. Muller, and H.-R. Buser. 2003. Caffeine, an anthropogenic marker for wastewater contamination of surface waters. *Environmental Science and Technology* **37:** 691–700.

———. 2006. Combined sewer overflows to surface waters detected by the anthropogenic marker caffeine. *Environmental Science and Technology* **40:** 4096–4102.

Burns, W. A., P. J. Mankiewicz, A. E. Bence, D. S. Page, and K. R. Parker. 1997. A principal-component and least-squares method for allocating polycyclic aromatic hydrocarbons in sediment to multiple sources. *Environmental Toxicology and Chemistry* **16:** 1119–1131.

Canuel, E. A., E. J. Lerberg, R. M. Dickhut, S. A. Kuehl, T. S. Bianchi, and S. G. Wakeham. 2009. Changes in sediment and organic carbon accumulation in a highly-disturbed ecosystem: The Sacramento–San Joaquin River delta (California, USA). *Marine Pollution Bulletin* **59:** 154–163.

Castle, D. M., M. T. Montgomery, and D. L. Kirchman. 2006. Effects of naphthalene on microbial community composition in the Delaware estuary. *FEMS Microbiology Ecology* **56:** 55–63.

Castro-Jimenez, J., and others. 2008. PCDD/F and PCB multi-media ambient concentrations, congener patterns and occurrence in a mediterranean coastal lagoon (Etang de Thau, France). *Environmental Pollution* **156:** 123–135.

Cetin, B., and M. Odabasi. 2005. Measurement of Henry's law constants of seven polybrominated diphenyl ether (PBDE) congeners as a function of temperature. *Atmospheric Environment* **39:** 5273–5280.

Cetin, B., S. Ozer, A. Sofuoglu, and M. Odabasi. 2006. Determination of Henry's law constants of organochlorine pesticides in deionized and saline water as a function of temperature. *Atmospheric Environment* **40:** 4538–4546.

Chiou, C. T., S. E. McGroddy, and D. E. Kile. 1998. Partition characteristics of polycyclic aromatic hydrocarbons on soils and sediments. *Environmental Science and Technology* **32:** 264–269.

Chiuchiolo, A. L., R. M. Dickhut, M. A. Cochran, and H. W. Ducklow. 2004. Persistent organic pollutants at the base of the Antarctic marine food web. *Environmental Science and Technology* **38:** 3551–3557.

Coates, J. D., R. T. Anderson, and D. R. Lovley. 1996. Oxidation of polycyclic aromatic hydrocarbons under sulfate-reducing conditions. *Applied and Environmental Microbiology* **62:** 1099–1101.

Coates, J. D., J. Woodward, J. Allen, P. Philp, and D. R. Lovley. 1997. Anaerobic degradation of polycyclic aromatic hydrocarbons and alkanes in petroleum-contaminated marine harbor sediments. *Applied and Environmental Microbiology* **63:** 3589–3593.

Cornelissen, G., O. Gustafsson, T. D. Bucheli, M.T.O. Jonker, A. A. Koelmans, and P.C.M. Van Noort. 2005. Extensive sorption of organic compounds to black carbon, coal, and kerogen in sediments and soils: Mechanisms and consequences for distribution, bioaccumulation, and biodegradation. *Environmental Science and Technology* **39:** 6881–6895.

Countway, R. E., R. M. Dickhut, and E. A. Canuel. 2003. Polycyclic aromatic hydrocarbon (PAH) distributions and associations with organic matter in surface waters of the York River, VA estuary. *Organic Geochemistry* **34:** 209–224.

Cousins, I., and D. Mackay. 2000. Correlating the physical–chemical properties of phthalate esters using the 'three solubility' approach. *Chemosphere* **41:** 1389–1399.

Dachs, J., R. Lohmann, W. A. Ockenden, L. Mejanelle, S. J. Eisenreich, and K. C. Jones. 2002. Oceanic biogeochemical controls on global dynamics of persistent organic pollutants. *Environmental Science and Technology* **36:** 4229–4237.

Daughton, C. G., and T. A. Ternes. 1999. Pharmaceuticals and personal care products in the environment: Agents of subtle change? *Environmental Health Perspectives* **107:** 907–938.

De Wit, C. A. 2002. An overview of brominated flame retardants in the environment. *Chemosphere* **46:** 583–624.

Dickhut, R. M., and K. E. Gustafson. 1995. Atmospheric inputs of selected polycyclic aromatic hydrocarbons and polychlorinated biphenyls to southern Chesapeake Bay. *Marine Pollution Bulletin* **30:** 385–396.

Dickhut, R. M., E. A. Canuel, K. E. Gustafson, K. Liu, K. M. Arzayus, S. E. Walker, G. Edgecombe, M. O. Gaylor, and E. H. MacDonald. 2000. Automotive sources of carcinogenic polycyclic aromatic hydrocarbons associated with particulate matter in the Chesapeake Bay region. *Environmental Science and Technology* **34:** 4635–4640.

Durhan, E. J., C. S. Lambright, E. A. Makynen, J. Lazorchak, P. C. Hartig, V. S. Wilson, L. E. Gray, and G. T. Ankley. 2006. Identification of metabolites of trenbolone acetate in androgenic runoff from a beef feedlot. *Environmental Health Perspectives* **114:** 65–68.

Eganhouse, R. P., D. L. Blumfield, and I. R. Kaplan. 1983. Long-chain alkylbenzenes as molecular tracers of domestic wastes in the marine environment. *Environmental Science and Technology* **17:** 523–530.

Eljarrat, E., A. De La Cal, D. Larrazabal, B. Fabrellas, A. R. Fernandez-Alba, F. Borrull, R. M Marce, and D. Barcelo. 2005. Occurrence of polybrominated diphenylethers, polychlorinated dibenzo-*p*-dioxins, dibenzofurans and biphenyls in coastal sediments from Spain. *Environmental Pollution* **136:** 493–501.

Fattore, E., E. Benfenati, G. Mariani, R. Fanelli, and E.H.G. Evers. 1997. Patterns and sources of polychlorinated dibenzo-*p*-dioxins and dibenzofurans in sediments from the Venice Lagoon, Italy. *Environmental Science and Technology* **31:** 1777–1784.

Ferguson, P. L., C. R. Iden, and B. J. Brownawell. 2001. Distribution and fate of neutral alkylphenol ethoxylate metabolites in a sewage-impacted urban estuary. *Environmental Science and Technology* **35:** 2428–2435.

Ferguson, P. L., R. F. Bopp, S. N. Chillrud, R. C. Aller, and B. J. Brownawell. 2003. Biogeochemistry of nonylphenol ethoxylates in urban estuarine sediments. *Environmental Science and Technology* **37:** 3499–3506.

Flaherty, C. M., and S. I. Dodson. 2005. Effects of pharmaceuticals on *Daphnia* survival, growth, and reproduction. *Chemosphere* **61:** 200–207.

Gardinali, P. R., and X. Zhao. 2002. Trace determination of caffeine in surface water samples by liquid chromatography–atmospheric pressure chemical ionization–mass spectrometry (LC-APCI-MS). *Environment International* **28:** 521–528.

Garrison, A. W. 2006. Probing the enantioselectivity of chiral pesticides. *Environmental Science and Technology.* **40:** 16–23.

Gustafsson, O., N. Nilsson, and T. D. Bucheli. 2001. Dynamic colloid–water partitioning of pyrene through a coastal Baltic spring bloom. *Environmental Science Technology* **35:** 4001–4006.

Hale, R. C., M. J. La Guardia, E. Harvey, and T. M. Mainor. 2002. Potential role of fire retardant-treated polyurethane foam as a source of brominated diphenyl ethers to the US environment. *Chemosphere* **46:** 729–735.

Hale, R. C., M. J. La Guardia, E. Harvey, M. O. Gaylor, and T. M. Mainor. 2006. Brominated flame retardant concentrations and trends in abiotic media. *Chemosphere* **64:** 181–186.

Harvey, R. G. 1997. *Polycyclic aromatic hydrocarbons.* Wiley-VCH, New York.

Hayes, L. A., K. P. Nevin, and D. R. Lovley. 1999. Role of prior exposure on anaerobic degradation of naphthalene and phenanthrene in marine harbor sediments. *Organic Geochemistry* **30:** 937–945.

Herberer, T. 2002. Occurance, fate, and removal of pharmaceutical residues in the aquatic environment: A review of recent research data. *Toxicology Letters* **131:** 5–17.

Hites, R. A. 2004. Polybrominated diphenyl ethers in the environment and in people: A meta-analysis of concentrations. *Environmental Science and Technology* **38:** 945–956.

Hofstetter, T. B., C. M. Reddy, L. J. Heraty, M. Berg, and N. C. Sturchio. 2007. Carbon and chlorine isotope effects during abiotic reductive dechlorination of polychlorinated ethanes. *Environmental Science and Technology* **41:** 4662–4668.

Hoh, E., and R. A. Hites. 2005. Brominated flame retardants in the atmosphere of the east-central United States. *Environmental Science and Technology* **39:** 7794–7802.

Horii, Y., K. Kannan, G. Petrick, T. Gamo, J. Falandysz, and N. Yamashita. 2005a. Congener-specific carbon isotopic analysis of technical PCB and PCN mixtures using two-dimensional gas chromatography–isotope ratio mass spectrometry. *Environmental Science and Technology* **39:** 4206–4212.

———. 2005b. Congener-specific carbon isotopic analysis of 18 PCB products using two dimensional GC/IRMS. *Bunseki Kagaku* **54:** 361–372.

Isosaari, P., H. Pajunen, and T. Vartiainen. 2002. PCDD/F and PCB history in dated sediments of a rural lake. *Chemosphere* **47:** 575–583.

Jacobs, M. N., A. Covaci, and P. Schepens. 2002. Investigation of selected persistent organic pollutants in farmed Atlantic salmon (*Salmo salar*), salmon aquaculture feed, and fish oil components of the feed. *Environmental Science and Technology* **36:** 2797–2805.

Jensen, K. M., E. A. Makynen, M. D. Kahl, and G. T. Ankley. 2006. Effects of the feedlot contaminant 17*a*-trenbolone on reproductive endocrinology of the fathead minnow. *Environmental Science and Technology* **40:** 3112–3117.

Jonsson, A., O. Gustafsson, J. Axelman, and H. Sundberg. 2003. Global accounting of PCBs in the continental shelf sediments. *Environmental Science and Technology* **37:** 245–255.

Kanaly, R. A., and S. Harayama. 2000. Biodegradation of high-molecular-weight polycyclic aromatic hydrocarbons by bacteria. *Journal of Bacteriology* **182:** 2059–2067.

Kanke, H., M. Uchida, T. Okuda, M. Yoneda, H. Takada, Y. Shibata, and M. Morita. 2004. Compound-specific radiocarbon analysis of polycyclic aromatic hydrocarbons (PAHs) in sediments from an urban reservoir. *Nuclear Instruments and Methods in Physics Research Section B: Beam Interactions with Materials and Atoms* **223-24:** 545–554.

Kelly, S. A., and R. T. Di Giulio. 2000. Developmental toxicity of estrogenic alkylphenols in killifish (*Fundulus heteroclitus*). *Environmental Toxicological Chemistry* **19:** 2564–2570.

Killops, S., and V. Killops. 2005. *Introduction to organic geochemistry*, 2nd ed. Blackwell, Oxford, UK.

Kvenvolden, K. A., F. D. Hostettler, J. B. Rapp, and P. R. Carlson. 1993. Hydrocarbons in oil residues on beaches of islands of Prince William Sound, Alaska. *Marine Pollution Bulletin* **26:** 24–29.

Lichtfouse, E., H. Budzinski, P. Garrigues, and T. I. Eglinton. 1997. Ancient polycyclic aromatic hydrocarbons in modern soils: C-13, C-14 and biomarker evidence. *Organic Geochemistry* **26:** 353–359.

Lindstrom, A., I. J. Buerge, T. Poiger, P. A. Bergqvist, M. D. Muller, and H. R. Buser. 2002. Occurrence and environmental behavior of the bactericide triclosan and its methyl derivative in surface waters and in wastewater. *Environmental Science and Technology* **36:** 2322–2329.

Lohmann, R., K. Breivik, J. Dachs, and D. Muir. 2007. Global fate of POPs: Current and future research directions. *Environmental Pollution* **150:** 150–165.

Lohmann, R., and K. C. Jones. 1998. Dioxins and furans in air and deposition: A review of levels, behaviour and processes. *Science of the Total Environment* **219:** 53–81.

Lohmann, R., E. Jurado, J. Dachs, U. Lohmann, and K. C. Jones. 2006. Quantifying the importance of the atmospheric sink for polychlorinated dioxins and furans relative to other global loss processes. *Journal of Geophysical Research-Atmospheres* **111,** D21303, Doi:10.102912005JD006923.

Macdonald, R. W., L. A. Barrie, T. F. Bidleman, M. L. Diamond, D. J. Gregor, R. G. Semkin, W.M.J. Strachan, Y. F. Li, F. Wania, M. Alaee, L. B. Alexeeva, S. M. Backus, R. Bailey, J. M. Bewers, C. Gobeil, C. J. Halsall, T. Harner, J. T. Hoff, L.M.M. Jantunen, W. L. Lockhart, D. Mackay, D. C. G. Muir, J. Pudykiewicz, K. J. Reimer, J. N. Smith, G. A Stern, W. H. Schroeder, R. Wagemann, and M. B. Yunker. 2000. Contaminants in the Canadian Arctic: 5 years of progress in understanding sources, occurrence and pathways. *Science of the Total Environment* **254:** 93–234.

Mackay, D., W. Y. Shiu, and K. C. Ma. 1992. *Illustrated handbook of physical–chemical properties and environmental fate for organic chemicals.* Lewis, Chelsea, MI.

Mandalakis, M., O. Gustafsson, C. M. Reddy, and X. Li. 2004. Radiocarbon apportionment of fossil versus biofuel combustion sources of polycyclic aromatic hydrocarbons in the Stockholm metropolitan area. *Environmental Science and Technology* **38:** 5344–5349.

Nash, J. P., D. E. Kime, L.T.M. Van der Ven, P. W. Wester, F. Brion, G. Maack, P. Stahlschmidt-Allner, and C. R. Tyler. 2004. Long-term exposure to environmental concentrations of the pharmaceutical ethynylestradiol causes reproductive failure in fish. *Environmental Health Perspectives* **112:** 1725–1733.

O'Malley, V. P., T. A. Abrajano, Jr., and J. Hellou. 1994. Determination of the ratios of individual PAH from environmental samples: Can PAH sources be apportioned? *Organic Geochemistry* **21:** 809–822.

Ogura, I., S. Masunaga, and J. Nakanishi. 2001. Atmospheric deposition of polychlorinated dibenzo-*p*-dioxins, polychlorinated dibenzofurans, and dioxin-like polychlorinated biphenyls in the Kanto region, Japan. *Chemosphere* **44:** 1473–1487.

Oros, D. R., D. Hoover, F. Rodigari, D. Crane, and J. Sericano. 2005. Levels and distribution of polybrominated diphenyl ethers in water, surface sediments, and bivalves from the San Francisco estuary. *Environmental Science and Technology* **39:** 33–41.

Padma, T. V. and R. M. Dickhut. 2002. Spatial and temporal variation in hexachlorocyclohexane isomers in a temperate estuary. *Marine Pollution Bulletin* **44:** 1345–1353.

Padma, T. V, R. M. Dickhut, and H. Ducklow. 2003. Variations in alpha-hexachlorocyclohexane enantiomer ratios in relation to microbial activity in a temperate estuary. *Environmental Toxicological Chemistry* **22:** 1421–1427.

Page, D. S., P. D. Boehm, G. S. Douglas, A. E. Bence, W. A. Burns, and P. J. Mankiewicz. 1996. The natural petroleum hydrocarbon background in subtidal sediments of Prince William Sound, Alaska, USA. *Environmental Toxicology and Chemistry* **15:** 1266–1281.

Paull, C. K., H. G. Greene, W. Ussler III, and P. J. Mitts. 2002. Pesticides as tracers of sediment transport through Monterey Canyon. *Geo-Marine Letters* **22:** 121–126.

Peeler, K. A., S. P. Opsahl, and J. P. Chanton. 2006. Tracking anthropogenic inputs using caffeine, indicator bacteria, and nutrients in rural freshwater and urban marine systems. *Environmental Science and Technology* **40:** 7616–7622.

Peters, K. E., C. C. Walters, and J. M. Moldowan. 2005. *The biomarker guide, I: Biomarkers and isotopes in the environment and human history*, 2nd ed. Cambridge University Press, New York.

Philp, R. P. 2007. The emergence of stable isotopes in environmental and forensic geochemistry studies: A review. *Environmental Chemistry Letters* **5:** 57–66.

Pies, C., T. A. Ternes, and T. Hofmann. 2008. Identifying sources of polycyclic aromatic hydrocarbons (PAHs) in soils: Distinguishing point and non-point sources using an extended PAH spectrum and *n*-alkanes. *Journal of Soils and Sediments* **8:** 312–322.

Plata, D. L., C. M. Sharpless, and C. M. Reddy. 2008. Photochemical degradation of polycyclic aromatic hydrocarbons in oil films. *Environmental Science and Technology* **42:** 2432–2438.

Reddy, C. M., N. J. Drenzek, N. C. Sturchio, L. Heraty, A. Butler, and C. Kimblin. 2002a. A chlorine isotope effect for biochlorination. *Geochimica et Cosmochimica Acta* **66:** A627–A627.

Reddy, C. M., T. I. Eglinton, A. Hounshell, H. K. White, L. Xu, R. B. Gaines, and G. S. Frysinger. 2002b. The West Falmouth oil spill after thirty years: The persistence of petroleum hydrocarbons in marsh sediments. *Environmental Science and Technology* **36:** 4754–4760.

———. 2002c. Radiocarbon as a tool to apportion the sources of polycyclic aromatic hydrocarbons and black carbon in environmental samples. *Environmental Science and Technology* **36:** 1774–1782.

———. 2004. Radiocarbon evidence for a naturally produced, bioaccumulating halogenated organic compound. *Environmental Science and Technology* **38:** 1992–1997.

Ribes, S., B. Van Drooge, J. Dachs, O. Gustafsson, and J. O. Grimalt. 2003. Influence of soot carbon on the soil–air partitioning of polycyclic aromatic hydrocarbons. *Environmental Science and Technology* **37:** 2675–2680.

Richardson, S. D. 2006. Environmental mass spectrometry: Emerging contaminants and current issues. *Analytical Chemistry* **78:** 4021–4045.

Richardson, S. D., and T. A. Ternes. 2005. Water analysis: Emerging contaminants and current issues. *Analytical Chemistry* **77:** 3807–3838.

Risebrough, R. W., W. Walker, T. T. Schmidt, B. W. Delappe, and C. W. Connors. 1976. Transfer of chlorinated biphenyls to Antarctica. *Nature* **264:** 738–739.

Rothermich, M. M., L. A. Hayes, and D. R. Lovley. 2002. Anaerobic, sulfate-dependent degradation of polycyclic aromatic hydrocarbons in petroleum-conta minated harbor sediment. *Environmental Science and Technology* **36:** 4811–4817.

Schwarzenbach, R. P., P. M. Gschwend, and D. M. Imboden. 1993. *Environmental organic chemistry*. Wiley, New York.

Schwarzenbach, R. P., P. M. Gschwend, and D. M. Imboden. 2003. *Environmental organic chemistry*, 2nd ed. Wiley, New York

Schwarzenbach, R. P., and others. 2006. The challenge of micropollutants in aquatic systems. *Science* **313:** 1072–1077.

Sinclair, E., and K. Kannan. 2006. Mass loading and fate of perfluoroalkyl surfactants in wastewater treatment plants. *Environmental Science and Technology* **40:** 1408–1414.

Slater, G. F. 2003. Stable isotope forensics: When isotopes work. *Environmental Forensics* **4:** 13–23.

Slater, G. F., H. K. White, T. I. Eglinton, and C. M. Reddy. 2005. Determination of microbial carbon sources in petroleum contaminated sediments using molecular C-14 analysis. *Environmental Science and Technology* **39:** 2552–2558.

Standley, L. J., L. A. Kaplan, and D. Smith. 2000. Molecular tracers of organic matter sources to surface water resources. *Environmental Science and Technology* **34:** 3124–3130.

Staples, C. A., P. B. Dorn, G. M. Klecka, D. R. Branson, S. T. O'Block, and L. R. Harris. 1998. A review of the environmental fate, effects, and exposures of bisphenol A. *Chemosphere* **36:** 2149–2173.

Stapleton, H. M., N. G., Dete, J. R. Dodder, C. M. Kucklick, M. M. Reddy, P. R. Schantz, F. Becker, B. J. Gulland, B. J. Porter, and S. A. Wise. 2006. Determination of HBCD, PBDEs and MeO-BDEs in California sea lions (*Zalophus californianus*) stranded between 1993 and 2003. *Marine Pollution Bulletin* **52:** 522-531.

Swartz, C. H., S. Reddy, M. J. Benotti, H. Yin, L. B. Barber, B. J. Brownawell, and R. A. Rudeland. 2006. Steroid estrogens, nonylphenol ethoxylate metabolites, and other wastewater contaminants in groundwater affected by a residential septic system on Cape Cod, MA. *Environmental Science and Technology* **40:** 4894–4902.

Ternes, T. A. 2001. Analytical methods for the determination of pharmaceuticals in aqueous environmental samples. *Trends in Analytical Chemistry* **20:** 419–434.

Vallack, H. W., D. J. Bakker, I. Brandt, E. Brorstrom-Lunden, A. Brouwer, K. R. Bull, C. Gough, R. Guardans, I. Holoubek, B. Jansson, R. Koch, J. Kuylenstierna, A., Lecloux, D. Mackay, P. McCutcheon, P. Mocarelli, and R.D.F. Taalman. 1998. Controlling persistent organic pollutants: What next? *Environmental Toxicology and Pharmacology* **6:** 143–175.

Van Den Berg, M., L. Birnbaum, T. Albertus, C. Bosveld, B. Brunström, P. Cook, M. Feeley, J. P. Giesy, A. Hanberg, R. Hasegawa, S.W. Kennedy, T. Kubiak, J. C. Larsen, F.X.R. van Leeuwen, A. K. Djien Liem, C. Nolt, R. E. Peterson, L. Poellinger, S. Safe, D. Schrenk, D. Tillitt, M. Tysklind, M. Younes, F. Wærn, and T. Zacharewski. 1998. Toxic equivalency factors (TEFs) for PCBs, PCDDs, PCDFs for humans and wildlife. *Environmental Health Perspectives* **106:** 775–792.

Venier, M., and R. A. Hites. 2008. Flame retardants in the atmosphere near the Great Lakes. *Environmental Science and Technology* **42:** 4745–4751.

Vetter, W., S. Gaul, and W. Armbruster. 2008. Stable carbon isotope ratios of POPs: A tracer that can lead to the origins of pollution. *Environment International* **34:** 357–362.

Wakeham, S. G., J. Forrest, C. A. Masiello, Y. Gelinas, C. R. Alexander, and P. R. Leavitt. 2004. Hydrocarbons in Lake Washington sediments: A 25-year retrospective in an urban lake. *Environmental Science and Technology* **38:** 431–439.

Wakeham, S. G., A. P. Mcnichol, J. E. Kostka, and T. K. Pease. 2006. Natural-abundance radiocarbon as a tracer of assimilation of petroleum carbon by bacteria in salt marsh sediments. *Geochimica et Cosmochimica Acta* **70:** 1761–1771.

Walker, S. E., and R. M. Dickhut. 2001. Sources of PAHs to sediments of the Elizabeth River, VA. *Soil and Sediment Contamination* **10:** 611–632.

Walker, S. E., R. M. Dickhut, C. Chisholm-Brause, S. Sylva, and C. M. Reddy. 2005. Molecular and isotopic identification of PAH sources in a highly industrialized urban estuary. *Organic Geochemistry* **36:** 619–632.

Wang, Z. D., S. A. Stout, and M. Fingas. 2006. Forensic fingerprinting of biomarkers for oil spill characterization and source identification. *Environmental Forensics* **7:** 105–146.

White, H. K., C. M. Reddy, and T. I. Eglinton. 2008. Radiocarbon-based assessment of fossil fuel-derived contaminant associations in sediments. *Environmental Science and Technology* **42:** 5428–5434.

WHO. 1979. *Environmental health criteria*, 9: *DDT and its derivatives*. World Health Organization, Geneva, Switzerland.

Yanik, P. J., T. H. O'Donnell, S. A. Macko, Y. Qian, and M. C. Kennicutt. 2003. Source apportionment of polychlorinated biphenyls using compound specific isotope analysis. *Organic Geochemistry* **34:** 239–251.

Yunker, M. B., R. W. MacDonald, D. J. Veltkamp, and W. J. Cretney. 1995. Terrestrial and marine biomarkers in a seasonally ice-covered arctic estuary: Integration of multivariate and biomarker approaches. *Marine Chemistry* **49:** 1–50.

Zander, M. 1983. Physical and chemical properties of polycyclic aromatic hydrocarbons. *In* A. Bjørstedh, ed., *Handbook of polycyclic aromatic hydrocarbons*. Marcel Dekker, New York.

Zencak, Z., J. Klanova, I. Holoubek, and O. Gustafsson. 2007. Source apportionment of atmospheric PAHs in the western Balkans by natural abundance radiocarbon analysis. *Environmental Science and Technology* **41:** 3850–3855.

Zhang, Q. H., and G. B. Jiang. 2005. Polychlorinated dibenzo-*p*-dioxins/furans and polychlorinated biphenyls in sediments and aquatic organisms from the Taihu Lake, China. *Chemosphere* **61:** 314–322.

Index